Lectures in Astrobiology I

Advances in Astrobiology and Biogeophysics

springeronline.com

This series aims to report new developments in research and teaching in the interdisciplinary fields of astrobiology and biogeophysics. This encompasses all aspects of research into the origins of life – from the creation of matter to the emergence of complex life forms – and the study of both structure and evolution of planetary ecosystems under a given set of astro- and geophysical parameters. The methods considered can be of theoretical, computational, experimental and observational nature. Preference will be given to proposals where the manuscript puts particular emphasis on the overall readability in view of the broad spectrum of scientific backgrounds involved in astrobiology and biogeophysics.
The type of material considered for publication includes:

- Topical monographs
- Lectures on a new field, or presenting a new angle on a classical field
- Suitably edited research reports
- Compilations of selected papers from meetings that are devoted to specific topics

The timeliness of a manuscript is more important than its form which may be unfinished or tentative. Publication in this new series is thus intended as a service to the international scientific community in that the publisher, Springer-Verlag, offers global promotion and distribution of documents which otherwise have a restricted readership. Once published and copyrighted, they can be documented in the scientific literature.

Muriel Gargaud
Bernard Barbier Hervé Martin Jacques Reisse (Eds.)

Lectures in Astrobiology

Volume I

With a Preface by Christian de Duve

and 40 Tables and 259 Figures Including Photos and Plates

Springer

Dr. Muriel Gargaud
Editor-in-Chief
Université Bordeaux 1
Observatoire Aquitain
des Sciences de l'Univers
2 Rue de l'Observatoire
33270 Floirac, France

Dr. Bernard Barbier
Centre de Biophysique Moléculaire
Rue Charles Sadron
45071 Orléans, France

Prof. Hervé Martin
Université Blaise Pascal
Laboratoire Magmas et Volcans
5 Rue Kessler
63038 Clermont-Ferrand, France

Prof. Jacques Reisse
Member of the Belgian Academy
of Sciences
Université Libre de Bruxelles (CP 165/64)
Department of Chemistry
Applied Sciences
Avenue F. Roosevelt, 50
1050 Brussels, Belgium

Library of Congress Control Number: 2004107506

ISSN 1610-8957

ISBN-10 3-540-22315-0 Springer-Verlag Berlin Heidelberg New York
ISBN-13 978-3-540-22315-3 Springer-Verlag Berlin Heidelberg New York

Springer is a part of Springer Science+Business Media
springeronline.com

Printed in Germany

Typesetting and production: LE-TEX Jelonek, Schmidt & Vöckler GbR, Leipzig
Cover design: Erich Kirchner, Heidelberg

Printed on acid-free paper 54/3141/YL - 5 4 3 2 1 0

"It seems plain and self-evident, yet it needs to be said: the isolated knowledge obtained by a group of specialists in a narrow field has in itself no value whatsoever, but only in its synthesis with all the rest of knowledge and only inasmuch as it really contributes in this synthesis toward answering the demand, 'Who are we?'"

Erwin Schrödinger

Foreword

Astrobiology, also known as bioastronomy or exobiology, refers to a vast area of scientific research. The formation of the solar system, its accretion and the formation of the planets, the origin of the molecules out of which living beings are made, the traces of present and past life within the solar system and elsewhere, as well as the search for extra-solar planets, are all part of astrobiology. And the above list is not exhaustive.

For obvious reasons, astrobiology is a field without the traditional barriers between astronomers, chemists, physicists, geologists, and biologists or between experimentalists and theorists, observers and those who model the observations. As such, a single researcher cannot possess all the knowledge necessary to be an "astrobiologist". One can even go a step further and say that while astrobiology clearly exists as a field of scientific research, there are no astrobiologists. Astrobiology exists at a higher level of organization where the knowledge is not that of an individual but that of a research community whose members all share the same interest for the fundamental questions concerning the emergence of life, its evolution, and how life is distributed on Earth and throughout the universe. Each person contributes in piecing together this vast puzzle through their knowledge and their experimental and theoretical tools.

As often, if not always, when treating questions dealing with the past or with a sort of "elsewhere" where one cannot go and that one can only study indirectly, we must be satisfied with plausible scenarios rather than clear proof or other certainties. In this way, the strategy of the astrobiologist is similar to that of an archeologist or a paleontologist. There exists, however, major differences between the path of a chemist interested in the origin of life, and thus in prebiotic chemical evolution, that of a biologist wanting to follow time back starting with current life, and that of a paleontologist searching for the traces of primitive life and its evolution and extinctions.

While paleontologists have some hard data at hand (fossils and other physical traces), the situation is very different for chemists, who are obliged to build a plausible scenario for the appearance of life based on hypotheses developed by specialists in other fields (composition of the primitive terrestrial atmosphere, addition of extraterrestrial organic material, etc.). For the most part, these hypotheses are unverifiable. The biologist, on the other hand, tries to use phylogenetic tools to find and understand LUCA, the first common ancestor who must have been preceded by other micro-organisms with no descendants.

Similarly, there is an important difference between the strategies of a geologist, expert in the transition between the Tertiary and Cretaceous periods, and the planetologist who would like to describe the Earth during the period of intense meteoritic bombardment. The former disposes of observations and measures (iridium content, sediment ashes, shocked quartz, etc.), which provide a reasonable explanation for the great biological Cretaceous Tertiary crisis caused by a major meteorite impact. The latter only has access to indirect data based on observations of lunar craters but also simulations, which are of course based on theoretical models.

Since every scientist has a limited area of expertise, the scenario that he/she proposes can only be validated by the constraints and parameters that he/she knows and masters. Such an individual strategy can thus lead to as many scenarios as there are researchers. A multidisciplinary approach has the advantage of subjecting each individual proposition to a much larger number of constraints. This naturally leads to the rejection of "weak" scenarios and to the emergence of more robust hypotheses. For example, it is pointless for a chemist to invoke the role of a prebiotic chemical reaction if the conditions needed for the reaction to occur are completely incompatible with the primitive Earth conditions determined by the planetologists. This simple example illustrates the importance of interdisciplinary discussion for all those who consider themselves to be astrobiologists. The CNRS summer schools such as Propriano in 1999 and 2003 and La Colle-sur-Loup in 2001 have contributed to strengthening the dialog within the French scientific community.

The goal of the first summer school, Exobio'99 in Propriano, was to provide participants with an objective image of what we know today about the early Earth conditions – the oceans, the proto-continents, the atmosphere, and even the climate – but also of what we know about the solar system during the first billion years of its history. Some stages in the chemical evolution that may have occurred on the young planet Earth, with a different solar radiation, less intense in the visible part of the spectrum but much more intense in the RX region were also discussed during the first summer school. The discussion then moves towards the biological evolution, the early stages of which are still very poorly understood. The problems related to the exploration of Mars and Titan were then addressed.

The second summer school, Exobio '01 in La Colle-sur-Loup, was more oriented towards the chemistry, molecular biology, biochemistry, and biological evolution of early Earth. Its main theme was the study of organisms referred to as extremophiles, which could provide information on the nature of the first unicellular organisms that populated the young oceans. Among the specific topics addressed were the autoformation of biological membranes, the possible origins of the homochirality of the constituents of living beings, the protometabolisms that may be inferred from the study of metabolisms, and the possible role of ribozymes before the emergence of catalysis by proteinic enzymes.

The texts that follow represent the first volume of the series "Lectures in Astrobiology" and are the result of the first two schools. The chapters were written for readers already familiar with the general topic of the origin of life and life "elsewhere" but not to the extent to which each specialist is in his own discipline. As such, they are meant as much for students as for established scientists seeking to broaden their horizons in the vast field of the origins of life. We hope these texts will initiate vocations and incite researchers and students specialized in one of the individual fields to join the broad forum of astrobiology. It is undeniable that the questions forming the basis of astrobiology are among the big questions that humanity has asked itself since its inception and which recent decades have attempted to answer; answers that seem more and more plausible although necessarily partial.

Acknowledgements

This book is the product of a multidisciplinary community, the members of which all question the knowledge from their disciplines at origin in order to build together the complex structure, which this area of research represents. One needs a very open mind as well as the ability to question ones ideas through the recent discoveries in other fields.

A group of international reviewers and ourselves have read the set of texts that follow. The final versions of these texts, after multiple rewritings and long discussions, sometimes required the opinions of five or six specialists.

We would therefore like to thank all of the authors who have accepted these remarks, criticisms, and multiple discussions warmly, but also all of the "specialist reviewers" who, through their expertise, have contributed to the general coherence of this work.

Last but not least, the editors call upon the reader's indulgence concerning some (or many!) misusages of the English language; English is not the mother tongue of the large majority of authors.

Muriel Gargaud
Bernard Barbier
Hervé Martin
Jacques Reisse

From *left* to *right*: Hervé Martin (geochemist), Muriel Gargaud (astrophysicist), Jacques Reisse (chemist), Bernard Barbier (biochemist)

Preface

The twentieth century will be remembered as the century of scientific revolutions. It started with the discoveries of physics, which revealed the fine structure of matter and the fundamental laws of nature. Then came cosmology, which traced the history of the Universe, from the Big Bang to the dizzying recession of myriad whirling galaxies. Finally, with the advances of biochemistry, cell biology, and molecular biology, life itself has disclosed its secrets.

These revolutions, in turn, have spawned new technologies – nuclear power, space travel, informatics, bioengineering – that could not even be conceived one century earlier; they have also opened new fields of inquiry that had been relegated before to the realm of the unknowable, objects only of gratuitous speculation or imaginative fiction. Among these new fields, the origin and evolution of life on Earth have become topics of intense theoretical and experimental research.

The latest offspring of this upheaval is exobiology, the science of extraterrestrial life, also known as astrobiology or bioastronomy. Of all branches of science, it is the most universal and all-encompassing, involving virtually every scientific discipline. It is also the emptiest, being so far without known object. No sign of life beyond our planet has yet been uncovered.

Whether or not its quest will one day be fulfilled, exobiology has already produced many valuable fruits and is bound to produce many more in the future. It has brought together and impelled physicists, chemists, cosmologists, astronomers, planetologists, geologists, paleontologists, biologists of all kinds, and other specialists who had until then labored each in the isolation of their individual disciplines to interact. It has stimulated many investigations that would otherwise have been performed with considerably less vigor, perhaps not at all. It has revealed a number of significant facts on the cosmic properties and interrelationships out of which life and mind emerged on our planet and may, perhaps, have done so elsewhere in our galaxy or in others. It has even alerted philosophers and theologians to a very real possibility that, only 400 years ago, was a heresy punishable by death. It has evoked new dreams in the collective imagination of humans who, ever since their distant ancestors started contemplating the skies, have asked the question: Is there life out there? Are there others like us elsewhere?

The present book is the outcome of a remarkable pluridisciplinary effort initiated by a group of French scientists and materialized already into two summer schools organized under the aegis of the CNRS, Exobio'99, in Propriano, Corsica (1999) and Exobio'01, in La Colle-sur-Loup, South of France (2001). The proceedings of these meetings are an invaluable source of information covering every aspect, from astronomy to molecular biology, likely to illuminate the exobiology problem. This exceptional documentation will now be generally available, thanks to the present work, in which most of the Exobio participants have summarized their contributions in a world-wide accessible form. As a participant myself, who, for personal reasons was unable to provide a chapter, I am particularly pleased and honored to have the opportunity to preface this truly unique compendium.

Christian de Duve

Contents

General Introduction

From the Origin of Life on Earth to Life in the Universe

André Brack

On Earth, life probably emerged in water, about 4 billion years ago, with the first chemical systems capable of self-reproduction and also capable of evolution. What did these primitive chemical systems look like? What materials did they use? Is it reasonable to expect similar systems to emerge on other celestial bodies? Recent joint efforts in astronomy, astrophysics, planetology, geochemistry, geology, biology and chemistry bring some preliminary answers to questions that have puzzled humans since Antiquity. Research is presently pursuing three main avenues: (i) the search for fossil traces of life in Archeaen sediments and for other imprints of life, (ii) the reconstitution, in a test tube, of artificial life capable of self-reproduction and evolution, (iii) the search for another example of natural life beyond the Earth.

1 The Search for Traces of Primitive Life and Other Imprints

1.1 Microfossils

The oldest fossils of micro-organisms have been found in greenstone belts in the Pilbara, Australia and in Barberton, South Africa (Schopf, 1993; Westall et al., 2001; see also Westall's Chap. 6, Part I). They are 3.4–3.5 billion years old. From a theoretical point of view, the earliest forms of life would most likely have had a chemolithoautotrophic metabolism, using inorganic materials as a source of both C and energy. Evidence for the presence of chemolithotrophs in the hydrothermal deposits from Barberton and the Pilbara comes from the N and C isotopic data (Beaumont and Robert, 1999). Considering the high temperatures on the early Earth, it is probable that the earliest micro-organisms were thermophilic. Structures resembling oxygenic photosynthetic micro-organisms such as cyanobacteria have been described and carbon isotope data have been interpreted as evidence for their presence (Schopf, 1993). However, these interpretations are strongly disputed (Brasier et al., 2002). Nevertheless, the fact that the micro-organisms inhabiting the Early Archaean environments of Barberton and the Pilbara formed mats in the photic zone suggests that some of them may already have developed anoxygenic photosynthesis (Westall et al., 2001).

1.2 Oldest Sedimentary Rocks

The oldest sedimentary rocks have been collected in southwest Greenland. The Isua sediments are 3.8 Ga old (Schidlowski, 1988), whereas the Akilia sediments formed 3.85 Ga ago (Mozsis et al., 1996). They testify to the presence of permanent liquid water on the surface of the Earth and to the presence of carbon dioxide in the atmosphere. They contain complex organic molecules. The isotopic signatures of the organic carbon found in Greenland metasediments provide indirect evidence that life may be 3.85 billion years old. Taking the age of the Earth as 4.5 billion years, this means that life must have began quite early in Earth's history. This isotopic evidence stems from the fact that the carbon atom has two stable isotopes, ^{12}C and ^{13}C. The $^{12}C/^{13}C$ ratio in abiotic mineral compounds is 89. In biological material, the process of photosynthesis gives a preference to the lighter carbon isotope and raises the ratio to about 92. Consequently, the carbon residues of previously living matter may be identified by this enrichment in ^{12}C. A compilation has been made of the carbon isotopic composition of over 1600 samples of fossil kerogen (a complex organic macromolecule produced from the debris of biological matter) and compared with that from carbonates in the same sedimentary rocks (Schidlowski, 1988). The enrichment in ^{12}C carbon encompasses specimens right across the geological time scale, including the Isua and Akilia rocks. This offset is now taken to be one of the most powerful indications that life on Earth was active nearly 3.9 billion years ago (Schidlowski, 1988; Mojzsis et al., 1996; Rosing, 1999). Although there are serious reservations concerning the first two studies (contamination by more recent fossilised endoliths, ^{12}C enrichment produced by thermal decomposition of siderite and metamorphic processes), the biogenicity of the ^{12}C enrichment measured by Rosing has been verified by stepped combustion (van Zuilen et al., 2002).

1.3 One-handedness of Life

One-handedness, also called homochirality (from the Greek *kheiros*, the hand; see Cronin and Reisse's Chap. 3, Part II), of the proteins synthesised via the genetic code is another characteristic of all living systems. Each carbon atom of the amino-acid skeleton occupies the centre of a tetrahedron. Except for the amino acid glycine, the four substituents of the central carbon atom are different. The carbon atom is therefore asymmetrical and it is not superimposable to its mirror image. Each amino acid exists in two mirror images called enantiomers (from the Greek *enantios*, opposed). All protein amino acids have the same handedness, they are homochiral. They are left handed and hence known as L-amino acids. Their right-handed mirror images are known as D-amino acids. L-amino acids engaged in a protein chain generate right-handed single-strand α-helices and asymmetrical multistrand β-sheet structures. Nucleotides, the building blocks of nucleic acids DNA and RNA, are also homochiral. Their nomenclature is more complex because each nucleotide possesses four asymmetrical carbon atoms. In this case, the geometry of a selected carbon atom was chosen. Following this mere convention, nucleotides are thus right handed. Right-handed nucleotides generally generate right-handed helical nucleic acids.

Is homochirality a mandatory requisite for life? Pasteur was probably the first person to realize that biological asymmetry could best distinguish between inanimate matter and life. Life that would simultaneously use both right- and left-handed forms of the same biological molecules appears, in the first place, very unlikely for geometrical reasons. Enzyme β-pleated sheets cannot form when both L- and D-amino acids are present in the same chain. Since the catalytic activity of an enzyme is intimately dependent upon the geometry of the chain, the absence of β-pleated sheets would impede, or at least considerably reduce, the activity spectrum of the enzymes. The use of one-handed biomonomers also sharpens the sequence information of the biopolymers. For a polymer made of n units, the number of sequence combinations will be divided by 2^n when the system uses only homochiral (one-handed) monomers. Taking into account the fact that enzyme chains are generally made up of hundreds of monomers, and that nucleic acids contain several million nucleotides, the tremendous gain in simplicity offered by the use of monomers restricted to one handedness is self-evident. Finally, if the biopolymers to be replicated were to contain L- and D-units located at specific sites, the replication process would have to be able to position the right monomers at the right place, but also to select the right enantiomer from among two species that differ only by the geometry of the asymmetrical carbon atoms. For example, the bacteria *Bacillus brevis* are able to synthesise *gramicidine A* constructed on a strict alternation of left- and right-handed amino acids. However, the biosynthesis of this peptide involves a set of complex and sophisticated enzymes that are homochiral. Life on Earth uses homochiral left-handed amino acids and right-handed sugars. A mirror-image life, using right-handed amino acids and left-handed sugars, is perfectly conceivable and might develop on another planet. Thus, homochirality can be a crucial signature for life.

Unfortunately, the direct clues that could help chemists to identify the molecules that participated in the actual emergence of life on Earth about 4 billion years ago have been erased by geological processes, such as plate tectonics and the permanent presence of running water (erosion), as well as by the unshielded solar ultraviolet radiation and by the oxygen produced by life.

2 Reconstructing Life in a Test Tube

Life emerged in water and the first self-reproducing molecules and their precursors were probably organic molecules built up with carbon atom skeletons. Liquid water participated in the early mechanisms of life as a solvent but also by exchanging hydrogen bonds with the organic molecules. Organic molecules were formed from gaseous molecules containing carbon atoms (methane, carbon dioxide, carbon monoxide), nitrogen atoms (nitrogen, ammonia), sulfur atoms (hydrogen sulfide, sulfur dioxide) and energy (electric discharges, cosmic and UV radiation, heat).

2.1 Primitive Earth Atmosphere

In 1924, the Russian biochemist Aleksander Oparin suggested that the small reduced organic molecules needed for primitive life were formed in a primitive atmosphere dominated by methane. The idea was tested in the laboratory by Stanley Miller (Miller, 1953). He exposed a mixture of methane, ammonia, hydrogen and water to electric discharges, to mimic the effects of lightning. Among the compounds formed, he identified four of the twenty naturally occurring amino acids, the building blocks of proteins. Since this historic experiment, seventeen natural amino acids have been obtained via the intermediate formation of simple precursors, such as hydrogen cyanide and formaldehyde. It has been shown that spark-discharge synthesis of amino acids occurs efficiently when a reducing-gas mixture containing significant amounts of hydrogen is used. However, the true composition of the primitive Earth atmosphere remains unknown. Today, geochemists favour a nonreducing atmosphere dominated by carbon dioxide. Under such conditions, the production of amino acids appears to be very limited. Strongly reducing environments capable of reducing carbon dioxide were therefore necessary.

2.2 Organic Synthesis

Deep-sea hydrothermal systems may represent likely reducing environments for the synthesis of prebiotic organic molecules (Baross and Hoffman, 1985; Holm and Andersson, 1995; 1998). The ejected gases contain carbon dioxide, carbon monoxide, sulfur dioxide, nitrogen, and hydrogen sulfide. For instance, high concentrations of hydrogen (more than 40% of the total gas) and methane have been detected in the fluids of the Rainbow ultramafic hydrothermal system of the Mid-Atlantic Ridge (Charlou et al., 1998). The production of hydrogen, a highly reducing agent favouring prebiotic syntheses, is associated with the hydration of olivine into serpentine and magnetite, a reaction known as "serpentinisation". Indeed, hydrocarbons containing 16 to 29 carbon atoms have been detected in these hydrothermal fluids (Holm and Charlou, 2001). Amino acids have been obtained, although in low yields, under conditions simulating these hydrothermal vents (Yanagawa and Kobayashi, 1992). According to Wächtershäuser (1998), primordial organic molecules formed near the hydrothermal systems; the energy source required to reduce the carbon dioxide might have been provided by the oxidative formation of pyrite from iron sulfide and hydrogen sulfide. Pyrite has positive surface charges and bonds the products of carbon dioxide reduction, giving rise to a two-dimensional reaction system, a "surface metabolism" (Wächtershäuser, 1998). Laboratory work has provided support for this new promising hypothesis (Huber and Wächtershäuser, 1997). Iron sulfide, hydrogen sulfide and carbon dioxide react under anaerobic conditions to produce hydrogen and a series of thiols, including methanethiol. Methanethiol and acetic acid have also been obtained from carbon monoxide, hydrogen sulfide, iron and nickel sulfides and catalytic amounts of selenium. Under specific conditions, thioesesters are formed that might have been the metabolic driving force of a thioester world, according to de Duve (de Duve, 1998). Hydrothermal vents are often disqualified (Miller and Bada, 1988) as efficient reactors for the synthesis of bio-organic molecules, because of the high temperature. However, the products that are synthesized in hot vents are rapidly quenched in the surrounding cold water, which may preserve those organics formed. When fluid containing glycine and copper chloride repeatedly circulated through the hot (225 °C) and cold (0 °C) chambers in a laboratory reactor that simulated a hydrothermal system, glycine peptides up to hexaglycine were obtained (Imai et al., 1999).

2.3 Delivery of Organics by Comets and Meteorites

Comets and meteorites may have delivered important amounts of organic molecules to the primitive Earth (see Despois and Cottin's Chap. 8, Part I). Nucleic acid bases, purines and pyrimidines, have been found in the Murchison meteorite (Stoks and Schwartz, 1982). One sugar (dihydroxyacetone), sugar-alcohols (erythritol, ribitol) and sugar-acids (ribonic acid, gluconic acid) have been detected in the Murchison meteorite but no ribose, the sugar moiety of ribonucleotides, themselves building blocks in RNAs (Cooper et al., 2001). Eight proteinaceous amino acids have been identified in one such meteorite, among more than 70 amino acids found therein. Cronin and Pizzarello found an excess of about 9% of L-enantiomers for some nonprotein amino acids detected in the Murchison meteorite (Cronin and Pizzarello, 1997). The presence of L-enantiomeric excesses in these meteorites points to an extraterrestrial process of asymmetric synthesis of amino acids, asymmetry that is preserved inside the meteorite. These excesses may help to understand the emergence of the biological asymmetry or one-handedness (see Cronin and Reisse's Chap. 3, Part II). The excess of the one-handed amino acids, as found in the meteorites, may result from the processing of the organic mantles of the interstellar grains from which the meteorite was originally formed. This processing could occur, for example, by the effects of circularly polarized synchrotron radiation from a neutron star, a remnant of a supernova. Strong infrared circular polarization, resulting from dust scattering in reflection nebulae in the Orion OMC-1 star-formation region, has also been observed (Bailey et al., 1998). Circular polarization at shorter wavelengths might have been important in inducing this chiral asymmetry in interstellar organic molecules that could be subsequently delivered to the early Earth (Bailey, 2001). Dust collection in the Greenland and Antarctic ice sheets and its analysis by Maurette (Maurette, 1998) show that the Earth captures interplanetary dust as micrometeorites at a rate of about 50–100tons per day. About 99% of this mass is carried by micrometeorites in the 50–500 micrometre size range. This value is much higher than the most reliable estimate of the normal meteorite flux, which is about 0.03tons per day. A high percentage of micrometeorites in the 50- to 100-micrometre size range has been observed to be unmelted, indicating that a large fraction traversed the terrestrial atmosphere without drastic heating. In this size range, the carbonaceous micrometeorites represent 80% of the samples and contain 2% of carbon, on average. This flux of incoming micrometeorites might have brought to the Earth about 10^{23} g of carbon over the period corresponding to the late heavy bombardment, when planetesimals, asteroids and comets impacted the Earth until −3.8 Ga, as inferred from the lunar craters. For comparison, this delivery represents more carbon than that engaged in the present surficial biomass, i.e. about 10^{18} g. These grains also contain a high proportion of metallic sulfides, oxides, and clay minerals that belong to various classes of catalysts. In addition to the carbonaceous matter, micrometeorites might also have delivered a rich variety of catalysts. They may have functioned as tiny chemical reactors when reaching oceanic water.

Micrometeorites, with a composition close to that of the most primitive meteorites, represent very primitive objects of the Solar System. The formation of some of their organic molecules could have occurred early in the protosolar nebulae. These molecules would have escaped the accretion phase within comets in the outer Solar System. Micrometeorites are probably witnesses of a chemical continuum, via the cometary grains, from the interstellar medium, where they form, to terrestrial oceans. Comets are the richest planetary objects in organic compounds known so far. Ground-based observations have detected hydrogen cyanide and formaldehyde in the coma of comets. In 1986, onboard analyses performed by the two Russian missions Vega 1 and 2, as well as observations obtained by the European mission Giotto and the two Japanese missions Suisei and Sakigake, demonstrated that the Halley comet shows substantial amounts of organic material. On average, dust particles ejected from the Comet Halley nucleus contain 14% of organic carbon by mass. About 30% of cometary grains are dominated by light elements C, H, O, and N, and 35% are close in composition to the carbon-rich meteorites. The presence of organic molecules, such as purines, pyrimidines, and formaldehyde polymers has also been inferred from the fragments analyzed by the Giotto Picca and Vega Puma mass spectrometers. However, there was no direct identification of the complex organic molecules probably present in the cosmic dust grains and in the cometary nucleus. Many chemical species of interest for exobiology have been detected in Hyakutake comet in 1996, including ammonia, methane, acetylene (ethyne), acetonitrile (methyl cyanide) and hydrogen isocyanide. In addition, the study of the Hale-Bopp comet in 1997 led to the detection of methane, acetylene, formic acid, acetonitrile, hydrogen isocyanide, isocyanic acid, cyanoacetylene, formamide and thioformaldehyde. It is possible, therefore, that cometary grains might have been an important source of organic molecules delivered to the primitive Earth. Comets orbit on unstable trajectories and sometimes collide with planets. The collision of Comet Shoemaker-Levy 9 with Jupiter in July 1994 gave a recent example of such events. Such collisions were probably more frequent 4 billion years ago, the comets orbiting around the Sun being more numerous. By impacting the Earth, comets delivered probably a substantial fraction of the terrestrial water (about 35% according to the estimation of Tobias Owen based on the relative contents in hydrogen and deuterium and cometary grains could have provided large amounts of organic molecules. The chemistry that is active at the surface of a comet is still poorly understood. Launched in March 2004, the cometary mission Rosetta will analyse the nucleus of a comet. The spacecraft will first study the environment of the comet during a flyby over several months and a probe will land to analyse the surface and the subsurface ice sampled by drilling.

2.4 Simulation Experiments

Ultraviolet irradiation of dust grains in the interstellar medium may result in the formation of complex organic molecules. The interstellar dust particles are assumed to be composed of silicate grains surrounded by ices of different molecules, including carbon-containing molecules. Ices of H_2O, CO_2, CO, CH_3OH, NH_3 were deposited at 12 K under a pressure of 10^{-7} mbar and irradiated with electromagnetic radiation representative of the interstellar medium. The solid layer that developed on the solid surface was analysed by enantioselective gas chromatography and mass spectrometry GC-MS. After the analytical steps of extraction, hydrolysis, and derivatization, 16 amino acids were identified in the simulated ice mantle of interstellar dust particles (Muñoz Caro et al., 2002). These amino-acid identifications confirmed the preliminary amino-acid formation obtained by Mayo Greenberg (Briggs et al., 1992). The chiral amino acids were identified as being totally racemic. Parallel experiments performed with ^{13}C-containing substitutes definitely exclude contamination by biological amino acids. The results strongly suggest that amino acids are readily formed in interstellar space.

Before reaching the Earth, organic molecules are exposed to UV radiation, both in interstellar space and in the Solar System. Amino acids have been exposed in Earth orbit to study their survival in space. The UV flux ($\lambda < 206$ nm) in the diffuse interstellar medium is about 10^8 photons $cm^{-2}\ s^{-1}$. In Earth orbit, the corresponding solar flux is in the range of 10^{16} photons $cm^{-2}\ s^{-1}$. This means that an irradiation over one week corresponds to 275 000 years in the interstellar medium. Amino acids, like those detected in the Murchison meteorite, have been exposed as free molecules and also as adsorbed molecules on clay minerals to space conditions in Earth orbit onboard the unmanned Russian satellites FOTON 8 and 11. Aspartic acid and glutamic acid not adsorbed on clay minerals were partially destroyed during exposure to solar UV. However, decomposition was prevented when the amino acids were embedded in clays (Barbier et al, 1998; Barbier et al., 2002). Amino acids and peptides have also been subjected to solar radiation outside the MIR station for 97 days as solid films as well as embedded into ground mineral material (montmorillonite clay, basalt and Allende meteorite). Different thicknesses of meteorite powder films were used to estimate the shielding threshold. After three months of exposure, about 50% of the amino acids were destroyed in the absence of mineral shielding. Peptides exhibited a noticeable sensitivity to space vacuum and sublimation effects were detected. Decarboxylation was found to be the main effect of photolysis. No polymerization occurred and no racemization (the conversion of L- or

D-amino acids into a racemic mixture) was observed. Among the different minerals used as 5-μm films, meteoritic powder offered the best protection whereas montmorillonite was least efficient. Significant protection from solar radiation was observed when the thickness of the meteorite mineral was 5 μm or greater.

2.5 Recreating the Chemistry of Primitive Life

By analogy with contemporary living systems, it is tempting to consider that primitive life emerged as a cellular object, requiring boundary molecules able to isolate the system from the aqueous environment (membrane). Also needed would be catalytic molecules to provide the basic chemistry of the cell (enzymes) and information-retaining molecules that allow the storage and the transfer of the information needed for replication (RNA). Starting from the small organic molecules possibly present in the primitive oceans, chemists have tried to reproduce the spontaneous formation of the three families of basic biological molecules in a test tube. They succeeded in reconstructing precursors of membranes (Deamer, 1998; see also Ourisson and Nakatani's Chap. 1, Part II) and of protein enzymes (Brack, 1993). Unfortunately, they have not yet been able to reconstruct precursors of RNA. Synthesis of sugars from formaldehyde leads to a highly complex mixture of compounds, among which ribose represents only a very tiny fraction. Prebiotic synthesis of short strands of RNA faces two main difficulties: synthesis of the first strand of RNA (especially ribose) and replication of small strands fed with racemic nucleotides. RNA analogs and surrogates have been studied. Considering the easy formation of hexose-2,4,6-triphosphates from glycolaldehyde phosphate, $CHO\text{-}CH_2O\text{-}PO_3H_2$, in a process analogous to the formose reaction, Eschenmoser and coworkers (Eschenmoser et al., 1999) synthesized polynucleotides containing ribose in the form of pyranose (pyranosyl-RNA or p-RNA) in place of the usual "natural" ribose, the furanose form found in RNA. p-RNAs form Watson–Crick-paired double helices that are more stable than RNA. Replication experiments have had marked success in terms of sequence copying but have failed to demonstrate template-catalysis turnover numbers greater than one. p-RNA seems to be an excellent choice as a genetic system if it can be demonstrated that the prebiotic synthesis of pyranosyl nucleotides is much easier than the synthesis of standard isomers.

Chemists are now tempted to consider that primitive replicating systems must have used simpler information-retaining molecules than biological nucleic acids or their analogs. They are looking for simple self-sustaining chemical systems capable of self-replication, mutation, and selection. It has been shown that simple molecules, unrelated to nucleotides, can actually provide exponentially replicating autocatalytic models. Beautiful examples of autocatalytic micelle growth have been published (Bachmann et al., 1992). However, these autocatalytic systems do not really store hereditary information and cannot therefore evolve by natural selection. These autocatalytic primitive living entities must have been simple and robust enough to survive the heavy bombardment and the nonshielded solar UV. Some investigations are also focusing on autocatalytic systems adsorbed on mineral surfaces (Luther et al. 1998; Orgel, 1998).

3 Search for Extraterrestrial Life

3.1 The Diversity of Bacterial Life as a Reference for Extraterrestrial Life

Life on Earth is based upon the chemistry of carbon in water. The temperature limits compatible with the existence of life are thus imposed by the intrinsic properties of the chemical bonds involved in this type of chemistry at different temperatures. Presently, the maximum temperature limit known for terrestrial organisms is around 113 °C for the deep-sea microbe *Pyrolobus fumarii* (Stetter, 1998). Important factors preventing life at temperatures well above 110 °C are the thermal instability of supramolecular systems involved in biological systems and the temperature dependence of membrane permeability. In some theoretical scenarios, life appeared at very high temperatures. This means that today's hyperthermophiles might be viewed as relics of the last common universal ancestor of all living beings. However, this hot origin of life hypothesis has been seriously disputed, based on the fact that RNA is very unstable at high temperatures (see Gribaldo and Forterre's Chap. 7, Part II). The most attractive hypothesis might be that life appeared in a moderately thermophilic environment, hot enough to boost catalytic reactions, but cold enough to avoid the problem of macromolecule thermal degradation (see Lopez-Garcia's Chap. 9, Part II). This hypothesis seems to be supported by gene-sequencing studies that suggest that hyperthermophiles appeared later than thermophiles (Galtier et al., 1999; Brochier and Philippe, 2002). Life is extremely diverse in the ocean at temperatures of 2 °C. Living organisms, especially micro-organisms, are also present in the frozen soils of Arctic and alpine environments. Antarctica has a wide range of extreme habitats, including microbial ecosystems developing in dry valley rocks. The lower limit for bacterial growth published in the literature is −12 °C, the temperature at which intracellular ice is formed.

Salt-loving organisms, known as halophiles, have been well studied. They tolerate a wide range of salt concentrations (1–20% NaCl) and some procaryotes, the extreme halophiles, have managed to thrive in hypersaline biotopes (salines, salted lakes). They are, in fact, so dependent on such high salt concentrations that they cannot grow (and may even die) at concentrations below 10% NaCl.

The chemistry of life on Earth is optimised for neutral pH. Again, some micro-organisms have been able to adapt to extreme pH conditions, from pH 0 (extremely acidic) to pH 12.5 (extremely alkaline), albeit maintaining their intracellular pH between pH 4 and 9. As with temperature, the intracellular machinery cannot escape the influence of pressure. However, there are organisms in the deepest parts of the ocean (pressure 1100 bar). The extreme pressure limit for life on Earth is unknown – environments of above 1100 bar have not been explored.

For a long time, it was believed that deep subterranean environments were sterile. An important recent development has been the recognition that bacteria actually thrive in the terrestrial crust. Subterranean micro-organisms are usually detected in subsurface oil-fields or in the course of drilling experiments. For example, recent research has demonstrated that microbes are present much deeper in marine sediments than was previously thought possible, extending to at least 750 m below the sea floor, and probably much deeper (Parkes et al., 2000). To depths of at least 432 m, microbes have been identified as altering volcanic glass. These data provide a preliminary, and probably conservative, estimate of the biomass in this important new ecosystem. It could amount to about 10% of the surface biosphere.

These discoveries have radically changed the perception of marine sediments and indicate the presence of a largely unexplored deep bacterial biosphere that may even rival the Earth's surface biosphere in size and diversity. Clearly, this discovery also has important implications for the probability of life existing in other planets of the Solar System and elsewhere.

3.2 The Search for Life in the Solar System

3.2.1 Life on Mars and the SNC Meteorites

Mapping of Mars by Mariner 9, Viking 1 and 2 and Mars Global Surveyor revealed channels resembling dry river beds. Odyssey's gamma-ray spectrometer instrument detected hydrogen, which indicated the presence of water ice in the upper metre of soil, in a large region surrounding the planet's south pole, where ice is expected to be stable (Boynton et al., 2002). The amount of hydrogen detected indicates 20 to 50 per cent ice by mass in the lower layer beneath the topmost surface. The ice-rich layer is about 60 cm beneath the surface at latitude 60°S, and approaches 30 cm of the surface at latitude 75°S. The inventory of the total amount of water that may have existed at the surface of Mars is difficult to estimate and varies from a total water depth of some metres to several hundred metres. It is generally considered that liquid water has been restricted to the very early stages of Martian history (see Bribing's Chap. 9, Part I). Mars possessed therefore an atmosphere capable of decelerating carbonaceous micrometeorites and chemical evolution may have been possible on the planet. The Viking 1 and 2 lander missions were designed to address the question of extant (rather than extinct) life on Mars. Three experiments were selected to detect metabolic activity such as photosynthesis, nutrition and respiration of potential microbial soil communities. Unfortunately, the results were ambiguous since, although "positive" results were obtained, no organic carbon was found in the Martian soil by gas chromatography-mass spectrometry. It was concluded that the most plausible explanation for these results was the presence at the Martian surface of highly reactive oxidants, like hydrogen peroxide, which would have been photochemically produced in the atmosphere. The Viking lander could not sample soils below 6 cm and therefore the depth of this apparently organic-free and oxidizing layer is unknown. Direct photolytic processes can also be responsible for the degradation of organics at the Martian surface. Although the Viking missions were disappointing for exobiology, in the long run the programme has proved to be extremely beneficial for investigating the possibility of life on Mars.

Prior to Viking, it had been apparent that there was a small group of meteorites, all of igneous (volcanic) origin, called the SNC meteorites (after their type specimens Shergotty, Nakhla and Chassigny) that had comparatively young crystallisation ages, equal to or less than 1.3 billion years. One of these meteorites, designated EETA 79001, was found in Antarctica in 1979. It had gas inclusions trapped within a glassy component. Both compositionally and isotopically, this gas matched, in all respects, the make up of the Martian atmosphere, as measured by the Viking mass spectrometer. The data provide a very strong argument that at least that particular SNC meteorite came from Mars, representing the product of a high-energy impact that ejected material into space. There are now thirty one SNC meteorites known in total (Jet Propulsion Laboratory, 2003). Two of the SNC meteorites, EETA79001 and ALH84001, supply new and highly interesting information (Brack and Pillinger, 1998). A subsample of EETA79001, excavated from deep within the meteorite, has been subjected to stepped-combustion. The CO_2 release from 200 to 400°C suggested the presence of organic molecules. The carbon is enriched in ^{12}C, and the carbon isotope difference between the organic matter and the carbonates in Martian meteorites is greater than that observed on Earth. This could be indicative of biosynthesis, although some as yet unknown abiotic processes could perhaps explain this enrichment. McKay and co-workers (McKay et al., 1996) have reported the presence of other features in ALH84001 that may represent a signature of relic biogenic activity on Mars, but this biological interpretation is strongly questioned (Bradley et al., 1997; Westall et al., 1998).

Because Mars had a "warm" and wet past climate, it should have sedimentary rocks deposited by running and/or still water on its surface as well as layers of regolith generated by impacts. Such consolidated sedimentary hard rocks ought therefore to be found amongst the Martian meteorites. However, no such sedimentary material has been found in any SNC meteorite. It is possible that they did survive the effects of the escape acceleration from the Martian surface but did not survive terrestrial atmospheric entry because of decrepitation of the cementing mineral. The 'STONE' experiment, flown by ESA, was designed to study precisely such physical and chemical modifications to sedimentary rocks during atmospheric entry from space. A piece of basalt (representing a standard meteoritic material), a piece of dolomite (sedimentary rock) and an artificial Martian regolith material (80% crushed basalt and 20% gypsum) were embedded into the ablative heat shield of Foton 12, which was launched on September 9 and landed on September 24, 1999. Such an experiment had never been performed before and the samples, after their return, were analysed for their chemistry, mineralogy and isotopic compositions by a European consortium. Atmospheric infall modifications are made visible by reference to the untreated samples. The results suggest that some Martian sediments could, in part, survive terrestrial atmospheric entry from space (Brack et al., 2002).

Even if convincing evidence for ancient life in ALH84001 has not been established, the two SNC meteorites (EETA79001 and ALH84001) do show the presence of organic molecules. This suggests that the ingredients required for the emergence of a primitive life were present on the surface of Mars. Therefore, it is tempting to consider that micro-organisms may have developed on Mars and lived at the surface until liquid water disappeared. Since Mars probably had no plate tectonics and since liquid water seems to have disappeared from the surface of Mars very early, the Martian subsurface perhaps keeps a frozen record of very early forms of life. NASA has planned a very intensive exploration of Mars and European and Japanese missions are also taking place. Exobiology interests are included, especially in the analysis of samples from sites where the environmental conditions may have been favourable for the preservation of evidence of possible prebiotic or biotic processes. The ESA Manned Spaceflight and Microgravity Directorate has convened a Exobiology Science Team to design a multiuser, integrated suite of instruments for the search for evidence of life on Mars. The priority has been given to the in situ organic and isotopic analysis of samples obtained by subsurface drilling (Brack et al., 1999). A first exobiology lander, called Beagle 2, was launched in June 2003, as part of the ESA Mars Express mission. It successfully separated from the orbiter but all altempts to contact it were unsuccessful, thus compelling the scientists to consider Beagle 2 as lost.

3.2.2 Europa

Europa appears as one of the most enigmatic of the Galilean satellites. With a mean density of about 3.0g cm^{-3}, the Jovian satellite should be dominated by rocks. Ground-based spectroscopy, combined with gravity data, suggests that the satellite has an icy crust kilometres thick and a rocky interior. The Voyager images showed very few impact craters on Europa's surface, indicating recent, and probably continuing, resurfacing by cryovolcanic and tectonic processes. Images of Europa's surface taken by the Galileo spacecraft show surface features – iceberg-like rafted blocks, cracks, ridges and dark bands – which are consistent with the presence of liquid water beneath the icy crust (Carr et al., 1998). Data from Galileo's near-infrared mapping spectrometer show hydrated salts that could be evaporites. The most convincing argument for the presence of an ocean of liquid water comes from Galileo's magnetometer. The instrument detected an induced magnetic field within Jupiter's strong magnetic field. The strength and response of the induced field require a near-surface, global conducting layer, most likely a layer of salty water (Kivelson et al., 2000; Zimmer et al., 2000).

If liquid water is present within Europa, it is quite possible that it includes organic matter derived from thermal vents. Terrestrial-like prebiotic organic chemistry and primitive life may therefore have developed in Europa's ocean.

If Europa maintained tidal and/or hydrothermal activity in its subsurface until now, it is possible that bacterial activity is still present. Thus, the possibility of an extraterrestrial life present in a subsurface ocean of Europa must be seriously considered. The most likely sites for extant life would be at hydrothermal vents below the most recently resurfaced area. To study this directly would require making a borehole through the ice in order to deploy a robotic submersible. On the other hand, biological processes in and around hydrothermal vents could produce biomarkers that would be erupted as traces in cryovolcanic eruptions and thereby be available at the surface for in situ analysis or sample return. Mineral nutrients delivered through cryovolcanic eruption would make the same locations the best candidates for autotrophic life.

3.2.3 Titan

Titan's atmosphere (see Raulin et al. Chap. 2, Part II) was revealed mainly by the Voyager 1 mission in 1980, which yielded the bulk composition (90% molecular nitrogen and about 1–8% methane). Also, a great number of trace constituents were observed in the form of hydrocarbons, nitriles and oxygen-containing compounds, mostly CO and CO_2. Titan is the only other object in our Solar System to bear any resemblance to our own planet in terms of atmospheric pressure (1.5 bar) and carbon/nitrogen chemistry. It represents, therefore, a natural laboratory to study the formation of complex organic molecules on a planetary scale and over geological times. The ISO satellite has detected tiny amounts of water vapour in the higher atmosphere, but Titan's surface temperature (94 K) is much too low to allow the presence of liquid water. Although liquid water is totally absent, the satellite provides a unique milieu to study, in situ, the products of the fundamental physical and chemical interactions driving a planetary organic chemistry. Titan also serves as a reference laboratory to study, by default, the role of liquid water in exobiology.

The NASA/ESA Cassini-Huygens spacecraft launched in October 1997 arrived in the vicinity of Saturn in 2004 and will perform several flybys of Titan, making spectroscopic, imaging, radar and other measurements. A descent probe managed by European scientists will penetrate the atmosphere and will systematically study the organic chemistry in Titan's geofluid. During 150 min, in situ measurements will provide detailed analysis of the organics present in the air, in the aerosols and at the surface (Lebreton, 1997; Raulin, 1998).

3.3 The Search for Life Beyond the Solar System

3.3.1 What Are We Looking for?

Apart from abundant hydrogen and helium, 114 interstellar and circumstellar gaseous molecules are currently identified in the interstellar medium (see Despois and Cotin's Chap. 8, Part I). It is commonly agreed that the catalog of interstellar molecules represents only a fraction of the total spectrum of molecules present in space, the spectral detection being biased by the fact that only those molecules possessing a strong electric dipole can be observed. Among these molecules, 83 contain carbon whereas only 7 contain silicon. Silicon has been proposed as an alternative to carbon as the basis of life. However, silicon chemistry is apparently less inventive and does not seem to be able to generate any life as sophisticated as the terrestrial carbon-based one. Can these molecules survive the violent accretion phase generating a stellar system? The origin and distribution of the molecules from the interstellar medium to the planets, asteroids and comets of the Solar System is presently at the centre of a debate based on isotope ratios. Some molecules could have survived in cold regions of the outer Solar System, whereas others would have been totally reprocessed during accretion. Whatever the case, the interstellar medium tells us that organic chemistry is universal. What about liquid water? New planets have been discovered beyond the Solar System. On October 6, 1995, the discovery of an extrasolar planet was announced. The planet orbits an eight-billion-year-old star called 51 Pegasus, forty-two light years away within the Milky Way (Mayor and Queloz, 1995). The suspected planet takes just four days to orbit 51 Pegasus. It has a surface temperature around 1000°C and a mass about 0.5 the mass of Jupiter. One year later, seven other extrasolar planets were identified. Among them, 47 Ursa Major has a planet with a surface temperature estimated to be around that of Mars –90 to –20°C) and the 70 Virginis planet has a surface temperature estimated of about 70–160°C. The latter is the first known extrasolar planet whose temperature might allow the presence of liquid water. As of July 2004, 123 exoplanets have been observed (Schneider, 2003).

3.3.2 How to Detect Extraterrestrial Life?

Isotope enrichment of ^{12}C and one-handedness are the two most remarkable signatures of biological terrestrial molecules. With the development of exploration planetary missions, almost any body within the Solar System can be subjected to in situ organic, isotope and mineral analysis. Rocks, in particular, may represent telltale biomarkers of biological activity on both macroscopic (stromatolites) and microscopic (microfossils of bacteria) levels.

Identifying extrasolar life will be more difficult. Extrasolar life will not be accessible to space missions in the foreseeable future. The formidable challenge to detect distant life must therefore be tackled by astronomers and radioastronomers. The detection of water and ozone (an easily detectable telltale signature of oxygen) in the atmosphere will be a strong indication but not an absolute proof (see Selsis et al.'s Chap. 10, Part I). Other anomalies in the atmospheres of telluric exoplanets (rocky Earth-like planets), such as the presence of methane, could be the signature of an extrasolar life. European astrophysicists are proposing the construction of an infrared interferometer made of five space telescopes to study the atmospheres of exoplanets. The mission, called Darwin-IRSI, is presently under study at ESA.

Finally, the detection of an unambiguous electromagnetic signal (via the SETI program) would obviously be an exciting event (SETI, http://www.seti-inst.edu/).

3.4 Panspermia, Interplanetary Transfer of Life

Although it will be difficult to prove that life can be transported through the Solar System, the chances for the different steps of the process to occur can be estimated. These include (1) the escape process, i.e. the removal to space of biological material that has survived being lifted from the surface to high altitudes; (2) the interim state in space, i.e. the survival of the biological material over timescales comparable with the interplanetary passage (3) the entry process, i.e. the nondestructive deposition of the biological material on another planet. The identification of some meteorites as being of lunar origin and some others as most probably being of Martian origin, shows that the escape from a planet of material ranging from small particles up to boulder-size, after the planet has suffered a high-energy impact, is evidently a feasible process. In that context, it is also interesting to note that bacterial spores can survive huge accelerations and shockwaves produced by a simulated meteorite impacts (Horneck et al., 2001).

In order to study the survival of resistant microbial forms in the upper atmosphere and free space, microbial samples have been exposed in situ aboard balloons, rockets and spacecraft (Horneck, 1999). The ESA Microgravity Programme has continued to support experiments of that type. A priori, the environment in space seems to be very hostile to life. This is due to the high vacuum, an intense radiation of galactic and solar origin, and extreme temperatures. In the endeavour to disentangle the network of potential interactions of the parameters of space, methods have been applied to separate each parameter and to investigate its impact on biological integrity, applied singly or in controlled combinations.

Space vacuum has been considered to be one of the factors that may prevent interplanetary transfer of life because of its extreme dehydrating effects. However, experiments in space have demonstrated that certain micro-organisms survive exposure in space vacuum for extended periods of time, provided they are shielded against the intense deleterious solar UV radiation. Most of the results concern the spores (a dormant form) of the bacterium *Bacillus subtilis.* Up to 70% of the bacterial spores survive short-term (e.g. 10 days) exposure to space vacuum. The chances of survival in space are increased if the spores are embedded in chemical protectants such as sugars, or salt crystals, or if they are exposed in thick layers. For example, 30% of *subtilis* spores survived nearly 6 years in space when embedded in salt crystals and 80% survived in the presence of glucose. Specimens of *Streptococcus mitis* survived two-and-a-half years inside a television camera landed on the Moon by the unmanned lunar lander Surveyor 3 and recovered by the astronauts of Apollo 12.

Solar UV radiation has been found to be the most deleterious factor of space, as tested with dried preparations of viruses, bacterial and fungal spores, with

DNA being the critical UV target for lethality. However, about 5% of a species of the extreme halophile *Haloarcula,* (a salt-loving bacteria), survived a two-week space environment during a Foton spaceflight. The radiation field of the Solar System is governed by components of galactic and solar origin. It is composed of electrons, alpha particles and cosmic heavy ions, the latter being the most ionising and therefore the most deleterious components. The heavy particles of cosmic radiation are conjectured as setting the ultimate limit on the survival of spores in space because they penetrate even heavy shielding. The maximum time for a spore to escape a hit by a heavy particle has been estimated to be 10^5–10^6 years. Considering the mean sizes and numbers of meteorites ejected from Mars and falling on Earth, radiative transfer models for galactic cosmic rays, and laboratory responses of *B. subtilis* spores and *Deinococcus radiodurans* cells to accelerated heavy ions, Mileikowsky et al. (2000) have calculated that transfer of viable microbes from Mars to Earth via impact ejecta is a possible event.

During the major part of a hypothetical journey through deep space, microorganisms are confronted with the 3-K cold emptiness of space. Laboratory experiments under simulated interstellar medium conditions point to a remarkably less damaging effect of UV radiation at these low temperatures. Treating *Bacillus subtilis* spores with three simulated factors simultaneously (UV, vacuum and a low temperature of 3 K), produces an unexpectedly high survival rate, even at very high UV fluxes. From these data, it has been estimated that, in the most general environment in space, spores may survive for hundreds of years (Weber and Greenberg, 1985).

ESA has initiated the development of the exposure facility EXPOSE, to be attached to the Columbus module of the International Space Station. This will allow the extensive study of the survival of bacteria in space.

4 Conclusion

Recent data collected from different disciplines have given important coherence to the search for life in the Universe that is summarised in Table 1.

More precisely,

1. Studies of the origin of life and the exploration of Mars tell us that the conditions, which probably allowed the emergence of life on Earth, were also present on Mars, 4 billion years ago.
2. There are perhaps hydrothermal systems producing organic prebiotic molecules in the ocean of Europa.
3. Terrestrial bacteria are capable of surviving extreme conditions (temperature, pH, pressure, high salinity, and deep sediments). They resist space conditions and are therefore capable of interplanetary transfer, thus favouring their dispersion in the Universe.

If life started on Earth with the self-organisation of a relatively small number of molecules, its emergence must have been fast and the chances for the appearance

Table 1. Life in the Universe

Terrestrial life as reference			Extraterrestrial life		
Relics/imprints of life	Origins of life	Limits of life	Life in the Solar System	Extrasolar life	Panspermia
• fossil bacteria • sediments • homochirality	• primitive oceans • ingredients • life in test tube	• temperature • high salinity • pH • pressure • deep sediments	• life on Mars and the SNC meteorites • life on Europa • Titan	• exoplanets • signatures of extrasolar life	• interplanetary transfer of life

of life on any appropriate celestial bodies are real. On the contrary, if the process required thousands of different molecules, the event risks being unique and restricted to the Earth. In addition to its immense cultural and societal appeal, the discovery of a second genesis of life, artificial or natural, would be strong support for the simplicity of the processes involved in the emergence of life. The exploration of Mars and Europa and the recent discovery of exoplanets, open real hope. Many scientists are convinced that bacterial life is not restricted to the Earth. Now, they have to prove it!

References

Bachmann P.A., Luisi P.L., and Lang J. (1992). Autocatalytic self-replicating micelles as models for prebiotic structures. *Nature*, **357**, 57–59.

Bailey J., Chrysostomou A., Hough J.H., Gledhill T.M., McCall A., Clark S., Ménard F, and Tamura M (1998). Circular polarization in star formation regions: Implications for biomolecular homochirality. *Science,* **281**, 672–674.

Bailey J. (2001). Astronomical sources of circularly polarized light and the origin of homochirality. *Origins Life Evol. Biosphere*, **31**, 167–183.

Barbier B., Chabin A., Chaput D., and Brack A. (1998). Photochemical processing of amino acids in Earth orbit. *Planet. Space Sci.*, **46**, 391–398.

Baross J.A. and Hoffman S.E. (1985). Submarine hydrothermal vents and associated gradient environment as sites for the origin and evolution of life. *Origins Life Evol. Biosphere*, **15**, 327–345.

Beaumont V. and Robert F. (1999). Nitrogen isotope ratios of kerogens in Precambrian cherts: a record of the evolution of atmospheric chemistry? *Precambrian Res.*, **96**, 63–82.

Boillot F., Chabin A., Buré C, Venet M., Belsky A., Bertrand-Urbaniak M., Delmas A., Brack A., and Barbier B. (2002). The Perseus exobiology mission on MIR: behaviour of amino acids and peptides in Earth orbit. *Origins Life Evol. Biosphere,* **32**, 359–385.

Boynton W.V. et al. (2002). Distribution of hydrogen in the near surface of mars: evidence for subsurface ice deposits. *Science,* **297**, 81–85.

Brack A. (1993). From amino acids to prebiotic active peptides: a chemical reconstitution. *Pure & Appl. Chem.*, **65**, 1143–1151.

Brack A., Baglioni P., Borruat G., Brandstätter F., Demets R., Edwards H.G.M., Genge M., Kurat G., Miller M.F., Newton E.M., Pillinger C.T., Roten C.-A., and Wäsch E. (2002). Do meteoroids of sedimentary origin survive terrestrial atmospheric entry? The ESA artificial meteorite experiment *STONE. Planet. Space Science,* **50**, 763–772.

Brack A. and Pillinger C. (1998). Life on Mars: chemical arguments and clues from Martian meteorites. *Extremophiles*, **2**, 313–319.

Brack A., Fitton F., and Raulin F. (1999). Exobiology in the Solar System & the search for life on Mars. ESA Special Publication SP 1231, ESA Pub. Noordwijk, The Netherlands.

Bradley J.P., Harvey R.P., and McSween H.Y. (1997). No "nanofossils" in martian meteorite, *Nature*, **390**, 454.

Brasier M.D., Green O.R., Jephcoat A.P., Kleppe A.K., van Kranendonk M., Lindsay J.F., Steele A., and Grassineau N. (2002). Questioning the evidence for Earth's oldest fossils. *Nature*, **416**, 76–81.

Briggs R., Ertem G., Ferris J.P., Greenberg J.M., McCain P.J., Mendoza-Gomez C.X., and Schutte W. (1992). Comet Halley as an aggregate of interstellar dust and further evidence for the photochemical formation of organics in the interstellar medium. *Origins Life Evol. Biosphere*, **22**, 287–307.

Brochier C. and Philippe H. (2002). A non-hyperthermophilic ancestor for bacteria. *Nature*, **417**, 244.

Carr M.H. et al. (1998). Evidence for a subsurface ocean on Europa. *Nature*, **391**, 363–365.

Charlou J.L., Fouquet Y., Bougault H., Donval J.P., Etoubleau J., Jean-Baptiste Ph., Dapoigny A., Appriou P., and Rona P.S. (1998). Intense CH_4 plumes generated by serpentinization of ultramafic rocks at the intersection of $15°20'$ N fracture zone and the Mid-Atlantic Ridge. *Geochim. Cosmochim. Acta*, **62**, 2323–2333.

Cooper G., Kimmich N., Belisle W., Sarinana J., Brabham K., and Garrel L. (2001). Carbonaceous meteorites as a source of sugar-related organic compounds for the early Earth, *Nature*, **414**, 879–883.

Cronin J.R. and Pizzarello S. (1997). Enantiomeric excesses in meteoritic amino acids. *Science*, **275**, 951–955.

De Duve C. (1998). Possible starts for primitive life. Clues from present-day biology: the thioester world, in *The Molecular Origins of Life: Assembling Pieces of the Puzzle*, (ed.) A Brack, p. 219–236, Cambridge University Press, Cambridge.

Deamer D.W. (1998). Membrane compartments in prebiotic evolution, in *The Molecular Origins of Life: Assembling Pieces of the Puzzle*, (ed.) A Brack, p. 189–205, Cambridge University Press, Cambridge.

Eschenmoser A. (1999). Chemical etiology of nucleic acid structure. *Science,* **284**, 2118–2124.

Galtier N., Tourasse N., and Gouy N. (1999). A non hyperthermophilic common ancestor to extant life forms. *Science*, **283**, 220–221.

Holm N. and Charlou J.L. (2000). Indications of abiotic formation of hydrocarbons in the Rainbow ultramafic hydrothermal system, Mid Atlantic Ridge, *Earth Planet. Sci. Lett.*, **191**, 1–8.

Holm N.G. and Andersson E.M. (1995). Abiotic synthesis of organic compounds under the conditions of submarine hydrothermal systems: a perspective. *Planet. Space Sci.*, **43**, 153–159.

Holm N.G. and Andersson E.M. (1998). Organic molecules on the early Earth: hydrothermal systems. in *The Molecular Origins of Life: Assembling Pieces of the Puzzle*, (ed.) A Brack, p. 86–99, Cambridge University Press, Cambridge.

Horneck G. (1999). European activities in exobiology in Earth orbit: Results and perspectives. *Adv. Space Res.,* **23**, 381–386.

Horneck G., Stöffler D., Eschweiler U., and Hornemann U. (2001). Bacterial spores survive simulated meteorite impact. *Icarus*, **149**, 285–193.

Huber C. and Wächtershäuser G. (1997). Activated acetic acid by carbon fixation on (Fe, Ni)S under primordial conditions. *Science*, **276**, 245–247.

Imai E.-I., Honda H., Hatori K., Brack A., and Matsuno K. (1999) Elongation of oligopeptides in a simulated submarine hydrothermal system. *Science*, **283**, 831–833.

Jet Propulsion Laboratory/NASA: Mars Meteorites (2003). http://www.jpl.nasa.gov/snc/

Kivelson M.G., Khurana K.K., Russel C.T., Volwerk M., Walker R.J., and Zimmer C. (2000). Galileo magnetometer measurements: a stronger case for a subsurface ocean at Europa. *Science*, **289**, 1340–1343.

Lebreton J.P. (ed.) (1997). Huygens: Science Payload and Mission. *ESA-SP 1177*, ESA Pub. Noordwijk, The Netherlands.

Luther A., Brandsch R., and von Kiedrowski G. (1998). Surface-promoted replication and exponential amplification of DNA analogues. *Nature*, **396**, 245–248.

Maurette M. (1998). Carbonaceous micrometeorites and the origin of life. *Origins Life Evol. Biosphere*, **28**, 385–412.

Mayor M.D. and Queloz D. (1995). A Jupiter-mass companion to a solar-type star. *Nature*, **378**, 355–359.

McKay D.S., Gibson E.K., Thomas-Keprta K.L., Vali H., Romanek C.S., Clemett S.J., Chellier X.D.F., Maechling C.R., and Zare R.N. (1996). Search for past life on Mars: possible relic biogenic activity in martian meteorite ALH 84001. *Science*, **273**, 924–930.

Mileikowsky C., Cucinotta F., Wilson J.W., Gladman B., Horneck G., Lindegren, Melosh J., Rickman H., Valtonen M., and Zheng J.Q. (2000). Natural transfer of viable microbes in space, Part 1: From Mars to Earth and Earth to Mars. *Icarus*, **145**, 391–427.

Miller S.L. (1953). The production of amino acids under possible primitive Earth conditions. *Science*, **117**, 528–529.

Miller S.L. and Bada J.L. (1988). Submarine hot springs and the origin of life. *Nature*, **334**, 609–611.

Mojzsis S.J., Arrhenius G., McKeegan K.D., Harrison T.M., Nutman A.P., and Friend C.R.L. (1996). Evidence for life on Earth before 3800 million years ago. *Nature*, **384**, 55–59.

Munoz Caro G.M., Meierhenrich U.J., Schutte W.A., Barbier B., Arcones Segovia A., Rosenbauer H., Thiemann W.H.-P., Brack A., and Greenberg J.M. (2002). Amino acids from ultraviolet irradiation of interstellar ice analogues. *Nature*, **416**, 403–405.

Orgel L.E. (1998). Polymerization on the rocks: Theoretical introduction. *Origins Life Evol. Biosphere*, **28**, 227–234.

Parkes R.J., Cragg B.A., and Wellsbury P. (2000). Recent studies on bacterial populations and processes in subseafloor sediments: A review. *Hydrogeol. J.*, **8**, 11–28.

Raulin F. (1998). Titan, in *The Molecular Origins of Life: Assembling Pieces of the Puzzle*, (ed.) A Brack, p. 365–385, Cambridge University Press, Cambridge.

Rosing M.T. (1999). ^{13}C depleted carbon microparticles in $>$ 3700 Ma seafloor sedimentary rocks from West Greenland. *Science*, **283**, 674–676.

Schidlowski M. (1988). A 3800-million-year isotopic record of life from carbon in sedimentary rocks. *Nature*, **333**, 313–318.

Schneider J. (2003). http://www.obspm.fr/encycl/f-encycl.html.

Schopf J.W. (1993). Microfossils of the early Archean Apex Chert: new evidence of the antiquity of life. *Science*, **260**, 640–646

SETI http://www.seti-inst.edu/

Stetter K.O. (1998). Hyperthermophiles and their possible role as ancestors of modern life. in *The Molecular Origins of Life: Assembling Pieces of the Puzzle*, (ed.) A Brack, p. 315–335, Cambridge Univ. Press, Cambridge.

Stoks P.G. and Schwartz A.W. (1982). Basic nitrogen-heterocyclic compounds in the Murchison meteorite. *Geochim. Cosmochim. Acta*, **46**, 309–315.

Van Zuilen M., Lepland A., and Arrhenius G. (2002). Abiogenic and biogenic graphite in the Isua Supracrustal belt. In *Isua Multidisciplinary Study Workshop*, Berlin, Jan 2002, Danish Geol. Surv. Copenhagen, p. 73–74.

Wächtershäuser G. (1998). Origin of life in an iron-sulfur world, in *The Molecular Origins of Life: Assembling Pieces of the Puzzle*, (ed.) A Brack, pp. 206–218, Cambridge University Press, Cambridge.

Weber P. and Greenberg J.M. (1985). Can spores survive in interstellar space? *Nature*, **316**, 403–407.

Westall F., Gobbi P., Mazzotti G., Gerneke D., Stark R., Dobrek T., and Heckl W. (1998). Combined SEM (secondary electrons, backscatter, cathodoluminescence) and atomic force microscope investigation of the carbonate globules in Martian meteorite ALH84001: Preliminary results, *SPIE, Instruments, Methods and Missions for Astrobiology*, **3114**, 225–233.

Westall F., De Wit M.J., Dann J., Van Der Gaast S., De Ronde C., and Gerneke D. (2001). Early Archaean fossil bacteria and biofilms in hydrothermally-influenced, shallow water sediments, Barberton greenstone belt, South Africa. *Precambrian Res.*, **106**, 93–116.

Yanagawa H. and Kobayashi K. (1992). An experimental approach to chemical evolution in submarine hydrothermal systems. *Origins Life Evol. Biosphere*, **22**, 147–159.

Zimmer C., Khurana K.K., and Kivelson M.G. (2000). Subsurface oceans on Europa and Callisto: Constraints from Galileo magnetometer observations. *Icarus*, **147**, 329–347.

Part I

The Early Earth and Other Cosmic Habitats for Life

First Part Introduction

Researches related to the origins of life can develop in two different directions: they can look towards the past and try to determine how life appeared on Earth, or they can explore present-day Universe in the hunt for life. Both ways proceed by using the same strategy: research of "favourable" conditions for life to appear. These conditions are the necessary prerequisites for prebiotic chemistry to occur. Of course, until now, nobody has been able to create life but, nevertheless, chemists determined the necessary conditions to generate amino acids, purine and pyrimidine bases and many other "small" molecules. They also know how to obtain (small) polypeptides and even how to create membranes.

Very often, many plausible prebiotic scenarios have been tested and, as several experimental procedures are able to lead to the same results, there is no unique solution. These different procedures permit minimal requirements to be established in terms of pressure, temperature, pH and reactants, which are necessary to obtain some of the main components (molecules as well as supramolecular systems) of living species. For instance, liquid water is a parameter that appears crucial for prebiotic chemistry and life development. Such a condition allows both thermodynamic and compositional constraints to be fixed. Of course, the basic elements for life (C, O, H, N, etc.) must also be available.

This is why planetologists develop intensive researches in order to determine if liquid water is or was present on planets and satellites of the Solar System (Mars, Europa ...). Moreover, the research of extrasolar planets becomes a very important and attractive challenge and the recent discovery of some extrasolar planetary atmospheres appears fundamental in the research of sites favourable for life development. For the Earth's and the other solar planets, the atmosphere origins (as well as the origin of water and C-, O-, N-, H-bearing molecules) are assessed to be endogenous or/and exogenous. Consequently, it becomes critical to determine the respective importance of endogenous and exogenous processes. In the exogenous case, matter is assumed to come from other parts of the Solar System and to have been transferred to Earth surface by comets, asteroids, meteorites and micrometeorites. These are the reasons why astrobiology is deeply interested not only in various aspects of cosmology and nucleosynthesis but also in the accretion process of the Solar System and more generally of planetary systems around stars. How elements and even molecules, initially present in the protosolar nebula, were redistributed and transformed during accretion is also a problem of major interest.

On the other hand, geological and biological investigations on Earth showed that autotrophic life can develop without light energy, as proved by life development associated to submarine black smokers. These hydrothermal vents are genetically linked to the internal Earth heat production and then to plate tectonics. This points to the possibility for life to develop not only because of solar energy but also due to planetary internal energy. On Earth, plate tectonics lead to a bimodal repartition on altitudes (ocean vs. continent). The possible discovery of a bimodal altitude distribution on Mars provides a strong argument in favour of an early plate-tectonic activity and also appears as a fundamental element for research of early life on this planet. Similarly, the possibility to have submarine hydrothermal vents on Europa, one of the satellites of Jupiter, is an exciting possibility. Evidence for the discovery of life in the Solar System or better on exoplanets, would demonstrate that the emergence of life does not result from the succession of highly improbable and complex events that occurred only on Earth. On the contrary, it would show that more simple processes are involved in the appearance of life, processes able to develop and to be reproduced in several places in the universe.

Looking to Earth past could appear easier as we are sure that at least one time in Earth history all conditions for life development were realised. Unfortunately, this is the only certainty we have and as we look farther into the past more the record of traces of life is tiny and ambiguous. Indeed, very old rocks (older than 3.8 Ga) are not very abundant and even if they are preserved they have been transformed by tectonic and metamorphic processes. The debate around Isua (3.85 Ga) traces of life is a perfect illustration of the difficulty to obtain a clear and unambiguous interpretation of geological records. Strategies are now being developed in order to clearly and unambiguously identify very old fossil traces of life.

In addition, some parameters such as atmosphere and also ocean compositions are drastically different today from what they have been on early Earth. Similarly, due to Earth cooling, terrestrial dynamics changed in the course of time, generating rocks such as komatiites that are no longer produced. Moreover, it is not clearly established when plate tectonics began to operate on Earth. Some scenarios for life development imply alternation of dry and aqueous periods, which means that emerged continents existed but we do not possess obvious evidences of emerged continents before 3.5 Ga. This does not indicate that they did not exist but it adds a greater level of uncertainty in the life-emergence scenarios.

This is why, in order to discuss the conditions prevailing on Earth at that time, researches develop towards the oldest preserved rocks. Unfortunately, rocks older than 4.0 Ga are still poorly known, so that modelling and simulation are still more or less the only ways to discuss and reconstruct the first 500 Ma of our planet.

1 The Formation of Solar-type Stars: Boundary Conditions for the Origin of Life?

Thierry Montmerle

1.1 The Ancestors of Solar-like Stars

A trivial, but necessary condition for life, as we know it on Earth, to appear, is that planets must form, and thus that "solar-type stars" must form. The lifetime of such stars must be counted in billions of years, since life took about one billion years to arise on Earth; their surface temperatures must be neither too hot, nor too cold, so that some "habitable zone" may exist inside their putative planetary systems. Prior to this, there must be enough circumstellar material to build, say, a Jupiter-like planet, or even a more massive one, as suggested by the hundred or more "exoplanets" known to date (see, e.g., J. Schneider's web page)[1]. Measured with respect to the mass of the Sun ($M_\odot$), the preplanetary circumstellar mass must therefore lie in the range $10^{-3}\,M_\odot$ (one Jupiter mass) to $10^{-2}\,M_\odot$.

Such definitions may look rather vague, but at least they will certainly include the "right" stars. For instance, there is a simple upper limit to the relevant stellar mass: beyond a surface temperature $T = 10\,000\,\mathrm{K}$, the stellar radiation heats and completely ionizes the environment, which does not seem very favourable for planets to form. The corresponding boundary spectral type is called "A", which means, on the main sequence, a star more massive than about 2–3 $M_\odot$. For low-mass stars (with $M < 1\,M_\odot$), the situation is less clear: although surface temperatures are lower than for the Sun (for which $T = 5800\,\mathrm{K}$), one can conceivably imagine a kind of "Earth" located near a small, "M"-type star (with mass $\sim 0.5\,M_\odot$, $T \sim 3000\,\mathrm{K}$), and glowing over its inhabitants with a nice dark reddish colour.

In what follows, we will thus concentrate on stars with masses $M < 2\,M_\odot$. This range can be observed easily, because these stars are the most numerous in the Galaxy (contrary to the bright, massive stars that illuminate beautiful nebulae like the Orion nebula M42). This advantage is partially offset by the fact that low-mass stars are rather dim, but thousands exist between 200 and 1500 light-years from the Sun, i.e., between regions like Ophiuchus or Chamaeleon, and regions like Orion. A lot of information can therefore be gathered on these stars.

The ancestors of the Sun and solar-like stars are the so-called "T Tauri stars" (named after their prototype star; see the review by Bertout 1989). Discovered through their variable luminosity during the 1940s by the American astronomer

[1] http://www.obspm.fr/encycl/encycl.html

A. Joy in the constellation Taurus, these stars were found to have unusual optical spectra, which display strong emission lines (indicating a tenuous gas), reminiscent of hot stars, while having all the other characteristics of cold stars. Because they were seen in the vicinity of dark clouds (sometimes illuminating them to form "reflection nebulae"), these faint stars had to be distant, thus be intrinsically much more luminous than field stars with the same spectral type. This important property could be explained in 1965 by the Japanese astrophysicist C. Hayashi, who calculated (analytically!) the collapse phases of stars taken as spherical gas masses of about $1\,M_{\odot}$. He showed that such stars had to be "pre-main sequence" stars, i.e., extract all their energy from gravitational contraction (and not from thermonuclear reactions), and be very luminous (> 10 solar luminosities) during a few million years because of their large radius (several solar radii), before eventually igniting in their cores thermonuclear reactions turning hydrogen into helium (Hayashi 1966).

Then nothing else happened on the astronomical front for the next ten years. It was thought that indeed the Sun went through a T Tauri phase during its early evolution, but the spectrum and the true nature of these stars remained a mystery. However, during the past twenty years, observations (spurring theoretical calculations and numerical simulations) underwent spectacular progress, with the advent of new technologies allowing a "multiwavelength" approach. This progress can be summarized in one major breakthrough: the discovery and characterization of circumstellar material around young stars. This point is of fundamental importance to our discussion: the existence of this material (which, as we shall see, takes many forms) is a prerequisite for planet formation, hence for the possible appearance of life.

1.2 "Young Stellar Objects" and Solar-type Protostars

1.2.1 Molecular Clouds and Molecular Outflows

A major step forward became possible at the beginning of the 1970s, when radioastronomy turned to millimetre waves, which are characteristic of the cold, molecular universe. The first observations were made in the mid-1970s, using the $1 \to 0$ transition of CO at 2.6 mm as a tracer of the H_2 molecule, which by itself does not emit millimetre waves. These observations immediately showed that the main gaseous component of "dark clouds" was molecular hydrogen, whereas the rest of the cold gas in our galaxy is mainly made of neutral hydrogen (HI), which emits the famous 21-cm line. The CO line indicated that these "molecular" clouds (as they came to be known), had a temperature of about 30 to 50 K, random velocities of a few km/s, and masses ranging from 10^4 to $10^6\,M_{\odot}$.

Almost immediately (in 1976; see the review by Lada, 1985), high-velocity components (50 km/s or more) were detected, and interpreted as "molecular outflows", most often "bipolar", i.e., having a "blueshifted" component (velocity oriented towards the observer), and the other "redshifted" (velocity oriented away from the observer), in a more or less symmetrical fashion. The centre of symmetry of such outflows would presumably be the exciting source, but most of the time nothing could be found in the optical. Sometimes, however, bright optical "jets" could be seen emanating from T Tauri-like stars (Fig. 1.1), aligned with the axis of the outflows (for a recent review, see Reipurth and Bally 2001). It was not immediately realized that this phenomenon was in fact the key to our understanding of the formation of solar-type stars.

Fig. 1.1. Example of a jet associated with low-mass young stars (HH111, © NASA & B. Reipurth). This spectacular image has been built from an optical image (*right*), and an infrared image (*left*), both taken with cameras aboard the Hubble Space Telescope (HST). From tip to tip, this jet is about 100 times as long as the diameter of the Solar System, and has been shown to be at the centre of a much larger complex of outflows, extending over $\sim$ 25 light-years. By comparison, the star closest to the Sun, the triple system α Centauri, is 4.3 light-years away. The source of the jet (which, in the IR, is symmetrical with respect to the optical), is deeply embedded in a molecular cloud and cannot be seen optically. The jet extremity on the right clearly displays a bow shock where the outflow material encounters the dense surrounding medium

1.2.2 T Tauri Stars

In an entirely different wavelength domain, the X-ray range, another important discovery would be made. Following the launch, in 1979, of the "Einstein Observatory" satellite, the first to carry special mirrors allowing to obtain X-ray images (with a spatial resolution of about $1'$), the observation of several nearby star-forming regions located about 150pc, or 500light-years away from the Sun (Chamaeleon, ρ Ophiuchi, Taurus-Auriga, etc.), revealed the existence of a new population of T Tauri stars, at least as numerous as the previously known population. These stars did not exhibit the spectral signature of "classical" T Tauri stars (like emission lines), in spite of being as luminous, but rather displayed the fairly common spectra of cool, giant stars. These stars were coined "weak-line" T Tauri stars.

In yet another wavelength domain, the infrared range, progress in detector technology allowed a decisive step to be taken. The observation of continuum spectra in the near- to mid-IR (about 1 to 10micrometres) showed that classical T Tauri stars displayed excess emission, which was not seen in weak-line T Tauri stars. This excess was interpreted around 1985 as resulting from the presence of a circumstellar dust disk, shedding in part its material on the central star (the "accretion disk"); then such a disk would be absent in weak-line T Tauri stars (see the review by Bertout 1989). This interpretation has been recently confirmed by direct imagery using the Hubble Space Telescope (Fig. 1.2).

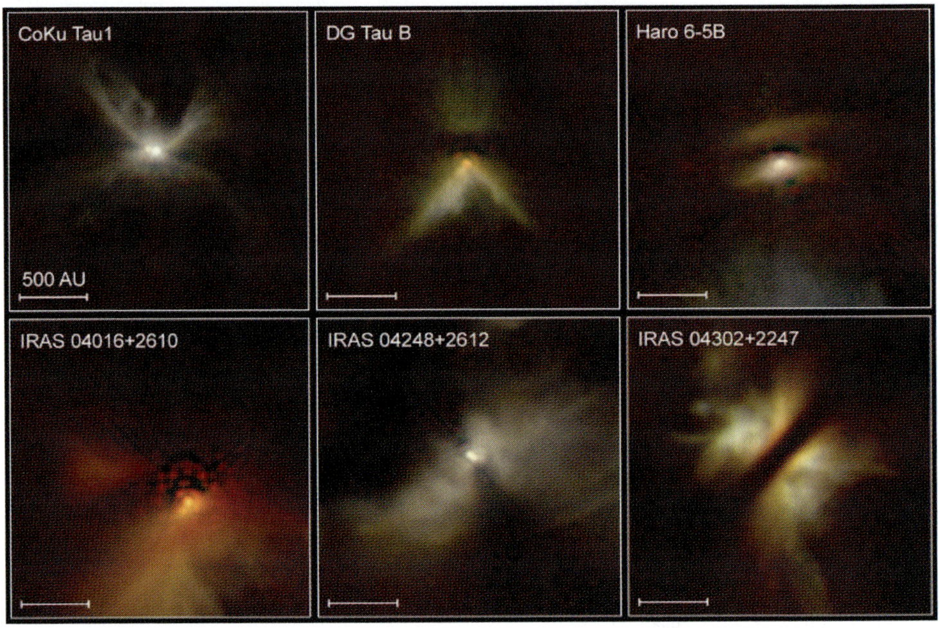

Fig. 1.2. Embedded young stars, observed in the IR by HST (©NASA 2000, and D. Padgett, W. Brandner, and K. Stapelfeldt). One clearly sees not only circumstellar disks oriented edge-on, but also outflows from a central source. Such disks are typically 10 times larger than the size of the present-day Solar System. The various outflow shapes probably result from their interaction with a circumstellar envelope

1.2.3 Protostars

The infrared domain also allowed the nature of the embedded sources of molecular outflows to be understood. These turned out to be either stars comparable to the classical T Tauri stars, surrounded by disks, or, in the case of an even more intense IR excess, by circumstellar envelopes: this was, in fact, the discovery of the first protostars. The best observations gave disk sizes on the order of ten times the size of the present-day Solar System (which has a radius of 50 times the Sun–Earth distance, i.e., the orbit of Pluto), and several hundred times that size in the case of the envelope.

However, the masses of disks and envelopes in the innermost regions cannot be computed from the IR spectrum. The reason is that in this case the IR radiation is "optically thick", which means that it comes from a "horizon", like the Sun, where the light comes from the surface but not from the inside. However, modelling of the IR spectrum shows that the optically thick part is located at less than 10 AU from the star in the case of disks, which are the hottest (roughly 100 K to 1000 K). In contrast, the millimetre domain is "optically thin", in other words the material is transparent and one can "see through" the whole material. In this case, the measured radiation flux is simply proportional to the mass. This domain corresponds to the outer, cooler regions of the disks. One can detect in particular the continuum emission of dust grains, which has been measured by telescopes such as the IRAM 30-m single-dish telescope, near Granada, Spain, using highly sensitive focal-plane bolometers. In this way, the "transparent" mass of disks (~ 0.001 to $0.1\,M_\odot$) and of envelopes (~ 1 to $10\,M_\odot$) has been measured (see for instance André and Montmerle 1994; Fig. 1.3). Then a new class of outflow source was discovered in 1993, which were invisible even in the IR, but detected

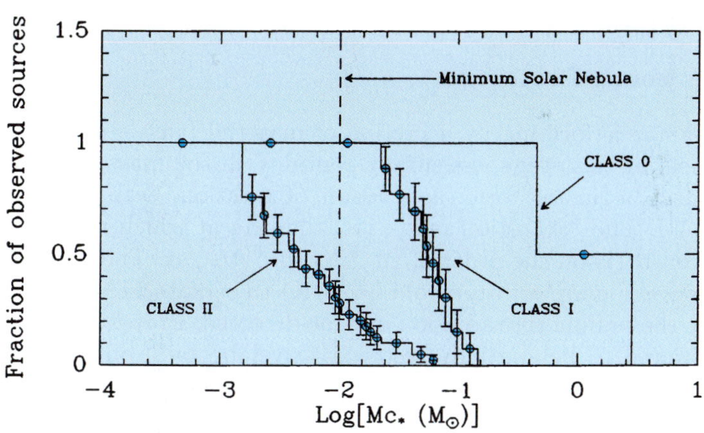

Fig. 1.3. Mass of circumstellar material around young stars in the ρ Oph cloud, estimated from mm observations (André and Montmerle 1994). This mass is that of envelopes in the case of Class 0 and Class I protostars, and of circumstellar disks in the case of Class II T Tauri stars (Class III T Tauri stars do not have disks). *Horizontal axis*: circumstellar masses in units of solar masses; *vertical axis*: number of sources relative to the total sample (about 100 sources here). Note that (i) the mass diminishes as the class number increases, which is explained by the fact that the circumstellar matter disappears to build the central star, and (ii) that only 30% of the Class II sources are more massive than the primitive solar nebula (which is basically the mass of Jupiter)

and resolved in the millimetre continuum (André, Ward-Thompson, and Barsony 1993), an indication that most of the mass was contained in the envelope, and very little in or around a central object. This stage would thus correspond to an even earlier phase than that found in other sources, since gravitational collapse would just be starting.

It was about time to attempt a "unified" interpretation of these various observations.

1.3 Stellar Evolution During the First Million Years

It is convenient to start with an IR/mm classification for young stars, because it is the only one available for embedded stars (which are optically invisible), while remaining usable when the stars are not embedded, i.e., are visible in the optical. Four classes have been defined, with a smooth transition from one class to the next. Roughly speaking, this classification is understood in terms of the early temporal evolution of a solar-type star, during which it eventually acquires its final properties.

The first stages of this evolution can be summarized as follows (Shu, Adams and Lizano 1987, André and Montmerle 1994, André, Ward-Thompson, and Barsony 2000; Fig. 1.4).

1.3.1 Protostars

"Class 0": Young Protostars

The central star is forming by accretion of material (gas + dust) from the envelope, which, at this stage, essentially contains all the mass of the future star. A central disk begins to form, as a result of rotation, with a radius of about 200 to 300 AU. The "accretion rate", i.e., the rate at which matter flows in towards the centre, is on the order of 10^{-5} to $10^{-6}\,M_{\odot}$/yr. The dust grains in the collapsing envelope are still too cold ($\sim$ 30 K) to radiate in the IR, even in the far-IR: only the millimetre radiation may be detected. Protostars in nearby dark clouds are found to be spatially resolved (envelope radii on the order of 1000 to 10 000 AU). Such sources are powering intense and well-collimated outflows (Bontemps et al. 1996). This phase lasts about 10^4 yr.

"Class I": Evolved Protostars

The envelope, now tenuous, has been almost completely used up to make the central star. Heated by the star, the envelope radiates strongly in the IR range, but less so in the mm range since its mass has decreased significantly in favour of the growing star. Molecular outflows are still present, but less focused and less collimated (Bontemps et al. 1996). Accretion takes place mainly via a circumstellar disk several hundred AU in radius. This stage lasts about 10^5 yr.

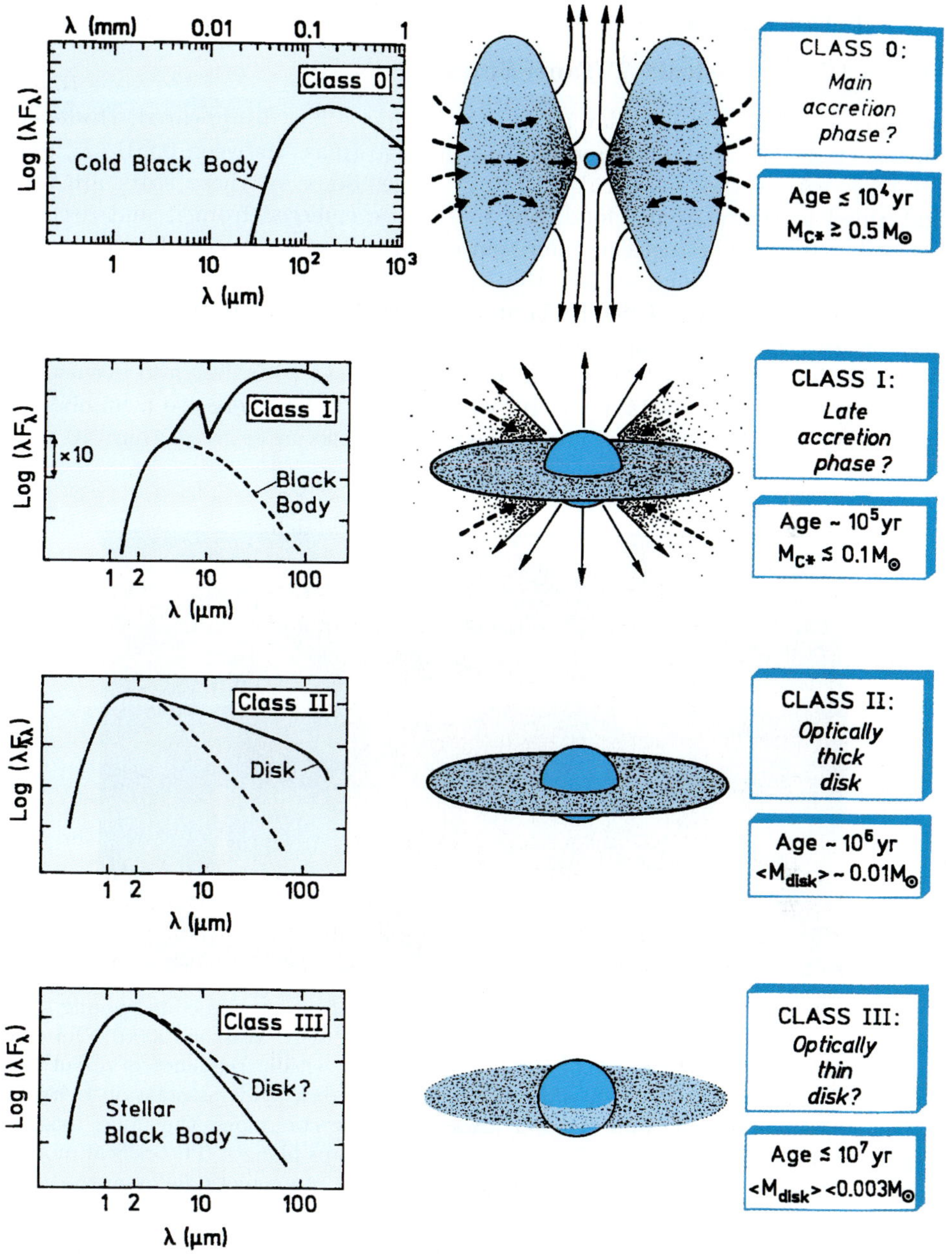

Fig. 1.4. The evolution of solar-type stars during the first few million years (André 1993). Note the increasing temperature as the star builds up (peaks of spectra on the *left*), from $\sim 30\,\mathrm{K}$ (protostellar envelope), to $\sim 3000\,\mathrm{K}$ (stellar photosphere)

1.3.2 T Tauri Stars

"Class II": "Classical" T Tauri Stars

At the end of the protostellar phase, the envelope has disappeared. However, an accretion disk remains, more or less massive (mass between $0.001\,M_\odot$ and $0.1\,M_\odot$): this is the "classical T Tauri" stage. Some of these stars are still embedded in their parent cloud, but most have emerged from it and become optically visible. The locus of their temperature–luminosity coordinates in the Herzsprung–Russell diagram (see the Appendix) as a function of mass is called the "birthline" (Stahler 1988, Palla and Stahler 1999). The interaction between the star and the disk is responsible for specific properties in their optical spectra (emission lines, etc.). The molecular outflows decrease with time and eventually disappear. This stage lasts a few 10^6 yr to a few 10^7 yr: there are a number of "old" classical T Tauri stars having kept their disks after more than 10 million years.

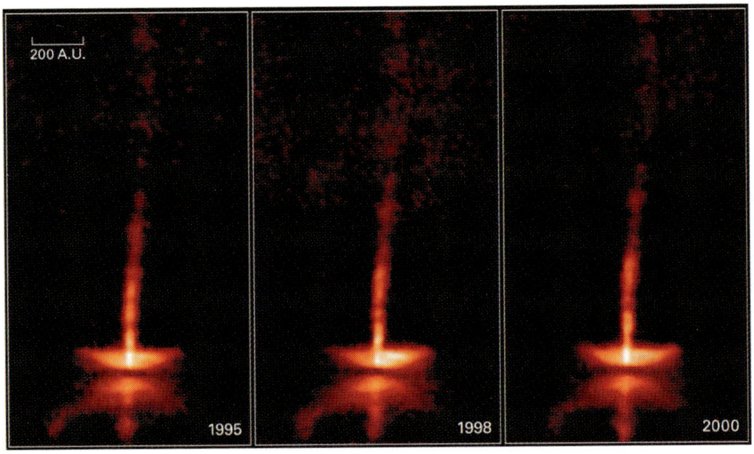

Fig. 1.5. Time variability of the HH30 *jet* emanating from an embedded young star, observed with HST between 1995 and 2000 (© NASA & A. Watson, 2000). One can clearly see knotty structures in the jet, leaving the star with a frequency of about one per month. One can also see an edge-on circumstellar disk. This association between accretion (onto the star) and ejection (away from the star along the polar axis) is typical of the formation stages of young stars and perhaps planets. Theoretical models explain the accretion–ejection coupling by the presence of magnetic fields in the close vicinity of the star and the inner disk

"Class III": "Weak-line" T Tauri Stars

The circumstellar disk has now disappeared. The reason for this is still poorly understood, but observations show that this disappearance takes place sooner or later, even for stars of the same mass. This transition must take place relatively rapidly (in less than 10^5 yr), because only a handful of "transition" T Tauri stars are known, i.e., surrounded by a ring rather than a disk (see, e.g., André et al. 1990). Molecular outflows have completely disappeared. This phase also lasts 10^6 yr to 10^7 yr or more: there are also "young" T Tauri stars aged less than 1 million yr without disks.

Therefore, a strong link is seen to emerge, from Class 0 to Class II, between disks and outflows, which is also suggested by direct imaging with HST (Hubble Space Telescope) (Fig. 1.5). Without going into details, current models rely on an "MHD" (magnetohydrodynamics) description, i.e., they invoke a strong role of magnetic fields to transfer along the polar axis a fraction of the material accreted via the disk. The main agent is the centrifugal force resulting from rotation. An estimated 10% of the accreted material is ejected in this way (more on this in Sect. 1.6 below). This is the important paradox of star formation: *to gain mass, mass must be lost.*

1.4 Importance for the Forming Solar System

1.4.1 Circumstellar Disks

The preceding section showed that the phenomenon most directly relevant for the formation of the Solar System is the existence of circumstellar disks, followed by their mysterious disappearance on timescales of 1 to 10 million years.

As we have seen, the existence of these disks is firmly established; thanks to millimetre interferometry, disk images may even be obtained, (including around double T Tauri stars as shown on Fig. 1.6). It can also be shown that the disks are in Keplerian rotation (i.e., each point rotates like a planet around the Sun), as the models had assumed previously (Simon, Dutrey, and Guilloteau 2000). In addition, single-dish observations of the mm dust continuum have shown that nearly 30% of circumstellar disks are more massive than Jupiter (André and Montmerle 1994, see Fig. 1.3).

The necessary conditions for planet formation, in terms of rotation and mass, appear therefore largely satisfied. But are they sufficient? The disappearance of disks, typically after a few million years, of course suggests that planets may be forming. But this is not yet proven observationally, because objects of size larger than a few mm, dust particles, pebbles, planetesimals and other aggregates are undetectable even in the nearest star-forming regions. On the other hand, accretion and/or ejection may also have exhausted the disk material.

Some optimism is, however, warranted if one considers that exoplanets have been found to exist (although so far the ones discovered are much more massive than the Earth), as do thin "second-generation" disks[2] around stars like

[2] These are very low-mass disks (about $10^{-6}\,M_\odot$, or 1/1000th of a Jupiter mass, or a few terrestrial masses), likely made of dust ejected by cometary-type objects orbiting the central star.

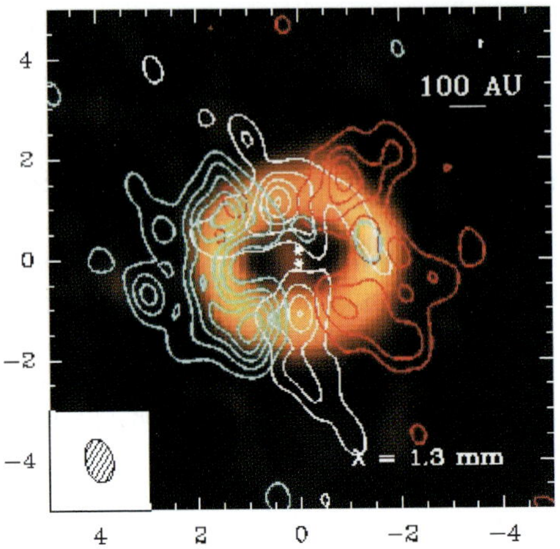

Fig. 1.6. The GG Tau binary system, another association between a disk and an outflow, this time seen nearly pole-on, observed in the mm range (© IRAM; Guilloteau, Dutrey, and Simon 1999). The disk, seen in the continuum emission of cold dust grains at 1.3 mm, is in fact circumbinary. By observing in addition the ^{13}CO molecule, one sees two opposite ("bipolar") outflows that move away (*red*) or towards (*blue*) the observer. The companion T Tauri stars GG Tau A and GG Tau B are indicated by asterisks at the centre of the image

β Pictoris, which has a mass of about $2\,M_{\odot}$. For the moment, the existence of all known exoplanets has been inferred from periodic stellar radial velocity variations, resulting from the so-called "reflex motion" that the planet exerts on the star, since the star–planet system, by Kepler's laws, must orbit around a common centre of gravity. This oscillation effect is maximized if (a) the planet is massive, and (b) if it is close to the star. This explains why, so far, most known exoplanets are massive (Jupiter mass or more) and close to the companion star. (In especially favourable cases, a smaller planet may be inferred from a tiny radial velocity modulation of the main effect.) On the basis of more than 100 known exoplanets, one now finds a $\sim$ 7% probability of detection of massive exoplanets close to solar-type stars. Since this is smaller than the 30% fraction of Jupiter-mass (or more) disks mentioned earlier, it probably implies that massive exoplanets may also exist far from the central star (as, in fact, does Jupiter itself), and are yet undiscovered because they do not induce an observably large radial velocity variation on the star. A recent discovery is that the probability of finding exoplanets rises significantly with the star's metallicity; the reasons for this are unclear, but it may be that planet formation is favoured by a higher metallicity (Santos et al. 2003).

However, problems remain. All the circumstellar disks known up to now are very large, typically 10 times larger than the size of the present-day Solar System (50 AU). But the spatial resolution currently available corresponds to several times this size at best, so that we are unable to see what is happening on such a small scale. This ignorance is an important missing piece of observational evidence, to bridge the gap between the "stellar" approach described above and the "planetary" approach practiced by Solar System astronomers. However, this gap should be filled within a few years, in particular with the availability of "ALMA", a giant millimetre interferometer under construction in the Chilean high plains,[3] as well as large (8–10-m class), ground-based telescopes and optical interferometers. Measuring the radial velocity distribution of disks down to very small distances from the central star ($< 10\,\mathrm{AU}$ at 150 pc) will allow accurate and direct measurements of their (dynamical) masses to be made. Also, one must take into account the role of the powerful jets and outflows, emitted perpendicular to the disks (see Sect. 2.1), which is ignored in current models for the first stages of planet formation. Finally, there remains the problem of binarity: we have mentioned above the existence of "circumbinary" disks (see Fig. 1.6). But we now know that over half of the stars are in reality binary systems (or "double stars"); the proportion reaches 80% in the case of young stars. This could totally change the evolution, in particular dynamical, of young planetary systems, for instance by ejecting planets out to large distances or cutting off the circumstellar disks. In fact, it is not impossible that the Sun may have been born in a (now dissolved) binary system. It is clear, however, that we are still far from having a coherent view of planet formation, even more so of the real fraction of stars expected to be surrounded by planets *now*.

1.4.2 High-energy Phenomena: X-ray Emission

A new factor, which has been underestimated or even unrecognized until recently, probably plays an important role, for star formation as well as for planet formation, and perhaps also for the origin of life. This is the ubiquitous X-ray emission from young stars, i.e., T Tauri stars and protostars (see Feigelson and Montmerle 1999 for a review). X-rays are mainly detected in the form of powerful flares, very similar to those seen in the images that the Yohkoh satellite send us every day of the Sun (see Fig. 1.7),[4] but much more intense (10^3 to 10^5 times stronger than on the quiet Sun).

In the course of these flares, the stellar X-ray luminosity L_X amounts to 10^{-4} to 10^{-3} times the total luminosity of the young star (called the "bolometric" luminosity L_{bol}), reaching a few per cent or more in exceptional cases. It can be shown that the "plasma" (a very hot, ionized gas, with a temperature of 10^7 to 10^8 K) that emits the X-rays is confined in very large magnetic loops (up to 2–3 stellar radii, i.e., about 10 solar radii for a young star). By analogy with the Sun, it is thought that the fast heating results from so-called "reconnection

[3] ALMA stands for "Atacama Large Millimetre Array". With 64 12-m antennas, this facility is the highest (alt. 5400 m) large man-made construction in history. It is managed by a US-European collaboration, soon to be joined by Japan.

[4] The X-ray emission from the Sun was discovered as early as 1949 by American physicists flying captured V2 rockets.

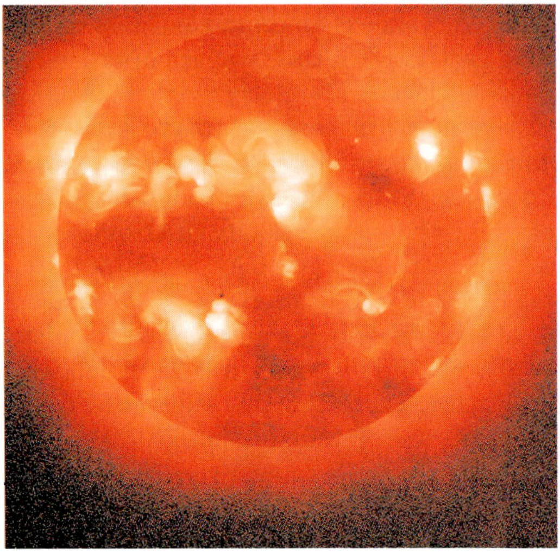

Fig. 1.7. The X-ray Sun, observed on a daily basis by the Yohkoh satellite, launched in 1991. These images (as well as other solar images) are available on the web (http://umbra.nascom.nasa.gov/images/latest.html). (Here, for instance, the date is 12/12/2000.) The *lighter areas* correspond to a very hot gas (plasma), heated to several million degrees and confined by magnetic field lines, anchored at the solar surface, as can be clearly seen here

events", in which magnetic field lines of opposite polarities get in contact in a "short-circuit", and suddenly release the magnetic energy previously stored in the course of various motions, for instance when magnetic footpoints are dragged by convective cells (e.g., Amari et al. 2000, Shibata and Yokoyama 2002). The plasma then cools radiatively in a few hours by the emission of X-ray photons.

Thanks to the ROSAT and ASCA X-ray satellites (from the 1990s), and more recently with the new generation of X-ray "space telescopes", Chandra and XMM, which combine high spatial resolution and high sensitivity, X-ray emission has been detected from hundreds of T Tauri stars, either concentrated in young stellar clusters like ρ Ophiuchi (Casanova et al. 1995, Grosso et al. 2000, Imanishi et al. 2001, Ozawa et al. 2004) or Orion (Feigelson et al. 2002a), or else more widely dispersed as in the Taurus-Auriga clouds (e.g., Stelzer and Neuhäuser 2001). It is precisely because of their X-ray emission that "weak-lined" T Tauri stars were discovered in the 1980s, in roughly equal numbers to the "classical" T Tauri stars, in other words effectively doubling the previously known population of T Tauri stars. For a given population of T Tauri stars (in the optical or in the infrared), the X-ray detection rate is essentially 100%, except for the less luminous ones (which are also the least massive), which may be below the X-ray sensitivity threshold. In other words, observations like the ones shown in Figs. 1.8 and 1.9 suggest that *all* T Tauri stars, with or without disks, emit

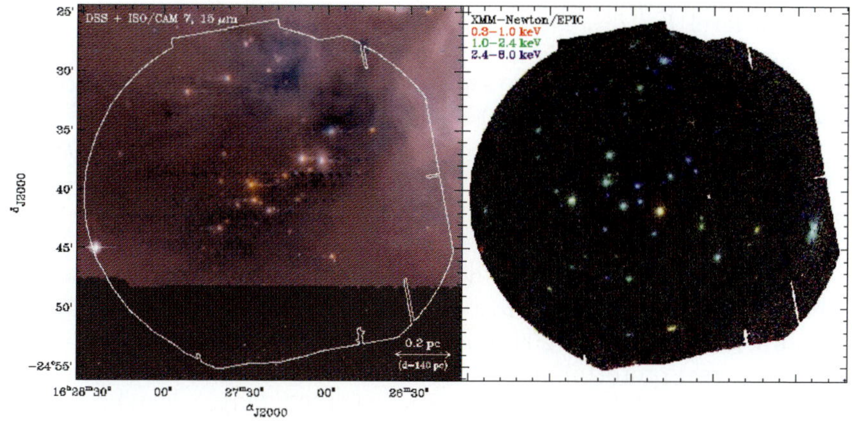

Fig. 1.8. Multiwavelength observations of the central region of the ρ Ophiuchi cloud (Ozawa, Grosso, and Montmerle 2004). The *background area* on the *left* figure is the mid-IR map obtained by the ISOCAM camera aboard the ISO satellite: the *bright spots* are identified with Class I protostars (located in the densest regions of the cloud; near the centre of the image), and with Class II and III T Tauri stars (outer sources). The approximately *circular white line* on the *left* is the field of view of the EPIC X-ray camera aboard the XMM satellite, corresponding to the image on the *right*. One sees an excellent correspondence between the IR sources and the X-ray sources. In particular, the Y-shaped "quartet" of protostars lying almost at the centre of the ISOCAM image is clearly visible in X-rays

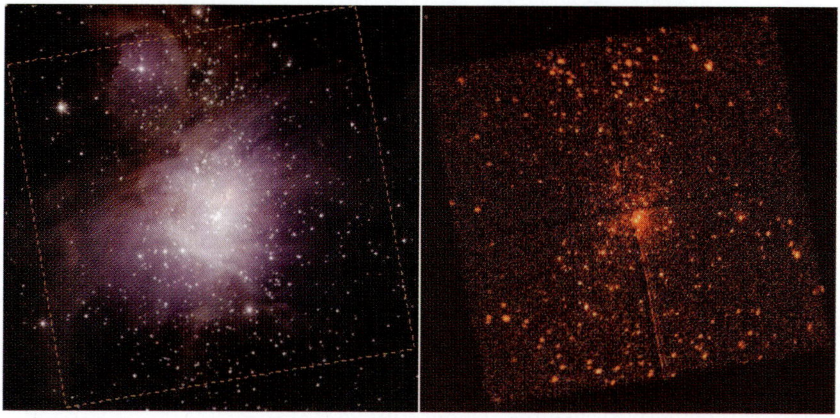

Fig. 1.9. The Orion nebula (M42), seen in the near-IR (2MASS survey, *left*) and in X-rays with Chandra (Feigelson et al. 2002a; *right*). As in the case of the ρ Oph cloud (Fig. 1.8), and in many other star-forming regions, there is an almost one-to-one correspondence between the IR and X-ray sources, most of which are T Tauri stars, and also more massive stars like the O stars of the Trapezium (centre of the image). This means that virtually all young stars emit X-rays

X-rays, at a level $L_X/L_{bol} \sim 10^{-4}$ to 10^{-3} with a fairly large dispersion (see, e.g., Grosso et al. 2000). This is, for instance, most vividly illustrated by the XMM image of the ρ Oph cloud (Fig. 1.8), or the Chandra image of the *Orion* Nebula (Garmire et al. 2000, Feigelson et al. 2002a) (Fig. 1.9).

As concerns protostars, whereas only a dozen were detected by ROSAT and ASCA (Feigelson and Montmerle 1999), they are now efficiently detected by Chandra and XMM. They all belong to clusters, and they all are of the "Class-I" type. Their X-ray luminosity tends to be higher than that of T Tauri stars by an order of magnitude, and their temperature also tends to be higher ($T_X > 5 \times 10^6$ K). A dozen Class I protostars have been detected in the ρ Oph cloud by Chandra (Imanishi et al. 2001), and more by XMM (Ozawa et al. 2004). In contrast, only two protostars had been detected by ROSAT, and only because they were displaying large flares (Grosso et al. 1997, Grosso 2001).[5]

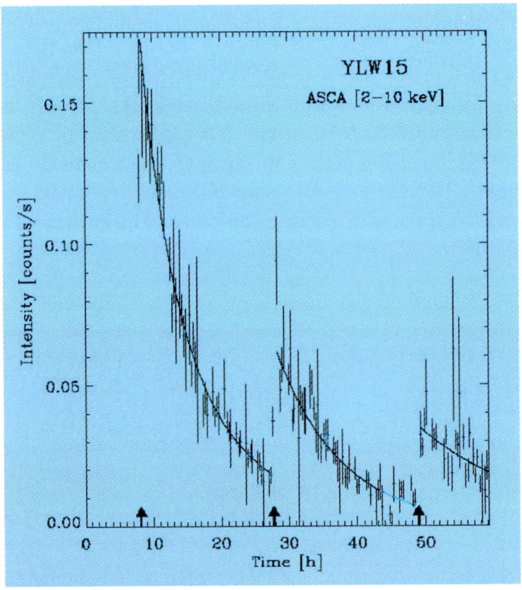

Fig. 1.10. The triple X-ray flare of the YLW15 Class I protostar in the ρ Oph cloud, observed with the ASCA satellite (Tsuboi et al. 2000). One clearly sees a quasiperiodic flaring phenomenon with a time interval of approximately 20 h (arrows). The curves are exponential fits to the flare decay by radiative cooling, from which various plasma parameters can be derived, like the temperature

[5] There has been a claim for the detection of two very young, Class 0 protostars with Chandra (Tsuboi et al. 2001), but this source is probably a double Class I source. Out of a sample of more than 20 Class 0 sources, none has been detected by Chandra or XMM. This is probably due to the very large extinction by the dense envelope, obscuring the central, growing star (Montmerle et al. 2004, in preparation).

On the basis of the X-ray light curves, the mechanism for X-ray flaring in protostars appears, in general, identical to that holding for T Tauri stars. However, in one, so far unique, case, the mechanism may be somewhat different. Indeed, a succession of three intense X-ray eruptions, separated by about 20 h each, could be observed with the ASCA satellite in YLW15, a Class I protostar of the ρ Oph cloud (Tsuboi et al. 2000), as shown in Fig. 1.10. To explain this phenomenon, Montmerle et al. (2000) proposed a model in which magnetic structures link the star to its accretion disk, in a fashion already invoked for neutron stars. The star is assumed to rotate faster than the inner accretion disk, a few stellar radii away, which results in shearing and magnetic reconnection, and thus X-ray flaring, at each turn, as shown in Fig. 1.11. This would be a particular case of a family of star-disk magnetic interactions expected by theory (e.g., Montmerle 2003).

Anyway, whatever the details of the emission mechanism, it is clear that the X-ray emission from Class I protostars is not a mere curiosity any longer, even if the detection rate is not as high as for T Tauri stars, which could be explained by the large extinction of these embedded objects. The bottom line is that we now know for sure that the X-ray emission of young, solar-type stars is almost

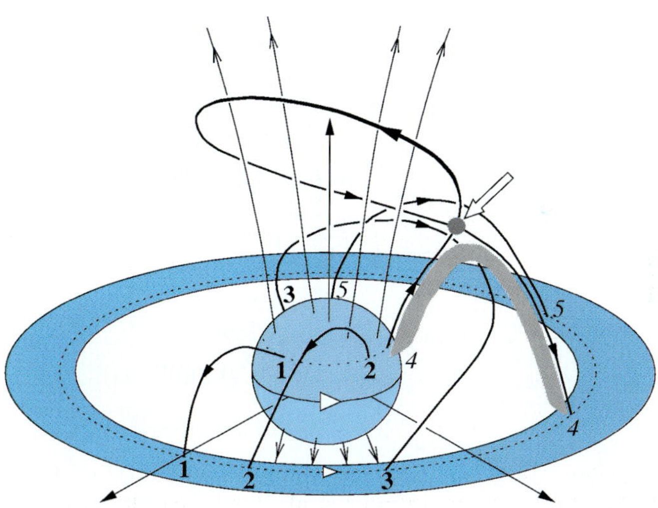

Fig. 1.11. Model of a star-disk magnetic interaction proposed by Montmerle et al. (2000) to explain the "triple flare" of YLW15. It is based on the standard explanation of solar flares – but here, instead of magnetic "footpoints" anchored at the solar surface, they are assumed to be anchored onto the star and on the inner part of the circumstellar disk. As a result of a rotation velocity difference between the star and the disk, the magnetic field lines are twisted around themselves, and after a full turn come into contact. This "short-circuit" (called "reconnection") liberates the magnetic energy stored during one stellar rotation. As rotation continues, other flaring events may occur, with a period close to the stellar period (here 20 h)

universal and intense (up to 1/10 of the total luminosity of the present-day Sun, or 10^5 times its X-ray luminosity), as early as 10^5 yr after gravitational collapse has started, and perhaps even earlier.

1.5 X-ray Irradiation: Ionization and Feedback Effects on Circumstellar Disks

In addition to providing us with information on the "magnetic state" of protostars and T Tauri stars (for instance relevant to ejection phenomena, or to the reconnection mechanisms), X-rays are important on another ground: they induce feedback effects on the dense surrounding circumstellar and interstellar material.

The first effect is the ionization of the accretion disk. Igea and Glassgold (1999) and Glassgold et al. (2000) have studied the case of a "standard" accretion disk, illuminated by a central X-ray source located a few stellar radii above the

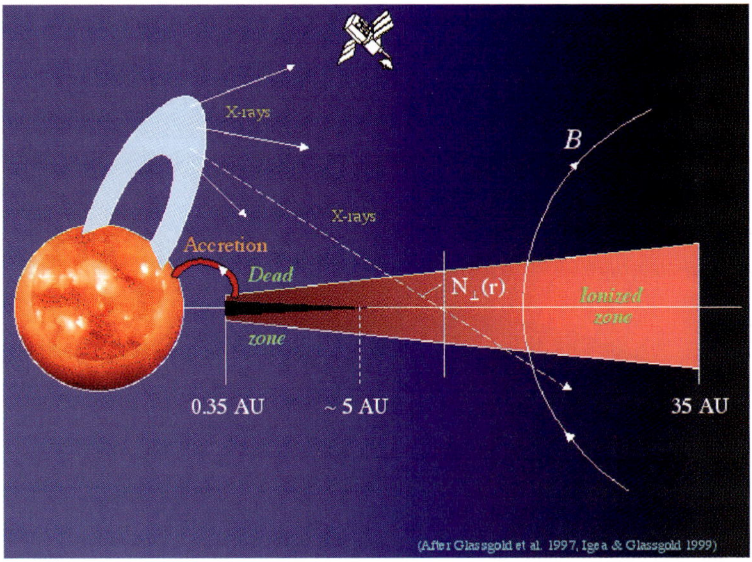

Fig. 1.12. Irradiation of a circumstellar disk by X-rays from a young star, based on the work of Glassgold, Najita, and Igea (1998). It turns out that the main quantity to measure the absorption of X-rays by the disk is the column density perpendicular to the disk ($N_{\perp}$). The ionization caused by X-ray irradiation results in coupling the disk material and magnetic fields. However, near the star, where the disk density becomes very high along the equator, X-rays cannot penetrate and the material is neutral ("dead zone"). These authors have suggested that this "protected" zone may be the location where the initial stages of planet formation take place. Elsewhere, the disk is turbulent, and this promotes an inwards flow of accretion towards the central star

equatorial plane, as suggested to some extent by observations (see Fig. 1.12). In such a disk, the density and temperature have a power-law dependence on the radius r, and the scale height is only a function of the local gravity at r (Hayashi, Nakazawa, and Nakagawa 1985); the free parameters are adjusted to the millimetre observations of disks (e.g., Beckwith et al. 1990, André and Montmerle 1994, Simon et al. 2000). The resulting ionization rate of the gas, $\xi_{\mathrm{e,X}}$ is then compared to that produced by galactic cosmic rays, $\xi_{\mathrm{e,CR}}$, which is always present in interstellar clouds, i.e., $\xi_{\mathrm{e,CR}} = 10^{-18}$ to 10^{-17} ionizations s^{-1}. The result, sketched in Fig. 1.12, is that the ionization of the accretion disk is dominated by X-rays, except in the densest equatorial regions where they cannot penetrate ("dead zone"), i.e., within a few AU of the central source. Everywhere else, the disk is partially ionized, and the ionization fraction $x_{\mathrm{e}} = n_{\mathrm{e}}/n_{\mathrm{H}}$ is roughly comparable to that of the interstellar medium ($x_{\mathrm{e}} \sim 10^{-7}$ or less)[6].

This novel way of considering the physical conditions of the accretion disk around the very young Sun (10^5 to 10^6 yr) should lead to a revision of the conditions for the "primitive solar nebula" adopted by Solar System experts. For instance, the link between the D/H ratios observed in comets and planets, and their implication on the origin of water (see Chap. 8 by Despois and Cottin) rests on an isotopic exchange in a neutral medium between interstellar ices, rich in deuterium, and protostellar hydrogen, whereas (except perhaps in the "dead zone") one should take into account a very weak ionization, comparable to that of the interstellar medium.

The ionization state of the disk has another important consequence: the coupling of matter with magnetic fields. Indeed, even very weakly ionized matter "sticks" to the magnetic field via collisions between neutral atoms and charged atoms (ions), a process called "ambipolar diffusion". When this process is coupled with the Keplerian motion of all atoms around the central object, it gives rise to a so-called "magnetorotational instability", discovered by S. Chandrasekhar and extensively studied by Balbus and Hawley (1991). This instability is invoked to explain the strong viscosity of accretion disks, and thus regulates accretion itself. In other words, X-rays exert a positive feedback on accretion: the higher the X-ray luminosity of the central object, the higher the ionization, and therefore the larger the accretion. The observation that a protostar like YLW15 (with its triple flare discussed above) has "low" and "high" X-ray emission states over several years, may indeed suggest a link between the intensity of X-ray emission and accretion (see discussion in Montmerle et al. 2000).

One interesting consequence is that such strong viscous coupling would not exist in the dead zone, which would then undergo accretion only if some other, nonmagnetic mechanism for an efficient viscosity were at work (which cannot be excluded). Note, however, that the very existence of an extended dead zone has been challenged by recent numerical computations, which take into account turbulent motions within the disk, and that show that the neutral,

[6] These results are however sensitive to the exact disk model used, and some disks may be less ionized than others (see Fromang et al. 2002).

dead zone volume tends to mix with the surrounding, ionized material: in the end, (weak) ionization probably dominates everywhere in the disk (Fleming and Stone 2003).

1.6 Disk Irradiation by Energetic Particles and "Extinct Radioactivities" in Meteorites

For the young Solar System, the X-rays may play another role. First, we know from solar observations that flares seen in X-rays also accelerate particles, i.e., electrons, protons and heavier nuclei, which induce so-called "spallation" nuclear reactions, that is, in-flight, low-energy collisions breaking the nuclei. Among the resulting fragments, some nuclei are produced in an excited state, others are radioactive: they are subsequently de-excited or decay, emitting γ-ray photons, such as those that were observed from the Earth in the solar photosphere by the SMM satellite (Murphy et al. 1991), and more recently by the RHESSI satellite (Lin et al. 2003). Again relying on the analogies between the magnetic activity of young stars and that of the Sun, one can then safely assume that particles must be accelerated in the flares of the young stars. This is in part confirmed by radio observations in the centimetre range, which show evidence for a non-thermal emission mechanism ("gyrosynchrotron"), thereby demonstrating that young stars are able to accelerate electrons to relativistic energies in their magnetic fields (see Güdel and Zucker 2000, and Güdel 2002). One thus concludes that the environment of Young stellar objects must be filled with high-energy particles, accelerated during flares, that are another source of "ionizing radiation" (in the nuclear sense of the term) for their surroundings, and in particular for the accretion disk, even though heavy particles like protons or α particles cannot be directly observed (see Feigelson et al. 2002b).

The importance of these high-energy particles has been recently demonstrated by considering the existence of "anomalies" in the isotopic composition of meteorites, related to "extinct radioactivities". More precisely, certain isotopes display an abundance excess over "cosmic abundances", which may result from the decay of parent radioactive isotopes. For instance, the natural isotope of magnesium is ^{26}Mg, which is normally produced as a result of stellar nucleosynthesis like most other atoms in nature. But inside some meteorites (chondrites) thought to be the most primitive material of the Solar System, one finds "aluminium-rich" inclusions (the normal aluminium isotope is ^{27}Al), which contain an excess of ^{26}Mg. The idea is that this excess ^{26}Mg is coming from the radioactive decay of ^{26}Al, which has a half-life of 7.5×10^5 yr (so that it is now "extinct"), according to the following reaction:

$$^{26}\mathrm{Al} \longrightarrow {}^{26}\mathrm{Mg} + e^+ + \gamma(1.8\,\mathrm{MeV})\,.$$

The problem is then to understand how this ^{26}Al was produced in the early Solar System, knowing that it can be produced in a wealth of nucleosynthetic sites (red giants, massive stars in their late stages of evolution known as Wolf–Rayet stars, novae, supernovae: see Diehl and Timmes 1998 for a review), and also by spallation reactions such as:

$$^{3}\mathrm{He} + {}^{24}\mathrm{Mg} \longrightarrow {}^{26}\mathrm{Al} + p$$

beyond a certain energy threshold for the incident particles (in general, a few MeV/n to a few tens of MeV/n). "Extinct radioactivities" in meteorites are measured in the form of ratios with a stable "daughter" nucleus: the ones that cannot be explained by continuous production in the Galaxy are ^{7}Be/^{9}Be, ^{10}Be/^{9}Be, ^{26}Al/^{27}Al, ^{41}Ca/^{40}Ca, ^{53}Mn/^{55}Mn, and ^{60}Fe/^{56}Fe.

The "simplest" case to interpret is that of beryllium. Indeed, ^{7}Be and ^{10}Be cannot be produced by stellar nucleosynthesis, but only by spallation reactions, for instance p+^{16}O or ^{3}He+^{16}O. ^{10}Be has a half-life of 1.5 Myr, while that of ^{7}Be is only 52 days. Therefore, the precise measurement of the ^{7}Be/^{9}Be and ^{10}Be/^{9}Be ratios in meteorites gives a direct indication of the level of irradiation of the young Solar System. These difficult measurements were successfully done by Chaussidon et al. (2002) and McKeegan et al. (2000), respectively, in the Allende meteorite, and it can be shown that the measured values, $^{7}\mathrm{Be}/^{9}\mathrm{Be} = 6 \times 10^{-3}$, and $^{10}\mathrm{Be}/^{9}\mathrm{Be} = 9 \times 10^{-4}$, can be accounted for by a high level of primitive irradiation, consistent with the high level of X-ray irradiation seen in young stars and protostars.[7]

The case of the other isotopic ratios is more complex, because the multiplicity of their possible creation sites and processes makes it impossible to relate the measured value to the irradiation mechanism alone. In contrast, one can calculate the ratios for a given level of the particle irradiation intensity. This calculation has been done by Gounelle et al. (2001) for the above isotopes, in the framework of a particular model allowing to do analytical calculations and using an extensive network of nuclear spallation reactions, and scaled to the X-ray luminosity of young stars. (The only exception is the ^{60}Fe isotope, which can be produced only in a neutron-rich environment such as supernovae, and thus cannot be explained in the framework of an irradiation model.) This model is the "unified" accretion–ejection MHD model for young stars (Shu et al. 1997, Shu et al. 2001). In this model, matter ejection takes place from a singular "X-point" (i.e., a ring in circular symmetry), where magnetic field lines converge. At this point, matter falls freely on the star (accretion), but a fraction flows "back" because of the centrifugal force, channelled along the field lines (ejection): this is the "X-wind" (Fig. 1.13). Along this ring, the strong magnetic pressure is assumed to relax by a fluctuating wind leading to successive reconnections. In turn, these reconnections give rise to X-ray emission, and then to particle acceleration.

[7] In a recent paper, Desch et al. (2003) invoke galactic cosmic-ray confinement within the Sun-forming molecular core to explain the ^{10}Be/^{9}Be ratio. In addition to the fact that it does not explain the ^{7}Be/^{9}Be ratio, this model appears somewhat ad hoc and does not explore the implications for the "standard" production of the Li, Be and B isotopes by galactic cosmic rays in the interstellar medium.

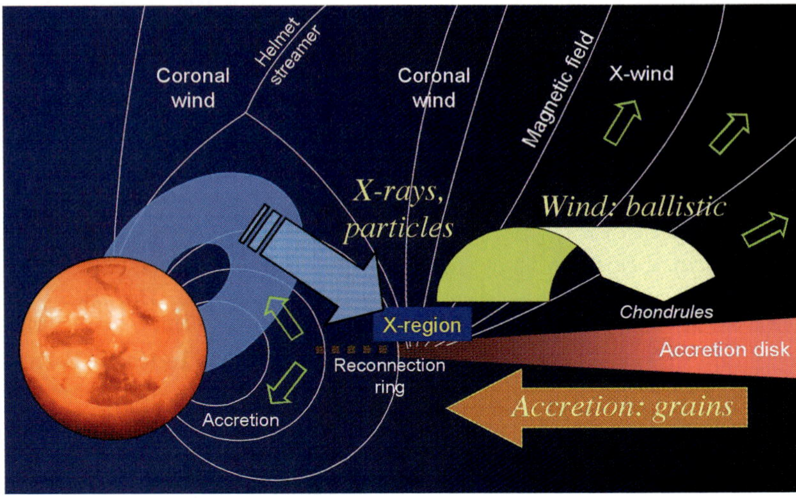

Fig. 1.13. *Background*: sketch of the accretion–ejection MHD model of Shu et al. (1997); see Fig. 1.12 for comparison. The large-scale structure of the magnetic field channels gas and dust both towards the central star (accretion) and outwards (ejection) under the effect of the centrifugal force. In this model, roughly 10–30% of the material is accreted, and the remainder is ejected in the form of a so-called "X-wind". In a complementary work, Gounelle et al. (2001) invoke various reconnection phenomena to generate flares and X-rays on the star or near the "X-ring", where inflow and outflow of material are separated by magnetic field lines. As a result, dust grains and premeteoritic material like chondrules are irradiated by energetic particles (protons and ^{3}He nuclei), creating various radioactive isotopes with lifetimes around 1 Myr or less, like ^{10}Be and ^{26}Al, now known as "extinct radioactivities". These grains are entrained by the outflowing material (X-wind), but are not tied to the magnetic field, hence "rain" back ballistically on the disk, far away ($\sim$ 1 AU or more) from the central star

This "X-wind" model is also a model for the streaming of matter, first flowing towards the central star in the equatorial plane of the disk, then lifted as a centrifugal wind along the outer magnetic field lines. Following these streaming motions, heavier bodies like dust grains "slide" towards the central star, then are entrained by the gas and "rain" back ballistically far out on the disk, because unlike the atoms they cannot "stick" to the magnetic field (see Fig. 1.13). As they go in, then out, dust grains undergo near the reconnection ring a quick melting and a strong irradiation, then cool far away from the central star, thus explaining the existence of refractory material in grains, which must have been processed at high temperature anyway in spite of their large distance from the Sun ($\sim$ 1 AU). Finally, the model assumes that the grains are spherical "minirocks", with a certain size distribution, made up of a refractory core surrounded by a mantle made of a mixture of refractory and volatile material. In this (admittedly in part ad hoc) way, specific spallation reactions dominate, and an overproduction of ^{41}Ca can be avoided (see Gounelle et al. 2001 for details).

Again adopting for the flares (here in the reconnection ring) a particle flux proportional to the X-ray flux, but including ^{3}He nuclei (frequently seen in so-called "impulsive" solar flares, where $^{3}\mathrm{He}/p \sim 0.3$) in addition to protons and α particles (helium nuclei), Gounelle et al. (2001, 2003) show that it is possible to *simultaneously* explain the production of ^{7}Be, ^{10}Be, ^{26}Al, ^{41}Ca and ^{53}Mn. The model also predicts a specific signature in the form of ^{50}V production. Measuring the $^{50}\mathrm{V}/^{51}\mathrm{V}$ ratio is exceedingly difficult, and the experiment is currently in progress (M. Gounelle, private communication). All in all, instead of invoking a separate explanation for each "extinct radioactivity", and at the cost of only a few free parameters, the model can explain 5 out of 6 isotopic ratios with the same mechanism. (Again, the sixth, ^{60}Fe, cannot be explained in this framework anyway.) The results are, however, highly dependent on the assumed grain characteristics, and this point is controversial among meteoreticists and astronomers: what would be a correct grain model? Also, is the "X-wind" model the best one to explain accretion and ejection? Are the assumed reconnections consistent with those inferred from X-ray observations? Does the reconnection ring configuration exist in MHD, etc.? In spite of many objections, this model, which could be called an "accretion–ejection-irradiation-transport" model is at present the most complete model available for the origin of extinct radioactivities in meteorites, and it undoubtedly yields encouraging results.

These results go even further than supporting a single "internal" explanation for the extinct radioactivities problem, because they highlight the necessary, and also probably sufficient, central role of the high-energy irradiation of the young Solar System. This model certainly downplays the possible influence of external factors, like a nearby supernova. In particular, the implication may be that the formation of the Solar System did not require a shock like that produced by a nearby supernova explosion. This hypothesis had been proposed as soon as the ^{26}Al extinct radioactivity was discovered in the Allende meteorite in the 1970s (see, e.g., Clayton 1985). Recent work, based on the galactic ^{26}Al emission as seen by the GRO satellite in the line of 1.809-MeV γ-rays, and observed in the vicinity of present-day star-forming regions, however, cannot exclude a small "pollution" of molecular clouds by nearby ^{26}Al nuclei (and/or ^{60}Fe) from massive stars and supernovae (Montmerle 2002).

1.7 The Origin of the Sun, and the Origin of Life

Modern observations have shown that solar-like stars (and presumably the Sun itself) are formed as a result of complex accretion and ejection processes, which take place on spatial scales dwarfing the present-day Solar System, and on typical timescales of order 1 Myr from the earliest, embedded stages (protostars) to the visible, disk phase ("classical" T Tauri stars). Whatever the actual physical mechanism(s), all theoretical models agree on the fact that magnetic fields play a central role in the formation of solar-like stars.

Hence, since its early youth ($\sim 10^5$ yr), and perhaps even since its formation, we know for sure that the Sun must have remained extremely magnetically active. If one adopts the X-ray luminosity of young stars as a scaling factor of this activity, one finds amplification factors with respect to the present-day Sun of order 10^4 to 10^5 at the protostar stage ($\sim 10^4$ yr), and 10^3 to 10^4 at the T Tauri stage ($\sim 10^6$ to 10^7 yr). This intense activity of all pre-main sequence solar-type stars is in a way confirmed for our own Solar System, since it may explain in a quantitative way the isotopic anomalies of "extinct radioactivities" in meteorites by a strong internal irradiation from the young Sun (see Feigelson et al. 2002b, Gounelle et al. 2003). It is no longer necessary to invoke external causes for the formation of the Solar System such as a supernova explosion. Also, as calculations show (and pending future observational evidence in the millimetre range), the accretion disks, which must be relevant for the "primitive solar nebula", should be everywhere partially ionized by the X-rays from the central star, except perhaps in the densest equatorial regions ("dead zone"), closer than 5 to 10 AU. The ionization fraction is comparable to that of the interstellar medium ($x_e \sim < 10^{-7}$), which could challenge the "neutral" chemistry of the primitive solar nebula adopted until now.

The active phase of T Tauri stars traced by X-rays continues throughout their lifetime, i.e., $\sim 10^6$ to 10^7 yr. For example, we know of X-ray emitting T Tauri stars, aged about 10 Myr, which are still surrounded by circumstellar disks and actively accreting (e.g., the η Cha group, see Lyo et al. 2003). But the eventual end of the T Tauri phase does not mean the end of the pre-main sequence evolution of solar-type stars. It will take many more tens of millions of years until the stars reach the main sequence. During this period, the X-ray activity declines, but remains high, as X-ray observations of open clusters of stars of different ages show (Pleiades, ~ 70 Myr, Hyades, ~ 800 Myr, etc.) (see Randich 2000; Fig. 1.14). For most irradiation phenomena, the key parameter is the "fluence", i.e., the integral of the product of the flux multiplied by the time interval. In the present case, the fluence of late phases is found to be comparable to that of the early phases (protostar, T Tauri). The "interplanetary media" of young stars are therefore subject to an intense bombardment by high-energy particles even in this late period. By comparison, the present-day Solar System, as well as planetary atmospheres, are subject to a very mild treatment by the Sun!

Thus, in a broad sense, we now possess an empirical knowledge of the formation and evolution of solar-like stars, before the main sequence phase starts as hydrogen ignition takes place. Some early phases, like the protostellar collapse, the details of the accretion process and the role of magnetic fields in the origin of jets and outflows, start to be understood from a theoretical standpoint. And we are fairly certain that X-ray irradiation, with its related disk ionization and particle-irradiation effects, must be important during the first million years of evolution. However, as discussed several times in this chapter, we still lack the constraints that high-angular resolution observations, i.e., on the scale of the in-

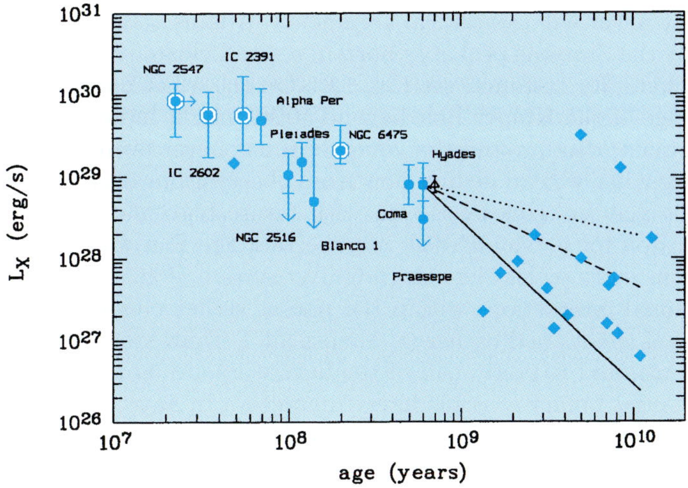

Fig. 1.14. Time evolution of the X-ray luminosity of solar-type stars, from the T Tauri phase (10^7 yr) up to now ($\sim 10^{10}$ yr), from X-ray observations of open clusters and field stars (Randich 2000). There is a decrease of a factor ~ 50 from the early stages to the main sequence (a few 10^8 yr, depending on the stellar mass). For field stars, which are roughly as old as the Sun, there is a large spread (*diamonds*), and the X-ray luminosity of the Sun ($\odot$) looks rather weak by comparison. The reason for this spread is not known; it may be due to one-to-one differences in the efficiency of the dynamo effect, in relation to the star's differential rotation. If one approximates the time evolution of the X-ray luminosity as a power law, the corresponding slopes would have indices –0.5, –1, –2, from *top* to *bottom*. As shown recently by the Chandra observations of Orion, the X-ray luminosity shows no correlation with rotation during the T Tauri phase, which might indicate a different type of dynamo mechanism during the earliest stages

ner Solar System (< 10 AU, say) could provide to theoretical models – planetary formation to begin with. We do not even know for sure whether giant planets or terrestrial planets form first, and on what timescale (10 Myr? 100 Myr?) (See the Chap. 2 by Petit and Morbidelli.) It is thus currently an open issue whether X-rays actually irradiate young planets, and for how long. But the situation is not hopeless, as new powerful telescopes are coming online in the next decade. Perhaps the greatest progress can be expected from the ground, in Chile to be precise, and from interferometric techniques (in the radio with ALMA, and in the optical/IR with the interferometric modes of the Very Large Telescope), allowing angular resolutions of order $0.01''$ (10 times better than now) to reach be reached, i.e., almost the Sun–Earth distance for the nearest young stars. We may then hope to "see" if and how planet formation takes place within circumstellar disks.

But another fundamental conceptual difficulty remains. Even if our knowledge of solar-like star formation were immensely better, we would still be faced with the difficulty of knowing what has actually happened to *our own* solar-like star, the Sun. Indeed, we do not know in what environment the Sun was born, and probably never will. Statistical arguments (Adams and Myers 2001) tend to indicate that the Sun was probably born in a small cluster of 10–100 stars (as in the ρ Oph cloud, for instance, see Fig. 1.8). On the other hand, the existence of the sharp edge of the Kuiper Belt beyond 50 AU might indicate that the primitive circumstellar disk around the young Sun has never been larger in the past, in which case what we can deduce now from observations of solar-like stars with 500 AU disks may be irrelevant, even with future high-resolution observations. The explanation for a "small" disk may be that the Sun was formed in a loose binary system more or less like the GG Tau system (Fig. 1.6), later dissipated by gravitational interactions within the parent stellar cluster. This idea is not crazy, since, as mentioned earlier (i) almost all T Tauri stars are found to be in binary systems, and (ii) only half of the field stars are binaries – implying that half of the initial binary systems have "divorced". In any case, we do not know yet if and how such a binary origin would have influenced the formation and evolution of planets, or the details of the irradiation by a binary X-ray source. Future numerical simulations will probably give us some clues.

Whatever the physical boundary conditions surrounding the birth and early history of the Sun, in the context of the problem of the origin of life the question is to know what has been the role of this intense irradiation by X-rays – and by the energetic particles that must be present along with them – emitted by the young Sun. On the time scale of the appearance of life on Earth ($\sim$ 1 Gyr), or even of the formation of the Earth ($\sim$ 100 Myr?), this intense activity looks like a brief "flash". But, like milk being "flash-pasteurized" by a brief exposure to high temperatures, this flash of magnetic activity can have irreversible consequences on some key molecules that could be present at that time. It is now clear that this is important for the isotopic composition of meteorites. It would even be interesting to compare the effects of such a flash with the radioactivity induced by the irradiation itself – which is going to contribute to the disk ionization for millions of years.

It is certainly very difficult today to make a quantitative estimate of the impact of the various high-energy irradiations on complex prebiotic molecules, or even on the D/H ratio used as a tracer of the origin of water on Earth, in the context of the new and/or speculative astronomical results reported above. One would need to have some idea of the primitive chronology of the appearance of water, and of life on Earth, to compare it to the X-ray luminosity, and by inference on the energetic particles, which is certainly premature now. But it is clear that this will necessarily have to be taken into account in the future, as precisely as possible.

Appendix: The Hertzsprung–Russell Diagram: Stellar Evolution at a Glance

The astronomers Einar Hertzsprung and Henry Norris Russell noticed very early on (towards the beginning of the 20th century) and independently, that stars tend to gather in specific parts of a diagram in which their luminosities are plotted as a function of their "effective" (surface) temperatures. Modern theory shows that stars evolve, and that they actually *move* as a function of time in such a diagram, widely referred to as the "Hertzsprung–Russell diagram", or "HR diagram" for short. This evolution generates an accumulation of stars where it is slow, and relative voids where it is rapid. A version of this diagram is shown on Fig. 1.15, based on 22 000 well-studied stars mostly drawn from a catalog based on data from the European satellite "Hipparcos", and also from the Gliese catalog of nearby stars.[8]

Figure 1.15 shows that the HR diagram is crossed by a thick, long, and wavy diagonal, running from hot, luminous stars, down to cool, low-luminosity stars. This diagonal is called the "main sequence", on which most stars lie. Note that, as is customary among astronomers, luminosities are indicated in solar units (on the vertical axis), and temperatures are increasing towards the left (on the horizontal axis). Figure 1.15 also shows the empirical division of stellar luminosities into "luminosity classes", from high (I) to low (V), as well as their "spectral types", indicated by letters (from O to M for historical reasons), which basically reflect their effective temperatures.

The theoretical works of Hans Bethe and collaborators showed, as early as 1938, that stars draw their luminosity from nuclear reactions in their cores. The "main sequence" is the locus where stars like the Sun are turning their hydrogen into helium in their centres, by means of a network of thermonuclear reactions (starting at a temperature of $\sim$ 15 MK). The main factor placing the stars on the main sequence is their mass. The stars on the upper left-hand side are the most luminous (up to $10\,000\,L_{\odot}$), i.e., the most massive (up to $100\,M_{\odot}$), while on the lower right-hand side lie the least luminous ones ($\sim 0.00001\,L_{\odot}$), which have a small mass (down to less than $0.1\,M_{\odot}$). The duration of the main sequence depends very much on the stellar mass: for massive stars it is only a few *million* years, while for the Sun it is about 10 *billion* years. The smallest stars stay on the main sequence "eternally", i.e., burn their hydrogen so slowly that this phase may take as long as hundreds of billions of years, i.e., much longer than the age of the present-day universe, which is estimated to be 12 billion years old. The present-day Sun is 4.5 billion years old, and thus lies half-way through its main sequence stage.

Figure 1.15 also shows a dense "wing" of stars to the right of the main sequence. This is the post-main sequence, or "giant" branch, into which stars evolve when they convert hydrogen into helium by a more energetic network of

[8] See http://www.anzwers.org/free/universe/hr.html

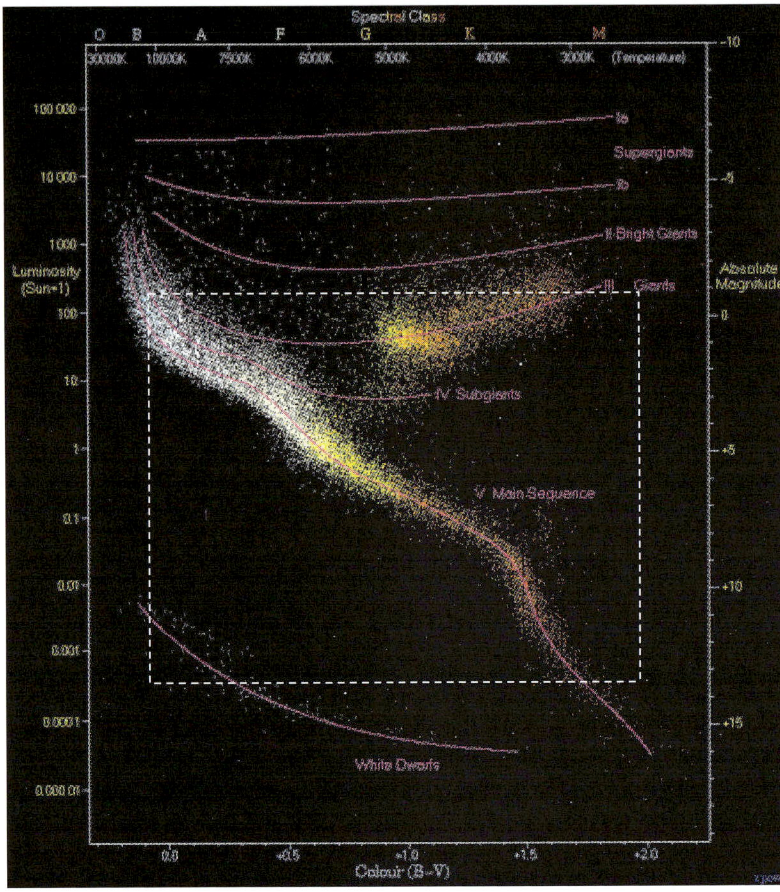

Fig. 1.15. The "Hertzsprung–Russell diagram". This diagram can be expressed in various units: absolute magnitude vs. colour or spectral type (directly from observations), luminosity vs. temperature (after the stellar type has been determined). The *long diagonal* is the "main sequence", on which stars turn hydrogen into helium via nuclear reactions. The *dotted rectangle* highlights the area of the "pre-main sequence" stages shown in Fig. 1.17 below

thermonuclear reactions (called the "CNO cycle"). The basic factor that makes these reactions happen is the increase in central temperature. As a result of this new energy source, the stars (mostly solar-type stars in this case) increase enormously in size. Subsequently, new thermonuclear reactions occur, in which helium burns and converts the stellar interior into carbon and oxygen during a few million years, while the envelope continues to expand. For instance, when it reaches that stage, 5.5 billion years from now, the Sun will become larger than the Solar System. In other words, it will engulf all the planets in a vast, expanding tenuous envelope, including the Earth: this is Doomsday as astronomers see it

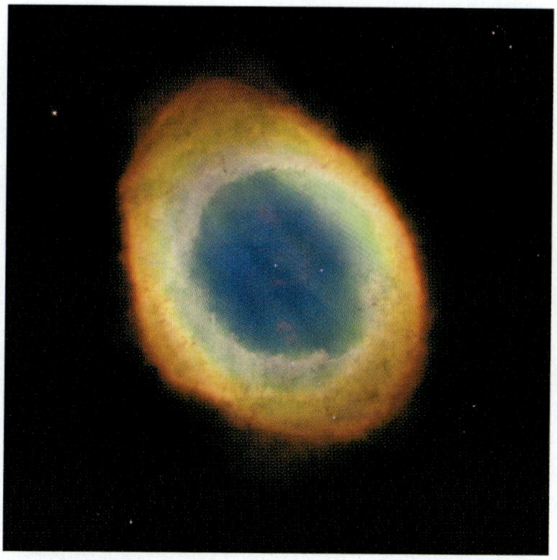

Fig. 1.16. Example of a "planetary nebula". This beautiful object (here in the constellation Lyra) is actually a dying star. The ring-like emission comes from an expanding envelope that may be much larger than the solar system, while the original, solar-type star shrinks to become a planet-sized "white dwarf" (on the picture, the *small dot* at the centre of the ring)

happening today in the so-called "planetary nebulae", such as the one located in the constellation Lyra shown on Fig. 1.16. In parallel, the core of the Sun-like stars shrinks to become a star in itself, called a "whitedwarf", a very hot (100 000 K), very small (planet-sized) and faint star (the small white dot at the centre of Fig. 1.16). White dwarfs slowly cool until they become dark after hundreds of billions of years.

One notable exception to this nuclear origin for stellar energy and evolution is a class of young stars called the "pre-main sequence" (PMS) stars, or "T Tauri stars" in the case of solar-like stars. Indeed, when they are young, stars draw their energy only from their slow contraction under the pull of gravity: *The PMS phase is a "non-nuclear" phase,* because at this stage the stars are not hot enough in their centres to ignite nuclear reactions. The PMS phase is very important for the problem of the origin of life because it is the phase during which planets may form.

T Tauri stars are too faint to be have been observed by Hipparcos, so none is included in Fig. 1.15. In contrast, Fig. 1.17 shows a dedicated HR diagram obtained for a cluster of low-mass young stars ($\sim 1\,M_\odot$), located in the constellation Chamaeleon, in the southern hemisphere, and drawn from another source[9] This HR diagram extends over a much smaller area than the "global"

[9] Simplified version, adapted from Lawson, Feigelson, and Huenemoerder 1996, *Monthly Not. R. Astron. Soc.*, **280**, 1071.

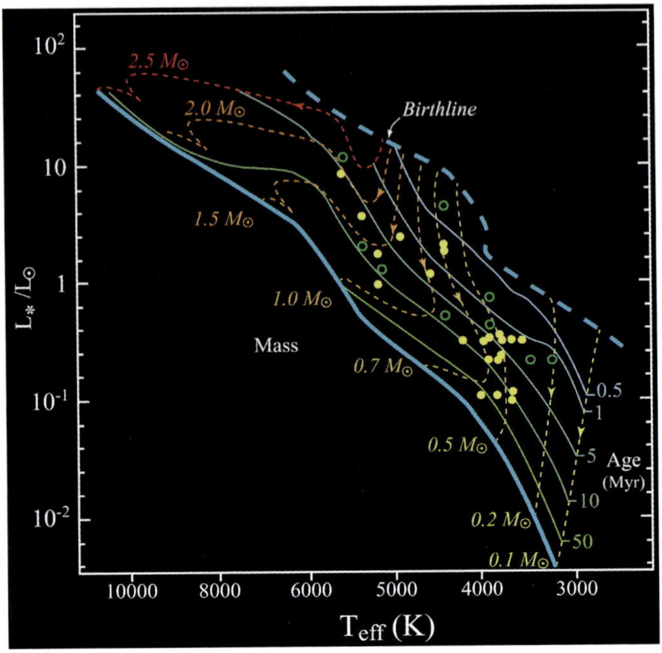

Fig. 1.17. The "Hertzsprung–Russell diagram" for solar-like pre-main sequence stars, also called "T Tauri" stars. By means of sophisticated models, one can assign an age and a mass to a star, once its luminosity and temperature are known. Stars with *filled dots* are surrounded by circumstellar disks, while stars with *open dots* are not. It is thought that disks may disappear because of planet formation. This diagram occupies the area indicated by the *dotted rectangle* in Fig. 1.15

HRD of Fig. 1.15. The dots represent the observational points, and the lines are so-called "evolutionary tracks" drawn from theoretical calculations of the evolution of PMS stars. Such tracks have been calculated by different authors[10], with slightly different results, and their intercomparison leads to uncertainties on stellar masses and ages that may reach 50% in some parts of the diagram. The so-called "birthline" shown on Fig. 1.17 is the locus where newly formed stars become optically visible.

Whereas the observational points are labeled in luminosities and temperatures, the theoretical tracks are labeled in *stellar masses* (dotted lines) and *ages* (continuous lines). The last continuous line indicates the start of the nuclear reactions turning hydrogen into helium ("zero age main sequence"). Using these tracks, one is able to assign (within some uncertainty) a theoretical mass and age to a star, from its observed luminosity and temperature. For instance, reading off the left panel, a T Tauri star with $L = 1.1\,L_{\rm sun}$ and $T = 4500$K

[10] e.g., D'Antona and Mazzitelli 1994, *Astrophys. J. Suppl.*, **90**, 467; Siess, Dufour, and Forestini 2000, *Astron. Astrophys.* **358**, 593.

has a mass of $1\,M_\odot$ and an age of 4.1 Myr. The present-day Sun is located at $M = 1M_\odot, L = 1\,L_\odot$ and $T = 5800\,$K.

The observational points in Fig. 1.17 are filled or open, depending on whether the stars have disks ("classical" T Tauri stars, filled points) or not ("weak-line" T Tauri stars, open points). One notes a concentration of filled points peaking near 3 Myr: in the present case, this is the estimated age of the Chamaeleon cluster (peak of star formation). At later ages, open points are more numerous, indicating that on average the disks disappear over timescales of order a few million years, although older disks are present in this particular cluster. A possible, but poorly understood, cause for this disappearance is the formation of planets.

References

Adams, F.C., Myers, P.C. 2001, *Astrophys. J.*, **553**, 744, Modes of Multiple Star Formation

Amari, T., Luciani, J.F., Mikic, Z., Linker, J. 2000, *Astrophys. J.*, **529**, L49, A Twisted Flux Rope Model for Coronal Mass Ejections and Two-Ribbon Flares

André, P. 1993, in The Cold Universe, Eds. T. Montmerle, C. Lada, F. Mirabel, J. Tran Thanh Van (Editions Frontières, Gif-sur-Yvette), p. 179, *Observations of Protostars and Protostellar Stages*

André, P., Montmerle, T. 1994, *Astrophys. J.*, **420**, 837, From T Tauri stars to Protostars: Circumstellar material and Young stellar objects in the ρ Ophiuchi cloud

André, P., Montmerle, T., Feigelson, E.D., Steppe, H. 1990, *Astron. Astrophys.* **240**, 321, Cold dust around Young stellar objects in the ρ Ophiuchi cloud core

André, P., Ward-Thompson, D., Barsony, M. 1993, *Astrophys. J.*, **406**, 122, Submillimeter continuum observations of ρ Ophiuchi A – The candidate protostar VLA 1623 and prestellar clumps

André, P., Ward-Thompson, D., Barsony, M. 2000, in *Protostars and Planets IV*, eds. V. Mannings, A.P. Boss, S.S. Russell (University of Arizona Press, Tucson), p. 59, From prestellar cores to Protostars: the initial conditions for star formation

Balbus, S.A., Hawley, J.F. 1991, *Astrophys. J.*, **376**, 214, A powerful local shear instability in weakly magnetized disks. I – Linear analysis. II – Nonlinear evolution

Beckwith, S.V.W., Sargent, A.I., Chini, R.S., Güsten, R. 1990, *Astron. J.*, **99**, 924, A survey for circumstellar disks around young stellar objects

Bertout, C. 1989, *Ann. Rev. Astron. Astrophys.*, **27**, 351, T Tauri stars – Wild as dust

Bontemps, S., André, P., Terebey, S., Cabrit, S. 1996, *Astron. Astrophys.*, **311**, 858, Evolution of outflow activity around low-mass embedded Young stellar objects

Butler, R.P., Vogt, S.S., Marcy, G.W., Fischer, D.A., Henry, G.W., Apps, K. 2000, *Astrophys. J.*, **545**, 504, Planetary Companions to the Metal-rich Stars BD -$10^0$3166 and HD 52265

Casanova, S., Montmerle, T., Feigelson, E.D., André, P. 1995, *Astrophys. J.*, **439**, 752, ROSAT X-ray sources embedded in the ρ Ophiuchi cloud core

Chaussidon, M., Robert, F., McKeegan, K.D. 2004, *35th Lunar and Planetary Science Conf.*, LPI, abstract 1568 [or LPI, **35**, 1568]

Clayton, D.D. 1985, *Nature*, **315**, 633, Aluminium clues to the formation of the Solar System

Diehl R., Timmes H. 1998, *Pub. Astr. Soc. Pacific*, **110**, 637, Gamma-ray line emission from radioactive isotopes in stars and galaxies

Desch, S.J.; Connolly, H.C., Jr. 2003, *Met. and Planet. Sci.*, **38**, Supplement, abstr. no. 5257, A cosmic-ray origin for CAI Beryllium 10

Feigelson, E.D., Montmerle, T. 1999, *Ann. Rev. Astron. Astrophys.*, **37**, 363, High-Energy Processes in Young stellar objects

Feigelson, E.D., et al. 2002a, *Astrophys. J.,* **574**, 258, X-ray emitting young stars in the Orion nebula

Feigelson, E.D., Garmire, G.P., Pravdo, S.H. 2002b, *Astrophys. J.*, **572**, 335, Magnetic flaring in the pre-main sequence Sun and implications for the early Solar System

Fleming, T., Stone, J.M. 2003, *Astrophys. J.*, **585**, 908, Local Magnetohydrodynamic Models of Layered Accretion Disks

Fromang, S., Terquem, C., Balbus, S.A. 2002, *Mounth. Not. Roy. Astr. Soc.*, 329, **18**, The ionization fraction in α models of protoplanetary disks

Garmire, G., et al. 2000, *Astron. J.*, **120**, 1426, Chandra X-Ray Observatory Study of the Orion Nebula Cluster and BN/KL Region

Glassgold, A.E., Feigelson, E.D., Montmerle, T. 2000, in *Protostars and Planets IV*, eds. V. Mannings, A.P. Boss, S.S. Russell (University of Arizona Press, Tucson), p. 429, Effects of Energetic Radiation in Young stellar objects

Gounelle, M., Shu, F.H., Shang, H., Glassgold, A.E., Rehm, K.R., Lee, T. 2001, *Astrophys. J.*, **548**, 1029, Extinct radioactivities and protosolar cosmic rays: self-shielding and light elements

Gounelle, M., Shang, S., Glassgold, A.E., Shu, F.H., Rehm, E.K., Lee, T., 2003, *LPI*, **34**, 1833, Early Solar System Irradiation and Beryllium-7 Synthesis

Grosso, N. 2001, *Astron. Astroph.*, **370**, L22, ROSAT-HRI detection of the Class I protostar YLW16A in the ρ Ophiuchi dark cloud

Grosso, N., Montmerle, T., Bontemps, S., André, P., Feigelson, E. D. 2000, *Astron. Astrophys.*, **359**, 113, X-rays and regions of star formation: a combined ROSAT-HRI/near-to-mid IR study of the ρ Oph dark cloud

Grosso, N., Montmerle, T., Feigelson, E.D., André, P., Casanova, S., Gregorio-Hetem, J., *Nature*, **387**, 56, An X-ray superflare from an infrared protostar

Güdel, M. 2002, *Ann. Rev. Astron. Astrophys.*, **40**, 217, Stellar Radio Astronomy: Probing Stellar Atmospheres from Protostars to Giants

Güdel, M., Zucker, A. 2000, in IAU Symp. 95 *Highly Energetic Physical Processes and Mechanisms for Emission from Astrophysical Plasmas* (ASP, San Francisco), p. 393, Gyrosynchrotron Emission from Stellar Coronae

Guilloteau, S., Dutrey, A., Simon, M. 1999, *Astr. Astrophys.*, **348**, 570, GG Tauri: the ring world

Hayashi, C. 1966, *Ann. Rev. Astron. Astrophys.*, **4**, 171, Evolution of Protostars

Hayashi, C., Nakazawa, K., Nakagawa, Y. 1985, in *Protostars and Planets II* (University of Arizona Press, Tucson), 1985, p. 1100, Formation of the Solar System

Hayashi, M.R., Shibata, K., Matsumoto, R. 1996, *Astrophys. J.*, **468**, L37, X-ray flares and mass outflows driven by magnetic interaction between a protostar and its surrounding disk

Igea, J., Glassgold, A.E. 1999, *Astrophys. J.*, **518**, 848, X-Ray Ionization of the Disks of Young stellar objects

Imanishi, K., Koyama, K., Tsuboi, Y. 2001, *Astrophys. J.*, **557**, 747, Chandra Observation of the ρ Ophiuchi Cloud

Lada, C.J. 1985, *Ann. Rev. Astron. Astrophys.*, **23**, 267, Cold outflows, energetic winds, and enigmatic jets around Young stellar objects

Lin, R.P., et al. 2003, *Astrophys. J.*, **595**, L69, RHESSI Observations of Particle Acceleration and Energy Release in an Intense Solar Gamma-Ray Line Flare

Lyo A.-R., et al. 2003, *MNRAS*, **338**, 616, Infrared study of the η Chamaeleon tis cluster and the longevity of circumstellar discs

McKeegan, K.D., Chaussidon, M., Robert, F. 2000, *Science*, **289**, 1334

Montmerle, T. 1997, in Proc. IAU Colloquium 161: *Astronomical and Biochemical Origins and the Search for Life in the Universe,* Eds. C. Moscovici, S. Bowyer, D. Werthimer (Bologna: Editrice Compositori), pp. 267–282, The infancy of solar-type stars and its relevance for the origin of life

Montmerle T. 2002, *New Astr. Rev.*, **46**, 8–10, 573, Irradiation phenomena in young solar-type stars and the early Solar System: X-ray observations and γ-ray constraints

Montmerle, T. 2003, *Adv. Sp. Res.*, **32**, 1067, The role of accretion disks in the coronal activity of young stars

Montmerle, T., Grosso, N., Tsuboi, Y., Koyama, K. 2000, *Astrophys. J.*, **532**, 1097, Rotation and X-ray emission from Protostars

Murphy, R.J., Ramaty, R., Reames, D.V., Kozlovsky, B. 1991, *Astrophys. J.*, **371**, 793, Solar abundances from gamma-ray spectroscopy – Comparisons with energetic particle, photospheric, and coronal abundances

Ozawa, H., Grosso, N., Mountmerle, T. 2004, *Astr. Astrophys.*, in press, The X-ray emission from Young Stellar Objects in the ρ Ophiuchi cloud eon as seen by XMM-Newton

Palla, F., Stahler, S.W. 1999, *Astroph. J.*, **525**, 772, Star Formation in the Orion Nebula Cluster

Randich, S. 2000, in *Stellar Clusters and Associations: Convection, Rotation, and Dynamos*, ASP Conf. Ser. 198, p. 401, Coronal activity among open cluster stars

Reipurth, B., Bally, J. 2001, *Ann. Rev. Astron. Astrophys.*, **39**, 403, Herbig-Haro flows: Probes of early stellar evolution

Santos, N.C., Israelian, G., Mayor, M., Rebolo, R., Udry, S. 2003, *A&A*, **398**, 363, Statistical properties of exoplanets. II. Metallicity, orbital parameters, and space velocities

Shibata, K., Yokoyama, T. 2002, *Astrophys. J.*, **577**, 422, A Herzsprung-Russell-like diagram for solar/stellar flares and coronae: Emission measure vs. temperature diagram

Shu, F.H., Adams, F.C., Lizano, S. 1987, *Ann. Rev. Astron. Astrophys.*, **25**, 23, Star formation in molecular clouds – Observation and theory

Shu, F.H., Shang, H., Glassgold, A.E., Lee, T. 1997, *Science*, **277**, 1475, X-rays and fluctuating X-winds from Protostars

Shu, F.H., Shang, H., Gounelle, M., Glassgold, A.E., Lee, T. 2001, *Astrophys. J.*, **548**, 1029, The origin of chondrules and refractory inclusions in chondritic meteorites

Simon, M., Dutrey, A., Guilloteau, S. 2000, *ApJ*, **545**, 1034, Dynamical Masses of T Tauri stars and Calibration of Pre-Main-Sequence Evolution

Stahler, S.W. 1988, *Astroph. J.*, **332**, 804, Deuterium and the stellar birthline

Stelzer, B., Neuhäuser, R. 2001, *Astron. Astrophys.*, **377**, 538, X-ray emission from young stars in Taurus-Auriga-Perseus: Luminosity functions and the rotation–activity–age relation

Tsuboi, Y., Imanishi, K., Koyama, K., Grosso, N., Montmerle, T. 2000, *Astrophys. J.*, **532**, 1089, Quasi-periodic X-Ray Flares from the Protostar YLW 15

Tsuboi, Y., et al. 2001, *Astrophys. J.*, **554**, 734, Discovery of X-Rays from Class 0 Protostar Candidates in OMC-3

2 Chronology of Solar System Formation

Jean-Marc Petit, Alessandro Morbidelli

The physical processes that contributed to the formation of the Solar System, and more particularly the inner Solar System (the terrestrial region inside Jupiter's orbit), are not yet understood. In particular, the time sequence of these processes is still confused. Many studies, approaching the problem from different points of view (star formation, high energy radiation physics, chemical evolution of the gaseous nebula, continuous and discrete medium dynamics, geophysics, short and long period radioactivity, etc.), have yielded substantial agreement on the fundamental phases of the evolution. However, the characteristic time scales inferred from these approaches agree only partially, and the transitions among the various phases of the accretion process are still obscure. Achieving a global understanding of the formation and evolution of the Solar System will require a multidisciplinary approach.

In the present chapter, we briefly present our current understanding and state its limitations. In the most favored scenario, the formation of the planets, or at least of their precursors the planetesimals, occurs just at the end of the Sun's formation. The main mechanism at work at this time is the accretion of material in the solar nebula, resulting in the formation of micron-size dust grains. These grains then collide with each other, growing larger objects named "planetesimals" that eventually reach a size of tens of kilometers, although in a way that is not yet fully understood. Subsequently a *runaway growth* occurs, in which the big planetesimals grow faster than the small ones, increasing the relative mass difference. This phase ends with the formation of the cores of the giant planets in the outer Solar System and of many planetary embryos and big asteroids in the inner Solar System. The giant planet cores rapidly accrete huge amounts of gas and become the giant planets. In the inner Solar System, the embryos formed during the runaway growth start colliding with each other because of the gravitational perturbation from the giant planets and eventually form the terrestrial planets: Mercury, Venus, Earth and Mars. Finally, a late delivery of material due to the impact of small bodies during their ejection from the Solar System may contribute to the composition of the outer layers of terrestrial planets.

2.1 Origin of the Proto-solar Nebula

The formation of the Sun and the proto-solar nebula is explained in great detail in Chap. 1, Part I by Montmerle. We just here review the main steps that set the stage for planet formation.

All begins with a giant molecular cloud of $10^6 M_\odot$ ($1 M_\odot$ = mass of the Sun), extending over 10^7 AU (1 AU = mean distance from the Earth to the Sun = 1.5×10^8 km), with an average particle number density of only $n_{\rm H} \simeq 10^4\,{\rm cm}^{-3}$. This cloud contracts under its own gravity, with magnetic and turbulent pressure slowing down the contraction. After about 10^7 yr, dense cores form with size, mass and density, respectively, on the order of 10^5 AU, $10^3 M_\odot$ and $n_{\rm H} \simeq 10^5\,{\rm cm}^{-3}$. The cores finally collapse to form the stars. The collapse can occur from the inside out, due to gravitational instability in the core, yielding a spontaneous star formation. In this scenario, the stars are isolated, with a large inter-star distance, as in the Taurus region. The collapse can also be induced by an external shock wave, in which case the stars are formed in clusters, and the inter-star distance is on the order of the envelope size, as in the Orion region (Pudritz, 2002).

At this point, the proto-star is hidden deep inside its envelope. Because of gravity, the envelope falls onto the central star, forming a circumstellar disk which keeps feeding new mass onto the star. The energy generated by the gravitational collapse is almost freely dissipated in a diffuse medium, so the contraction time is short, on the order of 10^5 yr. A significant fraction (10 to 30%) of the accreted matter is then ejected along the rotation axis in order to preserve the total angular momentum.

In the end, the star is fully formed and the envelope has disappeared. This is the exposed phase of the star. At this stage, however, the star becomes opaque to its own radiation.[1] The energy from the gravitational collapse is trapped, and the contraction becomes quasi-static, lasting 10^6–10^8 yr. For solar-type stars, this is known as the "T Tauri" phase. These stars are surrounded by massive disks of 0.003 to $0.1 M_\odot$ (3 to 100 times Jupiter's mass), mostly made of gas, which last for 10^6 to 10^7 yr while still accreting onto the star. The disks are often 10 times larger than our current Solar System and contain a large quantity of dust grains from which planetesimals start to form.

Finally, the gaseous component of the disk disappears, probably because of photo-evaporation due to the radiation produced by the central and neighboring stars. The star keeps contracting without a disk, increasing its temperature, to finally reach the "main sequence", where the temperature and pressure at the center of the star are high enough to initiate the nuclear fusion of hydrogen. This new source of energy manages to stop the gravitational contraction for time scales of order 10^9 to 10^{10} yr. This is the point reached by our Sun 4.5 Gyr ago (Prantzos and Montmerle, 1998).

[1] Photons produced in the core cannot freely and directly escape the star; one can only see the external layer.

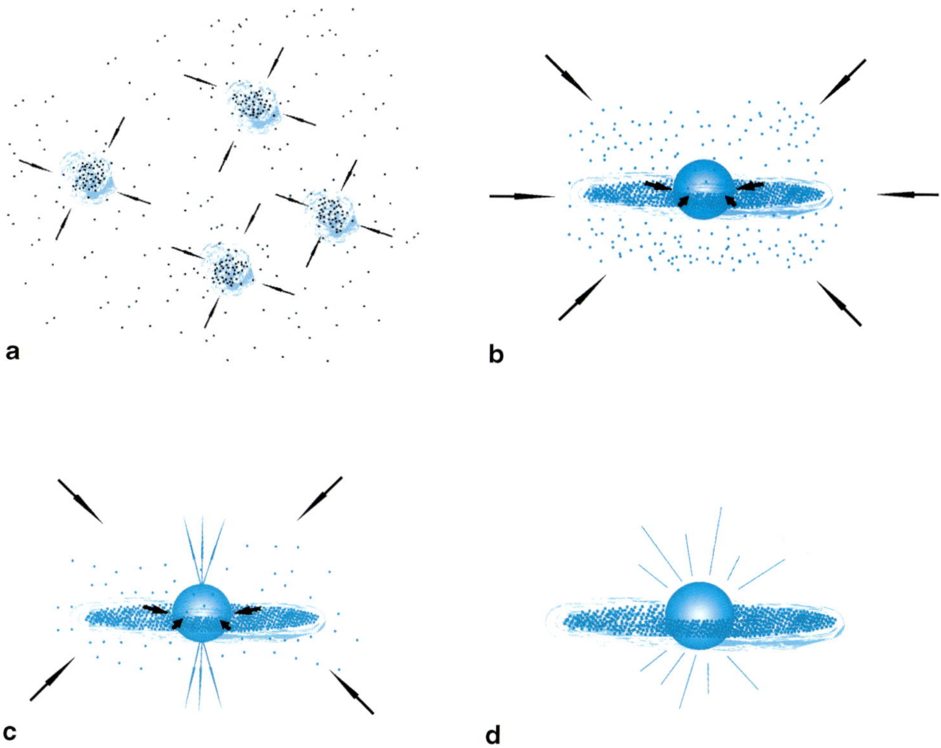

Fig. 2.1. The four stages of the star and stellar nebula formation process. (**a**) Cores form within molecular cloud envelopes on time scales of 10^7 yr. (**b**) A proto-star with a surrounding nebular disk forms at the center of a cloud core. It is hidden in the middle of its envelope. Accretion of the envelope onto the proto-star continues for 10^5 yr. (**c**) The proto-star becomes visible from the outside; this is the exposed phase. Because the proto-star is opaque to its own radiation, the contraction slows down and lasts for 10^6 to 10^8 yr. (**d**) The circumstellar disk that formed during the hidden phase survives for 10^6 to 10^7 yr

2.2 From Micron-size to Kilometer-size Bodies

In the proto-solar nebula, the most refractory materials condense first, gradually followed – as the local temperature drops – by more and more volatile elements.[2] The dust grains thus formed float in the gas. Collisions stick the grains together, forming fractal aggregates, possibly helped by electrostatic and magnetic forces. Other collisions then rearrange and compact the aggregates. When the grains

[2] The lower the condensation temperature of an element, the more volatile that element is. In a rough classification, an element is declared volatile if its condensation temperature is lower than that of water.

reach a size of approximately one centimeter, they begin to sediment rapidly onto the midplane of the disk in a time

$$T_{\rm sed} \sim \frac{\Sigma}{\rho_{\rm p}\Omega r} \sim \frac{\rho \bar{v}}{\rho_{\rm p}\Omega^2 r}, \tag{2.1}$$

where Σ is the nebula surface mass density, ρ is its volume mass density, $\bar{v}$ is the root mean square (r.m.s.) thermic excitation speed of gas molecules, $\rho_{\rm p}$ is the volume mass density of the particles, r is their radius and Ω is the local orbital frequency[3] of the gas (Goldreich and Ward, 1973; Weidenschilling, 1980). Assuming Hayashi's (1981) minimal nebula[4] at 1 AU (gas surface density of $6000\,{\rm g\,cm^{-2}}$, temperature of 600 K, particle volume mass density of $1\ {\rm g\ cm^{-3}}$), one gets

$$T_{\rm sed} \sim \frac{10^3}{r}\ {\rm y}, \tag{2.2}$$

where r is given in centimeters.

The growth from centimeter grains to kilometer-size planetesimals is still unexplained. Collisions between such bodies occur for speeds on the order of 10 m/s (a typical collision speed for Keplerian orbits with eccentricity $e \sim 10^{-3}$) and the physics is still poorly understood. Current theories predict the destruction of the grains at such speeds (Chokshi et al., 1993). However, recent experiments on collisions between micrometer-size grains (Poppe and Blum, 1997; Poppe et al., 2000) give a critical speed for accretion (speed below which grains accrete and above which they fragment) ten times larger than predicted by the theory. This may help solve the fragmentation paradox.

Grains are also subject to gas drag, making them fall on the central star. Because the gas is partially supported by its own pressure, it behaves as if it *feels* a central star of lower mass, and therefore it rotates more slowly than a purely Keplerian orbit at the same heliocentric distance. As solid particles tend to be on Keplerian orbits, they have a speed larger than that of the gas. This gas exerts a drag on the particles, the impact of which is given by the characteristic stopping time (that is, the time a particle would take to reach the speed of the gas if the drag force were constant), defined by:

$$T_{\rm s} = \frac{m\Delta V}{F_{\rm D}}, \tag{2.3}$$

[3] The gas as a whole orbits around the central star with a rotation frequency that depends on the mass of the star, the distance to the star and the pressure of the gas.

[4] The nebula profile is obtained by considering the mass of each planet, augmenting it by the mass needed to bring the observed material to solar composition, and spreading the resulting mass over the annulus ranging from half the distance to the previous planet to half the distance to the next planet. The minimal nebula has a resulting density that decays approximately as $r^{-3/2}$.

where m is the mass of the grain, ΔV is the difference between the particle velocity and the gas velocity, and $F_{\rm D}$ is the gas drag. For small particles, the Epstein gas drag law gives

$$T_{\rm s} \propto \frac{\rho_{\rm p} r}{\Sigma}, \tag{2.4}$$

while for big particles, Stokes' drag law gives

$$T_{\rm s} \propto \rho_{\rm p} r^2 \tag{2.5}$$

(Weideschilling, 1977). We then compare this time to the characteristic Keplerian time, $T_{\rm K} = 1/\Omega$. If $T_{\rm s} \gg T_{\rm K}$, the particle is almost decoupled from the gas. If $T_{\rm s} \ll T_{\rm K}$, it is strongly coupled to the gas, and it tends to move with the gas flow. The maximum effect occurs when $T_{\rm s} \sim T_{\rm K}$, which occurs for meter-size particles. Assuming the Hayashi minimal nebula, $\Delta V \sim 50$–$55\,{\rm m\,s^{-1}}$ at 1 AU from the Sun, the particle's radial speed induced by the gas, that is the drift speed toward the Sun, is then on the order of 10^3–$10^4\,{\rm cm\,s^{-1}}$ (Weideschilling, 1977). Therefore, meter-size particles should fall on the Sun in 5×10^2 to 5×10^3 yr. This is less than the collisional growth time, so that the particles should fall on the Sun before they can grow massive enough to decouple from the gas.

One way out of this problem is to have a density of solids larger than that of the gas. In this case, the growth time scale would be faster than the radial drift time scale. However, having disks with more dust than gas is not supported by observations or by current theories.

Another possibility is the existence of vortices in the proto-planetary disk, produced by turbulent viscosity of the nebula (Tanga et al., 1996). In this model, 70 to 90% of the particles are trapped in anti-cyclonic vortices. Once trapped, the particles do not continue to fall toward the Sun, but rather fall towards the vortex center. Falling time scales vary from a few tens of years to a few thousand years, depending mainly on the size of the particle and on the heliocentric distance (which increases all dynamical time scales). At the vortex center, particle dynamics are stable over the lifetime of the vortex. The particles' relative velocities are reduced (particles tend to follow the gas stream lines, so they all tend to have the same velocity) and the local density is increased, enhancing the accretion process. Therefore, vortices would help the accretion of kilometer-size planetesimals in two ways: by stopping sunward drift of meter-size bodies and by speeding up the accretion process due to the accumulation of the bodies at the vortex centers.

2.3 From Planetesimals to Planets

2.3.1 Formation of Planetary Embryos

After this troubled period, the gas nebula contains a disk of planetesimals with sizes ranging up to a few kilometers. However, the dynamics of accretion change because they become dominated by the effects of the gravitational attraction among the planetesimals, increasing the collisional cross-sections. A phase of runaway growth starts, during which the big bodies grow faster than the small ones, hence increasing their relative mass ratio (Greenberg et al., 1978). This process can be summarized by the equation:

$$\frac{\mathrm{d}}{\mathrm{d}t}\left(\frac{M_1}{M_2}\right) = \frac{M_1}{M_2}\left(\frac{1}{M_1}\frac{\mathrm{d}M_1}{\mathrm{d}t} - \frac{1}{M_2}\frac{\mathrm{d}M_2}{\mathrm{d}t}\right) > 0\,, \tag{2.6}$$

where M_1 and M_2 are, respectively, the masses of the "big" and "small" bodies and can be explained as follows.

Generally speaking, accretion is enhanced by a high collision rate, which occurs when the relative velocities are large, but also by large collisional cross-sections and gentle impacts, which occur when the relative velocities are low. Therefore, the relative velocity between the different planetesimal populations is the critical parameter that governs the growth regime.

At the beginning of the runaway growth phase the large planetesimals represent only a small fraction of the total mass. Hence, the dynamics are governed by the small bodies, in the sense that the relative velocities among the bodies are on the order of the escape velocity of the small bodies, $V_{\mathrm{esc}(2)}$. This velocity is independent of the mass, M_1, of the big bodies and is smaller than their escape velocity, $V_{\mathrm{esc}(1)}$. For a given body, the collisional cross-section is enhanced with respect to the geometrical cross-section by the so-called *gravitational focusing* factor:

$$F_{\mathrm{g}} = 1 + \frac{V_{\mathrm{esc}}^2}{V_{\mathrm{rel}}^2}\,, \tag{2.7}$$

where V_{esc} is the body's escape velocity and V_{rel} is the relative velocity of the other particles in its environment. Because $V_{\mathrm{rel}} \sim V_{\mathrm{esc}(2)}$, the gravitational focusing factor of the small bodies ($V_{\mathrm{esc}} = V_{\mathrm{esc}(2)}$) is of order unity, while that of the large bodies ($V_{\mathrm{esc}} = V_{\mathrm{esc}(1)} \gg V_{\mathrm{esc}(2)}$) is much larger. In this situation one can show that mass growth of a big body is described by the equation

$$\frac{1}{M_1}\frac{\mathrm{d}M_1}{\mathrm{d}t} \propto M_1^{1/3} V_{\mathrm{rel}}^{-2} \tag{2.8}$$

(Ida and Makino, 1993). Therefore, the relative growth rate is an increasing function of the body's mass, which is the condition for runaway growth.

The runaway growth stops when the mass of the large bodies becomes important (Ida and Makino, 1993) and starts to govern the dynamics. This occurs when:

$$n_1 M_1^2 > n_2 M_2^2\,, \tag{2.9}$$

where n_1 (n_2) is the number of big bodies (small bodies). In this case, $V_{\mathrm{rel}} \sim V_{\mathrm{esc}(1)} \propto M_1^{1/3}$ and, hence, $(1/M_1)(\mathrm{d}M_1/\mathrm{d}t) \propto M_1^{-1/3}$. The growth rate of the planetary embryos gets slower and slower as the bodies grow and the relative mass differences among the embryos also slowly reduce. In principle, one might expect that the small bodies catch up, narrowing their mass difference with the embryos. But, in reality, the now-large relative velocities prevent the small bodies from accreting with each other; the small bodies can only participate in the growth of the embryos.

The runaway growth phase happens throughout the entire disk, with time scales that depend on the local dynamical time (Keplerian time) and on the local density of available solid material. This density will also determine the maximum size and/or number of embryos when the runaway growth phase ends (Lissauer, 1987). Assuming a reasonable surface density of solid materials, the runaway growth process forms planetary embryos of lunar to Martian mass at 1 AU in 10^5–10^6 yr, with embryos separated by a few 10^{-2} AU. Beyond the so-called *snow line* at about 4 AU, where condensation of water ice (which occurs

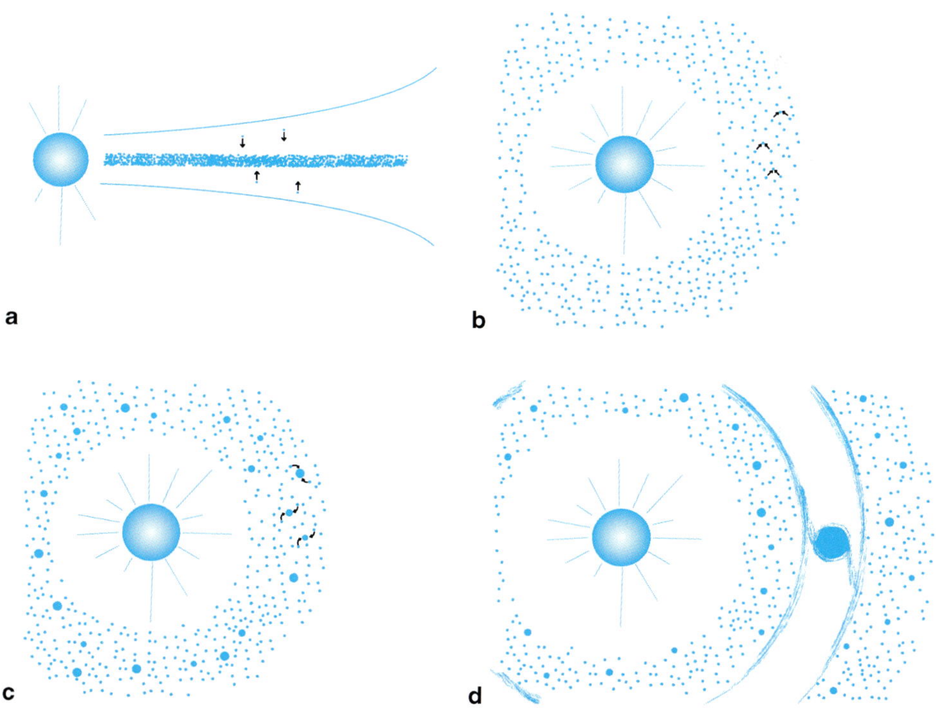

Fig. 2.2. Sketch of the accretion process in the early Solar System. (**a**) During the cooling of the nebula, dust grains condense. Through collisions, these grains accrete and reach centimeter size. They then *sediment* onto the central plane of the disk. (**b**) Once in the median plane, particles continue to grow, in an ordered manner, due to low collision speeds. (**c**) When bodies reach kilometer size, they begin to gravitationally attract their neighbors, with the biggest being more efficient than the smallest. This is the *runaway growth* phase. (**d**) When some embryos reach a large enough mass (a few Earth masses), they attract the nebular gas and form giant planets like Jupiter and Saturn

thanks to the low temperature) enhances the surface density of solid material, runaway growth could produce embryos as large as several Earth masses, $M_\oplus$, in a few million years (Thommes et al., 2003).

2.3.2 Formation of the Giant Planets

Observations and models of giant planet interiors give some important constraints on the composition and mass of the giant planets:

- Jupiter has a mass of $314\,M_\oplus$ and contains $\sim$ 10–45 $M_\oplus$ of heavy elements (elements other than hydrogen and helium; Guillot et al., 1997);
- Saturn has a mass of $94\,M_\oplus$ and contains $\sim$ 10–20 $M_\oplus$ of heavy elements;
- Uranus and Neptune have masses of 14 and 17 $M_\oplus$, respectively, of which only $\sim$ 1–2 $M_\oplus$ is hydrogen and helium.

To account for these constraints, the best current models for the formation of Jupiter and Saturn hypothesize a three-stage formation (Pollack et al., 1996):

1. The solid core accretes as explained in the previous section; beyond the snow line, the surface density of solid material is enhanced several times due to the presence of ice grains. This allows embryos to grow to about $10\,M_\oplus$ on a time scale of a million years (Thommes et al., 2003).
2. The accretion of the solid core slows down (see above), while a slow accretion of nebular gas begins due to the gravity of the core. The gas accretion continues at a roughly constant rate over many million years, until a total mass of 20 to 30 $M_\oplus$ is reached.
3. When the mass of the proto-planet reaches $\sim$ 20–30 $M_\oplus$, the gas gravitationally collapses onto the planet. The mass of the planet grows exponentially and reaches hundreds of Earth masses in $\sim 10^4$ yr.

The end of the exponential gas accretion is not yet understood. Most likely, the growth of the giant planets slows down when a gap is opened in the gaseous disk by the tidal forces from the planet, and stops when the nebula is dissipated (probably by photo-evaporation from the central star). The dissipation time of the nebula is poorly constrained both by models and observations, but should occur within 10 Myr. With this hypothesis, the Pollack et al. model yields an initial solid material surface density of

$$\sigma_{\mathrm{init}} \sim 10\,\mathrm{g/cm^2} \tag{2.10}$$

at Jupiter's location. Assuming a plausible surface density variation of $\sigma \propto a^{-2}$, where a is the distance to the Sun, the initial density of solid material at Saturn's location is $\sigma_{\mathrm{init}} \sim 3\,\mathrm{g/cm^2}$. The composition and formation time scale derived from this model for Saturn are in good agreement with observations.

For Uranus and Neptune, it is generally assumed that phases one and two of the previous model lasted more than 10 Myr and hence the nebula disappeared before the third phase of accretion could start. This would explain why these two planets accreted only a few Earth masses of gas. However, the formation of the cores of Uranus and Neptune is an open problem. Pollack et al. (1996) applied their model to Uranus. The initial surface density would be $\sigma_{\rm init} \sim 0.75\,{\rm g/cm^2}$. They assumed that the planetesimals needed to form Uranus were smaller (a few km to a few tens of km) than those needed to form Jupiter or Saturn (100 km radius); therefore, phase one lasted from 2 to 10 Myr, while phase two lasted past 10 Myr, thus accreting only a few Earth masses of gas. Extrapolating this to Neptune's location, they claimed that Neptune could still form in the required time. However, recent direct numerical simulations seem to imply that accretion by runaway growth requires that Uranus and Neptune (or at least their cores) formed very close to Jupiter and Saturn (Thommes et al., 2003), or even in between them (Thommes et al., 1999). Conversely, the formation of Uranus and Neptune from a system of smaller planetary embryos, similar to what happened for the terrestrial planets (see below), does not work, according to direct simulations performed by Levison and Stewart (2001). It is possible that the solution to the problem lies in the migration and excitation damping suffered by the embryos due to the tidal interaction with the gaseous disk.

The discovery of extrasolar giant planets of mass similar to that of Jupiter has forced a re-evaluation of a long-discarded giant-planet formation mechanism through a gravitational instability in the proto-planetary disk. Boss (1998) showed that a very massive disk ($> 0.1\,M_\odot$) may be cool enough to become gravitationally unstable. In about 10^3 yr, this would create giant gaseous protoplanets that are gravitationally bound and tidally stable and so should eventually form giant planets. This very short time scale can explain the presence of giant planets around young stars. New models of the interior of Jupiter suggest the possibility of a core mass that is considerably smaller than the one usually assumed ($10\,M_\oplus$ or more), or even of no core at all (Guillot et al., 1997). If this is the case, then only a gravitational instability mechanism could have formed Jupiter. Boss (2000) has shown that the instability can develop in a more conventional circumstellar disk of mass $\sim 0.1\,M_\odot$. However, this mechanism tends to form very big giant planets like Jupiter. It may not be able to form giant planets as low in mass as Saturn. If Uranus and Neptune were formed through this mechanism, then they must have lost their envelopes by some late process. Moreover, it is questionable whether asteroids and Jupiter trojans could have formed if Jupiter had accreted its current mass in the short time scale implied by the direct collapse mechanism (Kortenkamp and Wetherill, 2000; Kortenkamp et al., 2001). Even if these considerations cast serious doubts on the role of gravitational instability in the formation of our Solar System, it seems likely that this mechanism may have formed some of the recently discovered extrasolar giant planets (Boss, 2002).

2.3.3 Formation of the Terrestrial Planets and Primordial Sculpting of the Asteroid Belt

After a few 10^5 years of runaway growth, the embryos in the terrestrial planet region and in the asteroid belt region have lunar to Martian masses. They govern the local dynamics and start perturbing each other. The system becomes unstable, and the embryos' orbits begin to intersect (Chambers and Wetherill, 1998). Because of mutual close encounters, the embryos' dynamical excitation (increase of eccentricity and inclination[5]) moderately increases, and accretional collision among embryos starts to occur. The situation drastically changes when Jupiter and Saturn acquire their current masses. These two planets strongly perturb the dynamical evolution of the embryos in the asteroid belt region between 2 and 5 AU. The latter acquire a large dynamical excitation, begin to cross each other, and frequently cross the orbits of the embryos in the terrestrial planet region. The collision rate increases. Despite the high relative velocity, these collisions lead to accretion because of the large mass of the embryos.

The typical result of this highly chaotic phase – simulated with several numerical N-body integrations – is the elimination of all the embryos originally situated in the asteroid belt and the formation of a small number of terrestrial planets on stable orbits in the 0.5–2 AU region on a time scale of $\sim$ 100 Myr.

This scenario has several strong points:

- Planets are formed on well separated and stable orbits only inside 2 AU. Their number typically ranges from 2 to 4, depending on the simulations, and their masses are in the range of Mars to Earth mass (Chambers and Wetherill, 1998; Agnor et al., 1999)
- Quasi-tangent[6] collisions of Mars-mass embryos into the proto-planets are quite frequent (Agnor et al., 1999). These collisions are expected to generate a disk of ejecta around the proto-planets (Canup and Asphaug, 2001), from which a satellite is likely to accrete (Canup and Esposito, 1996). This is the standard, generally accepted scenario for the formation of our Moon.
- The accretion time scale of the terrestrial planets is about 100 Myr. This is compatible with several constraints on the chronology of accretion coming from geochemistry (Allègre et al., 1995). On the other hand, Hf–W chronology[7] seems to indicate that the formation of the Earth's core occurred within the first 30 Myr (Yin et al., 2002; Kleine et al., 2002; Schoenberg et al., 2002). This might suggest that Earth's accretion was faster than it appears in the simulations. However, if the cores of the embryos are not mixed with the mantles during the collisions – as indicated by SPH simulations[8] (Canup and Asphaug, 2001) – this time scale would measure the mean differentiation age of the embryos that participated in the formation of the Earth, not the time required for our planet to accrete most of its mass.

[5] The inclination is the angle between the invariant plane, i.e. the plane perpendicular to the total angular momentum of the Solar System, and the plane of the orbit.

[6] The relative velocity is almost parallel to the local horizontal plane at the impact point.

[7] Measurement of tungsten (W) isotope composition. ^{182}W is produced by isotopic decay of ^{182}Hf (hafnium) with a half-life of 9.4 Myr.

[8] Smooth Particle Hydrodynamics. See Benz (1990) for a general presentation of this method.

- All the embryos located beyond 2 AU are eliminated in two-thirds of the simulations (Chambers and Wetherill, 2001). They are either dynamically ejected from the Solar System, or they collide with the Sun, or they are accreted by the forming terrestrial planets.
- Simultaneously, the small planetesimals are subject to the combined perturbations of the giant planets and of the embryos (Petit et al., 2001). The dynamical excitation increases very rapidly (on a time scale of 1–2 Myr) and most of the small planetesimals are eliminated in a few million years, either by ejection from the Solar System or by collision with a growing planet or the Sun. In the asteroid belt (2–4 AU range) this leaves a remnant population of small bodies (the asteroids) on stable orbits with large eccentricities and inclinations, which contains only a very small fraction of the total mass initially in the region. This scenario explains the current mass deficit of the asteroid belt, the eccentricity and semi-major axis distribution of the largest asteroids (Fig. 2.3) and other more subtle properties of the asteroid belt population, such as the partial mixing of asteroid spectral types (Gradie and Tedesco, 1982). In the terrestrial planet region, a non-negligible fraction of the small planetesimal population reaches orbits with very high inclinations. These orbits are unstable but long-lived, so that this population decays slowly with time, with a half-life close to 100 Myr (Morbidelli et al., 2000b). This population could have played an important role in the so-called «Late Heavy Bombardment» of the terrestrial planets. The chronology of formation of the lunar basins, however, suggests that the cratering rate suddenly increased some 600 Myr after the formation of the planets. The origin of this cataclysm is unknown and cannot be due to the long-lived population of high-inclination planetesimals in the terrestrial region.

However, this scenario of terrestrial planet formation suffers from some weaknesses:

- The final orbits of the planets formed in the simulations are typically too eccentric and/or inclined with respect to the real ones. This could be due to the fact that the current simulations neglect the so-called phenomenon of "dynamical friction", namely, the effect of a large population of small bodies carrying cumulatively a mass comparable to that of the proto-planets. Dynamical friction should damp the eccentricities and inclinations of the most massive bodies.
- Obliquities of the terrestrial planets should have random values. However, in reality, only one planet (Venus) has a retrograde spin. Moreover, all planetary obliquities are compatible with an initial 0° obliquity, modified by the subsequent evolution in the framework of the current architecture of the planetary system (Laskar and Robutel, 1993).
- The planet formed in the simulations approximately at the location of Mars is typically too massive.

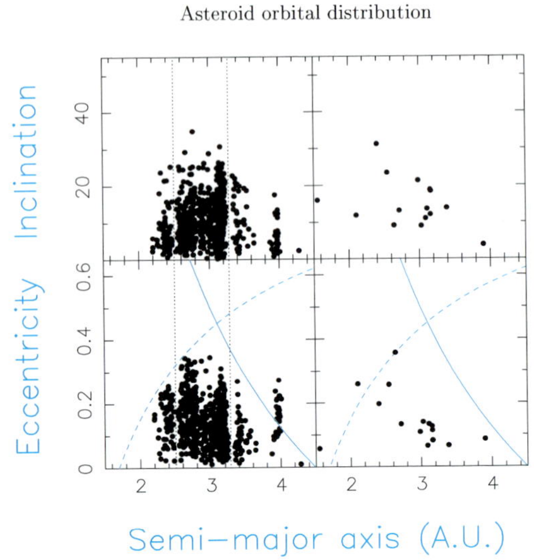

Fig. 2.3. *Left*: Inclination (*top*) and eccentricity (*bottom*) as a function of semi-major axis for real asteroids with diameter larger than 50 km (*solid line*: aphelion distance of 4.1 AU; *dashed line*: perihelion distance of 1.7 AU; this corresponds to the limits of the region outside of which orbits are unstable over the age of the Solar System due to interactions with Jupiter or Mars). *Right*: Final distribution of the surviving bodies in a simulation of the primordial sculpting of the asteroid belt (Petit et al., 2001). In this simulation, 2000 test particles initially on circular and coplanar orbits between 2 and 4 AU are integrated over 100 Myr under the gravitational perturbations exerted by the giant planets and by the embryos. Only 0.7% of the integrated particles survive at the end of the simulation. This model correctly reproduces the mass depletion of the asteroid belt and its large dynamical excitation. It can even produce asteroids in the Hilda region, with a semi-major axis of ~ 3.9 AU. The main concern is the almost complete depletion of the inner belt ($a < 2.5$ AU) due to the large orbital eccentricities of the planets formed in this model

2.3.4 Origin of Water on Earth

Simulations of terrestrial planet formation show that two-thirds of the planets accrete at least one embryo originally located in the outer asteroid belt (Morbidelli et al., 2000a). Carbonaceous meteorites thought to come from that region contain about 10% of their mass in water, so it is plausible to assume that embryos formed in the outer asteroid belt were comparably water-rich. In this way, the Earth could have accreted some 3–6×10^{25} g of water, or 10 to 20 times the amount of water currently on our planet (Morbidelli et al., 2000a).

On the other hand, small asteroids and comets would not have been important agents for the delivery of water to the Earth, because of mass and time scale reasons. Indeed, the planetesimals of cometary nature formed in the Jupiter-Saturn region were dynamically eliminated in the first 10^5 yr after the formation of these giant planets. At that time, planets in the terrestrial region had not yet formed from the population of planetary embryos. If water from these comets had already been accreted by the embryos, it would probably have been lost during the numerous subsequent giant impacts leading to the formation of the terrestrial planets. The comets originally in the Uranus–Neptune region might have reached the terrestrial planet region at much later epochs, during or after the latest stages of formation of the terrestrial planets. However, the collision probability of these objects with the Earth is tiny, so that even if $100\,M_\oplus$ of material existed in the form of comets, the water accreted by our planet would have been only 10% of that currently existing on Earth. This may explain why Earth's water has an isotopic composition very different from cometary water: comets would have contributed only a small fraction of the Earth's water budget. If several Earth masses of hydrated asteroids existed in the outer asteroid belt, these asteroids could have brought to the Earth roughly the current amount of water on the planet, according to the model of primordial depletion and excitation explained in the previous section. However, most of this water would have been delivered during the first 10^7 yr after the formation of Jupiter, namely, quite early during the history of the accretion of the Earth. It is therefore likely that most of this water was lost during the numerous giant impacts that occurred afterwards. In conclusion, from the dynamical viewpoint, the delivery of water by a hydrated embryo is the most plausible hypothesis (Morbidelli et al., 2000a). In this case, because much more water than necessary is delivered to the Earth, a large fraction can be lost during the giant impacts and still retain the Earth's current amount of water.

The presence of a large amount of water at some time during the Earth's formation (as predicted by the embryo delivery model) is compatible with recent geochemical models (Abe et al., 2000). However, some problems in this scenario appear when more careful attention is paid to chemical constraints. The first problem concerns the budget of siderophile elements in the Earth's mantle (Drake and Righter, 2002). Carbonaceous chondrites are characterized by a definite mass ratio between the siderophile elements and the water that they contain. Therefore, an embryo of carbonaceous chondritic nature would have brought to Earth siderophile elements in the same proportion as water. The amount of siderophile elements in the Earth's mantle is actually compatible with the delivery of the *current* amount of water, but not significantly more. This seems to be in conflict with the violent scenario of Earth formation and water accretion that results from the numerical simulations, because in this scenario one would expect that significant losses in the water budget occurred. The problem is currently unsolved. It is possible that the water-carrying embryo was partially differentiated, so that the ratio between siderophile elements and water in its mantle was lower than in the Earth-collected carbonaceous chondrites (debris of smaller undifferentiated bodies). If the core of the impactor joined the core of the proto-Earth without contaminating the mantle of the latter (as shown in SPH simulations), then a large amount of water could have been delivered together with a comparatively small amount of siderophiles. Another possibility is that the Earth underwent differentiation after the delivery of the water, depleting most of its mantle of the co-delivered siderophile elements. The chemistry of differentiation of water-rich bodies, however, is largely unknown. Some experts hypothesize that differentiation could not happen in the presence of water.

A second problem concerns the water's oxygen isotope composition, which is identical on the Earth and the Moon within measurement accuracy (Wiechert et al., 2001). All other bodies of the Solar System, from Mars to the asteroids, have a different oxygen isotope composition. In the modern scenario of the Moon's formation (Canup and Asphaug, 2001), the material that forms the proto-lunar disk is mostly of impactor origin. Therefore, the Earth and the proto-lunar impactor had to have the same striking similarity in oxygen isotope composition that we observe today on the Moon. This seems to be completely inconsistent with proto-planets growing by accretion of embryos from a large range of heliocentric distances, possibly up to the distance of the asteroid belt. A possible solution is that the amount of impactor material that ends up in the proto-lunar disk has been overestimated. In fact, to simplify the calculations, the SPH simulations used to model the giant impact assume a proto-Earth that is not spinning. As a consequence, to produce the angular momentum characteristic of the Earth–Moon system, the trajectory of the impactor is required to be quasi-tangent to the Earth's surface; this maximizes the amount of impactor mass in the disk. If the Earth's spin is assumed to be fast, then a more head-on impact could be used and would possibly enhance the Earth's contribution to the disk. Another more appealing possibility is that during the impact, or soon after the impact, when the temperatures reached silicate evaporation values, the oxygen isotope distributions of the proto-Earth and the impactor rapidly equilibrated (Chaussidon and Robert; Stevenson, private communications).

This discussion shows that the last word on the origin of the Earth has not yet been said, and that a final understanding will require much interdisciplinary work.

2.4 The Future

Our understanding of the formation and early evolution of our Solar System is still incomplete, although we can outline many of the steps. The formation of a proto-star from a molecular cloud seems to be broadly understood, as are the dynamics and the transport of different components through the Solar System. The main phases of giant and terrestrial planet formation can be modeled. In the coming years, much work will have to be invested in detailed study of the problems mentioned. Precisely dating the key events in the formation of the Solar System and establishing their chronological relationship will require a multidisciplinary approach. We summarize below the points that, in our opinion, are the most important to study in the near future.

- We need to improve our knowledge of the disk and the nebula, in particular their composition and the physico-chemical conditions in the early Solar System.
- Measuring the abundance of radioactive elements or of their daughter products gives chronological information on events lasting from less than one Myr to a few Gyr. However, this dating method assumes that the isotope composition of pertinent chemical elements was uniform throughout the Solar System. This should be the case if the radioactive elements were present in the molecular cloud prior to the collapse of the proto-solar nebula, often postulated as being produced during the supernovae explosion of the fastest evolving stars. However, the recent discovery that ^{10}Be and ^{7}Be were present in the early Solar System (Chaussidon et al., 2001, 2002) throws serious doubts on this view. These two radionuclei have extremely short half-lives, so that their production by a supernova would require an ad hoc synchronism between the supernova event and the Solar System's formation. It is more likely that these radionuclei formed in the vicinity of the Sun due to irradiation processes and then were disseminated through the proto-solar disk (Gounelle et al., 2001, 2003). In this case, there is no reason to believe that the resulting isotopic composition of beryllium was uniform throughout the solar disk. If the same is also true for other important radioactive elements, like ^{26}Al, most of the accretion chronology established so far would be invalidated. Therefore, theoretical studies as well as laboratory experiments will have to be carried out in order to determine the origin of short-period radioactive isotopes.
- Studies of short-period radioactive isotopes will allow us to precisely determine the formation time of the CAIs (*Calcium Aluminum Inclusions*) and chondrites with respect to solar and nebular evolution. This in turn will constrain the thermal evolution of the nebula on short time scales (seconds to minutes) as well as the energy sources necessary for their formation (initial heating, slow cooling).
- Current theories cannot account for the growth of bodies from millimeter-size grains to multi-kilometer planetesimals. Studies must tackle this question in two different ways: first from the point of view of collisional accretion of grains, to determine precisely the critical collision velocity from theory and experiments, and second from the point of view of particle-gas interaction, to determine the potential trapping of particles in vortices, the local relative velocities, the local surface density of particles and the radial velocity induced by gas drag.
- Although the first stages of giant planet formation seem rather well understood, we still have to understand how the exponential collapse of gas ends: is it due to the formation of a gap around the planet's orbit, or is it because of the dissipation of the nebula?
- Similarly, we will need to study in greater detail the interaction between proto-planets and nebulae: in our Solar System this interaction does not

seem to have generated a perceptible migration of our giant planets, whereas this phenomenon is invoked to explain the migration of planets over several AUs in various extra-solar systems.

- The differentiations of the Martian and Earth cores are supposed to occur around 15 Myr and before 30 Myr, respectively, according to the most recent interpretation of the Hf–W radioactive couple. A recent paper by Yoshino et al. (2003), studying the percolation of iron through the mantle to form the core, claims that the differentiation must occur in less than 3 Myr, everything else being just a mix of already differentiated bodies. A better knowledge of the physics of collisions and the destiny of the cores and mantles is required to solve this problem.
- The numerical simulation of terrestrial planet accretion needs to be improved, including a treatment for dynamical friction. Until the final orbits of the planets have eccentricities and inclinations comparable to the real ones and the mass of Mars is correctly reproduced, we cannot be sure we understand the terrestrial planet formation process. This is particularly true for the accretion time scale and the range of heliocentric distances from which the parent embryos came.
- The similarity in oxygen isotope composition between the Moon and the Earth challenges our understanding of terrestrial planet formation. Improved SPH simulations of the Moon-forming event will be required to better understand which fraction of the Moon's mass was originally in the impactor and which in the proto-Earth. The possibility that the oxygen isotope distributions of the proto-Earth and of the impactor rapidly equilibrated during/after the impact event needs to be investigated in detail.

Appendix:

A: Chemical and Isotopic compositions

In the previous chapter, we have described our understanding of the formation of the Solar System based on dynamical modelling and observations of seemingly similar systems. More information can be gathered from the physical and chemical system properties and, in particular, from its isotopic compositions.

Ordinary chondrites show a *zoned* formation (i.e preferential formation of bodies of a given chemical composition in certain regions rather than formation of all chemical compositions everywhere). Deficiency of some volatil elements results from condensation rather than later vaporisation : this implies a zoning at formation time. However, the continuity of the concentration curve implies a temporal evolution of the temperature versus heliocentric distance function. One notes a similar heterogeneity of isotopic concentrations, in particular ^{17}O and ^{18}O with respect to ^{16}O. Since isotopic compositions are homogeneized very rapidly, this seems to prevent the possibility of large scale turbulence.

Chondrites consist of three main components:
- CAIs (*Calcium Aluminium-rich Inclusions*)
- chondrules (Olivine, pyroxene)
- the matrix (HO, organic compounts, etc.).

Laboratory experiments on the formation of CAIs and chondrules tell us about the thermal evolution of the solar nebula. CAIs are formed at very high temperature; they are the oldest solids formed in the protosolar nebula and their structure requires that they cooled rapidly (in a few hours) in a very low density medium. Chondrules are the results of a gas-to-solid phase transition which generates some solid precursors, followed by heating to very high temperature (1900–2100 K) and a rather fast cooling (1400–1700 K). (Zanda, 2004). The radiation cooling of an isolated chondrule lasts only a few seconds and produces glass. The observed chondrules are a mixture of glass and cristal, which implies a cooling lasting at least 10 minutes and necessitates an external heat source. Moreover, the external layer of the chondrules is dusty, so the cooling must have occured in a dusty gas.

The heating and cooling times given here do not correspond to the entire nebula but rather to local events, probably due to shock waves.

B: Chronology from Radioactivity

For longer time scales, we resort to radioactivity studies. Radioactivities can be classified in two groups:

- *short term radioactivities*, (also called *extinct* radioactivities since the radioactive elements have totally decayed today) which, depending on their

half-lifes (see Tab. 2.1), tell us about phenomena lasting from a few hundred thousand years to a few years. They allow a precise relative dating.
- *long term radioactivities* are those for which the radioactive isotopes still exist, with half-lives ranging from 10^9 to 10^{11} years. They allow an absolute timing of Solar System events which however might not have by itself enough resolution to reveal the detailed chronology of the first nebular events.

Radioactive dating relies on a good knowledge of the sources of radioactive elements: an original molecular cloud enriched by sustained nucleosynthesis from other stars, or irradiation by the young Sun. In addition, one must assume a good isotopic homogeneisation of the medium. One then measures the time elapsed after isolating a body from a homogeneous medium with a significant chemical fractionation between the radioactive parent isotope and its stable daughter product. This is exactly what happens during chondrule formation. It will be nearly impossible to date the formation of primitive meteorites which occurs later since (i) the matrix results from the condensation of the gaz as a whole and hence show no strong fractionation and (ii) accretion in itself does not induce strong fractionation. But one can date the differentiated meteorites which experienced heating, and hence metal silicate and/or silicate/silicate fractionation. In the case of the Earth, the core differentiation, the primitive outgassing or extraction of the crust can be dated because of the large chemical fractionations they induce.

Using the Uranium-Lead chronometer (see Tab. 2.1) the absolute age of the oldest CAIs has been determined to be $4.567.2 \pm 0.6$ Gyr . This is conventionally taken as reference for the "age of the Solar System" though it only represents the age of the oldest solids found in meteorites and dated so far. It was also shown that internal thermal evolution of planetesimals (the parent body(ies) of the H-chondrites) occurred between 4.563 and 4.502 Gyr ago (Gopel et al., 1993) and that the end of the accretion, differentiation and atmosphere extraction of the Earth are about 0.1 Gyr younger than the Solar System age (Allegre et al., 1995).

To establish a precise chronology of the different events that led to the formation of the planets one must use the short term radioactive isotopes. From the $^{26}Al/^{26}Mg$ chronometer one can see that the formation of the chondrules was spread over 3 Myr after the formation of the first CAIs. Other short term chronometers show that the differentiation of the parent bodies of the iron meteorites and of the eucrites occurred over about 3–4 Myr. The differentiation (core-mantle) of Mars took about 10 Myr and that of the Earth took 29 Myr or more (Yin et al. 2002). However, as discussed above, the interpretation of the data on the differentiation chronology of the planets is subject of debate.

Ages given above assume that radioactive elements were homogeneously spread in the nebula and that their creation stopped at time $t = 0$. This can be true if those elements were created by the stellar and supernova environment of the molecular cloud. The end of radioactive elements creation would then correspond to the isolation of the nebula during the contraction of the cloud. However, the recent discovery of the extinct radioactivity of Be seems to contradict this point of view. Be is not created in supernovae, but rather by spallation[9], that is, by energetic irradiation. For this kind of radioactivity, one must move the origin of time to the epoch when our young Sun emitted a large amount of very energetic radiations. In addition, this could impair the homogeneity hypothesis.

Table 2.1. Lifetime of several short term (left column) and long term (right column) radiaoctive elements.

Parent-Daughter	Lifetime (10^6 years)	Pair	lifetime (10^9 years)
^{41}Ca-^{41}K	0.1	^{235}U-^{207}Pb	0.7
^{26}Al-^{26}Mg	0.7	^{40}K-^{40}Ar	1.2
^{10}Be-^{10}B	1.5	^{238}U-^{206}Pb	4.5
^{60}Fe-^{60}Ni	1.5	^{232}Th-^{208}Pb	14
^{53}Mn-^{53}Cr	3.7	^{176}Lu-^{176}Hf	35
^{107}Pd-^{107}Ag	6.5	^{187}Re-^{187}Os	46
^{182}Hf-^{182}W	9.4	^{87}Rb-^{87}Sr	49
^{129}I-^{129}Xe	16.4	^{147}Sm-^{143}Nd	101

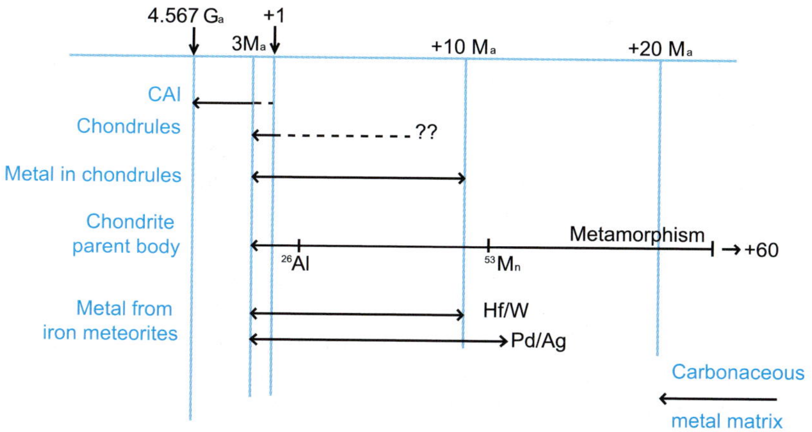

Fig. 2.4. Approximate chronology of formation of several components of the Solar System. Radioactive elements used for some of these dates are indicated

[9] Nuclear reaction during collision between nuclei. In the present case, this refers to the collision of energetic hydrogen or helium nuclei with grains in the nebula. The target nuclei are mostly carbon and oxygen

References

General

Albarede F. 2003. geochemistry : an introduction. Gordon and Breach.

Faure G. 1986. Principles of isotopes geology. John Wiley and sons.

Prantzos, N. and Montmerle, T. 1998. dans *Naissance, vie et mort des étoiles,* PUF, N° 330.

Specialised

Abe, Y., Drake, M., Ohtani, E., Okuchi, T. and Righter, K. 2000. Water in the early Earth. In *Origin of the Earth and the Moon,* (K. Righter and R. Canup, Eds.) Univ. of Arizona Press, Tucson, Arizona, USA.

Agnor, C.B., Canup, R.M. and Levison, H.F. 1999. On the Character and Consequences of Large Impacts in the Late Stage of Terrestrial Planet Formation. *Icarus* **142,** 219–237.

Allegre, C.J., Manhes, G. and Gopel, C. 1995. The age of the Earth. *Geochimica et Cosmochimica Acta* **59,** 1445–1456.

Benz, W. 1990. Smooth Particle Hydrodynamics: a Review. In *The Numerical Modeling of Nonlinear Stellar Pulsations,* (J.R. Buchler, Ed.) Kluwer Academic Publisher. 269–288.

Boss, A.P. 1998. Evolution of the solar nebula. IV. Giant gaseous protoplanet formation. *Astrophys. J.* **503,** 923–937.

Boss, A.P. 2000. Possible rapid gas giant planet formation in the solar nebula and other proto-planetary disks. *Astrophys. J.* **536,** L101–L104.

Boss, A.P., Wetherill, G.W. and Haghighipour, N. 2002. Rapid Formation of Ice Giant Planets. *Icarus* **156,** 291–295.

Canup, R.M. and Esposito, L.W. 1996. Accretion of the Moon from an Impact-Generated Disk. *Icarus* **119,** 427–446.

Canup, R.M. and Asphaug, E. 2001. Origin of the Moon in a giant impact near the end of the Earth's formation. *Nature* **412,** 708–712.

Chambers, J.E. and Wetherill, G.W. 1998. Making the Terrestrial Planets: N-Body Integrations of Planetary Embryos in Three Dimensions. *Icarus* **136,** 304–327.

Chambers, J.E. and Wetherill, G.W. 2001. Planets in the Asteroid Belt. *Meteoritics and Planetary Science* **36,** 3, 381–399.

Chaussidon, M., Robert, F., McKeegan, K.D. and Krot, A.N. 2001. Li, Be, B Distribution and Isotopic Composition in Refractory Inclusions from Primitive Chondrites: a Record of Irradiation Processes in the Protosolar Nebula. *Meteoritics and Planetary Science* **36,** A40.

Chaussidon, M., Robert, F. and McKeegan, K.D. 2002. Incorporation of Short-lived ^{7}Be in One CAI from the Allende Meteorite. *LPSC Abstrats* **33,** 1563.

Chokshi, A., Tielens, A.G.G.M., and Hollenbach, D. 1993. Dust Coagulation. *Astrophys. J.* **407,** 806–819.

Drake, M.J. and Righter, K. 2002. Determining the composition of the Earth. *Nature* **416,** 39–44.

Goldreich, P. and Ward, W.R. 1973. The formation of planetesimals. *Astrophys. J.* **183,** 1051–1061.

Gopel, C., Manhes, G.and Allegre, C.J. 1993. Constraints on the Thermal History of H Chondrites Deduced From the U-Pb Systematics of Phosphates. *Meteroritics* **28,** 354,

Gounelle, M., Shu, F.H., Shang, H., Glassgold, A.E., Rehm, K.E. and Lee, T. 2001. Extinct Radioactivities and Protosolar Cosmic Rays: Self-Shielding and Light Elements. *Astrophys. J.* **548,** 2, 1051–1070.

Gounelle, M., Shang, S., Glassgold, A.E., Shu, F.H., Rehm, E.K. and Lee, T. 2003. Early Solar System Irradiation and Beryllium-7 Synthesis. *LPSC Abstracts* **34,** 1833.

Gradie, J.C. and Tedesco, E.F. 1982. Compositional structure of the asteroid belt. *Science* **216,** 1405–1407.

Greenberg, R., Wacker, J.F., Hartmann, W.K. and Chapman, C.R. 1978. Planetesimals to planets: Numerical simulation of planetesimal evolution. *Icarus* **35,** 1–26.

Guillot, T., Gautier, D. and Hubbard, W.B. 1997. NOTE: New Constraints on the Composition of Jupiter from Galileo Measurements and Interior Models. *Icarus* **130,** 534–539.

Hayashi, C. 1981. Structure of the solar nebula, growth and decay of magnetic fields and effects of magnetic and turbulent viscosities on the nebula. *Prog. Theor. Phys. Suppl.* **70,** 35–53.

Ida, S. and Makino, J. 1993. Scattering of Planetesimals by a Protoplanet: Slowing Down of Runaway Growth. *Icarus* **106,** 210–227.

Kleine, T., Münker, C., Mezger, K. and Palme, H. 2002. Rapid accretion and early core formation on asteroids and the terrestrial planets from Hf–W chronometry. *Nature* **418,** 952–955.

Kortenkamp, S.J. and Wetherill, G.W. 2000. Terrestrial Planet and Asteroid Formation in the Presence of Giant Planets. I. Relative Velocities of Planetesimals Subject to Jupiter and Saturn Perturbations. *Icarus* **143,** 60–73.

Kortenkamp, S.J., Wetherill, G.W., Inaba, S. and Trilling, D.E. 2001. Asteroid Formation with a Pre-Existing Jupiter. *LPSC Abstracts* **32,** 1796.

Laskar, J. and Robutel, P. 1993. The chaotic obliquity of the planets. *Nature* **361,** 608–612.

Levison, H.F. and Stewart, G.R. 2001. Remarks on Modeling the Formation of Uranus and Neptune. *Icarus* **153,** 224–228.

Lissauer, J. 1987. Timescales for planetary accretion and the structure of the protoplanetary disk. *Icarus* **69,** 249–265.

Manhes, G. and Gopel, C. 1993. Thermal History of Ordinary Chondrites: Comparison and Evaluation of Chronological Tools. *Meteoritics* **28,** 3, 390.

Morbidelli, A., Chambers, J., Lunine, J.I., Petit, J-M., Robert, F., Valsecchi, G.B. and Cyr, K.E. 2000a. Source regions and time scales for the delivery of water to Earth. *Meteoritics and Planetary Science* **35,** 6, 1309–1320.

Morbidelli, A., Petit, J-M., Gladman, B. and Chambers, J. 2000b. A plausible cause of the Late Heavy Bombardment. *Meteoritics and Planetary Science* **36,** 3, 371–380.

Petit, J-M., Morbidelli, A. and Chambers, J. 2001. The primordial excitation and clearing of the Asteroid Belt. *Icarus* **153,** 338–347.

Pollack, J.B., Hubickyj, O., Bodenheimer, P., Lissauer, J.J., Podolak, M. and Greenzweig, Y. 1996. Formation of the Giant Planets by Concurrent Accretion of Solids and Gas. *Icarus* **124,** 62–85.

Poppe, T. and Blum, J. 1997. Experiments on preplanetary grain growth. *Adv. Space. Res.* **20,** 1595–1604.

Poppe, T., Blum, J. and Henning, T. 2000. Analogous experiments on the stickiness of micron-sized preplanetary dust. *Astrophys. J.* **533,** 454–471.

Pudritz, R. 2002. Clustered Star Formation and the Origin of Stellar Masses. *Science* **295,** 68–76.

Schoenberg, R., Kamber, B.S., Collerson, K.D. and Eugster, O. 2002. New W-isotope evidence for rapid terrestrial accretion and very early core formation. *Geochimica et Cosmochimica Acta* **66,** 17, 3151–3160.

Tanga, P., Babiano, A., Dubrulle, B. and Provenzale, A. 1996. Forming Planetesimals in Vortices. *Icarus* **121,** 158–177.

Thommes, E.W., Duncan, M.J. and Levison, H.F. 1999. The formation of Uranus and Neptune in the Jupiter-Saturn region of the Solar System. *Nature* **402,** 635–638.

Thommes, E.W., Duncan, M.J. and Levison, H.F. 2003. Oligarchic growth of giant planets. *Icarus* **161,** 431–455.

Weidenschilling, S.J. 1977. Aerodynamics of solid bodies in the solar nebula. *MNRAS* **180,** 57–70.

Weidenschilling, S.J. 1980. Dust to Planetesimals: Settling and Coagulation in the Solar Nebula. *Icarus* **44,** 172–189.

Wiechert, U., Halliday, A.N., Lee, D.-C., Snyder, G.A., Taylor, L.A. and Rumble, D. 2001. Oxygen Isotopes and the Moon-Forming Giant Impact. *Science* **294,** 345–348.

Yin, Q., Jacobsen, S.B., Yamashita, K., Blichert-Toft, J., Télouk, P. and Albarède, F. 2002. A short timescale for terrestrial planet formation from Hf–W chronometry of meteorites. *Nature* **418,** 949–952.

Yoshino T., Walter, M.J. and Katsura, T. 2003. Core formation in planetesimals triggered by permeable flow. *Nature* **422,** 154–157.

Zanda B. 2004. Chondrules. *Earth and Planetary Science Letters* **224,** 1–17.

3 The Origin and Evolution of the Oceans

Daniele L. Pinti

3.1 Introduction

Oceans play a key role in the evolution of life. The first organic molecules on Earth have been likely synthesized in aqueous solutions and primitive biota possibly survived near oceanic hydrothermal systems (Stetter, 1998; Holm and Andersson, 1998; Nisbet and Sleep, 2001). The oceans shielded organic molecules from the massive UV radiation (Cleaves and Miller, 1998) and protected living organisms from the heavy cometary and meteoritic bombardment of our planet (Sleep et al., 2001; Nisbet and Sleep, 2001).

Nonetheless, our knowledge of the origin and the evolution of the oceans is rather poor and an universally accepted model of formation for the terrestrial oceans is not yet available. The main reason is that the atmosphere–ocean system formed during the first 700 Ma of the history of the Earth (Fig. 3.1). During this period, called "Hadean" (from "Hadeus", the Greek god of Hell) and that extends from the Earth accretion to the end of the heavy meteoritic bombardment, 3.9 Ga (Ga = billion years), the geological record has been mostly wiped out by the intense tectonic activity of the young Earth.

The Hadean geological record is reduced to a handful of detrital zircons found in Western Australia (Wilde et al., 2001; Mojzsis et al., 2001). These zircons contain precious information on the presence of liquid water, very early in the history of the Earth, and their study is revolutionizing our geological view on the primitive Earth (Peck et al., 2001; Valley et al., 2002). Perhaps, liquid water occurred at the surface of the Earth 50 Ma after the end of the accretion. A few tens of millions of years later, the oceans may have reached the conditions of temperature, salinity and pH suitable for the survival of living organisms, probably extremophiles (Rothschild and Mancinelli, 2001), and this well before the end of the heavy meteoritic bombardment of the Earth. Occasionally, the impact of large asteroids could have boiled the ocean and momentarily sterilized the Earth (Nisbet and Sleep, 2001).

In this chapter, I will summarize the current state of knowledge on the origin of the oceans on the basis of theoretical models and the few geochemical and isotopic records that the primitive oceans have left in Precambrian rocks.

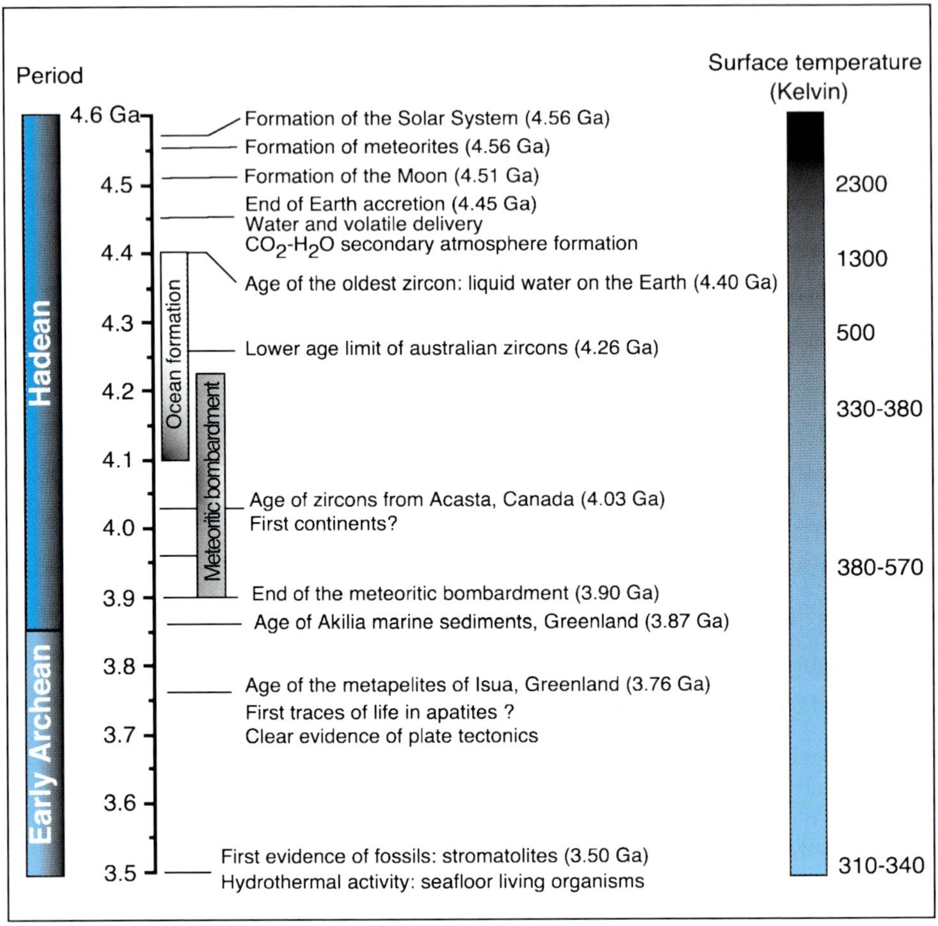

Fig. 3.1. Sketch showing the chronology of the first events that resulted in the formation of the oceans. Surface temperatures are indicative because the transition from uninhabitable hot to uncomfortably cold Hadean climates has been also suggested for the Hadean period (e.g. Sleep and Zahnle, 2001). The first traces of life in apatite at 3.8 Ga are graphite particles of possible organic origin found in Isua metasediments (Mojzsis et al., 1996). Recently, van Zuilen et al. (2002) questioned this result, suggesting an inorganic origin for the carbon. The age of Acasta orthogneisses are from Bowring and Williams (1999)

3.2 The Origin of Water

All theoretical models of formation of the oceans need a clear answer to a basic question: when and how water was delivered to our planet. However, the origin of water remains one of the most important subjects of debate and controversy in geosciences and astrophysics.

Knowing the timing of water delivery to the Earth may allow a choice to be made among the possible scenarios of the ocean formation proposed until now: (1) an early delivery of water during the accretion of the Earth, which implies that the oceanic water inventory was available since the beginning (Dauphas et al., 2000; Morbidelli et al., 2000; Robert et al., 2000); or (2) a continuous delivery of water through eons, which implies expanding oceans (Frank, 1986; Deming, 1999). Currently, the most accepted hypothesis is the former, but the extraterrestrial carrier of the water (chondritic vs. cometary) and the precise moment of the delivery (during the planetary growth or at the end of accretion), are still matters of debate (Owen, 1998; Delsemme, 1999; Morbidelli et al., 2000; Dauphas et al., 2000; Dauphas, 2003).

I would like to briefly discuss the long-standing hypothesis of a constant delivery of water throughout the history of the Earth, which implies an expanding ocean, if we assume that there is not a return of water to the mantle. It is mainly based on the theory of Frank et al. (1986) who showed, through a series of measurements on spacecrafts, that the Earth is annually hit by a large number of small cometary bodies. Their calculations indicated that about $2.2-8.5 \times 10^{21}$ kg of water has reached the Earth since its formation, if the influx rate is assumed constant. This is equivalent to three times the mass of water in the present-day oceans (1.4×10^{21} kg). More recent data of Frank and Sigwarth (1997) collected by the ultraviolet imagery of the POLAR spacecraft confirmed this finding. The cometary hypothesis is not universally accepted. Harrison (1999) showed, using models of the variations of the continental freeboard, that the constant addition of large quantities of water (comparable to the present-day ocean volume) throughout the history of the Earth is an unlikely process. His conclusion is also supported by geological evidence of a deep water table at the surface of the Earth, since the Archaean. The minimum water depth needed to form the 3.2 Ga Ironstone Pods in the Barberton greenstone belt is 982 m (de Ronde et al., 1997). Volcanic massive sulfides are also common in Archaean terrains. Some of them are analogous to sulfide deposits produced at present-day midocean ridges. To produce such a deposit, the ocean-floor pressure should be higher than the critical pressure, which is equivalent to an ocean depth of about 3 km (Harrison, 1999).

The universally accepted hypothesis is that the terrestrial water inventory (Table 3.1) was available soon after the Earth formation. A first hypothesis suggests that water and other volatiles degassed from the interior of the Earth, at the moment of its formation (Rubey, 1951). The second hypothesis suggests that a few planetary embryos accreted by the Earth at the final stage of its formation carried the bulk of water presently on Earth (Morbidelli et al., 2000). These planetary embryos may have had a chondritic composition and be originally formed in the outer asteroid belt (as the Trojan-class asteroids) (Morbidelli et al., 2000). Several authors challenged this interpretation, proposing alternative carriers of

Table 3.1. Concentration of water and D/H ratios in terrestrial and extraterrestrial reservoirs

Reservoirs	Mass or concentration of H_2O [kg]	D/H ($\times 10^{-6}$)	δD [‰ SMOW]	Ref.
- Whole Earth	$2.2–6.7 \times 10^{21}$	149–153	−40 to −20	1
- Primitive Earth	–	128–136	−180 to +130	1, 2
- Mantle	$5–50 \times 10^{20}$	143–149	−80 to −40	3
- Oceans	1.40×10^{21}	155.7	0	3
- Organic matter	1.36×10^{18}	135–145	−130 to −70	3
- Metamorphic rocks	3.60×10^{19}	140–146	−100 to −60	3
- Sedimentary rocks	2.32×10^{20}	143–145	−80 to −70	3
- PSN	–	21 ± 5	−865	4
- Carbonaceous chondrites[a]	6–22	128–181	−180 to +160	5–8
- Antarctic micrometeorites[a]	2–4	120–200	−229 to +285	9, 10
- Comets[a]	58–65	298–324	+900 to +1080	3, 4, 8

[a] H_2O concentration in wt%.
References: [1] Dauphas et al., 2000; [2] Déloule et al. 1991; [3] Lécuyer et al. 1998; [4] Robert 2001; [5] Boato 1954; [6] Robert and Epstein 1982; [7] Kerridge 1985; [8] Morbidelli et al. 2000; [9] Engrand et al. 1999; [10] Maurette et al. 2000.

water such as chondritic micrometeorites (Engrand et al., 1999; Maurette et al., 2000) or comets (Delsemme, 1999).

In a pioneering work on the origin of the oceans, Rubey (1951) tested different hypotheses. First, he argued that water derived from the weathering of the continental crust, which is known to contain large amounts of water, mostly in the form of hydrated minerals. However, using a simple mass balance, Rubey showed that the amount of water contained in the silicate rocks is insufficient to deliver water to the oceans. The continental crust has a mass of 2.4×10^{22} kg and it contains about 1% of H_2O (Krauskopf and Bird, 1995). Even assuming that all the continental crust has been weathered, only 10% of the terrestrial water inventory could have been delivered through this mechanism. Rubey considered another potential source of volatiles, i.e. volcanism. He observed that gas emissions from volcanoes are mainly composed of H_2O and CO_2 with minor amount of sulfates, nitrogen and rare gases. This corresponds to the volatile composition of the atmosphere, oceans and sediments. He argued that the degassing of the volatiles trapped during the Earth accretion, via the volcanism, could explain the formation of the atmosphere and the oceans.

The degassing of the Earth's interior has been largely demonstrated by noble gas studies (e.g., Allègre et al., 1986; Ozima and Podosek, 2001). During the first oceanographic cruises at Galapagos and EPR midocean ridges, at the end of the 1970s, an anomalous enrichment of the primordial isotope helium-3 was detected from deep water sampled at the axis of the ridge (Craig et al., 1975). The ratio between the primordial and the radiogenic isotope of helium, namely $^3He/^4He$ ratio, was higher than that expected in seawater, which should correspond to the atmospheric one. The degassing of primordial volatiles from the upper mantle was the cause of the observed enrichment of ^3He. The analyses of primordial noble gases in mantle rocks such as MORBs (midocean-ridge basalts), diamonds, xenoliths (e.g., Sarda et al., 1988; Staudacher et al., 1989; Ozima and Zashu, 1988; Poreda and Farley, 1992) and the discovery of extinct radionuclides, such as the Pu-I-Xe system (Butler et al., 1963; Staudacher and Allègre, 1982; Marty, 1989) has clearly demonstrated a catastrophic degassing that took place in the first 100 Ma after the Earth accretion. The observations of noble gas having a clear "solar" isotopic composition, such as neon, was another piece of evidence supporting the degassing of primordial gases trapped during the formation of the Earth from the solar nebula (Honda et al., 1991). It has been often considered, by comparison with noble gases, that the major volatiles (N, C, O, H) had the same origin.

The hypothesis of Rubey implies that the volatiles trapped during the formation of the Earth derive from the gas and dust of the protosolar nebula (PSN hereafter). However, the proximity of the Earth to the Sun implies high condensation temperatures that cannot allow incorporation of highly volatile elements such as N, C, O, H. Furthermore, the redox conditions that prevailed in this sector of the Solar System did not allow water to form, suggesting a source different from PSN for the oceans. This is confirmed by the isotopic signature of water and particularly by the isotopic ratio of hydrogen (D/H) measured in modern seawater (value of the standard mean ocean water, SMOW), which is equal to 155.7×10^{-6} ($\delta D_{SMOW} = 0‰$; Table 3.1). The D/H ratio is highly variable in the Solar System, but in a general way, it increases moving outward the Solar System due to a progressive enrichment of deuterium. This enrichment is produced by ion–molecule interactions in the interstellar medium (Robert et al., 2000). The D/H ratio in the PSN has been estimated to be $21 \pm 5 \times 10^{-6}$ using as reference the isotopic composition of helium in the Sun (Geiss and Gloecker, 1998) and the isotopic composition of molecular hydrogen in the high atmosphere of the giant planets (Gautier and Owen, 1983). The value of $21 \pm 5 \times 10^{-6}$ is more than seven times less than the D/H ratio measured in the oceans. Using appropriate models of the solar nebula, Drouart et al. (1999) showed that only planetesimals formed in the region of Jupiter–Saturn and in the outer asteroid belt could contain water with a D/H ratio similar to that of the oceans. Only two types of planetary bodies could be the source of this water: comets and hydrous carbonaceous chondrites (Morbidelli and Benest 2001).

Comets are bodies composed of ices (mainly H_2O and in minor amounts CH_3OH, CO and CO_2) together with particles of silicates, carbon and organic matter. The comets actually observed, such as Halley or Hale–Bopp, likely come from the Oort Cloud, a distant cloud of comet material located in the outer region of the Solar System, itself populated mostly by planetesimals that were originally in the Uranus–Neptune region and in the primordial Kuiper Belt.

Among the most primitive meteorites, carbonaceous chondrites contain a large amount of water (up to 22 g of H_2O per 100 g of rock; Kerridge, 1985). Currently, these meteorites are rare and only 4% of the falls on Earth corresponds to carbonaceous chondrites (Dauphas and Marty, 2001). However, hydrous carbonaceous chondrites seem to be a large reservoir of extraterrestrial material: the outer asteroid belt, located between Mars and Jupiter at 2 UA and that is the major source region of the meteorites arriving currently on Earth, seems to be dominated by carbonaceous chondrites; the lunar soil contains from 1 to 2% of such material; the Antarctic micrometeorites (AMM), that constitute the highest flux of extraterrestrial material actually reaching the Earth (40 000 tons per year), are hydrous carbonaceous chondrites (Engrand et al., 1999; Maurette et al., 2000).

To choose between these two potential candidates, we have a strong isotopic constraint, which is the ratio between deuterium and hydrogen (D/H ratio) in the water molecule. The values of the D/H ratios in the terrestrial and extraterrestrial reservoirs (PSN, AMM, carbonaceous chondrites, comets) are reported in Fig. 3.2 and Table 3.1. The water of the oceans has a D/H ratio of 155.7×10^{-6}, while the whole Earth has a D/H ratio of $149–153 \times 10^{-6}$ (Lecuyer et al., 1998). These values are close to those measured in the carbonaceous chondrites ($128–180 \times 10^{-6}$, average $149 \pm 6 \times 10^{-6}$; Boato, 1954; Robert and Epstein, 1982; Kerridge 1985; Robert, 2001) and those measured in the Antarctic micrometeorites AMM ($140–200 \times 10^{-6}$, average $154 \pm 16 \times 10^{-6}$, Engrand et al., 1999; Maurette et al., 2000). The D/H ratio in the water of comets has been measured only in Comet Halley ($316 \pm 34 \times 10^{-6}$, Eberhardt et al., 1995), in Comet Hyakutake ($290 \pm 100 \times 10^{-6}$, Bockelée-Morvan, 1998) and in Comet Hale–Bopp ($320 \pm 120 \times 10^{-6}$, Meier et al., 1998). The obtained values are 10–20 times larger than the D/H ratio of molecular hydrogen in the PSN and 2–3 times larger than the value found for modern seawater, suggesting that comets did not contribute significantly to the delivery of water to Earth.

A simple mass and isotopic balance shows that the amount of water delivered by comets on Earth is around 10% of the total (Dauphas et al., 2000). This is in agreement with independent estimates, based on the mean collision probability with Earth of comets coming from the giant planet region calculated by Morbidelli et al. (2000). This calculation indicates that $\sim 5 \times 10^{-5} M_T$ (M_T is the mass of the Earth $= 6 \times 10^{24}$ kg) of cometary material from the trans-Uranian region could have been accreted to Earth. Even assuming a 100% water composition of material and a 100% impact efficiency, this would imply the delivery of only 10–15% of the terrestrial water inventory.

The defenders of a cometary origin of water observed that the comets in which we measure the D/H ratio are "long-period" comets, probably formed in the Uranus–Neptune region or in the trans-Uranian region (Delsemme, 1999). Accepting the model of D/H enrichment in the PSN (Robert et al., 2000), it is understandable to find D/H ratios 10 times higher than that of the PSN in the outer regions of the Solar System. Owen and Bar-Nun (1995) and Delsemme (1999) suggested that comets delivering water on Earth were those formed in the Jupiter region. These comets would have formed at relatively high temperature (100 K) and consequently they would exhibit a lower D/H ratio due to exchange with protosolar hydrogen. Morbidelli et al. (2000) showed that the lifetime of comets from this region is extremely short (estimated lifetime of 1.5×10^5 yr), lowering the impact probability of these comets with an accreting Earth. They

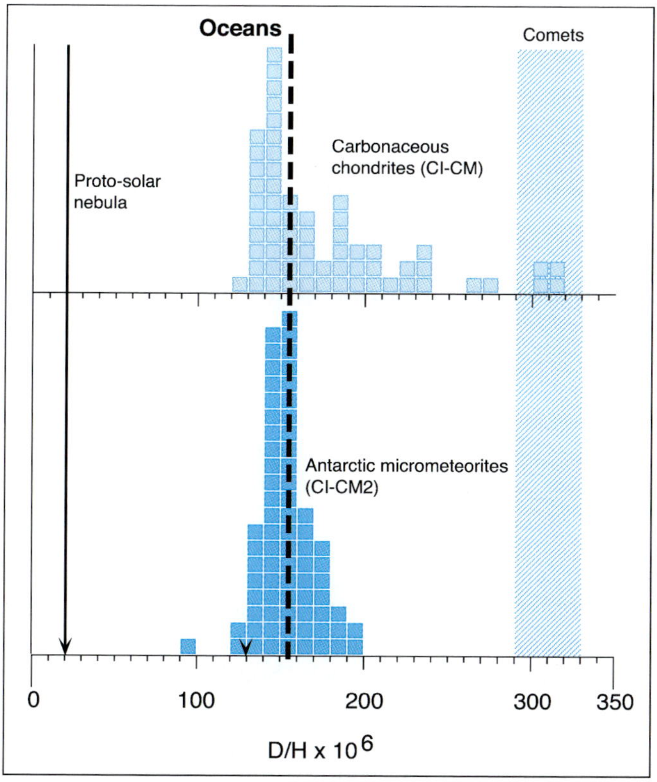

Fig. 3.2. Frequency distribution of the D/H ratios measured in carbonaceous chondrites and Antarctic micrometeorites compared to values for the PSN, Earth oceans and comets. Data: Déloule et al., 1991; Engrand et al., 1999; Maurette et al., 2000; Robert et al. 2000

concluded that the contribution of cometesimals to the terrestrial water inventory was negligible.

We can understand from these studies that the debate is far from settled. The accepted hypothesis is that 90% or more of the water has been delivered to Earth by small asteroids having a chondritic composition and coming from the outer asteroidal belt (Morbidelli et al., 2000). The timing of delivery is likely at the end of the accretion of the Earth (Dauphas, 2003). This is suggested by the presence of anomalous amounts of siderophile elements in the mantle (Dauphas and Marty, 2001).

The siderophiles are elements (such as those of the platinum group) that should have migrated to the core during the Earth primary differentiation, together with Ni-Fe alloy. However, a non-negligible amount of these elements is still present in the mantle and this can only be explained by a late addition, after the core formation and the mantle separation. The core formation is estimated at 50 Ma after the condensation of the PSN (Allègre et al., 1995). The relative abundance of siderophiles in the mantle is very similar to that of the primitive meteorites, such as the hydrous carbonaceous chondrites. Mass-balance calculations suggest that a meteoric flux equivalent to $4.5 \times 10^{-3}\,M_{\mathrm{T}}$ could explain the late delivery of siderophiles to the mantle (Dauphas and Marty, 2001). The same carbonaceous chondrite-like bodies, containing from 6 to 22 wt% of water (Table 3.1), could have delivered $1.6 - 6.0 \times 10^{21}$ kg of water to Earth, from 1 to 4 times the mass of the present-day oceans (1.4×10^{21} kg).

3.3 Formation of the Oceans: the Geological Record

If different views exist on the origin of water, there is a general consensus that the total oceanic water inventory was available soon after the end of the Earth accretion. The question is when did water condense on the surface to form stable oceans, habitable for life. The chronology of the ocean formation is mostly unknown and it depends on the physical parameters used for the different models of formation and the scarce geological records of the presence of liquid water at the surface of the Earth. The oldest marine sediments found so far are BIF (Banded Iron Formation) in a layered body of amphibolite and ultramafic rocks, crosscut by a quartz-dioritic dyke having a U-Pb age of 3.865 ± 11 Ma, at Akilia, West Greenland (Nutman et al., 1997). Other traces of ancient marine sediments have been found at the Isua Supracrustal Belt, Southern West Greenland and consist of metamorphosed pelagic sediments, probably turbidites, dated at 3.7 Ga (Rosing et al., 1996).

New geological evidence of liquid water comes from the recent U-Pb dating of individual detrital zircons ($ZrSiO_4$) from the Mt. Narryer quartzite and from Jack Hills metaconglomerate, Western Australia (Mojzsis et al., 2001; Wilde et al., 2001). Zircon is a common U-rich trace mineral in granitic rocks that preserves a detailed record of the magma genesis (Pidgeon and Wilde, 1998). Furthermore, the radioactive parent nuclides $^{235,238}U$ decay into the stable daughter products $^{206,207}Pb$. Both the parent nuclides and the daughter products can be preserved in zircons, even when a crystal has been removed from its host rock by weathering, transported as a detrital grain, deposited, hydrothermally altered and metamorphosed (e.g. Valley et al., 1994). This makes zircon a reliable geochronometer for very ancient rocks. The U-Pb ages obtained for the Mt. Narryer and Jack Hills zircons range from 4.28 to 4.40 Ga (Wilde et al., 2001). Because zircons are silicate minerals occurring in intermediate and acid rocks (such as granites), their presence strongly suggests that a differentiated continental crust existed less than 150 Ma after the end of the Earth accretion.

But the most intriguing feature of these zircons is their oxygen isotopic signature (Peck et al., 2001; Fig. 3.3). The $^{18}O/^{16}O$ ratios (denoted as $\delta^{18}O$) of the Jack Hills zircons range from values of 5.0‰, which is close to the primitive value of the Earth mantle ($\delta^{18}O = 5.3 \pm 0.3$‰, Valley et al., 1998) to $\delta^{18}O$ of 7.4 ± 0.7‰. This value can be explained only by a mixing between the mantle source of the zircons and a source enriched in ^{18}O.

The best explanation is that the mantle source of the Jack Hills zircons interacted, directly or indirectly, with surface waters. The products of weathering and low-temperature alteration (such as oceanic crust) have indeed higher $\delta^{18}O$ values (> 10‰, Valley et al., 2002). The $\delta^{18}O$ signature measured in the Hadean, Archaean and Proterozoic zircons shows a progressive increase towards values up to 10‰ (Fig. 3.3). This can be explained by an increase of the amount of supracrustal high-$\delta^{18}O$ material available for melting and assimilation, as the

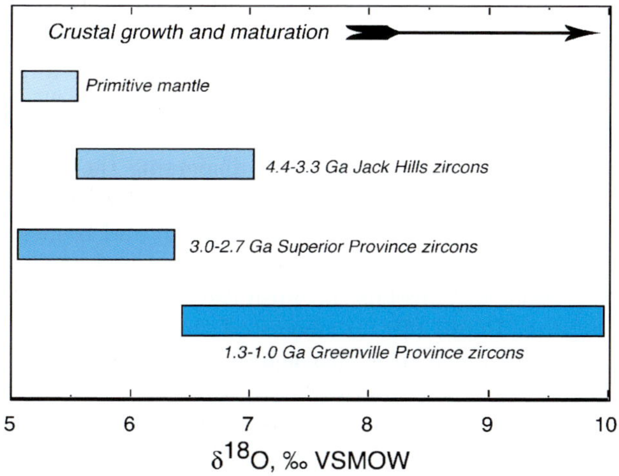

Fig. 3.3. Variations of the $\delta^{18}O$ measured in zircons of different ages. The progressive increase of the $\delta^{18}O$ with age is due to contamination of high $\delta^{18}O$-supracrustal material from a growing and maturing continental crust. Oxygen isotopic ratios are average values. Oxygen data for zircons: King et al., 1998; Peck et al., 2000, 2001. Data for primitive mantle: Valley et al., 1998

"wet" continental crust evolved and matured during the Precambrian (Peck et al., 2000). The presence of $\delta^{18}O$ of 7.4 ± 0.7‰ in a 4.404 ± 0.008-Ga old detrital zircon of Jack Hills could thus be the indirect and oldest record of interaction of a magmatic source with liquid water at the surface of the Earth (Peck et al. 2001).

There are several other lines of evidence that liquid water was present very early in the history of the Earth. Among others, we can cite the isotopic signatures of Neodymium (Nd) and Strontium (Sr) obtained in ancient minerals and rocks of ages ranging from 4.04 to 3.80 Ga. The Nd and Sr isotopic ratios suggest that between 10 and 15% of the volume of the present-day continental crust were formed at that period (Taylor and McLennan, 1995). This crust has a TTG (Trondhjemite-Tonalite-Granodiorite) composition, which is derived directly by partial melting of hydrated basalt (Martin and Moyen, 2002). This means that large volumes of water were available on Earth well before 4 Ga.

In the next section, I will illustrate the formation of the ocean based on the model proposed by Abe (1993) and particularly that of Sleep et al. (2001), which integrated the new evidence coming from the Australian zircons for extensive surface water at indeterminate temperature and long-lived continental crust, 4.40 Ga ago.

3.4 Formation of the Oceans: Chronology and Processes

The formation of the oceans can be summarized in four steps, here represented by the cartoons of Figs. 3.4.1 to 3.4.4. The first step is from the start of the Earth accretion (4.56 Ga ago) to the segregation of the core and the end of the interior degassing, estimated at 4.45–4.50 Ga (Allègre et al., 1995; Halliday, 2001). During this period, carbonaceous chondrites likely delivered water (and possibly organic molecules) to the Earth.

During impacts, the asteroids vaporized, dispersing the trapped volatiles in the atmosphere. A small part of them has been probably partitioned in the melt (the surface of the Earth was in a molten state) and re-gassed successively (Fig. 3.4.1). The total amount of water delivered to Earth was certainly higher than that in the present-day oceans, because the losses of volatiles were important at the beginning. Pepin (1991) suggested that at least 50% of the water delivered to Earth could have been dissociated in hydrogen and oxygen by the strong UV radiation that the Earth suffered. Further losses could have derived from impact erosion of the atmosphere by part of the arriving planetary bodies.

During the first chaotic moments of the Earth history, at about 4.50 Ga (Halliday, 2001), a collision with a planetoid having a size comparable with that of Mars affected our planet (Canup and Asfaugh, 2001; Fig. 3.4.1). The accretion of the debris formed the Moon. The energy of the collision has been estimated to be 4×10^{31} J (Sleep et al., 2001). This value, divided by the mass of the Earth, is 7×10^{6} J kg^{-1}, an energy comparable to that needed for the vaporization of

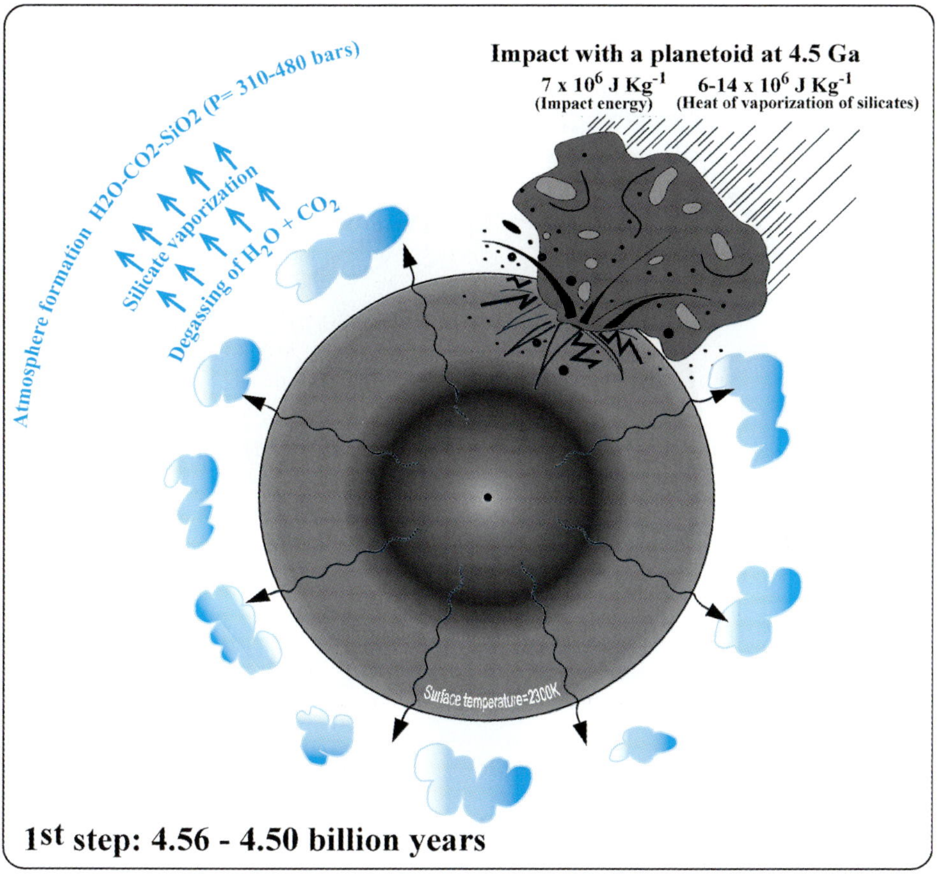

Fig. 3.4.1. Sketch representing the main phases of the formation of the terrestrial oceans

silicates at the surface of the Earth ($6-14 \times 10^6\,\mathrm{J\,kg^{-1}}$). After the collision, the temperature increased to 2300K ($\sim 2000^\circ$C), causing melting of the surface of the Earth. If some liquid water existed at that time, it was vaporized together with silicates (Sleep et al., 2001). The age of 4.50–4.45 Ga can be assumed as the initial time (T_0) of the ocean formation. The collision of this planetoid caused the formation of a dense atmosphere of gaseous silicates that rapidly cooled down and precipitated after a few thousand years (Sleep et al., 2001).

The residual atmosphere was constituted mainly of water vapor and CO_2 (nitrogen was close to 1% of the total volume and thus negligible, Kasting, 1993). If we assume that all the water of the present oceans (1.4×10^{21} kg H_2O) was in the atmosphere, the atmospheric partial pressure would be 270 bars. To this pressure, we must add the partial pressure of CO_2. The CO_2 in the primordial atmosphere was equivalent to the amount of carbon actually preserved in carbon-

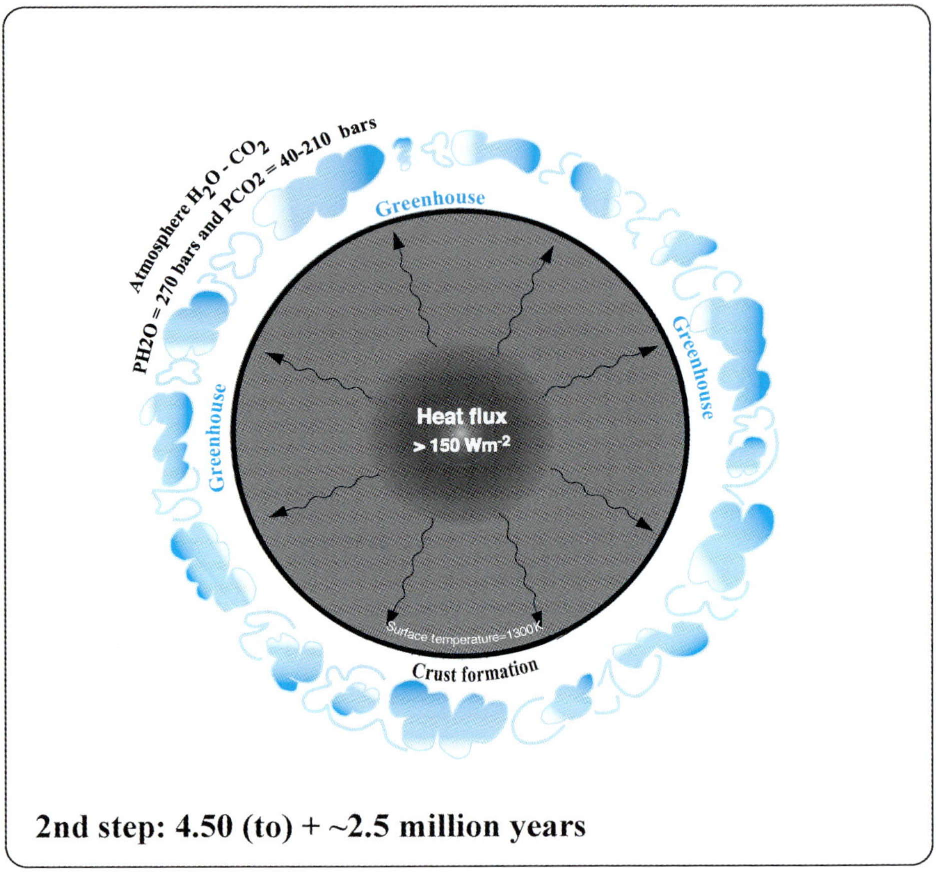

Fig. 3.4.2. Ibid

ates (3.65×10^{21} moles, Li, 2000), in continental sediments (1.12×10^{21} moles, Li, 2000) and in the biosphere (2.6×10^{17} moles, Faure, 1991). This amount would produce an atmospheric CO_2 partial pressure of 40 bars. Recently, Sleep et al. (2001) proposed that the total inventory of CO_2 in the present mantle (2.5×10^{22} moles, Zhang and Zindler, 1993) was concentrated in the primordial atmosphere. This amount is equivalent to a partial pressure of 170 bars. The primordial atmosphere could have been a dense mixture of 270 bars of H_2O and 40–210 bars of CO_2, an atmospheric composition somewhat similar to that of Venus.

A runaway greenhouse effect produced by the massive H_2O–CO_2 atmosphere dominates the second phase of formation of the oceans (Fig. 3.4.2). This runaway greenhouse maintained the surface of the Earth at the melting temperature, while the heat flow from the interior was greater than 150 W/m^2 (Abe, 1993). This is the critical surface heat flow that corresponds to the difference between

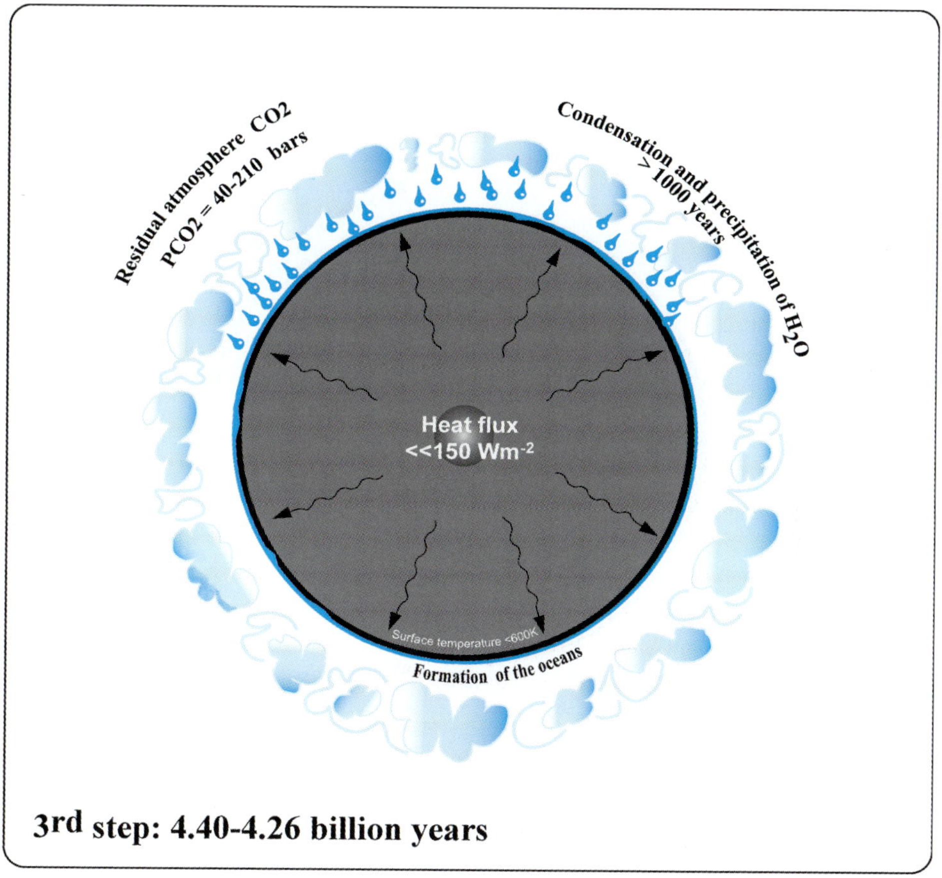

Fig. 3.4.3. Ibid

the critical greenhouse threshold and the heat supply by the Sun. For a surface heat flow lower than $150\,W/m^2$, the runaway greenhouse cannot be maintained longer and the surface of the Earth rapidly cooled down. In a few million years, the surface reached a temperature of around 1300 K, sufficiently cold to produce a solid rim of basaltic composition, which separated the atmosphere from the hot interior of the Earth. The surface of the Earth cooled down rapidly, allowing the condensation of water (Fig. 3.4.3).

The oceans formed by a sort of "universal deluge" due to the condensation of the atmospheric water vapor (Fig. 3.4.3). Abe (1993) suggested that the terrestrial oceans were produced in less than 1000 years, due to heavy rains. The raining rates were 7000 mm/year, which corresponds to 10 times the present-day raining rates at the tropical latitudes. Assuming an atmospheric pressure at the surface of several hundred bars, water starts to condense and precipitates at 600 K. The presence of stable liquid water at relatively low temperatures at

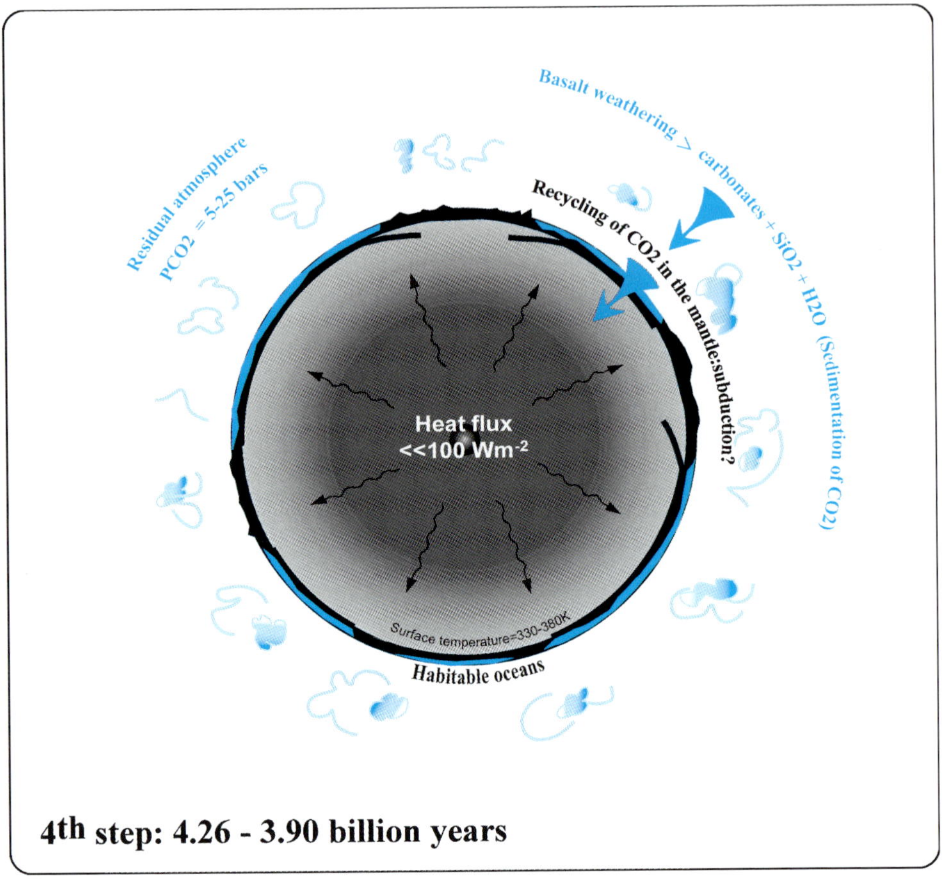

Fig. 3.4.4. Ibid

4.40–4.35 Ga, as suggested by the oxygen isotopic composition of zircons, indicates that the Earth cooling was a relatively rapid process (Valley et al., 2002). Temperatures lower than 600 K were probably available between 50 to 150 Ma after the formation of the Earth.

The neoformed oceans were too hot for the development of life. From our knowledge of extremophiles, the maximum temperature allowed for the survival of "hyperthermophiles" (heat-loving organisms) is from 80 to 110°C (Rothschild and Mancinelli 2001). When the terrestrial heat flux decreased to a value of 100 W/m^2, the greenhouse effect directly controlled the temperature of the oceans. Temperatures of 60 to 110°C canbe reached at equilibrium with a residual atmosphere of 5 to 25 bars of CO_2. If we assume an initial CO_2 partial pressure of 210 bars, this means that we must be able to recycle efficiently 185–205 bars of CO_2 into the mantle. Sleep et al. (2001) proposed the carbonation of the oceanic basaltic protocrust as the most efficient process of extraction of

CO_2 from the atmosphere. The most common reaction of carbonation is (Holland 1984):

$$CaSiO_3 + H_2O \rightarrow Ca^{2+} + 2HCO_3^- + SiO_2 \quad (3.1)$$

$$Ca^{2+} + 2HCO_3^- \rightarrow CaCO_3 + H_2CO_3 \quad (3.2)$$

The formed carbonates (3.2) could have been recycled to the interior of the Earth, by subduction (Fig. 3.4.4). This implies that plate tectonics was active since 4.4–4.3 Ga, which cannot be excluded but for which we have no direct evidence, because of the scarce geological record. Taking into account the instability of a young and thin protocrust, its partial melting caused by the impact of meteoritic bodies could have permitted a sort of crustal recycling. Zhang and Zindler (1993) showed that the present-day CO_2 recycling rate is 2.3×10^{13} mol/y. This means that from 40 to 210 bars of CO_2 can be removed from the atmosphere in a span of time ranging from 200 Ma to 1.0 Ga. The generation and recycling of oceanic crust was probably higher in the Hadean, because of the larger amount of heat to be evacuated from the interior of a very hot early Earth, helped by vigorous mantle convection (Nisbet et al., 1993; Pinti, 2002). Bickle (1986) estimated an Archaean plate generation 2–3 times higher than at modern MOR, which means a 2–3 times higher crustal recycling. Zhang and Zindler (1993) proposed a maximum initial recycling rate of 150×10^{12} mol/y. At this rate, the atmospheric CO_2 could have been recycled in a shorter time, between 30 and 160 Ma.

This means that at 4.3–4.2 Ga, the oceans were stable and cold enough to assure the survival of the first living communities, possibly hyperthermophiles. It is likely that the temperature of the oceans was not stable in this period but successive impacts of asteroids and comets may have locally increased temperatures up to 110 °C, or higher ($\geq$ 300 °C) causing the partial evaporation of the oceans and their recondensation (Nisbet and Sleep, 2001). Oceans were stable at the end of the Hadean (3.9 Ga) when the massive meteoritic bombardment of the Earth ended.

3.5 Chemical Composition of the Primitive Oceans

If the volume of the oceans has not likely changed in the last 4 billion years, its chemistry certainly has. Here, I present an overview of the evolution of the physical and chemical conditions that operated in the primitive oceans and where the first living organisms developed. For a complete treatment on the subject, the reader can consult the excellent book of Heinrich Holland "The Chemical Evolution of the Atmospheres and the Oceans" (Holland, 1984). The physical and chemical parameters of the oceans discussed here, which can have an impact on the development of life, are the temperature, the pH, the redox and the salinity. This is based both on thermodynamical modeling and, also, on the few traces that oceans left in the Precambrian marine sediments.

3.5.1 Temperature

In the last 4 billion years, the temperature of the oceans decreases from an initial value of 230°C (calculated at equilibrium with a massive CO_2 atmosphere, Sleep et al., 2001) to the present-day average of 20°C at the surface. The temperature decrease is linked to the progressive decline of the CO_2 amount in the atmosphere and the consequent weakening of the related greenhouse effect (Abe, 1993; Kasting, 1987, 1993; Tajika and Matsui, 1993; Fig. 3.5). The evolution of the ocean temperature can be compared with the geological record. A measurement of the past ocean temperatures could be that recorded by the isotopic ratio of the oxygen ($^{18}O/^{16}O$ expressed as $\delta^{18}O$) measured in cherts, a hard rock composed of very fine-grained silica (Knauth, 1994). Of particular interest is that $\delta^{18}O_{\text{water-rock}}$ in cherts decreases as the temperature of the ocean increases, due to isotopic fractionation during precipitation of silica in the ocean (Knauth and Lowe, 1978). Therefore, from the $\delta^{18}O$ values recorded by cherts of different ages we can provide a measure of the oceanic temperature at any time (Fig. 3.5). The variation of the $\delta^{18}O$ measured in the cherts of Archaean age suggests that these rocks precipitated in an ocean that was 50°C warmer than today. This result has been contested because it could simply reflect isotopic variations of the $\delta^{18}O$ of the ancient oceans. However, measurements of the $\delta^{18}O_{\text{SMOW}}$ of $0 \pm 3‰$ in Early Proterozoic seawater suggest that the isotopic signature of the ocean has not changed significantly during eons (Knauth and Robert, 1991). The problem with this paleothermometer is the doubtful origin of Archaean cherts. Most of them seem to be silica gel precipitated from hydrothermal fluids (Sugitani, 1992; Nijman et al., 1998; Pinti et al., 2001). Consequently, the temperature curve of Knauth and Lowe (1978) may reflect the temperature of silica precipitation from hot hydrothermal fluids, rather than that of ambient seawater.

An alternative method for measuring the past ocean temperatures could be the homogenization temperature of seawater inclusions preserved in ancient rocks. de Ronde et al. (1997) have shown that ancient seawater was preserved in bodies of iron oxide (Ironstone Pods) from the 3.5–3.2 Ga Barberton greenstone belt, South Africa. These rocks have been interpreted as deposits of Archaean seafloor hydrothermal vents (de Ronde et al., 1997). It is to be noted that Donald Lowe of Stanford University and Gary Byerly of Louisiana State University have recently contested this interpretation, showing that the Ironstone Pods are largely composed of goethite (a thermally unstable hydrated iron oxide mineral), deriving from quaternary dissolution of the Archaean siderite (Lowe and Byerly, 2003). They concluded that these pods are deposits of young subaerial springs and contain no record of Archaean environments. Although a remobilization of iron could have been produced during quaternary weathering, selected fluid inclusions by de Ronde et al. (1997) show primary textures and geochemical information can be considered genuine for Archaean times (F. Westall, personal communication).

If the fluids contained in the inclusions are preserved at different physical conditions of temperature and pressure (i.e. depth) than the original fluid, phase

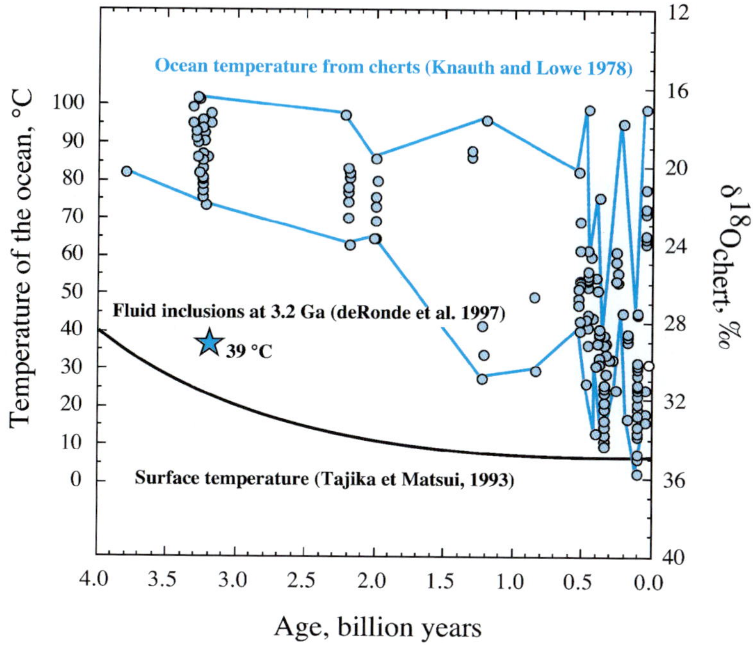

Fig. 3.5. Isotopic variations of oxygen (expressed as $\delta^{18}O$) measured in cherts of different ages compared to the ocean temperatures where the cherts precipitated. Data: Knauth and Lowe (1978). The lower curve represents the surface temperature variations as calculated by the model of Tajika and Matsui (1993). The *star* indicates the temperature at 3.2 Ga, obtained from homogenization temperatures of primary seawater inclusions from Ironstone Pods, South Africa (de Ronde et al., 1997)

separation will take place in inclusions. In the case of seawater, the change of temperature and pressure can produce the separation of a saline phase from the liquid. Heating or cooling down the sample homogenizes the phases contained in the inclusions and thus we can know the temperature of formation (Roedder, 1984). In the case of the Ironstone Pods, de Ronde et al. (1997) calculated an oceanic surface temperature of 39 °C, which is lower than the 70 °C proposed by Knauth and Lowe (1978) on the basis of the $\delta^{18}O$ in cherts (Fig. 3.5).

3.5.2 The pH and the Redox State

Archaean ocean was probably acid. This hypothesis is based on the fact that the primitive ocean was in equilibrium with an atmosphere mainly composed of CO_2. At the present time, the pH of the ocean is controlled by its equilibrium with the carbonates (dissolution and precipitation of $CaCO_3$) and the CO_2 fugacity in the atmosphere (Ottonello, 1997). The equation relating the CO_2 fugacity in the atmosphere and the pH of the ocean at equilibrium is:

$$10^{9.83} \cdot f_{CO_{2(g)}}^{-1} \cdot \left[H^+\right]^4 + \left[H^+\right]^3 - \left(10^{-7.82} \cdot f_{CO_{2(g)}} + 10^{-14}\right) \cdot \left[H^+\right] - 10^{-17.85} \cdot f_{CO_{2(g)}} = 0 \quad (3.3)$$

The curve of variation of CO_2 in the atmosphere, as calculated by Tajika and Matsui (1993), is reported in Fig. 3.6 together with the curve of variation of the pH in the ocean, calculated from (3.3). The ocean was probably acid until 2 Ga to become progressively basic until the present-day value of 8.2 (Fig. 3.6). The chosen model for the CO_2-level variation in the atmosphere predicts an initial value of 10 bars (Tajika and Matsui, 1993). In the Hadean, the CO_2 level could have been much higher (40–210 bars, Sleep et al., 2001) and the ocean pH directly controlled by the equilibrium of carbonation of the oceanic crust rather than by the precipitation of carbonates. More complicated thermodynamic models are thus needed to predict the pH of the Hadean ocean. Sleep et al. (2001) considered the carbonation reactions produced at equilibrium (3.1 and 3.2). The pH fixed by these reactions is controlled directly by the pCO_2. Sleep et al. (2001) calculated the theoretical pH from the reactions and consequently that of oceans at initial temperatures of 200–230 °C. The obtained pH ranges from 4.8 and 6.5, supporting the idea of an acid Hadean ocean.

Kempe and Degens (1985) proposed an alternative model for the Hadean ocean, based on their observations of the soda lakes of the east African rift.

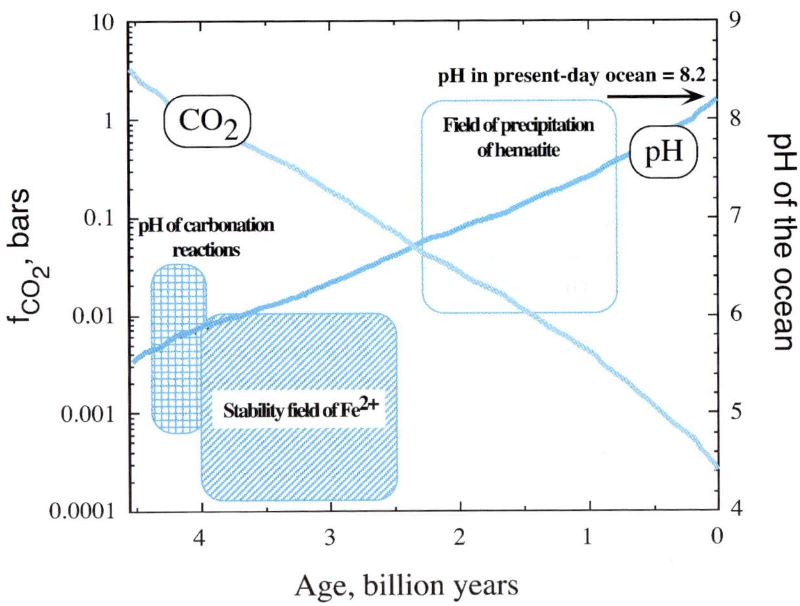

Fig. 3.6. Variations of the partial pressure of CO_2 in the atmosphere and change of pH of the ocean, at equilibrium. The different fields represent the variations of the pH of the ocean as obtained from different thermodynamic calculations. See text for more details

They suggested an early soda ocean, saturated in sodium carbonate (thermo-natrolite: $Na_2CO_3{\cdot}H_2O$). The resulting ocean would be alkaline, with pH ranging from 9 to 10.5. Thermo-natrolite could have been produced by reactions between sodium-rich rocks (such as those of the TTG, that constitute the primitive crust) and the CO_2 in the atmosphere:

$$2NaAlSi_3O_8 + CO_2 + 3H_2O \rightarrow Na_2CO_3 \cdot H_2O + Al_2Si_2O_5(OH)_4 + 4SiO_2 \quad (3.4)$$

$$3NaAl_3Si_3O_{10}(OH)_2 + CO_2 + 5H_2O \rightarrow Na_2CO_3 \cdot H_2O + 3Al_2Si_2O_5(OH)_4 \quad (3.5)$$

Sleep et al. (2001) argued against an early soda ocean, based on thermodynamic and mass balance. Reactions (3.4 and 3.5) at temperatures between 20 and 200°C take place only for pCO_2 comprised between 3 kbar up to 6 Gbar, while only 40 bars to 210 bars were probably available in the Hadean atmosphere. Mass balance indicates that the Na/Al ratio in the weathered crust should be of 1:1, while the most sodium-rich rocks, the magmatic suite TTG, has a Na/Al ratio of 1:2. In other terms, there was not enough sodium to saturate the ocean in $Na_2CO_3{\cdot}H_2O$.

The hypothesis of an Archaean acid ocean is supported by the large deposition, at the start of the Proterozoic era, of BIF, rocks composed by a succession of cherts, siderite ($FeCO_3$) and hematite (Fe_2O_3) (Morris, 1993; Isley, 1995). The amount of iron is between 30 and 45 wt%. The precipitation of most of BIF at the start of the Proterozoic can be explained by a progressive accumulation of Fe in its reduced form (Fe^{2+}) in the ocean during the Hadean and the Archaean (Holland, 1984). The Fe^{2+} ion is indeed soluble in water in an anoxic, reduced environment and for pH ranging from 0 up to 6. In an alkaline and oxygenated ocean, as modern seawater, the Fe^{3+} is precipitated as hydroxide $Fe(OH)_3$. The accumulation of iron during the Archaean and its precipitation as BIF in the Proterozoic is thus a clear indication of a major change in the pH and redox conditions of seawater and particularly the evolution from an anoxic and acid ocean to a basic and oxygenated one.

There are several other lines of evidence of the redox state of the ocean (Holland, 2002). The study of rare-earth elements in marine sediments shows this change due to the progressive oxygenation of the atmosphere and of the ocean (the Great Oxidation Event, Holland, 2002; Murakami et al., 2001; Yang et al., 2002). For example, Archaean microbialites (laminated limestones), found at the 2.52-Ga Campbellrand carbonate Platform, South Africa, show an excess of cerium (Ce) that cannot be found in the present-day microbialites (Kamber and Webb, 2001). Ce^{3+} is easily oxidized to Ce^{4+} in the presence of O_2 and is scavenged by iron hydroxides, as in the present-day ocean. The presence of a cerium positive anomaly in the Archaean marine sediments could thus be another sign of an anoxic ocean (Yang et al., 2002).

Other geochemical evidences of an anoxic ocean during the Archaean are the nitrogen isotopic signatures of kerogens (polymerized organic matter) preserved in ancient cherts (Beaumont and Robert, 1999; Boyd, 2001a; Pinti et al., 2001; Pinti, 2002). The average isotopic composition of N in present-day marine organic matter differs significantly from that of the atmosphere. The average value of the $\delta^{15}N$ in marine organic matter is +6‰(Peters et al., 1978; Boyd, 2001b), while that of the atmosphere is assumed to be 0‰. The difference between the $\delta^{15}N$ values in the organic matter and in the atmosphere is due to the isotopic fractionation during the process of denitrification of NO_3^- caused by anaerobic microorganisms (denitrifiers) (Peters et al., 1978).

In Early Archaean kerogens, the $\delta^{15}N$ ranges from close to 0‰ down to − 8.1‰ (Beaumont and Robert, 1999; Pinti et al., 2001, 2003). At the start of the Proterozoic, the $\delta^{15}N$ values in kerogens shift from negative values to positive values (Boyd and Philippot, 1998). This shift could well be related to the Great Oxidation Event (Holland, 2002). The oxygenation of the upper parts of the oceans at that time was almost certainly accompanied by a shift in the marine chemistry of N from NH_4^+-dominated to NO_3^--dominated.

In the absence of O_2 in the water column, NH_4^+ must have been the N-dominant containing species (Boyd 2001a). The NH_4^+ ion probably cycled efficiently between photosynthesizers, decomposers, and the water column. The most likely metabolic pathway for N in an anoxic ocean was N fixation or NH_4^+ uptake caused by bacteria, which could be able to produce the observed negative shift in the $\delta^{15}N$ values of Archaean kerogens (Beaumont and Robert, 1999; Pinti and Hashizume, 2001). The situation changed drastically when the concentration of O_2 in the water column increased. Nitrifying chemoautotrophs were then able to oxidize NH_4^+ to NO_2^- and subsequently to NO_3^-. These N-species were convertible to N_2 in anaerobic settings by denitrifiers, producing the positive shift of $\delta^{15}N$ observed in the present-day marine organic matter.

The hypothesis of an anoxic primitive ocean demands that all or nearly all of the sulfur in volatiles added to the ocean–atmosphere system in Archaean time was removed as a constituent of sulfides, mainly pyrite. This seems to be true. Anhydrite is known only as a replacement mineral before 2.3 Ga. The earliest significant anhydrite deposits are at 2.2 Ga. Barite is reasonably common in Archaean and early Proterozoic sediments. Its quantity is very small compared to that of contemporaneous pyrite, but its presence does indicate that SO_4^{-2} has been present in surface ocean water during the past 3.45 Ga. This is not surprising. Volcanic SO_2 reacts with H_2O at temperatures below 400 °C to give H_2S and H_2SO_4, without requiring the presence of oxygen (Holland, 2002).

3.5.3 Salinity

The salinity of the oceans derives from two distinct sources: weathering of the continental crust and oceanic hydrothermalism. In the present-day ocean, weathering is the dominant source of salinity, but during the Hadean and the Archaean periods, it was likely the oceanic hydrothermalism that was dominant (Derry and Jacobsen, 1988). The volume of the continents was lower than today (10–15% of the present-day volume) and the surface of the Earth was composed by a larger amount of small plates. Models suggest that the extension of the midocean ridges during Archaean was 10 times longer and production of new oceanic crust at MOR was 2–3 times higher than the present-day (Bickle, 1986). The input of dissolved species from hydrothermal activity at MOR and from interactions of hot fluids with oceanic crust was predominant during this period. The occurrence of large amounts of iron from the ridge crests, as a source of iron in BIF (Isley, 1995), is the evidence that the mantle buffered the chemistry of the ocean.

The Proterozoic era marks the end of an ocean salinity dominated by hydrothermal sources and the beginning of a salinity produced by the weathering of continents. Measuring the isotopic variation of the Sr and Nd ratios in the Archaean and the Proterozoic marine sediments have permitted observation of the change in the salinity sources of the ocean. The continental crust is enriched in rubidium (Rb) because this element is incompatible (which means that it tends to remain in the residual melt during magma crystallization, the melt forming the crust). The isotope ^{87}Rb decays to the isotope of mass 87 of strontium (^{87}Sr) and consequently, the isotopic ratio $^{87}Sr/^{86}Sr$ increases in the crust compared to the values observed in the mantle, which is correlatively depleted in Rb. Veizer and Compston (1976) measured the strontium isotopic ratios $^{87}Sr/^{86}Sr$ in Archaean carbonates and found values of 0.7025 ± 0.0015 to 0.7031 ± 0.0008. These values are close to that characteristic of the mantle during the Archaean (< 0.703). The present-day carbonates show strontium isotopic ratios higher than 0.715, indicating that the source of Sr in the ocean evolved with a progressive enrichment in ^{87}Sr due to the continental input (Veizer et al., 1989).

Isotopic studies of neodymium (Nd) and Sr in Archaean microbialites of the Campbellrand carbonate Platform, South Africa, allowed to constrain the ratio between the hydrothermal and the continental sources of salinity (Kamber and Webb, 2001). Figure 3.7 shows the isotopic ratios of Nd ($^{143}Nd/^{144}Nd$) and Sr($^{87}Sr/^{86}Sr$) measured in the 2.52-Ga microbialites. The Nd and Sr ratios show a mixing between the mantle and the continental crust reservoirs of Nd and Sr. From a simultaneous mass and isotopic balance, we can estimate the mantle and the continental flux ratio for the elements dissolved in the ocean at the transition between the Archaean and Proterozoic (2.5 Ga). The results suggest that the mantle flux of elements to the oceans was 3 times higher than the continental flux, while now the ratio between the mantle and the continental inputs is of 0.3 to 0.5 (Kamber and Webb, 2001).

The change in salinity sources, pH and redox state of the ocean has likely altered the relative concentration of dissolved species in ancient seawater. However, the salinity of the oceans was likely dominated, since the beginning, by halite (NaCl). A NaCl brine largely similar to modern seawater could have been produced by water–rock reactions at midoceanic ridges since Hadean (Sleep et al., 2001). Finding traces of preserved seawater in ancient rocks would be, of course, helpful for determining its chemistry. De Ronde et al. (1997) found primary inclusions at the 3.2-Ga Ironstone Pods of the Barberton greenstone belt, South Africa, containing two fluids mixed up. The first is seawater having a $NaCl–CaCl_2–H_2O$ composition. The second fluid is of hydrothermal ori-

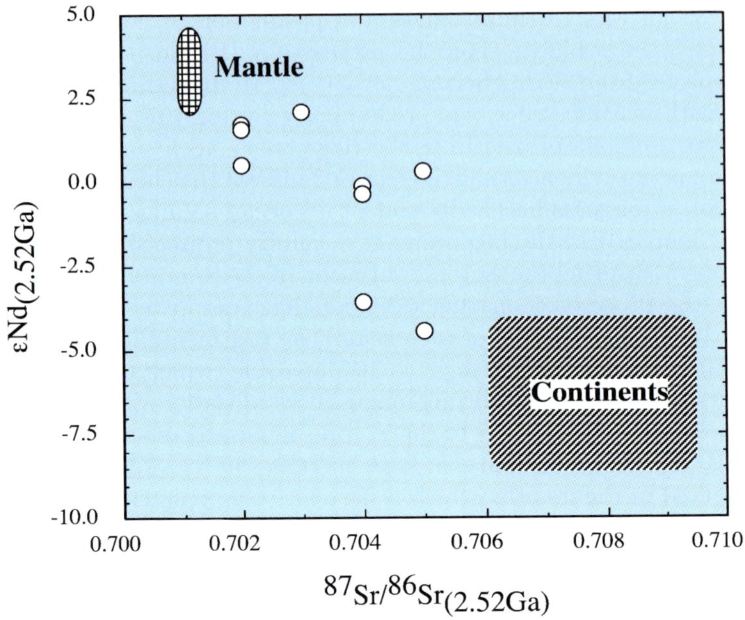

Fig. 3.7. Isotopic variations of Nd and Sr measured in 2.52-Ga microbialites from South Africa (Kamber and Webb, 2001). The variations can be explained by a mixing between a mantle and a continental input of dissolved species in the ancient ocean

gin and has a $CaCl_2$–$FeCl_2$–H_2O composition. Recent studies on primary fluid inclusions found in intrapillow quartz at the 3.5-Ga North Pole Dome, Western Australia, show the presence of ancient seawater having a NaCl–$CaCl_2$–H_2O composition similar to that found by de Ronde and his colleagues (Foriel et al. 2003).

The concentration of the major cations and anions dissolved in the Archaean fluids found in the Ironstone Pods is reported in Table 3.2 and Fig. 3.8. The Na/Cl ratio in the Archaean seawater is the same as the present-day ocean (0.858), but the total amount of sodium and chlorine is 165% higher than that of modern seawater. The concentrations of Ca^{2+} and Sr^{2+} are 22 and 50 times higher than the present-day ocean. The higher Ca and Sr concentrations could derive from reactions with carbonates (the seawater end-member fluid is a NaCl–$CaCl_2$–H_2O brine). The concentration of Mg^{2+} is mostly the same as that of modern seawater, while K^+ is 87% higher (Fig. 3.8). The concentration of SO_4^{2-} is lower than in modern seawater. This could be explained by reactions between calcium sulfates and hydrothermal fluids (de Ronde et al., 1997) or alternatively by reactions of these fluids with the oceanic crust (Bischoff and Dickson, 1991). However, these reactions would reduce considerably the Mg amount, which is not the case. Finally, the amount of halogens Br and I are 2.6 and 74 times higher, respectively, than what is observed in modern seawater.

Table 3.2. Concentrations of the main dissolved species in the Archaean ocean (in mmole/L). Data are from de Ronde et al. (1997)

Cl^-	Br^-	I^-	SO_4^{2-}	Na^+	K^+	Mg^{2+}	Ca^{2+}	Sr^{2+}
Archaean seawater								
920	2.25	0.037	2.3	789	18.9	50.9	232	4.52
Modern seawater								
556	0.86	0.0005	28.7	477	10.1	54.2	10.5	0.09

The Br/Cl ratio in the Archaean seawater is estimated at 2.5×10^{-3} (de Ronde et al., 1997), which is close to the mantle value of 2.9×10^{-3} (McDonough and Sun, 1995), but higher than what is measured in modern seawater of 1.54×10^{-3}. The I/Cl ratio in the Archaean seawater is 40×10^{-6} (de Ronde et al., 1997), which is higher than that of modern seawater (0.9×10^{-6}) but lower than that estimated for the mantle of 190×10^{-6}. Two hypotheses can explain the enrichment of halogens. The first is that the Archaean ocean chemistry was controlled by large-scale hydrothermal interaction between the oceanic crust/volcano-sedimentary pile and seawater. In other words, the ocean chemistry was buffered by the mantle, that is the source of I and Br. The second hypothesis assumes the role of the biosphere in the uptake of halogens from

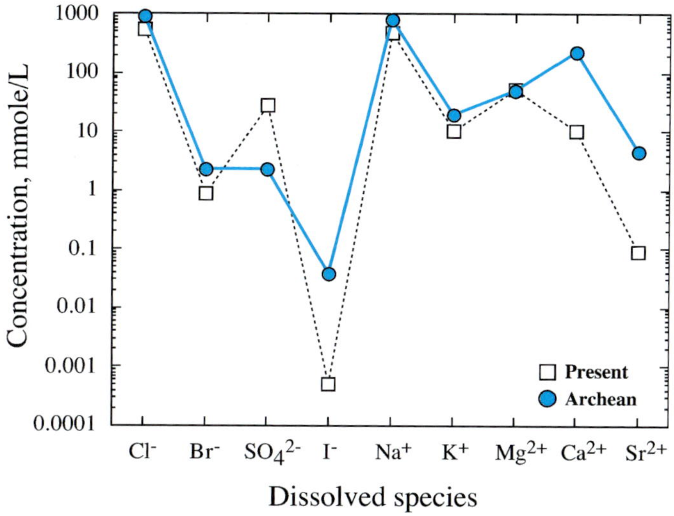

Fig. 3.8. Variations of the concentration of the main dissolved species in Archaean seawater and in modern seawater. Data from de Ronde et al. (1997)

seawater. In modern seawater, Br and I are depleted because they are fixed in plankton (they are metabolites for plankton). The burial of organic matter with sediments causes the scavenging of halogens from the ocean (Krauskopf and Bird, 1995). Channer et al. (1997) explained the enrichment of Br and I in Archaean seawater by two different hypotheses. The first is the presence of an ancient biosphere that interacted differently with halogens (absence of plankton or different metabolic pathways) and preserved halogens to be scavenged from the ocean. The second suggests that the carbon burial rate in the Archaean was lower than today. The carbon burial rate is indeed controlled by the rate of sedimentation, which in turn depends on the continental surface available for erosion. Smaller continents dominated the Archaean period and it is likely that sedimentation and carbon burial were reduced to a minimum value compared to the present-day (Des Marais et al., 1992), preserving high halogens concentrations in the water column.

3.6 Conclusions

The most recent models of ocean formation (Sleep et al., 2001) and recent geological findings (Mojzsis et al., 2001; Wilde et al., 2001) suggest that liquid water was present at the surface of the Earth soon after the accretion of our planet. The total oceanic water inventory was available since the beginning, delivered by hydrous carbonaceous chondrites and, in minor amount, by comets (Morbidelli et al., 2000; Dauphas and Marty, 2001; Dauphas, 2003). Probably 50 to 150 Ma later, the water vapor composing the secondary atmosphere of the Earth condensed to the surface to form the oceans (Abe, 1993). However, their temperature was still too high ($\geq 230\,°C$) to allow the survival of living organisms. Probably in a few tens of millions of years, the temperature of the oceans cooled down enough (80–110 °C) to allow the development of hyperthermophiles organisms, living on the oceanic floor or close to hydrothermal vents. At the beginning of the Archaean (3.9 Ga ago), the presence of marine sediments clearly indicates a stable oceanic environment. The ocean was probably more acid and anoxic than the modern one, being in equilibrium with a CO_2-dominated atmosphere (Kasting, 1991). The chemistry of the Archaean ocean was dominated by halite, as in modern seawater, but the total concentrations were higher, especially for elements such as halogens. This can be explained by an ocean chemistry buffered by the mantle, through large-scale hydrothermal interactions with the oceanic crust.

Acknowledgement

I thank the editors for inviting me to participate in this fascinating review of the early "life" of our planet. An anonymous reviewer, Jacques Reisse and handling editor Hervé Martin greatly improved the manuscript. Gilles Delaygue of

CEREGE gave valuable suggestions. Genevieve Roche is thanked for the color plates illustrating the birth of the oceans. Research work on Archaean is supported by French national funding CNRS-GDR Exobiology; CNRS-PNP; MAE. This is the LGMT contribution no. 45.

References

Abe, Y. (1993). Physical state of the very early Earth. *Lithos*, **30**, 223–235.

Allègre, C.J., Manhès, G., Gopel, C. (1995). The age of the Earth. *Geochim. Cosmochim. Acta*, **59**, 1445–1456.

Allègre, C.J., Staudacher, T., Sarda, P. (1986/87). Rare gas systematic: formation of the atmosphere, evolution and structure of the Earth's mantle. *Earth Planet. Sci. Lett.*, **81**, 127–150.

Beaumont, V., Robert, F. (1999). Nitrogen isotope ratios of kerogens in Precambrian cherts: a record of the evolution of atmosphere chemistry? *Precambrian Res.*, **96**, 63–82.

Bickle, M.J. (1986). Implications of melting for stabilisation of the litosphere and heat loss in the Archaean. *Earth Planet. Sci. Lett.*, **80**, 314–324.

Bischoff, J.L., Dickson, F.W. (1991). Seawater-basalt interaction at 200 °C and 500 bar. Implications for the origin of heavy-metal deposits and regulation of sewater chemistry. *Earth Planet. Sci. Lett.*, **25**, 385–397.

Boato, G. (1954). The isotopic composition of hydrogen and carbon in the carbonaceous chondrites. *Geochim. Cosmochim. Acta*, **6**, 209–220.

Bockelée-Morvan, D. et al. (1988). Deuterated water in comet C/1996 B2 (Hyakutake) and its implications for the origin of comets. *Icarus*, **193**, 147–162.

Bowring, S.A., Williams, I.S. (1999). Priscoan (4.00–4.03 Ga) orthogneisses from northwestern Canada. *Contrib. Mineral. Petrol.*, **134**, 3–16.

Boyd, S.R. (2001a). Ammonium as a biomarker in Precambrian metasediments. *Precambrian Res.*, **108**, 159–173.

Boyd, S.R. (2001b). Nitrogen in future biosphere studies. *Chem. Geol.*, **176**, 1–30.

Boyd, S.R., Philippot, P. (1998). Precambrian ammonium biogeochemistry: a study of the Moine metasediments, Scotland. *Chem. Geol.*, **144**, 257–268.

Butler, W.A., Jeffery, P.M., Reynolds, J.H., Wasserburg, G.J. (1963). Isotopic variations in terrestrial xenon. *J. Geophys. Res.*, **68**, 3283–3291.

Canup, R.M., Asphaug, E. (2001). Origin of Moon in a giant impact near the end of the Earth's formation. *Nature*, **412**, 708–712.

Channer, D.M.D.R., de Ronde, C.E.J., Spooner, E.T.C. (1997). The Cl-Br-I composition of ~3.23 Ga modified seawater: implications for the geological evolution of ocean halide chemistry. *Earth Planet. Sci. Lett.*, **150**, 325–335.

Cleaves, H.J., Miller, S.L. (1998). Oceanic protection of prebiotic organic compounds from UV radiation. *Proc. Natl. Acad. Sci. USA*, **95**, 7260–7263.

Craig, H., Clarke, W.B., Beg, M.A. (1975). Excess ^{3}He in deep water on the East Pacific Rise. *Earth Planet. Sci. Lett.*, **26**, 125–132.

Dauphas, N. (2003). The dual origin of the terrestrial atmosphere. *Icarus*, **165**, 326–339.

Dauphas, N., Marty, B. (2001). Inference on the nature and the mass of Earth's late veneer from noble metals and gases. *J. Geophys. Res.*, **107**, E 12, 1–7.

Dauphas, N., Robert, F., Marty, B. (2000). The late asteroidal and cometary bombardment of Earth as recorded in water deuterium to protium ratio. *Icarus*, **148**, 508–512.

de Ronde, C.E.J., Channer, D.M.D.R., Faure, K., Bray, C. J., Spooner, T. C. (1997). Fluid chemistry of Archaean seafloor hydrothermal vents: Implications for the composition of circa 3.2 Ga seawater. *Geochim. Cosmochim. Acta*, **61**, 4025–4042.

Deloule, E., Albarede, F., Sheppard, S.M.F. (1991). Hydrogen isotope heterogeneities in the mantle from ion probe analysis of amphiboles from ultramafic rocks. *Earth Planet. Sci. Lett.*, **105**, 543–553.

Delsemme, A.H. (1999). The deuterium enrichment observed in recent comets is consistent with the cometary origin of seawater. *Planet. Space Sci.*, **47**, 125–131.

Deming, D. (1999). On the possible influence of extraterrestrial volatiles on Earth's climate and the origin of the oceans. *Palaeogeogr. Palaeoclim. Palaeoecol.*, **146**, 33–51.

Derry, L.A., Jacobsen, S.B. (1988). The Nd and Sr evolution of Proterozoic seawater. *Geophys. Res. Lett.*, **15**, 397–400.

Des Marais, J., Strauss, H., Summons, R.E., Hayes, J.M. (1992). Carbon isotope evidence for the stepwise oxidation of the Proterozoic environment. *Nature*, **359**, 605–609.

Drouart, A., Dubrulle, B., Gautier, D., Robert, F. (1999). Structure and transport in the solar nebula from constraints on deuterium enrichment and giant planets formation. *Icarus*, **140**, 59–76.

Eberhardt, P., Reber, M., Krankowski, D., Hodges, R.R. (1995). The D/H and $^{18}O/^{16}O$ ratios in water from comet P/Haley. *Astron. Astrophys.*, **302**, 301–316.

Engrand, C., Deloule, E., Robert, F., Maurette, M., Kurat, G. (1999). Extraterrestrial water in micrometeorites and cosmic spherules from Antarctica: An ion microprobe study. *Meteorit Planet. Sci.*, **34**, 773–786.

Faure, G. (1991) *Principles and Applications of Inorganic Geochemistry: a Comprehensive Textbook for Geology Students*. Maxwell Macmillan International. New York, NY.

Foriel, J., Philippot, P., Banks, D., Rey, P., Cauzid, J., Somogyi, A. (2003). Composition of a 3.5 Gyr shallow seawater from the North Pole Dome, Western Australia. *Geochim. Cosmochim. Acta*, **67**, A100.

Frank, L.A., Sigwarth, J.B. (1997). Trails of OH emissions from small comets near Earth. *Geophys. Res. Lett.*, **24**, 2435–2438.

Frank, L.A., Sigwarth, J.B., Craven, J.D. (1986). On the influx of small comets into the Earth's atmosphere I. Observations. *Geophys. Res. Lett.*, **13**, 303–306.

Gautier, D., Owen, T. (1983). Cosmological implication of helium and deuterium abundances of Jupiter and Saturn. *Nature*, **302**, 215–218.

Geiss, J., Gloecker, G. (1998). Abundance of Deuterium and Helium in the protosolar Cloud. *Space Science Rev.*, **84**, 239–250.

Halliday, A.N. (2001). Earth science – In the beginning. *Nature*, **409**, 144–145.

Harrison, C.G.A. (1999). Constraints on ocean volume change since the *Archaean. Geophys. Res. Lett.*, **26**, 1913–1916.

Holland, H.D. (1984) *The Chemical Evolution of the Atmosphere and Oceans*. Princeton University Press. Princeton, NY.

Holland, H.D. (2002). Volcanic gases, black smokers, and the great oxidation event. *Geochim. Cosmochim. Acta*, **66**, 3811–3826.

Holm, N.G., Andersson, E.M. (1998) Hydrothermal systems. *In The Molecular Origins of Life. Assembing Pieces of the Puzzle*, ed. Brack, A., p. 86–99, Cambridge University Press. Cambridge, UK

Honda, M., McDougall, I., Patterson, D.B., Doulgeris, A., Clague, D.A. (1991). Possible solar noble gas component in Hawaiian basalts. *Nature*, **349**, 149–151.

Isley, A.E. (1995). Hydrothermal plumes and the delivery of iron to Banded Iron Formation. *J. Geol.*, **103**, 169–185.

Kamber, B.S., Webb, G.E. (2001). The geochemistry of late Archaean microbial carbonate: Implications for ocean chemistry and continental erosion history. *Geochim. Cosmochim. Acta*, **65**, 2509–2525.

Kasting, J.F. (1987). Theoretical constraints on oxygen and carbon dioxide concentrations in the precambrian atmosphere. *Precambrian Res.*, **34**, 205–229.

Kasting, J.F. (1991). Box models for the evolution of atmospheric oxygen: An update. *Palaeogeogr. Palaeoclim. Palaeoecol.*, **97**, 125–131.

Kasting, J.F. (1993). Earth's early atmosphere. *Science*, **259**, 920–926.

Kempe, S, Degens, E.T. (1985). An early soda ocean? *Chem. Geol.*, **53**, 95–108.

Kerridge, J.F. (1985). Carbon, hydrogen and nitrogen in carbonaceous chondrites: Abundances and isotopic compositions in bulk samples. *Geochim. Cosmochim. Acta*, **49**, 1707–1714.

King, E.M., Valley, J.W., Davis, D.W., Edwards, G.R. (1998). Oxygen isotope ratios of Archean plutonic zircons from granite-greenstone belts of the Superior Province: indicator of magmatic source. *Precambrian Res.*, **92**, 365–387.

Knauth, L.P., Lowe, D.R. (1978). Oxygen isotope geochemistry of cherts from the Onverwatch Group (3.4 billion years), Transvaal Group, South Africa, with implications for secular variations in the isotopic composition of cherts. *J. Geol.*, **41**, 209–222.

Knauth, P.L. (1994). Petrogenesis of chert. *Review Mineral.*, **29**, 233–258.

Knauth, P.L., Roberts, S.K. (1991) The hydrogen and oxygen isotope history of the Silurian-Permian hydrosphere as determined by direct measurement of fossil water. In *Stable Isotope Geochemistry: A Tribute to Samuel Epstein*, Vol. Special Publication No. 3, eds. Taylor, H.P., O'Neil, J.R., Kaplan, I.R., p. 91–104, The Geochemical Society. San Antonio; TX.

Krauskopf, K.B., Bird, D.K. (1995) *Introduction to Geochemistry*. McGraw-Hill. New York, NY.

Lecuyer, C., Gillet, P., Robert, F. (1998). The hydrogen isotope composition of seawater and the global water cycle. *Chem. Geol.*, **145**, 249–261.

Li, Y.-H. (2000) *A Compendium of Geochemistry; from Solar Nebula to Human Brain.* Princeton University Press. Princeton, NY.

Lowe, D.R., Byerly, G.R. (2003). Ironstone Pods in the Archean Barberton greenstone belt, South Africa: Earth's oldest seafloor hydrothermal vents reinterpreted as Quaternary subaerial springs. *Geology*, **31**, 909–912.

Martin, H., Moyen, J.-F. (2002). Secular changes in tonalite-trondhjemite-granodiorite composition as markers of the progressive cooling of Earth. *Geology*, **30**, 319–322.

Marty, B. (1989). Neon and xenon isotopes in MORB: implications for the earth-atmosphere evolution. *Earth Planet. Sci. Lett.*, **94**, 45–56.

Maurette, M., Duprat, J., Engrand, C., Gounelle, M., Kurat, G., Matrajt, G., Toppani, A. (2000). Accretion of neon, organics, CO_2, nitrogen and water from large interplanetary dust particles on the early Earth. *Planet. Space Sci.*, **48**, 1117–1137.

Meier, R., Owen, T.C., Matthews, H.E., Jewitt, D.C., Bockelée-Morvan, D., Biver, N., Crovisier, J., Gautier, D. (1998). A determination of the DH_2O/H_2O ratio in Comet C/1995 O1 (Hale–Bopp). *Science*, **279**, 842–844.

Mojzsis, S.L., Arrhenius, G., Friend, C.R.L. (1996). Evidence for life on Earth before 3,800 million years ago. *Nature*, **384**, 55–57.

Mojzsis, S.J., Harrison, M.T., Pidgeon, R.T. (2001). Oxygen-isotope evidence from ancient zircons for liquid water at the Earth's surface 4,300 Myr ago. *Nature*, **409**, 178–181.

Morbidelli, A., Chambers, J., Lunine, J.I., Petit, J.M., Robert, F. (2000). Source regions and timescales for the delivery of water to the Earth. *Meteorit. Planet. Sci.*, **35**, 1309–1320.

Morris, R.C. (1993). Genetic modelling for banded iron-formation of the Hamersley Group, Pilbara Craton, Western Australia. *Precambrian Res.*, **60**, 243–286.

Murakami, T., Utsunomiya, S., Imazu, Y., Prasad, N. (2001). Direct evidence of late Archean to early Proterozoic anoxic atmosphere from a product of 2.5 Ga old weathering. *Earth Planet. Sci. Lett.*, **184**, 523–528.

Nijman, W., de Bruijne, H., Valkering, M.E. (1998). Growth fault control of Early Archean cherts, barite mounds and chert-barite veins, North Pole Dome, Eastern Pilbara, Western Australia. *Precambrian Res.*, **88**, 25–52.

Nisbet, E.G., Cheadle, M.J., Arndt, N.T., Bickle, M.J. (1993). Constraining the potential temperature of the archaean mantle – a review of the evidence from komatiites. *Lithos*, **30**, 291–307.

Nisbet, E.G., Sleep, N.H. (2001). The habitat and nature of early life. *Nature*, **409**, 1083–1091.

Nutman, A.P., Mojzsis, S.J., Friend, C.R.L. (1997). Recognition of $\geq$ 850 Ma water-lain sediments in West Greenland and their significance for the early Archean Earth. *Geochim. Cosmochim. Acta*, **61**, 2575–2484.

Ottonello, G. (1997) *Principles of Geochemistry*. Columbia University Press. New York, NY.

Owen, T.C. (1998) The origin of the atmosphere. In *The Molecular Origins of Life*, ed. Brack, A., p. 13–34, Cambridge University Press. Cambridge, UK.

Owen, T.C., Bar-Nun, A. (1995). Comets, impacts, and atmospheres. *Icarus*, **116**, 215–226.

Ozima, M., Podosek, F.A. (2001) *Noble Gas Geochemistry*. 2nd edition. Cambridge University Press. Cambridge, UK.

Ozima, M., Zanshu, S. (1988). Solar-type Ne in Zaire cubic diamonds. *Geochim. Cosmochim. Acta*, **52**, 19–25.

Peck, W.H., King, E.M., Valley, J.W. (2000). Oxygen isotope perspective on Precambrian crustal growth and maturation. *Geology*, **28**, 363–366.

Peck, W.H., Valley, J.W., Wilde, S.A., Graham, C.M. (2001). Oxygen isotope ratios and rare earth elements in 3.3 to 4.4 Ga zircons: Ion microprobe evidence for high δ^{18}O continental crust and oceans in the Early Archean. *Geochim. Cosmochim. Acta*, **65**, 4215–4229.

Pepin, R.O. (1991). On the origin and ealy evolution of terrestrial planetray atmospheres and meteoritic volatiles. *Icarus*, **92**, 2–79.

Peters, K.E., Sweeney, R.E., Kaplan, I.R. (1978). Correlation of carbon and nitrogen stable isotope ratios in sedimentary organic matter. *Limnol. Oceanogr.*, **23**, 598–604.

Pidgeon, R.T., Wilde, S.A. (1998). The interpretation of complex zircon U-Pb systems in Archaean granitoids and gneisses from the Jack Hills, Narryer gneiss Terrane, Western Australia. *Precambrian Res.*, **91**, 309–332.

Pinti, D.L. (2002). The isotopic record of Archean nitrogen and the evolution of the early earth. *Trends in Geochemistry*, **2**, 1–17.

Pinti, D.L. Hashizume, K. (2001). N-15-depleted nitrogen in Early Archean kerogens: clues on ancient marine chemosynthetic-based ecosystems? A comment to Beaumont, V., Robert, F., 1999. *Precambrian Res.*, **96**, 62–82. *Precambrian Res.*, **105**, 85–88.

Pinti, D.L., Hashizume, K., Matsuda, J. (2001). Nitrogen and argon signatures in 3.8 to 2.8 Ga metasediments: Clues on the chemical state of the Archean ocean and the deep biosphere. *Geochim. Cosmochim. Acta*, **65**, 2301–2315.

Pinti, D.L., Hashizume, K., Philippot, P., Foriel, J., Rey, P. (2003). Nitrogen quest in Archean metasediments of Pilbara, Australia. *Geochim. Comochim. Acta*, **67**, A379.

Poreda, R.J., Farley, K.A. (1992). Rare gases in Samoan xenoliths. *Earth Planet. Sci. Lett.*, **113**, 129–144.

Robert, F. (2001) L'origine de l'eau dans le Système Solaire telle qu'elle est enregistrée par son rapport isotopique D/H. In *L'environnement de la Terre Primitive*, eds. Gargaud, M., Despois, D., Parisot, J.-P., p. 79–90, Presses Universitaires de Bordeaux. Bordeaux, France.

Robert, F., Epstein, S. (1982). The concentration and isotopic composition of hydrogen, carbon and nitrogen in the carbonaceous meteorites. *Geochim. Cosmochim. Acta*, **46**, 81–95.

Robert, F., Gautier, D., Dubrulle, B. (2000). The solar system D/H ratio: Observations and theories. *Space Sci. Rev.*, **92**, 201–224.

Roedder, E. (1984) *Fluid Inclusions.* Mineralogical Society of America. Washington D.C.

Rosing, M.T., Rose, N.M., Bridgwater, D., Thomsen, H.S. (1996). Earliest part of Earth's stratigraphic record: A reappraisal of the $>$ 3.7 Ga Isua (Greenland) supracrustal sequence. *Geology*, **24**, 43–46.

Rothschild, L.J., Mancinelli, R.L. (2001). Life in extreme environments. *Nature*, **409**, 1092–1101.

Rubey, W.W. (1951). Geologic history of seawater: An attempt to state the problem. *GSA Bull.*, **62**, 1111–1147.

Sarda, P., Staudacher, Th., Allègre, C.J. (1988). Neon isotopes in submarine basalts. *Earth Planet. Sci. Lett.*, **91**, 73–88.

Sleep, N.H., Zahnle, K. (2001). Carbon dioxide cycling and implications for climate on ancient Earth. *J. Geophys. Res.*, **106**, 1373–1399.

Sleep, N.H., Zahnle, K., Neuhoff, P.S. (2001). Initiation of clement surface conditions on the earliest Earth. *Proc. Natl. Acad. Sci. USA*, **98**, 3666–3672.

Staudacher, T., Allegre, C.J. (1982). Terrestrial xenology. *Earth Planet. Sci. Lett.*, **60**, 389–406.

Staudacher, T., Sarda, P., Richardson, S.H., Allègre, C.J., Sagna, I., Dmitriev, L.V. (1989). Noble gases in basalt glasses from a Mid-Atlantic Ridge topographic high at 14° N: geodynamic consequences. *Earth Planet. Sci. Lett.*, **96**, 119–133.

Stetter, K.O. (1998) Hyperthermophiles and their possible role as ancestors of modern life. In *The Molecular Origins of Life*, ed. Brack, A., p. 315–335, Cambridge University Press. Cambridge, UK.

Sugitani, K. (1992). Geochemical characteristics of Archean cherts and other sedimentary rocks in the Pilbara Block, Western Australia: evidence for Archean seawater enriched in hydrothermally-derived iron and silica. *Precambrian Res.*, **57**, 21–47.

Tajika, E., Matsui, T. (1993). Degassing history and carbon cycle of the Earth: From an impact-induced steam atmosphere to the present atmosphere. *Lithos*, **30**, 267–280.

Taylor, S.R., McLennan, S.M. (1995). The geochemical evolution of the continental crust. *Rev. Geophys.*, **33**, 241–265.

van Zuilen, M.A., Lepland, A., Arrhenius, G. (2002) Reassessing the evidence for the earliest traces of life, *Nature*, **418**, 627–630.

Valley, J.W., Chiarenzelli, J.R., McLelland, J.M. (1994). Oxygen isotope geochemistry of zircon. *Earth Planet. Sci. Lett.*, **126**, 187–206.

Valley, J.W., Kinny, P.D., Shulze, D., Spicuzza, M.J. (1998). Zircon megacrysts from kimberlite: Oxygen isotope variability among mantle melts. *Contrib. Mineral. Petrol.*, **133**, 1–11.

Valley, J.W., Peck, W.H., King, E.M., Wilde, S.A. (2002). A cool early Earth. *Geology*, **30**, 351–354.

Veizer, J., Compston, W. (1976). $^{87}Sr/^{86}Sr$ in Precambrian carbonates as an index of crustal evolution. *Geochim. Cosmochim. Acta*, **40**, 905–914.

Veizer, J., Hoefs, J., Ridler, R.H., Jensen, L.S., Lowe, D.R. (1989). Geochemistry of Precambrian carbonates: I. Archean hydrothermal systems. *Geochim. Cosmochim. Acta*, **53**, 845–857.

Wilde, S.A., Valley, J.W., Peck, W.H., Graham, C.M. (2001). Evidence from detrital zircons for the existence of continental crust and oceans on the Earth 4.4 Gyr ago. *Nature*, **409**, 175–178.

Yang, W., Holland, H.D., Rye, R. (2002). Evidence for low or no oxygen in the late Archean atmosphere from the ~2.76 Ga Mt. Roe #2 paleosol, Western Australia: Part 3. *Geochim. Cosmochim. Acta*, **66**, 3707–3718.

Zhang, Y.X., Zindler, A. (1993). Distribution and evolution of carbon and nitrogen in earth. *Earth Planet. Sci. Lett.*, **117**, 331–345.

4 Genesis and Evolution of the Primitive Earth Continental Crust

Hervé Martin

The oldest terrestrial sedimentary rocks recorded that life appeared very early on our planet probably since 3.87 Ga ago at Isua in Greenland (Schidlowski, 1988; Mojzsis et al., 1996; Rosing and Frei, 2004) and certainly before 3.5 Ga ago at Barberton in South Africa (Walsh, 1992; Westall et al., 2001) or at the North Pole barite deposit in north-western Australia (Shen and Buick, 2004). Life apparition, as well as organic molecule duplication, which is a prerequisite to life development necessitated specific physical and chemical conditions and environment. For instance, all proposed scenarios imply the presence of liquid water on the Earth surface (ocean, lake, pond, etc), whereas others envisage alternation of aqueous and dry periods (emerged continents). These conditions are classically realized in continental domains or at the continent/ocean interface. Consequently, in order to address the problem of the origin of life on our planet, it appears necessary, not only to know when liquid water condensed on the surface of the Earth to form oceans, but also to determine when the first continent formed and emerged. This last point leads to a discussion of:

- The primitive Earth thermal regimes that controlled both genesis and stabilization of continental masses.
- The dynamic and tectonic regimes, which, in association with thermal activity, can lead to the formation of mountain chains and consequently to the emersion of continents.

On the other hand, the chemical composition of the ocean (i.e. pH) is directly correlated to the mineralogical and chemical compositions of emerged continent. Indeed, weathering and alteration of surface rocks render some chemical elements (i.e. alkali) soluble, which, via rivers, are transported from continental crusts towards oceans, where they can accumulate. By this mean, the ocean composition is mostly controlled by the continent composition. From this point of view, detailed knowledge of the primitive continental crust composition appears necessary to discuss the origin of life. In addition, here too, knowledge of the primitive Earth tectonic regimes appears critical, as tectonic regimes create relief and thus generate potential energy necessary to transfer chemical elements from continents to oceans.

The aims of this chapter are to provide some constraints on these different aspects of early Earth dynamics. It will be divided into two parts: the first discusses the composition of the early continental crust and the mechanisms of its genesis; the second addresses the problem of its geodynamic as well as of its subsequent evolution. All these aspects will be envisaged in terms of a comparison with the present-day continental crust genesis and dynamics, which are well known and constrained.

4.1 Composition and Genesis of the Primitive Continental Crust

4.1.1 Introduction

The Earth continental crust has a low density ($d \sim 2.75$), thus contrarily to oceanic crust ($d \sim 3.1$), it cannot be extensively recycled or reincorporated into the mantle ($d \sim 3.3$). This is mainly this physical property that accounts for the preservation of crustal remnants as old as 4.0 to 4.4 Ga (4.0 to 4.4 billion years). Consequently, the continental crust is an archive that can be used to try to reconstruct the geological history of our planet. This history is classically divided into several main aeons (Fig. 4.1).

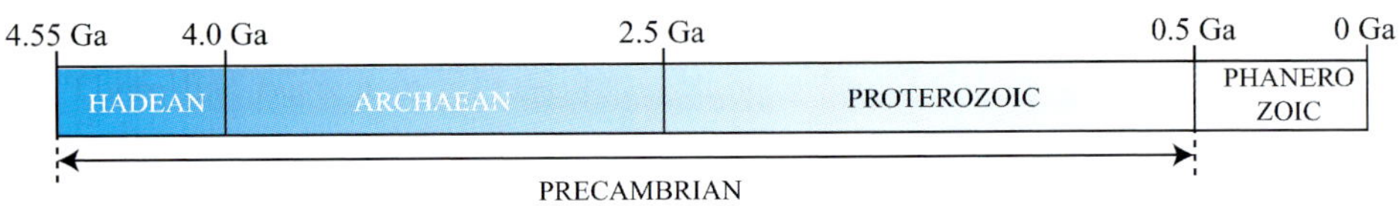

Fig. 4.1. Synthetic geological time scale

The Hadean extends from Earth accretion until the age of the oldest known rocks: the 4.03-Ga Acasta gneisses. In other words, except for zircon crystals from Australia, direct geological memory for this 0.55-Ga long period does not exist. The Archaean–Proterozoic boundary corresponds to a period of transition in Earth dynamics, indeed present-day-like geodynamic processes started to operate since Proterozoic times (2.5 Ga). These changes will be described and discussed in detail in this chapter. Finally, the Proterozoic–Phanerozoic transition is marked by the apparition and development of living beings with internal or external skeleton able to be preserved by fossilization.

Geographically, the early continental crust is widely distributed all over the world; on each continent it outcrops in very large domains. This appears even more obvious when the early continental crust only covered by a few kilometres of younger sediments is taken into account (Fig. 4.2).

After Earth accretion, the core rapidly differentiated from the mantle, a process that lasted less than 0.1 Ga and more probably less than 0.03 Ga (Kleine et al., 2002; Yin et al., 2002; Vityazev et al., 2003). On the contrary, the crust differentiated from the mantle by a slow and progressive mechanism that is still in progress today. Different models were proposed to account for juvenile continental crust growth. Figure 4.3 shows very important differences depending on the authors, however, all clearly point to the fact that 70 to 75% (volume) of the continental crust was generated and extracted from the mantle before 2.5 Ga. Consequently, the Archaean aeon appears as a period of intense magmatic activity that led to the extraction of about 3/4 of the continental crust from the mantle.

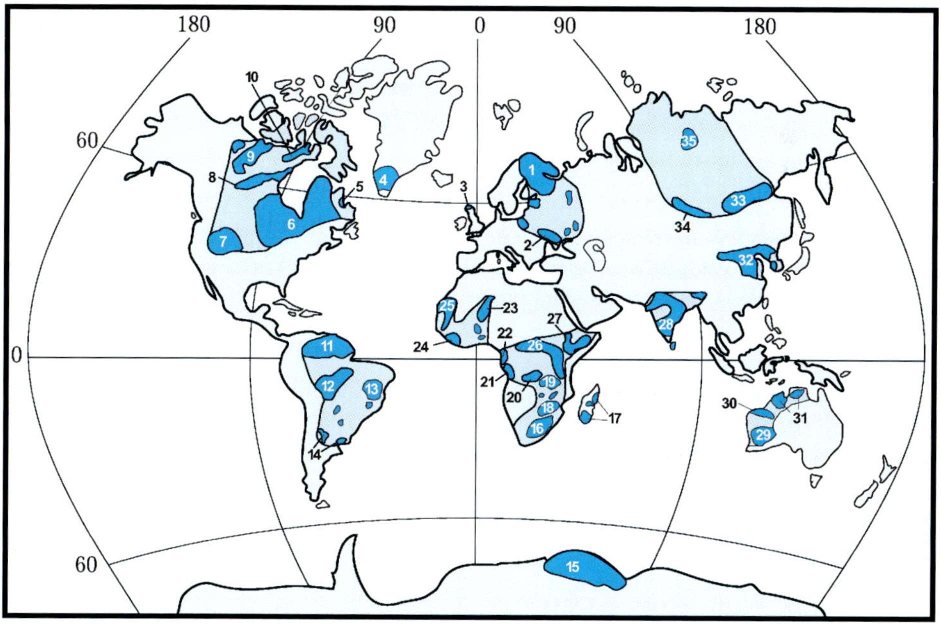

Fig. 4.2. Distribution of Archaean provinces (adapted from Condie, 1981; and Goodwin, 1991) Exposed Archaean terrains are in *dark blue*, and areas underlain by Archaean rocks are in *light blue*. (1) Baltic Shield; (2) Ukrainian Shield; (3) Scottish Shield; (4) Greenland Shield; (5) Labrador Shield; (6) Superior Province; (7) Wyoming Province; (8) Kaminak Group; (9) Slave Province; (10) Committee Bay Block; (11) Guiana Shield; (12) Guapore Craton; (13) São Francisco Craton; (14) Rio de la Plata and Luis Alves Massifs; (15) Napier Complex; (16) Kaapvaal Craton; (17) Madagascar Craton; (18) Zimbabwe Craton; (19) Zambian Block; (20) Kasaï Craton; (21) Chaillu Craton; (22) Cameroon N'tem Complex; (23) Tuareg Shield; (24) Man Shield; (25) Reguibat Shield; (26) Central Africa Craton; (27) Ethiopian Block; (28) Indian Shield; (29) Yilgarn block; (30) Pilbara block; (31) Litchfield, Rul Jungle and Nanambu Complexes; (32) Sino-Korean, Tarim and Yangtze Cratons; (33) Aldan Shield; (34) Baikal, Sayan and Yienisei fold belts; (35) Anabar Shield

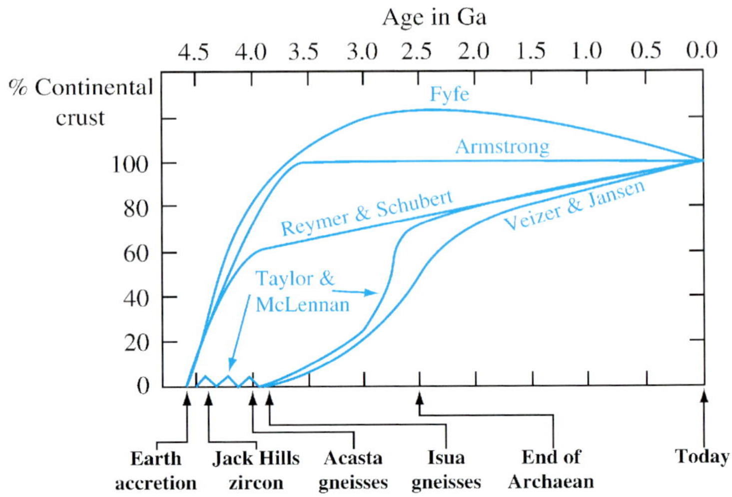

Fig. 4.3. Continental crust growth models after Taylor and McLennan (1985). Instead of big differences between them, all these models show that 70 to 75% of the juvenile continental crust has been generated and extracted from the mantle prior to 2.5 Ga. Curves are from Fyfe (1978); Veizer and Jansen (1979); Armstrong (1981); Reymer and Schubert (1984); Taylor and McLennan (1985)

4.1.2 Age of the Oldest Continental Crust

The oldest rocks so far discovered and forming huge outcrops ($\sim$ 3000 km^2) are exposed in Greenland (Table 4.1). They consist of magmatic rocks, now transformed into gneisses and called the Amitsôq gneisses; they were dated at 3.822 ± 0.005 Ga (Kinny, 1986). Associated volcanic and sedimentary rocks from Isua (Photo 4.1), yielded an age of 3.812 ± 0.014 Ga (Baadsgaard et al., 1984). Nearby, on Akilia island, a dyke cutting a Banded Iron Formation (BIF) gave an age of 3.872 ± 0.010 Ga, thus demonstrating that Akilia is still older than 3.872 Ga (Nutman et al., 1996; Nutman et al., 2002). Very recently, similar volcano sedimentary formation was discovered in Hudson Bay (Canada) at Porpoise Cove and dated at 3.825 ± 0.016 Ga (David et al., 2002). Both Isua and Akilia formations contain sedimentary rocks (BIF, quartzites) that constitute a direct proof that liquid water existed on the surface of the Earth at 3.87 Ga. In addition, these BIF contain organic matter with a δ^{13}C average isotopic constitution of -30 to -35‰ that could be interpreted as a biological signature (Mojzsis et al., 1996). Recently, U-rich sediments from the same locality were interpreted as indicators of oxidized ocean water resulting of oxygenic photosynthesis (Rosing and Frei, 2004). However, the reliability of these biological signatures is still subject to a very active controversy.

Table 4.1. Summary of the oldest crustal ages measured on terrestrial materials

Age in Ga	Formation	Country	Rock or Mineral
3.822	Amitsôq Gneisses	Greenland	Magmatic rock
3.825	Porpoise Cove	Canada	Magmatic rock
3.872	Akilia-Isua Gneisses	Greenland	Magmatic rock cutting BIF
4.030	Acasta Gneisses	Canada	Banded migmatite
4.300	Mt. Narryer	Australia	Zircon
4.404	Jack Hills	Australia	Zircon

At Acasta, in Northern Territories (Canada) smaller amounts ($20\,km^2$) of banded Archaean rocks are exposed. These small outcrops constitute the oldest continental crust so far discovered; it was generated at 4.030 ± 0.003 Ga (Bowring and Williams, 1999).

Until recently (2001) these rocks were the oldest known terrestrial materials, and the question was to determine whether a continental crust existed or not before 4.0 Ga. Nd isotopic composition of the oldest rocks provides indirect

Photo 4.1. Oldest known terrestrial sediments: the 3.865 Ga-old Isua gneisses in Greenland. (Photo G. Gruau)

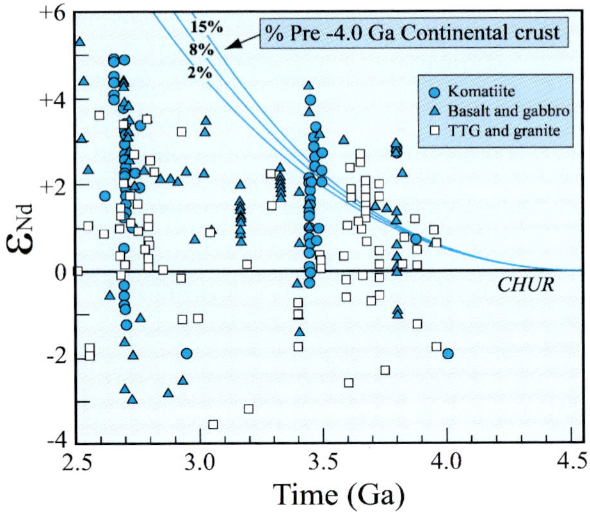

Fig. 4.4. $\varepsilon_{\mathrm{Nd}}$ vs. time diagram showing that even the oldest known rocks possess the $\varepsilon_{\mathrm{Nd}} > 0$ characteristic of mantle composition impoverished by extraction of continental crust prior to 4.0 Ga. *Blue curves* are theoretical mantle Nd isotopic composition assuming 2, 8 and 15% extraction of Hadean continental crust. Data are from McCulloch and Bennet (1993), Jahn (1997) and personal data. CHUR = chondritic uniform reservoir

evidence of a Hadean continental crust. Indeed, continental crust was extracted (fractionated) from the mantle, and consequently as the crust volume increased, the mantle composition changed.

The primordial Earth mantle is assumed to have the same Nd isotopic ratios as chondritic meteorites. In Fig. 4.4, $\varepsilon_{\mathrm{Nd}}$ represents the difference between the $^{143}\mathrm{Nd}/^{144}\mathrm{Nd}$ of a rock and that of chondrites; consequently a $\varepsilon_{\mathrm{Nd}} = 0$ will reflect a mantle source not affected by the extraction of the continental crust (CHUR = chondritic uniform reservoir), whereas $\varepsilon_{\mathrm{Nd}} > 0$ indicates a mantle impoverished by crust extraction; $\varepsilon_{\mathrm{Nd}} < 0$ would rather reflect crustal source or contamination. Obviously, even the oldest known rocks already display positive $\varepsilon_{\mathrm{Nd}}$, thus demonstrating that important volumes of continental crust were formed before 4.0 Ga. Simple calculations (McCulloch and Bennet, 1993) showed that about 10% of the volume of the present-day continental crust formed during the Hadean aeon. Similar studies performed on Hf (Vervoort et al., 1996) and Pb (Kamber et al., 2003) isotopes led to the same conclusion. The lack of known remnants of Hadean crust was considered to be due to the fact that this crust has been completely destroyed by intense meteoritic bombardment, or because, due to its relative small volume, it has been totally recycled into the mantle.

Very recently, zircon crystals, extracted from Jack Hills metaquartzites in Australia gave an age of 4.404±0.008 Ga (Wilde et al., 2001), which is the oldest age obtained on terrestrial material. The chemical characteristics of these zircon crystals demonstrate that they crystallized in granite-like magmas, thus establishing the existence of stable continental crust as early as 4.4 Ga ago. In addition, oxygen isotopes measurement ($\delta^{18}O = 8.5$ to 9.5‰) made on these minerals showed that their host magmatic rock source strongly interacted with liquid water (Mojzsis et al., 2001; Wilde et al., 2001). These data are extremely important as they demonstrate that: 1) granitic-like continental crust began to differentiate from the mantle very early in Earth history; 2) water condensed very rapidly after Earth accretion (150 Ma) such that oceans existed early thus creating necessary conditions for life development (see chapter by D. Pinti in this book).

4.1.3 Composition of the Early Continental Crust: Comparison with Modern Continental Crust

Archaean terrains are classically made up of 3 main lithologic units: a gneissic basement (~ 85% volume), volcano-sedimentary greenstone belts (~ 5% volume), (Photo 4.5) and late granitoids (~ 10% volume). Almost everywhere the gneissic basement seems to be the oldest unit on which greenstone belts deposited or into which granitoids intruded; consequently, it represents the early juvenile continental crust. In the field, it consists of grey gneisses, generally strongly deformed and metamorphosed (Photos 4.2, 4.3, 4.4). These gneisses contain quartz [SiO_2] + plagioclase [$NaAlSi_3O_8$–$CaAl_2Si_2O_8$] + biotite [$K(Fe,Mg)_3Si_3AlO_{10}$

Photo 4.2. Sete Voltas grey gneisses (TTG) Brazil. The coarse-grained rock (*left*) is 3.4 Ga old whereas the cross-cutting grey vein (*right*) is 3.17 Ga old (Photo H. Martin)

Photo 4.3. Peninsular gneisses (TTG) from Gurur (India) dated at 3.3 Ga. These grey gneisses are cut by younger basic veins (*black*) and granitic dykes (*white*). (Photo H. Martin)

Photo 4.4. 2.7-Ga old Naavala grey gneiss (TTG), Finland. This late Archaean rock appears more homogeneous and less deformed than its older equivalents (Photo H. Martin)

Photo 4.5. Pillow lavas from Kuhmo greenstone belt in Finland. These 2.65 Ga old lavas are typical of submarine emplacement. (Photo H. Martin)

$(OH)_2]$ ± hornblende $[Ca_2(Fe,Mg)_5Si_8O_{22}(OH)_2]$ + accessory minerals. Generally, alkali feldspar $[KAlSi_3O_8]$ is absent, such that their composition is that of tonalite (trondhjemite = Na plagioclase-bearing tonalite) and granodiorite (Fig. 4.5). By comparison, modern continental crust is granodioritic to granitic in composition: it contains alkali feldspar.

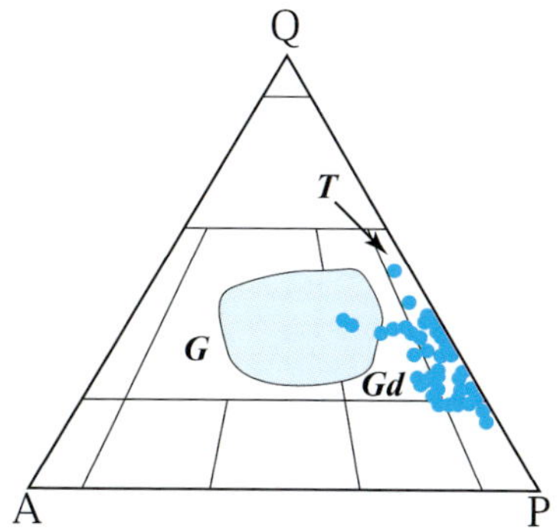

Fig. 4.5. Quartz–alkali feldspar–plagioclase (Q–A–P) modal diagram (Streckeisen, 1975). In such a triangle, a rock containing 100% quartz plots on the Q summit, whereas rocks without quartz plot on the A–P side (opposite to the Q apex); their position on this line depending on their A/P ratio. *Dark blue filled circles* = Archaean grey gneisses; *Pale blue field* = domain of composition of the modern continental crust; T = tonalite and trondhjemite; Gd = granodiorite; G = granite

Major elements even better discriminate modern and Archaean crusts. Grey gneisses are very homogeneous and have a constant chemical composition all over the world. In the normative An–Ab–Or triangle, they plot near the triple point between the trondhjemitic, tonalitic and granodioritic domains (Fig. 4.6), which is why they are generally referred to as TTG.

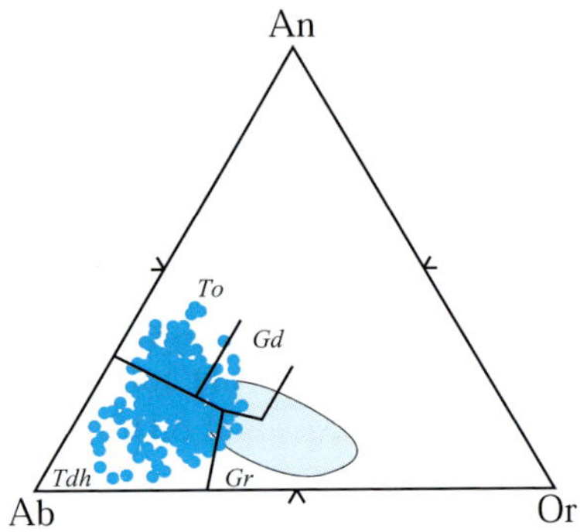

Fig. 4.6. Anorthite $[CaAl_2Si_2O_8]$–Albite $[NaAlSi_3O_8]$–Orthoclase $[KAlSi_3O_8]$ (An–Ab–Or) normative triangle (O'Connor, 1965; Barker, 1979). *Dark blue filled circles* = Archaean grey gneisses; *Pale blue field* = domain of composition of the modern continental crust; To = tonalite; Tdh = trondhjemite; Gd = granodiorite; Gr = granite

The K–Na–Ca triangle also allows to clearly distinguish between Archaean TTG, characterized by Na enrichment (Fig. 4.7A) and modern continental crust that is K-enriched and follows a classical calc-alkaline trend (Fig. 4.7B). It must

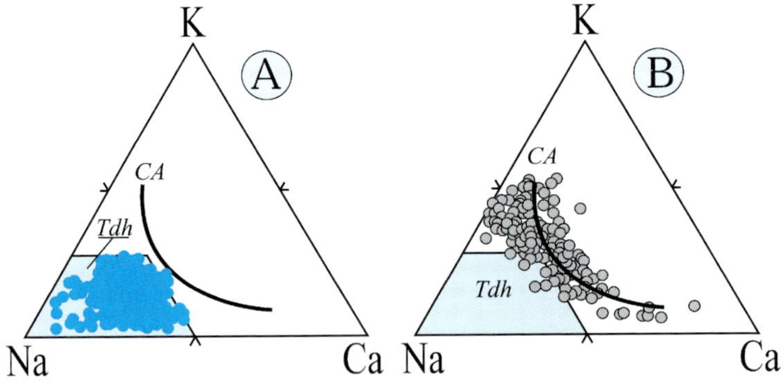

Fig. 4.7. K–Na–Ca triangle for Archaean TTG (**A**) and modern continental crust (**B**). TTG plot in the trondhjemitic field (Tdh) whereas their modern equivalents are K-enriched and follow a classical calc-alkaline differentiation trend (CA). The field of Archaean TTG (Tdh) is from Martin (1994)

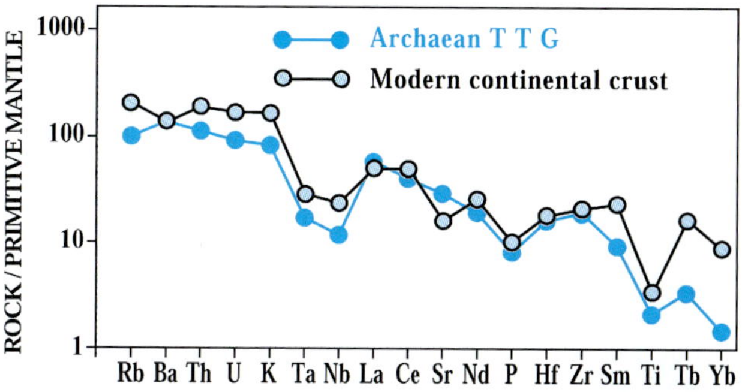

Fig. 4.8. Primitive mantle normalized Archaean TTG (*blue points*) and modern continental crust (*grey points*) average compositions. Compared to primitive mantle, all these elements are enriched in both Archaean and modern continental crusts; however, the latter is Rb, Th, U, Sm, Tb and Yb richer than Archaean TTG

also be noted that, contrarily to modern continental crust, Archaean TTG never show poorly differentiated (Ca-rich) terms. This can indicate that the source of TTG was different from that of modern crust.

Figure 4.8 shows the trace element contents of Archaean and modern continental crusts compared to the primitive mantle composition. All these elements have rock/primitive mantle ratios > 1, which means that during differentiation, they concentrated into the crust leaving a depleted mantle (see Rudnick and Gao, 2003). However, the degree of crust enrichment depends on the element, for instance, in modern continental crust Rb is about 150 times greater than in the primitive mantle whereas enrichment is only 10 times for Yb. Trace element patterns are also discriminative: compared to modern continental crust, Archaean TTG have lower contents in Rb, Th, U, K as well as in heavy rare earth elements (Sm, Tb, Yb) but are slightly richer in Sr (Fig. 4.8).

These differences appear more prominent and demonstrative when rare earth elements (REE) are taken into account (Fig. 4.9). Archaean TTG have high light REE contents ($La_N \sim 100$) whereas heavy REE contents are very low ($Yb_N \sim 2.5$) (The La_N and Yb_N values are called normalized values = concentration into the rock/concentration into chondrites), they do not possess a negative nor positive Eu anomaly (Fig. 4.9A). In modern continental crust La contents are high ($La_N \sim 100$) but heavy REE contents are also high ($Yb_N \sim 15$) and negative Eu anomaly is widespread (Fig. 4.9B). This difference in REE contents appears fundamental as it demonstrates that these two groups of rocks were generated from different sources and/or through different petrogenetic mechanisms (i.e. mechanisms such as partial melting, fractional crystallization, mixing, etc. that are actors of magma genesis).

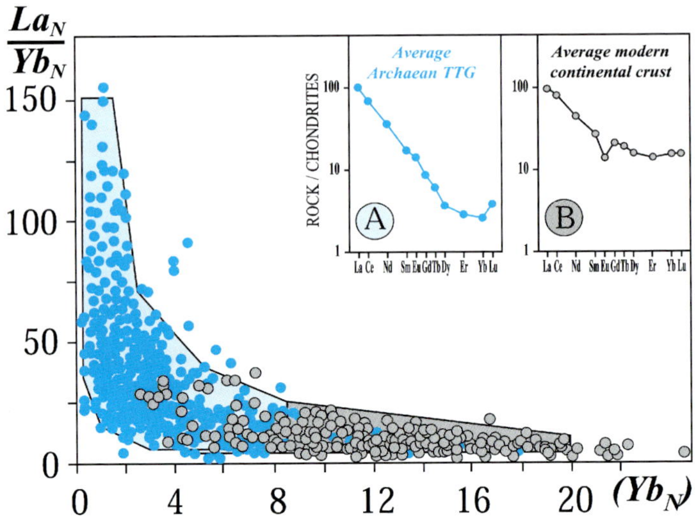

Fig. 4.9. Compiled $(La/Yb)_N$ ratios and Yb_N contents for Archaean TTG (*blue points*) and modern continental crust (*grey points*) from Martin (1986). A major change in continental crust composition clearly appears at the Archaean–Proterozoic boundary: TTG generated prior to 2.5 Ga are characterized by low Yb contents $(0.3 < Yb_N < 8.5)$ associated with strongly fractionated REE patterns (with high La_N/Yb_N) $(5 < (La/Yb)_N < 150)$; Continental crust younger than 2.5 Ga has high Yb content $(4.5 < Yb_N < 20)$ and moderately fractionated REE patterns $((La/Yb)_N \leq 20)$

4.1.4 Source of the Early Continental Crust

In most TTG, two main mechanisms are successively implied in magmatic-rock genesis: partial melting followed by fractional crystallization. This latter mechanism, which does not necessarily operate, usually consists in the extraction of high-density minerals from a lower density magma due to gravity forces. Geochemical diagrams plotting the logarithm of a compatible element (element that concentrates in crystals rather than in magmatic liquid) vs. the logarithm of an incompatible element (element that concentrates in magmatic liquid rather than in crystals) allows the role played by both fractional crystallization (subvertical trend, Fig. 4.10) and partial melting (subhorizontal trend, Fig. 4.10) to be clearly discriminated. For instance, in Fig. 4.10, TTG compositions plot along a vertical trend thus demonstrating that fractional crystallization played a role in their genesis. In addition, more accurate mathematical modelling shows that the degree of fractional crystallization remains low ($\ll 30\%$) and also allows the composition of the parental magma before that crystallization took place to be calculated.

Typically, TTG isotopic compositions $((^{87}Sr/^{86}Sr)i \sim 0.702$; $\varepsilon_{Nd} \sim 0)$ are very close to Archaean mantle values, thus precluding an origin of TTG

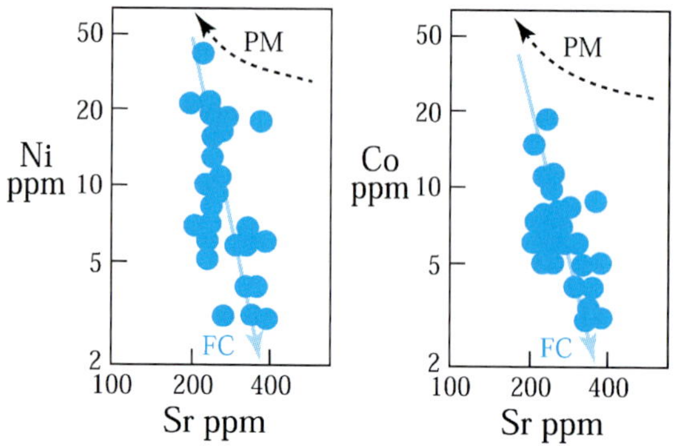

Fig. 4.10. Log (compatible element (Sr)) vs. Log (incompatible element (Ni, Co)) for TTG demonstrating that fractional crystallization can play a role in TTG genesis. In this diagram, partial melting (PM) draws *horizontal lines*, whereas fractional crystallization (FC) results in *vertical trends*. *Blue points* are Archaean TTG from Finland (Martin, 1987), they clearly follow a fractional crystallization trend and do not show any affinity for partial-melting mechanisms

by recycling (remelting) of an older continental crust. [$(^{87}Sr/^{86}Sr)i$ = initial $(^{87}Sr/^{86}Sr)$ = $(^{87}Sr/^{86}Sr)$ in the magma when it crystallized]. This corroborates the juvenile character of TTG and indicates that their parental magma had a direct or indirect mantle origin.

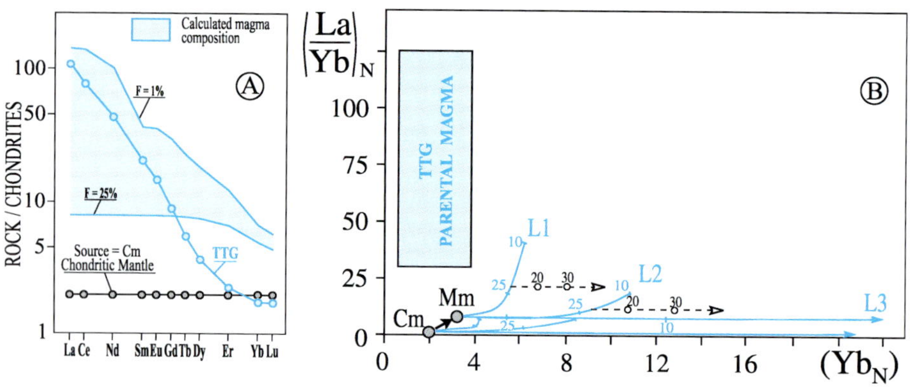

Fig. 4.11. (**A**) Modelling of chondritic mantle source melting (Martin, 1987). *Grey circles* = Chondritic mantle (Cm); *Blue circles* = TTG; *Pale blue field* = domain of REE patterns for liquids generated by 1 to 25% melting of the source. (**B**) $(La/Yb)_N$ vs. Yb_N diagram summarizing the different computed models of mantle melting. Both chondritic (Cm) and metasomatized (Mm) mantle compositions correspond to lherzolite containing 10% (L1; *inset* A) and 5% (L2) garnet or 5% spinel (L3). *Dotted lines* are olivine fractional crystallization trends at depth. *Blue numbers* indicate the degree of melting (F%) whereas *black* ones represent the degree of fractional crystallization. *Pale blue rectangle* = TTG parental magmas

- **Direct mantle origin:** magmas are generated by melting of a mantle peridotite. Mathematical modelling evidences that it is impossible to generate Archaean TTG by direct melting of primitive chondritic or metasomatized mantle (Fig. 4.11). A metasomatized mantle is a mantle that has been enriched in some elements (i.e. Rb, K, U, Th, LREE, etc.) by percolation of fluids through the peridotite.
- **Indirect mantle origin:** mantle melting generates a basaltic magma whose subsequent re-fusion gives rise to TTG. Before melting the basalt is metamorphosed into amphibolite. The same modelling, applied to an Archaean tholeiitic basalt (identical to oceanic crust basalt), shows that TTG can be produced by basalt melting only when garnet [$(Ca,Fe,Mg)_3Al_2(SiO_4)_3$] and hornblende [$Ca_2(Fe,Mg)_5(Si_8O_{22})(OH,F)_2$] are stable in the residue of melting (Fig. 4.12). This constraint is independent of both TTG age and geographical origin; it appears to be very important as it allows the temperature and pressure conditions at which amphibolite melting took place to be determined.

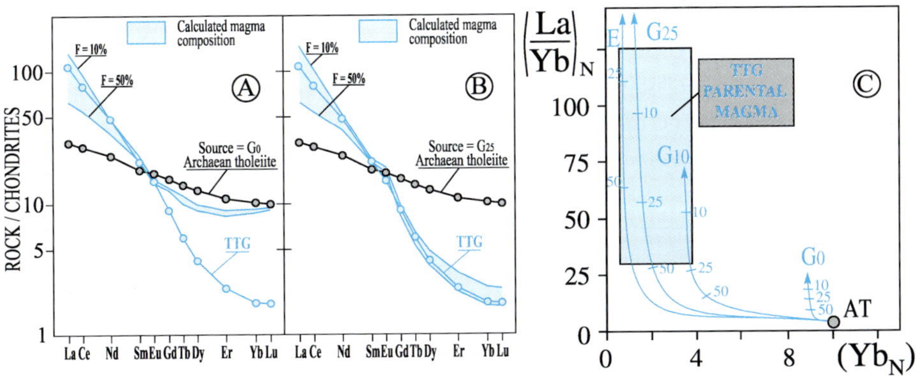

Fig. 4.12. Melting modelling for an Archaean tholeiite (basalt) metamorphosed into (**A**) garnet-free (G0) and (**B**) garnet-bearing (G25) amphibolite (Martin, 1987). *Grey circles* = Archaean tholeiite; *Blue circles* = TTG; *Pale blue field* = domain of REE patterns for liquids generated by 10 to 50% melting of the source. (**C**) $(La/Yb)_N$ vs. Yb_N diagram summarizing the different computed models of basalt melting. Source is an average Archaean tholeiite (AT) transformed into garnet-free amphibolite (G0; *inset* A); 10% (G10) and 25% (G25; *inset* B) garnet bearing amphibolite as well as into eclogite (E). *Pale blue rectangle* = TTG parental magmas

4.1.5 Mechanisms of Continental Crust Genesis

- **Genesis of the Modern Continental Crust**

Today, juvenile continental crust is extracted from the mantle in subduction zones. The oceanic crust, which is basaltic in composition is generated by melt-

ing of a dry mantle, consequently it contains anhydrous minerals such as olivine, pyroxenes and plagioclase. Immediately after its genesis, it interacts with hydrothermal water (hydrothermal activity is very important in ridge systems: white and black smokers) and also with ocean water, such that anhydrous minerals are transformed into hydrous minerals (talc, amphibole, serpentine, chlorite).

Consequently, in subduction zones, this is a water-rich oceanic crust that returns into the mantle. The average age of oceanic crust when it enters in subduction is 60 Ma (200 Ma maximum), consequently it had time to completely cool down (Fig. 4.13A), such that, from a thermal point of view, subduction corresponds to the introduction of a cold slab (oceanic crust) into a warm mantle. As a consequence, the mantle temperature decreases in the vicinity of the subducted slab, resulting in a strong deformation of isotherms along the Benioff plane (interface between the subducted slab and the overlying mantle wedge) (Fig. 4.13B).

During basalt partial melting, water plays a prominent role as it considerably lowers the solidus temperature: (solidus temperature = temperature of beginning of melting). For instance, Fig. 4.14 shows that at a pressure of 15 kbar an anhydrous basalt will begin to melt at about 1250°C whereas if 5% water is present, melting initiates at only 700°C. Today, in a subduction zone, the geothermal gradient along the Benioff plane is low (Fig. 4.14) and dehydration reactions in subducted basalt occur before the hydrous solidus temperature is reached. Consequently, the oceanic slab becomes dry and is unable to melt at low temperature; it remains totally solid when reintroduced into the mantle. Fluids (mainly water) liberated by dehydration reactions have a very low density when

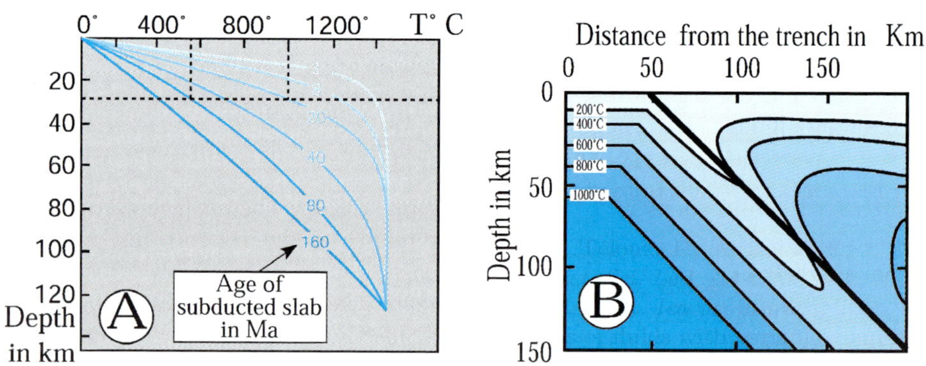

Fig. 4.13. (**A**) Depth vs. temperature diagram (after Parson and Sclater, 1977) showing that oceanic isotherm depends on the age of the subducted slab. For instance, at 30 km depth, for the same initial temperature, a 20-Ma old oceanic crust will be twice as hot as a 80-Ma old one. (**B**) *Schematic cross section* showing the thermal structure of a subduction where the slab is 50 Ma old (after, Delong et al., 1979). The *heavy line* represents the Benioff plane

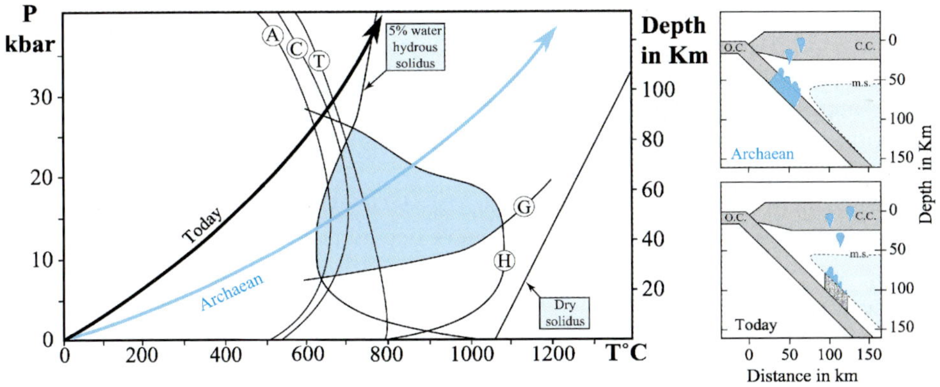

Fig. 4.14. Pressure vs. temperature diagram and schematic cross section of subduction zone showing the conditions of genesis of both primitive and modern continental crusts (after Martin, 1999). Today, geothermal gradients along the Benioff plane are low (*black arrow*) and the subducted oceanic crust dehydrates before reaching the temperature of its hydrous solidus (temperature of start of melting). The fluids due to slab dehydration rise up into the mantle wedge that they rehydrate and metasomatize, thus inducing its melting and giving rise to calc-alkaline magmas typical of modern juvenile continental crust. In Archaean times, geothermal gradients along the Benioff plane were greater (*blue arrow*) and the subducted slab reached temperature of hydrous melting before dehydration could took place. The subducted slab was able to melt in the stability field of both hornblende and garnet (*blue field*) thus generating TTG magmas. *P–T* diagram shows hydrous (5% water) and dry solidus curves for tholeiitic basalt (Wyllie, 1971; Green, 1982). Dehydration reactions correspond to destabilization of antigorite (serpentine) (A); chlorite (C) and talc (T). The *blue field* is the domain where a magma generated by basalt melting coexists with a residue containing both garnet (G) and hornblende (H). In the synthetic cross sections: O.C. = oceanic crust; C.C. = continental crust; m.s. = mantle hydrous solidus; *blue* domains correspond to place where magma is generated and emplaced; whereas *dotted* grey domains represent places where fluids pass through

compared to solid rocks and they will rise up towards the surface through the mantle wedge. Figure 4.13B shows that the mantle wedge remains hot, but that it cannot melt because it is dry. Water released by subducted slab dehydration, rehydrates mantle peridotite and consequently, it also considerably lowers its solidus temperature and induces its partial melting. In addition, oceanic crust dehydration fluids carry soluble chemical elements such as K, Rb, U, and La that modify (metasomatize) the mantle-wedge composition. In other words, modern juvenile continental crust is generated in subduction geodynamic environments, by melting of the mantle wedge, whose composition has been modified by fluids liberated by subducted slab dehydration. In such an environment, the residue of melting consists in olivine and pyroxenes: it does not contain garnet nor hornblende (amphibole).

- **Genesis of the Primitive Continental Crust**

Geochemical modelling showed that, in order to generate TTG, garnet and hornblende must have been stable mineral phases in the residue of melting of Archaean basalt. This implies that Archaean geothermal gradient crosscuts the blue field in Fig. 4.14 that is the *P–T* domain where the slab melt coexists with garnet and hornblende. Consequently Archaean geothermal gradients were greater than modern ones. Following such high geothermal gradients, the relative positions of dehydration reaction curves and hydrous basalt solidus are inverted: subducted oceanic slab reaches its hydrous solidus temperature before dehydration began and can melt at low temperature, giving rise to TTG magmas.

In summary, it appears that contrary to modern juvenile continental crust that is generated by metasomatized mantle wedge peridotite melting, primitive continental crust results in relatively shallow depth melting of subducted oceanic crust transformed into garnet amphibolite. Consequently, Archaean and modern juvenile continental crusts had different sources (oceanic crust basalt and mantle peridotite, respectively) and were derived through different mechanisms.

4.1.6 Test of the Model

- **Geochemical Modelling**

Geochemical modelling (Fig. 4.15) is in perfect agreement with the proposed model and clearly shows that TTG can be generated through garnet amphibolite melting, whereas modern juvenile continental crust composition is better accounted for by metasomatized peridotite melting.

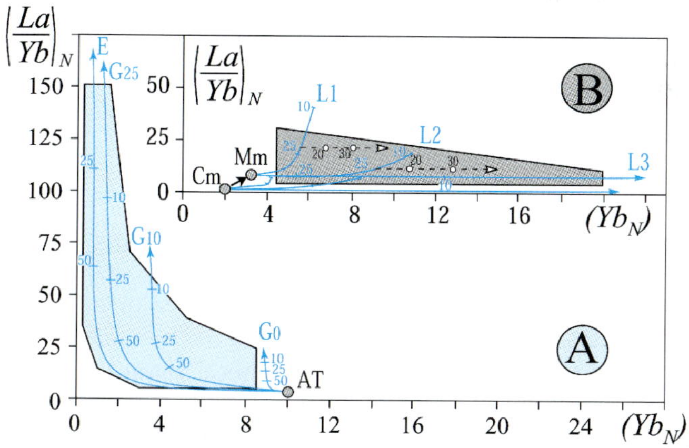

Fig. 4.15. $(La/Yb)_N$ vs. Yb_N diagram for Archaean (**A**) and modern (**B**) continental crusts. Computed models as well as symbols are the same as in Figs. 4.11B and 4.12C. There is a perfect agreement between model predictions and real rock compositions: Archaean crust is generated by garnet amphibolite melting, whereas modern crust is produced by metasomatized mantle-wedge peridotite melting

- **Experimental Melting**

Experimental melting has been performed for different pressure and temperature conditions. When water is present (hydrous melting), 20 to 30% melting of a basalt generates liquids that are trondhjemitic, tonalitic and granodioritic (TTG) in composition (Wolf and Wyllie, 1994; Rapp and Watson, 1995; Zamora, 2000; Rapp et al., 2003); if water is not available (dehydration melting) granodioritic to granitic magmas are produced (Prouteau et al., 1999). Trace-element patterns (especially REE) are identical to TTG ones only when garnet and hornblende are stable in the residue of melting (Fig. 4.16).

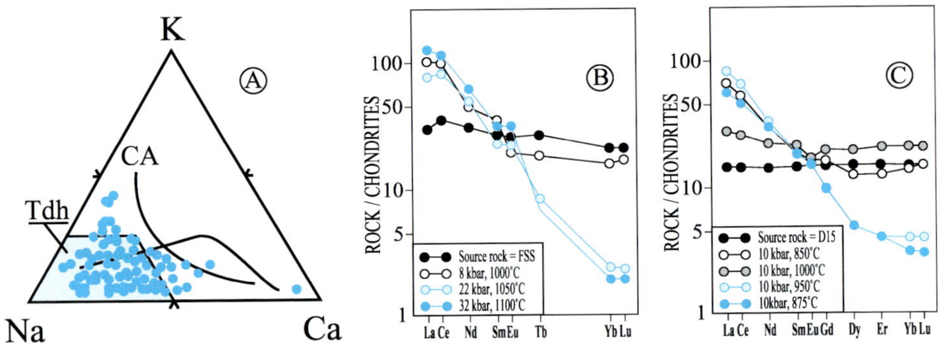

Fig. 4.16. (**A**) K–Na–Ca triangle showing the trondhjemitic composition of liquids obtained by experimental melting of basalts. *Blue field* = primitive continental crust; Trondhjemitic (Tdh) and classical calc-alkaline (CA) = modern juvenile crust, differentiation trends are also shown. (**B**) Chondrites normalized REE patterns for liquids generated by experimental basalt melting: FSS = tholeiite (Rapp et al., 1991); D15 = low-K tholeiitic amphibolite (Wolf and Wyllie, 1991). When residual garnet is absent (*open symbols*) magma HREE contents remain similar to source ones; when garnet is present (*blue symbols*), magma becomes strongly depleted in HREE

- **Modern Analogues of the Primitive Continental Crust**

Today, in modern subduction zones, the subducted oceanic crust is old, (average age = 60 Ma); it is cold and contributes to cooling of the mantle wedge, resulting in very low geothermal gradients along the Benioff plane. Figure 4.13A shows that the temperature of the subducted slab mainly depends on its age, which means of the time allowed for its cooling. For instance, at 30 km depth, for the same initial temperature, a 20-Ma old oceanic crust will be twice as hot as an 80-Ma old one. Consequently, if today, a very young oceanic crust is exceptionally subducted, high geothermal gradients are expected along the Benioff plane. Such a situation, necessarily limited in both time and space, should result in slab melting and in the genesis of magmas identical to primitive continental crust TTG.

This peculiar environment exists in a few places all around the Pacific ocean, but it is well exemplified in Patagonia, where the active Chile ridge itself is subducted under the South American plate (Fig. 4.17). To the North, the subducted Nazca plate is old (> 20 Ma) and the magmas produced are calc-alkaline, typical of the modern continental crust, with high Yb contents ($Yb_N > 6$). To the South, the subducted Antarctic plate is young (< 20 Ma) and the generated magmas, called adakites, have a composition different of classical calc-alkaline lavas; but they are perfectly similar to Archaean TTG. They are Yb-poor ($Yb_N \leq 6$) thus demonstrating that garnet and hornblende were residual phases during melting. Consequently, adakites cannot be generated by fusion of mantle wedge, but result from subducted slab melting.

A similar approach applied to all the subduction zones all over the world (Defant and Drummond, 1990) demonstrates that a good correlation exists between the age and consequently the temperature, of the subducted slab and the composition of the generated juvenile continental crust (Fig. 4.18): subduction of a young and hot slab generates TTG-like adakites, whereas subduction of an old and cold slab produces classical calc-alkaline magmas.

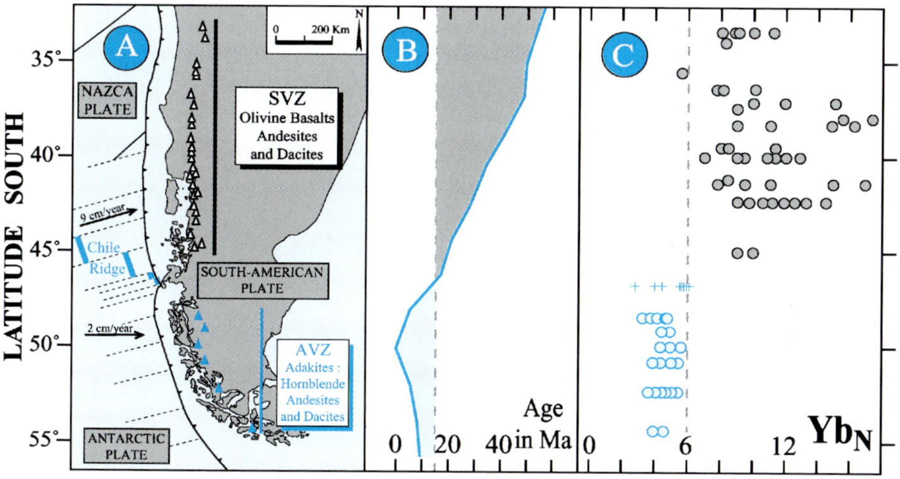

Fig. 4.17. (**A**) Schematic map of South Chile showing the triple junction between the Nazca, Antarctic and South-American plates, as well as the position of active volcanoes (SVZ = South volcanic zone; AVZ = Austral volcanic zone). (**B**) Age of the subducted oceanic crust located under the volcanic arc vs. latitude diagram showing that North of 45° latitude the Nazca plate is 20 to 50 Ma old whereas, to the South, the Antarctic plate is younger than 20 Ma. (**C**) Yb_N vs. latitude diagram. To the South, the subducted oceanic crust is young and the generated magmas have Archaean TTG compositions ($Yb_N < 5.5$; *blue crosses* and *circles*) typical of subducted slab melting. To the North, magmas are typical classical calc-alkaline ($8 < Yb_N < 20$; *grey circles*) and are generated by metasomatized mantle-wedge melting. (Martin, 1999)

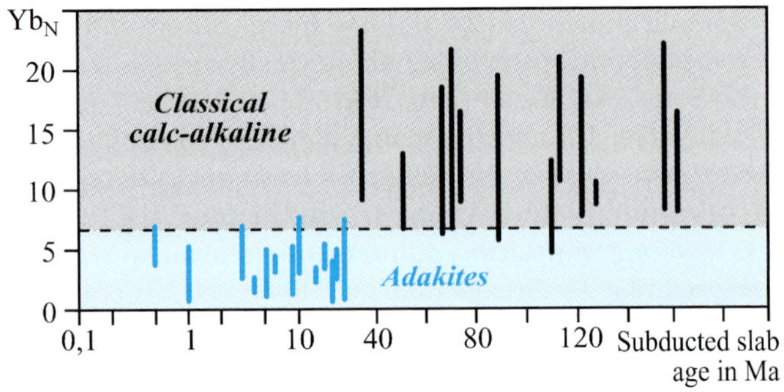

Fig. 4.18. Yb_N vs. subducted oceanic slab age (after Defant and Drummond, 1990). When $T < 30$ Ma, scale is logarithmic whereas it is linear for $T > 30$ Ma. *Horizontal bars* represent the range of variation of arc magma composition. A worldwide correlation exists between the age of the subducted oceanic crust and the composition of the subduction related magmas

4.1.7 Discussion

Geological data, geochemical modelling and experimental petrology demonstrate that Archaean TTGs were generated by partial melting of a hydrous basaltic source, leaving a garnet and amphibole-bearing residue. However, until recently, there was little agreement on the geodynamic setting in which this melting could have taken place. Two main environments were proposed (Fig. 4.19): 1) the basaltic material was previously underplated beneath a thickened crust (Arndt, 1992; Rudnick, 1995; Albarède, 1998) and melted due to mantle-plume activity; 2) basaltic rocks correspond to a subducted hot oceanic slab that melted instead of dehydrating (Martin, 1986; Condie, 1989; Rollinson, 1997; Albarède, 1998; Barth et al., 2002a; Barth et al., 2002b; Foley et al., 2002; Martin and Moyen, 2002; Kamber et al., 2003). Recently Kamber et al., (2002) and Kleinhanns et al., (2003) also proposed that TTG generated in a subduction environment but by extensive fractional crystallization of a basaltic magma.

As they cannot interact with ocean water, underplated basalts remain dry, consequently, their melting temperature must be very high (> 1200 °C). In addition, even if such high lower-crustal temperatures can be reached, amphibole will not be stable in melting residue and the generated magmas will be granitic instead of TTG (Prouteau et al., 1999). On the contrary, subducted basalts were already efficiently hydrated by interaction with hydrothermal and sea waters in a ridge system as well as on the ocean bottom.

On the other hand, modern analogues strongly militate in favour of the subduction model. Indeed, one of the more striking pieces of evidence in favour of the Archaean subduction model is that today, when Archaean-like thermal regimes are created in subduction environments, TTG-like magmas (adakites) are generated, whereas this kind of magmatism is totally unknown in association with plume systems. Recent works by Smithies et al. (2000) and Martin et al.

(2004) showed that the analogy between adakites and TTG is valid only when young TTG (3.0 Ga < TTG age < 2.5 Ga) are taken into account (see discussion in Sect. 4.3.2).

Figure 4.19 also clearly shows that after their genesis by subducted slab melting, TTG parental magmas have to cross the mantle wedge and will interact with it (see below). Underplated basalt melts will never cross any mantle slice such that no interaction would be possible.

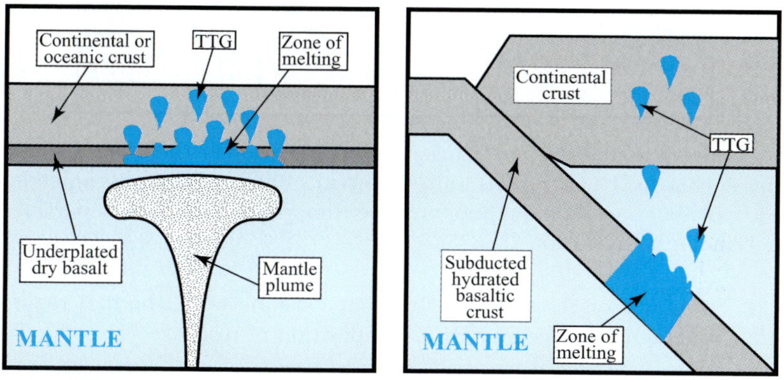

Fig. 4.19. Schematic diagrams summarizing the hypothesis of TTG genesis: melting of underplated dry basalt due to mantle plume activity (*left*); melting of a subducted hydrated basalt (*right*)

4.1.8 Summary – Conclusions

- The primitive juvenile continental crust is TTG in composition and has been extracted from the mantle through a 3-stage mechanism (Fig. 4.20):
  - Partial melting of the mantle generates a basaltic oceanic crust.
  - Melting of the basaltic crust transformed into garnet-bearing amphibolite gives rise to TTG parental magma.
  - Small extent fractional crystallization is able to produce the TTG differentiated suites.
- Comparison between primitive and modern juvenile continental crust petrogenetic mechanisms is summarized in Table 4.2.
- Study of modern analogues such as hot subduction of a young oceanic crust demonstrates that even today, in exceptional environments, it is possible to create the conditions of primitive juvenile continental crust genesis. These conditions are only realized locally and last a short period of time. In addition, this also demonstrates that the primitive continental crust can have formed in a geodynamic environment similar to modern subduction zones. Hot subduction, which is rare and exceptional today, was a normal and banal situation during the first half of our planet history.

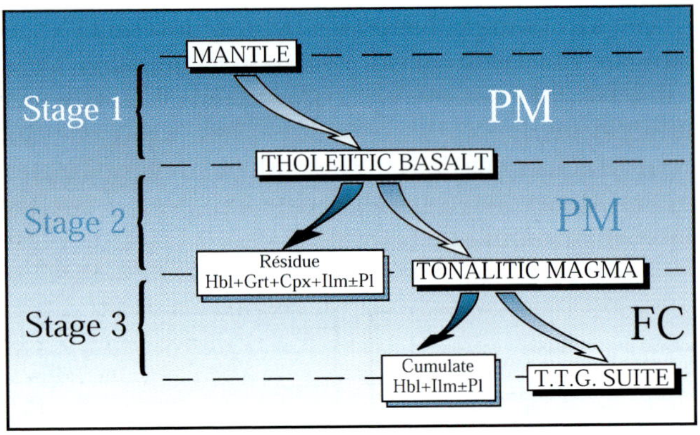

Fig. 4.20. Schematic diagram summarizing the succession of the different mechanisms implied in Archaean TTG genesis (after Martin, 1993). Hbl = hornblende; Grt = garnet; Pl = plagioclase; Cpx = clinopyroxene; Ilm = ilmenite; PM = partial melting; FC = fractional crystallization

Table 4.2. Summary of the relationships that exist between thermal regime in the subducted slab, magma source and REE composition of magmas

	Primitive continental crust	Modern continental crust
- Age of the subducted oceanic crust	Young (age < 25 Ma)	Old (25 Ma < age < 200 Ma)
- Temperature of the subducted oceanic crust	High = hot crust	Low = cold crust
- Geothermal gradient along the Benioff plane	25 to 30 °C km^{-1}	≤ 10 °C km^{-1}
- Subducted slab dehydration	No	Yes
- Subducted slab melting	Yes	No
- Magma source	Oceanic crust transformed into garnet amphibolite	Metasomatized mantle wedge
- Composition of the residue of melting	Garnet + hornblende ± plagioclase	Olivine + pyroxenes
- Magma composition	TTG	Classical calc-alkaline
- Yb_N	Low: $(0.3 < Yb_N < 8.5)$	High: $(4.5 < Yb_N < 20)$

The second part of this chapter will try to discuss the reasons why hot subduction was prominent during Archaean and what was the terrestrial dynamics during this period.

4.2 Evolution and Dynamic of the Primitive Continental Crust

4.2.1 Introduction: The Archaean Specificity

The primitive Earth is not only characterized by specific processes of continental-crust genesis, but several other features make it different from modern Earth. Among these differences the more obvious are the nature and abundance of Archaean rocks that can be summarized as follows:

- Some rocks are widespread in Archaean terrains whereas they are rare or nonexistent after 2.5 Ga: komatiites, Banded Iron Formations (BIF) (Photo 4.6), TTG.
- Others are abundant after 2.5 Ga and rare or unknown in Archaean terrains: andesites, magmatic per-alkaline rocks and eclogites.
- Some as high-Mg granodiorites (sanukitoids) are mainly known at the Archaean–Proterozoic boundary.

These differences are classically interpreted as reflecting Archaean thermal flux greater than today. This is well demonstrated by komatiites that are ultramafic (Fe–Mg-rich; Si-poor) lavas that cooled rapidly, as attested by their spinifex texture where olivine or pyroxene crystals instead of being massive, form long needles (Photo 4.7). These lavas exist only in Archaean terrains. Their petrologic study showed that they were produced by a high degree of mantle melting ($\geq 50\%$) and that they emplaced at temperatures ranging from 1525 °C

Photo 4.6. 2.7-Ga old Sandur banded iron formation (BIF) (India). On the right part is a detail view showing the alternation of quartz (*white*) and magnetite (*dark*) layers. (Photo H. Martin)

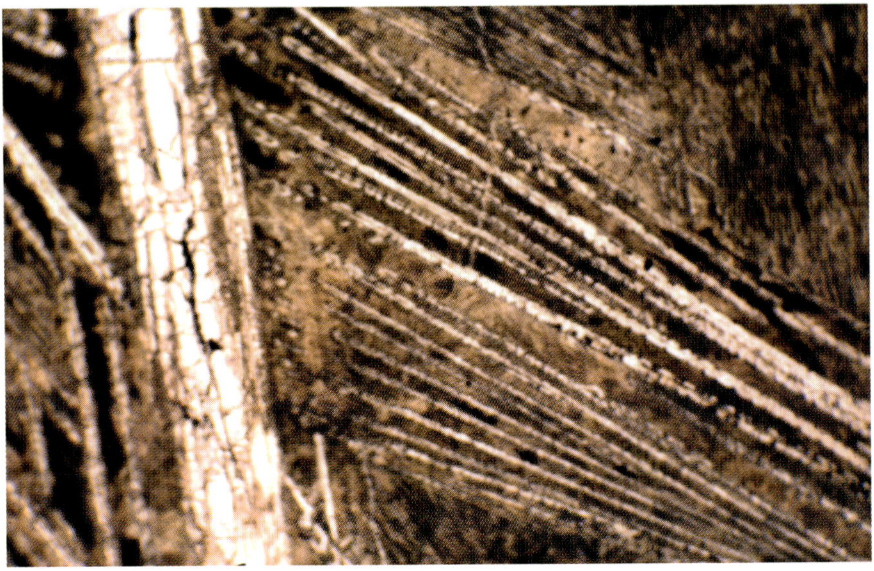

Photo 4.7. Spinifex textured olivine crystals in Abitibi (Canada) komatiites (2.7 Ga). These textures are typical of the high-temperature (1500 °C) Archaean ultramafic lavas (Photo H. Martin)

to 1650°C (Nisbet et al., 1993; Svetov et al., 2001). By comparison, modern basalts are produced today by only 25 to 35% mantle melting at temperatures of about 1250–1350°C. The depth of the komatiite source is still actively debated and estimations range from 200 to 600 km; the only certainty is that they were produced at great depth as proved by the diamond crystals that they contain (Capdevila et al., 1999). Diamond in komatiitic magmas also shows that their genesis took place below a continental lithosphere. Komatiites corroborate that the Archaean upper mantle temperature was greater during the first half of Earth history. Since its formation, the Earth has cooled such that after 2.5 Ga, it could not reach high temperatures and consequently was unable to produce high degrees of mantle melting, thus accounting for the disappearance of komatiites after Archaean times.

Different estimates consider that the early Archaean mantle temperature was 100 to 200°C greater than today. This conclusion is in perfect agreement with primitive continental crust studies that establish greater Archaean geothermal gradients. Scarcity or absence of eclogites, andesites and per-alkaline (K-, Na-rich; Ca–Al poor) magmas are also accounted for by the same cooling process. When Earth accreted it accumulated energy, such as: residual accretion heat; heat release by exothermic core-mantle differentiation; radioactive element (mainly U, Th, K) disintegration heat, etc. After 4.55 Ga this potential energetic stock is gradually consumed and consequently the Earth progressively cools (Fig. 4.21).

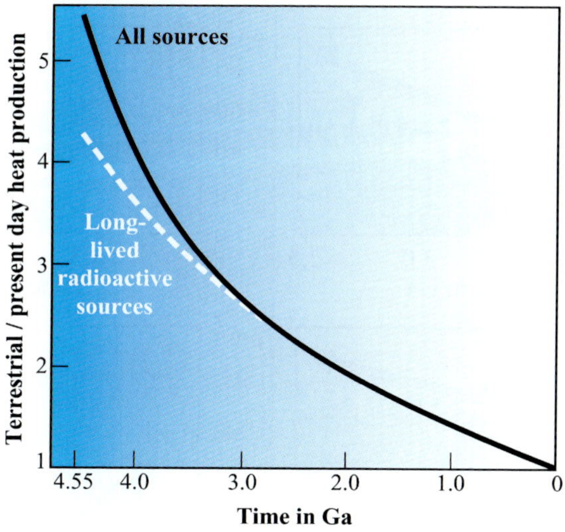

Fig. 4.21. Earth heat production vs. time diagram (after Brown, 1986)

4.2.2 Continental Crust and Earth Cooling

Earth progressive cooling is not only attested by komatiites, but it is also imprinted in the Archaean continental crust record. During TTG emplacement, from 4.0 to 2.5 Ga, Earth heat production decreased by at least a factor of 2 (Fig. 4.21), and consequently, in subduction zones, geothermal gradients along the Benioff plane also significantly changed. Figure 4.22 is a compilation of more than 1100 TTG analyses (Martin and Moyen, 2002), that shows MgO, Ni, (Na_2O + CaO) and Sr plotted against TTG emplacement age. For each period of time, TTG compositions scatter across a relatively ample range. This could be accounted for by either different degrees of partial melting of the basaltic source or subsequent fractional crystallization.

Geochemical modelling (Martin, 1994) shows that, during fractional crystallization, the main cumulate assemblage consists of hornblende and plagioclase with somewhat smaller amounts of accessory phases. During such a process, Mg and Ni remain in the cumulate (due to their very high crystal/liquid partition coefficients (Kd) for hornblende) and consequently, their content in the magma decreases during differentiation. The same conclusion can be drawn for Na_2O and CaO, as well as for Sr, whose low Kd values in hornblende are offset by their high partition coefficient in plagioclase.

In summary, fractional crystallization of plagioclase + hornblende results in a decrease of MgO as well as of Ni, (Na_2O + CaO) and Sr in the magma. Consequently, within each age bracket, the highest values of MgO, Ni, (Na_2O + CaO) and Sr can be considered as representative of the less differentiated and more primitive TTG parental magma. Thus, the upper envelope of the data set (arrows in Fig. 4.22) reflects the compositional variation of the primitive TTG parental

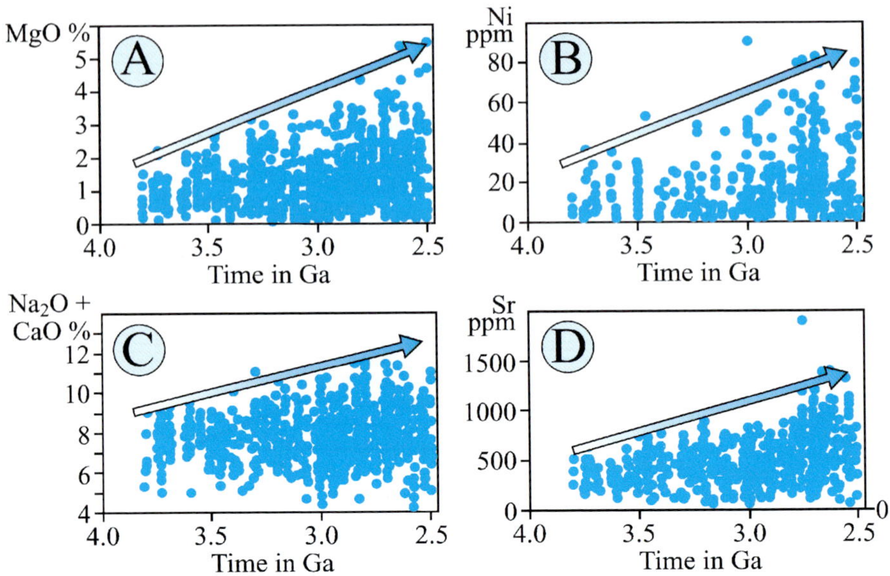

Fig. 4.22. Diagrams showing time-evolution of MgO, Ni, (Na_2O + CaO) and Sr content of the primitive TTG parental magmas from 4.0 to 2.5 Ga (after Martin and Moyen, 2002)

magmas with time, which can be interpreted in terms of melting conditions in the magma source as well as of possible interactions with the mantle. In all diagrams TTG parental magma show a temporal evolution: at 2.5 Ga TTG are significantly MgO-, Ni-, (Na_2O + CaO)- and Sr-richer than at 4.0 Ga.

Figure 4.23 is a MgO vs. SiO_2 diagram that not only shows the time-dependent MgO increase in TTG, but that also compares TTG composition with liquids produced by experimental melting of basalts. It clearly appears that for the same SiO_2 content experimental liquids are systematically MgO-poorer than TTG. Such differences have already been reported for adakites (Maury et al., 1996; Smithies, 2000; Prouteau et al., 2001). Based on experimental work, Rapp et al. (1999) demonstrated that, during their ascent through the mantle, slab melts undergo assimilation of olivine and react with mantle peridotite leading to the crystallization of orthopyroxene and garnet. This mechanism is able to lower SiO_2 and to significantly increase MgO and Ni contents of the magmas.

As high MgO and Ni contents in TTG can be regarded as reflecting slab melt/mantle interaction two main conclusions can be drawn:

1) The existence of mantle–melt interactions implies that the source of the melts is located at great depth, under a mantle slice. In other words, they are formed by melting of basalts under a significant mantle thickness, so that significant interactions can take place. This is a strong argument in favour of TTG genesis by melting of the subducted oceanic slab rather than under-

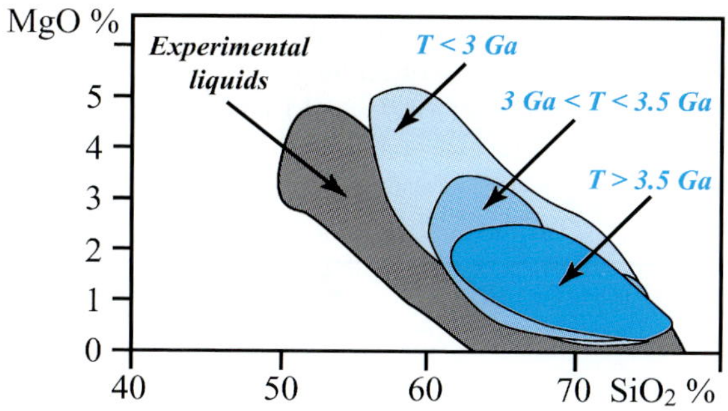

Fig. 4.23. MgO vs. SiO_2 diagram comparing composition of TTGs at different times (*blue fields*) with melts produced by experimental hydrous basalt melting (*grey area*). (after Martin and Moyen, 2002)

plated basalts. Indeed, unlike slab melting, the fusion of underplated basalts prevents felsic magmas from coming into contact with mantle peridotite and consequently precludes any interaction (Fig. 4.19).

2) The composition of the TTG parental magmas progressively evolves through time: it changes from low MgO and Ni at 4.0 Ga toward higher values at 2.5 Ga. This progressive change reflects increasing melt and mantle interactions: the efficiency of these interactions was greater at 2.5 Ga than at 4.0 Ga.

Both (Na_2O + CaO) and Sr are contained in huge amounts in plagioclase. Consequently, (Na_2O + CaO) and Sr contents of the TTG parental magma reflect the stability of plagioclase during melting: when plagioclase is stable in the residue, the magma is depleted in these elements, whereas when plagioclase is no longer stable magma is (Na_2O + CaO)- and Sr-rich. The temporal increase of (Na_2O + CaO) and Sr content in TTGs is thus interpreted as reflecting the declining role played by residual plagioclase in the genesis of TTG parental magmas from 4.0 to 2.5 Ga. In addition, Fig. 4.24 shows that plagioclase is only stable at relatively low pressure, such that the presence or absence of plagioclase in the residue of melting can be interpreted in terms of melting depth. Consequently, it appears that from 4.0 to 2.5, the depth of basalt melting progressively increased.

Figure 4.21 shows that Earth heat production was very high at 4.0 Ga such that very high geothermal gradients could be realized. Thus conditions of slab melting can be reached at shallow depth, within the stability field of plagioclase (Fig. 4.24). In the presence of residual plagioclase, Sr and (Na_2O + CaO) will remain in the residue of melting and thus will not enter the generated melts. On the other hand, melting of the slab at shallow depth implies a very thin mantle wedge over the slab melting-zone (Inset A in Fig. 4.24). Depending on crustal thickness and subduction dip, no overlying mantle wedge can even be envisaged (Smithies, 2000; Martin et al., 2004). Therefore, the ascending slab melts would either cross no mantle at all or only a very thin slice, thus reducing the probability of interaction. Moreover, the temperature of the mantle crossed by the slab melts would be low, thus diminishing the efficiency of possible reactions. This absence or scarcity of mantle–melt interactions would result in low MgO and Ni in TTG parental magmas.

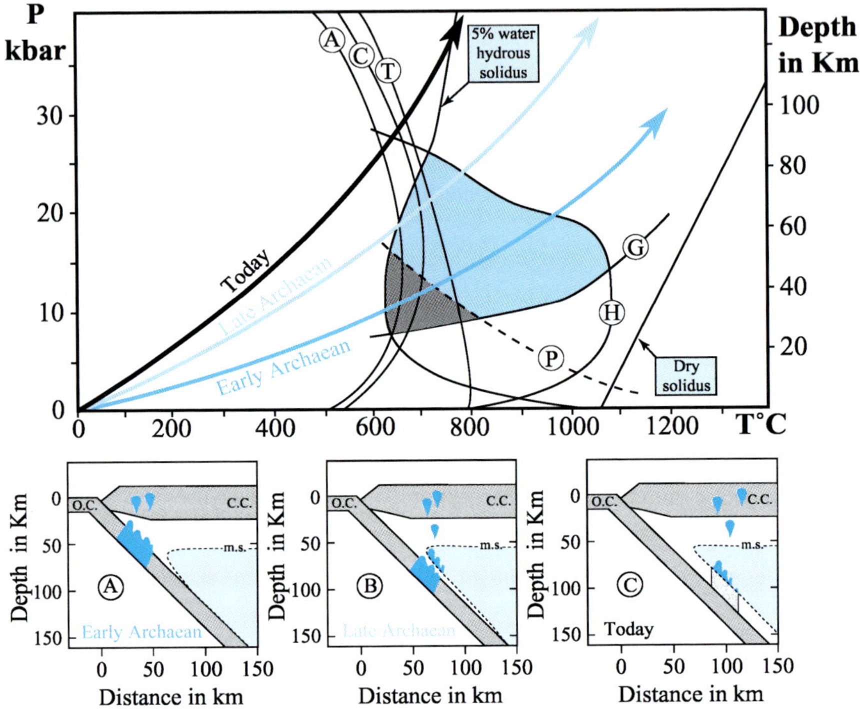

Fig. 4.24. Pressure vs. temperature diagram and synthetic cross section of subduction zones: in the early Archaean (4.0 Ga) geothermal gradients along the Benioff plane were very high, thus the subducted slab melted at a shallow depth and plagioclase was a residual phase. Because of the small thickness and the low temperature of the wedge, mantle–melt interactions were limited or absent. At 2.5 Ga, the Earth was cooler, geothermal gradients along the Benioff plane were lower such that slab melting occurred at greater depth without residual plagioclase. The overlying mantle wedge was thick and hot and interactions can occur between mantle and slab melts. Today, modern calc-alkaline magmas are generated by metasomatized mantle-wedge melting. *P–T* diagram shows hydrous (5% water) and dry solidus curves for tholeiitic basalt (Wyllie, 1971; Green, 1982). Dehydration reactions correspond to destabilization of antigorite (serpentine) (A); chlorite (C) and talc (T). Also labelled are the domains where a magma generated by basalt melting coexists with a residue containing both garnet (G) and hornblende (H) without (*blue field*) or with (*grey field*) plagioclase (P) In the synthetic cross sections: O.C. = oceanic crust; C.C. = continental crust; m.s. = mantle hydrous solidus; *dark blue domains* correspond to the place where magma is generated and positioned; *dotted grey domains* represents the place where fluids pass through

Between 4.0 and 2.5 Ga, the Earth's temperature decreased, leading to lower geothermal gradients along Benioff planes. Figure 4.24 indicates that, while a lower geothermal gradient would still allow hydrous slab melting, this process must occur at greater depth and outside the plagioclase stability field. In this case, (Na_2O + CaO) and Sr will not be retained by residual plagioclase and will enter the resulting melt. Inset B of Fig. 4.24 shows that the mantle overlying the melting zone is thicker and also warmer than for a shallow-depth melting. Because of this, ascending slab melts must cross an even thicker and hotter mantle peridotite slice, thus increasing the efficiency of interaction and yielding MgO- and Ni-rich magmas.

Still lower geothermal gradients (Fig. 4.24, inset C) prevent slab melting and rather favour slab dehydration. Aqueous fluids released by dehydration reactions ascend into the mantle wedge, which is then metasomatized and starts melting, giving rise to the typical calc-alkaline magmas.

In conclusion, it appears that the conditions of melting significantly changed from 4.0 to 2.5 Ga and that these changes are recorded in the TTG composition. As the Earth cooled, the depth of slab melting increased and as a result, slab melt and mantle–peridotite interactions were also augmented.

4.2.3 Archaean Tectonic

Today, mountain chains like the Alps or the Himalayas result in continental collision that gives rise to several tens of kilometre-sized overthrusting (Fig. 4.25). In other words, in collision environments, plate tectonics generates (and consequently is characterized by) great horizontal structures.

On the contrary, the more prominent figure in Archaean terrains consists of dome and basin structures that can be observed at all scales, from a few hundred kilometres to tens of kilometres in diameter. Generally, greenstone belts form basins stretched in between well-developed TTG domes (Fig. 4.26; Photo 4.8). Since Gorman et al. (1978), this mechanism is called sagduction and is considered as driven by gravity forces that result in vertical structures. This style of deformation that is well developed in Archaean terrains, persisted in some places during the lower Proterozoic, but is totally absent in Phanerozoic terrains. Here

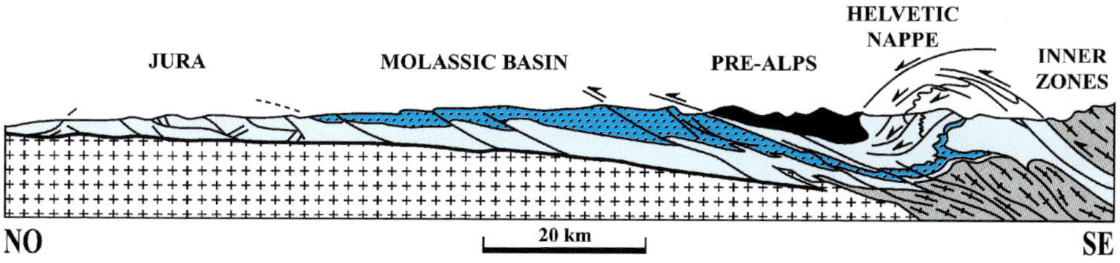

Fig. 4.25. Schematic E–W cross section of French Alps (after Merle, 1994) showing the great thrusts and the mainly horizontal character of this modern mountain chain

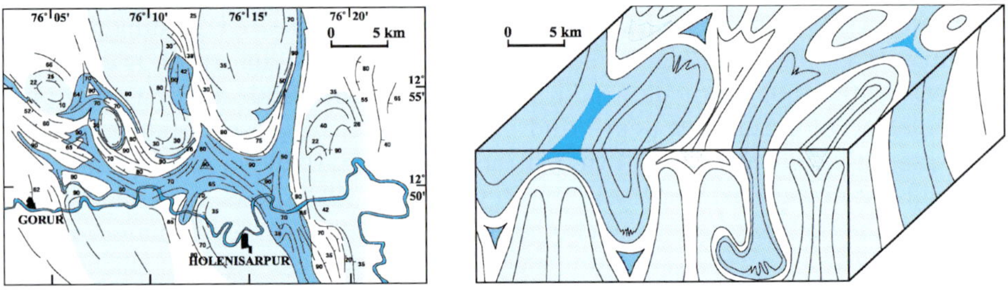

Fig. 4.26. Map (*left*) and diagram bloc (*right*) of the Holenarsipur area (S. India) showing the dome and basin structures typical of Archaean vertical tectonics. (after Bouhallier et al., 1995)

Photo 4.8. Satellite view of the Archaean Pilbara block (Australia) showing the sagduction structures: the greenstone belts (*dark green colour*) are located between TTG domes (*white-yellow colour*). (Picture width ~ 400 km). (Photo landsat)

too, this main change in tectonic style (from vertical to horizontal), seems to occur at about 2.5 Ga., even if the transition has been very slow.

Sagduction is gravity-driven tectonic movement that could be described as some kind of inverse diapirism. Indeed, when high density ($d = 3.3$) ultramafic rocks such as komatiites or even some iron-bearing sediments as BIF (Photo 4.6) are positioned over a low density ($d = 2.7$) continental crust made up of TTG, they generate a strong inverse density gradient. Gorman et al. (1978) calculated that a 5–7 km thick komatiitic lava layer was sufficient to initiate sagduction. More recently, Chardon et al. (1996) and Rey et al. (2003) showed that crustal heating from below can enhance the development of mechanical instabilities and diapirism. Sagduction structures are not only due to the down motion of high-density greenstones into the TTG basement but also to the concomitant ascent of the surrounding low-density TTG. Sagduction can create a depression in the central part of greenstones, a depression that can be filled by sediments (Fig. 4.27).

Sagduction efficiency is strongly dependent on the inverse density gradient. On Earth, the only high-density rocks able to position at the surface are komatiites and BIF, rocks that are only known in Archaean terrains. Indeed, after

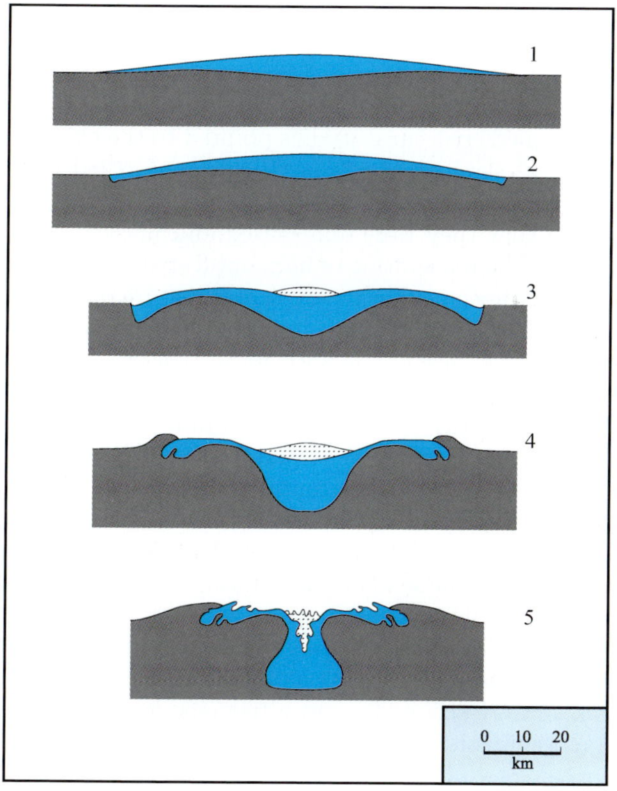

Fig. 4.27. Sagduction model as proposed by Gorman et al. (1978). High-density ultramafic rocks (*blue*) are emplaced over a low density continental crust (*grey*). Once started, the downward motion of ultramafic rocks creates in the centre of the structure a depression where sediments and volcanic rocks can deposit

2.5 Ga, Earth was so cold that mantle was not able to reach a degree of melting as high as 50%, necessary to generate komatiites. Today, most basalts are generated by degrees of melting of about 25% to 30%, giving rise to rocks whose density does not exceed 2.9 or 3.0. The resulting inverse density gradient is not great enough to allow sagduction initiation. As komatiites and BIF are restricted to Archaean times, vertical tectonics (sagduction) is also restricted to the primitive Earth crustal evolution.

For a long time, vertical tectonic style, typical of Archaean, has been considered as the only one able to operate on the primitive Earth. Consequently, several authors proposed that plate tectonics did not operate during Archaean times and that, due to high heat fluxes an Archaean lithosphere (considered only from the rheological point of view) never existed.

4.2.4 Archaean Plate Tectonic?

The existence and efficiency of plate tectonic processes has been strongly debated until recently. The predominance and omnipresence of vertical tectonic patterns as well as the lack of described Archaean oceanic crust relicts, constitute the main arguments against plate tectonic operating on the primitive Earth. However, recent detailed structural studies pointed to the existence of horizontal structures in almost all Archaean terrains (Bickle et al., 1980; de Wit et al., 1992; Blais et al., 1997; Choukroune et al., 1997). Horizontal structures are generally the older ones such that they were obliterated and masked by the more recent sagduction patterns. The recognition of horizontal structures is extremely important. Indeed, the most spectacular and typical structure of modern plate tectonics is continental collision, which builds mountain chains, generates crustal-sized overthrusting, resulting in large-scale horizontal structures (Fig. 4.25). Collision-related structures, not only demonstrate the existence of rigid plates on Earth surface, but also demonstrate their relative motion. Large-scale horizontal tectonic movement appears as proof that plate tectonics operate. During the 15 last years, overthrusts and great horizontal structures have been recognized in almost all Archaean terrains: in Finland (Blais et al., 1997), Greenland (Nutman and Collerson, 1991), South Africa (de Wit et al., 1992), Australia (Bickle et al., 1980; Bickle et al., 1993), Canada (Ludden et al., 1993; Choukroune et al., 1997), (Fig. 4.28).

Figure 4.29 shows great overthrusting structures from Finland in the Baltic shield. This place is of special interest as in the thrusting plane, from place to place, crop-out kilometre-size mafic and ultramafic lenses (Lentiira group) that are interpreted as remnants of oceanic crust. This conclusion is based on 3 sets of evidence:

- *Structural position*: The Lentiira group is exclusively located in the thrusting plane between two continental blocs (Kivijärvi and Naavala). This position, all along the suture is exactly the position where relics of oceanic crust (ophiolites) are expected and observed in Phanerozoic mountain chains.

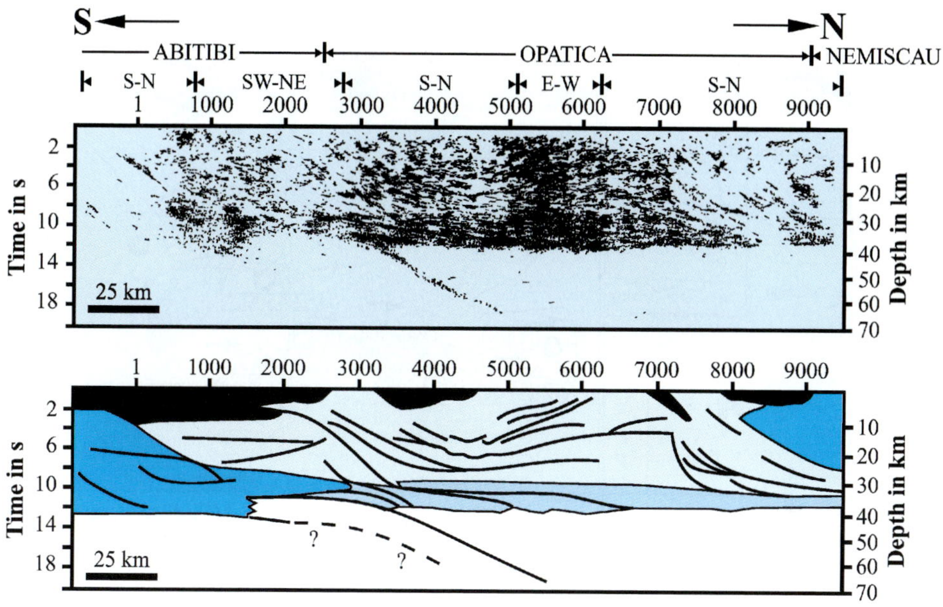

Fig. 4.28. Seismic cross section (*top*) and geological interpretation (*bottom*) of the 3.0–2.7 Ga old Archaean terrains of Abitibi and Opatica in Canada. Horizontal structures are not only restricted to continental crust, but are also imprinted in the mantle; they are interpreted as the result of collision between two crustal blocks (after Ludden et al., 1993; Choukroune et al., 1997)

- *Petrologic association*: The Lentiira group consists of an association of rocks (now slightly metamorphosed) that is identical to that observed in ophiolites and in modern oceanic crust.
- *Chemical composition*: Lentiira amphibolites have tholeiitic basalt compositions chemically identical to present-day midocean ridge basalts. They do not show any similarity with basalts generated in other geodynamic environment. Ultramafic rocks also have the same composition as ultramafic rocks from ophiolitic complexes.

These two pieces of evidence (horizontal tectonics and Archaean oceanic crust) are extremely important because they demonstrate that plate-tectonic-like processes operated since the beginning of Archaean times. A similar inference has been drawn from sedimentary (chert) record (Kato and Nakamura, 2003). This conclusion could appear inconsistent as it could imply that both horizontal and vertical tectonics operated during Archaean. In fact, it is now considered that these two mechanisms did not operate in the same time at the same place (Choukroune et al., 1997; Rey et al., 2003). As their modern equivalents, Archaean continental plates had rigid limits, which, when implied in a collision event, could produce large-scale overthrusting. On the contrary, the central part

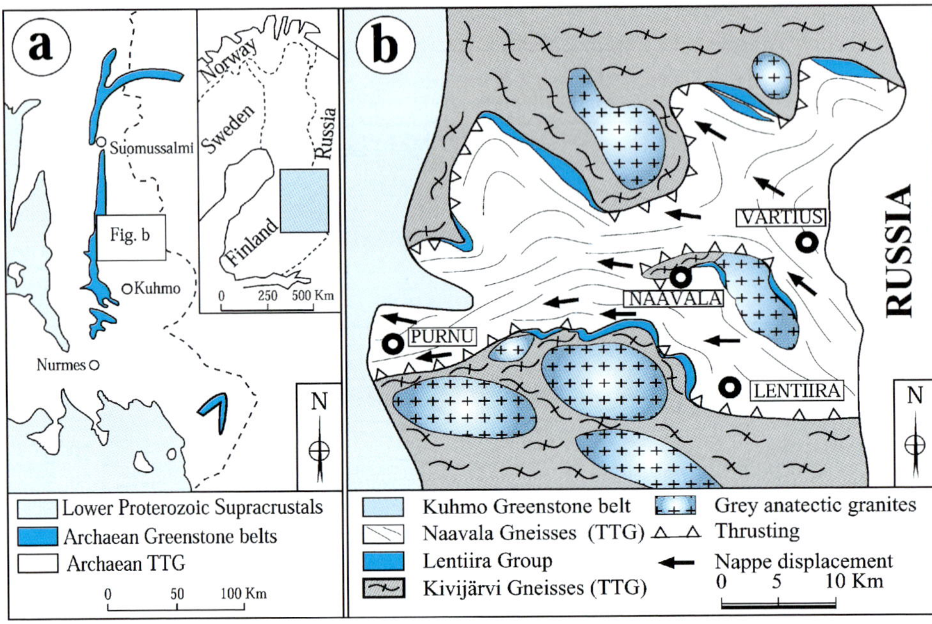

Fig. 4.29. (**a**) Schematic geological map of the western part of Archaean terrains from the Baltic shield (Blais et al., 1997); (**b**) Detailed structural map of the same area showing the 2.7 Ga Naavala TTG thrust over the 2.9-Ga old Kivijärvi TTG. Mafic and ultramafic lavas of the Lentiira group form huge lenses exclusively located in the thrusting plane, which is exactly the structural position expected for oceanic crust relicts

of continental plates, when reheated could give rise to sagduction and correlated vertical structures. It has been proposed that the source of reheating could have been mantle plumes positioned under continental plates (Rey et al., 2003). This hypothesis is attractive as it can also account for high-density komatiites genesis by mantle melting in the plume and for their emplacement over a low-density TTG continental crust, conditions required in order to initiate sagduction.

Consequently, it appears that the Archaean–Proterozoic transition did not consist of the change from vertical to horizontal tectonic, but rather to a regime where both horizontal and vertical tectonics coexisted before 2.5 Ga towards a regime where only horizontal tectonics were active after 2.5 Ga. As observed today, Archaean horizontal tectonics should have generated mountain chains.

4.2.5 Specificity of Archaean Plate Tectonics

If it is now widely admitted that plate tectonics operated during Archaean times, it is also obvious that its detailed modalities were different from present-day ones. Indeed, Earth heat production was more important on the early Earth than today (Fig. 4.21). On the other hand, studies of metamorphism showed that Archaean geothermal gradients into the central parts of Archaean continental crust were only slightly greater than today. This observation could seem contradictory with greater Archaean heat production but continental geothermal gradients only reflect the fraction of Earth internal heat evacuated by conduction. Convection is by far a more rapid and efficient mechanism of internal heat release and today most internal heat is evacuated by convection in midocean-ridge systems.

The large amounts of internal heat produced during Archaean times has necessarily been released, otherwise, accumulated heat should have resulted in melting the external part of our planet, which is not shown by geological record. As conduction is not efficient at all to evacuate internal heat, Archaean convection should have played this role and, as today, heat was released by ocean-ridge systems. As the amount of heat to evacuate was greater and as conduction has the same efficiency as today, it can be concluded that the excess of heat has been released through convective processes. The convection rate could have been slightly greater, but mainly the ridge length was significantly greater than today. The amount of released heat is correlated with the cubic square of the ridge length (Hargraves, 1986). As the Earth volume and surface did not significantly change since 4.5 Ga, a greater ridge length should result in smaller plates (Fig. 4.30).

These conclusions are strongly supported by a modern analogue that is the North Fiji basin (Lagabrielle et al., 1997). This basin is the place in the world where the greatest present-day heat flux has been measured ($> 240\,\mathrm{mW\,m^{-2}}$). As shown in Fig. 4.31, the North Fiji basin contains several active ridge segments delimiting small plates; there the active ridge length normalized to the basin surface, is 20 times greater than the average Pacific Ocean value. This obviously

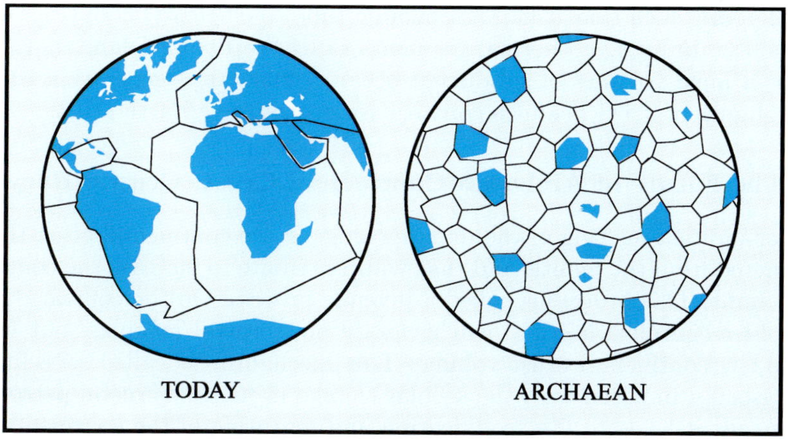

Fig. 4.30. Schema comparing modern plate size (*left*) to its supposed Archaean equivalent (*right*). The greater Archaean heat production resulted in a mosaic of plates smaller than today (after de Wit and Hart, 1993)

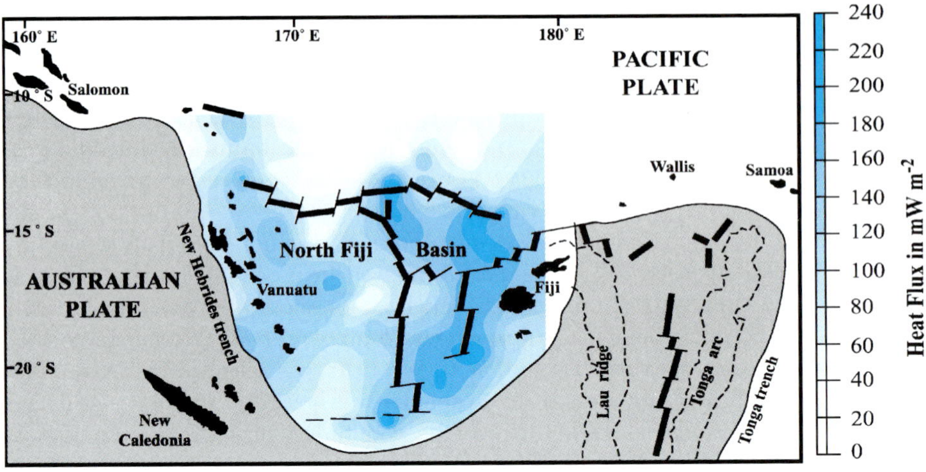

Fig. 4.31. Map of North Fiji basin (after Lagabrielle et al., 1997) showing the internal Earth heat flux in *blue*. The great length of ridge systems (in *black*) delimits several small lithospheric plates

demonstrates that a correlation exists between heat flux (evacuated internal heat), ridge length and plate size.

Archaean plates were smaller, and they also certainly have greater motion rate, such that oceanic crust entered more rapidly into subduction. Today, the average age of an oceanic plate when it enters in subduction is 60 Ma whereas it has been calculated that it was < 20 Ma (more probably about 10 Ma) during Archaean (Bickle, 1978). Consequently, during the first half of Earth history, not only was the oceanic crust hotter when it formed, but also when it subducted it was still young such that it had not enough time to efficiently cool (Fig. 4.13A). This resulted in high geothermal gradients along the Benioff plane that allowed melting of huge volumes of subducted oceanic crust, thus generating the enormous volumes of TTG magmas constitutive of primitive Earth crust.

4.2.6 The Future of Archaean Continental Crust: Crustal Recycling

Direct or indirect melting of the mantle can generate continental crust that will be extracted from the mantle and that will contribute to increase the continental crust volume. This process is called a juvenile process. On the contrary, remelting of older continental crust, will generate new crustal magmas, but will not increase the continental crust volume. This mechanism is called recycling. Recycling is superimposed onto the secular change in juvenile crustal petrogenetic mechanisms; it leads to the modification and maturing of the continental crust composition. Recycling can operate through two main mechanisms:

- Direct melting of TTG, that will generate granitic magma (K-richer than TTG), this can mainly occur in continental-collision environments.

- Weathering and erosion of TTG gives rise to sediments and sedimentary rocks. These rocks, when implied in orogeny are dragged to depth and metamorphosed, there they can undergo partial melting and generate granitic magmas.

The two recycling mechanisms finally result in the genesis of magma whose composition is controlled by the eutectic of the system, the composition of which is that of a granite. In such a system, TTG melting will generate a granite, but granite melting will also generate a granite and never a TTG. All recycling processes converge towards a modification of the continental crust composition that, in the course of time, becomes more granitic, more potassic (Fig. 4.32). This maturation is a unidirectional and irreversible evolution.

Before being able to recycle continental crust, it is necessary to create it or, in other words, to give it time to grow. This is why, during Archaean times, juvenile processes were prominent, whereas recycling appears as a subordinated process. Contrarily, after 2.5 Ga, the volume of continental crust is so great that recycling becomes prominent. The contrast between these two periods is magnified by the fact that, in the course of time, the juvenile crustal growth rate strongly decreased (Fig. 4.3).

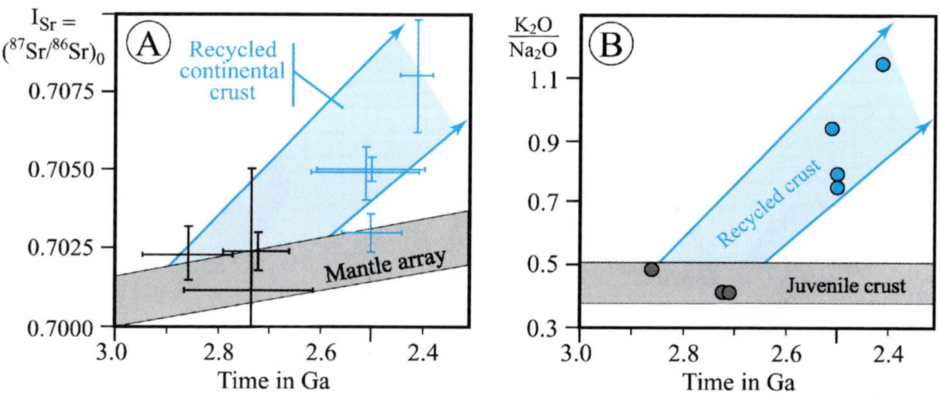

Fig. 4.32. I_{Sr} (**A**) and K_2O/Na_2O (**B**) vs. time diagram for the Archaean crust of the Baltic shield. Crustal evolution is characterized by an increase in isotopic initial ratios $^{87}Sr/^{86}Sr$ (I_{Sr}) and also by an augmentation of K_2O/Na_2O. *Grey fields* represent the composition of mantle-derived juvenile crust, whereas *blue domains* are those of crustal recycling

4.3 Some Open Questions

4.3.1 Episodic Crustal Growth

Since long ago, it has been established that crustal growth has been an episodic or, more precisely, an irregular process. Compilations of crustal ages (Condie,

1989; Condie, 1998, Photo 4.9) or estimates of continental growth rates (McCulloch and Bennet, 1993) (Fig. 4.33) clearly show that crustal growth proceeded by super-events (i.e. 3.8, 2.7, 1.8, 1.1 and 0.5 Ga, with possible minor events at 3.4 and 2.0 Ga).

Some authors consider that, because subduction is a continuous process, the episodic pattern of crust formation ages is a strong argument against crustal growth at converging boundaries (Albarède, 1998). In fact, geochronological data as well as Figs. 4.22 and 4.33 show that the crustal growth is not episodic and that juvenile continental crust has been generated during all long Archaean times, which in turn is an argument in favour of subduction. However, the same figures show that if crustal growth has been a continuous process through the whole Earth history, it has been affected by more or less regular periods of acceleration, of more intense crust generation rate. In other words, crustal growth has been a continuous mechanism but irregular in its efficiency. It must also be noted that periods of intense crust genesis are not restricted to Archaean times but are known until now, in perfect association with subduction processes.

Several authors consider that mantle-plume activity could be responsible for periodic intense crust production (Stein and Hofmann, 1994; Albarède, 1998). For instance, Albarède (1998) proposed that plume activity resulted in the emplacement of large oceanic plateaus, similar to the Mesozoic Ontong-Java plateau, where thick piles of plume basalts erupted on the oceanic floor. In this

Photo 4.9. Crystals of zircon extracted from Sete Voltas TTG (Brazil) and dated at 3.4 Ga. Length of longer crystals is about 0.1 mm. (Photo H. Martin)

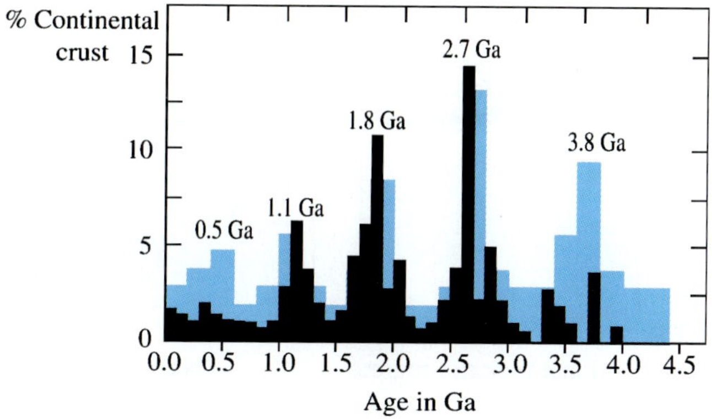

Fig. 4.33. Histograms showing the estimate rates of growth of continental crust averaged over 200-Ma intervals (after McCulloch and Bennet, 1993, *blue pattern*) and the distribution frequency (100 Ma intervals) of U-Pb zircon ages (after Condie, 1989; and Condie, 1998, *black pattern*). This diagram clearly shows that the growth of continental crust has been an irregular process

model, the continental crust is generated by magmatic processing of oceanic plateaus in subduction zones, or even possibly into the crust itself. This model that considers oceanic plateau basalt rather than oceanic crust basalt melting requires that an intense hydration of oceanic plateau basalts could occur before their magmatic processing. As pointed out by Windley (1998) descriptions of deep sections of Archaean oceanic plateaus are rare and when they exist their interpretation is subject to intense discussions as for the Belingwe greenstone belt from Zimbabwe (Kusky and Kidd, 1992; Bickle et al., 1994).

Condie (1998) proposed an alternative model also based on the Stein and Hofmann work (1994) (Fig. 4.34). He considers that as subduction is operating, the descending residual oceanic crust accumulates at the 660-km seismic discontinuity. When stored oceanic crust exceeds a threshold, it suddenly sinks into the mantle as a cold avalanche that reaches the mantle-core boundary (D″ layer). When cold slabs arrive at the D″ layer, they induce mantle-plume production. This catastrophic event results in some kind of mantle and thermal overturn. Plumes can partly generate oceanic plateaus, but Condie (1998) assumes that they mostly actively participate in heating the upper mantle, which increases the rate of oceanic-crust genesis, resulting in smaller and faster plates that enter more rapidly into subduction. Faster subduction of younger oceanic crust also results in increasing the rate of production continental crust (TTG). In addition, avalanche events can cause 'super' subduction zones above them, which can attract plates from great distances, thus aiding in the growth of a supercontinent over the place where the avalanche took place.

Moyen (1997) also pointed out cyclicity in continent growth and evolution, which he divided into 4 stages. The first stage consists of arc complexes, where continental crust (TTG) is generated by melting of a subducted slab. During

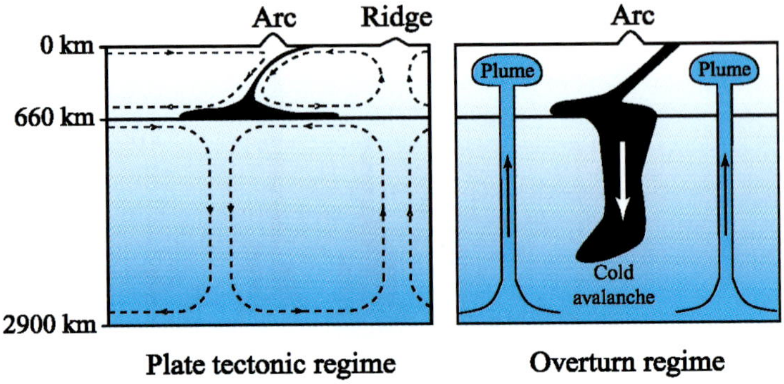

Fig. 4.34. Schematic cross sections of the Earth mantle, between (*left*) and during (*right*) overturn (after Stein and Hofmann, 1994). During period of "normal" mode, plate tectonics operates and the residual subducted oceanic slab accumulates at the 660-km boundary. During overturn, the 660-km boundary is broken, cold residual slabs sink into the lower mantle (cold avalanche) until the mantle-core limit. This process initiates the uprising of hot lower mantle plumes

a second episode, arcs accrete due to collision or collage that can be correlated to the cold avalanche as described by Condie (1998). The third stage consists of the reworking of the newly accreted crust. Indeed, genesis of juvenile continental crust also results in transferring and concentrating strongly incompatible elements such as U, Th and K from the mantle to the crust. These elements, which have radiogenic isotopes, contribute to heat the crust that can help in its melting and recycling. In addition, hot continental crust has lower viscosity and density, which favours sagduction mechanisms. Finally, the last episode is a quiescence period that can evolve towards a new cycle or towards cratonization and development of a thick and cold lithospheric keel.

4.3.2 Oceanic-crust Behaviour

Several authors (Nisbet, 1984; Vlaar, 1986; Arndt and Chauvel, 1991; Bickle et al., 1994) proposed that the Archaean oceanic crust could have been significantly thicker than today. Today, oceanic crust thickness ranges between 7 and 12 km, Archaean-crustal thicknesses of 20 to 22 km were proposed by Sleep and Windley (1982) and Bickle et al. (1994) and of at least 40 km by Arndt and Chauvel (1991). These conclusions are supported by the fact that the Archaean Earth heat production was greater than today, and consequently that greater volumes of magma were produced in ridge systems, these later resulting in thicker oceanic crust.

- **Archaean-mantle Temperature**

This parameter has been an important subject of discussion and debate. If everybody agrees that primitive Earth was hotter than today the temperatures proposed for the Archaean mantle vary in a wide range. For instance, Arndt and Chauvel (1991) consider a temperature difference of 200 to 400 °C, whereas Jarvis and Campbell (1983) conclude that this difference was not more than 100 °C. Based on hydrothermal records in BIF, Komiya (2001) considers that the Archaean mantle was 150 to 200 °C hotter. Most of the estimations point towards an Archaean mantle 100 to 200 °C hotter than today. Davies and Richard (1992) calculated that a 100 °C augmentation of the mantle temperature would decrease its viscosity by an order of magnitude. Consequently, the Rayleigh number for the mantle should have been 2 to 4 times greater, thus resulting in faster and more chaotic convection with smaller plates.

- **Oceanic-crust Thickness**

The basic assumption that leads to the conclusion that Archaean oceanic crust was thicker than today is that in ridge or plume systems, due to higher mantle temperature, the degree of mantle melting was greater and consequently greater volumes of magmas were generated. Models were calculated considering Archaean plates with geometry and characteristics similar to present-day ones. In such a system, especially in the Early Archaean, oceanic plates should have been several times thicker than today (Arndt and Chauvel, 1991; Bickle et al., 1994): 40 km and about 20 km thick for mantle temperature differences of 400 °C and 100–200 °C, respectively. However, as mentioned, for instance by Nisbet (1987) another possible way to accommodate greater magma production is to increase the rate of plate genesis and consequently of plate motion. Faster plate motion is able to evacuate a greater magma production without modifying crust thickness. As a consequence, faster plate production also implies faster subduction and Bickle (1978) was the first to propose that the average age of Archaean plates was 10–20 Ma when they enter into subduction, whereas it is of 60 Ma today. In other words, greater mantle temperature should result in thicker oceanic crust, only if the other parameters of mantle convection are supposed to remain unchanged. Modification of these parameters (i.e. convection rate) could significantly diminish or even cancel effects on crustal thickness.

- **Oceanic-crust Buoyancy and Subduction**

Today, after its genesis in ridge systems, oceanic crust progressively cools and consequently its density gradually increases. It becomes unstable after about 40 Ma and thus it can naturally sink into the mantle. An oceanic crust that enters into subduction only after 20 Ma will not be cold enough and consequently not dense enough to sink easily and spontaneously. The general buoyancy of a thick oceanic crust will lead to the same effect. Today, young oceanic crust can subduct (forced subduction) but it will result in low-angle or even in flat subduction (Gutscher et al., 2000). On the other hand, Archaean oceanic crust was underlain by a hotter and consequently less dense and viscous mantle. As proposed by Arndt and Chauvel (1991) a hotter mantle should result in a greater depth of adiabatic melting of plumes as well as in higher degrees of partial melting. In these conditions, komatiites can be generated. Komatiitic magmas are very dense and their addition to oceanic crust increases its whole density, such that it could "rapidly" acquire the negative buoyancy necessary to spontaneously subduct (Nisbet and Fowler, 1983). Geochemical data and modelling show that komatiites alone cannot be the source of continental crust. Indeed, experimental work of Foley et al. (2003) shows that melting of komatiites in subduction conditions would generate basalts and not continental crust. However, the subduction of an oceanic crust made up of an assemblage of komatiites and basalts, leads to the low-temperature melting of the hydrated basalt component, whereas komatiite would only melt at higher temperatures and remain in the residual crust. Such assemblages of komatiites and tholeiitic basalts are widespread in all Archaean greenstone belts all over the world. In addition, melt of TTG composition has a low density, whereas the residue of melting is very dense. So, after extraction of TTG magma, the density of the residual subducted slab significantly increases such that it will be able to pull down the whole oceanic lithosphere into the mantle. Martin and Moyen (2002) and Smithies et al. (2000; 2003) showed that Early Archaean TTG poorly or even never interacted with mantle-wedge peridotite; this observation, which indicates shallow melting of the subducted oceanic crust, is also interpreted as consistent with the forced flat subduction of oceanic crust (Smithies et al., 2003; Martin et al., 2004). These authors proposed that flat subduction was a dominant process in the Early Archaean.

Finally, addressing the Earth thermal budget, Nisbet (1987) showed that komatiites dissipate 30% more heat than the same volume of basalts, so they are also more efficient to evacuate Earth internal heat.

4.3.3 Archaean Mountains?

Today, stable continental lithosphere is rheologically characterized by a relatively thick fragile upper crust overlying a ductile lower crust and a resistant upper mantle. Implied in collision processes, continental crust gives rise to high-altitude mountain chains (i.e. Alps, Himalayas). As modern plate sizes are large, mountain chains are long. Due to higher heat production, the Archaean upper mantle was hotter, which could have modified its rheological properties making it less resistant to crustal load. Similarly, the continental crust was slightly hotter that its present-day equivalent, especially at plate boundaries. This leads to the conclusion that Archaean-lithosphere rheology was probably softer and more ductile than today. As a consequence, mountain chains were probably not able to reach or to maintain very high altitude for a long time period. Archaean mountains chains were certainly wider but not so long (due to smaller plate size) and less high than their modern equivalents. Metamorphic records in Archaean continental crust never shows pressures greater than 12.5 kbar equivalent to a maximum crustal thickness of 45 km (Percival, 1994). On the other hand, Nisbet (1987) considers that Archaean sedimentary facies demand that mountains existed. England and Bickle (1984) calculated that thick crust (and consequently mountains) may survive some ten million years when thickening is due to continental collision, whereas this is almost impossible if thickening results from magmatic accretion of sialic material.

In addition, it appears that collage of terrains was also an active process of crustal accretion during Archaean (Nutman et al., 1996). Contrarily to collision, this mechanism of juxtaposition of small continental blocks only produces a very moderate crust thickening and does not necessarily give rise to mountains.

4.3.4 Cool Early Earth and Late Heavy Bombardment

Exponential extrapolation of lunar meteorite impact rate to the Hadean is commonly interpreted in terms of an extremely intensive meteorite bombardment from 4.55 to 3.9–3.8 Ga on both the Moon and the Earth (Fig. 4.35). Since the 1960s (Donn et al., 1965) it has been considered that meteorite impacts could have played a prominent role in Earth crust genesis and evolution. Intense bombardment was a very strong argument to account for the lack of continental remnants older than 4.0 Ga. Arndt and Chauvel (1991) considered that bombardment could have mixed felsic and mafic components of the Early crust and also to have favoured plume development, and prevented lithosphere formation.

More recently it has been proposed that the 3.9–3.8 bombardment was not the logical suite of an exponential decrease of impact activity since the Earth

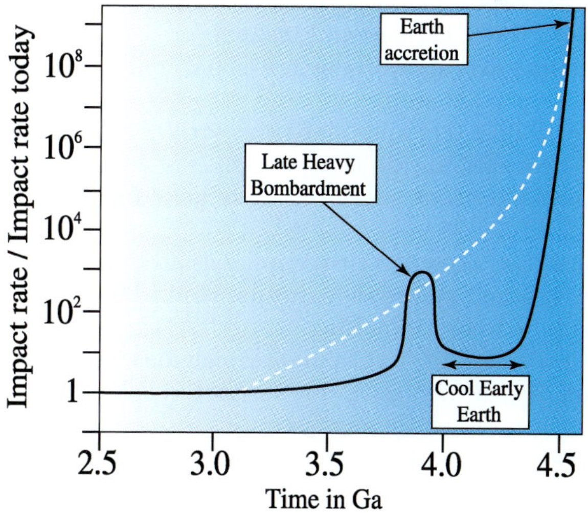

Fig. 4.35. Estimates of meteorite impact rate during Hadean and Archaean times (from Valley et al., 2002). The *dotted white line* represents the classical exponential decay of impact rate whereas the *black line* represents the cool early Earth with a late heavy bombardment model

accretion, but rather a catastrophic and sudden spike. Valley et al. (2002) developed the concept of cool early Earth as shown in Fig. 4.35. This model proposes that the impact rate remained more or less the same since Earth's accretion end, except during the 3.9–3.8 period where heavy bombardment took place (Ryder et al., 2000; Kring and Cohen, 2002). This late hypothesis accounts for the new constraints fixed by the discovery of Hadean zircons (Wilde et al., 2001) that indicated not only that continental crust existed at 4.4 Ga but also that it has been stable enough to undergo alteration and erosion processes such that zircon crystals could be preserved in sediments. This scenario is not easily imaginable under a constant heavy meteorite bombardment. In addition, oxygen isotope data on the same zircon crystals (Mojzsis et al., 2001) point to the existence of liquid water on the surface of the Earth at 4.4 Ga, which is also consistent with the "cool early Earth hypothesis".

4.4 Main Conclusions

- The oldest known continental crust is 4.03 Ga old (Acasta gneisses). Zircons from Australia, gave 4.404-Ga ages thus demonstrating that continental crust already existed for 4.4 Ga.
- All over the world, primitive continental crust has a very homogeneous composition that consists of tonalites, trondhjemites and granodiorites (TTG).
- TTG are generated from the mantle through a 3-stage process:
  - Mantle melting generates a basaltic crust similar to modern oceanic crust.
  - These basalts transformed into garnet-bearing amphibolite or eclogite, melt and give rise to the TTG parental magma.
  - This parental magma can evolve by fractional crystallization thus generating the TTG differentiated suite.
- Evidence of horizontal tectonics since the early Archaean demonstrate that plate tectonics operated during this period.
- TTG genesis probably took place in a subduction-like geodynamic environment.
- Subduction of a hot oceanic crust implies that the basalt hydrous solidus temperature is reached before dehydration could take place. Thus TTG are produced by subducted slab melting.
- The temporal evolution of TTG parental magmas shows that in the course of time, melting depth increased, due to Earth cooling and that interactions between slab melt and mantle increased.
- Archaean plates were smaller than today and probably moved at a greater rate, such that the average age of oceanic crust when it enters into subduction was < 20 Ma, whereas it is 60 Ma today.
- During Archaean, two tectonic regimes coexisted. At plate boundaries, similarly to modern collisional mountain chains, horizontal tectonics operated. In the plate central part, a vertical, gravity-driven tectonic could operate. This

latter was due to the emplacement of high-density materials (komatiites) over low-density continental crust (TTG). Disappearing of komatiites after 2.5 Ga also led to the disappearing of vertical tectonic.

- Horizontal tectonics allowed genesis of relief and mountain chains and consequently led to the continent immersion. However, due to higher geothermal gradients, which lower crustal viscosity, Archaean relief should have been lower than their modern equivalents.
- All changes observed through the Archaean–Proterozoic boundary (TTG vs. calc-alkaline; (horizontal and vertical) vs. only horizontal tectonics disappearance of komatiites; etc.) only result in a single process: progressive Earth cooling.
- After its genesis, primitive juvenile continental crust irreversibly evolved through recycling processes.

References

1. General

Books

The Archaean Crustal Evolution. 1994. K.C. Condie (ed.), Developments in Precambrian Geology. Elsevier, Amsterdam, 528 pp.

The Young Earth: An introduction to Archaean Geology. 1987 Nisbet, E.G.; Allen and Unwin, Boston, 402 pp.

The Continental Crust: Its Composition and Evolution. 1985. Taylor, S.R. and McLennan, S.M.; Blackwell Scientific Publications, Oxford, 312 pp.

Early Earth: Physical, Chemical and Biological Development. 2003. (C.J. Hawkesworth, C. Ebinge, C.J. Ebinger, C.M.R. Fowler, eds.), Geological Society of London; 352 pp.

Articles

Choukroune, P., Ludden, J.N., Chardon, D., Calvert, A.J., Bouhallier, H., 1997. Archaean crustal growth and tectonic processes: a comparison of the Superior province, Canada and the Dharwar craton, India. In *Orogeny Through Time* (J.P. Burg and M. Ford, eds.) *Geol. Soc. Spec. Publ.*, **121**: 63–98.

Defant, M.J. and Drummond, M.S., 1990. Derivation of some modern arc magmas by melting of young subducted lithosphere. *Nature*, **347**: 662–665.

Jahn, B.M., 1997. Géochimie des granitoïdes archéens et de la croûte primitive. In: R. Hagemann and M. Treuil (eds.), *Introduction à la Géochimie et ses Applications.* Editions Thierry Parquet.

Martin, H. and Moyen, J.-F., 2002. Secular changes in TTG composition as markers of the progressive cooling of the Earth. *Geology*, **30(4)**: 319–322.

Rapp, R.P. and Watson, E.B., 1995. Dehydratation melting of metabasalt at 8–32 kbar: implications for continental growth and crust-mantle recycling. *Journal of Petrology*, **36(4)**: 891–931.

2. Specialised

Albarède, F., 1998. The growth of continental crust. *Tectonophysics*, **296**: 1–14.

Armstrong, R.L., 1981. Radiogenic isotopes: the case for crustal recycling on a near-steady-state no-continental-growth Earth. *Philosophical Transactions of the Royal Society of London*, **A301**: 443–472.

Arndt, N.T., 1992. Rate and mechanism of continent growth in the Precambrian. In: S. Maruyama (ed.), *Evolving Earth Symposium*, Okazaki, pp. 38–41.

Arndt, N.T. and Chauvel, C., 1991. Crust of the Hadean Earth. *Bulletin of the Geological Society of Denmark*, **39**: 145–151.

Baadsgaard, H. et al., 1984. The zircon geochronology of the Akilia association and Isua supracrustal belt, West Greenland. *Earth and Planetary Science Letters*, **68**: 221–228.

Barker, F., 1979. Trondhjemites: definition, environment and hypothesis of origin. In: F. Barker (ed.), *Trondhjemites, Dacites and Related Rocks*. Elsevier, Amsterdam, pp. 1–12.

Barth, M.G., Foley, S.F. and Horn, I., 2002a. Partial melting in Archaean subduction zones: constraints from experimentally determined trace element partition coefficients between eclogitic minerals and tonalitic melts under upper mantle conditions. *Precambrian Research*, **113**: 323–340.

Barth, M.G., Rudnick, R.L., Carlson, R.W., Horn, I. and McDonough, W.F., 2002b. Re–Os and U–Pb geochronological constraints on the eclogite-tonalite connection in the Archaean Man Shield, West Africa. *Precambrian Research*, **118(3–4)**: 267–283.

Bickle, M.J., 1978. Heat loss from the Earth: constraint on Archaean tectonics from the relationships between geothermal gradients and the rate of plate production. *Earth and Planetary Science Letters*, **40**: 301–315.

Bickle, M.J., Battenay, L.F., Boulter, C.A., Groves, D.I. and Morrant, P., 1980. Horizontal tectonic interaction of the Archaean gneiss belt and greenstones, Pilbara block. *Contribution to Mineralogy and Petrology*, **84**: 25–35.

Bickle, M.J. et al., 1993. Origin of the 3500–3300 Ma calc-alkaline rocks in the Pilbara Archaean: isotopic and geochemical constraints from the Shaw batholith. *Precambrian Research*, **60**: 117–149.

Bickle, M.J., Nisbet, E.G. and Martin, A., 1994. Archaean greenstone belts are not oceanic crust? *Journal of Geology*, **102**: 121–138.

Blais, S., Martin, H. and Jégouzo, P., 1997. Reliques d'une croûte océanique archéenne en Finlande orientale. *Comptes Rendus de l'Académie des Sciences de Paris*, **325**: 397–402.

Bouhallier, H., Chardon, D. and Choukroune, P., 1995. Strain patterns in Archaean dome-and-basin structures: the Dharwar craton (Karnataka, South India). *Earth and Planetary Science Letters*, **135**: 57–75.

Bowring, S.A. and Williams, I.S., 1999. Priscoan (4.00–4.03 Ga) orthogneisses from northwestern Canada. *Contribution to Mineralogy and Petrology*, **134**: 3–16.

Brown, G.C., 1986. Processes and problems in the continental lithosphère: geological history and physical implications. In: N. Snelling (ed.), *Geochronology and Geological Record*. Geological Society of London Special Publication, London.

Capdevila, R., Arndt, N.T., Letendre, J. and Sauvage, J.-F., 1999. Diamonds in volcanoclastic komatiite from French Guyana. *Nature*, **399**: 456–458.

Chardon, D., Choukroune, P. and Jayananda, M., 1996. Strain patterns, décollement and incipient sagducted greenstone terrains in the Archaean Dharwar craton (south India). *Journal of Structural Geology*, **18**: 991–1004.

Choukroune, P., Ludden, J., Chardon, D., Calvert, A.J. and Bouhallier, H., 1997. Archaean crustal growth and tectonic processes: a comparison of the Superior province, Canada and the Dharwar craton, India. In: J.-P. Burg and M. Ford (eds.), *Orogeny Through Time*. Geological Society Special Publication, pp. 63–98.

Condie, K.C., 1981. *Archaean Greenstone Belts*. Elsevier, Amsterdam, 434 pp.

Condie, K.C., 1989. *Plate Tectonics and Crustal Evolution*. Pergamon, Oxford, 476 pp.

Condie, K.C., 1998. Episodic continental growth and supercontinents: a mantle avalanche connection? *Earth and Planetary Science Letters*, **163(1–4)**: 97–108.

David, J., Parent, M., Stevenson, R., Nadeau, P. and Godin, L., 2002. La séquence supracrustale de Porpoise Cove, région d'Inukjuak; un exemple unique de croûte paléo-archéenne (ca. 3.8 Ga) dans la Province du Supérieur. 23ème Séminaire d'information sur la recherche géologique, Ministère des ressources naturelles du Québec. (session 2).

Davies, G.F., 1992. On the emergence of plate tectonics. *Geology*, **20(11)**: 963–966.

de Wit, M.J. and Hart, R.A., 1993. Earth's earliest continental lithosphere, hydrothermal flux and crustal recycling. *Lithos*, **30**: 309–335.

de Wit, M.J. et al., 1992. Formation of Archaean continents. Nature, **357**: 553–562.

Defant, M.J. and Drummond, M.S., 1990. Derivation of some modern arc magmas by melting of young subducted lithosphere. *Nature*, **347**: 662–665.

Delong, S.E., Schwarz, W.M. and Anderson, R.N., 1979. Thermal effects of ridge subduction. *Earth and Planetary Science Letters*, **44**: 239–246.

Donn, W.L., Donn, B.D. and Valentine, W.G., 1965. On the early history of the Earth. *Geological Society of America Bulletin*, **76**: 287–306.

England, P.C. and Bickle, M.J., 1984. Continental thermal and tectonic regimes during the Archaean. *Journal of Geology*, **92**: 353–367.

Foley, S.F., Buhre, S. and Jacob, D.E., 2003. Evolution of the Archaean crust by delamination and shallow subduction. *Nature*, **421(6920)**: 249–252.

Foley, S.F., Tiepolo, M. and Vannucci, R., 2002. Growth of early continental crust controlled by melting of amphibolite in subduction zones. *Nature*, **417**: 637–640.

Fyfe, W.S., 1978. Evolution of the Earth's crust: modern plate tectonics to ancient hot spot tectonics? *Chemical Geology*, **23**: 89–114.

Goodwin, A.M., 1991. *Precambrian Geology*. Academic Press, 666 pp.

Gorman, B.E., Pearce, T.H. and Birkett, T.C., 1978. On the structure of Archaean greenstone belts. *Precambrian Research*, **6**: 23–41.

Green, D.H., 1982. Anatexis of mafic crust and high pressure crystallisation of andesite. In: R.S. Thorpe (ed.), *Andesites*. J. Wiley, New York, pp. 465–487.

Gutscher, M.-A., Spakman, W., Bijwaard, H. and Engdahl, E.R., 2000. Geodynamics of flat subduction: Seismicity and tomographic constraints from the Andean margin. *Tectonics*, **19(5)**: 814–833.

Hargraves, R.B., 1986. Faster spreading or greater ridge length in the Archaean. *Geology*, **14**: 750–752.

Jahn, B.M., 1997. Géochimie des granitoïdes archéens et de la croûte primitive. In: R. Hagemann and M. Treuil (eds.), *Introduction à la Géochimie et ses Applications*. Editions Thierry Parquet.

Jarvis, G.T. and Campbell, I.H., 1983. Archaean komatiites and geotherms: solution to an apparent contradiction. *Geophysical Research Letters*, **10**: 1133–1136.

Kamber, B.S., Collerson, K.D., Moorbath, S. and Whitehouse, M.J., 2003. Inheritance of early Archaean Pb-isotope viariability from long-lived Archarean protocrust. *Contribution to Mineralogy and Petrology*, **145**: 25–46.

Kamber, B.S., Ewart, A., Collerson, K.D., Bruce, M.C. and Mc Donald, G.A., 2002. Fluid-mobile trace element constraints on the role of slab melting and implications for Archaean crustal growth models. *Contribution to Mineralogy and Petrology*, **144**: 38–56.

Kato, Y. and Nakamura, K., 2003. Origin and global tectonic significance of Early Archaean cherts from the Marble Bar greenstone belt, Pilbara Craton, Western Australia. *Precambrian Research*, **125(3–4)**: 191–243.

Kinny, P.D., 1986. 3820 Ma zircons from a tonalitic Amîtsoq gneiss in Godthåb district of southern West Greenland. *Earth and Planetary Science Letters*, **79**: 337–347.

Kleine, T., Munker, C., Mezger, K. and Palme, H., 2002. A short timescale for terrestrial planet formation from Hf-W chronometry of meteorites. *Nature*, **418(6901)**: 949–952.

Kleinhanns, I.C., Kramers, D. and Kamber, B.S., 2003. Importance of water for Archaean granitoid petrology: a comparative study of TTG and potassic granitoids from Barberton Mountain and, South Africa. *Contribution to Mineralogy and Petrology*, **145**: 377–389.

Komiya, T., 2001. Secular change of the Composition and Temperature of the upper mantle. *AGSO-Geoscience Australia Record*, **37**: 54–56.

Kring, D.A. and Cohen, B.A., 2002. Cataclysmic bombardment throughout the inner solar system 3.9–4.0 Ga. *Journal of Geophysical Research*, **107(E2)**: 4-1–4-6.

Kusky, T.M. and Kidd, W.S.F., 1992. Remnants of an Archaean oceanic plateau, Belingwe greenstone belt, Zimbabwe. *Geology*, **20**: 43–46.

Lagabrielle, Y., Goslin, J., Martin, H., Thiriot, J.-L. and Auzende, J.-M., 1997. Multiple active spreading centers in the hot North Fiji basin (SW Pacific): a possible model for Archaean seafloor dynamics? *Earth and Planetary Science Letters*, **149**: 1–13.

Ludden, J., Hubert, C., Barnes, A., Mikereit, B. and Sawyer, E., 1993. A three dimensional perspective on the evolution of the Earth's largest Archaean crust. LITHOPROBE seismic reflection images in the southwestern Superior Province. *Lithos*, **30**: 357–372.

Martin, H., 1986. Effect of steeper Archaean geothermal gradient on geochemistry of subduction-zone magmas. *Geology*, **14**: 753–756.

Martin, H., 1987. Petrogenesis of Archaean trondhjemites, tonalites and granodiorites from eastern Finland: major and trace element geochemistry. *Journal of Petrology*, **28(5)**: 921–953.

Martin, H., 1993. The mechanisms of petrogenesis of the Archaean continental crust-comparison with modern processes. *Lithos*, **30**: 373–388.

Martin, H., 1994. The Archaean grey gneisses and the genesis of the continental crust. In: K.C. Condie (ed.), *The Archaean Crustal Evolution. Developments in Precambrian Geology*. Elsevier, Amsterdam, pp. 205–259.

Martin, H., 1999. The adakitic magmas: modern analogues of Archaean granitoids. *Lithos*, **46(3)**: 411–429.

Martin, H. and Moyen, J.-F., 2002. Secular changes in TTG composition as markers of the progressive cooling of the Earth. *Geology*, **30(4)**: 319–322.

Martin, H., Smithies, R.H., Rapp, R.P., Moyen, J.-F. and Champion, D.C., 2004. An overview of adakite, TTG and sanukitoid: relationships and some implications for crustal evolution. *Lithos*, (in press).

Maury, R.C., Sajona, F.G., Pubellier, M., Bellon, H. and Defant, M.J., 1996. Fusion de la croûte océanique dans les zones de subduction/collision récentes: l'exemple de Mindanao (Philippines). *Bulletin de la Société Géologique de France*, **167(5)**: 579–595.

McCulloch, M.T. and Bennet, V.C., 1993. Evolution of the early Earth: constraints from ^{143}Nd–^{142}Nd isotopic systematics. *Lithos*, **30**: 237–255.

Merle, O., 1994. *Nappes et Chevauchements. Enseignement des Sciences de la Terre.* Masson, Paris, 138 pp.

Mojzsis, S.J. et al., 1996. Evidence for life on Earth before 3,800 million years ago. *Nature*, **384**: 55–59.

Mojzsis, S.J., Harrison, M.T. and Pidgeon, R.T., 2001. Oxygen-isotope evidence from ancient zircons for liquid water at the Earth's surface 4,300 Myr ago. *Nature*, **409**: 178–181.

Moyen, J.-F., Martin, H. and Jayananda, M., 1997. Origine du granite fini-Archéen de Closepet (Inde du Sud): apports de la modélisation géochimique du comportement des éléments en traces. *Compte Rendus de l'Académie des Sciences de Paris*, **325**: 659–664.

Nisbet, E.G., 1984. The continental and oceanic crust and lithosphere in the Archaean: isostatic, thermal and tectonic models. *Canadian Journal of Earth Sciences*, **21**: 1426–1441.

Nisbet, E.G., 1987. The Young Earth: *An Introduction to Archaean Geology*. Allen and Unwin, Boston, 402 pp.

Nisbet, E.G., Cheadle, M.J., Arndt, N.T. and Bickle, M.J., 1993. Contraining the potential temperature of the Archaean mantle: A review of the evidence from komatiites. *Lithos*, **30**: 291–307.

Nisbet, E.G. and Fowler, C.M.R., 1983. Model for Archaean plate tectonics. *Geology*, **11**: 376–379.

Nutman, A.J. and Collerson, K.D., 1991. Very early Archaean crustal-accretion complexes preserved in the North Atlantic craton. *Geology*, **19**: 791–794.

Nutman, A.J., McGregor, V.R., Friend, C.R.L., Bennett, V.C. and Kinny, P.D., 1996. The Itsaq gneiss complex of southern Greenland; the world's most extensive record of early crustal evolution (3900–3600 Ma). *Precambrian Research*, **78**: 1–39.

Nutman, A.P. et al., 2002. 3850 Ma BIF and mafic inclusions in the early Archaean Itsaq Gneiss Complex around Akilia, southern West Greenland? The difficulties of precise dating of zircon-free protoliths in migmatites. *Precambrian Research*, **117 (3–4)**: 185–224.

O'Connor, J.T., 1965. A classification for quartz-rich igneous rocks based on feldspar ratios. *United States Geological Survey Professional Paper*, **525B**: 79–84.

Parson, B.A. and Sclater, J.G., 1977. An analysis of the variation of ocean floor bathymetry and heat flow with ages. *Journal of Geophysical Research*, **82**: 803–827.

Percival, J.A., 1994. Archeén high-grade metamorphism. In: K.C. Condie (ed.), *The Archaean Crustal Evolution. Developments in Precambrian Geology*. Elsevier, Amsterdam, pp. 357–410.

Prouteau, G., Scaillet, B., Pichavant, M. and Maury, R.C., 1999. Fluid-present melting of ocean crust in subduction zone. *Geology*, **27**: 1111–1114.

Prouteau, G., Scaillet, B., Pichavant, M. and Maury, R.C., 2001. Evidence for mantle metasomatism by hydrous silicic melts derived from subducted oceanic crust. *Nature*, **410**: 197–200.

Rapp, R.P., Shimizu, N. and Norman, M.D., 2003. Growth of early continental crust by partial melting of eclogite. *Nature*, **425**: 605–609.

Rapp, R.P., Shimizu, N., Norman, M.D. and Applegate, G.S., 1999. Reaction between slab-derived melts and peridotite in the mantle wedge: experimental constraints at 3.8 GPa. *Chemical Geology*, **160**: 335–356.

Rapp, R.P. and Watson, E.B., 1995. Dehydratation melting of metabasalt at 8–32 kbar: implications for continental growth and crust-mantle recycling. *Journal of Petrology*, **36(4)**: 891–931.

Rapp, R.P., Watson, E.B. and Miller, C.F., 1991. Partial melting of amphibolite/eclogite and the origin of Archaean trondhjemites and tonalites. *Precambrian Research*, **51**: 1–25.

Rey, P.F., Philippot, P. and Thebaud, N., 2003. Contribution of mantle plumes, crustal thickening and greenstone blanketing to the 2.75–2.65-Ga global crisis. *Precambrian Research*, **127(1–3)**: 43–60.

Reymer, A. and Schubert, G., 1984. Phanerozoic addition rates of the continental crust and crustal growth. *Tectonics*, **3**: 63–78.

Rollinson, H., 1997. Eclogite xenoliths in west African kimberlites as residues from Archaean granitoid crust formation. *Nature*, **389**: 173–176.

Rosing, M.T. and Frei, R., 2004. U-rich Archaean sea-floor sediments from Greenland – indications of $>$ 3700 Ma oxygenic photosynthesis. *Earth and Planetary Science Letters*, **217(3–4)**: 237–244.

Rudnick, R.L., 1995. Making continental crust. *Nature*, **378**: 571–577.

Rudnick, R.L. and Gao, S., 2003. The Composition of the Continental Crust. In: R.L. Rudnick (ed.), *The Crust. Treatise on Geochemistry.* Elsevier-Pergamon, Oxford, pp. 1–64.

Ryder, G., Koeberl, C. and Mojzsis, S.J., 2000. Heavy bombardment of the Earth at $\sim$3.85 Ga: The search for petrographic and geochemical evidence. In: R.M. Canup and K. Righter (eds.), *Origin of the Earth and Moon.* Arizona University Press, pp. 475–492.

Schidlowski, M., 1988. A 3,800-million-year isotopic record of life from carbon in sedimentary rocks. *Nature*, **333**: 313–318.

Shen, Y. and Buick, R., 2004. The antiquity of microbial sulfate reduction. *Earth-Science Reviews*, **64(3–4)**: 243–272.

Sleep, N.H. and Windley, B.F., 1982. Archaean plate tectonics: constraints and inferences. *Journal of Geology*, **90**: 363–379.

Smithies, R.H., 2000. The Archaean tonalite-trondhjemite-granodiorite (TTG) series is not an analogue of Cenozoic adakite. *Earth and Planetary Science Letters*, **182**: 115–125.

Smithies, R.H., Champion, D.C. and Cassidy, K.F., 2003. Formation of Earth's early Archaean continental crust. *Precambrian Research*, **127(1–3)**: 89–101.

Stein, M. and Hofmann, A.W., 1994. Mantle plumes and episodic crustal growth. *Nature*, **372**: 63–68.

Streckeisen, A., 1975. To each plutonic rock its proper name. *Earth Science Reviews*, **12**: 1–33.

Svetov, S.A., Svetova, A.I. and Huhma, H., 2001. Geochemistry of the komatiite-tholeiite rock association in the Vedozero-Segozero Archaean greenstone bet, Central Karelia. *Geochemistry International*, **39**: 24–38.

Taylor, S.R. and McLennan, S.M., 1985. *The Continental Crust: Its Composition and Evolution.* Blackwell Scientific Publications, Oxford, 312 pp.

Valley, J.W., Peck, W.H., King, E.M. and Wilde, S.A., 2002. A Cool Early Earth. *Geology*, **30**: 351–354.

Veizer, J. and Jansen, S.L., 1979. Basement and sedimentary recycling and continental evolution. *Journal of Geology*, **87**: 341–370.

Vervoort, J.D., Patchett, P.J., Gehrels, G.E. and Nutman, A.J., 1996. Constraints on early Earth differentiation from hafnium and neodymium isotopes. *Nature*, **379**: 624–627.

Vityazev, A.V., Pechernikova, G.V. and Bashkirov, A.G., 2003. Early accretion and differentiation of protoplanetary bodies and Hf-W chronometry, *Third International Conference on Large Meteorite Impacts*, Nördlingen (Germany), pp. 4035.

Vlaar, N.J., 1986. Archaean global dynamics. *Geologie en Mijnbouw*, **65**: 91–101.

Walsh, M.M., 1992. Microfossils and possible microfossils from the Early Archaean Onverwacht Group, Barberton Mountain Land, South Africa. *Precambrian Research*, **54**: 271–293.

Westall, F., de Wit, M.J., Van Der Gaast, S., De Ronde, C. and Gerneke, D., 2001. Early Archaean fossil bacteria and biofilms in hydrothermally-influenced, shallow water sediments, Barberton greenstone belt, South Africa. *Precambrian Research*, **106**: 93–116.

Wilde, S.A., Valley, J.W., Peck, W.H. and Graham, C.M., 2001. Evidence from detrital zircons for the existence of continental crust and oceans on the Earth 4.4 Ga ago. *Nature*, **409**: 175–178.

Windley, B.F., 1998. Tectonic models for the geological evolution of crust, cratons and continents in the Archaean. *Revista Brasileira de Geociencias*, **28**: 183–188.

Wolf, M.B. and Wyllie, P.J., 1991. Dehydration-melting of solid amphibolite at 10 kbar: textural development, liquid interconnectivity and applications to the segregation of magmas. *Contribution to Mineralogy and Petrology*, **44**: 151–179.

Wolf, M.B. and Wyllie, P.J., 1994. Dehydration-melting of amphibolite at 10 kbar: the effects of temperature and time. *Contribution to Mineralogy and Petrology*, **115**: 369–383.

Wyllie, P.J., 1971. The role of water in magma genesis and initiation of diapiric uprise in the mantle. *Journal of Geophysical Research*, **76**: 1328–1338.

Yin, Q. et al., 2002. A short timescale for terrestrial planet formation from Hf-W chronometry of meteorites. *Nature*, **418(6901)**: 949–952.

Zamora, D., 2000. Fusion de la croûte océanique subductée: approche expérimentaleet géochimique. Université Thesis, Université Blaise Pascal, Clermont-Ferrand, 314 pp.

5 Thermal Evolution of the Earth During the First Billion Years

Christophe Sotin

There is good evidence that life occurred on Earth during the first billion years of its history. Modelling the dynamics of the Earth at this period of time is critical to understand the conditions of the emergence of life. These conditions are the result of the coupling between the inner and outer envelopes of the Earth. Several processes such as volcanism, magnetic field and plate tectonics originate in the Earth's deep layers. They control the physical and chemical conditions of the outer layers (atmosphere, hydrosphere, and crust) where life appeared and developed. The goal of this chapter is to describe these internal processes and to present models for Earth's evolution. After a descriptive summary of our current knowledge of the Earth's deep interior, this chapter explains the mechanisms of heat transfer to the surface by subsolidus thermal convection, a process that drives the Earth's surface dynamics (volcanism and plate tectonics). The last part of this chapter addresses the Earth's magnetic field and how it prevents atmospheric escape and preserves the present atmosphere. Throughout this chapter, references to conditions existing on Earth-like planets are given to illustrate how the knowledge of these planets contributes to a better understanding of the history of our own planet.

5.1 Internal Structure and Dynamics of the Earth

5.1.1 Description of the Different Layers

Seismological data acquired during almost a century provide a radial description (1D) of the physical properties of the Earth (density, elastic constants, pressure). The model, known as PREM (*Preliminary Reference Earth Model*) was published in 1981 (Dziewonski and Anderson, 1981). It gives a more accurate description of the Earth's internal structure than the model proposed at the beginning of the 20th century (Lehmann, 1936 and review by Bolt, 1982), which determined the depth of the major discontinuities thanks to the reflection of seismic waves on these interfaces. The 'solid' Earth is divided into three main layers with distinct chemical compositions: the core, the mantle and the crust (Fig. 5.1).

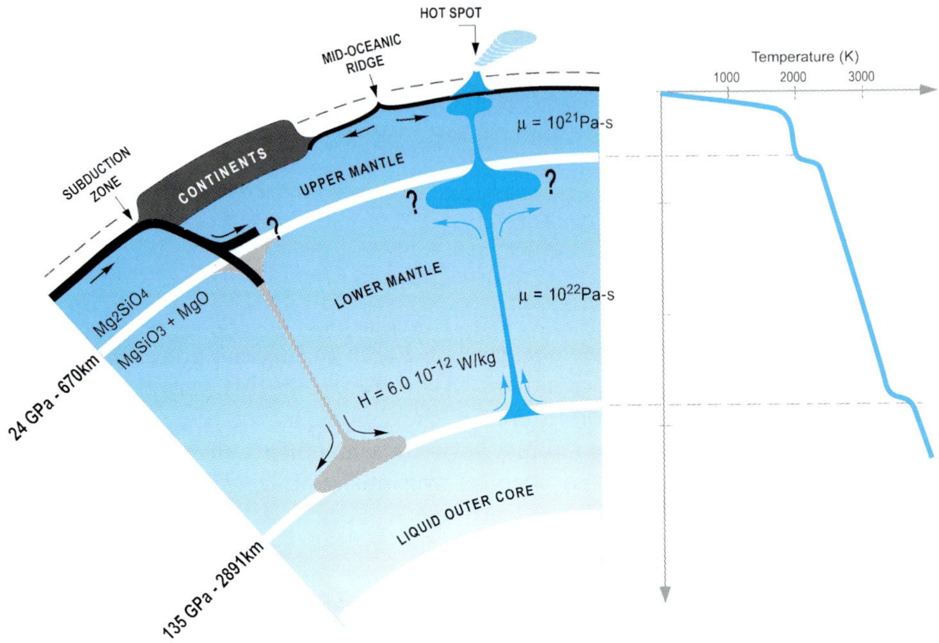

Fig. 5.1. Internal structure of the Earth. For sake of simplicity, iron is not identified in the different mineralogical compositions mentioned on the figure (One can assume that 10% of Mg is replaced by Fe). On the right is a simplified horizontally averaged temperature profile

5.1.1.1 The Core

The volume of the core is 16% of that of the whole solid Earth, whereas its mass is one third of the total mass of the Earth (6×10^{24} kg). Iron, nickel and a light element (presumably sulfur) are the main components of the core. Sulfur is proposed because it is compatible with liquid iron and it appears to be depleted in the crust and mantle compared to solar and meteoritic abundances. The core is divided into two domains: a solid inner core (1221 km in radius) and a liquid outer layer. Seismic velocities suggest that sulfur is present only in the outer liquid layer. This is confirmed by high-pressure laboratory experiments on the Fe–FeS system (see Sect. 5.3.2.2 and references therein). Motion of iron-rich liquid material in response to Coriolis forces, thermal convection due to the Earth cooling, and chemical differentiation related to the enrichment in sulfur as the inner solid core crystallizes (see Sect. 5.3.2.2), lead to the formation of electromagnetic currents responsible for the Earth's dynamo. The main component of the Earth's magnetic field is dipolar. The axis of the dipole rotates about the Earth's rotation axis but is parallel to it over geological time scales. Inversions of the magnetic field occur from time to time. These inversions are recorded in the oceanic crust and are the major argument to demonstrate the validity of plate tectonics (Vine and Matthews, 1963). If one understands that the Earth's dynamo originates in the liquid outer core, there is still a great deal of research necessary to explain its physical characteristics including the processes responsible for the reversals, the effect of chemical convection, and the importance of a solid inner core. More detailed description of the magnetic field (and its influence on the outer layers) is given in the last part (Sect. 5.3) of this chapter.

5.1.1.2 The Mantle

The mantle represents two third of the Earth's mass. It is composed of two layers: the upper mantle and the lower mantle (Fig. 5.1). The core–mantle boundary (CMB) is 2891 km deep. An envelope, named the D″ layer, is located at the CMB and its thickness lies between 100 km and 200 km as determined by seismic data. The composition of the D″ layer is not known. Seismic data suggest not only that this layer is a thermal boundary layer but also that it may have a different chemical composition from that of the lower mantle (Karato and Karki, 2001; Deschamps and Trampert, 2003). The origin of this differentiated layer is still controversial: is there contamination from the core or accumulation of subducted oceanic lithosphere?

The main chemical components of the mantle are: Si, Mg, Al, Fe, Ca, O. Earth mantle mainly consists in a rock called peridotite and made up of olivine [$(Mg, Fe)_2SiO_4$], pyroxenes [$(Mg, Fe)_2Si_2O_6$ and $Ca(Mg, Fe)Si_2O_6$] and an aluminous mineral: spinel [$MgAl_2O_4$] at pressures lower than 3 GPa (depth lower than 80 km) and garnet [$(Mg, Fe)_3Al_2Si_3O_{12}$] at higher pressure. The interface between the upper and the lower mantle is located at 670 km depth ($P = 23.8$ GPa) and is due to the transformation of olivine into perovskite [$(Mg, Fe)SiO_3$] and magnesowustite [$(Mg, Fe)O$]. Laboratory experiments have been carried out to characterize the Clapeyron curve of this transformation and show that it is endothermic ($\mathrm{d}P/\mathrm{d}T < 0$). This unusual characteristic implies that the transformation occurs at lower pressure (shallower depth) in upwelling plumes than in subducting lithosphere. This results in chemical buoyancy opposite to thermal buoyancy. The lower density of the hot plume, caused by its temperature above that of the surrounding olivine-rich mantle, may eventually be compensated by the higher density of perovskite and magnesowustite remaining in the upwelling plume. This effect may stop the plume at the lower/upper mantle interface (Turcotte and Schubert, 1982 or Sotin and Parmentier, 1989). This is invoked to explain the existence of two distinct chemical reservoirs, which produce the two different kinds of volcanism: the lower mantle for hot-spot volcanism and the upper mantle for midocean-ridge basalts. However, seismic tomography of the Earth's interior indicates that at least part of the cold upper mantle, the subduction zone where seismic velocities are greater than the horizontally averaged velocity, goes through this discontinuity and sink down to the CMB.

5.1.1.3 The Crusts

Most of the crust is formed by partial melting of deeper rocks (i.e. mantle peridotite); it consists in a thin envelope at the interface between the high-density deep interior where most of the geodynamic processes start (volcanism, plate tectonics, magnetic field, etc.) and the surface, where life exists. There are two kinds of crusts: the oceanic crust and the continental crust. The continental crust is at least 10% less dense than the oceanic crust. The thickness of the continental crust varies between a few km at rifting zones to more than 70 km under mountains chains; its average thickness is about 35 km, which is significantly thicker than the 6 km oceanic crust whose thickness does not seem to depend on spreading rates (Sotin and Parmentier, 1989). Due to their different density, their buoyancy is also contrasted such that their average altitude is different resulting in a bimodal altitude distribution on Earth's surface. If all the Earth's crust consisted of oceanic crust (a possible scenario during the first billion years of the Earth), then the whole Earth would be covered by seawater except at some of the largest hot spots, such as Hawaii for example.

Oceanic crust is formed by partial melting of the upper mantle during its adiabatic upwelling beneath the 60 000 km long midocean ridges. In the context of plate tectonics, the oceanic crust returns and is recycled within the mantle at subduction zones. The oldest known oceanic crust is about 180 Ma old.

Continental crust is formed by more complex mechanisms of partial melting. Because it is less dense than the oceanic crust it is more buoyant and only few continental crusts can be recycled within the mantle. When plate motion pushes two continental lithospheres one against another (India and Eurasia, for example), they collide; pile up such that mountains form with deep continental roots due to isostatic equilibrium.

5.1.1.4 Lithosphere–Asthenosphere Boundary

Earth envelopes can be defined not only based on their chemical composition but also on their rheological properties. The outer part of the Earth is rigid and is called the lithosphere; it includes the crust as well as the upper rigid part of the upper mantle. With depth, as temperature increases, rock deformation becomes controlled by creep, and then this envelope is called the asthenosphere. The 800 °C isotherm is generally considered as the limit between the lithosphere and the asthenosphere. At this temperature, peridotite has a low viscosity, enough to accommodate the stresses imposed by tectonic forces and thermal convection at geological time scales (tens of millions of years). This limit can also be defined by the seismic data: when heat transfer turns from conduction into convection (see Sect. 5.2.2.2), the temperature gradient rapidly decreases. As seismic velocities increase with decreasing temperature and increasing pressure, the change in thermal regime at the lithosphere–asthenosphere boundary is also marked by a rapid change in seismic-wave velocity. In the conductive outer layer, the high temperature gradient induces a decrease of the velocity profile. A low-velocity zone (LVZ) is therefore present at the interface between the conductive region and the adiabatic domain where the effect of pressure overcomes that of temperature leading to increasing seismic velocities. For people involved in convection models, the limit between lithosphere and asthenosphere can be defined as the top of the cold thermal boundary layer, a domain where cold thermal instabilities form (Sect. 5.2.1.1). Sometimes, the asthenosphere is limited to the LVZ. However, one can consider that the whole mantle affected by convection (or deforming by creep) belongs to the asthenosphere.

5.1.2 Internal dynamics of the Earth

5.1.2.1 Thermal Balance and Comparison with other Planets

Plate tectonics and hot spots are surface expressions of the Earth's internal dynamics, which are controlled by convective heat transfer. It is assumed that the planet started cooling just after the accretion period, 4.56 Ga ago because accretion processes had left a hot planet in a cold interplanetary space. The cooling rate is limited by the internal heating rate due to

- the decay of long-lived radiogenic elements present in the mantle and the crust (^{235}U, ^{238}U, ^{232}Th, ^{40}K),
- the latent heat of solidification of the inner core, a minor source of heat considering the small size of the inner core,
- the energy that sustains the magnetohydrodynamo and that is turned into heat by the Joule effect in the outer layer of the liquid core.

The natural decrease of the amount of available radiogenic elements should lead to a faster cooling today than in the past. However, this effect is counterbalanced by the temperature-dependent viscosity: the larger the temperature, the smaller the viscosity, and the more efficient the heat transfer. In the past, convection was more efficient (higher heat production) because the average temperature was greater. Today, convection is less efficient because the average temperature of the convecting mantle is lower (i.e. Sect. 5.2.2.3 and Fig. 5.6).

Earth heat flux has been measured at the surface of the Earth, mostly on continents, and has been modelled for the oceanic sea-floor (Turcotte and Schubert, 1982). About 42 TW (82 mW/m^2) come out of the Earth (1 TW = 1 tera W = 10^{12} W). This value is small compared to the extraterrestrial power received from the Sun (about 340 W/m^2). The 42 TW are produced by different sources: 8 TW by decay of radiogenic elements in the lithosphere (mostly in the continental crust), 24 TW by decay of radiogenic elements in the convective mantle), 4 TW by the core (latent heat and ohm-generated heat by the Joule effect in the outer layer of the liquid core) and only 6 TW by cooling of the Earth's mantle (equivalent to a cooling rate of 50 K/Ga).

Compared to other planets, the Earth is unique because of the existence of plate tectonics, a process that controls the dynamics and shape of its surface.

Mars and Venus are often called the Earth's sister planets because their density (and moment of inertia for Mars) is consistent with an internal structure similar to that of the Earth. Recent observations by spacecrafts (Magellan in 1989 for Venus or missions Viking, Mars Global Surveyor, Mars Odyssey for Mars) revealed that tectonic processes, similar to the Earth's plate tectonics, do not operate today on these two planets. Possible early plate tectonics on Mars is debated: the topographic dichotomy between the southern and the northern hemisphere (Plate 5.1) has been compared to a continental margin and the alignment of volcanoes on Tharsis Mons has been linked to volcanic chains above a hot spot (Sleep, 1994). However, the plate-tectonics theory for early Mars lacks additional arguments to fully convince the scientific community. Mars has a few volcanoes and most of them are located on Tharsis Mons (Plate 5.1). These volcanoes are not active. Some of them are amazingly big with elevations over 20 km. Crater counting on their slopes suggests that they may have been active some 100 Ma ago. Actually, recent (at geological time scales) volcanic activity is deduced from the study of SNC meteorites, which are supposed to come from Mars.

On Venus, crater counting indicates that the surface is young (between 500 Ma and 1 Ga) and homogeneous. The fact that the crater density is constant all over the surface is a strong argument against plate tectonics. It is assumed that a catastrophic event occurred 1 Ga ago, leading to complete resurfacing by volcanic rocks (e.g. Reese et al., 1999). This event is also likely to have modified the composition of the atmosphere. If radar images provided by the Magellan mission revealed the presence of many volcanic structures (Head et al., 1992), none of them seems to be active today. Indeed, the Magellan probe observed the planet over 3 years and has not detected a single volcanic event during that period of time.

The only body in the Solar System that has a terrestrial-like volcanic activity is Io. This satellite of Jupiter has about the same size as the Moon and it was a major surprise when telescopic observations followed shortly by Voyager images revealed Io's volcanic activity. There, the energy responsible for melting is tidal energy and not the energy stored during accretion (Io is too small), nor the radioactive decay of the radiogenic elements. During its 1.77 day orbit around Jupiter, Io's eccentricity, maintained by the Laplace resonance between Io, Europa and Ganymede, leads to differential gravitational forces by Jupiter. Dissipation of tidal energy depends on several parameters including viscosity and orbital frequency (Tobie et al., 2003). Although this heat source is quite different from those described above for the Earth, it provides enough heat to melt rocks and generate volcanism. Io heat flow has been estimated to be at least 20 times that of the Earth, somewhere between 2 and 10 W/m^2 (Matson et al., 2001).

Although Mars and Venus do not have active volcanoes, the interior of these planets is thought to convect. However, the convective domain would be overlaid by a thick conductive layer that would prevent the hot uprising material getting close enough to the surface such that adiabatic decompression is not sufficient

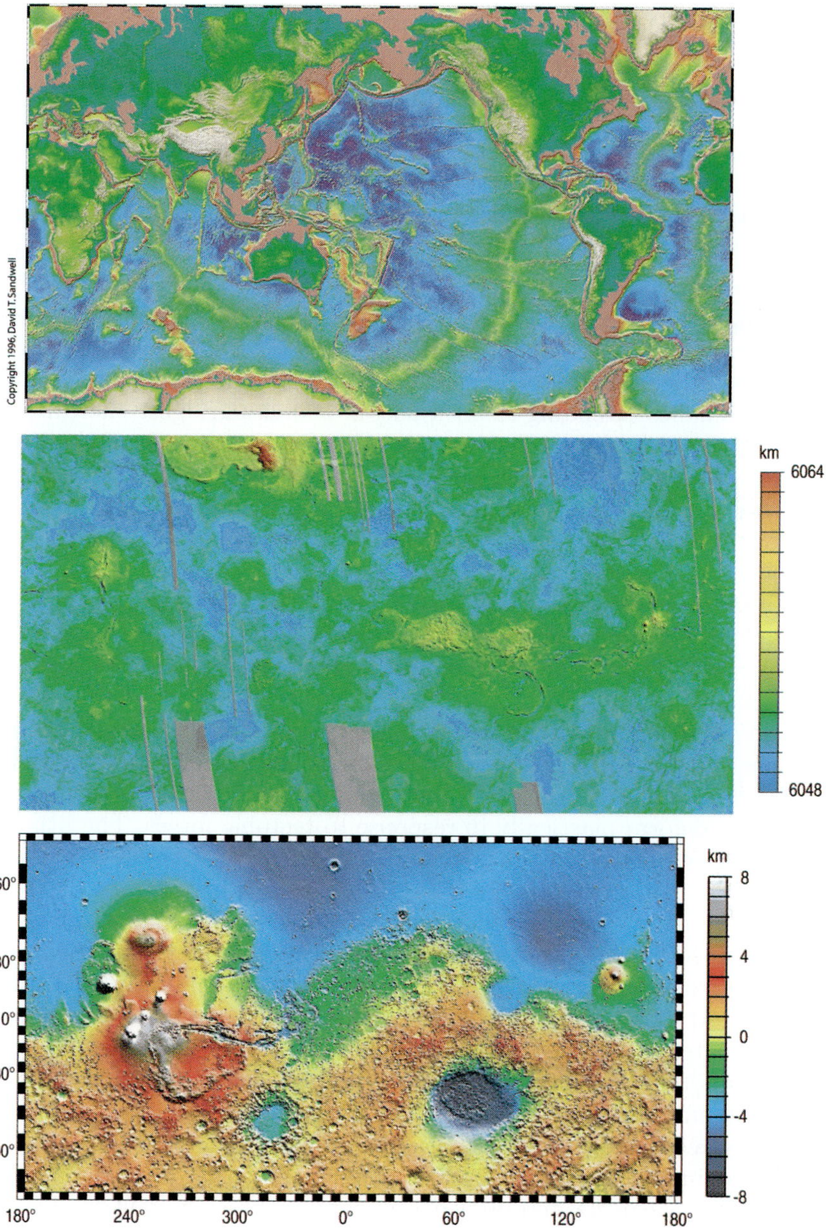

Plate 5.1. Topography of Earth, Venus and Mars. Largest variations occur on Mars where the summits of the volcanoes are more than 20 km above the mean radius of the planet. All three planets show large areas of low topography (oceanic floor for the Earth, the northern hemisphere for Mars and the plains on Venus). Mountains are present on both Earth and Venus (Aphrodite Terra on the equator and Ishtar Terra in the northern hemisphere) but there is no evidence of plate tectonics on Venus

to produce partial melting and consequently to built volcanoes. The relationship between the formation of the few volcanoes on both Mars and Venus is still debated and the link with the internal structure of their planets cannot be addressed without geophysical data such as seismic data. The Earth is unique because plate tectonics is a process that brings the hot material very close to the surface. A question, which is not yet solved, is why does plate tectonics occur and operate only on Earth. When did it start? Did it operate in the past on Mars and Venus?

5.1.2.2 Plate Tectonics

Plate tectonics theory developed in the 1960s after the discovery of the symmetric magnetic anomalies extending on both sides of midocean ridges. There are several components in the magnetic field of the Earth. The remnant magnetic field is recorded by a sedimentary rock or a volcanic rock at the time it formed. For volcanic rocks, minerals such as magnetite record a magnetic field parallel and proportional to the ambient magnetic field as its temperature decreases below the Curie point. Because the magnetic field reverses from time to time, the remnant magnetic field can point either in the same direction as the present magnetic field (normal polarity) or in the opposite direction (reversed polarity).

Oceanic crust is formed continuously at midocean ridges. Because Earth's surface remains constant, the oceanic crust is recycled into the mantle at subduction zones. The Earth's surface is divided into several plates (12 large and many more small ones), which are named the lithospheric plates: the rigid part of the Earth encompasses the crust and the upper part of the upper mantle that is too cold to deform plastically in response to tectonic forces. The lithospheric plates may have or may not have a continent. Their relative motion obeys Euler's theorem: the displacement of a rigid plate on a sphere can be described by a rotation about an axis that contains the centre of the sphere. The rotation can be described by the location of one pole (a pole is the intersection between the rotation axis and the surface of the sphere) and the angular velocity. The displacement of a plate relative to another one can be determined assuming one of the two plates to be fixed. To know the absolute displacement, the location of hot spots is taken as a reference frame on time scales of 200 Ma, the age of the oldest oceanic lithosphere. The displacement of plates on the Earth's surface can be easily reconstructed during these last 200 Ma, which is a short period of time compared to the age of the Earth (4.56 Ga). For periods older than 200 Ma, continent motion can be determined by analysing the remnant magnetic field of volcanic rocks but the reconstruction is more difficult since the oceanic part is missing.

A number of observations can be explained by plate-tectonics theory, including the location of seismic events at the border of two different plates, the formation of mountains, metamorphism, chain of volcanoes above hot spots, and the topography of oceanic sea floor. The sea floor subsides because the oceanic lithosphere progressively cools down such as its density increases as it moves away from the spreading centres. This simple cooling model well predicts both the topography and the geoid anomalies for lithosphere ages up to 130 Ma (e.g. Turcotte and Schubert, 1982). It also provides an expression for the heat-flux (q) that comes out of the oceanic lithosphere:

$$q = \frac{k(T_\mathrm{m} - T_\mathrm{s})}{\sqrt{\pi\kappa t}}, \tag{5.1}$$

where T_m is upper mantle temperature (on the order of 1350 °C), T_s is surface temperature, $k = 3\,\mathrm{W/m/K}$ is thermal conductivity, $\kappa = 10^{-6}\,\mathrm{m^2/s}$ is thermal diffusivity, and t is the age of the lithosphere ($t = x/u$ where x is the distance to the ridge axis and u is half the spreading rate). By integrating (5.1), one gets the total power that comes out of the oceanic lithosphere. At spreading centres, heat is transferred by hydrothermalism and therefore (5.1) does not apply. The amount of heat coming out of spreading centres is not yet known.

The displacement of lithospheric plates is the result of convection motions in the deep interior of the planet. The coupling between convection and plate tectonics is not yet well understood and is an active subject of research. So far, numerical models have a hard time to reproduce the complexity of the coupling. More specifically, it is difficult to reproduce the motion of the plates without prescribing weakness zones to simulate midocean ridges and subduction zones.

5.1.2.3 Hot Spots

The existence of hot spots is the second observation (after plate tectonics) that corroborates the existence of large and slow displacements in the Earth's mantle. Hot spots like Hawaii or La Réunion are the surface expression of big mantle plumes (diapirs) formed presumably at the CMB where hot thermal instabilities are generated. Due to plate tectonics, one thermal plume, which has a life of at least 200 Ma, generates several volcanoes, aligned along the direction of displacement of the plate (i.e. the Emperor chain in the Pacific ocean).

The basalts formed at midocean ridges (MORB) and those formed at hot spots come from the melting of two distinct reservoirs: the upper and lower mantle, respectively. One explanation, proposed in Sect. 5.2.1.2, is the endothermic transformation of olivine into perovskite and magnesowustite, which would prevent extensive mixing between the lower and the upper mantle. Laboratory experiments (Davaille, 1999) suggest that intraplate oceanic volcanism, more specifically the South Pacific superswell, which includes volcanoes of French Polynesia, may be related to chemical buoyancy. Compositional differences (i.e. iron content) between the lower and upper mantle can create density variations as large as those induced by temperature variations and would allow matter migration from the lower to the upper mantle. The origin of hot spots, their link with plate tectonics, their geometry (diameter of the tail, upwelling velocity, extension below the lithosphere, etc.) are still largely unknown.

5.1.2.4 Conclusion: Did Plate Tectonics Exist at the Beginning of the Earth's Evolution?

Our present knowledge of the relationships between mantle convection and plate tectonics is too weak to enable us to answer this question from a modelling point of view. In addition, the oldest known rocks are less than 4 Ga old and it is difficult to know whether plate tectonics already existed at that time. Plate tectonics is doubtless the surface expression of mantle convection. The velocity of plates (> 10 km/Ma) divided by the thickness of the mantle (2890 km) gives a deformation rate of about $10^{-16}\,\mathrm{ms}^{-1}$. This value is consistent with convective differential stresses around 10^5 Pa (1 bar) and a mantle viscosity of 10^{21} Pa s (i.e. Sect. 5.2.1.2). The remaining questions are: what triggered plate tectonics? What are the key parameters for plate tectonics to start (vigour of convection, viscosity profile in the lithosphere, presence of continents, etc.)? The following sections provide an update on thermal convection models based on recent studies that take into account the temperature dependence of viscosity.

5.2 Thermal Convection: the Driver of the Earth's Internal Dynamics

5.2.1 Basic Information About Thermal Convection

Thermal convection takes place within planets due to gravitational instabilities related to variations in temperature distribution. The uneven temperature distribution is caused by the fact that the planet is cooled from the outside and heated from the inside by the decay of long-lived radiogenic elements (or dissipation of tidal forces for Io) thus giving rise to a thermal gradient. Indeed, due to relatively low thermal diffusion rates in rocks, conduction appears as an inefficient mechanism to evacuate internal heat.

The volume of a mineral (V) depends on temperature (T) through the coefficient of thermal expansion ($\alpha = -(\mathrm{dLn}(V)/\mathrm{d}T)_P$). As this coefficient is small and can be considered as constant, density ($\rho = M/V$ where M is atomic mass) can be expressed as follows:

$$\rho = \rho_{\mathrm{ref}}\left[1 - \alpha\left(T - T_{\mathrm{ref}}\right)\right], \tag{5.2}$$

where ρ_{ref} is density of the mantle rock (peridotite) at a reference temperature (T_{ref}). Consequently, it appears that a cold peridotite is denser that its hot equivalent, therefore, peridotite is denser at its cold interface with the crust than at depth: this difference leads to gravity instability in the mantle. If the mantle viscosity (μ) is low enough for displacement to occur, then convective instability can take place and heat is transferred by convection. The vigour of convection is measured by the Rayleigh number (Ra), which is the ratio of the buoyancy force to the viscous drag:

$$\mathrm{Ra} = \frac{\alpha \rho g \Delta T b^3}{\kappa \mu}, \tag{5.3}$$

where g is gravity, ΔT is temperature difference between the top and the bottom of the mantle and b its thickness. Convection exists for Rayleigh numbers greater than a critical value that is about 10^3. For parameters relevant to the Earth's mantle, Rayleigh number ranges between 10^6 and 10^8, depending on the values chosen for the different parameters in (5.3), and whether we consider that the whole mantle is affected by a single convective process or if separate convection in the upper and lower mantle is assumed (Sect. 5.1.1.2). In the following paragraph, one type of convection pattern known as Rayleigh–Bénard convection is described. Although it cannot predict an accurate mantle temperature and heat flux for the Earth because it does not take into account (5.1) internal heating, (5.2) temperature dependence of viscosity and (5.3) lithospheric plate motions at the surface, it provides a clear and simple description of the convective process. The main interest of this description is to understand how one calculates the mean temperature of a convecting fluid, how one can estimate the heat transported by convection and how cold and hot plumes form.

5.2.1.1 Rayleigh–Bénard Convection: Theory and Limits of its Applicability for the Earth

Preliminary laboratory experiments were carried out at the beginning of the 20th century. A box filled with a fluid is limited by two horizontal plates at constant temperature. In these experiments, the viscosity of the fluid does not depend on temperature. The upper plate is maintained at room temperature whereas the lower one is hotter. Convective cells form and the material carries heat as it moves: cold material sinks and cools the lower part of the fluid, hot material rises and heats the upper part of the fluid (Fig. 5.2). The mean temperature of the convective fluid is the average between the temperatures at the top and the bottom of the box. The Nusselt number (Nu) is a measure of the efficiency of the convective transfer. It is equal to the ratio of the heat flux measured at the top of the convective fluid to the heat flux that would exist if there were no convection ($k\Delta T/b$). Laboratory experiments followed by numerical simulations (e.g. Sotin and Labrosse, 1999) show that the Nusselt number increases with Rayleigh number according to the following expression:

$$\mathrm{Nu} = \frac{q}{k\Delta T/b} = a\mathrm{Ra}^{1/3}, \tag{5.4}$$

where a is a constant that depends on boundary conditions (free slip or rigid boundaries) and on aspect ratio, which is the ratio of the length of the box to its height. One can note by using (5.3) and (5.4) that the convective heat flux (q) does not vary with the thickness of the box (b). The greater the value of the Rayleigh number, the more heat can be extracted by convective motion. Since viscosity increases with decreasing temperature, the efficiency of convective heat transfer decreases as a planet cools down.

The 2D geometry of the convective cells is maintained for Rayleigh numbers up to 10^7. Such a geometry can be easily compared with the Earth where upwelling sheets of hot material move upwards at midocean ridges and downwelling sheets of cold material move downwards at subduction zones (ignoring the angle of subducting slabs, which is not reproduced in these calculations). This analogy between a physical process and observations made Rayleigh–Bénard convection popular to explain the Earth's internal dynamics. However, the Earth exhibits several characteristics that are not compatible with Rayleigh–Bénard convection such as internal heating rate, cooling, temperature-dependent viscosity, plate tectonics, and spherical geometry. So Rayleigh–Bénard convection is a simple model

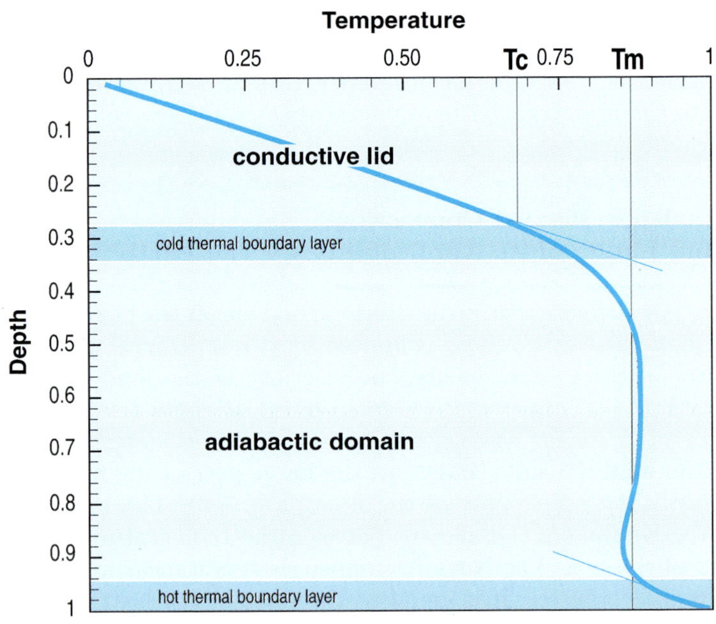

Fig. 5.2. Horizontally averaged temperature vs. depth for 2D thermal convection (from Deschamps and Sotin, 2000). The viscosity of the fluid depends very strongly on temperature. The different layers are: a conductive lid where viscosity is too high for plastic deformation to exist, a cold thermal boundary layer, a well-mixed convecting interior, a hot thermal boundary layer. In the well-mixed interior, the temperature gradient is adiabatic: in the present case it is isothermal because the effect of pressure is not taken into account. In the case of Rayleigh–Bénard convection (constant viscosity), the temperature of the well–mixed interior would be the mean of the upper and lower temperatures (0.5 in nondimensional units)

to explain and describe how a convection process can transfer heat but the temperatures and the heat fluxes deduced from this kind of convection pattern are different from those existing on Earth or on other planets.

The Earth's mantle is heated by the decay of long-lived radiogenic elements. Laboratory experiments and numerical simulations have described the 3D geometry of convection for such a fluid: a large number of cold plumes form at the top of the fluid (Plate 5.2), migrate downwards and vanish. Although one cannot predict where a plume forms, the number of plumes, the average temperature of the convective fluid and the size of plumes can be predicted by scaling laws. For a hot fluid cooled from above, the geometry is the same (Davaille and Jaupart, 1993).

In summary, the Earth's mantle is heated from within, cooled from above and is limited at the CMB by an isothermal hot surface. These conditions were taken into account for constant viscosity fluids (Sotin and Labrosse, 1999). The geometry resembles that of a fluid heated from within with some hot plumes,

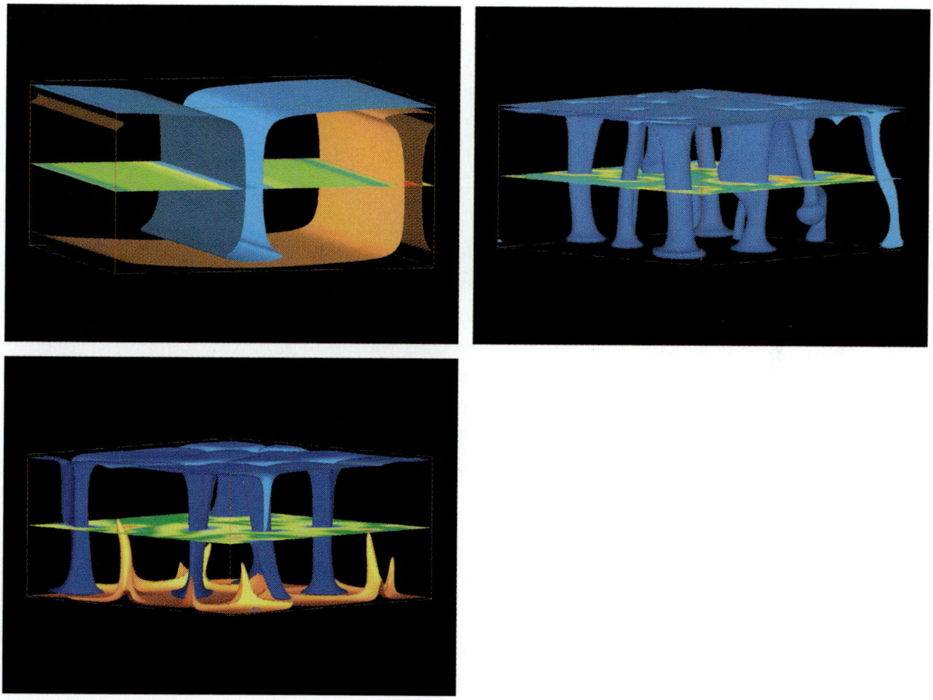

Plate 5.2. Thermal convection models showing three different patterns: Rayleigh–Bénard convection (no internal heating, fixed temperatures on top and bottom) with cold downwelling sheets and hot upwelling sheets, Rayleigh–Roberts convection (internal heating, fixed temperature on top, adiabatic gradient at the lower limit) with only cold plumes, and mixed heating (internal heating, fixed temperatures on top and bottom) with cold plumes and a few faint hot plumes

which form at the base of the layer and fade away before they get to the surface (Plate 5.2). In the following section, the viscosity of the Earth and the effect of temperature dependent viscosity are discussed.

5.2.1.2 The Earth's Viscosity

The main parameter that controls the efficiency of the convective process is viscosity. The larger the viscosity, the smaller the Rayleigh number (5.3) and the smaller the Nusselt number (5.4). The viscosity of the Earth's mantle is determined thanks to the postglacial rebound (e.g. Lambeck and Johnston, 1998). Models describing the postglacial rebound predict that the viscosity of the upper mantle is on the order of 10^{21} Pas and that the viscosity of the lower mantle is 5 to 50 times greater.

Viscosity (μ) is a parameter that controls the amount of strain rate ($\dot{\varepsilon}$) that a solid undergoes when being submitted to differential stresses ($\tau = \sigma - P$ where σ is stress and P is hydrostatic pressure) according to the relation (e.g. Turcotte and Schubert, 1982):

$$\tau = \mu\dot{\varepsilon}. \tag{5.5}$$

One important consideration is that the Earth's mantle is solid, as proved by the values of seismic velocities, but can be considered as a fluid at geological time scales (10s to 100s of Ma) because the small differential stresses (10^5 Pa) in the convective mantle can yield large deformation over long periods of time.

There has been a large number of laboratory experiments that aim at determining the strain rate of minerals and investigating the effect of different parameters including stress, temperature, grain size, water content, pressure, etc. Creep, which is the mechanism accommodating the deformation rate of a solid under constant differential stress, can be controlled by several mechanisms such as diffusion, dislocation, grain-boundary sliding, recrystallization, grain-boundary migration, etc. Depending on the conditions described above, one mechanism becomes predominant compared to the others. The creep rate is often written as an Arrhenius-type law:

$$\dot{\varepsilon} = A\tau^n \exp\left(-\frac{E + P\Delta V}{RT}\right), \tag{5.6}$$

where n is the stress exponent, E is the free activation energy, and ΔV is the activation volume. A material is said to be Newtonian if the stress exponent is equal to 1. In that case, viscosity does not depend on differential stress.

Laboratory experiments are carried out for large differential stresses for the deformation rate to be measurable on laboratory time scales (days, weeks). At such high stresses, the process that controls the creep behaviour is often the displacement of dislocations. But stresses induced by convective instabilities in the mantle are much smaller than the differential stresses reproduced in laboratory experiments. One obvious difficulty is to extrapolate laboratory measurements to natural conditions where the dominant mechanism can be diffusion or grain-boundary sliding. At low stresses, the physical processes that drive plastic deformation often yield a stress exponent close to 1. In that case, the rock has a Newtonian behaviour.

Mantle peridotites are made of minerals such as olivine and pyroxenes. Their viscosity is highly dependent on temperature: activation energy Q ($Q = E + P\Delta V$) ranging between 300 kJ/mole and 500 kJ/mole has been measured in laboratory experiments. The mean temperature of the upper mantle is about 1350 °C and variations of 100 °C yield viscosity variations of a factor of 10.

The fact that viscosity strongly depends on temperature (5.6) has been known for a long time. The role of temperature-dependent viscosity on the geometry and vigour of thermal convection has been investigated since the early 1990s by both laboratory experiments (Davaille and Jaupart, 1993; Manga and Weeraratne, 1999) and numerical models (Moresi and Solomatov, 1995, Grasset and Parmentier, 1998; Deschamps and Sotin, 2000; Choblet and Sotin, 2000). The other parameters (pressure, variable thermal expansion, variable thermal conductivity, etc.) are less influential on viscosity. Although their effect may not be negligible (e.g. Dumoulin et al., 1999), they will be ignored in the present study. The mantle peridotite is supposed to behave like a Newtonian body. Equations (5.5) and (5.6) provide the following expression for viscosity:

$$\mu = \frac{1}{A} \exp\left(\frac{Q}{RT}\right) . \tag{5.7}$$

Because the Earth is cooling down, Earth's viscosity is increasing and it is likely that the intensity of the convection process was much larger at the beginning of the Earth's evolution, 4.56 Ga ago.

5.2.2 Onset of Thermal Convection After Accretion

5.2.2.1 Temperature and Internal Structure of the Earth After Accretion

Models developed by Safronov (1958) assume that planets formed by accreting kilometre-sized (or smaller) planetesimals. Protoplanets grow quickly: in less than a few million years, their radius can be as large as several hundred to a few thousand kilometres. This timing is based on both recent models of planetary accretion in the inner Solar System (Weidenschilling et al., 1997) and geochemical measurements of extinct radioisotopes such as ^{26}Al in eucrites (e.g. Srinivasan et al., 1999). These bodies differentiated very quickly, in less than a few tens of millions years, as is suggested by isotopic studies of meteorites coming from a differentiated asteroid (e.g. Blichert-Toft et al., 2002).

The protoplanets grow by accreting asteroids and comets. As it impacts the planet and increases its volume and mass, this new material is heated because at least one part of the meteorite kinetic energy is transformed into heat. The velocity of the meteorite is greater than or equal to the escape velocity, which increases as the protoplanet grows. It can be shown that the temperature increases as the square of the radius of the growing planet. This kind of structure is gravitationally stable because the cold dense material is below the hot light material. Because accretion is a fast process for bodies up to 2000 km in radius (see above for the evidence), temperature cannot be transmitted by conduction on large distances ($x \cong (\kappa t)^{1/2} \cong 10\,\mathrm{km}$).

When the temperature is high enough, partial melting can develop (Fig. 5.3). Laboratory experiments suggest that two main processes occurred. First, if sulfur or oxygen are present, iron-rich liquid forms at low temperatures. At a pressure below 10 GPa, the melting temperature can be around 1100 or 1200 K, significantly below the melting temperature of peridotites, which is about 1350 K. It seems that for protoplanets about 1500 km in radius, a three-layer structure develops (Fig. 5.4): a cold undifferentiated core (temperature below the iron-alloy melting temperature) overlaid by an iron-rich liquid-pockets-bearing layer, and an outer layer containing both iron-rich melt and silicate melt. This outer layer may be what is generally referred to as magma ocean. Liquid iron may

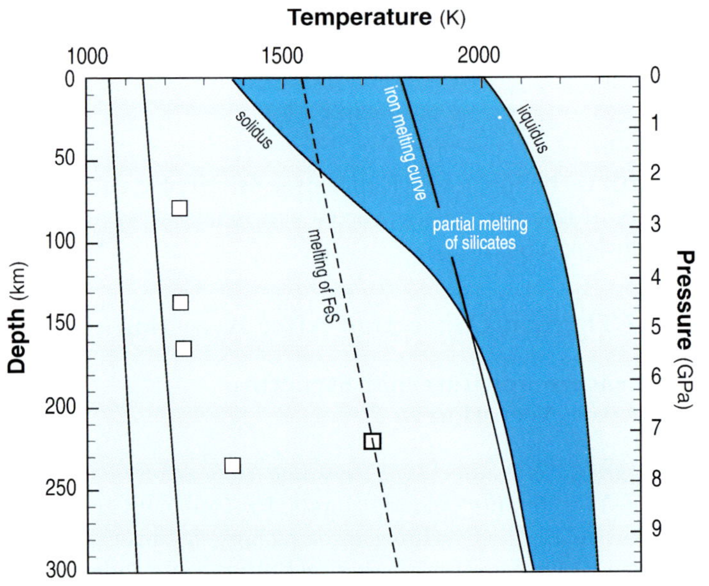

Fig. 5.3. Melting curves for different solids that may exist in the Earth. The liquidus and solidus curves of peridotites come from McKenzie and Bickle (1988). The *dark blue area* is the (pressure, temperature) domain of silicate partial melting. Melting curve of pure iron comes from Guillermet and Gustafson (1984). The *dashed curve* is the melting curve of FeS (Boehler, 1992). *Empty squares* are experimental values for the melting of Fe-FeS alloy (Usselman, 1975). *Straight lines* give melting curves (solidus) in the Fe-S-Ni-O system (Urakawa et al., 1987)

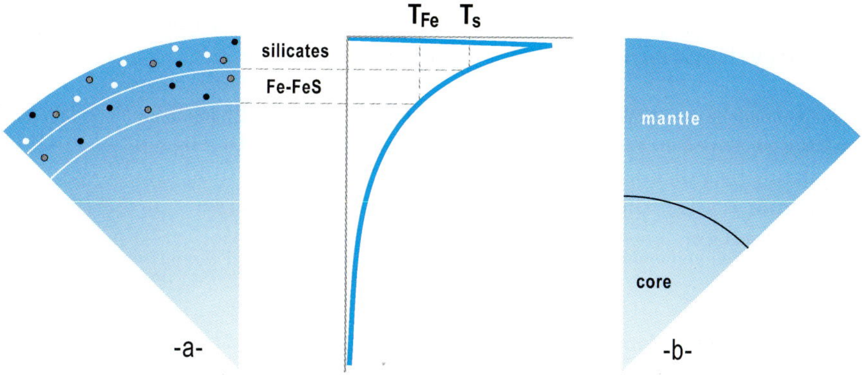

Fig. 5.4. Simple representation of the evolution of a protoplanet about 2000 km in radius. This very simple scenario is based on homogeneous accretion and knowledge of the melting temperatures of rocks and alloys. (a) Model of the internal structure of a protoplanet during accretion and before differentiation (temperature profile on the *right*); (b) Model of the internal structure of the protoplanet after core–mantle differentiation

eventually migrate to the centre of the planet because its density is higher than that of silicates. Lighter silicate melt migrates to the surface where it induces volcanism with important outgassing. Heat transfer is accommodated by this abundant volcanism. This kind of heat transfer is very different from what is known on Earth today but may resemble that of Io where Jovian tidal forces may heat up the shallow layers of the planet allowing partial melting and upwelling of the light silicate melt. The existence of a magma ocean is well documented for the Moon (e.g. Hartmann et al., 1986). Evidence is not so clear for the Earth because the present crust formed after this episode and because plate tectonics reprocessed the rocks that witnessed the early evolution of the Earth. The formation of an iron-rich liquid core must have been fast (less than 50 Ma) according to the ^{182}Hf-^{182}W chronometer (Halliday, 2000).

A process such as that described above would not be efficient to build planets as big as the Earth or Venus in less than 100 Ma (Weidenschilling et al., 1997). One possibility is that Venus and the Earth formed by accreting embryos whose size was similar to present-day Mars. One realistic model for the formation of the Earth–Moon system is that an object of Mars size impacted the Earth (e.g. Hartmann et al., 1986). Such a model is compatible with the idea that planets like Venus and Earth formed by accreting Mars-size bodies. Mars may be a sample of one of the 10 protoplanets that formed the Earth. During such impacts, iron cores, already formed in the 2000 km bodies, would merge to form an even bigger core. This tentative explanation is still lacking accretion models of differentiated bodies, although recent progress has been made for the Earth–Moon system (Canup, 2004, in press).

5.2.2.2 Thermal Convection in a Fluid Cooled From Above: Laboratory Experiments and Numerical Models

When a hot fluid is cooled from above, a cold thermal front propagates downward according to well-known laws of conduction. This problem was solved during the 19th century by Lord Kelvin. The temperature (T) of the fluid can be determined at each time t after the process started and at each depth z according to the following equation:

$$\frac{T - T_\mathrm{s}}{T_\mathrm{m} - T_\mathrm{s}} = \mathrm{erf}\left(\frac{z}{2\sqrt{\kappa t}}\right), \tag{5.8}$$

where T_s is surface temperature, T_m is temperature of the hot fluid, and κ is thermal diffusivity. This cooling increases the density of the outer layers because density increases with decreasing temperature (5.2). Gravitational instabilities can be triggered (Fig. 5.5) if the viscosity is small enough. Both laboratory experiments (Davaille and Jaupart, 1993) and numerical simulations (Moresi and Solomatov, 1995; Grasset and Parmentier, 1998; Choblet and Sotin, 1999) showed that the thermal instabilities occur in a thin layer called the thermal boundary layer (TBL), below a conductive lid where the viscosity is much too high (temperature much too low) for allowing deformation to occur (Fig. 5.2). The temperature difference across the TBL is proportional to a viscous temperature scale, which depends on the viscous characteristics (5.7) of the material. These different studies significantly modify the predictions made by the Rayleigh–Bénard convection pattern for both the temperature of the convective

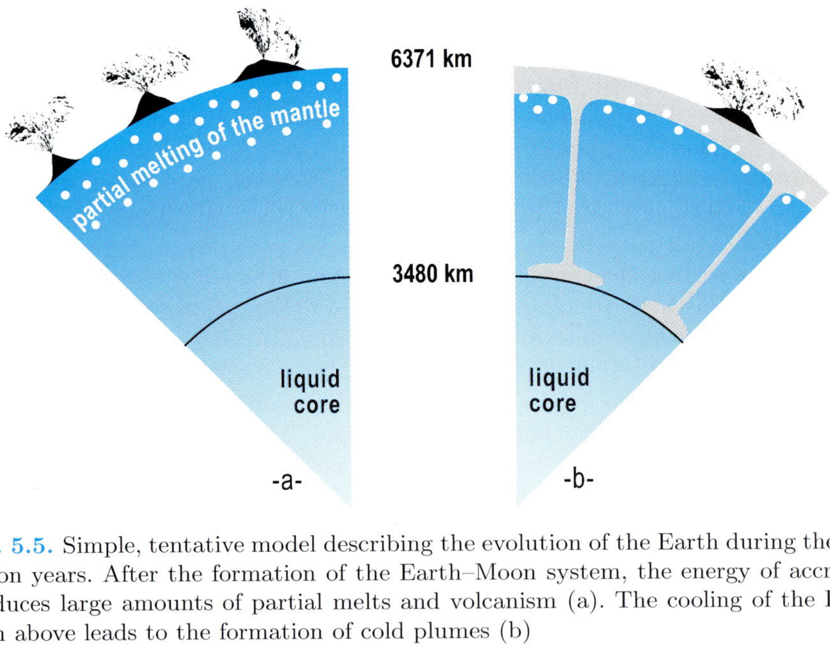

Fig. 5.5. Simple, tentative model describing the evolution of the Earth during the first billion years. After the formation of the Earth–Moon system, the energy of accretion produces large amounts of partial melts and volcanism (a). The cooling of the Earth from above leads to the formation of cold plumes (b)

fluid (T_m in Fig. 5.2) and the heat flux transferred by convection. The nondimensional temperature ($\theta = [T - T_s]/[T_1 - T_s]$ where T_1 is the temperature at the base of the mantle) predicted by Rayleigh–Bénard convection is 0.5, whereas these recent temperature-dependent viscosity models predict a much higher temperature (0.87 in the example shown in Fig. 5.2). On the other hand, the presence of a conductive lid on top of the convective layer (Fig. 5.2) limits the heat flux to a value much smaller than that predicted by Rayleigh–Bénard convection.

5.2.2.3 Application to the Earth: Cooling and the Onset of Convection

The rate of downward propagation of the cold front is controlled by thermal diffusivity ($\kappa = 10^{-6}\,\mathrm{m^2\,s^{-1}}$ for silicate rocks). Recent calculations (Choblet and Sotin, 2000) show that the time required for the onset of convection (first cold plume) is between some tens to some hundreds of Ma, depending on the value of the internal temperature. During this period, the mantle keeps heating up due to the decay of radioactive elements. Until thermal convection is setup, surface heat flux can be expressed by (5.1).

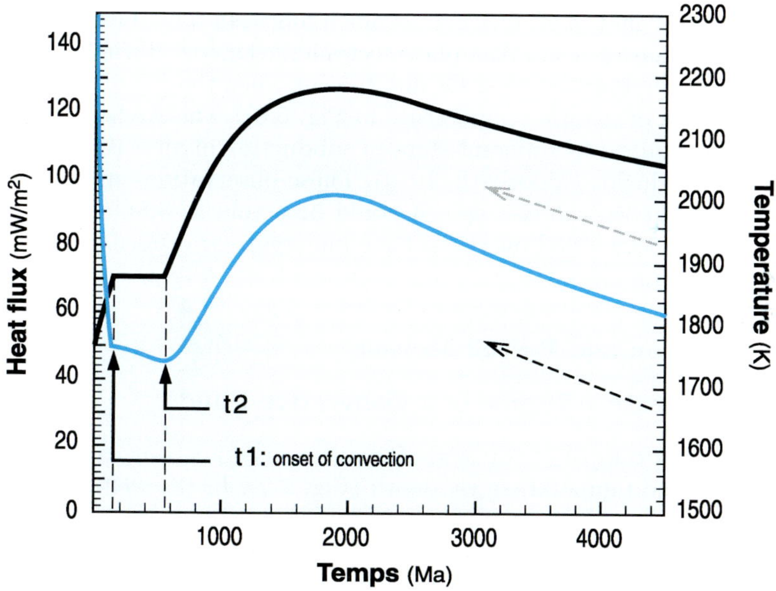

Fig. 5.6. Evolution of upper-mantle temperature (*black curve*) and surface heat flux (*blue curve*) vs. time. This model does not take into account plate tectonics and may be more appropriate for Venus or for Mars. The present-day heat flux ($80\,\mathrm{mW/m^2}$) is indicated by a *grey arrow* assuming that it has decreased during the last billions years. The present temperature is indicated by a *black dash arrow* assuming a cooling of 100 K/Ga. This figure shows the efficiency of plate tectonics to cool down a planet

Numerical experiments have been tested against laboratory experiments and were then applied for conditions closer to those existing on Earth (or other planets). After the onset of convection, there is a transient period (between t_1 and t_2 on Fig. 5.6). During that period, heat transfer is controlled by cold downwelling as described in Plate 5.2 and Fig. 5.5. Isotopic anomalies of ^{142}Nd in the 3.78 Ga old Isua supracrustal belt samples are better described by convection in a conductive-lid regime rather than by convection with plate tectonics (Boyet et al., 2003). After the transient period, scaling laws (e.g. Choblet and Sotin, 2000) can be used to predict evolution of heat flux and mantle temperature (Fig. 5.6). Note that the Earth cools down when the surface heat flux becomes greater than the power produced by the decay of the long-lived radioactive elements. According to Fig. 5.6, this event occurred about 1.8 Ga after accretion.

One important limitation of the present model is that it does not take into account plate tectonics in the Earth's evolution. But no model developed so far is able to simulate plate tectonics and thermal convection for temperature-dependent viscosity. One difficulty is that cooling is dominated by 3D structures, which impose 3D modelling and huge computer requirements to account for both the 3D structures and the temperature-dependent viscosity. Figure 5.6 shows that the present value of heat flux (mantle temperature) is higher (lower) than that predicted by the conductive-lid regime of mantle convection. This is consistent with the fact that plate tectonics is much more efficient than conductive-lid convection to transfer heat. Was plate tectonics triggered during the transient period by the cold downwellings? Or did it occur when the Earth started cooling down (maximum of mantle temperature in Fig. 5.6)? The Archaean greenstone belts can be explained by different steps of subduction of an oceanic crust under these Archaean shields (Moorbath, 1979). These observations suggest that subduction existed at least 2.5 Ga ago. It could be as old as 3.8 Ga if one believes that the oldest rocks found on Earth have undergone transformations linked to plate-tectonics processes.

5.2.3 Convection and Partial Melting

5.2.3.1 Temperature Profile of a Convective Fluid

If one takes the example of Rayleigh–Bénard convection, one can plot the horizontally averaged temperature vs. depth (Fig. 5.2). In the well-mixed domain, the temperature profile is adiabatic. If there is no pressure dependence then temperature is constant (T_m in Fig. 5.2). If pressure is important, then the temperature increase due to pressure can be written:

$$\frac{\mathrm{d}T}{\mathrm{d}P} = \frac{\alpha T}{\rho C_\mathrm{P}} \quad \text{or} \quad \frac{\mathrm{d}T}{\mathrm{d}z} = \frac{\alpha g T}{C_\mathrm{P}}. \tag{5.9}$$

If one assumes that the temperature at the base of the cold thermal boundary layer is 1350 °C, the temperature increase due to adiabatic compression down to the CMB can be as high as 2000 °C.

If the average temperature of the adiabatic domain is homogeneous in the vertical direction (Fig. 5.2) there are few horizontal temperature variations: the temperature in the centre of a cold plume is lower than in the centre of a hot plume. The temperature difference is on the order of the temperature difference across thermal boundary layers, around 100 K.

5.2.3.2 Temperature and Partial Melting of Peridotite: Where Is Partial Melt Located in the Mantle?

The melting temperature of dry peridotite increases with pressure and depth (Fig. 5.3). The experimental melting of mantle peridotite has been extensively studied in laboratories (e.g. McKenzie and Bickle, 1988). It shows that peridotite melting typically generates basalt magmas. The adiabatic temperature gradient is steeper than the dry solidus curve; consequently, mantle peridotite melts during its adiabatic decompression as it ascends towards the surface. Beneath a midocean ridge, analyses of basalt indicates that peridotite melting took place at about 50 km depth. If the mantle temperature is higher, as for instance in hot spots, the melting depth is also greater (100–120 km). As the early Earth was hotter, it can be considered that the depth of mantle melting was also greater at that time.

Another place of melting is located at subducted slabs where hydrated materials are entrained by subduction. The presence of water decreases the melting temperature of peridotite. Both the subducted oceanic crust and the hydrated mantle melt to form an andesitic magma.

5.2.3.3 Coupling Between Thermal Convection and Surface Processes (Volcanism and Tectonics)

Internal dynamics play an essential role not only in shaping the Earth's surface but also in the formation and evolution of the hydrosphere and atmosphere. Partial melting of the mantle followed by melt migration allowed volatile elements (H_2O, CO_2, He, etc.) to migrate to the surface and to be released in the atmosphere. Internal dynamics drive the formation and evolution of the atmosphere.

Internal dynamics is also a key element of the shape of the Earth. Above hot plumes, the edification of a volcano can lead to edifices several kilometres high (i.e. Hawaii, La Réunion). Collisions between continental portions of plates generate mountains. Continental crust is formed by hydrous melting of mantle peridotite or of subducted oceanic crust at the present time. Other processes such as melting at the base of a thick basaltic crust may have occurred during the first billion years of the Earth's evolution. Whatever process is involved, the continental crust is lighter than the oceanic crust. It is also thicker. Consequently, isostatic compensation makes the surface of continental crust higher than that of oceanic crust. As the volume of liquid water is small on Earth ($M_{H_2O} = 3 \times 10^{-4} M_{Earth}$), topographic variations create areas above sea levels and seashores, which may be important for the formation and evolution of life (Commeyras et al., this book).

Finally, volcanism is a potential energy source that can be used by organisms to survive in extreme environments. This is the case at midocean ridges where life develops around and close to black smokers. When the geologists dived to study midocean ridges, they discovered that life can develop without solar energy. It does not mean that life can form at midocean ridges (or where there is submarine volcanism) but that life can evolve and adapt itself to difficult conditions.

5.2.4 Conclusion: a Look at Other Planets

A different view is brought to us by the observation of other planets in the Solar System. The Earth-like body that evacuates the most internal energy seems to be Io. Heat is mostly released by hot-spot volcanism. The Galileo mission obtained a great deal of information about the Galilean satellites and Io in particular. It seems that the temperature of the lavas on Io is above the melting temperature of silicates and that these lavas are made of molten silicates. Some scientists believe that Io's volcanism may resemble that of the Earth some billions of years ago when the internal temperature was higher and the degree of melting greater. The lavas may be equivalent to what is known as komatiites on Earth.

Mars does not present any evidence of plate tectonics although mantle convection is predicted by thermal-evolution models. Models of thermal convection in the conductive-lid regime (no plate tectonics) give a surprising result when they are applied to Mars: if the concentration of long-lived radiogenic elements in Mars' mantle is similar to that of the Earth's mantle, then the temperature profile for Mars would lie in between the solidus and liquidus curves of peridotites at depth between 100 and 200 km. This means that a large volume of the Martian mantle may be partially molten. This model does not seem to make sense since no active volcanism has been detected since Mars has been observed. However, the data that would enable us to solve this question, are missing at the present time. It may be possible that very early in its history, Mars had plate tectonics, which allowed for a fast cooling of the planet (Sleep, 1994). This event may account for the topographic dichotomy between the northern and the southern hemispheres. But there is no evidence of both subduction zones and midocean ridges. Volcanoes on Tharsis Mons were active about 100 Ma ago and possibly later. They suggest that Mars' interior was hot enough to allow partial melting at shallow depths, such that melt could migrate and emplace on the surface. Data on the internal structure of Mars are critical to constrain its evolution models, compare it with the Earth's evolution and to better understand thermal evolution models, which drive the internal dynamics of planets.

On Venus, the lack of active volcanism does not exclude convection. Mountains located at the equator (Aphrodite Terra) and close to the North Pole (Ishtar Terra) suggest that horizontal displacements exist at some depth (underplating) to account for the large topographic variations, which may be explained by differences in the crustal thickness (Solomon et al., 1992).

These examples show that the relationship between internal processes (thermal convection) and surface features are difficult to establish. Observing other planets provides additional constraints that help to model these relationships. Understanding the link between thermal convection and plate tectonics is critical for the understanding of the Earth's evolution. The lack of plate tectonics on other Earth-like planets is an important aspect since models must work not only for the Earth but also for other planets and large satellites.

5.3 The Earth's Magnetic Field

5.3.1 Characteristics of the Magnetic Field

The magnetic dipole that accounts for about 90% of the Earth's magnetic field is not perfectly aligned with the Earth's rotation axis. This angle varies on human time scales. The magnetic field is also composed of non-negligible nondipolar components with magnitudes much smaller than that of the dipole. The remnant magnetic field recorded by magnetic minerals in magmatic rocks (i.e. when lavas crystallize and cool down at temperatures below the Curie temperature) demonstrates that the dipole has changed its orientation many times throughout Earth history. There have been about 40 major magnetic inversions during the last 100 Ma; their cause is not yet perfectly known and is still the subject of active research.

The magnetic field acts as a shield, which protects the planet against ionised solar particles by deflecting them away from the Earth. Without this magnetic shield, the high-energy solar ions would bombard and dissociate the molecules of the Earth's atmosphere. The light molecules, being faster, could more easily escape the Earth attraction. Such a process is also invoked to explain the thin atmosphere of Mars.

Bacteria and other organisms such as birds, turtles and humans use the magnetic field to orient themselves. There is a debate whether the magnetic field is required for life to develop on a planet's surface. Such a question is beyond the scope of the present chapter. During magnetic inversion, the magnetic shield is less efficient, or even not at all efficient, and the Earth's surface loses its protection against solar particles, which should be critical for life to survive. However, paleomagnetic data suggest that the duration of a reversal is so short (Coe et al., 1995) that living beings would survive it.

5.3.2 Magnetic Field and Core Dynamics

5.3.2.1 Convection in the Outer Core

The magnetic field is generated by the displacement of liquid iron in the outer core. Different forces (Coriolis force, thermal convection, electromagnetic forces) generate this motion. Some of the data available for the Earth's core are reported in Fig. 5.7 and a comprehensive discussion about the physical properties of the Earth's core can be found in Poirier (1994).

Numerical models describing the core dynamics are difficult to establish because the characteristics of the liquid outer core (its viscosity for instance) suggest that the thermal boundary layer could be as thin as a couple of metres. Therefore the grid sampling has to be on the order of 1 m to describe correctly the core dynamics and such a grid size is beyond the capabilities of present computers. To overcome this problem, hyperviscous models have been developed (Glatzmaier et al., 1995). In these models, the viscosity of liquid iron is magnified in order to allow for larger spacing between grid points. If these models do not simulate the real core, they still provide some hints on possible processes. For example, inversion of the magnetic field was obtained due to lateral variations of the heat flux imposed by the lower mantle. The relation between magnetic inversions and mantle dynamics such as the formation of hot spots, the presence of cold plumes or subducted slabs arriving at the CMB is still debated. Laboratory experiments are being set up to simulate self-sustained magnetohydrodynamics.

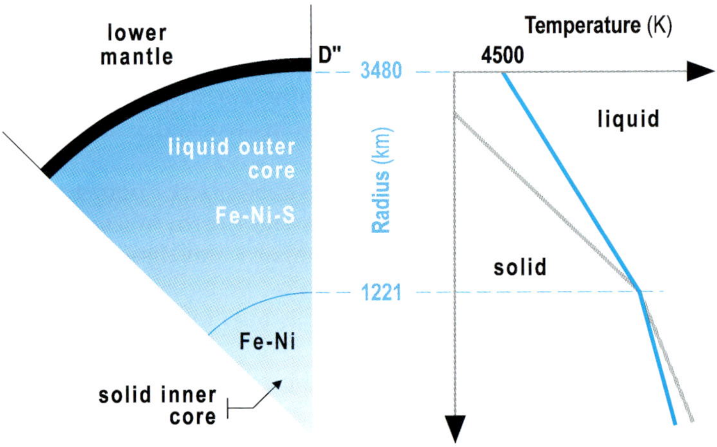

Fig. 5.7. Internal structure of the Earth's core. Temperatures are not well constrained because high-pressure experiments are still lacking. On the *right*, the temperature profile (*blue curve*) can be compared with the melting curve (*grey curve*). The slope of the melting curve is different in the solid core and in the outer liquid core because the outer liquid core gets enriched in light elements as the inner core freezes out. The freezing of the solid inner core is explained in the text

5.3.2.2 Solidification of the Inner Core

The currently favoured hypothesis is that, at the end of the Earth accretion period, the whole core was in a liquid state. The effectiveness of mantle convection controls the cooling rate of the core. As the core cooled from above, gravitational cold instabilities formed at the CMB. The number of cold plumes, which can form, is controlled by the amount of heat that the mantle can transfer. The global process looks simple until the temperature at the centre of the core becomes equal to the liquidus of the Fe–FeS (if S is the light element). Since Fe–FeS forms a solid mixture (Fig. 5.8), and if the amount of sulfur is smaller than at the eutectic composition, pure iron crystallizes and the remaining liquid is enriched in sulfur. The higher the sulfur content, the lower the liquidus temperature (Fig. 5.8); consequently, as solid core segregates, sulfur concentrates in the liquid core such that its crystallizing temperature (liquidus) decreases. In other words, the core must cool down even more as the solid core segregates. There are few laboratory experiments at pressures relevant to the Earth's core (135 GPa $< P <$ 350 GPa). But one can extrapolate the results obtained at smaller pressures up to 50 GPa (Boehllet 1986; Boehler et al., 1990). It seems that the solid core forms very slowly since only 4.5% of the core is solid at the present time (the radius of the solid core is 1220 km compared to 3480 km for the core). An alternative explanation is that the core started to cool down only very recently (Labrosse et al., 1997).

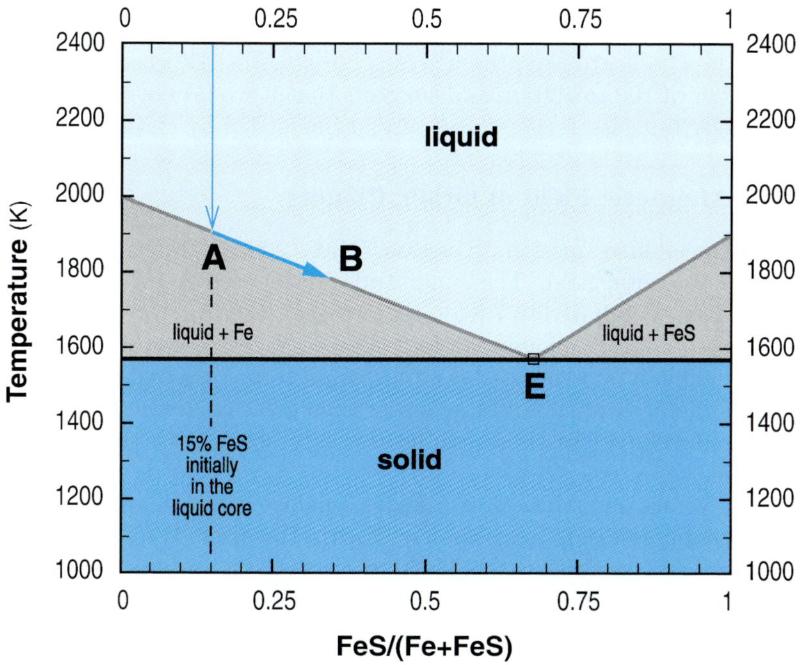

Fig. 5.8. Melting temperature of Fe–FeS alloy at P = 15 GPa according to experimental data (Fig. 5.3). For a given composition, the cooling of the fluid will form the first solid iron at point A. As the temperature decreases, the liquid gets enriched in sulfur (from A to B). E is the eutectic composition that may be eventually reached if the planet cools down sufficiently

Chemical convection may play a role in the dynamics of the liquid outer core. As the inner core crystallizes, sulfur is left over. As it is lighter than iron, it is redistributed in the liquid core. Consequently, the gravitational potential is changed and part of it is turned into heat by viscous dissipation. The presence of a solid core may also affect the geometry of iron circulation. If the solid core is too big, it would be difficult to have circulation of liquid iron parallel to the rotation axis of the Earth. The presence of a solid inner core may play an important role for the generation of a magnetic field. Is it required for a dipolar dynamo to exist? The answer is not yet known.

5.3.2.3 Core–Mantle Interaction

The cooling rate of the core is controlled by the cooling rate of the mantle. The heat flux at the CMB must be high enough to power the dynamo, the energy of the dynamo being transformed into heat by the Joule effect in the outer layers of the core. Typical values of 2 to 4 TW are often given in the literature (e.g. Labrosse, 1997). At the CMB, there is continuity of both temperature and heat flux. On the mantle side, the CMB can be considered as an isothermal boundary since the liquid nature of the core does not allow for large lateral temperature variations in the core.

Seismic data indicate the presence of a layer named D″. Its thickness would be on the order of 100 to 200 km and suggests that it would not be just a thermal boundary layer but also a chemical boundary layer fed by subducting slabs.

5.3.3 The Magnetic Field of Other Planets

Results of the Galileo mission (Kivelson et al., 2002) showed that Ganymede has its own magnetic field. The magnitude of the centred dipole is equal to $1.33 \times 10^{20}\,\mathrm{A\,m^2}$ (compared to $7.84 \times 10^{22}\,\mathrm{A\,m^2}$ for the Earth). The radius of the iron core is estimated at around 150 km. This small size could demonstrate that a magnetic field can be generated even for small iron cores and that other small bodies of the Solar System could generate their own magnetic field. Numerical models to be developed for the Earth have also to account for the magnetic field of Ganymede.

On both Venus and Mars, if a magnetic field exists, its amplitude is more than 4 orders of magnitude lower than on Earth. However, Mars Global Surveyor found magnetic anomalies in the Martian crust. This implies that Mars once had a magnetic field (Acuña et al., 1999). However the anomalies are difficult to interpret and the origin of the field (dyke intrusions, sills, hydrous alteration, etc.) is still under debate.

The planned future Mercury missions Messenger and Bepi Colombo (NASA and ESA, respectively) should provide additional information about another planet with a magnetic field. These data may help to understand how the magnetic field works and to address the problem of its existence during the first billion years of the Earth's history.

Conclusions

The internal structure of the Earth is well known thanks to seismic data. Laboratory experiments and numerical models continuously refine the knowledge of its internal dynamics. A number of challenging topics remain, including the relationship between plate tectonics and thermal convection, the fate of the magnetic field and the Earth's thermal evolution since the end of the accretion period. Plate tectonics have destroyed most of the rocks, thus erasing the geological record, which witnessed the first billion years of the planet. Clues about this period of time may thus come from the study of other planets such as Mars, Venus and the icy satellites of the giant planets. Comparison between the different planets helps to understand the physical and chemical conditions that prevailed when life first occurred on Earth.

Acknowledgement

Thanks go to the reviewers who corrected the first version of this paper. Illustrations were drawn by Alain Cossard.

References

Acuña, M.H. et al., Global distribution of crustal magnetization discovered by the Mars Global Surveyor MAG/ER experiment, *Science*, **284**, 790–793, 1999.

Blichert-Toft, J., Boyet, M., Télouk, P., Albarède, F., ^{147}Sm-^{143}Nd and ^{176}Lu-^{176}Hf in eucrites and the differentiation of the HED parent body, *Earth Planet. Sci. Lett.*, 167–181, 2002.

Boehler, R., The phase diagram of iron to 430 kbar, *Geophys. Res. Lett.*, **13**, 1153–1156, 1986.

Boehler, R., Melting of the Fe-FeO and the Fe-FeS system at high pressure: constraints on core temperature, *Earth Planet. Sci. Lett.*, **111**, 217–227, 1992.

Boehler, R., Von Bargen N., Chopelas A., Melting, thermal expansion, and phase transitions of iron at high pressures, *J. Geophys. Res.*, **95**, 21731–21736, 1990.

Bolt, B., *Inside the Earth*, Library of Congress Cataloging in Pubication Data, 191 pp. 1982.

Boyet, M., Blichert-Toft, J., Rosing, M., Storey, M., Télouk, P., Albarède, F., ^{142}Nd evidence for early Earth differentiation, *Earth Planet. Sci. Lett.*, **214**, 427–442, 2003.

Canup, R.M., Simulations of a Late Lunar-Forming Impact, *Icarus*, in press.

Choblet, G., Sotin, C., 3D thermal convection models of fluids with variable viscosity: comparison between secular cooling and volumetric heating, *Phys. Earth Planet. Int.*, **119**, 321–336, 2000.

Coe, R.S., Prevot, M., Camps, P., New evidence for extraordinarily rapid change of the geomagnetic field during a reversal. *Nature*, **374**, 687–692, 1995.

Davaille, A., Simultaneous generation of hot-spots and superswells by convection in a heterogeneous planetary mantle, *Nature*, **402**, 756–760, 1999.

Davaille, A., Jaupart, C., Transient high-Rayleigh number thermal convection with large viscosity variations, *J. Fluid Mech.*, **253**, 141–166, 1993.

Deschamps, F., Sotin, C., Inversion of 2D numerical convection experiments for a strongly temperature-dependent viscosity fluid, *Geophys. J. Int.*, **143**, 204–218, 2000.

Deschamps, F., Trampert, J., Mantle tomography and its relation to temperature and composition, *Phys. Earth Planet. Inter.*, **140**, 277–291, 2003.

Dumoulin, C., Douin, M-P., Fleitout, L., Heat transport in stagnant lid convection with temperature and pressure dependent Newtonian and non-Newtonian rheology, *J. Geophys. Res.*, **104**, 12 759–12 778, 1999.

Dziewonski, A.M., Anderson, D.L., Preliminary reference Earth model, *Phys. Earth Planet Int.*, **25**, 297–356, 1981.

Glatzmaier, G.A., Roberts, P., A three-dimensional self-consistent computer simulation of a geomagnetic reversal, *Nature*, **377**, 203–208, 1995.

Grasset, O., Parmentier, E.M., Thermal convection in a volumetrically heated, infinite Prandtl number fluid with strongly temperature-dependent viscosity: implications for planetary thermal evolution, *J. Geophys. Res.*, **103**, 18 171–18 181, 1998.

Halliday, A.N., Hf-W chronometry and inner Solar System accretion rates, in From dust to terrestrial planets (W. Benz et al., eds.), *Space Sci. Rev.*, **92**, 355–370, 2000.

Hartmann, W.K., Phillips, J., Taylor, G.J. (eds.), *Origin of the Moon*, Lunar and Planetary Institute, Houston, TX, 1986.

Head, J., Crumpler, L.S., Aubele, J., Guest J.E., Saunders, R.S., Venus volcanism: classification of volcanic structures, associations, and global distribution from Magellan data, *J. Geophys. Res.*, **97**, 13 153–13 197, 1992.

Karato, S.-I., Karki, B.B., Origin of lateral variation of seismic wave velocities and density in the deep mantle, *J. Geophys. Res.*, **106**, 21 771–21 783, 2001.

Labrosse, S., Poirier, J-P., LeMouel J-L., On the cooling of the Earth's core, *Phys. Earth Planet.* Int., **99**, 1–17, 1997.

Lambeck, K., Johnston, P., The viscosity of the mantle: evidence from analyses of glacial-rebound phenomena. In *The Earth's Mantle: Composition, Structure and Evolution*, 461–502, Jackson (ed.), Cambridge University Press, Cambridge, England, 1998.

Lehmann, I., P, Bureau Central seismologique international, Series A, *Travaux Scientifiques*, **14**, 88, 1936.

Manga, M., Weeraratne, D., Experimental study of a non-Boussinesq Rayleigh-Bénard convection at high Rayleigh and Prandtl numbers, *Phys. Fluid*, **11**, 2969–2976, 1999.

Matson, D.L., Johnson, T., Veeder, G.J., Blaney, D.L., Davies, A.G., Upper bound on Io's heat flow, *J. Geophys. Res.*, **106**, 33 021–33 024, 2001.

McKenzie, D., Bickle, M.J., The volume and composition of melt generated by extension of the lithosphere, *J. Petrology*, **29**, 625–679, 1988.

Moorbath, S., Les plus anciennes roches terrestres et la croissance des continents, In *La Dérive des Continents*, Pour la science, Belin (eds.), 1979.

Moresi, L.N., Solomatov, V.S., Numerical investigation of 2D convection with extremely large viscosity variations, *Phys. Fluids*, **7**, 2154–2162, 1995.

Poirier, J-P., Physical properties of the Earth's core, *C.R. Acad. Sci. Paris*, 318, 341–350, 1994.

Reese, C.C., Solomatov, V.S. and Moresi, L-N., Non-newtonian stagnant lid convection and magmatic resurfacing on Venus, *Icarus*, **139**, 67–80, 1999.

Safronov, V. S., The growth of terrestrial planets, *Vopr. Kosmog.* **6**, 63, 1958.

Sleep, N., Martian plate tectonics, *J. Geophys. Res.*, **99**, 5639–5655, 1994.

Solomon, S. et al., Venus tectonics: an overview of Magellan observations, *J. Geophys. Res.*, **97**, 13 199–13 255, 1992.

Sotin, C., Parmentier, E.M., Dynamical consequences of compositional and thermal density stratification beneath spreading centers, *Geophys. Res. Lett.*, **16**, 835–838, 1989.

Sotin, C., Parmentier, E.M., On the stability of a fluid layer containing a univariant phase transition: Application to planetary interiors, *Phys. Earth Planet. Int.*, **55**, 10–25, 1989.

Sotin, C., Labrosse, S., Thermal convection in an isoviscous, infinite Prandtl number fluid heated from within and from below: applications to the transfer of heat through planetary mantles, *Phys. Earth Planet. Int.*, **112**, 171–190, 1999.

Srinivasan, G., Goswani, J.N., Bhandari N., ^{26}Al in eucrite Piplia Kalan: plausible heat source and formation chronology, *Science*, **284**, 1348–1350, 1999.

Tobie, G., Mocquet, A., Sotin, C., Tidal dissipation within large icy satellites, submitted to *Icarus*, 2004.

Turcotte, D., Schubert, G., *Geodynamics*, John Wiley and Sons (eds.), 1982.

Urakawa, S., Kato, M., Kumazawa, M., Experimental study on the phase relations in the system Fe-Ni-O-S up to 15 GPa, In: M.H. Manghani and Y. Syono (eds.), *High Pressure Research in Mineral Physics*, 95–111, Terrapub, Tokyo, 1987.

Usselman, T.M., Experimental approach to the state of the core: Part 1. The liquidus relations of the Fe-rich portion of the Fe-Ni system from 30 to 1000 kb, *AM J. Sci.*, **275**, 278–290, 1975.

Vine, F.J., Matthews, D.H., Magnetic anomalies over ocean ridges, Nature, **199**, 947–949, 1963.

Weidenschilling, S. J., Spaute, D., Davis, D. R., Marzari, F., and K. Ohtsuki, Accretional Evolution of a Planetesimal Swarm: 2. The Terrestrial Zone, *Icarus*, **128**, 429–455, 1997.

6 The Geological Context for the Origin of Life and the Mineral Signatures of Fossil Life

Frances Westall

Life arose in a very different environment compared to that of the modern Earth. Unfortunately we have very little direct evidence of the conditions of the young Earth because of plate-tectonic activity, combined with massive resurfacing as a result the late stage (4.0–3.85 Ga) bolide impacts, have basically eliminated the first 500 million years of Earth history. Moreover, of the small amount of material remaining from the time interval between 4.0–3.5 Ga, the majority of the fragments have been severely altered by metamorphism, making deciphering of the early rock record difficult (but not impossible). The oldest, best-preserved supracrustal material is found in the 3.5–3.2 Ga old Barberton and Pilbara greenstone belts in South Africa and Australia, respectively. This is one billion years after the formation of the Earth. It is in sediments from the latter two belts that we find the first morphological signs of fossil life. From what information can be gleaned from this small database, I will present a resume in this chapter of the early evolution of the Earth in order to place the appearance and early evolution of life in its geological context. I will then discuss the biomineral signatures of early life and, finally, make brief reference to the value of these studies (geological context and biomineral signatures) in the search for extraterrestrial life.

The geological periods of interest for the appearance and early evolution of life include the Hadean Epoch (4.55–4.0 billion years ago [Ga – giga years]) that saw the formation of the Earth, its early geological evolution and, perhaps, even the origin of life; and the Early Archaean Epoch (4.0–3.2 Ga) from which we have the first fossil evidence of life (Fig. 6.1).

6.1 Geological Evolution of the Early Earth

Interpretations concerning the early evolution of the Earth are extremely difficult because of the lack rocks dating from the first 500 million years of Earth's history, as noted above. Information relating to the evolution of the early Earth is derived from the following sources:

- ancient zircon crystals that have been reworked and appear in younger rocks,
- ancient metamorphic terrains, the oldest of which occurs in Canada at Acasta and is about 4 Ga (Bowring and Housh, 1995; Bowring and Williams, 1999),

- ancient metamorphosed supracrustal rocks from Isua in S.W. Greenland (Appel and Moorbath, 1999),
- well-preserved supracrustal rocks younger than 3.5 Ga occurring in the Pilbara region of NW Australia and the Barberton Mountain Land in E South Africa,
- studies involving the modeling of hypothetical atmosphere/ocean, crust and mantle evolution scenarios, and
- information from other planets (geology, atmospheres, etc.) that is of relevance to the early Earth (comparative planetology). Given the meager quantity of actual data from the first billion years of Earth history, there is much debate concerning the interpretations that can be derived from the database.

The following is a brief overview of the main hypotheses regarding geologic, oceanic and atmospheric evolution.

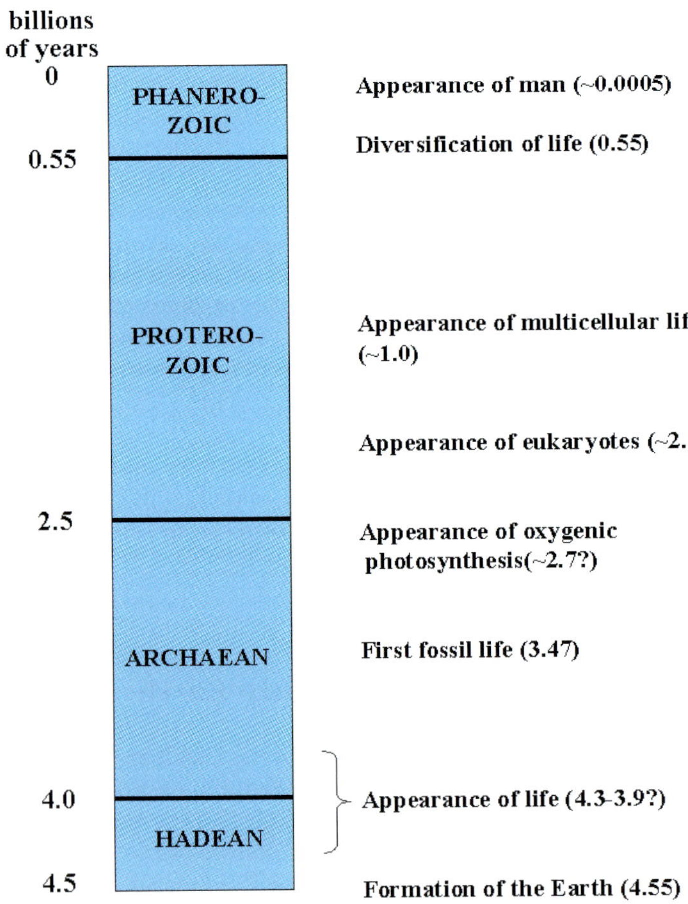

Fig. 6.1. Geological timescale

6.1.1 Crust

6.1.1.1 Early Fractionation of the Earth's Crust

For most of its history, at least since about 4.4–4.2 Ga, it appears that the early Earth has been a water-covered planet. This inference was made on the basis of the existence of zircon crystals that formed 4.4–4.2 Ga ago but now occur in younger sedimentary rocks in western Australia (Wilde et al., 2001; Mojzsis et al., 2001)[1]. Zircon crystals form by fractionation of magmatic crustal rocks that have been recycled (melted and fractionated) in the upper mantle by tectonic processes. Small amounts of zircon crystals can be formed in dry conditions, that is, without significant water: the crystals occurring in lunar crust are an example of zircons formed by this process (Nyquist and Shih, 1992). However, water is necessary for large amounts of fractionation, especially the kind of fractionation leading to the formation of granite-rich continental crust and, consequently, for the formation of large amounts of zircon crystals, which is the case on Earth (see Campbell and Taylor, 1983: No water, no granites – no granites, no continents[2]). Another indication that there was water on the surface of the Earth in Hadean times (4.5–4.0 Ga) comes from the analyses of the oxygen isotopes in the ancient zircon crystals. The results of these analyses indicate that the original magmatic source rock, that was melted in the mantle and then fractionated to produce the zircon crystals, had undergone low-temperature aqueous alteration by hydrothermal processes (Wilde et al., 2001). Although the interpretation that water must have been present on the Hadean Earth in large amounts has been questioned (e.g. Kamber et al., 2001; Halliday, 2001), the large number of Hadean age zircons occurring in younger crust does seem, nevertheless, to suggest that there was significant recycling and fractionation of water-rich oceanic crust to form continental crust. Further evidence for mixing of the crust and the upper mantle through recycling is the fact that the upper mantle appears to have reached its present state of oxidation by about 4.3 Ga (Tolstikhin and Marty, 1998).

The Hadean continents may have been represented by thickened areas of basaltic crust, similar to Iceland or the oceanic plateaux that occur today

[1] Radiogenic dating methods are used for dating material of this age. The radioactive "mother/daughter" element pairs used are those with long half-lives, such as ^{235}U–^{207}Pb and ^{238}U–^{206}Pb. These elements are abundant in highly resistant crystals, such as zircons ($ZrSiO_4$). They are so resistant to degradation that, although the rocks in which they originally formed have been completely eroded away, the ancient crystals remain as a component of younger sediments. They can be reworked time and time again, as in the example of 4.4 Ga old zircons occurring in the 3.0-Ga Narryer Gneiss Terrain of Western Australia (Wilde et al., 2001).

[2] Granites are magmatic rocks consisting in an assemblage of quartz, plagioclase and alkali-feldspar, as well as of variable amounts of micas; they are a fundamental component of continental crust. They formed by melting of hydrated crust, like the zircon crystals, which are very common in granites.

on oceanic crust. They would have contained granitoid bodies[3]. Some of this thickened oceanic crust may even have been subaerially exposed and weathered in the same way as Iceland protrudes from the Mid Atlantic Ridge system (Kramers, 2003). However, as noted above, there are no physical remnants of this ancient crust apart from zircon grains that now occur in younger rocks.

Why is there so little continental crustal material remaining from the Hadean and early Archaean Earth? A number of hypotheses have been put forward:

- major continental crustal formation started only after about 4.00 Ga (Kramers, 2003),
- only a small volume of continental material was initially produced (Kamber et al., 2001),
- the early continents were destroyed by subduction into the mantle because they were not underplated by lighter, more buoyant material, such as that representing the stabilising keel under modern continental crust (Kamber et al., 2001), and
- all earlier crust was completely mixed with basaltic lavas that were formed as a result of the massive impacts that occurred in the pre 3.85 Ga period and that resurfaced the planet (Arndt and Chauvel, 1991).

The geochemical information regarding early crustal formation is contradictory. Limited production of Hadean continental crust is indicated by the lack of depletion of certain elements in the mantle, according to measurements of the Nd and Hf isotopes and the Nd/Th ratios in one study (Kamber et al., 2001). However, modelling by Arndt and Chauvel (1991) suggests that large-scale mixing of the crust by large impacts would effectively have eliminated any isotopic evidence for the early formation of continental crust. Other studies show that more than 10% of the volume of the present-day crust was extracted before 4.0 Ga (McCulloch and Bennet, 1993; Jahn, 1997; Vervoort et al., 1996).

The oldest supracrustal[4] remnants (the $>$ 3.7 Ga Isua greenstone belt in Greenland (Fig. 6.2), and the 3.5–3.2 Ga greenstone belts of Barberton in South Africa and the Pilbara in Australia) appear to have survived tectonic recycling in the Early Archaean period because a stabilising keel underplated them. The formation of silica-rich granitic material (less dense and therefore more buoyant) in the younger portions of the Barberton and Pilbara greenstone belts (younger than 3.2 Ga), as a result of the continued fractionation of crustal material, also contributed to their survival (Martin, 1994; Eriksson, 1995). As noted

[3] Granitoids (Tonalite, Trondhjemites, Granodiorite and Granite) are rocks that belong to the same family as granites, they are quartz-rich but their relative proportions of plagioclase and alkali-feldspars are variable. Hadean and Early Archaean periods are characterised by genesis and emplacement of granitoids (TTG = Tonalite, Trondhjemite and Granodiorite).

[4] Supracrustal refers to rocks emplaced at the surface of the crust. They can be volcanic and/or sedimentary rocks in contrast with a simple igneous crust, such as the lunar mare.

above, without the presence of a stabilising keel, any fractionated Hadean crustal material would have been recycled back into the mantle through convection, as happened on the Moon (Kamber et al. 2001).

6.1.1.2 High Heat Flux on the Early Earth

The composition of the most ancient crustal rocks testifies to the high heat flux present on the early Earth. Ultramafic komatiitic basalts, typical of very hot mantle conditions (Arndt, 1994), are a common component of Early Archaean volcanic rocks, as are tonalite-trondjehmite granites. Moreover, hydrothermal activity in the Early Archaean appears to have been pervasive. This activity is manifested, for example, in the form of structurally controlled hydrothermal veins and springs (Nijman et al., 1999), as well as by the widespread replacement of volcanic and volcaniclastic protoliths by hydrothermal fluids, including veining and brecciation (Paris et al., 1985). Lowe and Byerly (1999) and Knauth and Lowe (2003) alternatively suggest that the hydrothermal activity on the early

Fig. 6.2. Ancient supracrustal rocks from the Isua greenstone belt in SW Greenland. (**a**) Basalt pillow lava with the pillows (e.g. *arrow*) elongated due to deformation. (**b**) Pillow lava breccia with the fragmented basalt pillows (*arrow*) cemented by hydrothermal chert. (**c**) Alternating chert (C) and amphibolite (A) layers. (**d**) Conglomerate with pebbles elongated due to deformation

Earth was not such a pervasive phenomenon but that the warm ocean waters were, on the other hand, strongly influenced by hydrothermal effusions.

The aspect of heat dissipation on a hotter Earth with a thicker crust has provoked a number of controversies. More vigorous recycling would be one method of heat dissipation. In terms of tectonic regimes, Lowe (1999) concludes that Early Archaean tectonics were very similar to those of the present day, on the basis of volcanic bimodality (basic-felsic composition), as well as evidence for horizontal crustal deformation and strike-slip faulting, as occurs during subduction.

6.1.2 Oceans and Atmospheres

6.1.2.1 The Origin of the Volatiles

The Earth's first atmosphere was probably a reducing mixture of H_2, H_2S, CO_2, H_2O and rare gases. However, a comparison of the abundances of rare gases still present in the atmosphere today with the cosmic abundances of these gases indicates that the Earth must have lost its primordial atmosphere (Kasting, 1993). This first atmosphere was probably lost through erosion by the solar wind as well as by impact erosion, for instance, by the massive impact with a Mars-sized planet that led to the formation of the Moon. The first atmosphere was replaced by a slightly reducing to neutral atmosphere of CO_2, H_2O, and H_2S (with minor amounts of other gases such as CH_4, CO, N_2) (Walker, 1985; Kasting, 1993; Nisbet, 1995). The second atmosphere was produced by outgassing, as well as by volatiles imported via comets, meteorites and micrometeorites. Study of the noble gas Xe suggests that by 4.35 Ga, the amount of H, C and N in the surface reservoirs of the Earth was much the same as it is today (Kramers, 2003). However, this latter study also indicates that a significant part of the present atmosphere is a remnant of loss, rather than the product of later degassing.

There has been much discussion concerning the origin of the Earth's volatile elements (see also chapter by Pinti, this volume). Drake and Righter (2002) have summarised the main hypotheses as follows:

- The planetesimals from which the Earth accreted were essentially dry and most of the volatiles were brought in during the Hadean (the "late veneer") via volatile-rich carbonaceous chondrite meteorites (e.g. Morbidelli et al., 2000).
- Wet accretion with little or no exogenous volatile source (e.g. Drake and Righter, 2002). Dauphas et al. (2000) and Robert (2001) estimate that between 50–90% of the volatiles originated from these "wet" planetesimals.
- A predominantly cometary origin for the volatiles (e.g. Delsemme, 1998; Owen and Bar-Nun, 2000). One problem with this theory is that the isotopic ratio of D/H that has been measured in (only) three comets is twice that of the Earth's oceans (Robert, 2001). Owen and Bar-Nun (2000) suggest, however, that the cometary component would only have represented about 30% of the volatile source (this estimate is only 10% according to Dauphas et al., 2000).

A further hypothesis proposed by Maurette et al. (2001) is that most of the volatiles were imported by micrometeorites in the $< 500\,\mu m$ size region. The D/H ratios of the micrometeorites fit perfectly to those in the present-day oceans (Maurette et al., 2000, 2001).

The early history of the Earth was strongly influenced by late-stage cataclysmic impact events. Two scenarios have been proposed, both based on comparison with the lunar cratering record but with different interpretations. One hypothesis posits a gradual decline in the intensity and size of impactors between the time of formation of the Earth and about 3.8 Ga (Hartmann et al., 2000). The second hypothesis proposed that the late-stage impacts were concentrated in a period of $20-200$ my between 4.0–3.85 Ga (Ryder et al., 2000; Kring and Cohen, 2000; Valley et al., 2002). One of the pre 3.85 Ga impactors produced the giant Aitken Basin at the south pole of the Moon. It has been calculated that between zero and 6 extraterrestrial bodies (bolides) of the size of that which produced the Aitken Basin, or even larger, probably hit the Earth in the Hadean. At a minimum, impacts of this size would have evaporated the upper 200 m of the Earth's oceans and, at maximum, they would have completely evaporated all the surface volatiles and sterilised the surface of the Earth, (Sleep et al., 1989). Geochemical analysis of the remains of the impactors from the Apollo landing sites on the Moon indicate that they originated from differentiated asteroids, but not those that are particularly carbon rich, nor from comets. These conclusions appear to support the accretion of a wet Earth with a "late veneer" addition from carbonaceous chondrite meteorites or comets, or they could support the Maurette hypothesis of micrometeorite importation. A further implication of the hypothesis regarding a short, relatively late "cataclysm" is that the Earth was very likely to have been resurfaced by the attendant destruction and volcanic activity (Kring and Cohen, 2002). The fact that there are no supracrustal remnants older than 3.75–3.8 Ga may reflect a period of pre-existing global resurfacing (N.B. these rocks do contain, however, geochemical evidence for the presence of pre-existing evolved crustal rocks).

6.1.2.2 Faint Young Sun Paradox

An important point to take into consideration with respect to the composition of atmospheric gases and the presence of liquid H_2O at the Earth's surface is the hypothesis that the early Sun was about 25% weaker than it is today (Sagan and Chyba, 1997). In order to compensate for the weaker radiation from the Sun so that water could remain liquid at the surface of the Earth, either the concentration of CO_2 in the atmosphere had to have been much higher (0.2 bars), or the atmosphere had to have contained another greenhouse gas, such as CH_4 (Nisbet and Sleep, 2001; Pavlov et al., 2001). However, if the CH_4 originated from volcanic outgassing, this would indicate reduced upper-mantle conditions, but this was clearly not the case (Delano, 2001). Kasting et al. (2001) hypothesise, on the other hand, a microbial origin (from methanogenic bacteria) for the methane, at least during the Late Archaean. This hypothetical hydrocarbon smog could have protected the Earth's surface (and the biota) from UV radiation.

6.1.2.3 Fate of the CO_2

It appears that the atmosphere of the early Earth was rich in CO_2. If this was the case, what happened to the CO_2? The conventional theory is that the gas, in the form of H_2CO_3, was removed by reaction with exposed continental rocks to produce carbonate-alteration products that were, in turn, transported to the sea through rain and river runoff for final burial in deep-sea sediments. Ultimately, the latter would be removed into the mantle by subduction of the sediment-coated oceanic crust. However, substantial amounts of subaerial continental landmasses are necessary for this weathering process. As noted above, on the young Earth, if they did exist, continental landmasses were few and far between. Moreover, the older portions of the 3.5–3.2 Ga Barberton and Pilbara greenstone belts clearly show that much of the (proto) continental landmass was actually submerged. Certainly, subaerial weathering of volcanic islands and exposed portions of the incipient "midocean ridges" would have contributed to the removal of CO_2 from the atmosphere by weathering. This would have been especially efficient given the high ambient temperatures ($> 85\,^{\circ}$C, Kasting, 1993; 50–80 °C Knauth and Lowe, 2003) and the highly reactive nature of the volcanic rocks. However, in the absence of large extents of exposed landmass, probably the most important sink for CO_2 was provided by subaqueous carbonate alteration (carbonatisation) of newly formed oceanic basalt. Dissolved CO_2 would undergo chemical reaction with the vast amount of freshly formed volcanic material produced by the hotter, and volcanically more active, Earth. This phenomenon has been described from the Early Archaean greenstone belts in the Pilbara, Barberton and Isua (de Wit and Hart, 1993; Rosing et al., 1996). The carbonates produced by this process would have been recycled together with the oceanic crust, thus leaving little record. Grotzinger and Kasting (1993) had suggested that the lack of carbonate deposits in the Early Archaean formations may have been due to a lack of Ca ions in solution, the available Ca ions being taken up by sulfate in the chemical hierarchy of precipitation. However, large deposits of marine sulfates are not known from the Early Archaean rock record.

6.1.2.4 Oxygen on the Early Earth

At present, the amount of free oxygen in the atmosphere is about 21%. The amount of free oxygen in the atmosphere during the earliest part of Earth's history is, however, much debated. Ohmoto (1997, 1999), for instance, proposes that there were significant levels of oxygen on the early Earth. Some oxygen would have been produced in the upper atmosphere as a result of photodissociation of H_2O molecules; indeed, the presence of banded iron formation (BIFs) deposits consisting of the oxidised, ferric form of iron, attests to the presence of a certain amount of oxygen. The observational data, however, point overwhelmingly to an anoxic atmosphere: neither oxidised detrital minerals nor palaeosols have been observed in the Archaean terrains (Rye et al., 1995; Rasmussen and Buick, 1999a,b), although there is a contested description of oxidised palaeosols from a 2.6 Ga old horizon in South Africa (Watanabe et al., 2000). Sulfur isotope studies (Farquhar et al., 2000) as well as trace element investigations of marine deposits also support a reducing environment in the Archaean (van Kranendonk et al., 2003). Kasting (1993) estimates that oxygen levels for the early Earth were $< 1\%$ present atmospheric levels (PAL). The low levels of atmospheric O_2 would imply that there was no ozone layer. This would have meant high levels of UV radiation at the surface of the Earth with important consequences for life forms within its sphere of influence.

The later rise in O_2 levels in the atmosphere appears to be related to two, simultaneous phenomena:

- the sequestration into the mantle by subduction of organic-rich sediments deposited on extensive continental platforms, and thus the removal of C (and consequently CO_2) from the volatile-rich outer layer at the surface of the planet (Lindsay and Brasier, 2000), and
- a rise in O_2 in the atmosphere related to the biogenic production of O_2 by oxygenic photosynthesising microbes (cyanobacteria). This basically means that, as tectonic processes removed the CO_2 from the atmosphere, it was replaced by O_2 produced by the newly evolved cyanobacteria. Until recently it had been thought that cyanobacteria had already evolved in the Early Archaean (e.g. Schidlowski, 1988; Schopf, 1993; Mozjsis et al., 1996; Rosing, 1999). However, for reasons explained in more detail below, it appears that this hypothesis was based on erroneous interpretations regarding the microfossils and isotope data. The oldest morphological cyanobacterial fossils from the Transvaal in South Africa date back to ~ 2.9 Ga (Noffke et al., 2003), whereas the oldest chemical fossil evidence, in the form of hopanoid derivatives of cyanobacterial hopanes, occur in 2.7 Ga carbonaceous shales from the Hammersley Basin (Summons et al., 1999; Brocks et al., 2003).

6.1.2.5 Oceanic Conditions

Estimates of the pH conditions of the early ocean range from neutral (Holland, 1984) to slightly acidic due to the high pCO_2 in the atmosphere (and consequently in the oceans) (Grotzinger and Kasting, 1993; Sleep et al., 2001), and even sodic (Kempe and Degens, 1985). Mass-balance calculations, however, suggest that the ocean had a pH between 4.8 and 6.5 (Sleep et al., 2001).

Fluid-inclusion studies provide a mixed indication for salinity. Most studies suggest higher salinities in the Archaean (de Ronde et al., 1994; Knauth, 1998), although some results suggest salinities similar to those of today (e.g. Appel et al., 2003). Ocean salinity appears to have been dominated by halite since the Archaean. Pinti (this volume) provides an elegant review of the present state of understanding regarding oceanic conditions in the Hadean/Archaean.

6.1.3 Bolide Impacts and the Origin of Life/Early Life

Another environmental aspect of relevance to the origin and evolution of life is that of bolide impacts, already mentioned above in connection with the import of volatiles to the Earth. Comparison with the lunar cratering record suggests that the flux of bolide impacts was particularly high in the pre 3.85 Ga period (Hartmann, 2000; Ryder et al., 2000). The great influx of bolides would probably have had severe consequences for early life, possibly destroying its first, tentative steps if life had already developed by 4.0 Ga (Maher and Stevenson, 1988; Sleep et al., 1989). Alternatively, if the impacts caused only the elimination of the upper layers of the oceans, early life-inhabiting ecosystems, such as deep-sea hydothermal vents, would have been protected and may have survived the impacts. On the other hand, if life had either not developed prior to 4.0 Ga, or if it had been completely destroyed during the bolide impacts, these impacts could have had a positive effect on the origin of life: the massive amount of energy released during the bombardment and in the period immediately afterwards would have provoked intense hydrothermal activity, thus providing sites where life could have (re)appeared (Baross and Hoffman, 1985; Holm, 1992; Kring and Cohen, 2002). Evidence for continued bolide activity comes from the finding of layers of spherulites of extraterrestrial origin in both the Barberton and the Pilbara greenstone belts (Lowe and Byerly, 1986; Byerly et al., 2002; Kyte et al., 2003).

This chapter will not address the origin of life, but it is worth noting that Maurette (1998) suggests the possibility that scoriacious micrometeorites could have acted as microreactors in which important prebiotic molecules could have formed. Whatever its origin and whenever its appearance, life was well developed and well distributed by 3.47 Ga, as we will see below.

6.2 Potential Early Habitats for Life

Once life arose, early Earth provided a relatively wide range of potential habitats (Nisbet, 1995; Nisbet and Sleep, 2001). Changes in the availability of certain habitable niches would have occurred as the Earth itself continued to evolve. During the Hadean the crust would have been predominantly subaqueous. Larger amounts of landmass on top of continental crust appeared during the Early to Middle Archaean. Given the limited rock record from the Early Archaean, it is difficult to estimate the rate of colonisation of the various environmental niches. Already the oldest microfossil remains from the Barberton and Pilbara greenstone belts document that life was widespread in shallow water areas and had invaded the intertidal zone, possibly even becoming subaerial (see below). The main habitats may be summarised as follows:

- The deep-ocean environment would have included hydrothermal vents (both hot black smokers and perhaps lower-temperature white smokers), volcanic rock surfaces (including fractures), the sediment-covered ocean floor, and possibly even the subsurface sediment. The sediments would have been locally and distally derived. *In situ* or autochthonous sediments could have been formed by the alteration products of the oceanic basalts (carbonates and clays) and chemical precipitations due to hydrothermal exhalations. Distal or allochthonous sediments would have been derived from sedimented volcanic ash fall and maybe turbiditic input from tectonically unstable, sedimented slopes.
- The open ocean provided a planktonic environment including colonisation of floating particles.
- Around the flanks of exposed hot-spot volcanoes, emerged portions of basaltic crust, and the small exposures of continental crust, there would have been: (a) shallow water basins and/or lagoons that hosted lava surfaces, sediment surfaces (Fig. 6.3), hot springs, as well as planktonic environments; (b) an intertidal environment with the same types of substrate (i.e., sediment/rock surfaces, hot springs); and (c) a subaerial environment with hot springs, rock surfaces, regolith surfaces and maybe pools of standing water would also have potentially been available for organisms capable of surviving in those habitats.
- Work over the last decade has shown that there is an important microbial niche in the deep subsurface (Gold, 1992; Frederickson and Onstott, 1996). Perhaps this niche was already colonised in the Early Archaean period (equivocal C isotope data from hydrothermal veins could be interpreted in this sense, e.g. Ueno et al. 2001). It has even been suggested that the deep habitat could have been the locus for the origin of life (Gold, 1992).

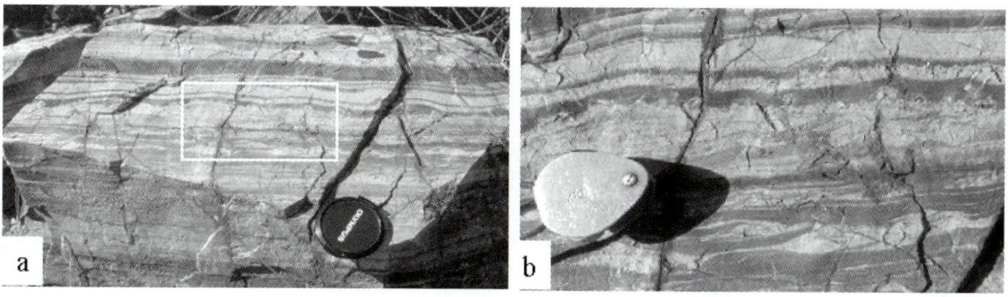

Fig. 6.3. Shallow-water sediments from the Pilbara greenstone belt, N.W. Australia. (**a**) The rock is a silicified volcaniclastic tuff. The *light* layers represent coarse-grained ash and the *darker* layers fine-grained ash. (**b**) Sedimentary structures such as ripple marks are present, indicating deposition in shallow waters within the influence of wave action

6.3 Early Archaean Fossil Record

Evidence for (microbial) life in ancient rocks can be revealed using a number of methods. They include:

- Isotopic analysis of elements such as C, S, N, and Fe. Different isotopic species of these elements can be fractionated during microbial metabolism, for instance the photosynthetic pathway leads to an increase in the ratio of ^{12}C to ^{13}C. However, in dealing with rocks as old as those from the Early Archaean epoch (older than 3.3 Ga), account needs to be taken of the effects of changes in the original signal due to geological processes over time, such as diagenesis and metamorphic alteration (Robert, 1988).
- Study of the degradation products of the original organic matter making up the organisms (chemical biomarkers). Organic matter is broken down into increasingly smaller components with time due to biogenic degradation, hydrolysis and oxidation, not to mention the effects of heating during metamorphism. The end product of this breakdown, kerogen, consists of resistant macromelecules, which can be transformed into pure graphite during high-grade metamorphism. Some of the most common methods used in the study of organic biomarkers include HPLC, GCMS, Raman and IR spectroscopy, and ToF-SIMs.
- The study of morphological fossils. Organic matter readily chelates mineral ions in solution that subsequently polymerise around and within the organic template to form an authentic reproduction of the original structure. These structures can be observed using a variety of microscopic methods and even macroscopic study in the case of large-scale development of microbial mats.
- Trace-element investigations. This new field of investigation takes as its starting point the fact that organic material can act as a trap for trace elements (e.g. van Kranendonk et al., 2003).

This discussion of the earliest record of life on Earth will concentrate mainly on the morphological evidence for early life but will also take into account carbon, sulfur, and nitrogen isotope data, as well as rare biochemical investigations. A brief review of the processes of microbial fossilisation and the criteria used in the identification of fossilised microbes is given in Sect. 6.4.

6.3.1 The Isua Greenstone Belt

The oldest rocks that could contain a record of life are the > 3.75 Ga old metasediments occurring in the Isua greenstone belt of SW Greenland (Appel and Moorbath, 1999). Carbon isotope analyses of some of the rocks have been interpreted to indicate evidence for microbial fractionation of carbon (Schidlowski, 1988; Mojzsis et al., 1996; Rosing, 1999). Some of the measurements were made on bulk rock samples, ground to a powder, and some on *in situ* measurements of graphite. However, recent ($<$ 8000-y old), endolithic cyanobacteria and fungal hyphae occur (sometimes as fossils) in these rocks (Westall and Folk, 2003) (Fig. 6.4). Obviously the presence of fossilised and nonfossilised endoliths poses a contamination problem with respect to measurements made not only on bulk samples, but also on *in situ* measurements, if steps have not been taken to eliminate this possibility.

Stepped combustion analysis is one method of showing up a contaminated carbon-isotope signal in ancient rock samples, at the same time as bringing to light the original, endogenous signal (e.g. van Zuilen et al., 2002). Furthermore, it has been demonstrated that graphite in the Isua rocks can be produced by thermal decomposition of siderite (iron carbonate) and need not necessarily be of biogenic origin (van Zuilen et al., 2002; Lepland et al., 2002).

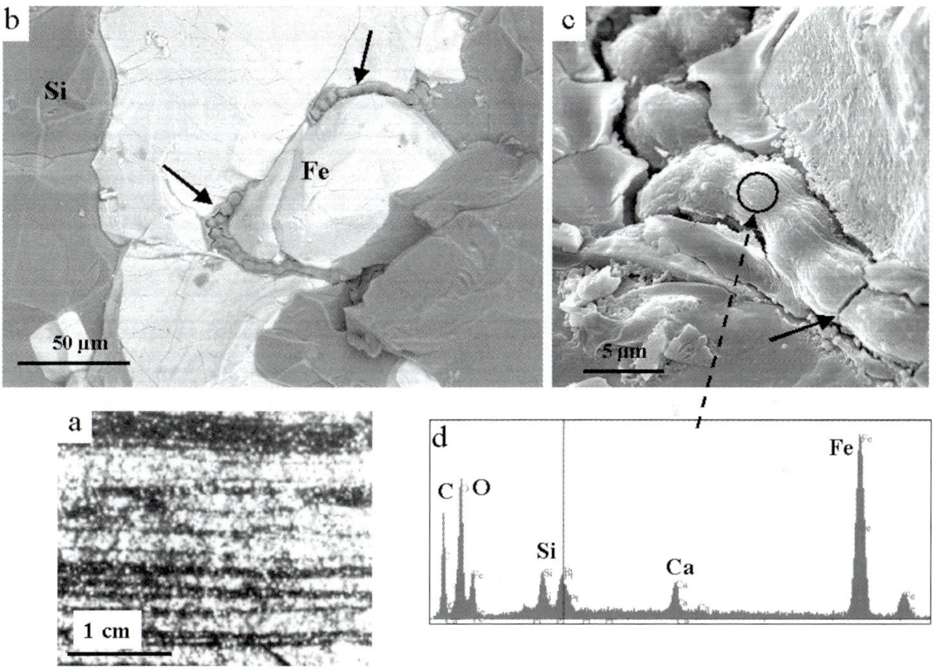

Fig. 6.4. Recent, fossilised endolithic bacteria (**b**,**c**) in a banded iron formation (BIF) (**a**) from Isua (> 3.75 Ga). (**a**) BIF sample with alternating magnetite-rich layers (*dark*) and chert-rich layers (*light*). (**b**) Backscatter image of the interface between a dark magnetite rich layer and a light chert-rich layer. The chert (Si, i.e. SiO_2), now metaquartzite appears dark because of its low atomic weight, whereas the magnetite (Fe, i.e. $Fe^{2+}Fe^{3+}{}_2O_4$) appears light because of its high atomic weight. Endolithic micro-organisms (*arrows*), in this case, cyanobacteria inhabit cracks in the rock as a refuge from the severe climatic conditions in Greenland. (**c**) Cracks (*arrow*) between individual cells in this cyanobacterial chain demonstrate that these micro-organisms have been fossilised. The *circle* is the area analysed by EDX. (**d**) Although the organisms have been fossilised, they still contain a trace of carbon according to the EDX analyses

The high metamorphic grade of these rocks makes it unlikely that morphologically identifiable microbial remains could still be found (see Appel et al., 2003). However, previous micropalaeontological investigations described structures resembling microbial fossils from banded iron formation (BIF) samples, which were interpreted as being indigenous to the sample, i.e. formed at the same time as the sedimentary rock. A recent re-evaluation of the work of Pflug and Jaeschke-Boyer (1979) and Pflug (2001) has shown that the so-called microfossil *Isuasphaera Isua* is certainly a metamorphic phenomenon (Appel et al., 2003; see also Bridgewater et al., 1981; and Roedder, 1981). Other microfossils, such as *Appelella ferrifera* (Robbins, 1987; Robbins et al., 1987; Robbins and Iberall, 1991) most probably represent recent endolithic contamination (see Fig. 6.4) (Westall and Folk, 2003).

6.3.2 The Barberton and Pilbara Greenstone Belts

6.3.2.1 Microfossils

The oldest, best-preserved sediments containing *bona fide* morphological evidence of endogenous fossil life, are the 3.5–3.2 Ga old volcaniclastic and chemical sediments from the Barberton greenstone belt in South Africa and the Pilbara greenstone belt in NW Australia. These sediments are very well preserved and document widespread evidence for life in shallow water to littoral (intertidal), and perhaps even subaerial, environments (Walsh, 1992; Walsh and Westall, 2003; Westall and Walsh, 2003; Westall et al., 2001a; Westall et al., 2002; Westall, 2003). The microfossils identified include (Fig. 6.5):

- Filaments falling into two size ranges: small filaments $< 0.25\,\mu m$ in width and up to several tens of micrometers in length, and larger filaments up to $2.5\,\mu m$ wide and about $100\,\mu m$ long;
- Vibroid-shaped structures (curved rods), $2–3.8\,\mu m$ in length and $1\,\mu m$ in width;
- Rods $1\,\mu m$ in length;
- Spherical to oval structures with size modes centered around $0.4–0.5\,\mu m$, $0.8–1.0\,\mu m$, and about $1.3\,\mu m$.

These structures are interpreted as fossil microbial filaments, vibroids, rods and coccoids, respectively, on the basis of criteria outlined in the next section.

6.3.2.2 Microbial Mats

The original organisms created fine, delicate mats on the surfaces of sediments deposited in shallow-water environments (Fig. 6.5a,b), or thicker, more robust mats formed in more extreme, partially subaerial to subaerial, littoral environments (Fig. 6.6). Embedded evaporite minerals and dessication cracks in the more robust mats testify to subaerial exposure. The mats are often macroscopically visible as laterally extensive, fine, crinkly laminae, or as domed structures (tabular and domal stromatolites respectively, Fig. 6.7).

Fig. 6.5. Silicified microfossils from shallow water, volcaniclastic sediments in the Pilbara greenstone belt (3.46 Ga) from rock shown in Fig. 6.3. (**a**) Thin section showing development of a delicate biofilm (crinkly brownish layer shown by *white arrows*) at the interface between two sediment layers (the large crystal is a piece of silicified pumice). (**b**) SEM detail of the biofilm layer in (**a**) showing a consortium of microbial filaments (*large arrow*), rod- and coccoid-shaped microfossils (*small arrows*), associated with a polymer film of extracellular polymeric substances (EPS) that coat the adjacent grains. (**c**) Chains of coccoidal bacteria (*arrows*) exhibiting a deflated, rugged appearance. This is due to cell lysis (loss of their cell contents) before fossilisation. (**d**) Deflated, twisted microbial filament. All the microfossils still contain carbon

6.3.2.3 Preservation

The mats and microfossils from the Early Archaean formations in Barberton and the Pilbara were preserved mainly by replacement by silica (silicification), although in one case a mat was replaced by both calcium carbonate (calcified) and silica. The pervasive silicification was due to the high concentration of silica in the Archaean ocean waters (there were no siliceous micro-organisms to remove the silica – modern oceans are undersaturated with respect to silica owing to the presence of these organisms), as well as to the frequency of hydrothermal activity pouring silica-rich fluids into the environment. Both the fossilised mats

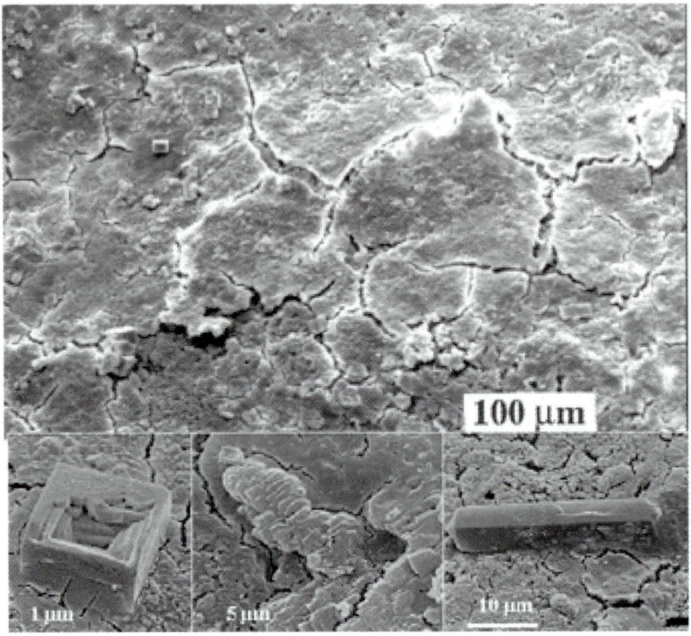

Fig. 6.6. SEM image of a robust microbial mat formed in a (partially?) subaerial littoral environment. The *top photograph* shows a cracked, dessicated, thick film. The lower images are details of minerals that precipitated onto the surface of the mat: from *left* to *right*: carrobite, calcite and aragonite (?). The first two are typical of evaporite environments

and the micro-organisms still preserve a certain amount of organic carbon in their structures (e.g. Fig. 6.8). They are thus described as being carbonaceous or permineralised, the latter term referring to the permeation of the organic matter by a mineral. As a consequence of the contrast between the dark, carbonaceous fossils and the transparent, embedding silica, the mats, some of the larger microbial filaments, and decimicrometer-sized colonies of coccoidal fossils may be visible in petrological thin section (20–30 µm thick, polished slices of rock) as permineralised fossils (Fig. 6.8). Most of the microfossils, on the other hand, contain volumetrically very little remnant carbon (N.B. an average living bacterial cell, such as *Escherichia coli*, only contains about 2.8×10^{-13} g carbon).

Since the silicified microfossils are also encased in silica (chert), and since they are generally extremely small (0.5–1 µm for the cocci, 0.1–0.5 µm wide by some tens of µm long for the filaments, and 2–3.8 µm for the vibroids), these structures are invisible as individuals in thin section (except, as noted above, where they occur as larger colonies) and require special treatment, such as delicate etching in the fumes of HF (Westall, 1999) in order to be observed with a high-resolution scanning electron microscope.

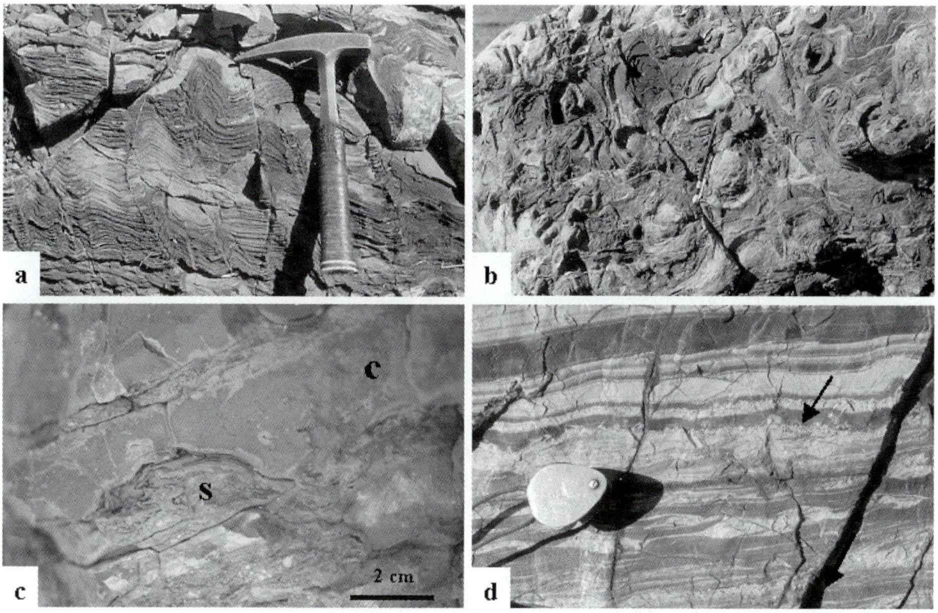

Fig. 6.7. Tabular and dome-shaped stromatolites from the Pilbara greenstone terrain. (**a**,**b**) Cross section and plan view of dome-shaped stromatolites. (**c**) Section through a dome-shaped stromatolite (*s*) that collapsed under the weight of the precipitated silica (*c*), as shown by the contorted layering on one side of the structure. The collapse of the layers within the stromatolite indicates plastic deformation of originally soft layers. (**d**) Volcaniclastic ashes deposited in a shallow water basin. Delicate microbial mats (tabular stromatolites, see Fig. 6.5) formed at the surfaces of these subaqueously deposited sediments (*arrow*)

6.3.2.4 Biogenic Isotopes

C isotope values for samples from the Early Archaean formations in Barberton and the Pilbara range from −14‰ to −42‰ (Hayes, 1983; Walsh, 1989; Strauss and Moore, 1992; Walsh and Lowe, 1999; Westall et al., 2001; Ueno et al., 2001; Brasier et al., 2002). Although these values (−14‰ to −42‰) are consistent with a biologically produced signal, Hayes (1983) and Robert (1989) showed that post depositional processes could alter the original isotopic record. Moreover, Brasier et al. (2002) hypothesise that abiotic Fischer–Tropf synthesis could produce such a fractionation of the carbon-isotope signal. Isotope investigations on these rocks are also subjected to contamination problems, especially considering the fact that most reliable measurements are obtained from samples with a relatively high bulk C content.

As with the carbon-isotope data, there are contradictory interpretations of the sulfur isotopes. Shen et al. (2001) propose that sulfur-reducing bacteria produced the sulfur-isotope ratios measured in Early Archaean sulfides hosted in

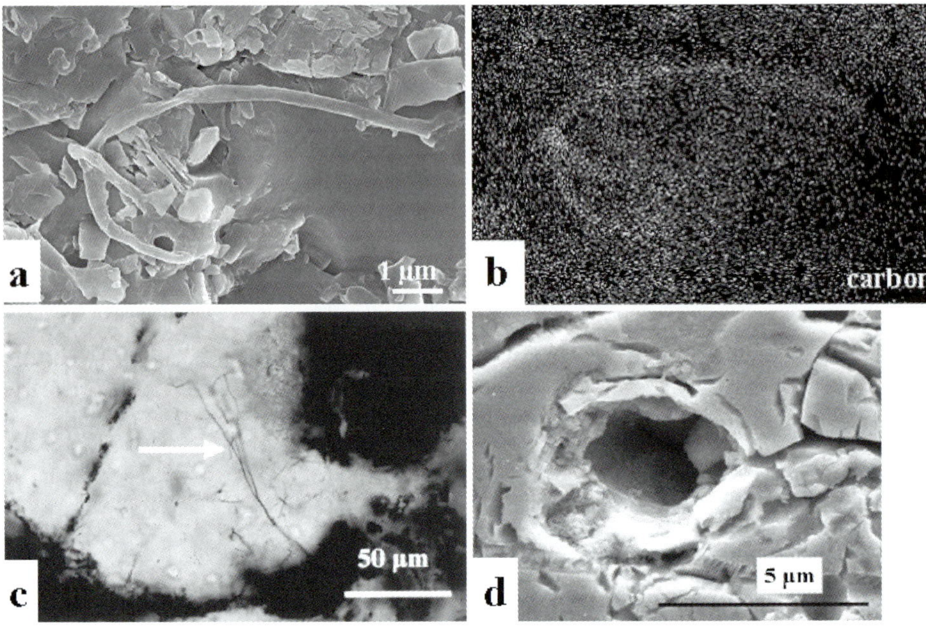

Fig. 6.8. Microfossils viewed in different modes: (**a**) filament observed with a high-resolution SEM; (**b**) EDX map of the same filament demonstrating its carbonaceous composition (the embedding mineral is chert, SiO_2); (**c**) light microscope observation of a large, carbonaceous filament embedded in translucent quartz; (**d**) SEM view of a cross section through a large filament embedded in chert showing the granular nature of the permineralised cell wall

sedimentary barite rocks in the Pilbara. Runnegar (2002), on the other hand, suggest that the S isotope values in the Early Archaean rocks are of purely abiogenic origin. Other studies of N isotopes in Early Archaean hydrothermal and kerogenous cherts in the Pilbara by Beaumont and Robert (1999), Pinti and Hashizume (2001) and Pinti et al. (2001) suggest the presence of chemolithotrophic bacteria.

6.3.2.5 Kerogen and Chemical Biomarker Studies

The amount of remnant carbon (kerogen) associated with these microfossiliferous cherts is very small, ranging normally from 0.01–0.5% and occasionally up to 1% (Hayes et al., 1983; Walsh, 1989; Westall et al., 2001a). The kerogen in these rocks is relatively mature, in the sense that they have lost their functional groups. Raman spectroscopy has been used to document the carbonaceous nature of certain "filaments" in Early Archaean cherts from the Pilbara (Schopf et al., 2002; Brasier et al., 2002), although there is much discussion as to the interpretation of their respective results. Schopf et al.

(2002) believed that the kerogen is of biogenic origin, whereas Brasier et al. (2002) contend that it is abiogenic. The oldest reliable chemical biomarkers were obtained from 2.7 Ga oil shales from the Hammersley range in Australia (Summons et al., 1999; Brocks et al., 2003). These investigations document the presence of hopanoids and steranes in the highly carbonaceous sediments.

6.3.3 Inferences Regarding the Early Archaean Microbiota

Although it is impossible to deduce metabolic behaviour from purely morphological fossil remains, other methods can be used to make some inferences regarding the kind of organisms that these Early Archaean microfossils might represent. They include:

- analysis of the microenvironments in which the microfossils occur,
- the isotope signal,
- a comparison with modern micro-organisms,
- theoretical considerations relating to early life.

6.3.3.1 Environment

The Early Archaean environment appears to have been an "extreme" environment based on the geological context outlined above. In the first place, the surface seems to have been relatively hot and strongly influenced by volcanic and hydrothermal activity. The ocean water was probably saltier than it is at present and may even have been slightly acidic (N.B. there are counter hypotheses to these inferences, see above). Furthermore, the atmosphere was most likely anoxic. Consequently, unless the UV radiation was filtered by an organic smog, as proposed by Lovelock (1988) and Pavlov et al. (2001), any life in shallow water to exposed areas would have been bathed in high levels of UV radiation. (N.B. Higher levels of CH_4 in the atmosphere due to intense volcanic activity would also have caused a "dirty" atmosphere that would have prevented much of the UV radiation penetrating to the surface.)

6.3.3.2 Organism type

– *Chemolithotrophs, Sulfate-Reducing Bacteria*

Based on a knowledge of modern microbial metabolisms, as well as an understanding of the carbon and energy sources that were probably present on the early Earth, the earliest forms of life would most likely have had a chemolithoautotrophic metabolism, in which inorganic materials are used as a source of both C and energy (see Nisbet, 1995, for an eloquent hypothesis regarding the evolution of early life). Evidence for the presence of chemolithotrophs in the hydrothermal deposits from Barberton and the Pilbara comes from the N and C isotopic data (e.g. Beaumont and Robert, 1999; Isozaki et al., 1999; Pinti and Hashizume, 2001; Pinti et al., 2001; Ueno et al., 2001). Heterotrophic (chemo-organotrophic) organisms using fermentation or anaerobic respiration would have evolved from the chemolithotrophs. Based on S-isotope measurements of early Archaean cherts, Shen et al. (2001) proposed that bacteria using the relatively evolved sulfate-reducing metabolism had already evolved.

– *Anoxygenic Photosynthesisers*

Had some of the Early Archaean organisms already developed photosynthesis (the use of light energy)? Nisbet (1995) proposes a couple of pathways by which bacteria could have developed the ability to harvest light energy, an energetically far more favorable metabolic pathway than chemotrophy. One of Nisbet's (1995) scenarios hypothesises that motile bacteria around a hydrothermal vent could have developed thermosensors (e.g. bacteriochlorophylla), thus preadapting them to "photosynthesise using volcanogenic H_2S in shallow water". In the second scenario he suggests that carotenoid pigments could have provided early micro-organisms in shallow-water environments with protection from harmful UV radiation and "may have functioned as light-harvesting pigments in early photosynthesis".

The fact that the micro-organisms inhabiting the Early Archaean environments of Barberton and the Pilbara formed mats, some of which exhibit upward directed growth (Hofmann et al., 1999; Westall et al., 1999, 2002) in shallow water to exposed environments, suggests that some of them may already have developed photosynthesis. Early studies based on carbon-isotope data as well as morphological resemblance suggested that many of the microfossils from the Early Archaean represented oxygenic cyanobacteria, (e.g. Schidlowski, 1988; Walsh, 1992; Schopf, 1993; Mojzsis et al., 1996; Rosing, 1999). However, this interpretation can be questioned for the following reasons:

- the $-26‰$ $\delta^{13}C$ value found in the ancient sediments, which is supposed to be typical for the rubisco pathway in cyanobacteria (Schidlowski, 1988), is also produced by anoxygenic, thermophilic chemolithotrophic micro-organisms (Reysenbach and Cady, 2001);
- some of the cyanobacteria-like microfossils are artefacts of kerogen deposition in hydrothermal veins (Brasier et al., 2002);
- although there are many microfossils in the numerous microbial mats formed in the photic zone that I have studied, there are no cyanobacteria-like microfossils (they are organisms that, furthermore, should be more readily preservable than other bacteria because their characteristic thick, polysaccharide-rich sheaths are highly susceptible to fossilisation);
- there is no mineral evidence for oxygenic conditions associated with the microbial mats formed in the photic zone. The evidence for bacteria and microbial mats in Early Archaean, sunlight-bathed environments suggests, however,

that they could have been formed by anoxygenic-photosynthesising organisms (Westall et al., 2002).

– *Organism Type: Conclusions*

Given the sedimentological and geochemical information regarding the environmental conditions, the isotopic data, and the fact that the microfossils found to date occur in microbial mats or colonies in shallow water to (partially) subaerial habitats, close to or in juxtaposition to hydrothermal activity, we may hypothesise that:

- The Early Archaean organisms must have been thermophilic anaerobes, given the relatively high environmental temperatures;
- They were adapted to living in water with salinities slightly higher than the present day and certainly the organisms that constructed mats in evaporitic environments must have been at least halotolerant if not halophilic;
- Those organisms living in extremely shallow water to subaerial environments must have been relatively UV resistant (the micro-organisms may have developed gene-repair mechanisms and, in any case, the thick polymer layers of the microbial mats, with their outer layers of dead cells, would also have served as protection (see Cockell, 2000);
- They probably included xerophilic organisms (i.e. organisms that were able to survive in dry environments (some of the microbial mats show evidence of dessication cracks and evaporite minerals are embedded in their surfaces);
- Some of the micro-organisms may have been anoxygenic photosynthesisers (but not oxygenic, cyanobacteria-like organisms).

6.4 The Fossilisation of Bacteria

The existence of these very ancient examples of terrestrial life, indeed, the earliest morphological fossil examples of life on Earth, calls for an explanation of the processes by which such soft-bodied, delicate structures can be preserved. Examples of fossil bacteria and their exopolymeric substances (EPS) are known from throughout the 3.5 Ga old fossil rock record (Westall, 1999; Westall et al., 2000).

Functional groups, such as carboxyl, hydroxyl or phosphate in the cell wall or in the degraded cytoplasm within the organism chelate metal ions in solution. The nucleated mineral ions thus replace the organic template (Westall et al., 1995; Fig. 6.9), trapping the organic molecules in their matrix in nonoxidising environments. Minerals that are known to replace bacteria and their EPS, include silica (chert),

calcium carbonate (calcite, aragonite), calcium phosphate (apatite and other species), iron carbonate (siderite), iron oxide (haematite) and iron sulfide (pyrite). The original bacteria may have been alive or may have exhibited varying stages of deterioration prior to being fossilised. If they are not so badly

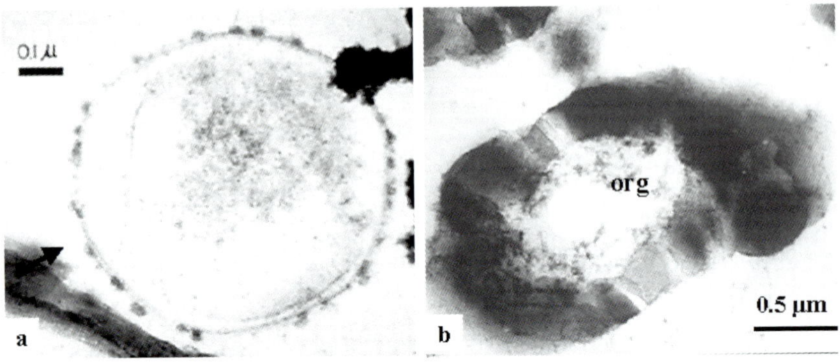

Fig. 6.9. The process of fossilisation as demonstrated by the experimental fossilisation (silicification) of bacteria. (**a**) TEM observation of the chelation of mineral ions (the small, dark aggregates) to functional groups on the cell envelope. (**b**) Further nucleation and polymerisation of the mineral leads to the formation, in this case, of a crust around the bacterium. Sometimes the fossilising mineral may replace the whole organism. In this example the remnant, degraded organic matter (org) remains trapped inside the crust

degraded as to be completely unrecognisable, a certain amount of degradation may actually aid identification. For example, Fig. 6.5b–d show filamentous and coccoidal microfossils that have a partially deflated appearance owing to the fact that they were dead and had lysed[5] before fossilisation.

Fossil bacteria are identified on the basis of morphological, colonial and biogeochemical criteria (Westall, 1999; Westall, 2003). Morphological characteristics include, size, shape, outer cell envelope texture, cell division, evidence for cell lysis, and flexibility (filamentous bacteria) (Fig. 6.10). Bacteria seldom occur as isolated individuals. Rather, they invariably occur in colonies of a few to up to millions of individuals. Thus, various aspects of the microbial colony can be used to aid identification. These include (i) the fact that more than one species, often with different morphologies, may coexist together in a consortium (e.g. Fig. 6.10), and (ii) that bacteria almost invariably produce copious quantities of EPS, forming fine to very robust and complexly structured biofilms or microbial mats, which can incorporate detrital particles as well as authigenically precipitated minerals (Fig. 6.6) (Westall et al., 2000). These microbial mats may produce macroscopically visible constructions, such as laterally extensive, finely wavy, planar laminations (tabular stromatolites) or hummocky to dome-shaped stromatolites (e.g. Fig. 6.7). If carbonaceous matter is still associated with the microfossils, then isotopic analyses may, with caution, be used to infer microbial fractionation (e.g. Hayes et al., 1983; Beaumont and Robert, 1999; Shen

[5] Upon death of the organism, an enzyme is automatically released that starts to break down the cell wall, letting the cytoplasm or internal cell contents escape (a process known as cell lysis). This produces a "burst or deflated balloon" effect that, to my knowledge, is impossible to imitate by nonbiogenic structures, mineral or inorganic.

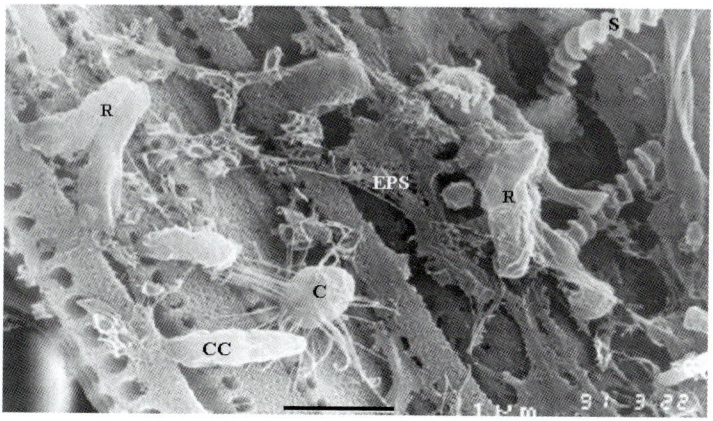

Fig. 6.10. The fossilisable characteristics of bacteria include morphological and colonial criteria, as well as biochemical signals. This SEM image of a colony of modern bacteria, composed of different species with different morphologies, illustrates some of the colonial and morphological aspects of bacteria that could be preserved in the fossilised state. The bacteria include a coccoid (C), a chain of coccoids within a common outer envelope, showing cell division (CC) (N.B. compare with the fossilised example in Fig. 5c), rods (R) exhibiting a wrinkled outer surface due to cell lysis (compare with Fig. 6.5c), and spiral morphologies (S). The EPS associated with the bacteria is in a highly degraded state.

et al., 2001; Pinti et al., 2001). Likewise, analysis of the degradation products of microbial biomolecules can point to a biogenic origin and, sometimes, a specific type of micro-organism (e.g Summons et al., 1999; Brocks et al., 2003). Other methods have been recently incorporated into the battery of tests to determine organic geochemical biogenicity including, time-of-flight-secondary-ion-mass-spectrometry (ToF-SIMS) (Toporski et al., 2000), Raman spectroscopy (Wynn-Williams and Edwards, 2000; Schopf et al., 2002), and atomic force microscopy (Kempe et al., 2003).

There are inherent drawbacks to any methods of identification when used in isolation. For instance, abiotic molecules can produce fine films and globular structures that, at first glance, imitate those produced by micro-organisms and their EPS. However, the abiotic organics cannot reproduce the totality of features, nor the complexity found in microbes and their biofilms or mats. Likewise, certain minerals form spherical or oval shapes, sometimes in clusters, which can look like microbial colonies. Again, it is the totality of features, which aids distinction of abiogenic structures from biogenic phenomena. With respect to biogeochemical analyses, similar problems arise. Moreover, the problem of younger "contamination" by endolithic organisms can be serious, as noted above (Westall and Folk, 2003).

Another important factor to take into account is the geological context and the specific environment of deposition of the rock. For instance, whereas one would not necessarily expect to find bacteria inside hot (> 120 °C) hydrothermal veins, or hot basalt rocks (i.e. lava flows), bacteria can penetrate cracks in the cooled lavas, or between mineral particles, as well as colonising the surfaces of lavas extruded into water (Fisk et al., 1998). In conclusion, the only way to avoid misinterpretations is to use as much evidence as possible from different methods of analyses, starting with sound geological field work in order to place the potentially fossiliferous rocks into a well-constrained environmental context. For identifying morphological fossils, much observational experience with both modern microbes and fossilised examples is an additional necessity.

6.5 Conclusions and Perspectives

The search for information pertaining to the early evolution of the Earth, and the concomitant appearance and evolution of life on Earth, is fraught with difficulties, not only because of the lack of rock material of a suitable age. (Mars would be ideal for such research since there has been no wholesale destruction of the oldest rocks by tectonic processes, as occurred on Earth.) Nevertheless, the meager rock record contains evidence for widespread life in shallow-water to subaerial environments by about 3.5 Ga. Life took the form of filamentous, vibroid-shaped, rod and coccoidal micro-organisms that formed delicate biofilms in subaqueous habitats and robust biofilms in exposed, littoral environments. Sedimentological and geochemical data regarding the environment, together with the isotopic data, suggest that the variety of micro-organisms represented by the fossils were UV resistant, thermophilic anaerobes that also included halophiles and xerophiles. Possibly anaerobic photosynthesis had appeared by this time, although there is no evidence for the development of aerobic photosynthesisers, such as cyanobacteria.

In terms of the fossil record, very careful analysis of the potentially biogenic microstructures needs to be undertaken in order to be able to distinguish between abiogenic bacteriomorphs (organic or inorganic, *in situ* or introduced during sample preparation) and true bacterial fossils. The hazard of younger contamination of older, fossiliferous material is especially a problem with carbon-rich material. The development of sensitive *in situ* methods of analysis will aid determination of the biogenetic potential of a particular structure. The challenge facing us now is the development of such techniques, which will be able to provide reliable information on a very small amount of remnant carbon in very small structures.

Looking into the future, such developments are of vital importance in preparing laboratory techniques for the analysis of Martian samples brought back to Earth via sample-return missions. The amounts of material available to individual laboratories will be extremely small and, given the controversy surrounding the interpretations of possible signs of life in Martian meteorite ALH84001 (McKay et al., 1996), a battery of *in situ* methods of analysis is necessary.

Acknowledgement

Many colleagues have given me invaluable help in my search for understanding the geological context of early life. I especially would like to thank Maarten de Wit, Wouter Nijman, Martin van Kranendonk, and François Robert (who cleared up my confusion about the import of volatiles and early atmosphere formation). Maud Walsh, Hervé Martin, and Jacques Riesse are thanked for their very helpful comments on the manuscript. My husband Daniele Marchesini has been my irreplaceable field companion in Barberton and the Pilbara. I have also been helped by many fruitful discussions with Maud Walsh regarding the microbial fossils. The CNRS, Danish Geological Survey, NRC, LPI and NAI (David McKay) all provided financial support for field and laboratory work.

References

Appel, P.W.U., Moorbath, S. (1999). Exploring Earth's oldest geological record in Greenland. *Eos*, **80**, 257–264.

Appel, P.W.U., Rollinson, H., Touret, J.L.R. (2001). Remnants of an Early Archean (> 3.75 Ga) sea-floor, hydrothermal system in the Isua greenstone belt. *Precambrian Res.*, **112**, 27–49.

Appel, P.W.U., Moorbath, S., Touret, J. (2003). Isuasphaera Isua revisited. *Precambrian Res.* **176**, 173–180.

Arndt, N. (1994). Archean komatiites, in *Archean Crustal Evolution*, ed. K.C. Condie, p. 11–44, Elsevier, Amsterdam.

Arndt, N., Chauvel, C. (1991). Crust of the Hadean Earth. *Bull. Geol. Soc. Denmark*, **39**, 145–151.

Baross, J.A., Hoffman, S.E. (1985). Submarine hydrothermal vents and associated gradient environment as sites for the origin and evolution of life. *Origins Life Evol. Biosph.*, **15**, 327–345.

Beaumont, V., Robert, F. (1999). Nitrogen isotope ratios of kerogens in Precambrian cherts: a record of the evolution of atmospheric chemistry? *Precambrian Res.*, **96**, 63–82.

Bowring, S.A., Housh, T. (1995). The Earth's early evolution. *Science*, **269**, 1535–1540.

Bowring, S.A., Williams, I.S., 1999. Priscoan (4.00–4.003 Ga) orthogneisses from northwestern Canda. *Contrib. Mineral. Petrol.*, **134**, 3–16.

Brasier, M.D., Lindsay, J.F. (1998). A billion years of environmental stability and the emergence of eukaryotes: New data from northern Australia. *Geology*, **2**, 555–558.

Brasier, M.D., Green, O.R., Jephcoat, A.P., Kleppe, A.K., van Kranendonk, M., Lindsay, J.F., Steele, A., Grassineau, N. (2002). Questioning the evidence for Earth's oldest fossils. *Nature*, **416**, 76–81.

Bridgewater, D., Allart, J.H., Schopf, J.W., Klein, C., Walter, M.R., Baarghoorn, E.S., Strother, P., Knoll, A.H., Gorman, B.E. (1981). Microfossil-like objects from the Archaean of Greenland: a cautionary note. *Nature*, **289**, 51–53.

Brocks, J.J., Love, G.D., Snape, C.E., Logan, G.A., Summons, R.E., Buick, R. (2003). Release of bound aromatic hydrocarbons from late Archean and Mesoarchean kerogens via hydropyrolysis. *Geochim. Cosmochim. Acta.*, **67**, 1521–1530.

Byerly, G.R., Lowe, D.R., Wooden, J., Xie, X. (2002). An Archean Impact Layer from the Pilbara and Kaapvaal Cratons, *Science*, **297**, 1325–1327.

Campbell, I.H., Taylor, I.R. (1983). No water, no granites – no granites, no continents. *Geophys. Res. Lett.*, **10**, 1061–1064.

Cockell, C.S. (2000). The ultraviolet history of the terrestrial planets – implications for biological evolution. *Planet. and Space Sci.*, **4**, 203–214.

Dauphas, N., Robert, F., Marty, B. (2000). The late asteroidal and cometary bombardment of Earth as recorded in water deuterium to protium ratio. *Icarus*, **148**, 508–512.

Delano, J.W. (2001). Redox history of the Earth's interior since $\sim$ 3900 Ma: Implications for prebiotic molecules. *Origins Life Evol. Biosph.*, **31**, 311–341.

Delsemme, A. (1998). Cosmic origin of the biosphere, in *The Molecular Origins of Life*, ed. A. Brack, pp. 100–118, Cambridge Univ. Press, Cambridge.

De Ronde, C.E.J., de Wit, M.J., Spooner, E.T.C. (1994). Early Archaean ($>$ 3.2 Ga) iron-oxide-rich, hydrothermal discharge vents in the Barberton greenstone belt, South Africa. *Geol. Soc. Amer. Bull.*, **106**, 86–104.

De Vries, S.T., Nijman, W. (2001). Environmental conditions and hydrothermal systems: Buck Ridge Chert Complex, Barberton, SA, in *Int. Archean Symp.* eds. K.F. Cassidy, J.M. Dunphy, M.J. van Kranendonk, M.J.), p. 224–226, AGSO-Geosc. Aust., 2001/37, Perth.

De Wit, M.J., Hart, R.A. (1993). Earth's earliest continental lithosphere, hydrothermal flux and crustal recycling. *Lithos*, **30**, 309–336.

Drake, M.J., Righter, K. (2002). Determining the composition of the Earth. *Nature*, **416**, 39–44.

Fisk, M.R., Giovannoni, S.J., Thorseth, I.H. (1998). Alteration of oceanic volcanic glass: Textural evidence of microbial activity. *Science*, **281**, 978–980.

Eriksson, K.A. (1995). Crustal growth, surface processes and atmospheric evolution of early Earth, in *Early Precambrian Processes*, eds. M.P. Coward, A.C. Ries, Geol. Soc. Lond., Sp. Pub. **95**, 11–26.

Farquahr, J., Bao, H., Thiessen, M. (2000). Atmospheric influence of Earth's earliest sufur cycle. Science, 289, 756–758.

Fredrickson, J.K., Onstott, T.C. (1996). Microbes deep inside the earth. *Sci. Am.* **27(4)**, 42.

Gold, T. (1992). The deep, hot biosphere. *Proceedings of the National Academy of Sciences* **89**, 6045–6049.

Grotzinger, J.P., Kasting, J.F. (1993). New constraints on Precambrian ocean composition. *J. Geol.*, **101**, 235–243.

Halliday, A.N. (2001). In the beginning. *Nature*, **409**, 144–145.

Hartmann, W.K., Ryder, G., Dones, L., Grinspoon, D. (2000). The time-dependent intense bombardment of the primordial Earth/Moon system, in *Origin of The Earth and Moon* Eds. R.M. Canup and K. Righter, 493–512, University of Arizona Press.

Hayes, J.M. (1983). Geochemical evidence bearing on the origin of aerobis, a speculative hypothesis, in *Earth's Earliest Biosphere* ed. J.W. Schopf, 291–301, Princeton Univ. Press, Princeton.

Hayes, J.M., Kaplan I.R., Wedeking K.W. (1983). Precambrian organic chemistry, preservation of the record. In *Earth's Earliest Biosphere* ed. J.W. Schopf, 93–134, Princeton Univ. Press. Princeton.

Hofmann, H.J., Grey, K., Hickman, A.H., and Thorpe, R.I. (1999). Origin of 3.45 Ga coniform stromatolites in Warrawoona Group, Western Australia. *Geol. Soc. Am. Bull.*, **111**, 1256–1262.

Holland, H. D. (1984). *The Chemical Evolution of the Atmosphere and Oceans.* Princeton Univ. Press, Princeton.

Holm, N.G. (1992). Why are hydrothermal systems proposed as plausible environments for the origin of life? *Origins Life Evol. Biosph.*, **22**, 5–14.

Isozaki, Y., Ueno, Y., Yurimoto, H., Maruyama, S. (1999). 3.5 Ga kerogen-rich silica dykes from North Pole area, Western Australia: seafloor biosphere in the Archean? *AGU Fall Meeting Abst.* B42B–20.

Jahn, B.M., 1997. Géochimie des granitoides archéens et de la croûte primitive. In: R. Hagemann and M. Treuil (Eds.), *Introduction à la Géochimie et ses Applications.* Editions Thierry Parquet.

Kamber, B.S., Moorbath, S., Whitehouse, M.J. (2001). The oldest rocks on earth: time constraints and geological controversies, in *The Age of the Earth: from 4004 BC to AD 2002,* eds. C.L.E. Lewis, S.J. Knell, Geol. Soc. Lond. Sp. Pub. **190**, 177–203.

Kasting, J.F. (1991). Box models for the evolution of atmospheric oxygen: An update. *Paleogeogr. Paleoclimat., Paleoecol.*, **97**, 125–131.

Kasting, J.F. (1993). Earth's early atmosphere. *Science*, **259**, 920–926.

Kasting, J.F., Pavlov, A.A., Siefert, J.L. (2001). A coupled ecosystem-climate model for predicting the methane concentration in the Archean atmosphere. *Origins Life Evol. Biosph.*, **31**, 271–285.

Kempe, A., Schopf, J.W., Altermann, W., Kudryavtsev, A.B., Heckl, W.M., 2002. Atomic force microscopy of Precambrian microscopic fossils. *Proc. Natl. Acad. Sci.*, **99** (14): 9117–9120.

Kempe, S., Degens, E.T. (1985). An early soda ocean? *Chem. Geol.*, **5**, 95–108.

Knauth, L.P. (1998). Salinity history of Earth's early ocean. *Nature*, **395**, 554–555.

Knauth, L.P., Lowe, D.R. (2003). High Archean climatic temperature inferred from oxygen isotope geochemistry of cherts in the 3.5 Ga Swaziland Supergroup, South Africa. *Geol. Soc. Am. Bull.*, **115**, 566–580.

Knoll, A.K. (1992). The early evolution of eukaryotes: a geological perspective. *Science*, **256**, 622–627.

Kramers, J.D. (2003). Volatile element abundance patterns and an early liquid water ocean on Earth. *Precambrian Res.*, **126**, 379–394.

Kring, D.A., Cohen, B.A. (2002). Cataclysmic bombardment throughout the inner solar system 3.9–4.0 Ga. *J. Geophys. Res.*, **107**, 10.1029/2001JE001529.

Kyte, F.T., Shukolyukov, A., Lugmaor, G.W., Lowe, D.R., Byerly, G.R. (2003). Early Archean spherule beds: Chromium isotopes confirm origin through multiple impacts of projectiles of carbonaceous chondrite type. *Geology*, **31**, 283–286.

Lepland, A., Arrhenius, G., Cornell, D. (2002) Apatite in early Archean Isua supracrustal rocks, southern West Greenland: its origin, association with graphite and potential as a biomarker. *Precambrian Res.*, **118**, 221–241.

Lindsay, J.F., Brasier, M.D. (2000). A carbon isotope reference curve for c. 1700 to 1575 Ma, McArthur and Mount Isa Basins, northern Australia. *Precambrian Res.*, **99**, 271–308.

Lindsay, J.F., Brasier, M.D. (2002). Did global tectonics drive early biosphere evolution? Carbon isotope record from 2.6 to 1.9 Ga carbonates of Western Australian basins. *Precambrian Res.*, **114**, 1–34.

Lovelock, J.E. (1988). *The Ages of Gaia*. W.W. Norton, New York.

Lowe, D.R. (1999). Geologic evolution of the Barberton greenstone belt and vicinity, in *Geologic evolution of the Barberton greenstone belt, South Africa*, eds. D.R. Lowe, G.R. Byerly, *Geol. Soc. Am. Spec. Paper*, **329**, 287–312.

Lowe, D.R., Byerly, G.R. (1986). Early Archean silicate spherules of probable impact origin, South Africa and Western Australia, *Geology*, **14**, 83–86.

Maher, K.A., Stevenson, D.J. (1988). Impact frustration of the origin of life. *Nature*, **331**, 612–614.

Martin, H. (1994). The Archean grey gneisses and the genesis of the continental crust. In: K.C. Condie (Ed.), *The Archean crustal evolution*. Developments in Precambrian Geology. Elsevier, Amsterdam, pp. 205–259.

Maurette, M. (1998). Carbonaceous micrometeorites and the origin of life. *Origins Life Evol. Biosphere.*, **28**, 385–412.

Maurette, M., Duprat, J., Engrand, C., Gounelle, M., Kurat, G., Matrajt, G., Toppani, A. (2000). Accretion of neon, organics, nitrogen and water from large interplanetary dust particles on the early earth. *Planet. Space Sci.*, **48**, 1117–1137.

Maurette, M., Matrajit, G., Gounelle, M., Engrand, C., Duprat, J. (2001). La matière extraterrestre primitive et les mystères de nos origines, in *L'Evironnement de la Terre Primitive*, eds. M. Gargaud, D. Despois, J.-P. Parisot, p. 99–127, Presses Univ. Bordeaux, Bordeaux.

McCulloch, M.T., Bennet, V.C. (1993) Evolution of the early Earth: constraints from ^{143}Nd-^{142}Nd isotopic systematics. *Lithos*, **30**, 237–255.

McKay, D.S., Gibson, E.K., Thomas-Keprta, K.L., Vali, H., Romanek, C.S., Clemett, S.J., Chillier, X.D.F., Maedling, C.R., Zare, R.N. (1996). Search for past life on Mars: possible relic biogenic activity in Martian meteorite ALH84001. *Science*, **273**, 924–930.

Morbidelli, A., Benest, D. (2001). Evolution primordiale du sytème solaire interne et origine de l'eau, in *L'Evironnement de la Terre Primitive*, eds. M. Gargaud, D. Despois, J.-P. Parisot, 91–98, Presses Univ. Bordeaux, Brodeaux.

Morbidelli, A., Chambers, J., Lunine, J., Petit, J.M., Robert, F., Valsecchi, G.B., Cyr, K.E. (2000). Source regions and timescales for the delivery of water to the Earth. *Meteoritics and Planet. Sci.*, **35**, 1309–1320.

Mojzsis, S.J., Arrhenius, G., Keegan, K.D., Harrison, T.H., Nutman, A.P. Friend, C.L.R. (1996). Evidence for life on Earth before 3,800 million years ago. *Nature*, **384**, 55–59.

Mojzsis, S.J., Harrison, T.M., Pidgeon, R.T. (2001). Oxygen-isotope evidence from ancient zircons for liquid water at the Earth's surface 4,300 Myr ago. *Nature*, **409**, 178–181.

Nijman, W., de Bruijne, K.H., Valkering, M. (1999). Growth fault control of Early Archaean cherts, barite mounds and chert-barite veins, North Pole Dome, Eastern Pilbara, Western Australia. *Precambrian Res.*, **95**, 247–274.

Nisbet, E.G. (1995). Archaean ecology: a review of evidence for the early development of bacterial biomes, and speculation on the development of a global-scale biosphere, in *Early Precambrian Processes*, eds. M.P. Coward, A.C. Ries, 27–52, Geological Soc. London, Oxford.

Nisbet, E.G., Sleep, N.H., (2001). The habitat and nature of early life. *Nature*, **409**, 1083–1091.

Noffke, N., Hazen, R., Nhleko, N. (2003). Earth's earliest microbial mats in a siliclastic marine environment (2.9 Ga Mozaan Group, South Africa). *Geology*, **31**, 673–676.
Nyquist, L.E., Shih, C.Y. (1992). The isotopic record of lunar volcanism. *Geoch. Cosmochim. Acta*, **56**, 2213–2243.
Ohmoto, H. (1997). Evidence in pre-2.2 Ga paleosols for the early evolution of atmospheric oxygen and terrestrial biota. *Geology*, **24**, 1135–1138.
Ohmoto, H. (1999). Redox state of the Archean atmosphere: Evidence from detrital heavy minerals in ca. 3250–2750 Ma sandstones from the Pilbara craton, Australia: Comment. *Geology*, **27**, 1151–1152.
Owen, T.C. (1998). The origin of the atmosphere, in *The Molecular Origins of Life*, ed. A. Brack, 13–34, Cambridge Univ. Press, Cambridge.
Owen, T.C., Bar Nun, A. (2000). Volatile contributions from icy planetesimals, in *Origin of the Earth and Moon*, eds. R.M. Canup, K. Righter, p. 459–471, University of Arizona Press, Arizona.
Paris, I., Stanistreet, I.G., Hughes, M.J. (1985). Cherts of the Barberton greenstone belt interpreted as products of submarine exhalative activity. *J. Geol.*, **93**, 111–129.
Pavlov, A.A., Kasting, J.F., Brown, L.L., Rages, K.A., Freedman, R. (2001). Greenhouse warming by CH_4 in the atmosphere of early earth. *J. Geophys. Res.*, **105**, 11 981–11 990.
Pflug, H.D. (2001). Earliest organic evolution. Essay to the memory of Bartholomew Nagy. *Precambrian Res.*, **106**, 79–91.
Pflug, H.D., Jaeschke-Boyer, H. (1979). Combined structural and chemical analysis of 3,800-Myr-old microfossils. *Nature*, **280**, 483–486.
Pinti, D.L. (2004) The origin and evolution of the oceans. This volume.
Pinti, D.L., Hashizume, K. (2001). ^{15}N-depleted nitrogen in Early Archean kerogens: clues on ancient marine chemosynthetic-based ecosystems? *Precambrian Res.*, **105**, 85–88.
Pinti, D.L., Hashizume, K, Matsuda, J.-I., (2001). Nitrogen and argon signatures in 3.8–2.8 Ga metasediments: Clues on the chemical state of the Archean ocean and the deep biosphere. *Geochim. Cosmochim. Acta*, **65**, 2301–2315.
Rasmussen, B., Buick, R. (1999a). Redox state of the Archean atmosphere: Evidence from detrital heavy minerals in ca. 3250–2750 Ma sandstones from the Pilbara craton, Australia: *Geology*, **27**, 115–118.
Rasmussen, B., Buick, R. (1999b). Redox state of the Archean atmosphere: Evidence from detrital heavy minerals in ca. 3250–2750 Ma sandstones from the Pilbara craton, Australia: Reply. *Geology*, **27**, 1152.
Reysenbach, A.-L., Cady, S.L. (2001). Microbiology of ancient and modern hydrothermal systems. *Trends in Microbiol.*, **9**, 79–86.
Robbins, E.I. (1987). Appelella ferrifera, a possible new iron-coated microfossil in the Isua Iron Formation, southwestern Greenland, in *Precambrian Iron Formations.* eds. P.W.U. Appel, G.L. LaBerge, p. 141–154, Theophrastes, Athens.
Robbins, E.I., LaBerge, G.L., Schmidt., R.G. (1987). A model for the biological precipitation of Precambrian Iron-Formations – B. Morphological evidence and modern analogs in *Precambrian Iron Formations.* eds. P.W.U. Appel, G.L. LaBerge, p. 97–139, Theophrastes, Athens.
Robbins, E.I., Iberall, A.S. (1991). Mineral remains of early life on Earth? On Mars? *Geomicrobiol. J.*, **9**, 51–66.

Robert, F. (1988). Carbon and oxygen isotope variations in Precambrian cherts. *Geochim. Cosmochim. Acta*, **52**, 1473–1478.

Robert, F. (2001). L'origine de l'eau dans le système solaire telle qu'elle est enregistrée par son apport isotopique D/H, in *L'Evironnement de la Terre Primitive*, eds. M. Gargaud, D. Despois, J.-P. Parisot, p. 79–90, Presses Univ. Bordeaux, Bordeaux.

Roedder, E. (1981). Are the 3,800 Myr-old Isua objects microfossils, limonite-stained fluid inclusions, or neither? *Nature*, **293**, 459–462.

Rosing, M.T. (1999). 13C depleted carbon microparticles in $>$ 3700 Ma seafloor sedimentary rocks from West Greenland. *Science*, **283**, 674–676.

Rosing, M.T., Rose, N.M., Bridgewater, D., Thomsen, H.S. (1996). Earliest part of Earth's stratigraphic record: reappraisal of the $>$ 3.7 Ga Isua (Greenland) supracrustal sequence. *Geology*, **24**, 43–46.

Runnegar, B. (2002). Archean sulfates from Western Australia: implications for Early Archean atmosphere and chemistry. Goldschmidt Conf., Davos, August, 2002# 3859.

Ryder, G., Koeberl, C., Mojzsis, S.J. (2000) Heavy bombardment on the Earth at $\sim$ 3.85 Ga, The search for petrographic and geochemical evidence, in *Origin of the Earth and Moon*, (eds.) R.M. Canup, K. Righter, p. 475–492, University of Arizona Press, Arizona.

Rye, R., Kuo, P.H., Holland, H.D. (1995). Atmospheric carbon dioxide concentrations before 2.2 billion years ago. *Nature*, **378**, 603–605.

Sagan, C., Chyba, C. (1997). The early sun paradox: organic shielding of ultraviolet-labile greenhouse gases. *Science*, **276**, 1217–1221.

Schidlowski, M. (1988). A 3,800-million-year isotopic record of life from carbon in sedimentary rocks. *Nature*, **333**, 313–318.

Schidlowski, M., Hayes, J.M., Kaplan, I.R. (1983). Isotopic inferences of ancient biochemistries: carbon, sulfur, hydrogen and nitrogen, in *Earth's Earliest Biosphere,* ed. J.W. Schopf, p.149–186, Princeton Univ. Press, Princeton.

Schopf, J.W. (1992). Paleobiology of the Archean, in *The Proterozoic Biosphere*, eds. J.W. Schopf and C. Klein, p. 25–42, Cambridge Univ. Press, Cambridge.

Schopf, J.W. (1993). Microfossils of the Early Archean Apex Chert: new evidence of the antiquity of life. *Science*, **260**, 640–646.

Schopf, J.W., Walter, M.R., (1983). Archean microfossils: new evidence of ancient microbes, in *Earth's Earliest Biosphere,* ed. J.W. Schopf, 214–239, Princeton Univ. Press, Princeton.

Schopf, J.W., Packer, B.M., (1987). Early Archean (3.3 billion to 3.5 billion-year-old) microfossils from Warawoona Group, Australia. *Science*, **237**, 70–73.

Schopf, J.W., Kudryavtsvev, A.B., Agresti, D.G., Wdowiak, T.J., Czaja, A.D. (2002). Laser-Raman imagery of Earth's earliest fossils. *Nature*, **416**, 73–76.

Shen, Y., Buick, R., Canfield, D.E. (2001). Isotopic evidence for microbial sulfate reduction in the early Archaean era. *Nature*, **410**, 77–81.

Sleep, N.H., Zahnle, K.J., Kasting, J.F., Morowitz, H.J. (1989). Annihilation of ecosystems by large asteroid impacts on the early Earth. *Nature*, **342**, 139–142.

Sleep, N.H., Zahnle, K., and Neuhoff, P. S. (2001). Initiation of clement surface conditions on the early Earth, *Proc. Nat. Acad. Sci. USA*, **98**, 3666–3672.

Strauss, H., Moore, T.B. (1992). Abundances and isotopic compositions of carbon and sulfur species in whole rock and kerogen samples, in *The Proterozoic Biosphere: A Multidisciplinary Study* eds. J.W. Schopf, C. Klein, p.709–798, Camdridge Univ. Press, Cambridge.

Summons, R.E., Jahnke, L.L., Hope, J.M., Logan, J.H. (1999). 2-Methylhopanoids as biomarkers for cyanobacterial oxygenic photosynthesis. *Nature* , **400**, 554–557.
Tolstikhin, I.N., Marty, B. (1998). The evolution of crustal volatiles: a view from helium, neon, argon and nitrogen modelling. *Chem. Geol.*, **147**, 27–52.
Toporski, J.K.W., Steele, A., Westall, F., Avci, R., Martill, D.M., Mckay, D.S. (2001). In-situ biomarker detection using TOF-SIMS and high resolution electron microscopy imaging of an exceptionally well preserved bacterial biofilm from the 28 million year old Enspel formation. *Geoch. Cosmoch. Acta*, **66**, 1773–1791.
Ueno, Y., Maruyama, S., Isozaki, Y., Yurimoto, H. (2001). Early Archeans (ca. 3.5 Ga) microfossils and 13C-depleted carbonaceous matter in the North Pole area, Western Australia: Field occurrence and geochemistry, in *Geochemistry and the Origin of Life*, eds. S. Nakashsima, S. Maruyama, A. Brack, and B.F. Windley, 203–236, Universal Acad. Press, Tokyo.
Valley, J.W., Peck, W.H., King, E.M., Wilde, S.A., (2002). A Cool Early Earth. *Geology*, **30**, 351–354.
Van Kranendonk, M., Webb, G.E., Kamber, B.S. (2003) Geological and trace element evidence for a marine sedimentary environment of deposition and bogenicity of 3.45 Ga carbonates in the Pilbara, and support for a reducing Archean ocean. *Geobiology*, **1**, 91–108.
Van Zuilen, M., Lepland, A., Arrhenius, G. (2002). Reassessing the evidence for the earliest traces of life. *Nature*, **418**, 627–630.
Vervoort, J.D., Patchett, P.J., Gehrels, G.E., Nutman, A.J. (1996). Constraints on early Earth differentiation from hafnium and neodymium isotopes. *Nature*, **379**, 624–627.
Walker, J.C.G. (1985): Carbon dioxide on the early Earth. *Origins of Life*, **16**, 117–127.
Walsh, M.M. (1989). *Carbonaceous cherts of the Swaziland Supergroup, Barberton Mountain Land, Southern Africa.* PhD thesis, Louisiana State University.
Walsh, M.M. (1992). Microfossils and possible microfossils from the Early Archean Onverwacht Group, Barberton Mountain Land, South Africa. *Precambrian Res.*, **54**, 271–293.
Walsh, M.M., Lowe, D.R. (1999). Modes of accumulation of carbonaceous matter in the early Archaean: A petrographic and geochemical study of carbonaceous cherts from the Swaziland Supergroup, in Geologic evolution of the Barberton greenstone belt, South Africa, eds. D.R. Lowe, G.R. Byerly, *Geol. Soc. Am Spec. Paper*, **329**, 115–132.
Walsh, M.M., Westall, F. (2003). Archean biofilms preserved in the 3.2–3.6 Ga Swaziland Supergroup, South Africa, in *Fossil and Recent Biofilms,* eds. W.E. Krumbein, T. Dornieden, M. Volkmann, Kluwer, Amsterdam, 307–316.
Watanabe, Y., Martini, J.E.J., Ohmoto, H. (2000). Geochemical evidence for terrestrial ecosystems 2.6 billion years ago. *Nature*, **408**, 574–578.
Westall, F. (1999). The nature of fossil bacteria. *J. Geophys. Res.*, **104**, 16 437–16 451.
Westall, F. (2003). Stephen Jay Gould, les procaryotes et leur évolution dans le contexte géologique. *Paléovol*, **2**, 485–501.
Westall, F., De Wit, M.J., Walsh, M.M., Folk, R.L., Chafetz, H., Gibson, E.K. (1999). An Early Archean, organic carbon-rich microbialite (3.3–3.4 Ga) from the Barberton greenstone belt, South Africa. *Intl. Soc. Study of the Origin of Life (ISSOL)*, San Diego, Abstr.
Westall, F., Folk., R.L. (2003) Exogenous carbonaceous microstructures in Early Archaean cherts and BIFs from the Isua greenstone belt: Implications for the search for life in ancient rocks. *Precambrian Res.*, **126**, 313–330.

Westall, F., Walsh, M.M. (2003) Fossil biofilms and the search for life on Mars. In *Fossil and Recent Biofilms* (eds. W.E. Krumbein, T. Dornieden, and M. Volkmann, Kluwer, Amsterdam, 447–466.

Westall, F., Boni. L., Guerzoni, M.E. (1995). The experimental silicification of microorganisms. *Palaeontology.*, **38**, 495–528.

Westall, F., Steele, A., Toporski, J., Walsh, M., Allen, C., Guidry, S., Gibson, E., Mckay, D., Chafetz, H. (2000). Polymeric substances and biofilms as biomarkers in terrestrial materials: Implications for extraterrestrial materials. *J. Geophys. Res.*, **105**, 24 511–24 527.

Westall, F., De Wit, M.J., Dann, J., Van Der Gaast., S., De Ronde., C., Gerneke., D. (2001). Early Archaean fossil bacteria and biofilms in hydrothermally-influenced, shallow water sediments, Barberton greenstone belt, South Africa. *Precambrian Res.*, **106**, 93–116.

Westall, F., Brack, A., Barbier, B., Bertrand, M., Chabin, A. (2002). Early Earth and early life: an extreme environment and extremophiles – application to the search for life on Mars. *ESA Spec. Pub.* , **518**, 131–136.

Wilde, S.A., Valley, J.W., Peck, W.H., Graham, C.M. (2001). Evidence from detrital zircons for the existence of continental crust and oceans on earth 4.4 Gyr ago. *Nature*, **409**, 175–178.

Wynn-Williams, D.D., Edwards, H.G.M. (2000). Proximal analysis of regolith habitats and protective biomolecules in situ by laser Raman spectroscopy: Overview of terrestrial Antarctic habitats and Mars analogs. *Icarus*, **144**, 486–503.

7 Lake Vostok, Antarctica: Exploring a Subglacial Lake and Searching for Life in an Extreme Environment

Jean Robert Petit, Irina Alekhina, Sergey Bulat

7.1 Lake Vostok and Ice Core Data in a Nutshell

7.1.1 Generalities

The recently discovered subglacial lakes in Antarctica are icy-water-geologic complex systems with unique characteristics. More than 100 lakes have been detected (Siegert et al., 1996, 2001), some of which are very large (up to $\sim$14 000 km^2). The high pressure, excess gas, low temperature, absence of solar energy and isolation of subglacial lakes from surface biota for thousands and perhaps millions of years make them extreme environments that may host life different from that found on the Earth's surface.

The discovery of icy Europa, a Jovian satellite covered by 10- to 100-km thick ice with a possible subglacial ocean (Carr et al., 1998, Pappalardo et al., 1999), has revived the unanswered question on the possible existence of extraterrestrial life (Chyba and Phillips 2001). In this context, the subglacial lakes of Antarctica, buried under more than 3000 m of ice, provide a possible analogue of the hydrosphere of such icy environments and have stimulated technological projects related to the development of robot exploration systems.

The subglacial Lake Vostok, (Lake Vostok hereafter) named after the nearby Russian station above the lake, is the most documented subglacial lake to date. It is the largest and deepest known subglacial lake and likely the oldest, making it one of the most attractive water bodies for geological, geophysical, glaciological and biological studies (see e.g. Kennicutt 2001). At Vostok, the ice is 3750 m thick and has been drilled down to 3623 m for paleoclimate studies. The ice drilling did not enter the lake water but penetrated 84 meters into ice formed by refreezing of the lake water (Jouzel et al., 1999). These 84 m of so-called accretion ice have provided samples that provide information on the composition of the lake water, opening an unexpected window into this unknown environment.

In this chapter, we first briefly describe the general features of the lake and the surrounding area based on the results of a geophysical survey. We then summarise results from the study of the first 3310 m of the ice core and the climatic record over the last 400 000 years as well as the mean chemical composition of the glacier ice. We will take a close look at questions concerning the origin of the lake, the formation of the accretion ice, the energy and mass balance and the biological content of the lake.

From certain thermodynamic properties of the water-ice equilibrium, a scenario for the water cycle and accretion-ice formation is proposed. The heat and mass balances are established and the renewal time of the lake water is deduced. A thermohaline-type circulation is likely present in the lake and no anomaly in geothermal flux is required to form the lake or to maintain it. This information and the energy balance should provide useful constraints for future more complex 3D numerical modelling.

The isotope composition of the ice is used to constrain the location of the melting area as well as to suggest a possible lake input from hydrothermals.

Finally, we present the state-of-the-art in relation to biological research. Biological studies of Antarctic ice are very scarce in general and for the studies of Vostok accretion ice, the published results are controversial. Indeed the very low biological content makes direct contamination (forward-contamination) a critical issue that is very difficult to overcome. From the most recent studies for which chemical and biological contamination is better documented and controlled, Vostok lake ice is very clean regarding chemical and biological content and the lake water appears primarily lifeless. Only a few bacterial phylotypes have been identified in accretion ice that are not listed as forward-contaminants of the laboratory environment or drilling and handling operations. Indeed, the chemical and biological contents of Lake Vostok are very different from those of open lakes. Lake Vostok is likely ultraoligotrophic and its probable excess of oxygen represent a potential harsh constraint. However, some niches are suggested, within the deep faults at the bottom or in sediments close to fault mouths, in which chemoautotrophic micro-organisms are likely protected and fed into the lake by hydrothermal fluids boosted by sporadic local or distant seismic events. They could be integrated in the accretion ice by a particular ice-bedrock configuration upstream of Vostok.

In the vicinity of ocean rifts, black smokers may supply life chains and create a “ biomagmatic environment”. At the bottom of Lake Vostok there is no molten rock rise, but byproducts from distant tectonic events may supply biological activity or create a “biotectonic environment”. This represents an interesting alternative scenario for primary production, for such extreme environments. Finally, the multidisciplinary approach developed for the search for life in Lake Vostok ice samples followed some guiding principles that could be applicable for the search for life in other extreme environments or for the study of extraterrestrial samples.

7.1.2 Lake Vostok: Present Knowledge and Some Open Questions

Lake Vostok, is the largest subglacial water body (14 000 km^2) among more than 100 lakes that have been detected in Antarctica (Siegert, personal communication). Subglacial lakes have been identified by characteristic strong mirror-like and very flat radio echo sounding reflectors (e.g. Siegert et al., 1996). Subglacial lake mapping requires a high-resolution grid, and in some cases lake contours can be obtained from accurate mapping of the ice-sheet surface relief from altimetry satellite imagery (Ridley et al., 1993, Rémy et al., 1999).

The ice-sheet surface is smoothed by the glacier cover but nevertheless it reflects subglacial relief. In the presence of a large lake, the glacier surface is a very flat area over hundreds of km^2, which is the case for the lake near the station of Vostok (Fig. 7.1). The Russian station of Vostok (78°28'S, 106°48'E, altitude 3488m a.s.l., −55.5°C mean annual temperature) was established in 1957 at the south geomagnetic pole during the International Geophysical Year. By coincidence, this station is located at the southern end of the lake (Fig. 7.1). The ice thickness at this location is 3750m and deep drilling in the ice began in the 1960s for geophysical and paleoclimate studies. The deepest continuous ice core (3623m) was obtained in 1998.

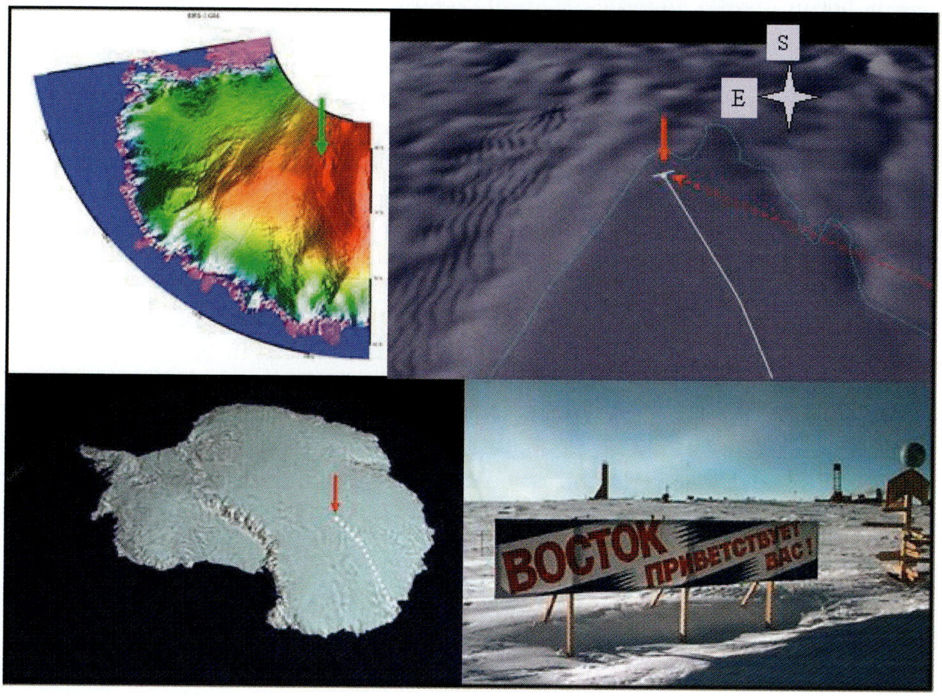

Fig. 7.1. Antarctica and Lake Vostok as seen from space and from the surface. Top left: Topographic map of a sector of Antarctica as deduced from ERS-1 satellite data (from Rmy et al, 2000). The colours are related to the surface relief, with the Lake Vostok area (*arrow*) appearing in a relatively uniform colour. Top right: Synthetic imagery of the southern part of Lake Vostok (Radarsat image, Canadian agency). The boundary of the subglacial lake with bedrock and the track used by the tractor train that supplies the station are indicated. The *dotted red arrow* from right to left represents the ice flow line near Vostok station (*solid arrow*), located 5km from the southeastern lake shore. Bottom left: A synthetic image of Antarctica. Bottom right: The welcome sign at Vostok station, the memorial geomagnetic pole and the drilling towers in the background.

The lake is crescent shaped and extends northward from Vostok station for more than 250 km (Siegert et al., 2001). Its width reaches 60 km, while the water bathymetry estimated from recent seismic studies indicates a water trough as deep as 1200 m, however, with an overall mean water thickness of 400 m (Masolov et al., 2001). The lake area is comparable in size to Lake Ontario or the island of Corsica in the Mediterranean sea, or roughly half the size of lake Baïkal. The water volume is 5600 km^3, i.e. 30 times the volume of Lake Geneva.

The glacier overlays the lake and flows mainly eastward over the lake width with a southeast movement observed locally in the Vostok area ($\sim$ 130° azimuth). From south to north, there is a slight surface slope and the glacier is 50 m higher at a distance of 250 km from Vostok. At this location, the ice thickness is 600 m greater (4250 m) than at Vostok, i.e. about 10 times the surface elevation difference, ensuring the isostatic equilibrium of the floating tongue. The glacier/lake interface therefore has a slope opposite to that of the glacier surface, deepening northward from the Vostok area by about 600 m into the lake (Siegert et al., 2001).

Accretion ice was mapped from an aerogeophysical survey. The lake ice is mostly located in the southern area and was detected as far as 165 km from Vostok (Bell et al., 2002), representing more than 50% of the lake area. The average ice thickness is about 295 m (Bell et al., 2002), larger than the 220-m thickness detected at the Vostok site (Jouzel et al., 1999). The lake ice accretes under the glacier base and is exported by glacier movement. To compensate, an inflow of water and a melting area would be required in the northern area (Siegert et al., 2001). The lake water is therefore slowly renewed and the glacier is expected to supply the lake water with ice impurities (salt, dust) and gases (O_2, CO_2, etc.) but also with micro-organisms from the surface.

We know relatively little about such subglacial bodies and even lack a qualitative empirical description of the processes that occur within the lake, e.g. the location of the melting area, the processes leading to the freezing of the lake water, the link between melting and freezing rates and the available geothermal heat and observed geophysical data. Previous models designed to study ice-sheet dynamics and date ice sequences used the presence of water as a constraint for basal melting (Ritz 1992). Therefore it was a surprise when the deepest part of the Vostok core was extracted and identified as lake ice. Also, the renewal time of the lake water has been estimated at values from 4500 years to more than 100 000 years (Kapitsa et al., 1996, Jean Baptiste et al., 2001, Bell et al., 2002). Estimation of the glacier flow from radio echoes and the modelling of the water circulation suggested melting and freezing rates up to several decimetres per year (Siegert et al., 2000). Indeed the lake system is practically isothermal and is virtually a closed system. Therefore most of the energy exchanges are dominated by the latent heat of melting and freezing. A rate of 10 cm per year represents an equivalent heat flux of about 1 W/m^2 that is $\sim$ 20 times the expected geothermal flux. The energy balance of the whole lake has to be considered. Other questions remain unanswered regarding how and when the lake was formed. Did the lake exist before the main Antarctic glaciations or did it result from ice melting due to an anomalous geothermal heat flow? Is the lake a collector of a vast subglacial hydraulic system? How is the heat redistributed within the lake? What is the chemical composition of the water and its salinity?

Concerning the presence of life in Lake Vostok, after the current discoveries and documentation of life in various extreme environments on Earth, it is reasonable to expect that Lake Vostok may contain life, and micro-organisms or footprints of their activity should then be found in the accreted ice. This will, however, be difficult to prove or disprove. Indeed, the biological content of ice from Antarctica is still undocumented, studies are scarce, not due to any lack of scientific interest, but rather because polar ice is chemically very clean and the biological signature is very small. The search for evidence of any biological activity in the ice is a major challenge as it can be easily confused with forward-contamination.

To prevent the borehole from collapsing during drilling and core extraction, the hole is filled with a fluid to counterbalance the ice pressure. At Vostok a low-grade kerosene based mixture was used and the fluid covers the entire ice-coring system as well as the ice cores. Fortunately the ice is not permeable to fluids and only the outer part of the core is contaminated. Special decontamination procedures are used to remove the outer part of the core. Antarctic ice contains very low concentrations of chemical impurities and insoluble dust at a level of around $10^{-5}\,\mathrm{g\,L^{-1}}$ (or 10 ppb) and the internal part of the ice must be processed in dust-free clean areas. Standardised procedures are now available and are used on a routine basis to assess the chemical content of the ice for major ions. On the other hand, the biological content of the ice is unknown and may vary greatly. The published results for the accreted ice raise a number of questions (Vincent, 2000) and the associated chemistry is questionable. In addition, assuming that the lake is seeded by germs from the surface biota, it is uncertain whether the micro-organisms would be capable of resisting the water veins and the oxygen dissolved in the ice matrix (Lipenkov and Istomin 2001) as they are buried in the glacier. Moreover, once released in the lake after a million years or so, could they continue to thrive in the harsh environment of the lake, with no lag for adaptation and possibly no capacity to adapt because of their damaged genome. The lake represents an extreme environment, with subzero temperatures, no light, high pressure and high oxygen and nitrogen contents supplied by melt water depleted for organic carbon and mineral salts. The likelihood of biological findings therefore needs to be discussed and assessed with respect to the specific constraints of such an environment.

7.1.3 Climate Record and Ice Chemical Properties

7.1.3.1 The Climate Record

The paleoclimatic record deduced from analysis of the first 3310 m of the ice core was the first long record from ice covering the last 400000 years (Petit et al.,

1999). The past temperature, greenhouse gas content (CO_2, CH_4) and concentrations of aerosols of marine and continental origin have been reconstructed (Fig. 7.2).

The variations of the temperature at the surface of the ice sheet are deduced from the stable isotope composition of the ice. The profile displays an overall amplitude of about 12°C. Slightly different estimates of climatic changes (up to 18°C) have been proposed (Salamatin et al., 1998). From the mean time separating each warm period (interglacial), the record suggests the presence of cycles with a periodicity of about 100000 years. Glacial periods dominate the records, representing almost 90% of the cycle duration. The present warm period, referred to as the Holocene, started circa 12000 years ago and appears to be a very stable period with respect to other interglacials. The overall climate

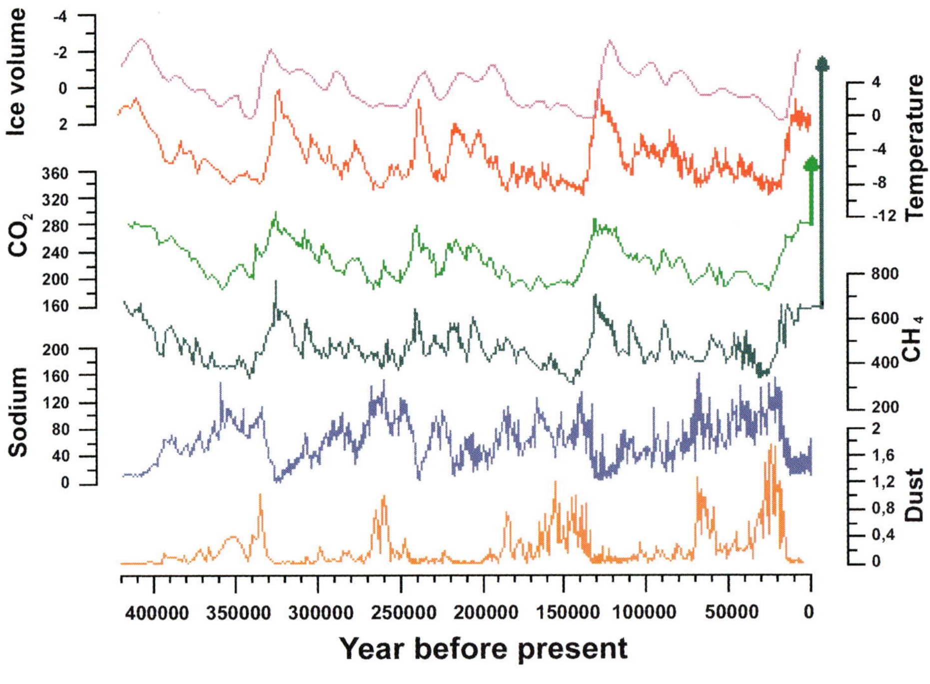

Fig. 7.2. The climatic record over the last 400000 years deduced from the first 3310 m of the Vostok ice core (adapted from Petit et al, 1999). From top to bottom:
-Global ice volume (in relative units) as deduced from the marine sediment record.
-Temperature (difference with the present surface temperature) deduced from the stable isotope composition of the ice.
-Records of greenhouse gases: CO_2 (ppmv) and CH_4 (ppbv) as deduced from entrapped air bubbles. Note the recent increase up to the present level for CO_2 and CH_4, reflecting anthropogenic activity since the 1850's.
-Profile of sodium concentration (ppb), representative of marine aerosols.
-Profile of continental dust concentration (ppm).

record displays a "saw tooth" structure with temperatures gradually decreasing from the interglacial period to reach the minima of the glacial period. This was followed by a more rapid deglaciation, taking place in just a few millennia.

Spectral analysis of the record indicates the presence of three major periodicities of about 100000, 40000 and 20000 years. These characterise orbital geometry variations and the Earth's movement around the Sun (i.e. orbital eccentricity, obliquity and precession) that trigger large climatic variations by modulating the solar energy received by the Earth according to latitude and season. The Vostok record confirms the astronomical theory of paleoclimate that is also supported by marine records (e.g. Imbrie, 1992).

Impurities in the Antarctic atmosphere include the small fraction of primary aerosols emitted by the ocean and the continents as well as those resulting from gas to particle conversion or, for the most recent period, anthropogenic activity. Sodium (Na) characterises sea spray while dust, for which the mass is mostly represented by particles with sizes between 0.8 and 3 μm, originated mostly from the continents and arid zones. During glacial periods, the colder atmosphere was also drier than during interglacials, reducing the water cycle and precipitation rate and also the capability of the atmosphere to be naturally washed out by precipitation. This is of importance because it significantly increases the residence time of aerosols and their ability to be disseminated worldwide. For glacial periods, the measured sodium concentration is 3 to 4 times higher than throughout the Holocene period, an effect partly due to the reduced precipitation and longer residence time in the atmosphere, but also to a more cyclonic activity around Antarctica. These factors, taken together, more than account for the greater sea-ice extent.

For glacial periods, the dust record shows concentrations of up to 750ppb, 50 times greater than for interglacial periods (15 ppb). Again a drier atmosphere and more efficient meridional transport toward Antarctica are suggested as main factors to which a source effect is added by the significant extension of continental aridity. The geochemical composition of the dust (isotopes of Sr and Nd) made it possible to determine their geographic origin, mostly in southern South America and in particular Patagonia (Basile et al., 1997, Delmonte et al., 2004). This area is very sensitive to climate change because it is under the influence of sea ice extension in the Drake passage. Moreover, the Southern Andes, on which ice caps can develop, acted as a meteorological barrier to precipitation. Together, this leads to intense periglacial erosion in the mountains, efficient transport of sediment to the plains by seasonal runoff and intense aeolian surface aggradation of the outwash plain during the dry seasons.

Air bubbles entrapped within the ice provide atmospheric air samples that are naturally and continuously collected during snow and firn densification. Between the different glacial and interglacial periods, greenhouse gases contents varied similarly as did the temperature. For CO_2, the concentrations varied from 180 to 300ppmv (parts per million by volume) and for CH_4, from 360 to 700ppbv (parts per billion by volume). For the recent period, since 1850, CO_2 and CH_4

have both steadily increased and have reached the highest concentrations ever observed for the past 400,000 years, 360 ppmv and 1700 ppbv, respectively. Such changes are the consequence of anthropogenic activity with the use of fossil fuels for energy as well as the intensification of agricultural activity.

Over the last 400000 years preceding this recent period, the record shows clear correlations between temperature and greenhouse gases, indicating that these gases participated and contributed to the amplitude of the temperature variations. During the last deglaciation, the very small radiative forcing change due to greenhouse gases (CO_2 but also CH_4 and other gases such as N_2O, etc.) was amplified by positive reactions from the climate system: decrease of sea ice, increase of water vapour, increase of surface albedo, etc. Finally, up to 50% of the global temperature change, which was about 4 °C from the glacial to interglacial period, may be due to the resulting greenhouse gas effect. Such results and amplification effects are now also supported by climate simulations using general circulation models (GCM). Information derived from the ice-core records makes it possible to describe the natural changes of the climate, and is useful to test climatic models with scenarios from the past (e.g. glacial period). The estimate of climate sensitivity to greenhouse gases as deduced from the past and from GCM is of importance for predicting future global warming and the environmental changes due to the almost inescapable doubling of the atmospheric CO_2 by the end of this century.

7.1.3.2 Chemical Composition of the Glacier Ice

Antarctic ice contains impurities and salts at very low concentrations. This results from the natural distillation process of the water cycle and the very distant oceans and continents that are the major sources. Antarctic ice contains 0.3 to 0.5 $mg\,L^{-1}$ of total salts, i.e. 100 to 1000 times less than the content of commercial mineral water (100 to 2500 $mg\,L^{-1}$, Table 7.1). Chemical analysis of the ice requires special equipment and sensitive techniques. Ice-core samples must first be decontaminated by removing the outer part. The inner part is then analysed in a dust-free laboratory. The use of clean rooms and deionised ultra-filtered pure water are prerequisites for ice chemistry as in the fields of medicine or microelectronics.

Melt water from Antarctic ice is acidic (pH $\sim$ 5.7), mainly due to the very low contribution of continental dust and carbonates and the excess sulfate ions such as sulfuric acid from the atmosphere. Aerosols from sea spray contribute most of the Na^+, Ca^{2+}, Mg^{2+} and Cl^- ions found in the ice. Dust from the continents contributes Ca^{2+}, Mg^{2+}, K^+, $NO_3{}^-$ and Fe^{3+} ions. The sulfates are dominated by sulfuric acid that is the end product of the atmospheric oxidation of sulfur components such as dimethyl sulfur (DMS), produced from the biological activity of the marine phytoplankton. On the other hand, sporadic volcanic events inject large amounts of ash and gases into the atmosphere. Ash layers are sometimes visible in ice cores. The gases, mostly represented by SO_2, also contribute to

Table 7.1. Mean composition (mg per litre or ppm) of commercial mineral water and Vostok ice (from Legrand et al., 1988). Dry residues for ice are given as the sum of the soluble salts only. All water names are registered trademarks ®. (1) Mineral water enriched with magnesium and calcium sulfate, (2) Mineral water with low salt content

	pH	Dry residue	Ca^{2+}	Mg^{2+}	K^+	Na^+	HCO^{3-}	Cl^-	$SO_4{}^{2-}$	NO^{3-}
Hepar (1)	7	2580.0	555.0	110.0	4.0	14.0	403.0	11.0	1179.0	2.9
Contrex (1)	7.2	2125.0	486.0	84.0	3.2	9.1	403.0	8.6	1187.0	2.7
Vals	7.2	2021.4	45.2	21.3	32.8	453.0	1403.0	27.2	38.9	< 1
Evian (2)	7.2	309.0	78.0	24.0	1.0	5.0	357.0	4.5	10.0	3.8
Valvert (2)	7.3	201.0	68.0	2.0	0.2	1.9	204.0	4.0	18.0	4.0
Volvic (2)	7	109.0	9.9	6.1	5.7	9.4	65.3	8.4	6.9	6.3
Vostok mean ice	**5.7**	**0.455**	**0.026**	**0.010**	**0.004**	**0.086**	**0.000**	**0.095**	**0.187**	**0.047**
Vostok recent ice	**~5.7**	**0.200**	**0.004**	**0.004**	**0.002**	**0.033**	**0.000**	**0.014**	**0.127**	**0.018**
Hepar/Vostok		5670	21 346	11 000	1000	162	na	115	625	145
Volvic/Vostok		239	380	610	1425	109	na	88	36	13 4

sulfuric-acid formation in the atmosphere and to ice acidity. Sulfuric-acid inputs modify the electrical conductivity of the ice and volcanic events are detectable by a nondestructive electrical method often used to establish the ice core chronology.

Antarctic ice is very low in carbon from both mineral and organic origin. It does not contain carbonates. Some carbonates may be present in soils and in dust at the time of emission from the continents, however, they are likely neutralised by sulfuric acid during transport to Antarctica. Carbon is present in CO_2 and CH_4 gases entrapped in the air bubbles of the ice. As ice contains $\sim 8\%$ atmospheric air with 180–280 ppmv of CO_2 and 600–1300 ppbv of CH_4, it represents only 10 ppb of carbon (Anklin et al., 1995). Organic acids from vegetation and biological activity are also found in Antarctic ice (Legrand and Saigne 1988, Legrand et al., 1993) as traces of carboxylic acids (acetic, formic, oxalic, etc.) or methane sulfonic acid (MSA). Altogether this represents only 1 ppb of organic carbon or $\sim 10^{-7}\,\mathrm{mole\,C\,L^{-1}}$. Assuming that continental dust emissions are from soils containing about 1 per mil of carbon from humus and vegetation residues, a maximum of 1000 ppb of dust concentration as observed during the glacial period would contribute to only 1 ppb of organic carbon in ice ($10^{-7}\,\mathrm{mole\,C\,L^{-1}}$).

7.1.4 Accretion Ice

On the Vostok ice core, below 3310 m depth, the horizontal layering is disturbed by the glacier dynamics. The climatic record is discontinuous and difficult to interpret. At 3539 m depth there is a sharp transition with changes in several parameters of the ice. The first change concerns the visible sediment inclusions of millimetre size (Fig. 7.3) that occur randomly (from 0 to 20 inclusions ob-

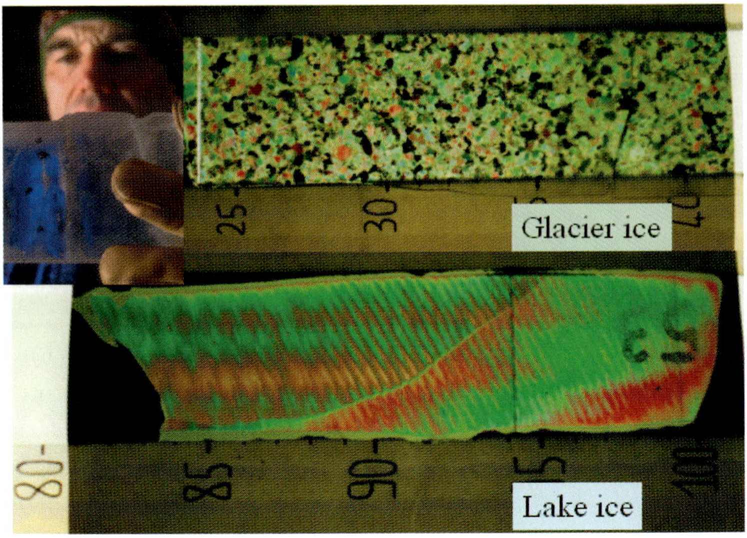

Fig. 7.3. Accretion ice samples. Top left: Sediment inclusions observed in a Vostok ice core sample at 3607 m depth (photograph: L. Meydard/CNRS). Thin ice sections of ice core observed with polarised light. Top: Polycrystalline glacier ice. Bottom: Accreted ice (3553 m depth) showing a single 32 cm long crystal. Colour irregularities are due to a rough ice surface.

served for each metre of the ice-core increment), the second is a change in the size of the ice crystals from a few millimetres in the overlying glacier layers to one decimetre (Fig. 7.3) and even up to one meter, and the third is a drastic decrease in the electrical conductivity of the ice (Fig. 7.5). Laboratory studies also indicate that the ice does not contain gases and the stable isotopic composition (deuterium) of ice displays a sharp transition in just 30 cm and a shift of about 10‰ (see Sect. 7.3 for notation) with respect to the overlying glacier ice (Figs. 7.4 and 7.5). From 3539 m to 3623 m depth, the isotope composition is almost constant, representing another significant difference with respect to the variable climatic record (Fig. 7.4).

The isotope composition change was first used to support the formation of basal ice by a freezing process (Jouzel et al., 1999; Souchez et al., 2000), assuming that the lake water had an isotope composition equivalent to that of the Vostok core. During freezing, isotope fractionation occurs and the ice is enriched in heavy isotopes with respect to the water. Further considerations suggest that the lake-water composition is not well defined because the melting zone is about 250 km from Vostok in the northern area of the lake, and the observed enrichment of the accreted ice could be explained primarily by an origin effect (see Sect. 7.3.1).

The very low air content of the basal ice likely provides the best straightforward evidence that the deep ice was formed by a slow freezing process. The

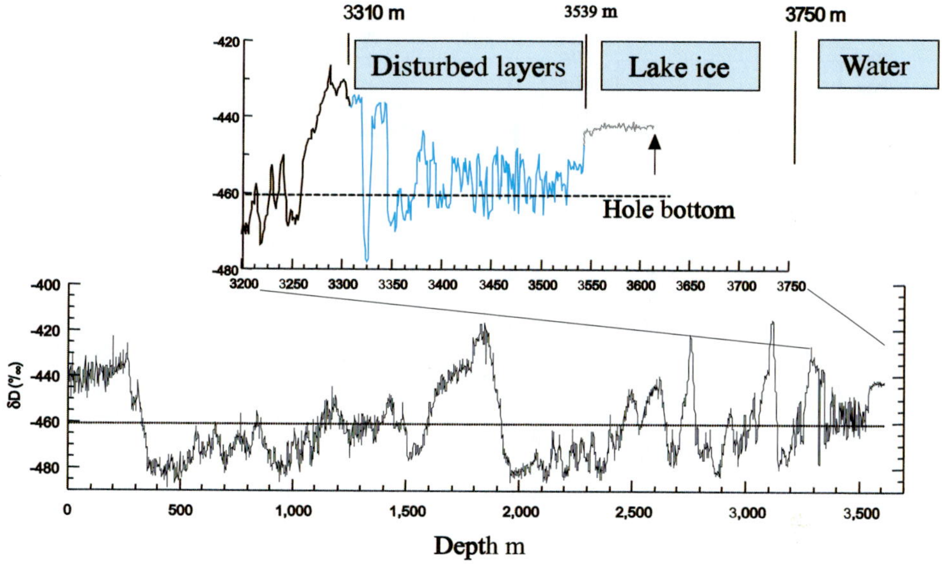

Fig. 7.4. Down core profile of the stable isotope concentration (Deuterium) of the Vostok ice core. From the surface to 3310 m, layering is preserved and the last 4 climatic cycles can be recognised. For deeper ice and down to 3539 m, the climatic record is disturbed by ice deformation and bedrock influences. At 3539 m depth, a sharp transition occurs at the beginning of the accretion ice and ice composition becomes almost constant. Drilling operations were stopped at 3623 m depth. The water interface is at 3750 m (adapted from Jouzel et al., 1999).

glacier ice, down to 3538 m depth, contains $\sim$ 8% of air, observable as bubbles for the first 500 m and as air hydrates (Lipenkov, 2000) for deeper ice. The air content of the lake ice is 100 to 1000 times lower than glacier ice and is close to the detection limit of the apparatus (Souchez et al., 2002). When ice forms, its lattice does not accept impurities and expels most ions and gases. Melting and refreezing is a technique used in the laboratory to extract air from glacier ice for analysis.

The unusual size of the ice crystals and the very low electrical conductivity support the proposed freezing process. The size of the crystals is large, similar to those commonly found in lake ice. In addition, the very low electrical conductivity suggests that impurities have been rearranged within the lattice, suggesting a process involving an equilibrium with water. Furthermore, thermodynamic considerations for the lake and its interface imply that the accretion ice must have been formed when the glacier/water interface was inclined. Due to the variable ice thickness and therefore the variable pressure at the interface, the melting point changes and the deeper interface must melt while the lake water freezes in the upper levels (see Sect. 7.1.5 and Fig. 7.8).

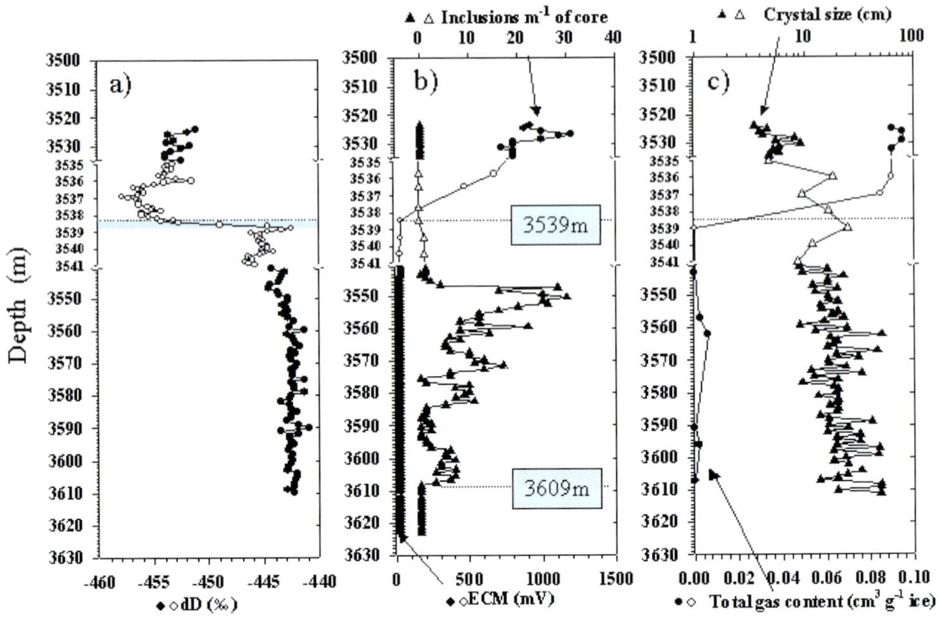

Fig. 7.5. Ice property changes as observed near the glacier-accretion ice boundary at 3539 m depth on the Vostok ice core. a) Deuterium content (in δD ‰); b) Inclusion number concentration (number per metre of ice increment: m^{-1}) and electrical conductivity measurement (ECM) of the ice (given in mV or in relative units) c) Ice crystal size (cm) and total air content of the ice ($cm^3 g^{-1}$) (adapted from Souchez et al., 2003).

At the Vostok site, the total accretion ice is about 220 m thick (from 3539 to 3750 m). It can be divided in two main parts:

- From 3539 to 3609 m depth: 71 m referred to as "*accretion ice* I", containing visible inclusions from sediments or from bedrock.
- From 3609 to 3623 m: 14 m of clear ice without inclusions ("accretion ice II"). It is assumed that this ice remains similar down to the interface with the lake at 3750 m. The total thickness would be 141 m.

The ice sheet flows from west to east and the flow line passing by Vostok originates at Ridge B, an area 300 km upstream. In this area, a deep ice core has been drilled for paleoclimate studies (Dome B ice core). Along the flow line passing by Vostok, the ice sheet velocity is about 3 ± 0.8 m/a (Bell et al., 2002). As accretion starts as soon as the glacier comes afloat on the lake, the first 71 m of accretion ice I starts to form in a shallow water bay upstream of Vostok (see Fig. 7.6). Radar and seismic surveys indicate that the bay is 11 km wide (Bell et al., 2002) and opens into the lake in the north (see Fig. 7.7). The ice sheet floats over the shallow bay, then overlays a bedrock rise close to the bay (Fig. 7.6).

The presence of visible inclusions in accretion ice I suggests that the basal ice was in contact with bedrock or with sediments. The inclusions are millimetre-

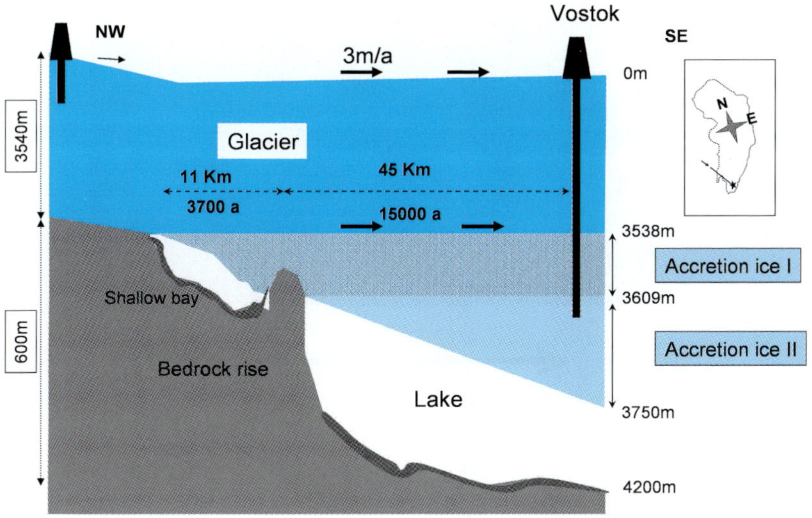

Fig. 7.6. Sketch of the cross section of the ice sheet and the lake along the flow line passing by Vostok. The ice sheet flows from the Ridge B area, about 300 km from Vostok, where deep drilling (Dome B) has been performed. The glacier comes afloat in the lake in a shallow bay and then overlays a bedrock rise 11 km from the shore. The two types of accretion ice are shown: accretion ice I contains visible sediments while the layer of accretion ice II is clear. The ice sheet floats for ∼ 56 km eastward over the lake before reaching Vostok. The mean velocity of the ice (∼ 3 m/a) can be used to estimate transit times and mean accretion rates (see the text).

sized aggregates formed by small clay particles (micrometre sized), sometimes surrounding a submillimetre-sized mineral (possibly quartz). Chemical analysis indicates the presence of calcium, magnesium and sulfates in large amounts as well as sea salt (sodium chloride) (Montagnat et al., 2000, Souchez et al., 2000, Cullen and Baker 2002, De Angelis et al., 2004). In particular, sulfates of calcium (Cullen and Baker 2002) and magnesium have been observed, suggesting gypsum or sediments from soils formed in an environment subject to strong evaporation (De Angelis et al., 2004). The lake water is not saturated with salts, and the high concentration of soluble sea salts within discrete inclusions suggests they have been preserved from dissolution, possibly by clay coating, and/or included in the accretion ice by a rapid freezing process.

In summary, in accretion ice I, major chemical elements are clearly enriched with respect to the glacier composition in discrete layers containing sediment inclusions, with ion concentrations up to 1000 ppb for sulfates, ∼ 600 ppb for chloride, ∼ 500 ppb for sodium, and ∼ 200 ppb for calcium and magnesium. On the other hand, accretion ice II, as well as some of the discrete layers without sediments in accretion ice I, are depleted with respect to the glacier ice. The sulfates (11 ppb), chloride (22 ppb), sodium (14 ppb), calcium (4 ppb) and magnesium (1.4 ppb) mostly represent sea salt and calcium-magnesium sulfate (De

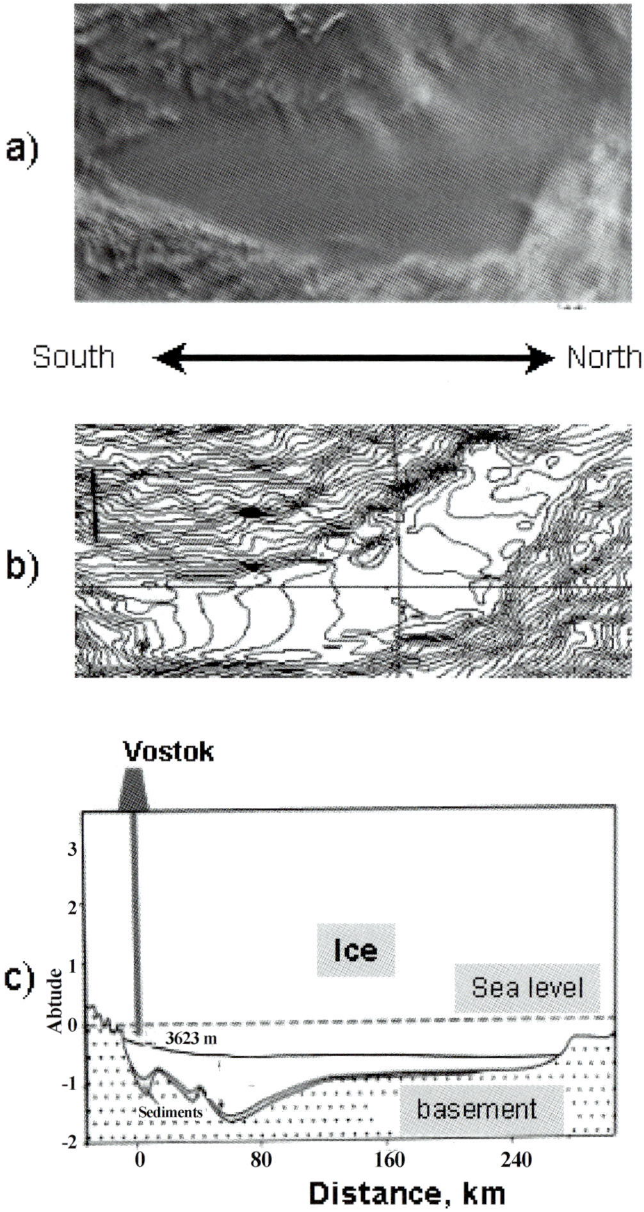

Fig. 7.7. Satellite and seismic information for Lake Vostok. (a) Synthetic Radasat imagery (Canadian Space Agency) The arrow represents a distance of 250 km. (b) Ice sheet surface topography (from Rmy et al., 2000). (c) North-south cross section of the glacier and lake as deduced from seismic and radar studies (adapted from Masolov, et al., 2001). The mean water thickness is 400 m but reaches 1200 m in a trough. The drilling tower marks the location of Vostok station.

Angelis et al., 2004). In all, the $\sim$ 60ppb of salts is almost 10 times lower than for recent snow and is likely representative of undisturbed accretion ice or deep lake conditions. During freezing, ions are not fully integrated in the ice lattice. Souchez et al., (2000) discussed the distribution of these ions between ice and water during freezing. From experiments, a reasonable range for the distribution coefficient (i.e. ice to water ion concentration ratio) is from 0.0008 to 0.0028, giving a lake salinity between 0.03 and 0.08‰. Such values correspond to mineral water with a low salt content.

When floating over the lake, basal friction is negligible and the glacier tongue moves as one piece. The basal and surface ice velocity are similar (i.e. $\sim$ 3m/a). We deduce the 71m of accretion ice I was formed in about 3700 years, giving an apparent mean accretion rate of 19mm/a. However, this value is representative of accretion only for the shallow bay area since a subglacial relief located 11km from the shore may cause the accretion ice I to thicken.

Accretion ice II starts to form once the glacier tongue is freely floating over the deep lake. In the southern area, the water depth is more than 600m and, after the bay relief, no bedrock rise has been detected along the flow line to Vostok. On the ice core, from 3609m to 3750m depth ($\pm$10m), accretion ice II represents 141m ($\pm$10m). The apparent accretion rate for deep lake conditions would be $\sim 9.5 \pm 2.7$mm/a, using 45km ($\pm$3km) as the distance between the bedrock rise and the Vostok site.

7.1.5 Lake Setting and Possible History of Lake Vostok

The east Antarctic continent is an old shield that separated from the "Gondwana" supercontinent, along with Australia, Africa and South America, some 200 millions years ago. It has been suggested that the lake is cradled in an aborted rift zone that likely extended 1500km northward to Prytz Bay, into which the huge Lambert glacier flows (Leichenkov et al., 2002). A recent intensive aerogeophysical survey (gravimetric, magnetic, radio echo soundings) and the deep (down to 40km) crust-layer teleseismology soundings, suggest that a continental-sheet thrust, probably of Proterozoic age, moved backward to close the rift, overlapping a passive continental margin. The depression in which the lake is cradled likely resulted from a subsequent reactivation of the fault system, reopening the rift (Studinger et al., 2002, 2003). Several recorded seismic events suggest that the fault system is still active, but the direction of movement (compression or extension) has not been determined (Studinger et al., 2003). The depression in the bedrock resembles a bath tub 250km long by 50km wide with a mean basement 500 to 600m lower than the average surrounding bedrock. The basement of the southern edge of the lake (under 3600m of ice) is almost at sea level, while the northern edge, under 4200m of ice, is 600m below sea level. This gives the two sides of the lake basin a slope inclined northward (Fig. 7.8). As the ice from Ridge B flows perpendicularly to the sides, the ice sheet comes afloat and forms a glacier/lake interface that is inclined and permanently regenerated.

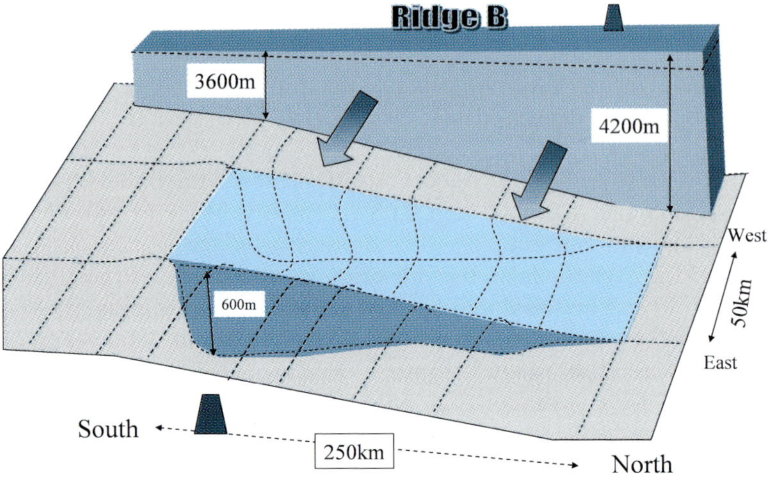

Fig. 7.8. Sketch of the Lake Vostok basin and the overlying glacier. The ice sheet flows from Ridge B and the glacier-water interface is inclined northward, similarly to the bedrock topography. The surface of the floating ice sheet is inclined in the opposite direction and a 50 m elevation difference compensates for the greater ice thickness (Kapitsa et al., 1996). One drilling tower indicates the Dome B drilling location on Ridge B and the other the Vostok site at the southern end of the lake.

In the southern area, a deeper gully (1200 m) was detected at the lake bottom just 30 km north of Vostok. The bottom of the lake is $\sim$ 1600 m below mean sea level (Masolov et al., 2001). This depression suggests that the subglacial trough was not formed from glacial scouring alone, but also from tectonic phenomena. If so, the bedrock depression existed before the Antarctic glaciations and, under such conditions, a protolake was likely already present there 34 millions year ago (Barett 1996). Sea salts and gypsum found in excess in sediment inclusions within the accreted ice (De Angelis et al., 2004) support the idea of an evaporating pool filled with sea water.

Once formed in the bedrock, the bath tub also formed a collector for precipitation and runoff. The site is located 1400 km from the ocean and prior to the quaternary glaciations, the climate was cold and arid in the vicinity of the geographic South Pole. The reduced precipitation and probable evaporation associated with the intense solar radiation contributed to the concentration of salts and possibly the formation of evaporite sediment. Therefore, a small protolake likely existed and because of the depth of the pool, the carving by the young Antarctic ice sheet was not sufficient to sweep out the sediments. The protolake may have been covered by ice, operating similarly to some present lakes in the dry valleys of Antarctica, fed by runoff from the summer season and concentrating additional salts. For example, Lake Vanda is covered by ice and the deep-water layers are highly saline and at a temperature of +24°C, as a result of the solar heating through the ice and the stratification of the deep layers.

A high-saline "protolake" likely existed at Vostok before the Antarctic glaciations. Then once the ice sheet overlaid the region, the salts were dissolved. The volume of the lake increased and once the cavity geometry was large enough, it reached a steady state depending on the mean temperature at the glacier/water interface (see Sect. 7.1.5).

Presently, accretion ice forms in the southern part of the lake and is exported out of the lake area by glacier movement. If steady state is assumed, melting must occur in another part of the lake (northern area) to renew the lake water. As salts are excluded when water freezes, we can estimate the theoretical contribution to lake salinity assuming that the lake water was initially fresh (glacier melt water) and the lake volume constant with time. This contribution would give a water salinity of about 100 ppm (0.10‰) assuming $300 \times 10^{-6}\,\mathrm{g\,l^{-1}}$ as the mean salt content of the melting ice and 300 renewal cycles over the last ~ 30 millions years. With 0.10‰ salinity, the lake water would correspond to low-salt mineral spring water of the "Volvic" type (see Table 7.1), but without carbonates, and acidic with a slight excess of sulfuric acid. A salinity of ~ 0.02 to 0.06‰ is suggested from the chemical analysis (see Sect. 7.1.4) of accretion ice II.

As for salts, gases are excluded from the ice lattice during freezing, and they accumulate in the lake. Glacier ice contains 8% air (STP) and after 300 renewal cycles, the lake may contain as much as 21 and 5 litres STP (or 26 and 8 g) of nitrogen and oxygen per litre of water, respectively. Due to the pressure and temperature of the lake, some of the gases likely form air hydrates (clathrates). Nevertheless, the lake water may remain saturated (~ 0.5 to 1 g O_2 L^{-1}) or at least highly enriched in dissolved gases (up to 50 times for oxygen) with respect to an open lake (Lipenkov and Istomin 2001, McKay et al., 2003).

7.1.6 Ice–Water Equilibrium

The glacier floats over the lake and the hydrostatic pressure is balanced. With respect to the southern end, the greater ice thickness and buoyancy of the northern end are compensated by a higher ice sheet surface elevation. The temperature of the ice sheet increases from $-55\,^{\circ}\mathrm{C}$ at the surface up to melting point at the base, in response to the geothermal heat flux. The upward heat flux through the ice (F in $\mathrm{W/m^2}$) is given by:

$$F = -K \times \mathrm{d}\Theta/\mathrm{d}Z \tag{7.1}$$

where K is the thermal conductivity ($2.2\,\mathrm{W\,m/^{\circ}C}$ for ice at $0\,^{\circ}\mathrm{C}$) and $\mathrm{d}\Theta/\mathrm{d}Z$ the vertical temperature gradient in the ice (positive downward). In Antarctica, the expected value for the mean heat flux is from 45 to $55\,\mathrm{mW/m^2}$ (Ritz, 1992). The temperature gradient in the ice sheet depends mainly on the ice dynamics driving the downward advection of cold ice. In the steady state, for a given ice thickness, the mean accumulation rate and surface temperature are the most important parameters (Ritz, 1992; Salamatin et al., 2003). At the bedrock/ice interface, the ice layers are warmed and they melt as soon as the geothermal heat flux exceeds the heat loss. Water may seep and accumulate in a bedrock depression, possibly forming a lake. The lake volume will increase until the heat flux at the interface reaches a new equilibrium.

To melt or to freeze 1 mm of ice in one year, an equivalent heat flux of 9.78 mW/m^2 (positive or negative, respectively) is required. For instance, in alpine glaciers, the entire ice mass is at the melting point and in equilibrium with water. A geothermal heat flux of 80 mW/m^2 melts 8 mm of basal ice annually. As a practical implication, the continuous streams of water produced by this melting at high altitude in the Alps are sometimes used to supply hydroelectric power plants in the valleys.

The melting of very deep ice layers of the Antarctic ice sheet was suggested theoretically already in the 1960s (e.g. Robin, 1955). However, for a given area, it is very difficult to predict melting and the formation of subglacial lakes using numerical models because many key parameters are not well known (Ritz, 1992). In addition to the bedrock topography, ice dynamics, geothermal heat, ice deformation and the vertical velocity of the ice, the layer history and glacier dynamics must be taken into account. Furthermore, the model should be integrated over a very long time. A small difference in the heat equilibrium and temperature profile may result in an ice sheet with a basal ice layer at a subfreezing temperature, frozen on the bedrock, instead of a water layer with a thickness of one hundred meters.

The temperature for water–ice equilibrium depends on a number of parameters including the interface curvature, salt concentration, pressure and gas content. Neglecting the gas content, it can be expressed by the relationship (Raymond and Tusima 1979):

$$T_f = T_0 - 2 \times \alpha_c / C + \beta_s \times s - \beta_p \times P \tag{7.2}$$

where T_f is the temperature in °C, T_0 is the triple point of water (0.01 °C by convention), α_c the mean effect of the interface curvature (C), β_s a coefficient for the chemical effect of impurities at a concentration of s and β_p the Clausius–Clapeyron coefficient for the pressure effect.

The curvature effect depends on crystal size. When two different ice crystals are in contact, the smaller crystal gradually disappears to the benefit of the larger one in order to minimise the total energy of the pair. This property drives the growing process of ice crystals in glaciers.

Chemical impurities decrease the melting point temperature (cryoscopy effect). For sea salt, β_s is $-$ 0.0573 °C psu^{-1} (psu: practical salinity unit or per mil). This property is used for example to melt snow or prevent ice formation on roads during winter by salting, as long as the temperature is not too cold.

For subglacial lakes, the water salinity is driven by the temperature of the interface, or, if water circulates, by the average lake temperature. In both cases, the water temperature is the result of an equilibrium between cooling by the upper glacier and warming by geothermal heat. If we consider a saline lake that accumulated in a bedrock depression before Antarctic glaciation; Once the lake is covered by the ice sheet, if the surface temperature, glacier dynamics and geothermal heat flux impose a temperature of $-11\,°C$ at the interface, a lake could exist but with 18% salinity, assuming sodium chloride. With a salinity of 0.1‰ or 1‰, the interface temperature will be 0.005 or $0.05\,°C$ lower than the freezing point for pure lake water.

The pressure lowers the melting point temperature of the ice by $0.0074\,°C/bar$, or $7.2 \times 10^{-4}\,°C/m$ of water or $6.7 \times 10^{-4}\,°C/m$ of ice (ice density is 0.923 for deep ice at Vostok). At Vostok, the glacier thickness is 3750m in the southern area and 4200m in the northern area, giving 337 and 377bars (33.7 and 37.7MPa) for the pressure and $-2.5\,°C$ and $-2.8\,°C$ for the temperature of the ice/water interface (pure water case) respectively. Though this $0.3\,°C$ difference is small and associated with the inclined interface, it is sufficient to drive the heat exchanges and initiate water circulation within the lake. This will force the system toward a horizontal water/ice interface and toward temperature homogenisation.

When applied to Lake Vostok, anticipating water circulation within the lake, one may assume the mean temperature of the lake water is close to $-2.65\,°C$, corresponding to the mean temperature of the interface. If the water is slightly saline or contains gases, all temperatures are shifted to a more negative value but the difference between the high and low levels of the interface would remain the same. Thus the mean water temperature divides the lake into two different regions (Fig. 7.9):

- In the lower half of the lake (northern area), the mean water temperature ($-2.65\,°C$) should be systematically warmer than the glacier interface (from -2.65 to $-2.8\,°C$). The heat from the water will therefore be transferred to the interface and the ice melts.
- In the upper part of the lake (southern area), the opposite phenomenon occurs. The mean water temperature ($-2.65\,°C$) appears to be super-cooled with respect to the interface temperature (-2.5 to $-2.65\,°C$). The lake water freezes and the ice accretes.

Finally, for a subglacial lake, an inclined interface is not thermodynamically stable and leads to the ice melting in the deep part and the water freezing in the higher levels. For Vostok, as soon as satellite information showed an inclined interface with a level difference of about 600m between the northern and southern area (Ridley et al., 1993, Kapitsa et al., 1996), the presence of accretion ice under the Vostok drilling site could be expected. Indeed, in January 1998, drilling operations at Vostok reached 3623m and lake ice was finally identified by laboratory stable isotope measurements in March 1998. Subsequently, theoretical aspects were considered and the phenomenon compared to the evolution of the unstable vertical sides of an ice cube floating in a glass of pure water (H. Miller, personal communication).

Energy transfers in the lake are mostly the result of water circulation, which is in turn driven by water-density differences. Under atmospheric pressure, the

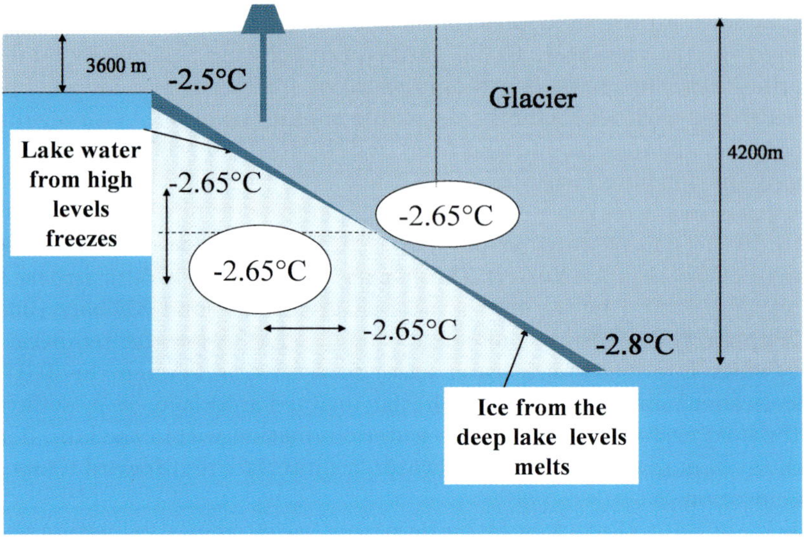

Fig. 7.9. Sketch of the temperature of the water-ice interface across Lake Vostok. Due to mixing by water circulation, the mean lake water temperature should be close to the average interface temperature (e.g. $-2.65\,^{\circ}$C). In the low levels of the lake, the water is warmer than the interface temperature and the ice melts. Conversely, in upper lake levels, the water appears to be super-cooled with respect to the interface temperature and it freezes.

density of fresh water has a maximum at $4\,^{\circ}$C. This value decreases with pressure as does the freezing point of water, but with a steeper gradient (see Souchez et al., 2000 and references herein). For pressure higher than 290 bars ($\sim$ 3170 m of ice), the water freezes before it reaches its maximum density. For Lake Vostok conditions, the volume expansion of water ($\mathrm{d}V/V$) is positive and the density decreases as temperature increases: the value is $1.8 \times 10^{-5}\,^{\circ}\mathrm{C}^{-1}$ for fresh water (Wuest and Carmak 2000) and $3.85 \times 10^{-5}\,^{\circ}\mathrm{C}^{-1}$ for sea water (Jenkins and Bombosch 1995). Between the high and the low level of the lake interface, the $0.3\,^{\circ}$C difference in temperature would give a 0.6 to 1.2×10^{-5} difference for the water density.

Water density is sensitive to salinity. The expansion coefficient is negative (shrinking) and is -7.86×10^{-4}/psu for sea water (Jenkins and Bombosch 1995). With a salinity difference of 0.1‰, the density changes by $\sim 8 \times 10^{-5}$, i.e. ~ 10 times more than the change due to the temperature effect. In Lake Vostok, glacier ice melting produces fresh water in the lower part of the inclined interface and the density difference should result in upwelling. Under the ice shelves of Antarctica, floating on the ocean over more than 600 km, melting of the deep ice produces a fresh water current under the tilted interface, a phenomenon referred to as "ice pumping" (Lewis and Perkin, 1986). In Lake Vostok, an analogous phenomenon likely occurs but with a lower magnitude.

7.2 Empirical Model for Water Cycle and Energy Balance

7.2.1 Underlying Concepts

Today, the ice melts in the northern area of the lake and the water refreezes in the southern area. The lake represents an open system but due to the limited ice and water exchanges, it can be considered as an almost isolated system and most of the available energy comes from geothermal heat. As the glacier floats, there is no dragging force and the heat produced from the ice deformation is negligible. The melting of the ice and the freezing of the water mobilise the energy as latent heat. Due to the small vertical temperature gradient at the base of the glacier and close to the interface ($0.02\,°\mathrm{C\,m^{-1}}$), the heat input needed to warm the glacier ice before melting and the heat released by the export of the accretion ice is very small. As the overall lake is isothermal and does not accumulate heat, the geothermal heat is almost totally transmitted to the glacier through the interface, as is the case for an overall simple conduction effect. The lake can be thought of as a thermal machine using the water phase changes and water circulation to transfer energy.

At Vostok, the accretion rate is about 9.5mm/a, corresponding to a latent heat of $92\mathrm{mW/m^2}$ to be released. If the freezing is entirely due the cooling by overlying ice, $92\mathrm{mW/m^2}$ should be diffused through the glacier. In fact, only a heat flux of $\sim 44\mathrm{mW/m^2}$ diffuses upward as deduced from borehole measurements at Vostok. A remaining part of the latent heat ($\sim 48\mathrm{mW/m^2}$) therefore enters the lake and must be released through the ice elsewhere than in the Vostok area in order to maintain the isothermal state of the lake. The formation of ice within the lake water represents an efficient way to furnish heat to the lake water, assuming some water may reach the supercooled state. This is the case under ice shelves where small ice discs called frazil ice are formed within fresh water (low density) plumes from deep ice melting, rising within the sea water (higher density) along the inclined ice/water interface. Similar phenomena likely occur within Lake Vostok.

The scenario for the lake processes can be based on those occurring under ices shelves (Jenkins and Bombosch 1995), with the process leading to frazil formation within the water layer seeping under the tilted interface of ice shelves (Souchez et al., 2000). Here, we also suggest that frazil increases in density and consolidates into accretion ice via a sintering process similar to that described for wet snow densification (Colbeck 1973, 1980, Raymond and Tusima 1979). We also use available information from a geophysical survey and observations from the accretion ice analyses (Souchez et al., 2000). The scenario is built up in 10 main steps.

1. In the northern area, at 4200m depth, the temperature of the glacier/lake interface is $-2.8\,°\mathrm{C}$. The main lake water mass, which is slightly saline and

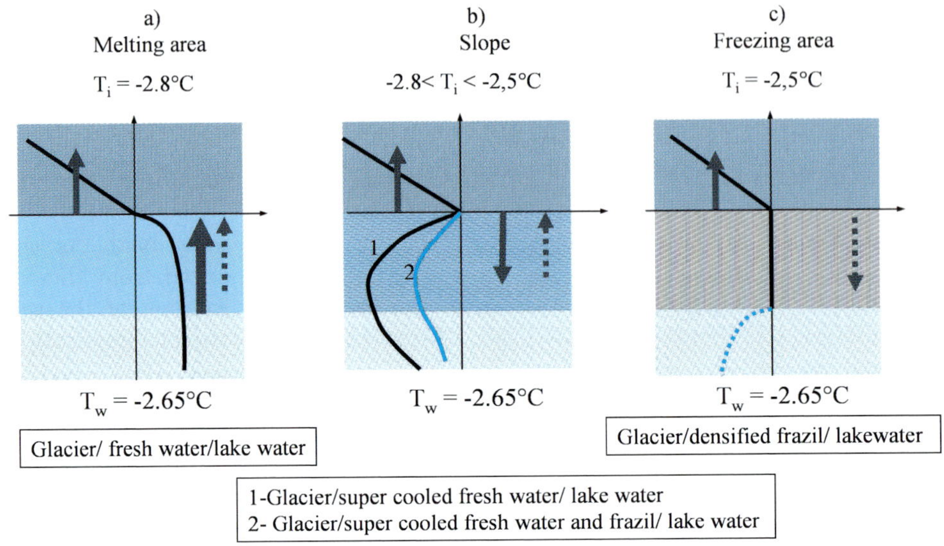

Fig. 7.10. Schematic temperature profiles through the glacier-lake interface (adapted from Jenkins et Bomboch, 1995). The vertical scale represents a few centimetres or decimetres; and the temperature scale is a few hundred $^\circ C$. Solid and dashed arrows indicate heat and salt fluxes respectively.

a) Melting area: the interface temperature (Ti) is $-2.8\,^\circ$C while the mean water temperature Tw is $-2.65\,^\circ$C. Ice melts and heat and salt fluxes are toward the interface.
b) Slope area: Ti is between -2.8 and 2.5 °C. The fresh water layer rising and seeping along the slope is cooler than the pressure-melting point. The heat flux is from the interface toward the lake, and when frazil ice forms, the latent heat is released to fresh water and to the lake 1) initial temperature profile, 2) after frazil formation.
c) Accretion area: the water frazil mixture accumulates in high levels of the lake and is transformed into ice. All the mass is at $Ti = -2.5\,^\circ$C, latent heat is diffused through the glacier and salts are expelled.

at a temperature of $-2.65\,^\circ$C, carries the energy. One part is transferred to the interface and the ice melts. The temperature profile through the interface is shown in Fig. 7.10a. The heat and salt fluxes are toward the interface. As the interface is warmer than the glacier, part of the heat diffuses through the glacier.

2. The fresh-water layer with a density lighter than lake water ($\sim 8 \times 10^{-5}$ if lake salinity is 0.1‰) does not mix immediately with the lake water: the salt diffusion is lower than the heat diffusion. Fresh-water buoyancy causes the layer to seep along the tilted interface. Under the ice shelves, melting is significant since the ocean may provide considerable amounts of energy and the produced fresh water layer may be as thick as 100 m at a distance of 400 km from the main melting area (Jenkins and Bombosch 1995). For Lake

Vostok, we may expect a thickness only ranging between a few decimetres and one metre.

3. As the fresh-water layer moves up along the interface, pressure decreases and the corresponding freezing (melting) point increases (see (7.2)). Soon the fresh-water layer is cooler than the interface and supercooled with respect to the local pressure melting point. With a temperature deviation of 0.01 to 0.1 °C, frazil ice forms. The density of the water and frazil mixture decreases (ice density ~ 0.9) so that the mixture will rise and reach upper lake levels. Concerning the exchanges of energy, the temperature of the interface is fixed and is systematically warmer than the temperature of the seeping fresh water layer (Fig. 7.10b). Note that the heat diffuses from the interface toward the lake, transferring the latent heat from frazil formation to the lake water.
4. Frazil concentration increases within the seeping layer. When ice crystals are in contact, they may grow in order to minimise the surface energy (Colbeck 1980). The larger ones grow from the smaller ones, exchanging water molecules and energy. In such an isothermal transformation, the total amount of ice remains constant (Raymond and Tusima 1979).
5. The frazil–water mixture accumulates in the high level of the lake and comes into contact with the glacier at the equilibrium temperature of −2.5 °C (Fig. 7.10c). The evolution of the system is again driven by thermodynamics and moves toward the formation of compact ice. This is done by the sintering of the ice crystals and the freezing of the interstitial water, a phenomenon similar to the sintering of wet snow (Raymond and Tusima 1979, Colbeck 1980). The efficiency of transformation into ice depends on the capability of the system to export the latent heat released from interstitial water freezing.
6. The export of the latent heat through the glacier requires physical contact between packed frazil crystals and the glacier, since conduction through the water is less efficient. This condition implies a minimum density for the water–frazil mixture. Here we assume the frazil crystals accumulate as stable randomly packed spheres, each having six neighbours to ensure static stability. This concept is similar to the approach developed for the densification of snow (Anderson and Benson 1963). Simple cubic packing is assumed, corresponding to one sphere at each corner of a cube. The density or occupancy coefficient is given by the ratio between the sphere volume and the volume of the cube with six faces tangent to the sphere. In this case the density is $\pi/6$ or $\sim 52.3\%$ (Anderson and Benson 1963).
7. A freezing front moves from the glacier boundary downward. The presence of the compacted frazil just below the glacier retains some of the interstitial water and prevents it from sinking as it cools, increasing its density. The large ice crystals from the glacier will compete with the frazil crystals. It is therefore expected that the large glacier crystals ultimately integrate frazil crystals in their lattice.
8. When the frazil–water mixture is transformed into ice, the lake/glacier interface represents the warmest point and the latent heat can diffuse toward

the glacier (Fig. 7.10c). This heat flux therefore corresponds to the latent heat released by freezing of the interstitial water (or host water).

9. Accreted ice forms slowly and the freezing front expels salts, impurities as well as gases from the ice lattice toward the water. The water density increases and the water sinks from the high levels of the lake allowing recycling of the salt and part of the heat.
10. Primary energy is provided by geothermal heat. The behaviour of a deep-water lake is likely complex. If the lake water is fresh, vertical convection and plumes may develop in a few days (Wuest and Carmak 2000). The deep layers may become stratified, similar to what is observed in some lakes in the dry valley in Antarctica. Indeed, as Lake Vostok is very old, the salts have widely diffused and deep layers are likely subject to convection. On the other hand, at the glacier/lake interface, stratification is maintained by the continuous production of fresh water.

7.2.2 Sketch of the Hypothetical Water-circulation Pattern

Based on the water property changes and neglecting deviation by the Coriolis force, we suggest a simplified water-circulation pattern in Lake Vostok represented by a two-loop system (Fig. 7.11) sharing a common intermediate water body. The first loop develops in the upper part of the lake and is a thermohaline loop in which temperature and salt vary. The loop is started by fresh-water production in lower levels at a temperature of $-2.8\,^{\circ}$C, the layer seeps along the interface in which frazil forms and the temperature gradually increases up to about $-2.5\,^{\circ}$C, in the upper levels. There, accretion ice consolidates and salts, impurities and gases are rejected. A warm and salty downward water loop follows at $-2.5\,^{\circ}$C. It recycles and mixes the heat and salts into the intermediate water. The intermediate water is then further pushed toward the melting area.

The second loop is likely purely thermal. The intermediate water (temperature $-2.65\,^{\circ}$C) carries the heat toward the melting area, causing the ice to melt. The salinity of the layer close to the interface will be influenced by the fresh-water production and some of lake water will be dragged upwards. As the heat diffuses more rapidly than the salt, the underlying water layer will be cooled down just to the interface temperature of the area (i.e. $\sim -2.8\,^{\circ}$C) causing the water to slowly sink to the lake bottom. There, the water is warmed by geothermal heat and upwells toward the intermediate lake water that is kept at a constant temperature of about $-2.65\,^{\circ}$C.

The intermediate water enters the melting area, carrying the heat. It will be cooled by $0.08\,^{\circ}$C, i.e. from $-2.65\,^{\circ}$C to $-2.73\,^{\circ}$C, the latter value being the mean temperature of the interface in the melting area. A volume of $1000\,\mathrm{cm}^3$ of intermediate water would carry about 80 cal, corresponding to the energy required to melt 1 g of ice or produce $1\,\mathrm{cm}^3$ of fresh water. Therefore the flow of the intermediate water should be 1000 times greater than the flow of the fresh water loop.

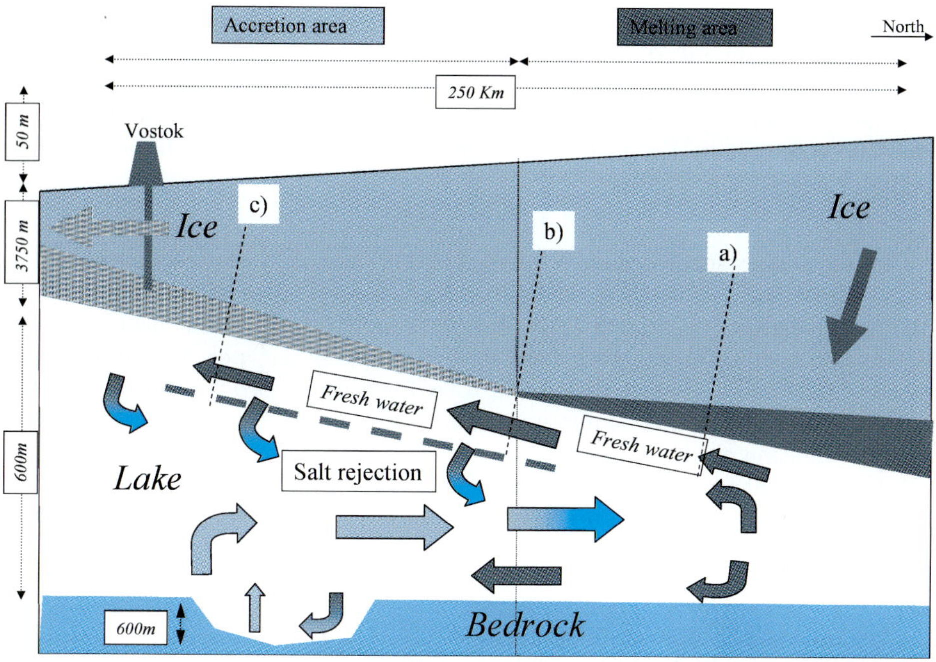

Fig. 7.11. Sketch of the expected water circulation pattern within Lake Vostok. Fresh water is produced in the lower levels of the lake. A fresh water layer upwells and seeps along the interface. Accreted ice forms in the upper levels of the lake and salts are rejected. The upper water layer becomes denser and sinks, recycling the salt and a part of the latent heat toward intermediate water. The intermediate water carries the heat toward the melting area. One loop cools and sinks to the lake bottom. Geothermal heating allows the deep lake water to rise and maintain the intermediate water at the same temperature. a) b) and c) are locations of the interface for which the temperature profiles are described in Fig. 7.10.

7.2.3 Energy Balance of the Lake and Water-renewal Time

In the absence of a lake, the geothermal heat would be transmitted to the glacier by simple conduction. The lake collects the geothermal heat that it finally restores to the glacier. Neglecting the energy exchanges linked to ice advection and export and the other possible losses within the lake, we can state that the geothermal heat will be totally used for melting the glacier ice and then fully transmitted to the glacier as accretion ice is formed. Mass balance will be in steady state when the volume of glacier ice supplying the lake is compensated by the export of an equivalent volume of accreted ice out of the lake.

The latent heat of accretion ice is partly released into the glacier by the freezing of the interstitial water in the accretion area and partly released into the lake water as the latent heat of frazil formation. As the lake is isothermal, and if we assume the accretion phenomenon requires a proportion of frazil that remains constant with time, the latent heat from frazil formation must be diffused out of the lake. From the interface properties (see Sect. 7.2.2 and Fig. 7.10) this latent heat can only be diffused through the glacier in the melting area.

The total energy available from geothermal heat is given by the product of the heat flux (G) by the lake surface area (S). The total ice volume (V_g) that melts or accretes is:

$$V_g = S \times G/L_f \tag{7.3}$$

Using a lake surface area S of $\sim$ 14 000 km^2 and a geothermal heat flux G of 46 mW/m^2, a likely value for the Vostok area (Siegert and Dowdeswell, 1996) and deduced hereafter (see Table 7.2), and L_f, the latent heat for water or ice and corresponding to a heat flux of 9.78 mW/m^2 per mm of ice, V_g is $\sim 70 \times 10^{+6}$ m^3/a. This corresponds to a fresh-water production of about $\sim$ 2.2 m^3/s, implying a flow of about 2200 m^3/s for the intermediate water moving into the melting area.

The renewal time of the lake water (T_r) can now be deduced. If "H" is the mean water depth, we obtain:

$$T_r = H \times L_f/G \tag{7.4}$$

Giving $\sim$ 80000 years for Lake Vostok ($H = 400$ m).

The mean melting (V_{mml}) and mean freezing rate (V_{mfr}) over the entire lake are:

$$V_{mml} = V_{mfr} = G/L_f \tag{7.5}$$

With a geothermal heat flux of 46 mW/m^2, about 4.75 mm of glacier ice could be melted annually. Conversely the formation of 4.75 mm of accretion ice per year releases an equivalent heat flux of 46 mW/m^2 into the glacier. Accretion and melting area of the lake are physically separated: one corresponds to the deep part of the glacier lake interface, the other to the upper level of the lake. The effective melting rate (V_{ml}) and accretion rate (V_{fr}) should depend on their respective area. Adopting "m" as the lake part for melting, or more precisely the area without accretion, and "$(1-m)$" as the part for accretion, the melting and freezing rates can be written:

$$V_{ml} = G/(L_f m) \tag{7.6}$$

$$V_{fr} = G/(L_f(1-m)) \tag{7.7}$$

When the melting area and freezing area are equivalent ($m = 0.5$), both the accretion rate and melting rate are 9.5 mm/a. This value appears consistent with the rate we deduced for accretion ice II (see Sect. 7.1.3).

7.2.4 Application to Heat Fluxes and Mass-balance Velocity

The ice melting is assumed to use all the geothermal heat available, represented by "$G \times S$". In the accretion area, the heat flux that diffuses through the glacier represents the latent heat released by the freezing of interstitial water. This is given by the expression $G \times S \times (1 - F)$, in which F is the frazil portion. The heat flux at the base of the interface in the accretion area is therefore:

$$Q_1 = G \times (1 - F)/(1 - m) \tag{7.8}$$

The total latent heat of frazil formation is equivalent to: $G \times S \times F$ and is released into the lake water. This energy will be recycled and finally diffused through the glacier only in the melting area. The corresponding upward heat flux through the interface in the melting area is:

$$Q'_2 = GF/m \tag{7.9}$$

The average heat flux through the interface is equivalent to the geothermal flux G (i.e. $Q_1 \times (1 - m) + Q'_2 \times m$).

The contribution of ice advection to the heat fluxes (Q') can be estimated. At Vostok, near the interface, the temperature gradient in ice is $\sim 0.02\,°\mathrm{C\,m^{-1}}$. A total of 141 m of accretion ice II was formed over a 15000-year period, and was cooled by 1.4 °C on average. This represents a heat flux of $0.78\,\mathrm{mW/m^2}$. The upward heat flux through the glacier in the accretion area is:

$$Q_0 = Q_1 + Q' \tag{7.10}$$

Similarly, in the melting area, a heat flux (Q'') enters an ice volume from the glacier that is warmed before melting. Therefore the heat flux diffused through the glacier in the melting area is:

$$Q_3 = Q'_2 - Q'' \tag{7.11}$$

If melting and freezing rates are similar, $m \sim 0.5$, and we may adopt $Q' \sim Q'' \sim 0.78\,\mathrm{mW/m^2}$.

At Vostok, the heat flux through the ice (Q_0) is $44\,\mathrm{mW/m^2}$, assuming $2.2\,\mathrm{W/m/°C}$ for the thermal conductivity of the ice and $0.02\,°\mathrm{C/m}$ for the borehole temperature gradient. We deduce $Q_1 \sim 43\,\mathrm{mW/m^2}$ from (7.10). If the accretion and melting areas are equivalent ($m = 0.5$), and using $F \sim 0.52$, we can deduce the geothermal heat $G \sim 46\,\mathrm{mW/m^2}$ (from (7.8)), $Q'_2 \sim 49\,\mathrm{mW/m^2}$ and $Q_3 \sim 48\,\mathrm{mW/m^2}$.

The total energy directed toward the melting area is represented by the total geothermal energy ($G \times S$) plus the latent energy released by frazil formation ($G \times S \times F$). The total equivalent heat flux arriving in the melting area is given by:

$$Q_4 = G(1 + F)/m \tag{7.12}$$

Figure 7.12 is a sketch of the heat fluxes for Lake Vostok using $m = 0.5$. In the melting area, $Q_4 \sim 141\,\mathrm{mW/m^2}$ and therefore is 3 times larger than the geothermal heat flux. An equivalent heat flux of $92\,\mathrm{mW/m^2}$ allows melting of 9.5 mm of glacier ice annually, while $48\,\mathrm{mW/m^2}$ is diffused through the glacier.

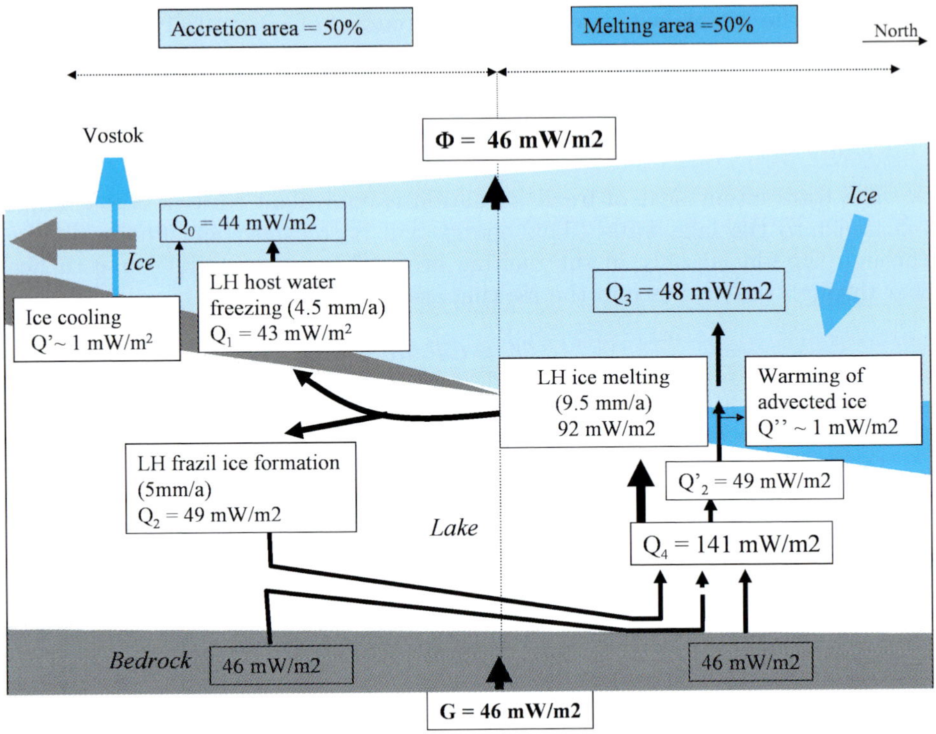

Fig. 7.12. Sketch of the heat balance and heat cycle for Lake Vostok. Melting and accretion area are taken as half the lake area ($m = 0.5$). Accretion and melting rate are 9.5 mm/a and represent an equivalent flux of latent heat (LH) of 92 mW/m^2. Accretion ice is formed by frazil ice (5 mm/a or 49 mW/m^2) and host water (4.5 mm/a or 43 mW/m^2). Latent heat from host water freezing is released in the accretion area (e.g. Vostok area). The latent heat of frazil ice formation (Q_2) is recycled by water circulation and finally diffused through the glacier but only in the melting area (Q'_2). G is the geothermal heat flux carried toward the melting area by water circulation Q_4 is the total equivalent heat flux arriving in the melting area. Q' and Q" are the heat flux from ice advection and ice export respectively (see text). Q_0 is the heat flux measured in the Vostok borehole.

Bell and coworkers (Bell et al., 2002) used radio echo sounding and detected 200 to 300 m thick accretion ice (295 m on the average). Accretion ice is detected on 21 out of 36 tracks performed every 8 km and covering the lake, extending up to 165 km northward from Vostok. Given the crescent shape of the lake area, the accretion ice should be present over 53% of the lake area. By deduction, the melting area or the area without accretion should represent 47% of the lake.

Table 7.2 shows a sensitivity calculation for the heat and the mass balance of Lake Vostok, according to plausible values of Q_0, F and accretion area ($1-\mathrm{m}$). Note that the heat flux value through the glacier in the melting area is always

Table 7.2. Sensitivity of renewal time of lake water, heat fluxes, melting and accretion rates according to observations and hypothesis. Parameters (*second row*) are defined in the text. Values used in Fig. 7.12 correspond to those from the third row

• Heat Flux at Vostok (measured $\sim 45\,\mathrm{mW/m^2}$)	Q_0	44	46	48	46	48	45
• % of frazil (packing of spheres $\sim 52.35\%$)	F	0.53	0.52	0.52	0.52	0.52	0.5235
• Accretion surface area (observations $\sim 53\%$)	$1-m$	0.50	0.52	0.52	0.54	0.54	0.57
• Mean geothermal heat flux ($\mathrm{mW/m^2}$)	G	46	49	51	51	53	53
• Heat flux through glacier in melting area (north lake)	Q'_2	48	52	55	57	59	64
• Mean accretion rate (observed 9.5 mm/a for ice II)	V_{fr}	9.5	9.6	10.1	9.6	10.1	9.5
• Mean melting rate (mm/a)	V_{ml}	9.5	10.4	10.9	11.3	11.8	12.6
• Heat flux injected in the melting area ($\mathrm{mW/m^2}$)	Q_4	141	155	162	168	176	187
• Ice volume to melt or to freeze ($10^{+6}\mathrm{m^3/a}$)	V_{g}	68	72	76	75	78	78
• Renewal time of the lake (ka)	T_{r}	85	80	76	77	74	71
• Mass balance velocity of the glacier (m/a)	v	1.39	1.49	1.55	1.54	1.61	1.61

greater than the heat flux at Vostok, whatever the chosen parameters. While the glacier ice is thicker in the melting area (~ 4200 m) with respect to accretion area (~ 3540 m) the heat flux and the temperature gradient at the base of the glacier are very sensitive to the vertical advection of the ice and therefore to the mean surface snow accumulation rate. The greater heat value is consistent with a higher accumulation rate in the northern area (30 to 40 mm/a) than in the Vostok area (20 mm/a). This is confirmed by a modelling approach of the glacier for the melting area (Salamatin et al., 2003).

A volume of about $70 \times 10^{+6}\,\mathrm{m^3}$ of accreted ice is formed annually under the ice sheet and an equivalent volume has to be exported out of lake by glacier movement. As accretion ice is 295 m thick and detected over 165 km along the eastern shore, the cross-sectional area is $48\,\mathrm{km^2}$. A horizontal velocity (v) of ~ 1.5 m/a is required to export $70 \times 10^{+6}\,\mathrm{m^3/a}$. A surface velocity of 3 ± 0.8 m/a was measured near Vostok. However, there, the ice flow is likely disturbed as suggested by its azimuth (130° instead of 90°) and not representative of the whole southern area of the lake. Indeed, from radar interferometry over the area upstream of Vostok, the surface velocity was deduced for two transects and was shown to gradually increase from 0 at Ridge B to 1.5 m/a on the western shore of Lake Vostok (Kwok 1998, Siegert and Kwok 2000). This latter agrees well with the balance velocity of the lake, supporting the energy approach developed above.

7.3 Some Implications of the Isotope Composition of Accreted Ice

The isotope composition of the water is measured either by the D/H ratio (deuterium content from HDO water to the hydrogen from H_2O water) or from the $^{18}O/^{16}O$ ratio of $H_2{}^{18}O$ and $H_2{}^{16}O$, respectively. The isotope ratios are usually expressed as a deviation (per mil) from a standard of the mean composition of ocean water (SMOW) given for $(D/H)_{standard}$ as 155.76 ppm and for $(^{18}O/^{16}O)_{standard}$ as 2005.2 ppm. We use

$$\delta D = \{(D/H)_{sample}/((D/H)_{standard} - 1)\} \times 1000 \quad (7.13)$$

$$\delta^{18}O = \{(^{18}O/^{16}O)_{sample}/((^{18}O/^{16}O)_{standard} - 1)\} \times 1000 \quad (7.14)$$

and for the isotope ratio (C)

$$C = 1 + \delta D/1000 \quad \text{or} \quad C = 1 + \delta^{18}O/1000 \quad (7.15)$$

Deuterium and ^{18}O water isotopes have found applications in such diverse fields as cloud physics, paleoclimate studies, hydrology and chemistry. Due to different molecular masses, the physical properties of heavy water are slightly different from those of normal water and fractionation occurs during the condensation phase of the water cycle. Applied to Lake Vostok, no fractionation occurs upon melting but when water freezes, the heavy isotope preferentially integrates the condensate phase (e.g. accretion ice) while the lighter isotope tends to stay in the water. This can be expressed using the general relationship:

$$C_{ice} = \alpha \times C_{water} \quad (7.16)$$

For ice–water equilibrium, the fractionation coefficient (α) is 1.0208 for deuterium and 1.003 for ^{18}O.

7.3.1 Constraint on the Geographical Location of the Melting Area

In steady state, the lake volume (V_L) is constant ($dV_L = 0$) and the volume of ice that melts annually and supplies the lake (Q_{in}) is balanced by an equivalent volume that accretes and moves out of the lake (Q_{out}). This also applies for the stable isotope composition of the lake water that should be to constant with time (t). Therefore, the composition of the ice that melts (C_{in}) should equal to that of the accretion ice (C_{out}) that is exported.

For a given constituent, the balance of the number of molecules (ΔN) in a given time (Δt) is:

$$\Delta N = V_{\rm L} \times \Delta C = Q_{\rm in} \times C_{\rm in} \times \Delta t - Q_{\rm out} \times C_{\rm out} \times \Delta t \tag{7.17}$$

If we introduce the renewal time of the water (τ), we can write:

$$V_{\rm L} = Q_{\rm in} \times \tau = Q_{\rm out} \times \tau \tag{7.18}$$

This gives:

$$Q_{\rm in} \times \tau \times \Delta C = (Q_{\rm in} \times C_{\rm in} - Q_{\rm out} \times C_{\rm out}) \times \Delta t \tag{7.19}$$

$$\Delta C = (C_{\rm in} - C_{\rm out}) \times \Delta t/\tau \tag{7.20}$$

Applied to water isotope molecules we obtain:

$$C_{\rm out} = \alpha \times C_{\rm water} \tag{7.21}$$

and after rearrangement of (7.20)

$$\Delta C_{\rm water}/(1 - \alpha \times C_{\rm water}/C_{\rm in}) = C_{\rm in} \times \Delta t/\tau \tag{7.22}$$

After integration, this becomes:

$$C_{\rm water} = C_{\rm in}/\alpha + (C_0 - C_{\rm in}/\alpha) \times \exp(-\alpha \times t/\tau) \tag{7.23}$$

where C_0 corresponds to the lake composition at the initial time.

After a duration of three times the renewal time of the lake water (i.e. $\sim$ 250 000 years for Lake Vostok), a steady sate is established (Fig. 7.13). The lake composition becomes constant and the isotope composition of accretion ice is equivalent to that of the melting ice ($C_{\rm water} = C_{\rm in}/\alpha = C_{\rm out}/\alpha$).

For deuterium, the isotope compositions measured on the Vostok ice core are:

- $\delta D_{\rm H} = -439.0$‰ for the Holocene period (the last 10000 years)
- $\delta D_{\rm out} = -442.5$‰ ± 0.8‰ for the accretion ice
- $\delta D_{\rm mean} = -460.9$‰ for the mean composition of the ice over the last 400000 years.

As the lake has likely existed for more than 400000 years, we may assume it is close to its steady state. The accretion ice composition is significantly different (by $\sim$ 18‰) from the mean value for glacier ice, indicating that the flow line passing by Vostok cannot be the main melting area.

Indeed, along the northward route from Vostok, surface snow composition is enriched in heavy isotopes and a geographical variation is observed (Qin Dahe et al., 1994). The whole lake area is under the influence of the same continental climate, and both the altitude and mean annual temperature are similar at all locations. The observed changes are likely due the variable influences of air

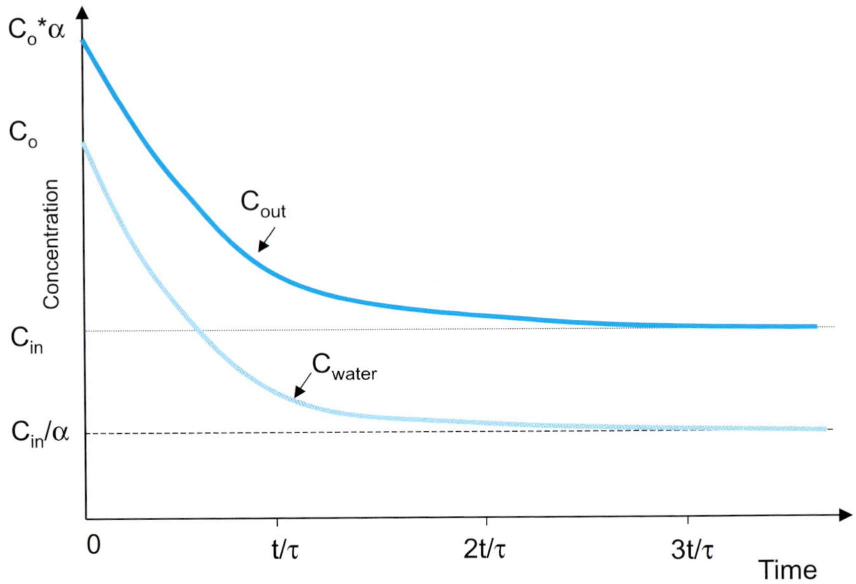

Fig. 7.13. Concentration versus time for the isotope composition of a lake (C_{water}) and for accretion ice (C_{out}) assuming the lake volume is constant. The lake, with an initial composition C_0, is supplied by a constant water composition C_{in}. After 3 times the renewal time (τ) $C_{water} = C_{in}/\alpha = C_{out}/\alpha$ (see text).

masses, each having a different origin, thermal history and mean pathway in the atmosphere. This is similar to an orographic effect (Qin Dahe et al., 1994) or to boundaries between air masses.

The glacier that overlies the lake flows westward from the Ridge B area, located 300 km upstream of Vostok, where a deep ice core (Dome B) has been drilled (Fig. 7.15). From the Komsomoskaya ice core, collected at a site located 550 km north of Vostok, the isotope composition is – 399.1‰ (average over the last 10000 years). This value is 40‰ higher than the Vostok value, giving an apparent geographic gradient of 7.27‰/100 km. Moreover, three surface-snow traverses allowed the collection of snow samples every 50 or 100 km: two from Komsomolskaya to Vostok (Petit et al. 1991, Qin Dahe et al., 1994) and one from Komsomolskaya to Dome B (Vaikmae, personal communication). Similar geographic gradients have been obtained: 8.21 and 7.24‰/100 km for the first two traverses and 7.7‰/100 km for the third.

The geographic gradient of isotope composition likely persisted throughout the past glacial and interglacial periods. Indeed, four deep drilling sites (Dome B, Dome C, Komosmoslkaya, and Vostok) let us describe the climate over the last 15 to 30 thousand years in the lake region. The four isotope profiles are consistent and display very similar amplitude changes for the deglaciation and last glacial periods, indicating a consistent and homogenous climate variation over the East

Antarctic plateau (Jouzel et al., 2001). As a consequence, the isotope composition record over the last 400000 years from any site in the vicinity of Vostok likely mimics the climatic record from Vostok with simply a shift corresponding to a constant value. Moving north from Vostok, this shift depends on the distance (D), and for the mean isotope value (δD) it can be written:

$$\delta D_{\text{mean}} = -460.9‰ + 0.077 \times D \tag{7.24}$$

where 0.077‰/km is the mean geographical gradient adopted for deuterium and −460.9‰ the mean value for the Vostok site. The area for which the mean isotope composition corresponds to the mean value of the accretion ice ($\delta D_{\text{out}} = \delta D_{\text{in}} = -442.5‰$), is located at a distance of $\sim 240 \pm 25$ km from Vostok and should be the centre of the glacier melting area (Fig. 7.14).

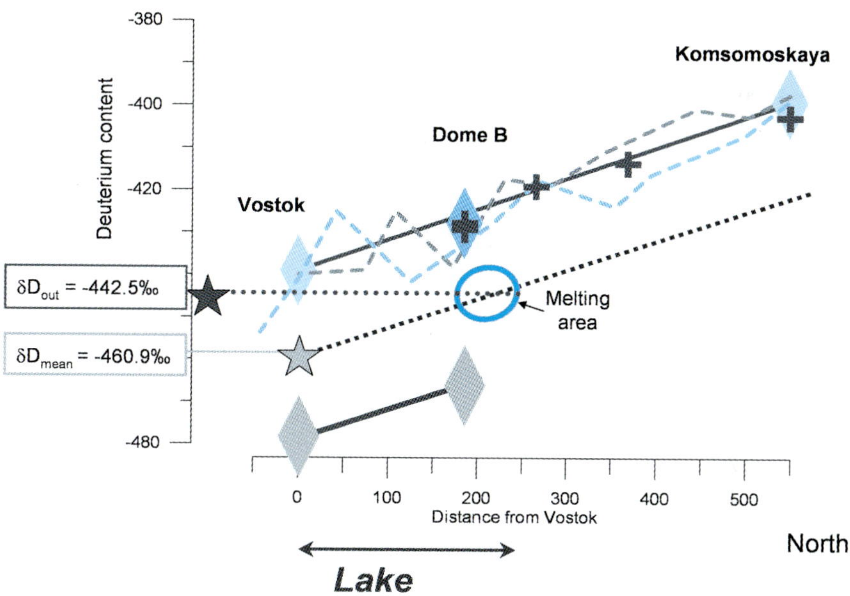

Fig. 7.14. Geographic variation of the isotope composition (‰) of the snow and ice near the surface and from deep cores in the vicinity of Lake Vostok. Diamonds refer to the mean isotope values for Vostok, Dome B and Komsomolskaya deep ice cores respectively: the three *upper diamonds* are the mean values for the Holocene and the two *lower diamonds* are for the last glacial period. Surface snow composition (mean values over the last 10 years) along routes from Vostok to Komsomolskaya (Qin Dahe et al., 1994, Petit et al., 1991, *dashed line*) and from Dome B to Komsomolskaya (R. Vaikmae, personal communication, crosses) are also shown. The two solid lines refer to the mean trend for the Holocene (top) and last glacial period (bottom). Over this area, we assume that the mean isotope composition of the ice (δD_{mean}) follows the same overall mean trend (dotted line). For the area located ∼ 240 km25 km from Vostok (circle), the composition of the ice that supplies the lake is equivalent to that of accretion ice (δD_{out}).

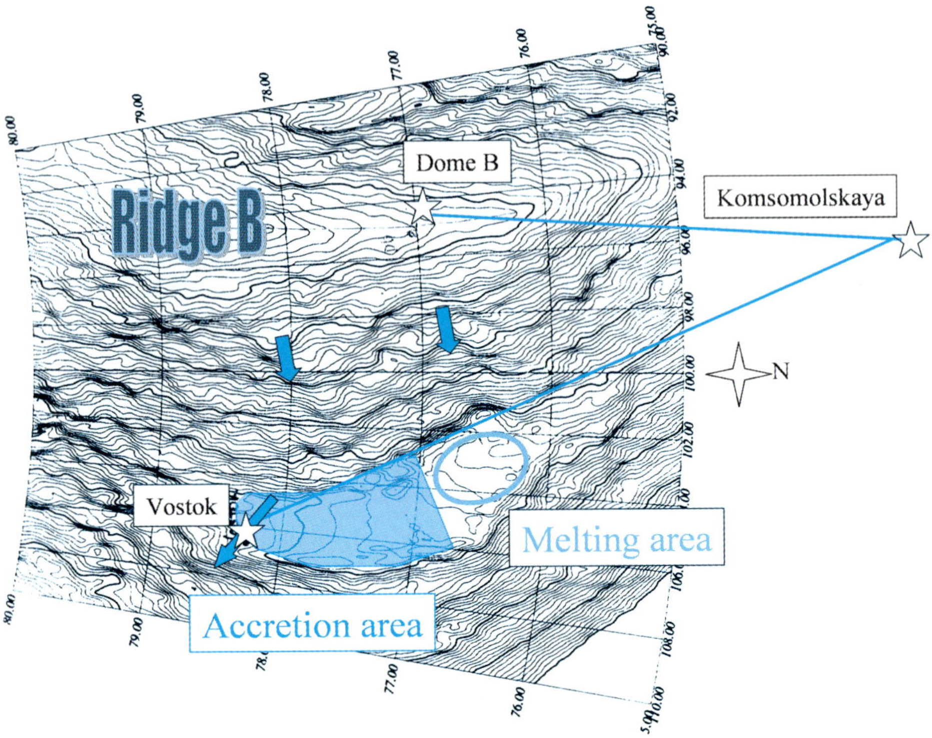

Fig. 7.15. Accretion area and expected melting area of Lake Vostok. Surface topography is from Rmy et al. (2000). Dome B, Komsomolskaya and Vostok are drilling sites, linked by traverse routes. The arrows (azimuth $\sim 90°$ East) indicate the ice flow according to the mean slope (from Siegert et al, 2001). The ice velocity increases from 0 m/a at Ridge B to 1.5 m/a at the boundary of Lake Vostok (Siegert and Kwok, 2000). At Vostok, the mean ice velocity is $\sim$ 3 m/a (Bell et al, 2002) but the azimuth of the flow line is $\sim$ 130° East, likely due to a local effect. On the lake, the shaded area corresponds to the zone (up to 165 km) where accretion ice was detected (Bell et al, 2002). The melting area is represented by the circle 240 km from Vostok.

This result is consistent with the interface properties causing the deep glacier ice to melt (see Sect. 7.1.5). In addition, Gorman and Siergert (1999) observed penetration of radio echo soundings $\sim$ 20 m into the lake water in the northern area (77 °S, 105 °E), implying that the lake water is very pure. Indeed, this may correspond to a location of an active fresh-water production area. Moreover, the water thickness gives an idea of the thickness of the fresh-water layer that does not mix with the lake water.

Finally, the geographical variation of the isotope composition restricts the melting area to a rather small zone. This excludes the lake as the collector centre for a large subglacial hydraulic system extending well outside the actual lake contour.

7.3.2 The δD–$\delta^{18}O$ Relationship

Let us consider the balance for the two water isotopes (deuterium and ^{18}O) throughout the water cycle in Lake Vostok, from glacier ice to water to accretion ice. For water, when two phases are in equilibrium, the delta value for deuterium is 8 times greater than that for ^{18}O, as a result of the molecular mass dependence of the fractionation effect: $(D–H)/H = 1$ and $(^{18}O–^{16}O)/^{16}O = 1/8$ (e.g. Criss 1999). The meteoric water line (MWL) defines the δD and $\delta^{18}O$ behaviour of precipitation on a global scale, with a slope of 8 and a residue of $\sim$ 10 ($\delta D = 8 \times \delta^{18}O + 10$) (Craig 1961, Jouzel et al., 2000).

The deuterium excess is defined as $d = \delta D - 8 \times \delta O$ (Dansgaard 1964) and is associated with the different behaviour of the two isotopes when the phases are not in equilibrium. For example, evaporation over the ocean represents a net export of water vapour and a nonequilibrium process determined by the diffusion properties of the isotope molecules within air. This is represented by a kinetic fractionation coefficient that is added to the fractionation coefficient. The deuterium excess is used as a proxy of this effect for the atmospheric water cycle.

For East Antarctica, the deuterium excess of snow precipitation varies between sites and with the climate primarily in response to the different conditions of vapour formation over the ocean (temperature, relative humidity, wind) and to a lesser degree to the variable thermal histories of the air masses (Vimeux et al., 1999). Along the route northward from Vostok, the deuterium excess for recent snow decreases from 18‰ at Vostok to 14‰ at Komsomolskaya at 600 km, to about 2‰ near the coast, 1400 km away (Qin Dahe et al., 1994). At Vostok, for the last 400000-year record, including the glacial and interglacial periods, the overall δD and $\delta^{18}O$ values are aligned with a slope of 7.86 and the average deuterium excess is 14‰. At Dome B, the mean deuterium excess is $\sim$ 11‰ over the last 30000 years that include the last glacial period (Vimeux et al., 2001, M. Stievenard, personal communication).

For steady-state lake conditions, the deuterium excess of the ice that supplies the lake should be equivalent to that of accretion ice. The mean δD and $\delta^{18}O$ composition of accretion ice is $-$ 442.5‰ and $-$ 56.3‰, respectively, and the deuterium excess is 7.5‰. On a δD–$\delta^{18}O$ plot (Fig. 7.16), values of the accretion ice are clustered on the right of the meteoritic line for Vostok as well as on the right of the Dome B meteoritic line, representing the area that supplies the lake. The mean value for accretion ice should then intersect a hypothetical precipitation line from sites as far as 800 km north of Vostok, well outside the lake area and therefore unrealistic. With a deuterium excess of 7.5‰ instead of 11‰, accretion ice appears enriched by 0.5‰ in ^{18}O with respect to the inflow ice (Fig. 7.16).

This weak decrease in the excess of deuterium is unlikely due to a kinetic fractionation effect, because an opposite trend is expected, and the lake would be continuously depleted in ^{18}O and no longer in steady state. On the other hand, the ^{18}O enrichment may result from isotope exchange with rich ^{18}O such as the

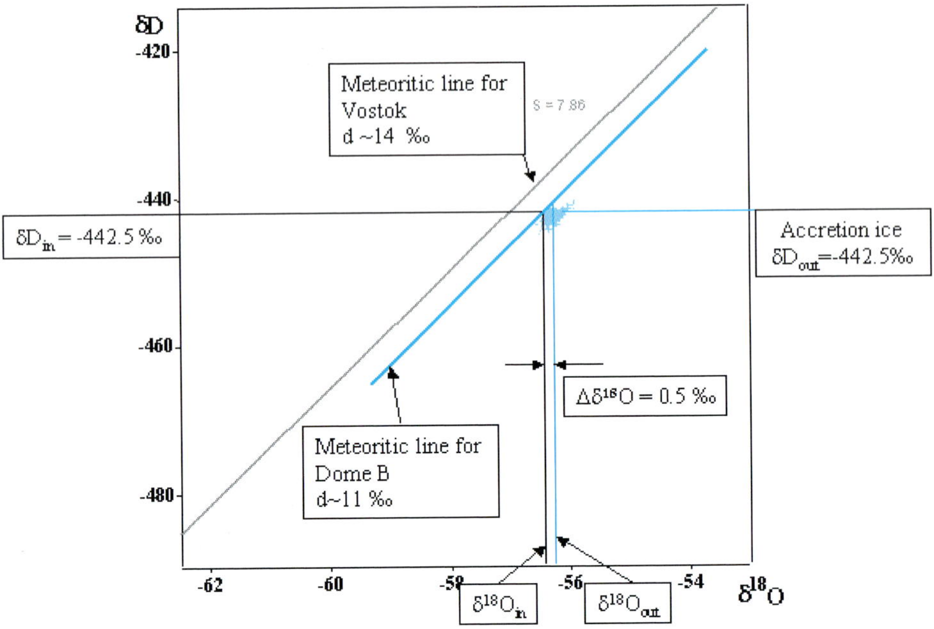

Fig. 7.16. δD-δ^{18}O plot for accretion ice composition and for the meteoritic lines of the Vostok (from Jouzel et al., 1999) and Dome B areas (as deduced from Vimeux et al., 2001, and M. Stievenard, personal communication). The mean accretion ice composition clusters on the right of the Dome B meteoritic line and appears to be enriched in ^{18}O with respect to the composition of the ice that supplies the lake ($\delta^{18}O_{in}$).

dissolved atmospheric oxygen or interaction with the bedrock or a contribution of hydrothermal water.

Atmospheric oxygen is enriched (+23.5‰) as a result of the preferential use of light oxygen by the vegetation and animals, producing the so-called "Dole effect" (Dole et al., 1954). Assuming the isotope exchange could take place between the lake water and the atmospheric oxygen accumulated within lake water, the 0.5‰ change in water composition (−56.3‰) requires a 6 per mil contribution from oxygen with a composition of +23.5‰. The atmospheric oxygen molecule abundance with respect to the concentration of oxygen molecules in ice is only 10^{-5} and therefore too low for a continuous enrichment.

The formation water is saline and encountered in oil fields when drilling, or as saline surface springs in regions undergoing tectonic compression (Criss, 1999). It is derived from old trapped sea water, and is found in the pore volume of rocks in sedimentary basins, where the temperature is 30 to 150°C. It may have undergone extensive interaction with rocks allowing reactions and exchanges with ^{18}O-rich carbonates or silica. This may be the case for Lake Vostok assuming it was an open lake during its early stage of formation, concentrating salts or evaporites. Similarly, in hydrothermal areas, spring water is known to be enriched in ^{18}O with respect to precipitation supplying the system (Clayton and Steiner, 1975, Craig, 1963). The "oxygen shift" with respect to the meteoritic water line (e.g. the δD–$\delta^{18}O$ line for the composition of the precipitation supplying the basin) results from isotope exchanges with ^{18}O-rich rocks at high temperature during the transfer. This may be the case for Lake Vostok, for which deep faults are expected within the bedrock, into which water may seep and cause warming down to great depths. Moreover, seismotectonic activity has been detected in the vicinity of the lake (Studinger et al., 2002). As faults move during seismic events, energy is liberated promoting reactions and isotope exchanges between water and bedrock as well as the flushing of water through faults. The fractured bedrock that surrounds the lake represents a huge reservoir of enriched ^{18}O from silicates, for example, which may potentially supply the lake water for a long time. The existence of a warm reservoir in relation to the lake or a hydrothermal contribution to the lake is also suggested from biological findings (see Sect. 7.4).

Seismotectonic activity is independently suggested by the excess of ^{4}He found in the accretion ice (see Sect. 7.4.3.1 and Fig. 7.19), reflecting possible degassing from the bedrock (Jean Baptiste et al., 2001, 2003).

7.4 Biological studies

7.4.1 Previous studies

Aeolian dust transported to the East Antarctic plateau originates mostly from the southern part of South America, in particular Patagonia (Basile et al., 1997, Delmonte et al., 2004). Germs from soils may be also transported and spread over East Antarctica and Vostok or the Ridge B area. Once included into the ice layers, they are buried in the glacier and possibly subsequently released into the lake when the layers melt. The probability of the lake being effectively seeded by micro-organisms from the surface biota will depend on the capability of the germs (microbial or spores) to resist aeolian transport, then cold ice temperatures, pressure and grinding within the glacier for a million years or so, and finally the ultralow organic carbon contents, oxygen excess and darkness in the lake environment. On the other hand, prior to glaciation of the Antarctic continent, Lake Vostok was likely an open lake with its own lifeforms that may have evolved and adapted to subglacial conditions.

In the 1980s, the Russian researcher Sabit S. Abyzov was the first to investigate the biological contents of ice cores from Vostok station, applying the classical methods of culturing and microscope observations (Abyzov 1993, Abyzov et al., 2001). The outer part of the cores was discarded using a sterile system that melts and collects only the inner part of the ice. From these studies, various yeasts, fungi and bacteria have been observed and some have been grown by culturing. While the success of culturing decreased with the ice-core depth, results were obtained down to samples taken at 2200m depth, suggesting metabolic re activation after 150000 years of "anabiosis" or dormancy.

These pioneering studies were the first in this field, however, their results must be treated with caution. Indeed, from ice-chemistry investigations, we have learnt a great deal over the last twenty years about the very low impurity contents of polar ice and in particular contamination issues, the need to work in dust-free facilities and to use highly sensitive equipment and preconcentrations techniques as well as artificial pure deionised water, together with an analytical protocol that includes the processing of blanks (false or sham samples). These techniques are now used widely and routinely by the ice-core community of chemists, making it possible to reliably determine the chemical composition of the polar ice (e.g. Boutron et al., 1994, Legrand and Mayewski 1997). For biological investigations, it is difficult to accept the conclusions of studies not carried out under dust-free and sterile conditions. To date, the biological contents of the glacier ice from Vostok (especially from deep horizons) remains to be confirmed.

The accretion ice samples provided an opportunity to investigate the lake biocontents and three main studies have recently been published (Karl et al., 1999; Priscu et al., 1999; Christner et al., 2001); all these studies detected micro-organisms. Cells have been observed at concentrations from 100 to 25000 cells per ml of melt water (Karl et al., 1999; Priscu et al., 1999), the development of germs from direct plating has been obtained (Christner et al., 2001) and the activity detected (Karl et al., 1999). *DNA* amplification from nonconcentrated ice samples was achieved, implying that a substantial number of cells are present in the water samples and the lake as well (Priscu et al., 1999; Christner et al., 2001). These micro-organisms found and identified by cloning and sequencing show little diversity and are similar to those found on the Earth's surface and/or near human-influenced environments (Priscu et al., 1999). The total organic carbon concentration (TOC) of water samples were found to range from 80 to 500 ppb. Due to the partitioning from ice formation, the TOC contents of Lake Vostok water should be comparable to some other open lakes and supply heterotrophic micro-organisms with nutrients (Karl et al., 1999; Priscu et al., 1999). In summary, the biological results support the seeding scenario by the glacier as well as the ability of micro-organisms from common environments to adapt to Lake Vostok extreme environments.

This may appear to be credible and at least represents progress in biological studies. However, the chemical results published along with the biological findings raise several questions from ice geochemists. First, the concentrations of the observed cells have the same magnitudes as those found in insoluble dust of comparable size, e.g. 1000–10 000 per g for particles $> 0.8\,\mu m$. Dust is measured in water samples by Coulter counter and observed by electron microscopy (Gaudichet et al., 1988M; Petit et al., 1999).

When a water sample from Antarctic ice is observed under an optical or electron microscope (using 0.4 or $0.2\,\mu m$ porosity), particles of mineral dust (size $0.8–3\,\mu m$) are most abundant while organic material and micro-organisms are very rare. This is contrary to the case for ice samples from the Alps or the Andes that are often rich in micro-organisms. For example, an Andean glacier sample soon develops micro-organisms easily detectable by the Coulter technique if left for a few days in the lab at ambient temperature and exposed to day light. For this reason, Andean samples should be processed soon after melting and separately from Antarctic ice samples to prevent cross-contamination. For Antarctic ice, bacterial development within water samples is very rare and when it occurs, contamination or insufficient care in cleaning tools and beakers are the most likely causes.

Several attempts have been made to reproduce the culturing and DNA amplification experiments and tests by performing corresponding tests on the accreted ice samples from Vostok. The decontamination procedure was the same as used for chemical analysis and was performed under a sterile hood. All the platings, under various environmental conditions, including those specific to psychrophiles, were unsuccessful (A.M. Gounot, unpublished and personal communication, S. Rogers, personal communication). For DNA, a series of amplifications were also nonconclusive and/or the false samples were positive (M. Blot, D. Faure, unpublished). From filter and microscopic preparations, we observed micro-organisms only once (Fig. 7.17), but for an ice sample that was too small to apply the decontamination procedure in full.

Furthermore, the published values for the bulk size of total organic carbon, from 80 to 500 ppb (Karl et al., 1999; Priscu et al., 1999), do not correspond to the expected values of 1 to 10 ppb from organic acid measurements or from an estimate based on dust concentration (see Sect. 7.1.3.2). Such low concentrations are difficult to measure in the small water samples ($\sim$ 10 g) generally available for the accretion ice and special techniques must be applied (see Sect. 7.4.2). Finally, TOC concentrations as high as 80 to 500 ppb are unlikely to represent the concentration in ice. The drilling fluid used for Vostok deep ice coring is low-grade kerosene that adheres to the ice core and is very difficult to remove, giving the ice-core samples a distinct odour. The high TOC values may be the fact of an insufficient decontamination procedure and the presence of some remaining drilling fluid. We know now that drilling fluid is not sterile and contains micro-organisms (see below), casting doubt on the published biological results given the lack of controls.

By the end of 2003, no independent confirmation of chemical and biological findings for Vostok ice had yet been published, another indication of the dubious nature of the already published results.

7.4.2 Recent Studies

As already reviewed, previous studies by microscopic observation, organic chemistry, DNA amplification, metabolic activity detection and culturing have suggested that Lake Vostok may host life (Karl et al. 1999; Priscu et al., 1999; Abyzov et al., 2001; Christner et al., 2001). However, a major uncertainty in

Fig. 7.17. View of the clean room lab, operator protection and some handling and washing operations used in ice core decontamination at LGGE in Grenoble. Bottom left: A microphotograph view of contamination of an accreted ice sample by the drilling fluid. Due to the small size of the ice sample, application of the full decontamination process was not possible. Once filtered and coloured by DIAPI, sediment inclusions (clays mineral) appear in yellow while microorganisms (such as cocci) appear in fluorescent green.

Lake Vostok geomicrobiological studies is the extent of ice-core contamination by the drilling fluid and overall handling (Vincent 1999, 2000). Due to the very low biomass in Vostok ice samples (Bulat et al., 2002, 2003; Petit et al., 2003), the search for life within them proved to be difficult due to the high probability of contamination similar to forward-contamination when searching for life in very poor media in terms of planetary protection policies (Space Studies Board 2000, Rummel 2001).

In this respect, we present here the results of a study we recently conducted using a series of chemical and biological controls to validate our findings. We have used 16S ribosomal gene sequencing to estimate the bacterial contents of Lake Vostok accretion ice samples at 3551 m and 3607 m, formed 20–15 ka ago. In addition, one sample of glacier ice at 3001 m, with an age of about 300 ka (Petit et al., 1999), was taken for comparison. The accretion-ice horizon contained small visible inclusions of sediment trapped when the glacier moved across a shallow depth bay upstream of the Vostok site (Jouzel et al., 1999; see also Figs. 7.6 and 7.19).

7.4.2.1 Decontamination Procedure and Chemical Controls

Decontamination proved to be a critical issue. Stringent decontamination techniques currently used in ice-core chemistry allow reliable measurements of major ions as well as trace elements (Legrand and Mayewski 1997; Boutron et al., 1994) present in ice for central Antarctica at concentration levels of 10^{-9} and 10^{-12} g/g, respectively. In our study we applied the rigorous ice-chemistry-based decontamination procedure for ice-core samples to produce the most reliable chemical composition for major ions and dissolved organic carbon (DOC). Major ion and light carboxylic acid concentrations are sensitive to gaseous contamination from the lab environment (Legrand et al., 1993) and served as a first control of our decontamination procedures applied to the samples prior to molecular biological studies. For samples that satisfactorily passed the chemical control, we implemented numerous and comprehensive biological controls to avoid misleading results.

Briefly, the ice-core decontamination procedure was as follows. Ice cutting and surface cleaning were performed in a cold room (−15 °C). Then pretreated ice samples were rinsed three times with pure low DOC (< 2 ppb C) water and melted in clean sterile polycarbonate jars at room temperature. Melt water was further concentrated up to 2000 times using filtering units equipped with 3 Kd membranes removing all matter larger than a 10 base-pair DNA. Core rinsing and melt-water treatment was carried out within a class 100 laminar hood (EU-ISO-5) in a class 10000 (EU-ISO-7) dust-free laboratory (Fig. 7.17), and the laboratory personal used surgical supplies (sterile cloths, gloves, masks and shields). Surfaces and tools were cleaned with chemical decontamination solutions.

Note that the ice samples were prepared in a different building than the one used for DNA extraction and PCR analyses to avoid any impact of PCR products on samples. The class 10000 dust-free laboratory was never used for DNA work.

For the part of the Vostok ice core deeper than 3520 m (including basal glacier ice and both accretion ice I and II), an initial reference profile of major ion and light carboxylic acid concentrations of glacial and accretion ice samples revealed very low contents of NH_4^+ (0.1–4 ppb), NO_2^- (< 0.1 ppb), acetate (0.5–2 ppb) and formate ions (< 0.5 ppb), consistent with those previously found in other Antarctic cores (Legrand and Saigne 1988). The concentration of DOC, influenced by the presence of kerosene from the drilling fluid, was shown to be low and similar in both glacial (24 ± 14 ppb) and accretion (17 ± 7 ppb C) ice. We consider this data to be consistent with the low carbon contribution from the light carboxylic acids and the values lie well below those previously reported for accretion ice (80–500 ppb C) (Karl et al., 1999, Priscu et al., 1999).

7.4.2.2 Biological Controls and Documentation of a Contamination Database

The chemistry-based decontamination procedure that we followed is reliable for ice-core chemical analyses (e.g. major ions). However, core rinsing even with ultrapure water may be not sufficient to remove all biological aliens (e.g., Kawai et al., 2002; Kulakov et al., 2002). In addition, commercial DNA extraction and PCR enzymes, reagents and disposable plastic articles suitable for use under low-biomass conditions are not always certified DNA-free (e.g., Grahn et al., 2003; Corless et al., 2000; Millar et al., 2002). For this reason, we had to develop our own contaminant database in order to decipher potential contaminants within the ice signals.

The five types of control for contaminant cells and foreign DNA inspected with PCR and DNA sequencing were as follows: (i) sham (negative) DNA extraction including reagents, spin filters and capture tubes; (ii) negative PCR–PCR reagents including DNA polymerases that is known to generate PCR products with no template DNA added (Corless et al., 2000 and references herein); (iii) ice-core ultrapure wash water; (iv) original Vostok drilling fluid from different depths in borehole; (v) laboratory environment – dust collected in laboratories in which DNA was extracted.

For all five controls, corresponding clone libraries for the V3 region of 16S rDNA were constructed and a total of 215 expected-size clones were sequenced and analysed. In this way, the core of the contaminant database was established. It is currently updated with each new ice sample treated. In addition, any known contaminant clones deposited in public databases (e.g., by Norman Pace) or published elsewhere (e.g., Tanner et al., 1998; Cisar et al., 2000; Kawai et al., 2002; Kulakov et al., 2002) must be employed.

7.4.2.3 Ice-core Findings with Respect to Contaminant Database: Building Likelihood Criteria

For ice samples with mineral inclusions, DNA extraction was performed directly from sediment material., It was performed from concentrated melt water for all others.

To amplify bacterial and archaeal 16S ribosomal RNA genes, a seminested PCR with broad-range primers was implemented. The resulting clones were only 405–430 bases in length because it was impossible to amplify full-sized 16S rRNA genes using different DNA polymerases and primers. This may be indicative of very small amounts of DNA or its ancient status (Cooper and Poinar 2000) due to possible damage, for instance, by oxygen excess content in the lake (Lipenkov and Istomin 2001). Note that the DNA was extracted from highly concentrated melt water and we were nevertheless forced to use a very sensitive seminested PCR scheme to be able to generate signals. One-round PCR did not work while PCR inhibitor tests performed with *E. coli* genomic DNA added to original ice DNA samples were always negative. Preliminary control experiments performed by *E. coli* cell counting followed by PCR extinction showed that the detection limit of our PCR schemes was equivalent to 5–8 cells per ml of melt water.

A total of 121 clones from both accretion (101 clones) and glacial (20 clones) ice samples were analysed. The ice clones showed high redundancy resulting in the recognition of twenty 16S rDNA phylotypes, of which 16 were seen in accretion ice and three also in glacier ice. We will discuss below the question of accretion-ice findings only.

Most phylotypes from accretion ice had close Genbank homologues (>98%), i.e. were identified (Stackebrandt and Goebel 1994) or assigned to known sequences, but three were distinctly different from their nearest neighbours (<96% similarity).

Bacterial taxa identified by 16S ribosomal DNA sequences were indexed on the basis of specially developed criteria. The contaminant database was very useful but not sufficient by itself (Table 7.3). Since the biological contents of the Lake Vostok are hard to predict, a degree of confidence was introduced in the table allowing us to discard obvious cases with the highest scores as contaminants, while those with the lowest scores were referred to as likely contaminants.

Table 7.3. Indexing contaminant criteria for bacteria recorded in Vostok ice core. The keys are also used in Table 7.4

Key score[a]	Regulation
A^{+5}	Taxa recorded in the contaminant database
B^{+4}	Taxa proved to be contaminants (e.g., Tanner et al., 1998, Cisar et al., 2000, Kawai et al., 2002, Kulakov et al., 2002)
C^{+3}	Aliphatic and aromatic hydrocarbons degrading bacteria
D^{+3}	Human/animal/plant saprophytes, commensals and pathogens (and those associated with human activity including food, etc.)
E^{+1}	Taxa revealed in waste water, activated sludge, contaminated aquifer etc. (linked to human activity)
F^{+1}	Taxa revealed in other (physically separate) glaciers[b]

[a] Confidence: $^{+1}$ – lowest score, $^{+5}$ – highest score.

[b] Tropical regions where taxa have been found in mountain glacier samples (Christner et al., 2001) do not supply Antarctica with aeolian dust. Southern South America does. Moreover the Antarctic ice sheet does not presently contribute to the transfer of the above surface biota into the lake (refer to our data on glacier ice findings) and the high oxygen content estimated for present day lake water is assumed to be harmful to common-air surface biota.

As a result, thirteen of the 16 phylotypes in accretion ice were indexed as contaminants. Two extra phylotypes were regarded as likely contaminants because of their uncertain status (Table 7.4). And only one 16S rDNA phylotype (12 clones; two sequence types differing by one nucleotide substitution) successfully passed our contaminant database and other contaminant screening criteria (Table 7.4).

Note that no bacterial species passing the contamination test was found in either the glacial and the accretion ice suggesting a small chance of microbes transiting from the 1-Ma old (Salamatin et al. 2003) glacier melt water into the lake and then being refrozen in accretion ice (Siegert et al., 2001). The fact that until now we have no confident findings from glacier-ice samples supports this view (Bulat et al., unpublished), which could be due to the presence of oxygen dissolved in the ice matrix in surprisingly large amounts (about 1.0 O_2 mg L^{-1}) (Lipenkov et al., 2002, Ikeda et al., 2000). Such an oxygen concentration should destroy any DNA (Lindahl 1993) and likely also in the glacier ice with time.

Table 7.4. Bacterial 16S rDNA clones recovered from Vostok ice core that passed contamination test

Clones[a] recovered in ice sample: #3607 inclusion	#3607b polycrystal	Closest neighbours in GenBank	Phylotype status
	2 (AF532060)	**Beta-*Proteobacteria*** • 100% *Hydrogenophilus thermoluteolus* TH-1 (AB009828) • 99.5% *Hydrogenophilus hirschii* yel5a (AJ131694) • 99.1% *Hydrogenophilus thermoluteolus* TH-4 (AB009829)	Lake Vostok indigenous
	2 (AF532058)	**Beta-*Proteobacteria*** • 97% rDNA clone pACH94 (AY297809) • 96% *Rhodocyclus sp.* R6 (AJ224937), rDNA clones (AF204245,AF502232, (AF204244)	Linked to E* Likely contaminant
3 (AF532061)		**OP11 Candidate division** • 91.7% rDNA clone C011 (AF507682) • 82–83% rDNA clones of OD1 candidate division (AY193189, AY193185)	Linked[b] to A1* and E*[c] Likely contaminant

* Refer to Table 7.3.
[a] Differs by 1–5 nucleotides within a phylotype.
[b] Both, ice clones and contaminant 'A1' clone, showed the same low level similarity to soil bacterial rDNA clones of one location (Dunbar et al., 2002; Kuske et al., 1997; Swaty et al., 1998).
[c] Contaminated aquifer (N. Pace, unpublished).

7.4.2.4 Reliable Biological Findings

The only phylotype that successfully passed our contaminant database and other contaminant screening criteria (Table 7.4) was the extant thermophilic bacterium *Hydrogenophilus thermoluteolus* (beta-Proteobacteria) that is encountered until now only in hot springs in the Izu district (Japan) (Goto et al., 1977; Hayashi et al., 1999) with closely related species *H. hirschii* known in the hot springs of Yellowstone National Park, USA (Stohr et al., 2001) (Fig. 7.18). A sequence of rDNA ascribed to the genus *Hydrogenophilus* was recently recovered by Lopez-Garcia et al. (2003) from hydrothermal sediment at the Mid-Atlantic ridge. It is potentially capable of growing into a chemolithoautotroph by oxidising H_2 and reducing CO_2(Goto et al., 1977; Hayashi et al., 1999; Stohr et al., 2001). In addition to the absence of *Hydrogenophilus sp.* in our contaminant database as well

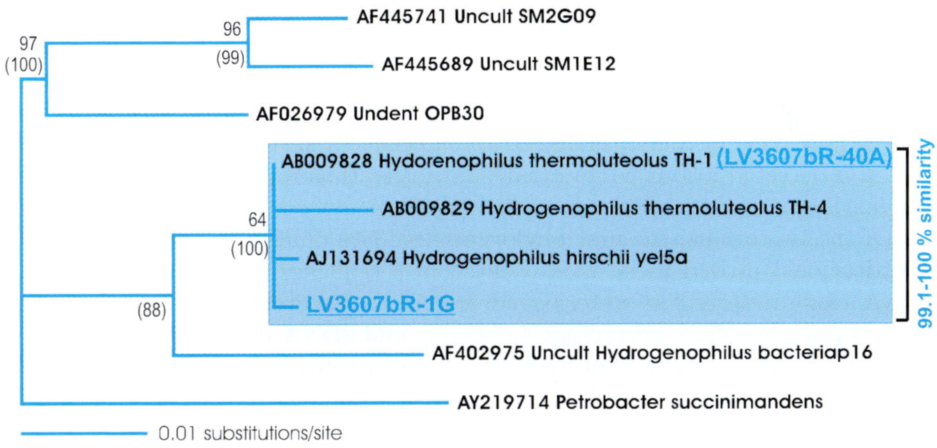

Fig. 7.18. Phylogenetic analysis of the bacterial rDNA clone LV3607bR-40A (a 428 nucleotide mask of unambiguously aligned positions) recovered from accretion ice sample #3607b. Bootstrap values of more than 50% are shown at the nodes (in brackets - support for full length sequences). Ice clones are in bold and underlined print. Species limits are depicted with transparent boxes. Clone 3607bR-40A displays a sequence identical to AB009828. Neighbouring sequences were determined with the BLAST (Altschul et al. 1997) program from public databases (GenBank). Sequences were aligned using the Clustalw program (Higgins et al. 1994). Phylogenetic trees were estimated using the maximum parsimony, minimum evolution (LogDet/paralinear distance parameter correction) and maximum likelihood criteria in PAUP (version 4.0b10) (D. Swofford, Sinauer Associates). The trees obtained were analysed by a Kishino-Hasegawa test in order to select the best, which were consequently optimised under maximum likelihood criteria. Bootstrap values were obtained by the maximum-parsimony method on 100 replicates with random-order sequence additions. For phylogeny reconstruction, aligned sequence blocks were fully cleaned from deletions or insertions, unassigned bases and ambiguous local alignments. Species limits depicted in a tree were defined using 3% sequence divergence (0.03 substitutions per site) (Stackebrandt and Goebel 1994).

as from other known contaminant data (Table 7.3), the limited distribution of the closest relatives of this species on a global scale (Izu district, Japan and Yellowstone, USA) provides an additional indication supporting its noncontaminant status.

Note that the thermophilic bacterium *Hydrogenophilus thermoluteolus* was detected in a sample with crystal boundaries (#3607b) while analysis of extra monocrystal samples with (#3607) and without sediment inclusions (#3607c) gave no reliable signal (Table 7.4). Because the bacterium we have found is a relevant thermophilic species with an optimal growth temperature around 50 °C (Goto et al., 1977; Hayashi et al., 1999), we consider its occurrence in ice to be an indication of long-term microbe concentration at the crystal boundaries by a freezing process rather than microbes thriving in ice veins as suggested (Price 2000).

7.4.2.5 The Unclassified Biological Findings

As mentioned, the above three *16S rDNA* phylotypes recovered in Vostok ice-core samples could not be identified because of low similarity with sequences in Genbank. One of them, related to *Chondromyces sp.* of delta-Proteobacteria, was also detected in our contaminant database, thus representing obvious contamination.

Of the two others, the first phylotype (2 clones differing by three nucleotide substitutions) proved to be most related in a sequence to one uncharacterised rDNA clone detected in waterlogged archaeological wood (97% similarity) as well as polyphosphate-accumulating strains and rDNA clones (96% similarity) of *Rhodocyclus* group of phototrophic beta-Proteobacteria affiliation (Crocetti et al., 2000) (Table 7.4) while demonstrating no confident phylogenetic clustering with them.

Most phototrophic beta-Proteobacteria of the *Rhodocyclus*, *Rubrivivax* and *Rhodoferax* genera are mesophilic freshwater bacteria with optimal growth temperatures between 25–30 °C. The *Rhodocyclus*, *Rhodoferax* and *Rubrivivax* species are facultative anaerobic bacteria that preferably grow under anoxic conditions by anoxygenic photosynthesis. All species grow photoheterotrophically and most species are able to grow photoautotrophically with hydrogen providing the reducing power for CO_2 fixation. At the same time they are all able to grow aerobically to microaerobically in the dark by oxygen-dependent respiration (Imhoff 2003). However, because these bacteria are common in stagnant water exposed to light and enriched in organic compounds and nutrients (Imhoff 2003), it would appear to be unrealistic to consider the phylotype recovered in ice as presenting lake biota. For this reason, it was referred as likely contaminant (Table 7.4).

The second uncertain phylotype (3 clones differing by two nucleotide substitutions each) showed rather distant relationships with uncultured bacteria of OP11 and OD1 Candidate Divisions and phylogenetically clustered with an rDNA clone from hot, dry, nutrient-poor sandy-loam soil nearby the Sunset Crater (Arizona, USA) (91.7% sequence similarity) (Dunbar et al., 2002; Kuske et al., 1997; Swaty et al., 1998) (Table 7.4). All the closest neighbours from GenBank are uncultured representatives of OP11 and OD1 Candidate Divisions and indistinctly phylogenetically related to each other (at the division level) and the ice phylotype as well, thus making this finding unresolved in terms of its taxonomy, features and living conditions.

However, our contaminant database (ice-core wash water dataset) contains one phylotype that demonstrated a low-score sequence similarity to the same set of soil bacterial clones (83–85% and namely 81% for rDNA clone C011 in Table 7.4) (Sunset Crater, Arizona, USA) along with OD1 clones (84–85% and namely 84% for rDNA clone AY193189) sampled in a contaminated aquifer (N. Pace, unpublished) while showing no evident phylogenetic relationship to the ice phylotype (79% sequence similarity). Nevertheless, because we are now increasingly aware of possible bacterial contamination, the incidence of a contaminant phylotype that shares the same set of soil bacterial clones with the ice phylotype may indicate a contaminant status for the latter (Table 7.4).

Meanwhile, considering two indecisive phylotypes, the latest, of OP11 Candidate Division, has a better chance of originating from the ice. Due to the largely uncharacterised nature of microbial habitats in the biosphere, environmental versatility of closely related species and ubiquity of some microbial general predictions about physiology based solely on 16S rDNA similarity, especially low similarity, must be interpreted, if at all, very cautiously.

7.4.2.6 Deliberations on Biological Findings in Lake Vostok Ice

Various bacteria previously discovered in Vostok accretion ice (Karl et al., 1999; Priscu et al., 1999; Abyzov et al., 2001; Christner et al., 2001) and identified by terminal restriction fragment length polymorphism (Priscu et al., 1999) and 16S rRNA gene sequencing (Christner et al., 2001) have suggested the presence of life in Lake Vostok. These studies may be valid, but results suffer from the absence of chemical controls or, when available, the ice chemistry is questionable (see Sect. 7.4.1).

In our approach, we have recognised up to now only one bacterium (12 clones; two sequence types) that successfully passed stringent contaminant criteria and, hence, can be considered as relevant to the accretion ice contents. It was found in one sample at 3607-m depth and represents the extant thermophilic facultative chemolithoautotroph *Hydrogenophilus thermoluteolus* of beta-Proteobacteria found until now only in hot springs. No reliable findings were detected in the sample at 3551m, and all other phylotypes detected (total 16 phylotypes) are presumed to be contaminants. This indicates that contamination, mostly from the laboratory environment (dust) as well as human sources, even after rigorous decontamination procedures, is still an important source of foreign bacteria. This fact along with the inability to generate PCR products in one round of PCR, indirectly suggests a very low number of cells and/or DNA amounts in the Vostok accretion ice.

Our findings are of special interest because the bacterium we discovered represents a genuine thermophilic species. Taking into consideration the limits of any PCR approach (e.g., PCR bias in estimating species composition and diversity (Dunbar et al., 2002; Lueders and Friedrich 2003) and despite the fact that we do not know enough about the correlation between rRNA sequence type and phenotype (while we obtained 100% sequence similarity for the region studied), we regard our finding of *Hydrogenophillus thermoluteolus* as an indication for the geothermal environment beneath Lake Vostok. It is likely that in such an environment, Archaea may be present (e.g., Barns et al., 1996). However, as in previous studies (Priscu et al., 1999), we have found no indication of their presence in the Lake Vostok ice samples.

7.4.3 Hydrothermal Environment in Lake Vostok?

7.4.3.1 Arguments from Geophysics and Ice Geochemistry

The geothermal environment beneath Lake Vostok is not intuitive for such a site and somewhat puzzling. The absence of ^{3}He-isotope enrichment in the accretion ice with respect to the glacier ice excludes the contribution of mantle-derived hydrothermal fluids or black smokers (Jean Baptiste et al., 2001; Jean Baptiste et al., 2003). However, three indicators lend support to geothermal activity, not necessarily mantle derived.

A first indicator is the geological setting. The lake is up to $\sim$ 1200 m deep and lies in a depression within the thick continental basement: its sharp shores (Masolov et al., 2001) and likely U-shaped cross-section imply deep faults (Fig. 7.19) where water can seep to great depth, warm up ($\sim$ 20 °C/km) and rise, causing hydrothermal circulation.

The second indicator is the long-term tectonic activity. The observed earthquakes from a fault in the vicinity of the lake have been used to support the possible introduction of energy into the lake, possibly fueling hydrothermal circulation (Studinger et al., 2003a,b). Past fault activity is supported by ^{4}He degassing. A three-fold ^{4}He excess in the $\sim$ 15 kyr-old accretion ice is observed with respect to the overlying glacier ice (Jean Baptiste et al., 2001; Jean Baptiste et al., 2003). ^{4}He, a byproduct of uranium series decay, remains entrapped within the rock matrix but could be released under tectonic activity. Bottom-lake sediments may also be the source of ^{4}He, but in this case the down core ^{4}He contents of accreted ice would increase with depth in response to the upward He diffusion within the ice. In fact the opposite trend is observed, and ^{4}He concentration is higher in the upper accretion layers, decreasing with depth (Jean Baptiste et al., 2003). This suggests that a ^{4}He source is likely located upstream of Vostok, and therefore in the shallow depth bay (Fig. 7.19). The excess of ^{4}He found in $\sim 15-20$ ka-old accretion ice suggests that tectonic activity was already present at that time.

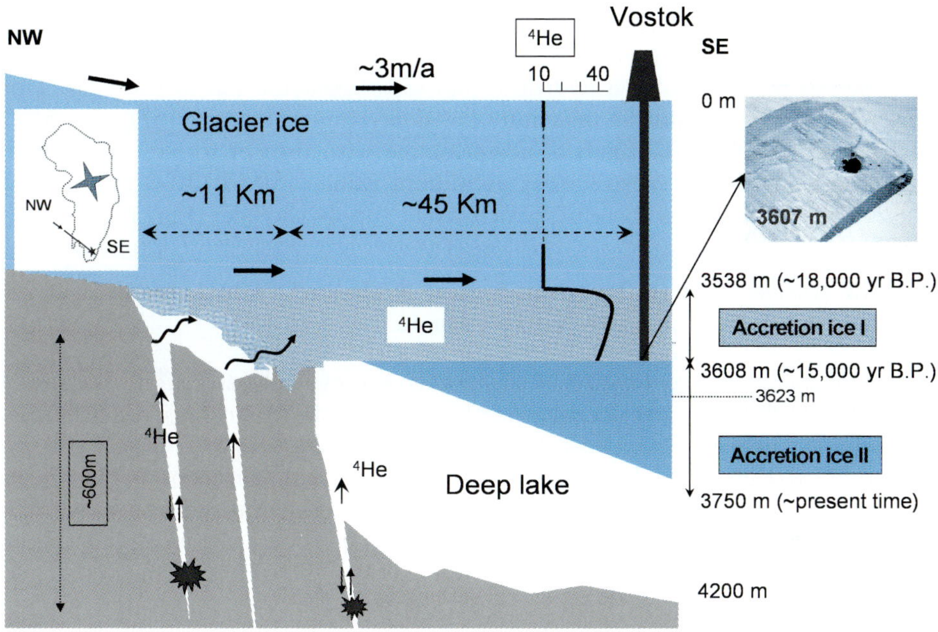

Fig. 7.19. Refined sketch of the glacier and lake basement along the Vostok ice flowline. The rock basement includes escarpments where deep faults allow water to seep in depth and circulate. Fault activity and ^{4}He degassing from the rocks are depicted by an explosion symbol and vertical arrows, respectively. A sketch of down core ^{4}He concentration (Jean Baptiste et al., 2001) shows a constant value for glacial ice and a change by a factor of 3 at the glacier-accretion ice boundary. In accretion ice, a decreasing trend is observed from 38.5 nmole kg^{-1} at 3551 m depth to 35.5 nmole kg^{-1} at 3609 m depth suggesting a ^{4}He contribution from a shallow depth area upstream of Vostok. On top right, a sediment inclusion is shown within the ice at 3607 m depth.

Thirdly, hydrothermal area spring water is known to be enriched in ^{18}O isotopes with respect to precipitation supplying the system (Craig 1963; Clayton and Steiner 1975). This 'oxygen shift' with respect to the meteoritic water line (MWL) may results from the isotope exchanges with ^{18}O-rich rocks and silicates at high temperature (Craig 1963). Lake Vostok is continuously renewed and its turnover time (τ) is $\sim$ 80 ka (see Table 7.2 and Sect. 7.2.4), small with respect to the age of the Lake ($\sim$ 1 Ma or more). Assuming a mass balance, the input (i.e. melting ice) equals the output (i.e. accreted ice exported by glacier movement) and the equilibrium holds for the isotope composition of ice and water (i.e. deuterium δD and oxygen δ^{18}O). The main melting area is located $\sim$ 240 km north of Vostok (see Sect. 7.3.1), where the glacier flows down the slope from Ridge B and is the thickest over the lake (Bell et al., 2002). This scenario satisfies the deuterium balance (Souchez et al., 2003, and see Sect. 7.3.1). However, when compared to the MWL of the ice core from Dome B (Vimeux et al., 2001), a site located upstream of the melting area, accretion ice composition appears to be enriched in $\delta^{18}O$ by $\sim 0.5‰$ (see Sect. 7.3.2 and Fig. 7.16), which can be regarded as an imprint of a hydrothermal water contribution to the lake. This enrichment may explain the unusually low deuterium excess ($d = \delta D - 8 \times \delta^{18}O \sim 7.5‰$) (Jouzel et al., 1999) of accretion ice with respect to ice composition from the Dome B and Vostok area (minimum value $\sim$ 10 to 13‰) (Vimeux et al., 2001 and references therein), which likely supply the lake (Fig. 7.16).

7.4.3.2 Assessment of Tectonic-fueled Niches or "Biotectonic" Environments in the Lake Vostok Basin

The geological setting, long-term seismotectonic evidence from ^4He degassing and the '^{18}O shift' of accretion ice, together support the existence of a geothermal environment beneath Lake Vostok.

Provided that Lake Vostok is a vestige of an old continental-rift system (Leitchenkov et al., 2003), or a collision zone of continental-sheet thrusts (Studinger et al., 2003b), seismotectonic activity has been operating locally for a long time. Local and distant tectonic events could activate fluid circulation (Dziak and Johson 2002) and promote chemical reactions between rocks and water. Among chemicals, hydrogen could be released (Wakita et al., 1980) and this powerful reducer could be consumed by micro-organisms (Chappelle et al., 2002). Our finding of *Hydrogenophilus thermoluteolus* in Lake Vostok accretion ice containing sediments is consistent with such a scenario.

A very low organic carbon and nutrient content measured in accretion ice is more favourable to sustaining chemoautotrophic rather than heterotrophic biota. Our finding of *Hydrogenophillus thermoluteolus*, a well-known thermophilic facultative chemolithoautotroph inhabiting hot springs, can be regarded as proof of such a biota. In addition, we consider its recognition exclusively from an ice interval that contains sediment inclusions as an indicator of a geothermal environment underneath the lake. Such organisms are unlikely to thrive in the open lake, where the high concentration of dissolved oxygen should be a significant constraint for at least heterotrophic life (Mikell et al., 1986). This suggests that the bacteria like *H. thermoluteolus* perhaps live deep in the 'hot' basement faults filled with sediments. It can also be speculated that such faults were colonised long ago – soon after their formation and before the onset of the first Antarctic glaciation around 34 Myr ago (Barrett 2003).

The hydrothermal plumes boosted by seismotectonic events can flush out bacteria and sediments from faults towards their vents. Some faults are likely open in the shallow bay upstream of Vostok, where the frazil ice accumulates and is rapidly transformed into ice as it is compressed by glacier transport against the relief rise closing the bay. Soluble salts (i.e. sodium chloride, calcium and magnesium sulfate) display discrete peaks with high levels of concentration within accretion ice I. These salts do not originate from the lake water but rather from ancient evaporite deposits (De Angelis et al., 2003, and submitted) likely below the lake sediments. The peak concentration supports the scenario involving rapid transfer and integration into the accreted ice, preventing dissolution in water. This could apply to mineral sediments with cells or DNA, thus preventing their possible damage (oxidation) by contact with the lake water.

7.5 Conclusions and Future work

7.5.1 Geophysical and Geochemical Aspects

We should first point out the paleoclimatological interest of ice cores from polar areas. The continuous natural sampling of the atmosphere and imprisonment within air bubbles is a unique property. Available records presently cover several thousand years and new projects are planned. The European Epica project (European Ice Core in Antarctica), funded by 10 nations and bringing together more than 60 scientists, plans to drill two new deep ice cores: one at Dome C and the other in the Droning Maud land area, with the aim of extending the paleoclimate record beyond four climatic cycles and of detailing the last glacial period. The ice core will also provide new material for the search for life in Antarctic ice, which remains to be documented.

Exploration of subglacial lakes will be improved and the lake bodies further documented by remote sensing and aerogeophysical surveys. The accretion ice recovered from the Vostok ice core represents the first window to the subglacial lake environment and likely represents the best proxy for the composition of the water. Further drilling and intrusion into the lakes for direct sampling or/and deployment of in-situ laboratories appears necessary to obtain a better understanding of the processes occurring in subglacial lakes. Such studies may be able to resolve the debate on fundamental issues concerning the water content and the history of lakes. However, to fulfil this ambitious goal, new clean drilling techniques need to be developed to prevent chemical and biological contamination by equipment and from the surface and human-influenced environment. This is the most critical issue, especially given that only 130 m of ice remains to be drilled above Lake Vostok and that the lake appears to be virtually germ free. Conversely, sampling techniques must also be developed that will ensure planetary protection similarly to those developed for the return of extraterrestrial samples. Since 1995, several international conferences and workshops on Antarctic subglacial lakes have been organised and a group of SCAR (Scientific Committee on Antarctic Research) experts have been assigned to study the topics linked to subglacial lake exploration (Kennicutt, M.C. 2001, and http://salegos-scar.montana.edu).

From the considerations we have developed above, the subglacial Lake Vostok likely existed prior the Antarctic glaciation as an open lake, or maybe as a swamp with water evaporating. The lake likely contained concentrated sea salt during this early stage, or at least sea water. This may partly account for the high concentration of several evaporites like salts found in sediment inclusions within the accreted ice samples (Cullen and Baker 2002; De Angelis et al., 2003). Spikes of soluble salt within accreted ice may be also associated with discrete tectonic events that may flush the salts through the fault and possibly inject them into the lake or directly into the accreted ice I through the fault mouths located in the shallow bay upstream of Vostok. As the glacier moves, it may drag the frazil ice crystals against subglacial relief, causing the sintering of frazil ice crystals and the rapid formation of accreted ice I.

As the ice cap was covering Antarctica, the lake volume likely increased and the expected salinity of the present-day lake is likely close to that of spring water. Further chemical analysis of accretion ice II would be relevant to this purpose (Souchez et al., 2000, 2003). Aside from needing favourable bedrock topography, lake formation depends on the temperature at the base of the ice sheet. This results in the energy balance between geothermal heat and the ice-sheet dynamics. Since this balance is too complex to be established accurately by numerical modelling and needs to be integrated over a very long time, prediction of lake formation will remain a difficult task. To date, direct detection by remote sensing or by radio echo sounding and aerogeophysical survey are the most effective methods, providing data to be integrated in modelling.

The glacier/lake interface is of prime importance. In Lake Vostok, it is tilted and, as a physical property applicable to other lakes, this leads to ice melting in the deep part and accretion-ice formation in upper levels. The slight difference in salinity between the glacier melt and lake water causes the newly formed fresh water to seep and rise along the interface, become supercooled and freeze. Frazil ice is likely to form within the seeping layer. It further accumulates and transforms into compact ice in the upper levels of the lake. Water circulation in the lake may have a thermohaline loop that should develop along the interface. Deep in the lake, a thermal loop is suggested, but to satisfy the heat transport its overall flow should be 1000 times higher than the thermohaline loop. Such circulation is obviously overly simplistic, but at least it is consistent with the observed accreted ice formation in the southern area. Further 3D numerical modelling, as already developed (Wuest and Carmak 2000; Mayer et al., 2003), will be pertinent in this respect.

From geometrical considerations, the frazil ice may account for about 52% of the accretion ice, while the remaining part is produced by subsequent freezing of the interstitial lake water in which the frazil is bathing. The melting and freezing areas are physically separated and represent about 47 and 53% of the lake area, respectively. The overall mean accretion and melting rates over the entire lake are about 10 mm/a and the renewal time of the water is about 80 ka. A total volume of $70 \times 10^6\,\mathrm{m}^3$ melts annually and an equivalent volume accreted after being exported out of the lake area. The mean balance velocity of the accretion area is ~1.5 m/a, which appears consistent with model results for the ice sheet upstream of the lake (Siegert and Kwok 2000). Future work involving the ice-sheet survey, surface-velocity measurements and temperature profiles in bore holes from selected areas over the lake would help to improve our knowledge of the mass and energy balance of the lake system and modelling (Salamatin et al., 2003).

Accretion ice is likely formed by the slow downward freezing front from the glacier. This slow process allows the formation of ice crystals with lattices with low concentrations of defects (Montagnat et al., 2000), an interesting material to study. The ice impurities, salts and gases, are mostly rejected during freezing and they accumulate in the lake. For salts, assuming the lake overturned 300 times since its formation while maintaining a constant water volume, the contribution of glacier salts to lake salinity is almost negligible ($\sim 0.1‰$). For gases, however, the lake likely accumulated up to $726\,\mathrm{g\,L^{-1}}$ for nitrogen and $8\,\mathrm{g\,L^{-1}}$ for oxygen in total. Due to the high pressure and temperature of the lake, part of the gases likely formed air hydrates (clathrates), nevertheless lake water may still be highly gas saturated (Lipenkov and Istomin 2001). The lake water likely contains an excess of oxygen with respect to open lakes, representing an additional obstacle for micro-organisms in the lake as well as for those from the surface biota that could be transported through the ice and freed in the lake.

7.5.2 Some Lessons for Future Research in Molecular Biology

The recent studies attempting to estimate the microbial contents of Lake Vostok accretion ice clearly show that, in the case of low biomass, the forward-contamination of samples represents a very difficult problem. It seems that Lake Vostok accretion ice as a whole is essentially germ free, indicating that the water body itself probably hosts only highly sparse life, if any, unless the lake water loses its biological content during the accretion process. An intercomparison recently performed in two independent laboratories indicates no reliable findings in several samples from deepest accretion ice II (Bulat, unpublished; S. Rogers, personal communication). It may seem very surprising that a giant water body such as Lake Vostok could be sterile, however, some desert areas without any apparent traces of life have already been found (Navarro-Gonzáles et al., 2003).

Of 16 bacterial phylotypes initially recovered from the accretion ice, only one proved relevant to the lake environment, while the 15 others are presumed to be contaminants. The best way to avoid contamination in future studies would be to use stringent ice-chemistry-based decontamination procedures and comprehensive biological controls including creation of a contemporary contaminant database as a prerequisite to identifying and categorising sources of extraneous cells and nucleic acids.

Seismotectonic influences exist in Lake Vostok (Studinger et al., 2003a, b) and probably sustain the geothermally fueled biological environment beneath the lake. Such influences could provide an alternative explanation for biological primary production to those based on black smokers (McCollom 1999).

The results of the present study have distinct implications, first concerning the search for lifeforms within Antarctic ice cores that may provide a million years of history on the evolution of micro-organisms. Secondly, this work is relevant to the field of exobiology and the search of life on Jovian icy moons and Mars, where low microbial biomass is expected, thus making forward-contamination by the spacecraft's terrestrial bioload a very crucial issue (Chyba and Phillips 2001; Space Studies Board 2000). Specifically, the ice decontamination protocol developed as well as the contaminant database established in the present study represent "lessons" for exobiology applications.

Concerning the search for life in subglacial environments on Earth (Siegert 2000, 2001) or on the Jovian satellite Europa (Carr et al. 1998; Pappalardo et al.,

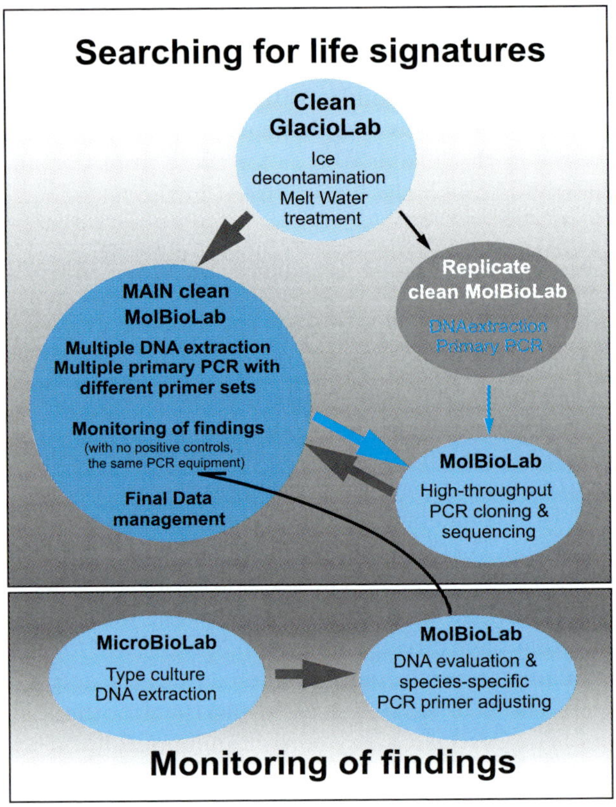

Fig. 7.20. An outline of a network for reliable molecular biological analysis of ice cores. Two areas (Searching for life and Monitoring of findings) are not just physically separated, but no exchange by products (only by information) is allowed. Abbreviations mean the following laboratories: GlacioLab – Glaciology, MolBioLab – Molecular Biology, MicroBioLab – Microbiology. Replicate clean laboratory refers here to the laboratory in which the ice samples with interesting findings are restudied by DNA extraction and primary PCR amplification after requesting sample aliquots from GlacioLab. Coloured arrows denote the following product/information flows: *red* - ice meltwater samples ready-to-extract-DNA, *blue* – primary PCR products ready-to-clone, *black* – data on sequences and PCR conditions and *green* – type culture genomic DNA.

1999), which has many similarities with Lake Vostok, our recommendations can be summarised as follow:

- Apply stringent ice decontamination procedures to achieve chemistry and trace DNA analysis standards.
- Document the biological contents of various environments (including humans) in contact with ice samples, to form contaminant databases.
- Use relevant biological tools (microscopy, culturing, PCR with both broad-range gene-specific and random primers or DNA-hybridisation) carefully to cover all known and expected biodiversity (provided that the life is DNA based).
- Verify microbial findings through their possible metabolic (redox) profiles taking into account known physical and chemical features of icy environments (e.g. in Lake Vostok – no light, temperature near freezing point, low DOC and other nutrient contents, excess oxygen and nitrogen).

In addition, we highly recommend that a network be established for studies on the microbial contents of ice cores, similar to the one we have already established for work in the field of molecular biology (Fig. 7.20). Such a network is also in line with ancient DNA research recommendations (Cooper and Poiner 2000).

In general, the reliability of biological ice-core findings increases when: (i) ice-chemistry procedures meet the necessary standards for low concentration measurements in polar ice; (ii) molecular biology procedures meet the recommendations for ancient DNA research as well as the practice of trace DNA analysis for forensic purposes; (iii) biological findings are different from those in the various environments in contact with the ice core; (iv) biological findings are discussed with respect to the expected conditions of subglacial environments. A multidisciplinary approach is therefore an obvious requirement for such studies.

Acknowledgement

We would like to dedicate this text to the memory of M. Blot (deceased in September 2002). He was an enthusiastic colleague who initiated biological studies of Antarctic ice in France.

We would like to thank M. Gargaud for editing assistance and for her fervent support and suggestions.

Vostok ice-core samples were obtained from a joint project involving Russia, France and the USA. We acknowledge RAE, IPEV and OPP NSF for logistic support and the SPMI drilling team for field work.

We thank E. Coursange, M. de Angelis, S. Greilich, M. Schock, D. Faure, A.M. Gounot, A. Richaulme, J. Martins, P. Taberlet, D. Prieur, F. Rémy, R. Souchez, R Vaikmae, F. Vimeux and M. Stievenard for assistance and helpful discussions. We thank J. Reisse, A. Salamatin and M. Siegert for helpful comments on the manuscript. We thank S. Rogers for conducting comparison tests on accretion ice.

The biological research is a contribution to Project 4 of the Subprogram "Study and research of the Antarctic", FTP "World Ocean" of the Russian Federation. The research was supported by the CNRS (Geomex), CNES, and a Balzan donation by C. Lorius. I.A. and S.B. were funded by the CNRS (SDU and SDV) and by Joseph Fourier University (Grenoble France).

References

Abyzov, S. S., Microorganisms in the Antarctic ice, *In Antarctic Microbiology*, Wilay-diss, Inc., N.Y., 265–295, 1993.

Abyzov, S. S., Mitskevitch, I.N., Poglazova, M.N., Barkov, N.I., Lipenkov, V.Y., Bobin, N.E., Kudryashov, B.B., Pashkevich, V.M., Ivanov, M.V., Microflora in the basal strata at Antarctic ice core above the Vostok lake, *Advances in Space Research*, **28**, 701–706, 2001.

Altschul, S. F., T.L. Madden, A.A. Schaffer, J. Zhang, Z. Zhang, W. Miller, D.J. Lipman, Gapped BLAST and PSI-BLAST: a new generation of protein database search programs, *Nucleic Acids Res.*, **25**, 3389–3402, 1997.

Anderson, D. L., and C.S. Benson, *The Densification and Diagenesis of Snow*, 391–411 pp., The MIT Press, Cambridge, Mass, 1963.

Anklin, M., J.-M. Barnola, J. Schwander, B. Stauffer, B. Raynaud, Processes affecting the CO_2 concentrations measured in Greenland ice, *Tellus*, **47 B**, 461–470, 1995.

Barns, S. M., C. F. Delwiche, J.D. Palmer, N. R. Pace, Perspectives on archaeal diversity, thermophily and monophyly from environmental rRNA sequences, *Proc. Natl. Acad. Sci.* USA, **93**, 9188–9193, 1996.

Barrett, P., Antarctic paleoenvironment through Cenozoic times, *Terra Antarctica*, **3**, 103–119, 1996.

Barrett, P., Cooling a continent, *Nature*, **421**, 221–223, 2003.

Basile, I., F.E. Grousset, M. Revel, J. R. Petit B.E. Biscaye, N.I. Barkov, Patagonian origin of glacial dust deposited in East Antarctica (Vostok and Dome C) during glacial stages 2, 4 and 6, *Earth and Planetary Science Letters*, **146**, 573–579, 1997.

Bell, R. E., M. Studinger, A. Tikku, G.K.C. Clarke, M.M. Gunter, C.C. Meertens, Origin and fate of lake Vostok water frozen to the base of the east Antarctic ice sheet, *Nature*, **416**, 307–310, 2002.

Boutron, C. F., J.P. Candelone, and S. Hong, Past and recent changes in the large scale tropospheric cycles of lead and other heavy metals as documente in Anatrctic and Greenland snow and ice: a review, *Geochem. Cosmoschem. Acta*, **58**, 3217–3225, 1994.

Bulat, S., I.A. Alekhina, V.Y. Lipenkov, G. Leitchenkov, D. Raynaud, and J.R. Petit Limitations for life in lake Vostok, Antarctica, Geophys. *Res. Abstract*, **5**, European Geophysical Society, Abstract 03288, 2003.

Bulat, S. A., I.A. Alekhina, M. Blot, J.R. Petit, D. Waggenbach, V.Y. Lipenkov, D. Raynaud, and V.V. Lukin, Thermophiles microbe signatures in Lake Vostok Antarctica, *Eos Trans*, **83 (19)**, B021–A09, 2002.

Carr, M. H., M. J. S. Belton, C. R. Chapman , M. E. Davies, P. Geissler, R. Greenberg, A. S. McEwen, B. R. Tufts, R. Greeley, R. Sullivan et al., Evidence for a subsurface ocean on Europa, *Nature*, **391**, 363–365, 1998.

Chappelle, F. H., O'Neill, K., Bradley, P.M., Methé, B.A., Ciufo, S.A., Knobel, L.L. and Lovley, D.R., A hydrogen-based subsurface microbial community dominated by methanogens, *Nature*, **415**, 312–315, 2002.

Christner, B., E. Mosley-Thompson, L. Thompson, and J. Reeve, Isolation of bacteria and 16S rDNAs from Lake Vostok accretion ice, *Environmental Microbiology*, **3**, 570–577, 2001.

Chyba, C. F., Phillips, C.B., Possible ecosystems and the search for life on Europa, *Proc. Natl. Acad. Sci. USA*, **91**, 801–804, 2001.

Cisar, J. O., Xu, D. Q., Thompson, J., Swaim, W., Hu, L. and Kopecko, D. J., An alternative interpretation of nanobacteria-induced biomineralization, *Proc. Natl. Acad. Sci. USA*, **97**, 11511–11515, 2000.

Clayton, R. N., and A. Steiner, Oxygen isotope studies of the geothermal system at Wairakei, New Zealand, *Geochem. Cosmochem. Acta*, **39**, 1179–1186, 1975.

Colbeck, S. C., Thermodynamics of snow metamorphism due to variation in curvature, *Journal of Glaciology*, **26**, 291–301, 1980.

Colbeck., S. C., *Theory of Metamorphism of Wet Snow*, pp. 11, CREEL, Hanover, New Hampshire, 1973.

Cooper, A. P., Poinar, H.N., Ancient DNA: Do It Right or Not at All., *Science*, **289**, 1139, 2000.

Corless, C.E., M. Guiver, R. Borrow, V. Edwards-Jones, E.B. Kaczmarski, A.J. Fox, Contamination and sensitivity issues with a real-time universal 16S rRNA PCR, *J. Clin. Microbiol.* **38**, 1747–1752, 2000.

Craig, H., Isotopic variations in meteoritic waters, *Science*, **133**, 1702–1703, 1961.

Craig, H., *The Isotopic Geochemistry of Water and Carbon in Geothermal Areas*, 1–53 pp., Consiglio Nationale dell Ricerche, Laboratorio di Geologia Nucleare, Pisa, 1963.

Criss, R. E., *Principle of Stable Isotope Distribution*, 254 pp., Oxford University Press, New York, Oxford, 1999.

Crocetti, G. R., P. Hugenholtz, P.L. Bond, A. Schuler, J. Keller, D. Jenkins, L.L. Blackall, Identification of polyphosphate-accumulating organisms and design of 16S rRNA-directed probes for their detection and quantitation, *Appl. Environ. Microbiol.*, **66**, 1175–1182, 2000.

Cullen, D., and I. Baker, Observation of sulfate crystallites in Vostok accretion ice, *Material Characterization*, **48**, 263–269, 2002.

Dansgaard, W., Stable isotope in precipitation, *Tellus*, **16**, 436–468, 1964.

De Angelis, M., M.H. Thiemens, J. Savarino, and J.R. Petit, Contribution of an ancient evaporitic-type reservoir to Lake Vostok chemistry, *Geophys. Res. Abstract*, **5**, European Geological Society, Abstract 05627, 2003.

De Angelis, M., J.R. Petit, J. Savarino, R. Souchez, and M.H. Thiemens, Contribution of an ancient evaporitic-type reservoir to lake Vostok chemistry, *Earth Planet Sci. Lett.*, **222**, 751–765, 2004.

Delmonte, B., I. Basile-Doelsch, J.R. Petit, V. Maggi, R.M. Revel, A. Michard, E. Jagoutz, and F.E. Grousset, Comparing the Epica and Vostok dust records during the last 220,000 years: stratigraphical correlation and origin in glacials periods, *Earth Science Reviews*, **66**, 63–87, 2004.

Dole, M., G.A. Lane, D.P. Rudd, and D.A. Zaukelies, Isotopic composition of atmopsheric oxygen and nitrogen, *Geochem. Cosmochem. Acta*, **6**, 65–78, 1954.

Dunbar, J., S.M. Barns, L.O. Ticknor, C.R. Kuske, Empirical and theoretical bacterial diversity in four Arizona soils, *Appl. Environ. Microbiol*, **68**, 3035–3045, 2002.

Dziak, R. P., and H.P. Johson, Stirring the oceanic incubator, *Science*, **296**, 1406–1407, 2002.

Gaudichet, A., M. de Angelis, E. Lefevre, and J.R. Petit, Mineralogy of insoluble particles in the Vostok Antarctic ice core over the last climatic cycle (150 kyr), *Geophys. Res. Lett.*, **15**, 1471–1474, 1988.

Gorman, M. R., and M.J. Siegert, Penetration of Antarctic lakes by VHF electromagnetic pulses: information on the depth and electrical conductivity of basal water bodies, *J. Geophys. Res.*, **104**, 29331–29320, 1999.

Goto, E., T. Kodama, Y. Minoda, Isolation and culture conditions of thermophilic hydrogen bacteria, *Agric. Biol. Chem.*, **41**, 685–690, 1997.

Grahn, N., M. Olofsson, K. Ellnebo-Svedlund, H-J. Monstein, J. Jonasson Identification of mixed bacterial DNA contamination in broad-range PCR amplification of 16S rDNA V1 and V3 variable regions by pyrosequencing of cloned amplicons., *FEMS Microbiology Letters*, **219**, 87–91, 2003.

Hayashi, N. R., T. Ishida, A. Yokota, T. Kodama, Y. Igarashi, Hydrogenophilus thermoluteolus gen. nov., sp. nov., a thermophilic, facultatively chemolithoautotrophic, hydrogen-oxidizing bacterium, *Int. J. Syst. Bacteriol.*, **49**, 783–786, 1999.

Higgins, D., J. Thompson, T. Gibson, J.D. Thompson, D.G. Higgins, T.J. Gibson, CLUSTAL W: improving the sensitivity of progressive multiple sequence alignment through sequence weighting, position-specific gap penalties and weight matrix choice, *Nucleic Acids Res.*, **22**, 4673–4680, 1994.

Ikeda, T., A.N. Salamatin, V.Ya. Lipenkov, T. Hondoh, Diffusion of Air Molecules in Polar Ice Sheets. in *Physics of Ice Core Records*, edited by Hondoh, T., pp. 393–421, Hokkaido University Press, Sapporo, 2000.

Imbrie, J., On the structure and origin of major glaciation cycles. I linear responses to Milankovich forcing, *Paleoceanography*, **7**, 701–738, 1992.

Imhoff, J. F., The phototrophic beta-Proteobacteria. in *The Prokaryotes: An Evolving Electronic Resource for the Microbiological Community*, edited by Dworkin, M., et al., pp. release 3.13 http://link.springer-ny.com/link/service/books/10125/. Springer-Verlag, New York, 2003.

Jean Baptiste, P., J.R. Petit, V.Y. Lipenkov, D. Raynaud, and N.I. Barkov, Constraints on hydrothermal processes and water exchange in Lake Vostok from helium isotopes, *Nature*, **411**, 460–462, 2001.

Jean Baptiste, P., J.R. Petit, D. Raynaud, J. Jouzel, and S. Bulat , Helium signature and seismotectonic activity in Lake Vostok, *Geophys. Res. Abstract*, **5**, European Geological Society, 2003.

Jenkins, A., and A. Bombosch, Modeling the effect of frazil ice crystals on the dynamics and thermodynamics of ice shelf water plumes, *Journal Geophysical Research*, **100**, 6967–6981, 1995.

Jouzel, J., G. Hoffmann, R.D. Koster, and V. Masson, Water isotopes in precipitation: data/model comparison for present-day and past climates, *Quaternary Science Reviews*, **19**, 363–379, 2000.

Jouzel, J., V. Masson, O. Cattani, S. Falourd, M. Stievenard, B. Stenni, A. Longinelli, S.J. Jonhsen, J.P. Steffenssen, J.R. Petit, J. Schwander, R. Souchez, and N.I. Barkov, A new 27 ky high resolution East Antarctic climate record, *Geophys. Res. Lett.*, **28**, 3199–3202, 2001.

Jouzel, J., J.R. Petit, R. Souchez, N.I. Barkov, V.Y. Lipenkov, D. Raynaud, L. Stievenard, N.I. Vassiliev, V. Verbecke, and F. Vimeux, More than 200 meters of lake ice above subglacial Lake Vostok, Antarctica. *Science*, **286**, 2138–2141, 1999.

Kapitsa, A., J.F. Ridley, G. d. Q. Robin, M.J. Siegert, and I. Zotikov, Large deep freshwater lake beneath the ice of central Antarctica, *Nature*, **381**, 684–686, 1996.

Karl, D. M., D.F. Bird, K. Bjökman, T. Houlihan, R. Shakelford, and L. Tupas, Microorganisms in the accreted ice of Lake Vostok, Antarctica, *Science*, **286**, 2144–2147, 1999.

Kawai M., M., E., Kanda, H., Yamaguchi, N., Tani, K. and Nasu, M., 16S ribosomal DNA-based analysis of bacterial diversity in purified water used in pharmaceutical manufacturing processes by PCR and denaturing gradient gel electrophoresis, *Appl. Environ. Microbiol.*, **68**, 699–704, 2002.

Kennett, J. P., K.G. Cannariato, I.L. Hendy, and R.J. Behl, Carbon isotopic evidence for methane hydrate instability during Quaternary interstadials, *Science*, **288**, 128–133, 2000.

Kennicutt, M. C., (Ed.) Subglacial lake exploration: workshop report and recommendations, 2001.

Kulakov, L. A., M.B. McAlister, K.L. Ogden, M.J. Larkin, J.F. O'Hanlon, Analysis of bacteria contaminating ultrapure water in industrial systems, *Appl. Environ. Microbiol.*, **68**, 1548–1555, 2002.

Kuske, C. R., S.M. Barns, J.D. Busch, Diverse uncultivated bacterial groups from soils of the arid southwestern United States that are present in many geographic regions, *Appl. Environ. Microbiol*, **63**, 3614–3621, 1997.

Kwok, R., Implications of ice motion over lake Vostok, paper presented at Lake Vostok: a curiosity or a focus for interdisciplinary study?, Washington D.C., 1998.

Legrand, M., M. de Angelis, F. Maupetit, Field investigation of major and minor ions along Summit (central Greenland) ice cores by ion chromatography, *J. Chromatography*, **640**, 251–258, 1993.

Legrand, M., C. Lorius, N.I. Barkov, and V.N. Petrov, Atmospheric chemistry changes over the last climatic cycle (160,000 yr) from Antarctic ice, *Atmospheric Environment*, **22**, 317–331, 1988.

Legrand, M., and P. Mayewski, Glaciochemistry of polar ice cores: a review, *Review of Geophysics*, **35**, 217–243, 1997.

Legrand, M., and C. Saigne, Formate, acetate and methanesulfonate measurements in antarctic ice: some geochemical implications, *Atmospheric Environment*, **22**, 1011–1017, 1988.

Leitchenkov, G. L., V.N. Masolov, V.V. Lukin, S.A. Bulat, R.G. Kurinin, V.Ya. Lipenkov, Geological nature of subglacial Lake Vostok. In *Geophysical Research Abstracts*, paper presented at EGS - AGU - EUG Joint Assembly, European Geophysical Society 2003, Abstract 03433, Nice, France, 2003.

Lewis, E. L., and R.G. Perkin, Ice pumps and their rates, *J. Geophys. Res*, **91**, 11756–11762, 1986.

Lindahl, T., Instability and decay of the primary structure of DNA, *Nature*, **362**, 709–715, 1993.

Lipenkov, V., V.A. Istomin, S.A. Bulat, D. Raynaud, and J.R. Petit, An estimate of the dissolved oxygen concentration in subglacial lake Vostok, *EOS Trans. AGU*, **83 (19)**, Spring Meet Suppl., 587–588, 2002.

Lipenkov, V. Y., Air bubbles and air-hydrates crystals in the Vostok ice core. in *Physics of Ice Core Records*, edited by Hondoh (ed), T., pp. 327–358, Hokkaido University Press, Hokkaido, 2000.

Lipenkov, V. Y., and V.A. Istomin, On the stability of air hydrate crystals in subglacial lake Vostok, *Mater. Glyatsiol. Issled.*, **91**, 138–149, 2001.

Lopez-Garcia, P., S. Duperron, P. Philippot, J. Foziel, J. Susini and D. Moreira, Bacterial diversity in hydrothermal sediment and epsilonproteobacteria dominance in experimental microcolonizers at the Mid-Atlantic Ridge, *Environ. Microbiol.*, **5**, 961–976, 2003

Lueders, T., and M.W. Friedrich, Evaluation of PCR amplification bias by Terminal Restriction Fragment Length Polymorphism analysis of small-subunit rRNA and mcrA genes by using defined template mixtures of methanogenic pure cultures and soil DNA extracts., *Appl. Environ. Microbiol.*, **69**, 320–332, 2003.

Masolov, V. N., V.V. Lukin, A.N. Shermetiev, and P. S.V., Geophysical investigations of the subglacial lake Vostok in eastern Antarctica, *Doklady Earth Science*, **379A**, 734–738, 2001.

Mayer, C., and M.J. Siegert, Numerical modelling of ice sheet dynamics across the Vostok subglacial lake, central East Antarctica, *J. Glaciology*, **46**, 197–205, 1999.

McCollom, T. M., Methanogenesis as a potential source of chemical energy for primary biomass production by autotrophic organisms in hydrothermal systems on Europa., *J. Geophys. Res.*, **104**, 30729–30742, 1999.

McKay, C. P., K.P. Hand, P.T. Doran, D.T. Andersen, and J.C. Priscu, Clathrate formation and the fate of noble and biologically useful gases in Lake Vostok, Antarctica, *Geophys. Res. Lett.*, **30**, 1702, doi: 1710.1029/2003GL017490, 2003.

Mikell, A. T., Jr., B.C. Parker, E.M. Gregory, Factors affecting high-oxygen survival of heterotrophic microorganisms from an Antarctic Lake, *Appl. Environ. Microbiol.*, **52**, 1236–1241, 1986.

Millar, B. C., J. Xu, J.E. Moore, Risk assessment models and contamination management: implications for broad-range ribosomal DNA PCR as a diagnostic tool in medical bacteriology, *J. Clin. Microbiol.*, **40**, 1575–1580, 2002.

Montagnat, M., P. Duval, P. Bastie, B. Hamelin, O. Brissaud, M. de Angelis., J.R. Petit, and V.Y. Lipenkov., High crystalline quality of large single crystals of subglacial ice above Lake Vostok (Antarctica) revealead by hard X-ray diffaction, *C.R. Acad. Sci., Earth and Planetary Sciences*, **333**, 419–425, 2001.

Navarro-Gonzáles, R., Rainey, F.A., Molina, P., Bagaley, D.R., Hollen, B.J., de la Rosa, J., Small, A.M., Quinn, R.C., Grunthaner, F.J., Cáceres, L. et al., Mars-like soils in the Atacama desert, Chile, and the dry limit of microbial life., *Science*, **302**, 1018–1021, 2003.

Pappalardo, R. T., Belton, M. J.S., Breneman, H.H., Carr, M.H., Chapman, C.R., Collins, G.C., Denk, T., Fagents, S., Geissler, P.E., Giese, B. et al., Does Europa have a subsurface ocean? Evaluation of the geological evidence, *J. Geophys. Res.*. **104**, 24015–24055, 1999.

Petit, J. R., Blot, M. and Bulat, S., Le lac Vostok : à la découverte d'un environnement sous-glaciaire et de son contenu biologique. in *Les Traces du Vivant*, edited by M. Gargaud, D. D., J-P. Parisot and J. Reisse, pp. 275–317, Presses Universitaires de Bordeaux, Pessac, 2003.

Petit, J. R., J. Jouzel, D. Raynaud, N.I. Barkov, J.M. Barnola, I. Basile, M. Bender, J. Chappellaz, M. Davis, G. Delaygue, M. Delmotte, V.M. Kotlyakov, M. Legrand,

V.Y. Lipenkov, C. Lorius, L. Pepin, C. Ritz, E. Saltzman, and M. Stievenard, Climate and atmospheric history of the past 420,000 years from the Vostok ice core, Antarctica, *Nature*, **399**, 429–436, 1999.

Petit, J. R., J.W.C. White, N.W. Young, J. Jouzel, and Y.S. Korotkevich, Deuterium excess in recent Antarctic snow, *J. Geophys. Res.*, **96**, 5113–5122, 1991.

Price, P. B., A habitat for psychrophiles in deep Antarctic ice, *Proc. Natl. Acad. Sci. USA*, **97**, 1247–1251, 2000.

Priscu, J. C., E.E. Adams, B. Lyons, M.A. Voytek, D.W. Mogk, R.L. Brown, C.P. McKay, C.D. Takacs, K.A. Welch, C.F. Wolf, J.D. Kirshtein, and R. Avci, Geomicrobiology of subglacial ice above Lake Vostok, Antarctica, *Science*, **286**, 2141–2143, 1999.

Qin Dahe, J.R. Petit, J. Jouzel, and M. Stievenard, Distribution of stable isotopes in surface snow along the route of the 1990 international Transantarctica expedition, *Journal of Glaciology*, **40**, 107–118, 1994.

Raymond, C., and K. Tusima, Grain coarsening of water saturated snow, *Journal of Glaciology*, **22**, 83–105, 1979.

Rémy, F., P. Schaeffer, and B. Legresy, Ice flow physical process derived from ERS1 high-resolution map of Greenland and Antarctica, *Geophys. Int. J.*, **139**, 6645–6656, 1999.

Ridley, J. F., W. Cudlip, and S. Laxon, Identification of sub glacial lakes using ERS-1 radar altimeter, *Journal of Glaciology*, **39**, 625–634, 1993.

Ritz, C., Un modèle thermo-mécanique d'evolution pour le basin glaciaire antarctique Vostok-glacier Byrd: sensibilité aux valeurs des parameters mal connus., Univ. Joseph Fourier Grenoble I, Grenoble, 1992.

Robin, G. d. Q., Ice movement and temperture distribution in glaciers and ice sheets, *J. Glaciology*, **39**, 625–634, 1955.

Rummel, J. D., Planetary exploration in the time of astrobiology: protecting against biological contaminations., *Proc. Natl. Acad. Sci.*, **98**, 2128–2131, 2001.

Salamatin, A. N., V.Y. Lipenkov, N.I. Barkov, J. Jouzel, J.R. Petit, and D. Raynaud, Ice-core age dating and palaeothermometer calibration based on isotope and temperature profiles from deep boreholes at Vostok Station (East Antarctica), *J. Geophys. Res.*, **103**, 8963–8977, 1998b.

Salamatin, A. N., J.R. Petit, and V.Y. Lipenkov An estimate of LV isolation time from a sensitivity experiment for the melting area, *EoS, Geophys. Research Abstract*, **5**, Abstract 08277, 2003.

Siegert, M., and R. Kwok, Ice-sheet radar layering and the development of preferred crystal orientation fabrics between lake Vosk and Ridge B, central East Antarctica, *Earth and Planetary Science Letters*, **179**, 227–235, 2000.

Siegert, M. J., M.R. Dowdeswel, and N.F. Gorman, An inventory of Antarctic subglacial lakes, *Antarctic Science*, **8**, 281–286, 1996.

Siegert, M. J., and M.R. Dowdeswell, Spatial variations in heat at the base of the Antarctic ice sheet from analysis of the thermal regime above subglacial lakes, *J. Glaciology*, **42**, 501–509, 1996.

Siegert, M. J., J.C. Ellis-Evans, M. Tranter, C. Mayer, J. Petit, A. Salamatin, and J.C. Priscu, Physical, chemical and biological processes in Lake Vostok and other Antarctic subglacial lakes, *Nature*, **414**, 603–609, 2001.

Souchez, R., P. Jean Baptiste, J.R. Petit, V.Y. Lipenkov, and J. Jouzel, What is the deepest part of the Vostok ice core is telling us?, *Earth-Science Reviews*, **60**, 131–146, 2002.

Souchez, R., J.R. Petit, J. Jouzel, M. DeAngelis, and J. Tison, Re-assessing lake vostok's behaviour from existing and new ice core data, *Earth and Planetary Science Letters*, **217**, 163–170, 2003.

Souchez, R., J.R. Petit, J. Tison, Jouzel J., and Verbeke V., Ice information in subglacial lake Vostok, central Antarctica, *Earth and Planetary Science Letters*, **181**, 529–538, 2000.

Space Studies Board, *Preventing the Forward Contamination of Europa,*. Natl. Acad. Sci. Press, Washington, DC, 2000.

Stackebrandt, E., and B.M. Goebel, Taxonomic note: A place for DNA-DNA reassociation and 16S rRNA sequence analysis in the present species definition in bacteriology, *Int. J. Syst. Bacteriol.*, **44**, 846–849, 1994.

Stohr, R., A. Waberski, W. Liesack, H. Volker, U. Wehmeyer, M. Thomm, Hydrogenophilus hirschii sp. nov., a novel thermophilic hydrogen- oxidizing beta-proteobacterium isolated from Yellowstone National Park, *Int. J. Syst. Evol. Microbiol.*, **51**, 481–488, 2001.

Studinger, M., R.E. Bell, G.D. Karner, A.A. Tikku, J.W. Holt, D.L. Morse, T.G. Richter, S.D. Kempf, M.E. Peters, D.D. Blankenship, R.E. Sweeney, and V.L. Rystrom, Ice cover, landscape setting and geological framework of lake Vostok, east Antarctica, *Earth and Planetary Science Letters*, **205**, 195–210, 2003a.

Studinger, M., R.E. Bell, G.D. Karner, A.A. Tikku, V. Levin, C.A. Raymond, and A. Lerner Lam, Lake Vostok, from a continental margin to a subglacial lake, *EOS Trans. AGU*, **83** (19) Sprint Meet., 587–588 2002.

Studinger, M., G.D. Karner, R.E. Bell, V. Levin, C. Raymond, and A.A. Tikku, Geophysical models for the tectonic framework of the Lake Vostok region, East Antarctica, *Earth and Planetary Science Letters*, **216**, 663–677, 2003b.

Swaty, R. L., C.A. Gehring, M. Van Ert, T.C. Theimer, P. Keim, T.G. Whitham, Temporal variation in temperature and rainfall differentially affects ectomycorrhizal colonization at two contrasting sites, *New Phytol.* **139**, 733–739, 1998.

Tanner, M. A., B.M. Goebel, M.A. Dojka, N.R. Pace, Specific ribosomal DNA sequences from diverse environmental settings correlate with experimental contaminants, *Appl. Environ. Microbiol.*, **64**, 3110–3113, 1998.

Vimeux, F., V. Masson, J. Jouzel, J.R. Petit,s E.J. Steig, M. Stievenard, R. Vaikmae, and J.W.C. White, Holocene hydrological cycle changes in the Southern hemisphere documented in East Antarctic deuterium excess records, *Climate Dynamics*, **17**, 503–513, 2001.

Vimeux, F., V. Masson, J. Jouzel, M. Stievenard, and J.R. Petit, Glacial-interglacial changes in ocean surface conditions in the Southern Hemisphere, *Nature*, **398**, 410–413, 1999.

Vincent, W. F., Icy life on a hidden lake, *Science*, **286**, 2094–2095, 1999.

Vincent, W. F., De la vie dans le lac Vostok? in *La Recherche*, pp. 14–16, 2000.

Wakita, H., Y. Nakamura, I. Kita, N. Fujii, and K. Notsu, Hydrogen release: new indicator of fault activity, *Science*, **210**, 188–190, 1980.

Wuest, A., and E. Carmak, A priori estimates of mixing and circulation in the hard to reach water body of lake Vostok, *Ocean Modelling*, **2**, 29–49, 2000.

8 Comets: Potential Sources of Prebiotic Molecules for the Early Earth

Didier Despois, Hervé Cottin

"Of Ice and Men..."

Why should we be interested in comets when studying the origin of life on Earth? First, comets are rich in water and carbon, two essential constituents of terrestrial life; part of Earth's water and carbon might be of cometary origin. But comets might also have brought Earth one step further on the way to the emergence of life.

Early last century, Chamberlin and Chamberlin proposed that infalling carbonaceous chondrite meteorites could have been an important source of terrestrial organic compounds (Chamberlin and Chamberlin 1908). Oró was the first in 1961 to propose from observations of carbon- and nitrogen-containing radicals in cometary comae that comets may have played a similar role (Oró 1961):

> "I suggest that one of the important consequences of the interactions of comets with the Earth would be the accumulation on our planet of relatively large amounts of carbon compounds which are known to be transformed spontaneously into amino acids, purines and other biochemical compounds."

It is now quite obvious, from observations and laboratory experiments, that comets are important reservoirs of a wide variety of organic compounds.

Comets will be considered in this chapter mainly as potential reservoirs of carbon and organic molecules for the early Earth. We present first their general characteristics, then the chemical composition of cometary matter as deduced from observations and constrained by laboratory experiments, the origin and evolution of this matter – and of comets themselves – the various cometary contributions to the early Earth, and the main future of ground-based and space developments for comet science. We conclude with a list of important questions for which answers should and will be sought in the forthcoming years.

The book "Comets and the Origin and Evolution of Life", edited by Thomas et al. (1997), is a basic reference on the topic. Comets have been discussed as possible sources of prebiotic molecules by Delsemme (2000), Greenberg (e.g. 1982, 1993, 1998) and Chyba and Sagan (1997). Despois et al. (2002) and Crovisier (2004) have discussed recently the observed molecular content of comets in this perspective, while Whittet (1997) and Ehrenfreund et al. (2002) relate more generally astrochemistry to the origin of Life. Recent general reviews on cometary science are Irvine et al. (2000), Festou et al. (1993a,b), and Bockelée-Morvan et al. (2004); the book Comets II (in press) edited by Festou, Keller and Weaver will present up-to-date information on many aspects of comets. In this book, Petit and Morbidelli (2004) survey the Solar System formation, and discuss the origin of water on Earth. Chapters of the French edition of this book (Gargaud et al., 2001, 2003), not included in this English edition, also contain much information on related topics only briefly alluded to here: build up and evolution of the Earth atmosphere (Marty; Selsis and Parisot), build up of the hydrosphere and D/H ratio (Robert), micrometeorites (Maurette). Recent related papers in English are Kasting (1993, 2003), Dauphas et al. (2000), Robert (2002), and Maurette (1998, 2000).

8.1 General Description of Comets

"Comets are small bodies of the Solar System, made of a mixture of ices and dust."

8.1.1 The Cometary Nucleus

The definition above describes in fact a specific part of comets, the cometary nucleus. Comets are most of the time far from the Sun and reduced to this nucleus, a solid body named a "dirty ice snow-ball" by Whipple (1950); it is only when comets get close enough to the Sun (less than a few UA) that they develop their other more spectacular and better known components – coma and tail – described below. Comet 1P/Halley[1] is the first of the only three comets whose nuclei have been closely observed by a space probe (the other two are 19P/Borrelly, observed from 3400 km by the Deep Space 1 spacecraft in September 2001, and P/Wild 2 observed by the Stardust spacecraft from 236 km in January 2004). In 1986, several probes were sent to observe comet Halley, among which the European Space Agency probe "Giotto" passed within approximately 500 km from the comet nucleus. The pictures taken by Giotto (Fig. 8.1) revealed an irregular-shaped body $8 \times 8 \times 16$ km by size, whose volume is roughly equivalent to a sphere of 10 km in diameter (see Newburn et al., 1991). The bright comet C/1996 B2 Hyakutake, which appeared in 1996 and passed rather close to the Earth (~ 0.1 AU), had a relatively small nucleus, about 3 km in diameter. Another comet, C1995 O1/Hale–Bopp, was detected for the first time during summer 1995 and became quite visible in spring 1997; although passing ten times further from the Earth, it has been more spectacular than Hyakutake due to its much larger diameter, approximately 45 km (see Weaver and Lamy 1999).

[1] 1P/Halley is the official complete name of the comet, according to the International Astronomical Union – see ref. CBAT for a description of the naming rules; the familiar names (e.g. "comet Halley") are also used in this chapter and should be considered as abbreviations.

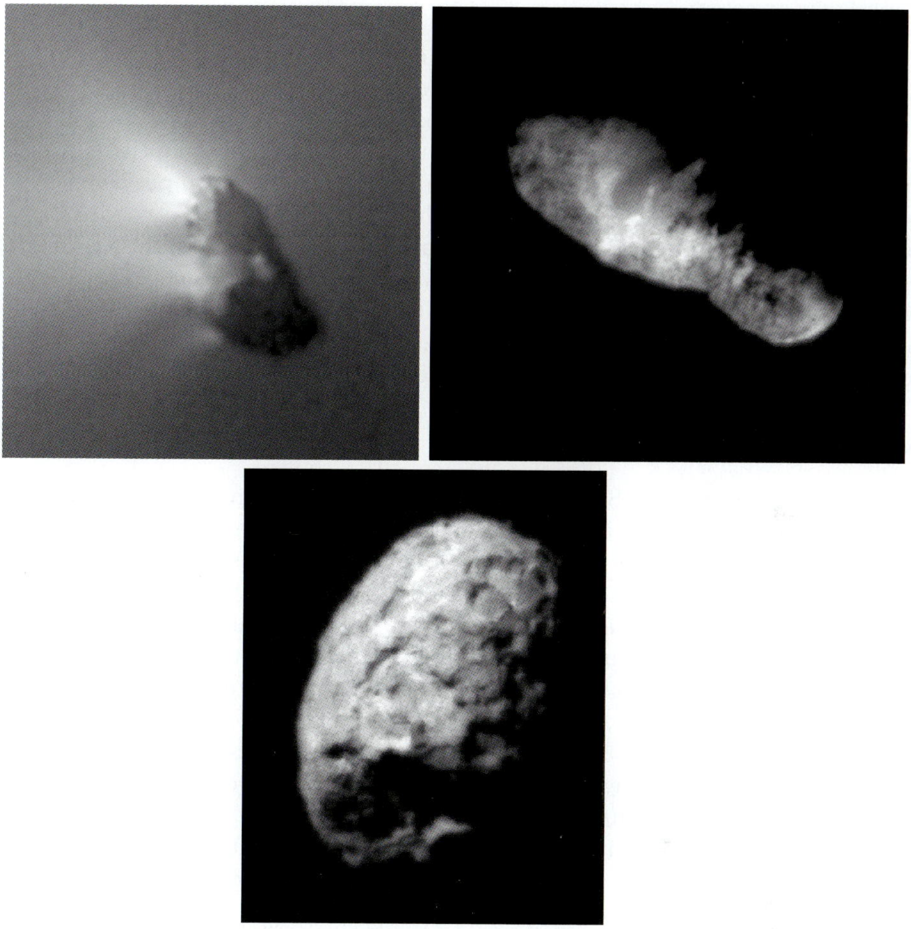

Fig. 8.1. (**a**) The nucleus of comet Halley, as seen by the ESA probe Giotto in 1986. (© ESA and MPIfA) (**b**) the nucleus of comet P/Borrelly observed by NASA probe Deep Space 1 in 2001 (© NASA) (**c**) the nucleus of comet P/Wild 2 recently observed (Jan. 2004) with a resolution of 20 m by Stardust, the NASA probe that will return dust samples collected during the flyby (© NASA)

The density of cometary nuclei is difficult to measure precisely: some estimates are as low as $0.2\,g/cm^3$ or as high as $1.8\,g/cm^3$. However, most values are in the range $0.5–0.9\,g/cm^3$ (e.g. Ball et al., 2001). This is lower than the density of water ice, and would confirm the image of the "dirty snowball". Using this density, the mass of comet Halley is about $10^{18}\,g$ (1000 billion tonnes).

The cohesion of cometary nuclei is low at least in some cases, as shown by comet D/1993 F2 Shoemaker-Levy 9: when the comet passed near Jupiter, the tidal forces let the nucleus split up into a swarm of more than 20 major pieces, which later crashed onto the planet (see Noll et al., 1996). The fact that cometary nuclei have irregular shapes shows that these bodies were never molten – otherwise they would be shaped according to the equilibrium figure of a liquid body: a sphere or, if the body rotates, an ellipsoid.

What are the ices and dust mentioned in the definition of a comet made of? Cometary ices are water ice mostly, mixed with smaller quantities of other species like methanol (CH_3OH), CO, CO_2, ... – the detailed composition is given below (Sect. 8.2.1 and Table 8.1). As for the dust grains, they are made of small particles of carbon, organic matter or silicates. The total mass of dust is roughly comparable to the mass of ices, within a factor of 10: some comets are "dusty", while other are "gaseous" (e.g. Rolfe and Battrick, 1987). Despite the presence of a large amount of ice inside the nucleus, the surface itself appears quite dark: for comet Halley, the proportion of reflected light (called *albedo*) is only 4%.

8.1.2 Comet Motion

Comets move around the Sun following orbits that are in most cases ellipses (the Sun being at one focus); in some cases the orbit may be parabolic or slightly hyperbolic (e.g. Festou et al., 1993a). The eccentricity is generally high, and often near 1 (parabola) or even slightly higher (hyperbola). For comparison, the eccentricity of a circle is zero, and that of the quasicircular orbit of the Earth is 0.017 (at present). The comet speed in the region of the Earth's orbit (i.e. at one AU from the Sun) is about 45 km/s (or 162 000 km/h). Combining (vectorially) this speed with Earth's own motion around the Sun ($\sim$ 30 km/s) leads to relative speeds ranging between 15 and 75 km/s. This confers a large kinetic energy to cometary particles entering the Earth's atmosphere and leads to dramatic effects when the nucleus itself impacts the Earth.

Comets are classified into various families according to their orbits. Long-period comets are defined as having an orbital period above 200 years. Among the long-period comets, one commonly calls "new comets" those for which aphelion (point of the trajectory furthest away from the Sun) is located beyond approximately 1000 AU and for which eccentricity is close to 1. This is not an official, well-defined, comet category; as these comets have periods above 10 000 years (and up to tens of Myr), no historical record of previous passages exists, and they appear "new" to us. Their orbits are very elongated ellipses, which, when they pass close to the Sun, look very much like parabolas. Comets with periods below 200 years are called short-period comets. Among those, one finds the *Halley family*, with periods from 20 to 200 years, and random inclinations of their orbital plane with respect to the ecliptic, and the *Jupiter family*, with periods of approximately 6 years, and small inclinations to the ecliptic.

8.1.3 Comet Reservoirs: Oort Cloud and Kuiper Belt

To explain the present number of observed comets (despite the rapid sublimation of the ice (see Sect. 8.1.5) and the origin of the various orbital families, the existence of two comet reservoirs is required.

The classical or *outer Oort cloud* (Oort, 1950) is a spherical reservoir situated between ~ 20 000 and 100 000 AU; the inner radius is loosely defined (see below), whereas the outer radius corresponds to half the distance to the nearest stars, beyond which an object can no longer be considered gravitationnally linked to the Sun. Oort Cloud comets are thought to be icy planetesimals scattered by the giant planets, originating mostly from the Uranus–Neptune zone and to a lesser extent from the Jupiter–Saturn zone.

Due to stellar encounters, or to the effect of the gravitational potential of the Galaxy, some of these objects are regularly sent back towards the planet zone with nearly parabolic orbits ("new" comets). The possible quantitative importance of a more tightly bound *inner Oort cloud* (3000–20 000 AU), has been recently underlined (Fernandez, 1997, Levison et al. 2001) considering the very probable option that the Solar System and its Oort cloud formed in the dense stellar environment of a young open star cluster. Formation scenarios of the Oort cloud (e.g. Duncan et al., 1987) are still very uncertain, at least quantitatively. The total number of comets in the Oort cloud could be on the order of 10^{12}–10^{14} depending on the model and the size distribution, while the total mass estimate ranges from 1 to 250 M_{Earth}.

While the Oort cloud hypothesis explains quite well the long-period comet distribution, numerical simulations (Fernandez 1980) have stressed the importance of another reservoir, the *Kuiper Belt,* to explain the number of short-period comets. The existence of this belt was proposed first by Leonard (1930) and Edgeworth (1943), (the history of the concept, including the Kuiper (1950) paper and the later role of Cameron and Whipple, is discussed in Green 1999). The region has the shape of a thick torus and contains many small icy bodies beyond the orbit of Neptune. Since the discovery of the first Kuiper Belt object (KBO) by Jewitt and Luu in 1992, much observational and theoretical work has been undertaken (see Morbidelli and Brown (2004), Jewitt (2002) and de Bergh (2004) for recent reviews on the dynamical and physical properties of KBOs). Collisions in the Kuiper belt are thought to produce fragments, which are later scattered by Neptune and the other planets towards the inner Solar System, first as Centaurs, then as Jupiter Family comets. From the ~ 800 objects now observed (summer 2003) a total population of 70 000 bodies larger than 100 km in diameter is estimated, and 1 to 10 billion bodies larger than 1 km. Several families, with probably different origin, are recognized in the Kuiper Belt (see Morbidelli and Brown 2002 and Jewitt 2002 for details). A subpopulation, the plutinos, is in a collision-avoiding resonant orbit with Neptune, and encompasses Pluto itself (and its satellite Charon). Pluto is considered as the largest KBO (with $D \sim 2500$ km), and several other large objects are known (Quaoar, $D \sim 1300$ km; Charon, $D \sim 1200$ km; Varuna, $D \sim 900$ km).

The formation of these reservoirs has been accompanied by substantial scattering of comets towards the zone of the terrestrial planets; these comets possibly supply carbon and organics to their surfaces (see Sect. 8.5) but probably make only a limited contribution to Earth's water (see Chap. 2, Part I by Petit and

Morbidelli). The composition of the comets coming from the two reservoirs is expected to differ more or less strongly according to the various formation models. Understanding the formation scenario of comets is vital for any considerations on the ubiquity (or scarcity) of terrestrial planets favourable to life development elsewhere in the Universe.

8.1.4 The Active Comet

When a comet approaches the Sun, it enters the active phase. A coma develops around the nucleus, which consists of a roughly spherical halo of neutral gas and dust particles. The radius of the coma depends on the chemical species, and ranges between less than a thousand to more than one million km (e.g. Crovisier and Encrenaz, 2000) (Fig. 8.2). What we call here "radius" is indeed a characteristic length describing the progressive decrease of the spatial density of particles. Beyond this coma, an atomic hydrogen cloud is produced by the photodissociation of a number of species, mainly H_2O and OH; this cloud can reach 1 to 10 million km. Starting from the nucleus and the coma, three types of cometary tails develop: a plasma tail, a dust tail, and a tail of neutral sodium

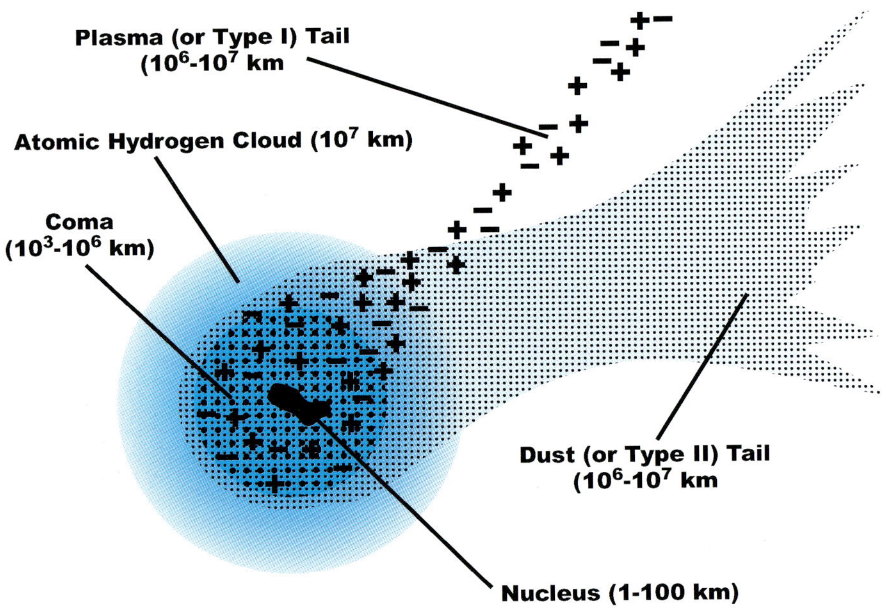

Fig. 8.2. Schematic view of an active comet. The Sun is (approximately) in the direction opposite to the ion tail, whose shape is due to the interaction with the rapid ($\sim 400 - 800$ km/s) charged particles of the solar wind. The dust particles that were previously released from the nucleus are slowly pushed away by the pressure of the solar photons, which, combined with the comet motion (here toward the left) explains the shape of the dust tail (see e.g. Festou et al., 1993b)

atoms – the latter clearly seen recently in observations of comet Hale–Bopp (Cremonese et al., 1997).

The plasma tail is made of (atomic and molecular) ions and electrons. It has a characteristic blue colour due to the CO^+ ions, a very rectilinear shape and is almost perfectly opposed to the Sun's direction. It is produced by the interaction of the solar wind – itself made of protons and charged particles – with the ions produced in the coma (Fig. 8.3). The yellow-white dust tail consists of small grains that have typically a diameter between 1 and 10 micrometres, and that reflect and scatter the light of the Sun. Its shape depends on the balance between the gravitational force from the Sun, and the radiation pressure due the photons of the solar radiation.

An antitail is seen in some comets, looking like a narwhal tusk pointing toward a direction opposite to the other tails. It is in fact always present, but not always visible. It reveals the presence of an accumulation of particles in the orbital plane of the comet. These particles are larger than those of the normal dust tail, and have thus been ejected at low speed from the cometary nucleus (they are less easily lifted by the gas flow) and are less sensitive to solar-radiation pressure (their mass/surface ratio is larger). By a projection effect, these particles appear as a spike at the head of the comet, when the Earth happens to cross the comet orbital plane. This feature usually appears only on the Sun side, but

Fig. 8.3. Image of a comet with well-developed tails (Hale–Bopp). The plasma tail (thin, rectilinear and bluish) is clearly recognized from the dust tail (diffuse, curved and white-yellow). (© Photo R. Lauqué)

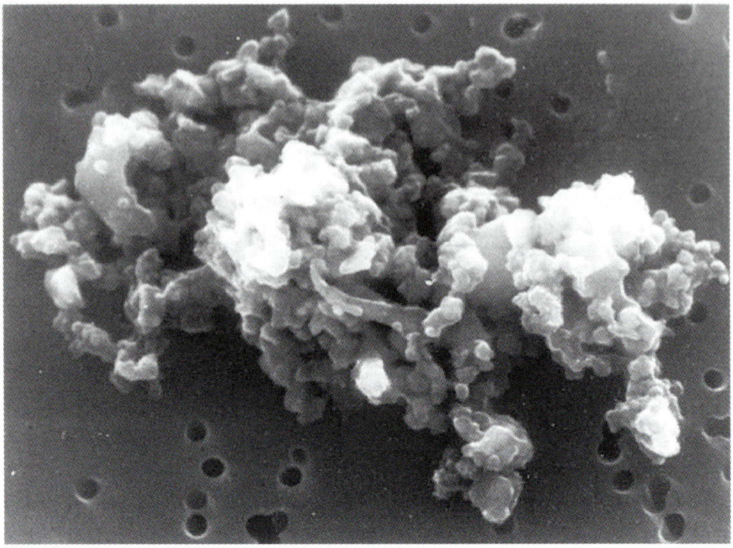

Fig. 8.4. Brownlee particle (interplanetary dust collected in the stratosphere). For the fluffiest of these particles, the origin is probably cometary

can sometimes also be distinguished as a thin line in the antisolar direction, superposed to the normal dust tail. The "neck-line" recently observed on comet P/Churyumov-Gerasimenko is an antitail (Weiler et al., 2004).

Cometary dust particles disperse more or less rapidly into the interplanetary space, according to their shape, nature and initial velocity. Some of them cross the path of the Earth; if large enough they produce meteors. One can get an idea of cometary dust particles by looking at particles collected in the Earth stratosphere ("Brownlee particles", or stratospheric interplanetary dust particles – SIDP). Many of these particles have a large surface to mass ratio and a very irregular, and somewhat fractal-like structure (Fig. 8.4). All collected particles are, however, not as irregular: one finds also among SIDPs more compact particles, similar to small blocks of clay. It is tempting – but perhaps too simplistic – to attribute to the most irregular particles a cometary origin, whereas the most compact ones would come from asteroids.

The third cometary tail, the sodium tail, made of neutral sodium atoms, was discovered for the first time in the Great Comet C/1910 A1, and reobserved more recently in comets C/1957 P1 (Mrkos) and Hale–Bopp. Sodium is pushed away in the antisolar direction by the radiation pressure, in a way similar to what happens for dust particles.

8.1.5 Gas and Dust Production

The weakest H_2O production rates currently detected in comets are of 10^{26} molecules/s (3kg/s) when the comet is favourably situated. In comet Hale–Bopp at perihelion, the production rate reached up to 10^{31} molecules/s, which corresponds to 300 tons/s. These rates imply that a bright comet typically loses a 1-m thick sheet at each perihelion passage. Thus for a nucleus of about a few tens km in diameter, one expects the comet to disappear after a few thousand passages close to the Sun. This implies that today short-period comets ($P < 200$ yr) cannot have occupied their present orbit since the beginning of the Solar System.

The expansion speed of most molecules in the coma is about 1 km/s. Photodissociation reactions take place, for example $H_2O \rightarrow H + OH$, followed by $OH \rightarrow O + H$. For the photodissociation fragments, the speeds can reach several km/s (8 and 21 km/s in the case of H produced from H_2O). The matter ejected by comets is almost equally distributed in mass between volatiles (mainly H_2O) and dust – within a factor 10. It is possible that the dispersion of the gas/dust ratio between comets represents a variation in the initial composition of the cometary nucleus; alternatively, it can also be an effect of evolution after many passages close to the Sun.

The dark surface observed on comet Halley's nucleus may consist of an accumulation of dust, and/or of complex refractory organic molecules. This crust may be broken by the explosion of gas pockets heated under the crust: this would enrich the coma in dust particles, and appear as an outburst of the comet.

8.1.6 Remote Activity, Outbursts and Split Comets

The activity of comets strongly depends on the distance to the Sun. It has long been thought that cometary activity developed only when the comet was closer than 4 AU from the Sun, because the "dirty snow ball" model of the nucleus suggested that water, the major volatile, controlled the emission of other molecules and of solid particles. The sublimation of H_2O ice is important only at heliocentric distances smaller than 4 AU and drives the coma development for such distances. However, a preperihelion cometary activity was detected in comet Hale–Bopp as far as 7 AU from the Sun. This activity has been attributed to the sublimation of carbon monoxide. CO is indeed present in large quantities with respect to H_2O, from 5% to above 20% – depending on the authors, the measurement techniques and the circumstances (type of comet, distance to the Sun). CO outgassing can, by a process similar to H_2O outgassing, lift dust particles and create a coma (Bockelée-Morvan and Rickman, 1999). After its passage at perihelion, comet Hale–Bopp was still outgassing CO as far as 14 AU.

Cometary matter is usually divided into two components: "refractories" and "volatiles"; the word "refractories" applies to species that do not sublimate. This classification is sometimes refined, introducing "hypervolatiles" (Ar, N_2, ...), and "semirefractories" (e.g. organic refractories, as opposed to more refractory species like silicates). Note here that water, traditionally ranking as the major volatile species in comets, behaves at 7 AU like a refractory.

A peculiar comet, 29P/Schwassmann-Wachmann 1, which follows an almost circular orbit at 6 AU from the Sun (i.e. somewhat beyond Jupiter), harbours intense and unpredictable outbursts; here also, CO is at the origin of this activity. Comet Halley showed an activity outburst after its perihelion passage, when it was already 14 AU from the Sun. Because of the lack of spectroscopic information, one does not know for sure if this outburst was also due to CO; on the pictures, only the development of a coma was noticed.

Breakup of cometary nuclei (see Boehnhardt 2002) is not uncommon. Some comets break into a few pieces (C/1975 v1 West), some into many pieces (57P/du Toit-Neujmin-Delporte in 2002), some even vanish (C/1999 S4 LINEAR). 90% of the Sun-grazing comets observed by the SOHO satellite belong to same orbital family: Kreutz family (Lamy and Biesecker 2002). They may have their origin in the splitting of one large comet. Depending on their size, the fragments have a lifetime between hours and years. The rate of splitting events of periodic comets has been evaluated to 0.01/yr/comet (Chen and Jewitt 1994). This makes splitting an important mass-loss factor, comparable to ordinary activity, and an important cause of comet death. While tidal forces close to Jupiter were clearly responsible for the breakup of Shoemaker-Levy 9, the other cases of splitting can occur at any place along the orbit, generally far from any planet. Other explanations (beside tidal forces – when applicable) involve thermal stress and gas explosions, possibly linked to runaway exothermic transformation of amorphous to crystalline ice, rotational breakup due to centrifugal force (Jewitt 2002) and fragmentation due to a collision with another small Solar System object.

Splitting events are an important source of information on the internal structure of comets. During the complete splitting of C/1999 S4 LINEAR, Bockelée-Morvan et al., (2001) detected no change in the relative molecular abundances, which pleads in favour of a rather homogeneous composition of the nucleus. Encounters with split cometary nuclei are a likely explanation for impact-crater chains seen on giant planet satellites, and that probably occurred on Earth too.

8.1.7 Nucleus Modelling: Outgassing and Internal Temperatures

Detailed modelling of the physical processes involved in nucleus formation, evolution and outgassing is a key tool to address important issues such as the pristine character of cometary matter, and the relation between abundances in the nucleus and those measured in the coma (see Sect. 8.2.1.4). Important phenomena and comet parameters include: ice sublimation; porosity of the cometary matter, which decreases from small to large objects, contributing to the density variation of icy bodies (from $\sim 0.5\,g/cm^3$ for small comets to $2\,g/cm^3$ for Pluto); heat transport by sublimated gases in the pores, which usually dominates solid thermal conductivity; radioactive heating by long-period elements like ^{235}U, ^{238}U, ^{40}K; disruptive and shattering collisions between comets; amorphous to crystalline ice transition, with possible release of trapped gas; surface processing by solar and galactic photons and particles. In the formation phase, heat from the short-period element ^{26}Al and heat due to accretion itself may be critical, but their role and magnitude vary greatly from one model to another. The difference between the production rates of CO (highly volatile) and H_2O as a function of distance to the Sun is well reproduced by models (e.g. Prialnik 1999, 2002, Enzian 1999, Coradini et al., 1997a,b, Capria 2002, Benkhoff 1999, Huebner and Benkhoff 1999a); a review of those models is presented by Huebner et al., (1999b). However, several other phenomena take place that lead to the observed abundance in the coma; they make the link between nucleus and coma abundances for some species more complex (HNC, H_2CO, ...), whereas not preventing in other cases a simple connection, as shown by species whose release closely follows that of H_2O, without significant effects of differential sublimation (see Sect. 8.2.1.4 and Fig. 8.5).

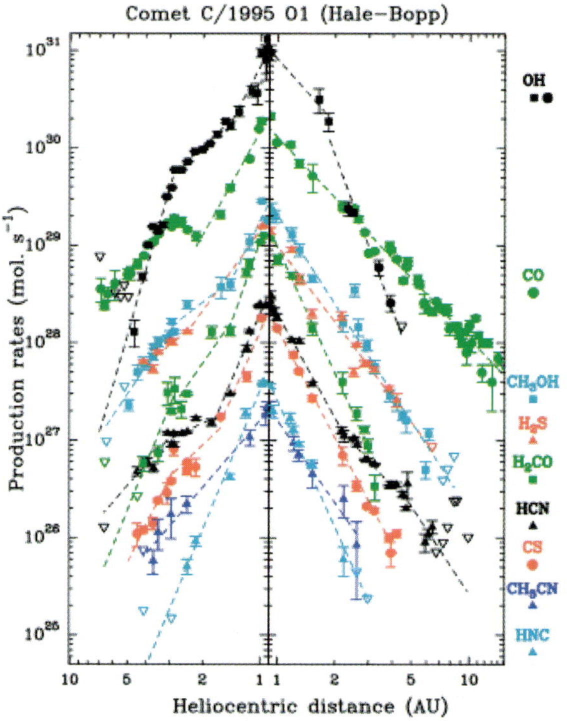

Fig. 8.5. This diagram displays the production rate Q_X of various cometary molecules as a function of heliocentric distance r_H (Biver et al. 2002). In this logarithmic scale Q_X shows, in general, a linear trend, corresponding to a power law in r_H. Whereas some species clearly depart from the mean behaviour (HNC, H_2CO, CO, CS – see discussion in text), the variation of the remaining species (e.g. CH_3OH, HCN, H_2S, CH_3CN) are roughly parallel ($\sim r_H^{-2}$), which sets a limit on the importance of differential sublimation effects for these species

8.1.8 Internal Temperature and the Case for Liquid Water

A crucial issue for prebiotic chemistry and exobiology is the maximum temperature reached by cometary matter during its history. Prialnik and Podolak (1999) consider that comets form rapidly, which leads to a key role of ^{26}Al decay in heating the interior of that comet. The maximal temperature reached somewhere in the nucleus ranges from low (35–90K for a small 1-km radius comet with low/ high (0.1/0.9) porosity) to 270–210K for a 100-km radius object. A very different situation is found in McKinnon (2002), who modelled the thermal history of large icy bodies in the outer solar nebula (Kuiper Belt). The accretion there is slow enough for ^{26}Al to have mostly decayed, and collisional accretion is the main heat source. The temperature remains below 100K in most places for radii below 100km, but reaches more than 250K in the case of KBO Varuna ($R \sim 400$km).

The prospect for high temperature inside the nucleus raises the question of possible liquid water inside comets or Kuiper Belt objects. Liquid water is a key parameter for chemical evolution toward prebiotic molecules, such as amino acids more complex than glycine. Therefore liquid water events on comets might allow the synthesis of more molecules than those one would expect in "dry" comets. From the model quoted above, liquid water seems excluded at present times in the bodies of modest size ($<$ 100km diameter) reaching the zone of terrestrial planets. Conditions were more favourable in the past but only a very limited parameter range would allow liquid water to be present for some time (Podolak and Prialnik 1997); in the case of KBOs, the body should be among the largest (Pluto?).

8.2 Chemical Composition of Comets as Deduced from Observations

8.2.1 Volatiles

8.2.1.1 Observations

Many observations and new discoveries concerning cometary volatiles were made recently, particularly thanks to the development of new instruments (large millimetre-wavelength radio telescopes, among which the IRAM 30-m dish on Pico Veleta and the Plateau de Bure interferometer have proven especially useful; infrared instruments, either ground-based or space-borne: Keck Telescope with NIRSPEC, and ISO satellite) and with the presence of very active comets: Hyakutake in 1996 and Hale–Bopp in 1997 (Bockelée-Morvan 1997, Despois 1999, Crovisier 1999, Bockelée-Morvan et al., 2000). Ten years before, the passage of comet Halley in 1985–86 had already brought considerable progress, from ground-based observations as well as from space probes, in particular Giotto and Vega 1 and 2 (Crovisier and Schloerb 1991, Encrenaz and Knacke 1991).

The majority of new detections of parent molecules (i.e. molecules coming directly from the nucleus) have been performed through radioastronomical observations. This technique provides very safe identifications of the molecular species through very high resolution study ($R = \lambda/\Delta\lambda \sim 10^7$ or better) of their rotational spectrum. Molecular abundances from radio data are usually considered to be accurate within 10–20% (calibration uncertainties), provided the excitation model for the molecule is correct. When comparing various techniques (UV, IR, radio), abundances agree usually well (within a factor of 2 at most). Among the species detected for the first time in Hale–Bopp, formamide (NH_2CHO) – a species already complex for astronomers! –already bears some resemblance to glycine NH_2-CH_2-COOH.

8.2.1.2 Molecular Abundances in Cometary Ices

Molecules detected in comets are given in Table 8.1; abundances – by molecule number – are expressed in per cent with respect to H_2O. The majority of the values are from Hale–Bopp observations. The next columns indicate, where available, the comet-to-comet variation range, and the number of comets observed (detected + undetected; Biver et al., 2002). The meaning of these abundances is discussed below (Sect. 8.2.1.4).

The most abundant species, with abundances around 10% of H_2O, are carbon monoxide CO (from 1 to 30%, see below), carbon dioxide CO_2 (5%) and methanol (1–5%). At the $\sim$ 1% level one finds the first sulfur-bearing species, H_2S, formaldehyde H_2CO (its exact abundance is hard to determine, due to the extended source problem discussed below), and the first nitrogen-bearing species: ammonia NH_3 (0.6%). Then, three hydrocarbons that are observable only in the infrared: methane CH_4 and ethane C_2H_6(0.6%) and acetylene C_2H_2 (0.1–0.5%). With 0.25%, ethylene glycol CH_2OHCH_2OH is the second alcohol. One also finds at these levels other sulfur species like SO_2 (0.1%), OCS and CS_2, and HCN (0.2%). SO (0.2–0.8%) is probably mostly a product of SO_2 photodissociation (estimated itself around 0.1%). The least abundant currently detectable species have a relative abundance of approximately 10^{-4}. Among them, isocyanic acid HNCO, formic acid HCOOH, the disulfur S_2, methylformate $HCOOCH_3$, HNC (see below), methyl cyanide CH_3CN and cyanoacetylene HC_3N (in comparable abundances), thioformaldehyde H_2CS and formamide NH_2CHO.

8.2.1.3 A Few Important Cases

- *Newly Found Species*

Bockelée-Morvan et al., (2000, note), and Bockelée-Morvan and Crovisier (2002) report the identification of acetaldehyde CH_3CHO, with an abundance of about 0.02% relative to H_2O from a reanalysis of comet Hale–Bopp data. Recently, CH_2OCH_2OH (ethylene glycol – commonly "antifreeze") has been identified

Table 8.1. Relative production rates of molecules in the coma, expressed by molecule number and in % relative to water ice (adapted from Despois et al., 2002). This rates are believed to be good tracers of molecular abundances in cometary ices, except for some species, like CO, H_2CO, HNC (see text). The values are for comet Hale–Bopp (Bockelée-Morvan et al., 2000, Crovisier et al., 2004a,b), except when indicated Hya. (Hyakutake) or I–Z (Ikeya–Zhang). When the molecule has been measured in several comets a range (Biver et al., 2002) is given

Molecule	Abundance (H_2O = 100)	Intercomet variation	Detected comets + upper limits	Notes
H_2O	100			
H_2O_2	< 0.03			
CO	23	< 1.7 – 23	5 + 4	Var.
CO_2	6 (various)			
CH_4	0.6			
C_2H_6	0.6			
C_2H_2	0.2			
C_4H_2	0.05?(I–Z)			Prelim.
CH_3C_2H	< 0.045			
CH_3OH	2.4	< 0.9 – 6.2	15 + 2	
H_2CO	1.1	0.13 – 1.3	13 + 2	Extend.
CH_2OHCH_2OH	0.25			
HCOOH	0.09			
$HCOOCH_3$	0.08			
CH_3CHO	0.025			
H_2CCO	< 0.032			
c-C_2H_4O	<0.20			
C_2H_5OH	< 0.10			
CH_2OHCHO	<0.04			
CH_3OCH_3	< 0.45			
CH_3COOH	< 0.06			
NH_3	0.7			
HCN	0.25	0.08 – 0.25	24 + 0	
HNCO	0.10			
HNC	0.04	< 0.003 – 0.035	5 + 2	Extend.
CH_3CN	0.02	0.013 – 0.035	4 + 0	
HC_3N	0.02			

Table 8.1. (continued)

Molecule	Abundance (H_2O = 100)	Intercomet variation	Detected comets + upper limits	Notes
NH_2CHO	0.015			
NH_2OH	< 0.25			
HCNO	< 0.0016			
CH_2NH	< 0.032			
NH_2CN	< 0.004			
N_2O	< 0.23			
Glycine I	< 0.15			
C_2H_5CN	< 0.010			
HC_5N	< 0.003			
H_2S	1.5	0.12 – 1.5	11 + 3	
OCS	0.4			Extend.
SO	0.3			Extend.
CS_2(from CS)	0.17	0.05 – 0.17	9 + 0	
SO_2	0.2			
H_2CS	0.02			
S_2	0.005 (Hya)			
CH_3SH	< 0.05			
NS	0.02			Parent?
PH_3	< 0.16			
NaOH	< 0.0003			
NaCl	< 0.0008			

Notes (see text for discussion):
- Extend.: extended source – in this case the abundance depends somewhat on the model adopted for the spatial distribution of the source
- Var.: CO shows large variability in a given comet along the orbit, and from comet to comet
- C_4H_2 values are preliminary
- NS is probably a photodissociation product from an unknown parent.
- CS_2 abundance indirectly deduced from CS abundance
- CO_2 measured to be 20% in Hale–Bopp at a distance of 2.9 UA by the ISO satellite
- Glycine is a compound with a very low volatility. Nondetection in the gaseous phase does not imply absence from the refractory component. Glycine I refers to one of the two most stable conformers of the molecule.

from the same archive spectra (Crovisier et al., 2004b); surprisingly, its abundance (0.25%) is higher than the upper limit for ethanol (0.1%), which may be an important constraint on the formation processes of cometary molecules. C_4H_2 has been possibly identified in the near infrared (NIR) spectrum of comet C/2002 C1(Ikeya–Zhang) (Magee–Sauer 2002). NS ($x \sim 0.02\%$), a radical, probably a dissociation product, has been detected in comet Hale–Bopp (Irvine et al., 2000b); its parent is not yet known: HNCS, HNSO, NH_4SH have been suggested.

- *Supervolatiles*

Very volatile ("supervolatile") species, like Ar, N_2, H_2, CO are important to set an upper limit to the highest temperature ever reached by the cometary nucleus and thus its degree of thermal processing. Bockelée-Morvan and Crovisier (2002) report doubts on the detection of Ar and N_2 (from N_2^+). H_2 appears to be a product of water dissociation, not a nuclear species. To date, the most volatile species are thus CO and CH_4.

- *Glycine Upper Limit in Volatiles*

Table 8.1 includes upper limits from radio observations of the coma (Bockelée-Morvan and Crovisier 2002). The limit ($x < 0.15\%$ or ~ 0.004 by weight or 6×10^{-5} mol/g) on glycine (the simplest amino acid, and one of the 20 that are the building blocks of proteins) is not very stringent, due to the intrinsic weakness of its radio lines. Furthermore, this limit concerns only the fraction of glycine molecules that could be released from the nucleus due to ice sublimation. Note that pure solid glycine itself is a refractory compound ($T_{\rm subl}$ is ~ 500 K for 1 atm pressure) and that thermal destruction occurs at a similar temperature. In the CM carbonaceous chondrite meteorite Murchison, glycine abundance (here with respect to refractories) is about 10^{-7} mol/g (Cronin and Chang 1993).

8.2.1.4 Are Coma Abundances a True Image of Ice Composition in the Nucleus?

On a plot of their production rates according to the distance to the Sun (Fig. 8.5 "Christmas tree plot" Biver et al., 2002), a majority of individual molecules vary in a similar way, which suggests that most of the cometary material escapes as a whole, without strong differentiation between chemical species. A few species (most notably CO, H_2CO, and HNC) harbour, however, a peculiar behaviour, and require additional phenomena to be taken into account: differential sublimation, extended source, and coma chemistry.

- *Differential Sublimation*

The great volatility of CO induces an important variation of the CO/H_2O ratio with heliocentric distance. CO can not only move through the micropores present in the ice, but even through the structure of crystalline ice itself; as a result, the surface layers of the nucleus can, according to cases, be either depeleted or enriched in CO. This phenomenon has been modelled in detail (see Sect. 8.1.7).

• *Extended Sources*

Some coma molecules do not appear to be emitted directly from the nucleus, but are more probably secondary products of photo- or thermal degradation of larger species, or products of chemical reactions in the coma. Thus they appear to come from an extended source surrounding the nucleus. Bockelée-Morvan and Crovisier (2002) summarise different indicators of extended source (spatial distribution, line shape, production behaviour with distance to the Sun). The best established cases are H_2CO and CO. Recent observations indicate that SO, CS, OCS and HNC harbour extended source characteristics. Photodissociation is likely to explain the case of SO (from SO_2) and CS (from a yet uncertain parent, maybe CS_2). HNC may be partly chemically produced (see below). The origin of H_2CO is not only in the nucleus, but also in the coma, where it is released from a "carrier": maybe organic CHON grains, maybe polymers (poly-oxymethylene – POM, which has been detected after experimental simulations of cometary ice analogs), maybe both. The degradation of polyoxymethylene (POM) has been studied in the laboratory and is found to be a possible source for H_2CO (Cottin et al., 2001, 2004). Various CN-bearing compounds including hexamethylenetetramine (HMT, see Fig. 8.13) and poly-HCN are possible sources of extended CN radical in the coma (Fray et al., submitted) in addition to the photodissociation of HCN.

• *Coma Chemistry*

HNC is a case for which coma reactions (other than photodissociation) can affect molecular abundances. The peculiar behaviour of the production rate of this species in comet Hale–Bopp, with its strong increase when the comet approached the sun, leads to the proposal of its production from HCN through chemical reactions in the coma as an explanation. Rodgers and Charnley (2001a) shows that reactions between HCN and high-speed ("suprathermal") hydrogen (from H_2O dissociation) could explain HNC production in comet Hale–Bopp. This process, however, is not efficient for smaller comets such as comet Hyakutake, for which the origin of HNC is still an open question (Irvine et al., 2003, Rodgers and Charnley 2001, Charnley et al., 2002a,b). It should be noted that in interstellar clouds, and thus perhaps also in protosolar nebulae, HNC and HCN are both present, and that HNC is stable in ice at sufficiently low temperature. Therefore, the presence of some HNC in the nucleus is not excluded.

Are some other minor species due to reactions in the coma? Rodgers and Charnley (2001b) show it cannot be the case for the detected species $HCOOCH_3$ and HCOOH, which are then expected to be directly released from the nucleus.

8.2.2 Grains

8.2.2.1 Nature of Grains

The chemical composition of the grains of comet Halley has been analyzed in situ by mass spectrometry (instrument PUMA on board the Vega 1&2 spacecraft, and instruments PICCA, IMS and NMS on board Giotto spacecraft – Altwegg et al., 1999). When grains are ejected from the nucleus, any embedded ice sublimes within the first few kilometres. Therefore, we will only consider the refractory (nonvolatile) component of grains in this section. Vega spacecraft have collected particles between 80 000 and 120 000km from the nucleus. About 5000 particles were analyzed, with mass ranging from 5×10^{-17} to 5×10^{-12} g (which means a size ranging from 0.04 to 2 µm if one assumes a density equal to 1). The total mass that was analyzed was only a few nanograms. Their elemental composition shows that the particles are made of a mixture of silicates and organic material. In comet Halley's coma, when Giotto made its measurements, the ratio between silicates and organic matter was found to be between 1 and 2. When organic matter is very abundant, the grains are called "CHON" grains (i.e., compounds of C, H, O and N). According to some models, the small mineral particles might be embedded in a matrix of carbonaceous matter (Greenberg 1982); however, mass spectrometer measurements show the existence of pure silicate an d pure organic grains, beside grains of mixed composition (Fomenkova, 1999).

Figure 8.6 shows the production of grains in mass and number for comet Halley as derived by Crifo (Crifo 1995, Crifo and Rodionov 1997) from in situ measurements presented in McDonnell et al. (1991). It must be noted that data up to 1 mm are direct in situ measurements, whereas higher values are extrapolations (relevant within an order of magnitude of uncertainty).

The temperature of grains can be derived from observations, but the best estimations are reached through calculations based on hypotheses about their shape and composition (Crifo 1988). Two extreme temperatures for each grain

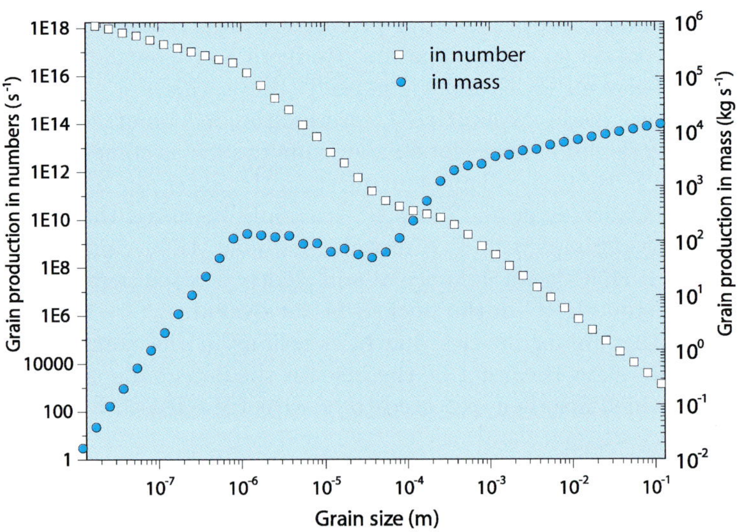

Fig. 8.6. Dust-grain production from comet Halley nucleus assuming a density of $1\,\mathrm{g\,cm^{-3}}$ (Cottin et al., 2004)

size are achieved, depending on whether the grain is made of olivine (low temperature) or amorphous carbon (high temperature). Olivine represents the silicate nucleus of grains with a very low absorbance in the visible; it is a lower temperature limit for grains. Amorphous carbon (aC) is representative of dark material, and stands as an upper limit for a strongly absorbing organic component of grains. As grain composition is poorly known, one cannot pretend to better constrain their temperature than between those two extremes. Grains temperatures are presented on Fig. 8.7 for a spherical shape; these temperatures are reached within the first few tens of kilometres from the nucleus (Crifo 1991). For porous or nonspherical grains the temperature tends to be higher.

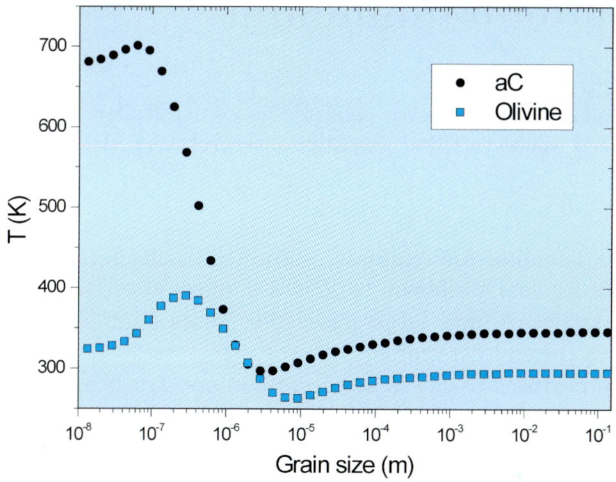

Fig. 8.7. Grain temperature as a function of size and composition (Cottin et al., 2004)

8.2.2.2 Organics in Grains

The most fruitful works related to the organic composition of grains have been reported in Kissel and Krueger 1987, and Krueger and Kissel 1987). To interpret the PUMA's mass spectra, they consider that all sudden dissipation processes near solid surfaces after an impact are governed by the same rules of molecular ion formation (Fig. 8.8). Such impacts occur when a dust particle hits the mass spectrometer's target. This hypothesis yields a good agreement between their predictions and other observations for small molecules (e.g. HCN, CH_3CN).

For larger molecules the problem is more complex, as they cannot survive the impact process without fragmentation. In addition to the difficulty of assembling correctly the fragments to reconstitute the original molecule, the nature of the fragments themselves is uncertain, several fragments having the same mass (to the resolution of the spectrometer). Thus the identification by these authors

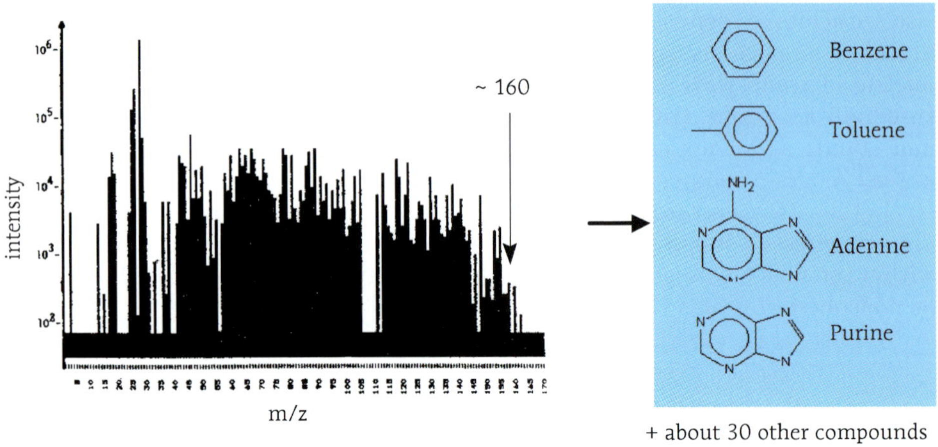

Fig. 8.8. Cumulated mass spectra collected in comet Halley by instrument PUMA on board VEGA-1 spacecraft and tentative interpretation. Adapted from Kissel and Krueger (1987)

of purines and adenine, for example, should be considered as very tentative. However, what is clearly shown by PUMA mass spectra is that compounds with complex structure and large molecular masses (at least with m/z up to $160\,\mathrm{g\,mol^{-1}}$) are present on grains.

First interpretations of the heavy-ion mass spectra from the PICCA instrument led Huebner (1987) to conclude the presence of polyoxymethylene (POM) in comet Halley. After further analyses, Huebner et al., (1989) discussed the presence of POM derivatives such as sulfur copolymers. As already stated in this chapter, the presence of POM in comets is coherent with the detection of extended sources of H_2CO in the coma. Indeed, this molecule could be the degradation product of POM. The authors also suggest the presence of HCN polymers such as polyaminocyanomethylene (PACM: $(NH_2–C–CN)_n$), even though these polymers are very difficult to detect because in the coma $HCN/H_2CO \sim 0.1$ and because from thermodynamic considerations N-containing polymers should be less volatile than O containing polymers (Krueger et al., 1991). The presence of HCN/H_2CO copolymers could also be suggested on comets because these two molecules polymerize very easily by themselves.

Later, however, Mitchell et al., (1992) showed that the mass spectrum pattern that led Huebner to announce the detection of polyoxymethylene is only characteristic of a mixing of molecules composed of C, H, O and N atoms (CHON molecules). Therefore PICCA mass spectra do not imply the presence of POM, but only of a large diversity of CHON molecules, among which POM could be present. The most we can say, if we consider laboratory simulation results presented latter in this chapter and the detection of an extended source of formaldehyde, is that the presence of POM on cometary nuclei is possible.

Beside polymers, heteropolymers and large CHON molecules, other compounds have been suggested: pure carbon grains, PAHs (polycyclic aromatic hydrocarbons, i.e. aromatic cycles linked together), very "branched" aliphatic hydrocarbons (Fomenkova 1999). PAH detection from in situ UV spectroscopy in Halley was announced (Moreels et al., 1994); PAHs were not seen in the ISO spectra of Hale–Bopp, possibly due to the lower temperature of the comet at the time of these observations (Crovisier 1999). A weak residual feature at 3.28 μm was, however, seen by Bockelée-Morvan et al., (1995) in various comets, and especially the dustier ones, which might be due to PAHs; this would correspond to a relative abundance to water 1.5 to 10×10^{-6}, at least 100 times below that deduced from UV data. A list of compounds probably and possibly detected in comet Halley grains by mass spectroscopy is reported in Table 8.2.

Table 8.2. Organic molecules inferred from mass spectra of gas and dust particles in comet Halley. The confidence levels of detections by MS are established as follows: – Confirmed: Molecule also detected by remote observations; – High: Molecule not detected by remote observations but present after laboratory irradiations of cometary ice analogs; – Medium: Molecule detected only by mass spectroscopy with a good confidence level according to the authors; Low: Molecule only inferred by mass spectroscopy with a low confidence level according to the authors

Molecule	Family	Mass spectrometer	Confidence level	Ref.
Hydrocyanic Acid	C–N–H	PUMA, PICCA	Confirmed	1, 2, 3
Methyl cyanide	C–N–H	IMS	Confirmed	7
Acetonitrile	C–N–H	PUMA, PICCA	Confirmed	1, 2, 3
Methanol	C–O–H	NMS	Confirmed	6
Formaldehyde	C–O–H	PUMA, PICCA	Confirmed	3
Formic acid	C–O–H	PUMA	Confirmed	1, 2
Acetaldehyde	C–O–H	PUMA, PICCA	Confirmed	1, 3
Ammonia	N–H	PUMA	Confirmed	2
Isocyanic acid	C–N–O–H	PUMA	Confirmed	1, 2
Ethane	C–H	NMS	Confirmed	8
Acetylene	C–H	NMS	Confirmed	8
Acetic acid	C–O–H	PUMA	High	1
Polyoxymethylene	C–O–H	PICCA	High (itself or derivatives)	4, 5
Ethene	C–H	NMS	High	8
Iminoethane	C–N–H	PUMA	Medium	1, 2
Aminoethene	C–N–H	PUMA	Medium	1, 2
Pyrroline	C–N–H	PUMA	Medium	1

Table 8.2. (continued)

Molecule	Family	Mass spectrometer	Confidence level	Ref.
Pyrrole	C–N–H	PUMA	Medium	1, 2
Imidazole	C–N–H	PUMA	Medium	1
Pyridine	C–N–H	PUMA	Medium	1, 2
Pyrimidine	C–N–H	PUMA	Medium	1, 2
Ethyl cyanide	C–N–H	IMS	Medium	7
Pentyne	C–H	PUMA	Low	1
Hexyne	C–H	PUMA	Low	1
Butadiene	C–H	PUMA	Low	1
Pentadiene	C–H	PUMA	Low	1
Cyclopentene	C–H	PUMA	Low	1
Cyclopentadiene	C–H	PUMA	Low	1
Clyclohexene	C–H	PUMA	Low	1
Cyclohexadiene	C–H	PUMA	Low	1
Benzene	C–H	PUMA	Low	1
Toluene	C–H	PUMA	Low	1
Propanenitrile	C–N–H	PUMA	Low	1
Iminomethane	C–N–H	PUMA	Low	1
Iminopropene	C–N–H	PUMA	Low	1
Purine	C–N–H	PUMA	Low	1, 2
Adenine	C–N–H	PUMA	Low	1, 2
Polyaminocyanomethylene	C–N–H	PICCA	Low	5
Methanolnitrile	C–N–O–H	PUMA	Low	1
Methanalimine	C–N–O–H	PUMA	Low	1
Aminomethanol	C–N–O–H	PUMA	Low	2
Aminomethanal	C–N–O–H	PUMA	Low	2
Oxyimidazole	C–N–O–H	PUMA	Low	1
Oxypyrimidine	C–N–O–H	PUMA	Low	1
Xanthine	C–N–O–H	PUMA	Low	1

1 (Kissel and Krueger 1987); 2 (Krueger and Kissel 1987); 3 (Krueger et al., 1991); 4 (Huebner 1987); 5 (Huebner et al., 1989); 6 (Eberhardt and Krankowsky 1995); 7 (Geiss et al., 1999); 8 (Altwegg et al., 1999)

8.2.2.3 Cometary Silicates

The nature of other refractory components is constrained by infrared observations of comets. Identification is achieved through a comparison of observations with laboratory spectra of different kinds of silicates and possibly carbonaceous grains, and it should be noted that there is no unique solution to the fitting problem. It appears that grains are made of a mixture of crystalline and amorphous silicate (Brucato et al., 1999, Crovisier et al., 1997, Harker et al., 2002, Hayward and Hanner 1997, Nuth et al., 2002, Wooden et al., 2000).

Crystalline forsterite (a magnesium-rich olivine) has been detected thanks to the ISO infrared satellite (Crovisier et al., 1997). Ground-based telescopes have enabled a study of the infrared spectrum around 10 μm, and determined the ratio pyroxene/olivine. According to Harker et al., (2002) amorphous silicates make up about 70% (by mass) of the submicrometre-sized ($\leq 1\,\mu m$) silicate dust grains in Hale–Bopp's coma, and are distributed roughly equally between olivine and pyroxene composition, and between Fe-rich and Mg-rich species; the remaining 30% are predominantly Mg-rich crystalline silicates with a crystalline

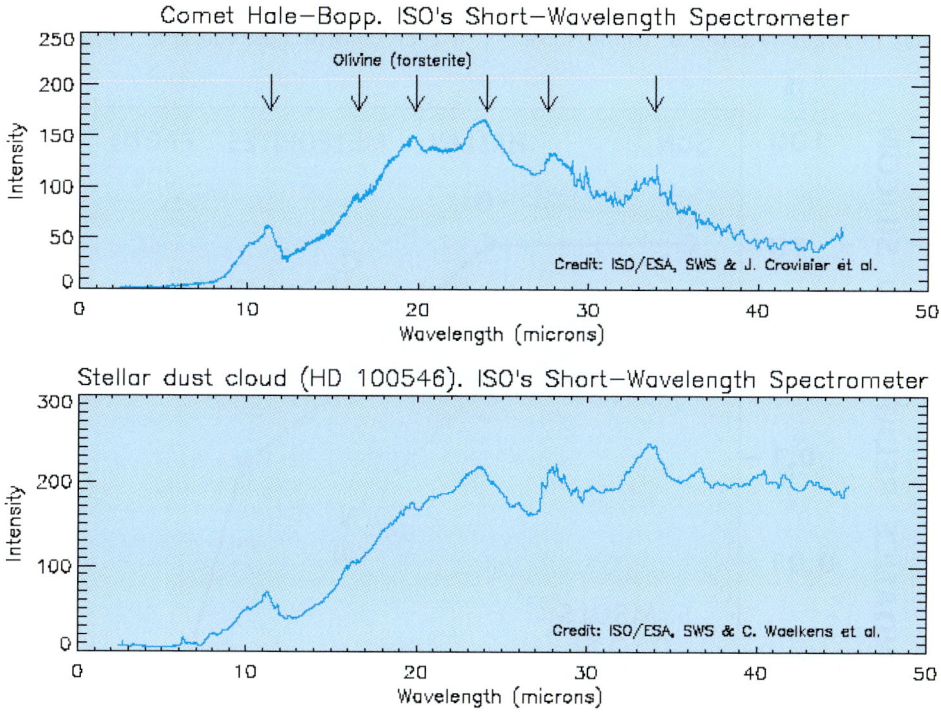

Fig. 8.9. Comparison of the infrared spectrum of Hale–Bopp obtained by ESA's Infrared Space Observatory ISO (Crovisier et al., 1997) with that of a young star of type AeBe (e.g. Malfait et al., 1998). The features in the comet spectrum lead to the identification of a crystalline silicate, fosterite, a magnesium-rich olivine. The AeBe stars are slightly more massive than the Sun, but the dust that surrounds them and that is responsible for this IR emission is considered to be similar to dust in the protosolar nebula. In HD100456, the similarity is even more striking and points towards a high crystalline mass fraction and small particle size ($< 10\,\mu m$) in both cases (Bouwman et al., 2003)

olivine:orthopyroxene ratio about 4 to 5 : 1 that seems to vary slightly with heliocentric distance. Figure 8.9 shows a comparison between the infrared spectrum of Hale–Bopp and circumstellar matter around a young star. However, data of such a high quality is only available for one comet (Hale–Bopp).

8.2.3 Elemental and Isotopic Composition

8.2.3.1 Elements

The composition in elements that can be deduced by combining dust and gas measurements (Geiss 1988) leads to abundances normalized to silicon close to solar abundances for carbon and oxygen. However, one notes a depletion (underabundance) of hydrogen of about 600, due to the high H_2 volatility and a depletion of 2 for nitrogen, which is explained in the same way by the fact that nitrogen is often in the form of N_2, a very volatile gas, too (Fig. 8.10). The

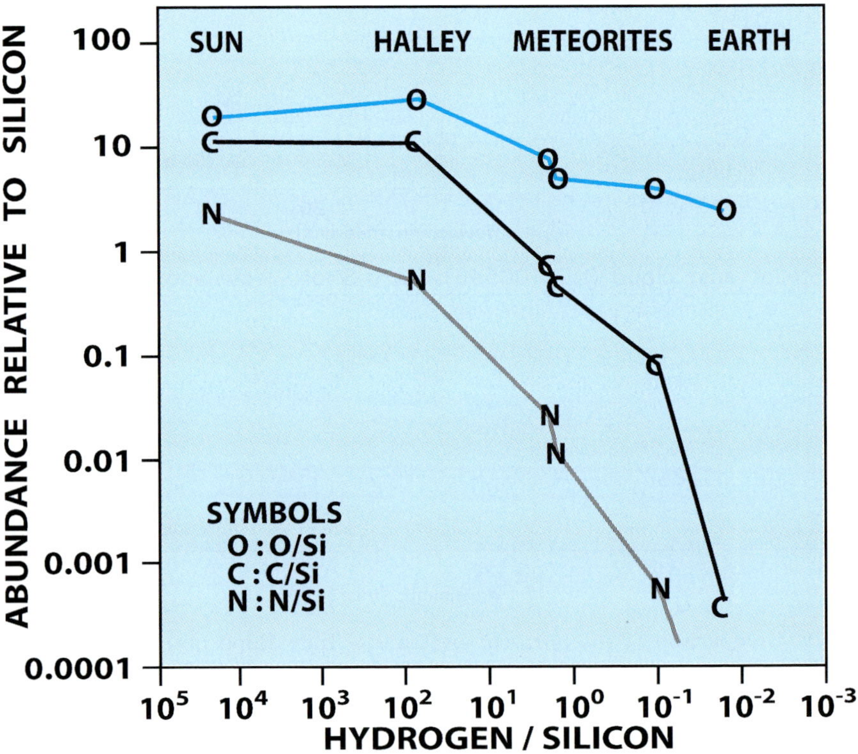

Fig. 8.10. Comparison of elemental abundances in the Sun, comet Halley, and (C1, C2) meteorites (Geiss 1988). The element abundance ratios in P/Halley are uncertain by a factor of about two (excluding the variability and uncertainty in the gas/dust ratio)

gas:dust ratio derived in the case of Halley is 1:2 in mass (Jessberger and Kissel 1991); note that the computation of the elemental composition for the whole comet uses this ratio and is quite sensitive to it.

In Fig. 8.10, the elemental composition of comets is also compared with that of carbonaceous chondrites (Geiss 1988). Elements of the series Mg–Al–Ca, as well as oxygen have comparable abundances within a factor of 2. On the other hand, in comet Halley, abundances of carbon, hydrogen, nitrogen and of any other element appearing mostly in volatile compounds are larger than in carbonaceous chondrites. This leads us to say that cometary matter is more "pristine" than that of carbonaceous chondrites, in the sense that it underwent less heating and thus was less processed.

8.2.3.2 Isotopic Abundances

Much of the information on the origin of cometary matter and on the importance of the cometary contribution to the Earth's water content derives from the study of isotopic abundances, which have been recently reviewed by Altwegg and Bockelée-Morvan (2003). The D/H ratio in particular is very important data to investigate the origin of cometary matter (e.g. Bockelée-Morvan et al., 1998, Mousis et al., 2000, Robert 2002) and its delivery to the Earth (Dauphas et al., 2000; see Sect. 8.5.3). It was measured for two species: HDO/H_2O ($D/H \sim 3 \times 10^{-4}$) and DCN/HCN ($D/H \sim 2.3 \times 10^{-3}$); upper limits on the order of 10^{-2} or below have also been set for deuterated formaldehyde, methanol, ammonia and H_2S. Molecules appear enriched in deuterium: as can be seen in Fig. 8.11, D/H in water for the 3 comets for which it has been measured is twice the ocean SMOW value, and an order of magnitude higher than the protosolar value of 2.5×10^{-5}. This enrichment is however not as high as in the interstellar gas (where multiply deuterated species like D_2CO and even ND_3 have been observed). Measurements of the O, C, N and S isotopes display, in general, abundances very close to solar abundances (e.g. Jewitt et al., 1997, Lecacheux et al., 2003), which confirms that the comets are Solar System objects. This does *not* contradict the assumption of an interstellar origin of the cometary matter discussed below. Indeed, the Sun and the Solar System as a whole were formed from a homogeneous clump of interstellar matter and it is from the same fragment of IS matter that the comets were made.

Some carbonaceous grains analyzed by in situ mass spectroscopy in Halley (Jessberger and Kissel 1991) showed a very high ratio $^{12}C/^{13}C$, rising up to 5000, whereas the normal ratio in Solar System objects is 89. These small grains would have been formed in the envelope of certain stars known as AGB stars (asymptotic giant branch) and would then have been injected into the interstellar medium. They thus would have been mixed with the interstellar cloud fragment that condensed to form the protosolar nebula, then later incorporated as such in the matter of forming comets. As this high departure from solar $^{12}C/^{13}C$ values is concentrated in very small grains, which do not represent a large quantity

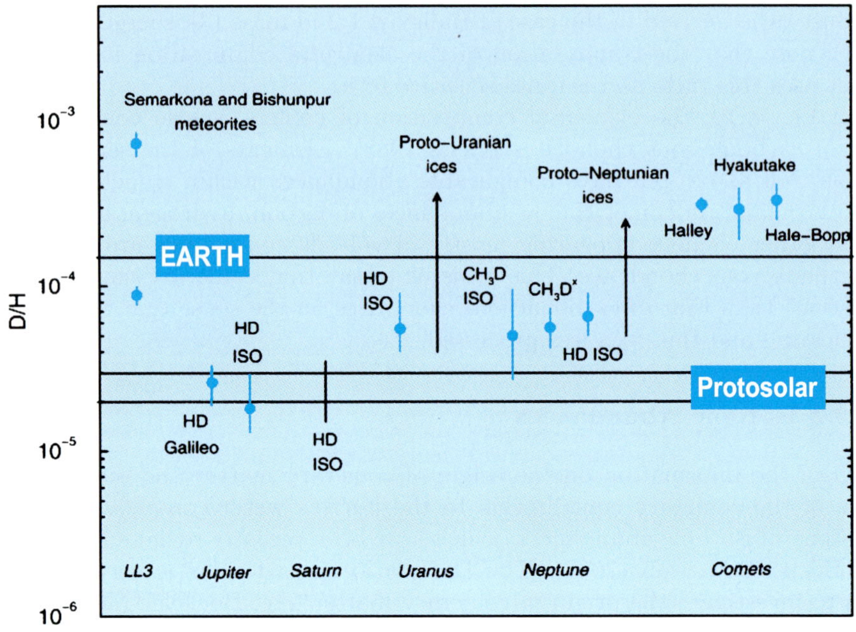

Fig. 8.11. D/H ratios in the Solar System. For comets, only measurements of D/H in water are shown (from Hersant et al., 2001)

of carbon, the average value of the cometary $^{12}C/^{13}C$ ratio remains close to the solar value, namely approximately 89. An intriguing measurement has been made in CN (Arpigny et al., 2003) suggesting the existence of parents of the CN radical other than HCN – possibly solid organic compounds – with strongly nonsolar N isotopic ratios.

8.2.4 Are All Comets Similar?

Abundance variations are observed from one comet to another, in particular for CO (from less than 1% to more than 30% according to comets and to the distance from the Sun) and CH_3OH (1 to 5%; e.g. Bockelée-Morvan 1997). This composition is deduced from gas abundances measured in the coma, making assumptions on the spatial distribution of the molecules, the limited chemistry in the coma, and the sublimation process.

That all comets are not similar already has observational evidence, as seen from dust content and daughter species C_2, CN (A'Hearn et al., 1995, Rolfe and Battrick 1987). Column 3 in Table 8.1 shows the range spanned by the observed molecular abundances in a sample of comets (Biver et al., 2002). Different compositions are expected, to some extent, between comets coming from the Oort cloud and the Kuiper Belt due to their different formation zone (respectively Jupiter–Neptune versus mostly transneptunian, see Sect. 8.4 and Morbidelli and

Table 8.3. Molecules detected during experimental simulations of cometary ice analogs. Italic letters refer to molecules actually detected in comets. (t) means tentative detection only in the analogs. Amino acids (alanine, AIB, ... except glycine) were detected after acid hydrolysis of the room temperature residue. Updated from Cottin et al., 1999

Hydrocarbons:
CH_4
C_2H_2, C_2H_4, *C_2H_6*
C_3H_8, C_4H_{10}
C_5H_{10}, C_5H_{12}
C_6H_{12}, C_6H_{14}
C_7H_{16}

Amides:
NH_2CHO
CH_3CONH_2
$HOCH_2CONH_2$
$NH_2(CO)_2NH_2$
$HOCH_2CH(OH)CONH_2$

Amines:
$HOCH_2CH_2NH_2$
$HCNH(NH_2)$
Diaminopyrrole
Diaminofurane
Triaminopropane
$(CH_2)_6N_4$ (HMT)

Ketones:
CH_3COCH_3
$HOCH_2COCH_3$
$HOCH_2CH_2COCH_3$
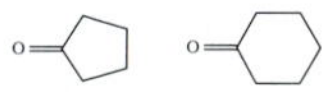

Carboxylic acids:
$HCOOH$
CH_3COOH (t)
$HOCH_2COOH$
$HOCH_2CH(OH)COOH$
$HOCH_2CH_2COOH$
$NH_2COCOOH$

Aromatic Compounds:
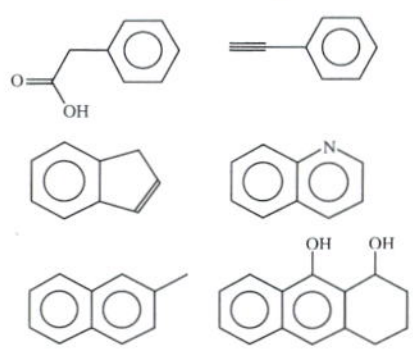

Alcohols:
CH_3OH
CH_3CH_2OH
$HOCH_2CH_2OH$
$HOCH_2CH(OH)CH_2OH$
$C_4H_8(OH)_2$
C_5H_9OH (t)
$C_5H_{11}OH$

Amino Acids:
NH_2CH_2COOH (Glycine)
$NH_2CH(CH_3)COOH$ (Alanine)
$CH_3CH_2CH(NH_2)COOH$ (α ABA)
$CH_3CH(NH_2)CH_2COOH$ (β ABA)
$(CH_2NH_2)(CH_3)CHCOOH$ (AIBA)
Sarcosine
Ethylglycine
Valine, Proline, Serine
Aspartic acid
Diaminopropanoic acid
Diaminobutyric acid
Diaminopentanoic acid
Diaminohexanoic acid

Aldehydes:
H_2CO
CH_3OCH_2CHO (t)

Ethers:
$CH_3OCH_2OCH_3$ (t)
$C_3H_6O_3$ (Trioxane) (t)
$(\text{-}CH_2\text{-}O\text{-})_n$ (POM)

Esters:
$HCOOCH_3$
CH_3COOCH_3
$CH_3CH_2COOCH_3$

Others: *CO*, *CO_2*, C_3O_2, H_2O_2, H_2CO_3, N_2H_4, *$HNCO$*, NH_2CONH_2, $NH_2CONHCONH_2$

Brown 2004). Observational evidence has been claimed for variations in the C_2H_6/H_2O and CH_4/H_2O ratios (Mumma et al., 2002, and Gibb et al., 2003, respectively), but the statistics is still limited.

8.3 Laboratory Simulation of Cometary Matter

8.3.1 Experimental Simulations

Only volatile molecules have been detected so far in cometary atmospheres, as no direct cometary sample has ever been analyzed (except for IDPs, whose origin cannot, however, be uniquely assigned to comets). For a better insight into more complex, less volatile material, one has to turn to experimental laboratory work. The principle of such experiments is the following: from observations of the most abundant species in comas and in the interstellar medium, one can infer the probable composition of the nucleus ices. A gaseous sample of the key species is deposited under vacuum on a cold substrate and irradiated during or after deposition by UV photons or charged particles. Condensed ices are sometimes simply warmed up slowly without irradiation. When the sample is warmed up for analysis a refractory organic residue remains on the substrate as the volatiles sublimate (Fig. 8.12). Greenberg has called this residue "yellow stuff". Bernstein et al., (1997) have shown that the organic residue is formed only when the initial mixture of ices contains polar molecules such as CH_3OH and NH_3.

The diversity of organic compounds synthesized is remarkable but their identification is seldom exhaustive. Table 8.3 is a simplified list of all the detected compounds. The simplest compounds such as CO, CO_2, H_2CO and CH_4 are

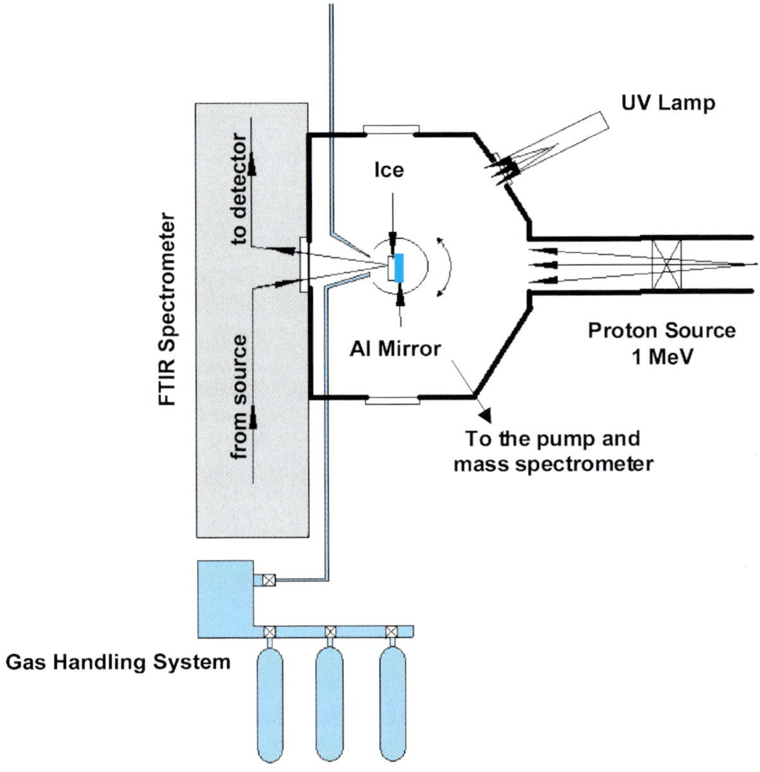

Fig. 8.12. A typical experimental setup allowing the irradiation (by UV and/or energetic protons) of cometary ice analogs made by deposition of a gas mixture on a rotating aluminium mirror cooled down to 10 K in a cryostat. The ice evolution can be analysed in situ by infrared reflection spectroscopy, and the volatiles released during warming, by mass spectrometry. The room-temperature residue can be collected for further analysis such as GC-MS (gas chromatography coupled to mass spectrometry), HPLC (high-performance liquid chromatography) and many others. From Hudson and Moore (1999)

detected in almost all the experiments, if the irradiated ice contains the appropriate elements. For more complex molecules, it depends on the ice composition and the nature of the energy source.

In addition to chemical transformations, it must be mentioned that experiments on the trapping of gases during ice condensation suggest that these processes play an important role in determining the composition of the ice and could lead to important enrichment or depletion between the gaseous and solid phases (Notesco and Bar-Nun 1996, Notesco and Bar-Nun 1997, Notesco et al., 1997).

8.3.2 Energy Deposition

Three kinds of energetic processing occur on icy coated dust grains in interstellar clouds (potentially precometary ices – see Sect. 8.4) or in the outer layers of comet ices in the Solar System:

In interstellar clouds, icy coated dust particles are subjected to processing by:

- *Charged particles*: Galactic cosmic rays.
- *UV-photons*: Lyman α photons from neighboring stars (in the diffuse outer regions of a cloud), or UV-photons induced by galactic cosmic rays (in the inner regions of dense clouds).
- *Thermal processes*: Cycling between the cold centre of a dense cloud and its warmer diffuse outer regions.

In the Solar System, the outer layers of comets undergo the same processes:

- *Charged particles*: Galactic cosmic rays, mainly in the Kuiper belt and the Oort cloud. This process has the largest effect on the outer few metres of the nucleus.
- *UV-photons*: Solar-UV, mainly in the inner Solar System when the comet is close to the Sun. This process would affect the outer few micrometres of the nucleus. Also during the Solar System formation, in the external layers of the disk, when solar UV luminosity was much higher than today.
- *Thermal processes*: During the formation of the Solar System (depending on the region in which the comet accretes), and in the inner Solar System (when the comet approaches perihelion).

Due to the diversity of environments involved, constraining the degrees to which different processes affect cosmic ices is a highly convoluted problem. Differences between the products synthesized during processing, according to the energy sources, could give information on the history of cometary matter and comets. Investigations are in progress to address this question.

8.3.2.1 UV Irradiation

UV irradiation is performed using a flowing-hydrogen discharge lamp (powered by a microwave cavity) delivering mainly Lyman α photons (122 nm) and a broad band of photons centred at 160 nm (see Allamandola et al., 1988, and Cottin et al., 2003) for a detailed description). The irradiated ices comprise common cometary small molecules but the initial abundances of CH_3OH, NH_3 and/or CO relative to H_2O are usually higher than those deduced for present-day comets and displayed in Table 8.1. During these experiments a wide variety of organic compounds have been identified. From an initial mix of H_2O: CO: NH_3 (ratio = 5: 5: 1), glycine, the simplest amino acid, acetamide, glyceramide, and many other molecules have been detected by GC-MS (Briggs et al., 1992). Analysis by MS-MS on the organic residues formed, leads to the detection of heavier compounds: several cyclic molecules and PAHs (Greenberg and Mendoza-Gomez 1993). The composition of the heaviest part of the residue is still unknown but an elemental composition based on the overall structure of the mass spectra is given by Greenberg and Li (1998) (C: O: N: H = 1: 0.06: > 0.001: 1.1).

Among the molecules synthesized after such irradiations of ices, one of them is of great interest. Bernstein et al., (1995) have identified abundant hexamethylenetetramine (HMT – $C_6H_{12}N_4$) in the refractory residue. This compound has exobiological implications since its acid hydrolysis products are amino acids (Wolman et al., 1971). Typically, for an initial composition of H_2O: CH_3OH: CO: NH_3 (10: 5: 1: 1), the organic residue at 300 K contains HMT ($\sim$ 60%), ethers and POM-like polymers ($\sim$ 20%), ketones and amides ($\sim$ 20%). 1/5 of the carbon and 1/2 of the nitrogen from the initial ice composition remain in the refractory part (Bernstein et al. 1995). Thus a large fraction of HMT is formed (60% of the residue) whereas only 5% NH_3 is present in the ice before irradiation. A scheme of HMT production is shown in Fig. 8.13. Formaldehyde is produced by methanol UV oxidation. It then reacts with ammonia to produce methylimine and its trimer: hexahydro–1,3,5–triazine. Successive reactions with formaldehyde and ammonia result in the formation of HMT (Bernstein et al., 1995). Methanol plays a key role and it has been shown by ^{13}C isotopic substitution that it is the source of HMT's carbon. The production of HMT and some HMT family molecules has been studied in great detail by Muñoz Caro (2003).

HMT photolysis produce nitriles (HCN and RCN) and isonitriles (RNC) (Bernstein et al 1994, Cottin et al 2002). Knowing this, it is interesting to note that it has been shown that the CN radical might have an extended source in comet Halley (Klavetter and A'Hearn 1994). According to these authors, this source could be large molecules present on grains because the observed HCN cannot explain the amount of CN in comets. Other observed CN-containing compounds (HNC, CH_3CN, HC_3N) are only minor products and cannot explain the discrepancy. Thus having been produced in the interstellar ices, as we have just seen, HMT could act as a parent (HMT $\rightarrow$ CN) or "grandparent" (HMT $\rightarrow$ RCN $\rightarrow$ CN) molecule for the CN extended source. As yet, there has been no observation of HMT in comets. If present, its detection in the IR would be masked by the Si–O and C–O vibration bands that are in the same region as the strongest HMT infrared signatures. Direct detection of HMT should be

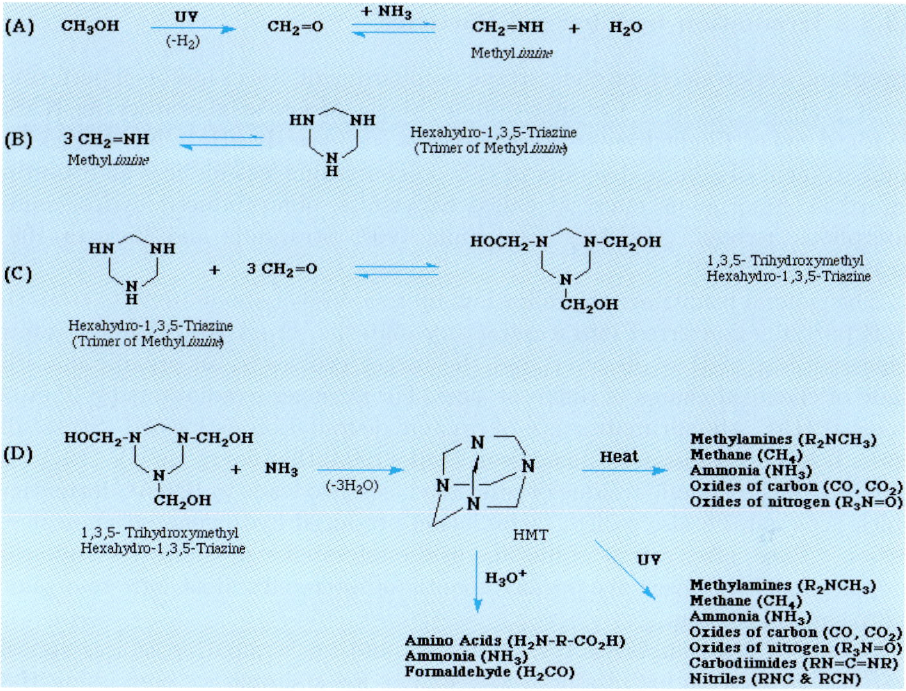

Fig. 8.13. Hexamethylenetetramine (HMT) chemistry. From (Bernstein et al., (1995))

feasible by in situ measurements with the GC-MS on board the lander of the Rosetta mission.

Bernstein et al (2002) and Muñoz Caro et al (2002) have detected a great number of amino acids (such as glycine, alanine, sarcosine, valine, proline, serine, etc.) in residues obtained after UV irradiation of ice mixtures made of H_2O: NH_3: CH_3OH: HCN and H_2O: CH_3OH: NH_3: CO: CO_2, respectively. Unhydrolyzed residues produce only a trace of glycine whose detection has already been reported by Briggs et al (1992) without any liquid water introduced to the analysis protocol. The detection of the other amino acids requires an acid hydrolysis of the residue in very strong conditions (HCl $\geq 6M$ and $T \geq 100°$C). Therefore it is not clear to date if 1) amino acids are present themselves in the laboratory residues, and henceforth in cometary ices, or if "only" amino acids precursors are synthesized, and 2) if the residues' processing (acid hydrolysis) is relevant to any chemistry that could have turned the amino acids' precursors imported by cometary impacts in the primitive oceans of the early Earth into actual amino acids. Note that the hydrolysis conditions would break a peptidic bond; some dipeptides have been found in the extraterrestrial organic matter of the Yamato-791198 and Murchison meteorites, but only glycine-glycine dimers and with a low (10^{-4}) abundance relative to glycine (Shimoyama and Ogasawara 2002).

8.3.2.2 Irradiation by Charged Particles

Important work concerning the particle bombardment of ices has been performed by Strazzulla's team in Catania and in Marla Moore's laboratory at NASA Goddard Space Flight Center. The particles used are H^+, He^+, N^+ or Ar. The bombardment of a large diversity of carbon-containing ices induces an evolution toward an amorphous material called Strazzulla "ion-produced hydrogenated amorphous carbon" (IPHAC) (Strazzulla 1997; Strazzulla and Baratta 1991; Strazzulla et al 1991).

The general results are the following: up to a dose of about 10 eV/C atom the ice is partially converted into a refractory material. From 10 to 25 eV/C atom, a massive loss of H is observed and the target evolves to an organic material made of chemical chains of different sizes. For stronger irradiations ($\geq$ 25 eV/C atom) IPHAC, the ultimate state of organic degradation, is formed (Strazzulla 1997). It has been shown by Jenniskens et al. (1993) that energetic UV (10 eV) irradiation of the organic residue of processed ices also leads to IPHAC formation, which then can be also called: "irradiation-produced hydrogenated amorphous carbon". Thus, after a typical lifetime in the interstellar medium, UV radiation and/or particles convert the organic mantle of interstellar dust into amorphous hydrogenated carbon.

A set of data comparing UV photolysis and ion irradiation of ices showed that the yield of major products was similar for a simple ice containing H_2O and CO_2 (Gerakines et al 2000), showing that the ice chemistry seems to be quite similar whether induced by photons or charged particles. Indeed, HMT production has been reported when interstellar ice analogs containing CH_3OH and NH_3 are irradiated with protons (Cottin et al., 2001), the same as with photons. The main difference is that UV photons (typically at 122 nm) only affect a few tenths of a micrometer in water-dominated ice, whereas protons can reach and alter the ice composition down to a few metres depth.

Chemical differentiation is more noticeable when molecules not dissociated by photons with a wavelength of 122 nm and above are involved in the chemical processes. This is the case for CO and N_2. Different results are obtained if pure CO is photolyzed or proton irradiated (Gerakines and Moore 2001). Likewise $N_3{}^+$ is detected when ices containing N_2 are proton irradiated, but not after photolysis (Hudson and Moore 2002). It seems that it is more the energy level than the way it is deposited (UV or charged particles) that matters for the chemistry.

We note that in these kinds of experiment, C_3O_2, which is sometimes evoked as an extended source of CO, has been detected after irradiation of ices containing CO or CO_2 (Brucato et al. 1997; Moore et al 1991). Kobayashi et al (1995) and Kasamatsu et al., (1997) were the first to report amino-acid production from irradiated ices. After an irradiation by 3-MeV protons of mixtures containing water, ammonia and a carbon-containing molecule (carbon monoxide, methane or propane), they detected by HPLC several amino acids: glycine, and for the first time in a cometary simulation, alanine, aminobutiric acid and aminoisobutiric acid. These new detections were not made directly from the organic residue, but after an acid hydrolysis in water. Likewise for UV irradiation, only traces of glycine were found during the analysis of unhydrolyzed residues.

8.3.2.3 Thermal Processing of Ices

Polyoxymethylene and associated molecules and polymers have been detected when several mixtures containing formaldehyde, instead of being irradiated, were warmed slowly to room temperature (Schutte et al., 1993a, Schutte et al., 1993b).

There are many differences between the organics detected with or without UV processing of ices. Without irradiation, HMT is not detected, which is quite surprising as H_2CO and NH_3 readily react in the gaseous phase to form HMT (Bernstein et al., 1995; Walker 1964). Likewise, ketones, amides or esters, easily synthesized under irradiation, are quite rare in those thermal experiments. It seems that UV photons provide enough energy to surmount the energy barrier for formation of these molecules. Without UV, POM's production is favoured since it requires less energy.

8.3.3 Relevance and Importance of Laboratory Simulations

This section can not be concluded without discussing the relevance of experimental simulations. Of course, an irradiation of a few hours can not reproduce millions of years or more of slow evolution with complex heterogeneous chemistry in an interstellar environment that will never be completely reproduced in the laboratory. Nevertheless, in the 3.4-μm region, infrared spectra of methane and butane mixtures after a particle irradiation present a very good fit with the observations of dust particles in the diffuse interstellar medium (i.e. highly processed material), and even with spectra of residues from the Murchinson meteorite (Pendleton et al., 1994). The same results have been obtained with residues of H_2O: CO: NH_3: $CH_4/C_2H_2/CH_3OH$, which have been exposed to direct solar UV radiation on the EURECA space station (Greenberg and Li 1997, Greenberg et al., 1995). These are also highly processed materials. Thus Strazzulla's IPHAC appears to be similar to the refractory mantle of dust grains in the harsh conditions of the diffuse interstellar medium. The less-processed mantle formed in molecular clouds is almost certainly composed of the large range of molecules detected after experimental simulations, and the abundances of characteristic molecules like HMT or POM depends on the history of the grain and the relative contribution of the different energy sources: UV and proton irradiation lead to HMT, thermal processing to POM-like polymers. The importance of such simulations is underlined if one considers that data drawn from these simulations are necessary for the preparation of space missions to comets. A good illustration of this point is the selection and calibration of chromatographic columns for the COSAC experiment (ROSETTA mission – ESA) that requires an anticipation of the nature of the molecules to be searched for (see Sect. 8.6).

8.4 Origin and Evolution of Cometary Matter

8.4.1 Origin of Cometary Matter

We have already seen (Sect. 8.1.3) how icy planetesimals populated the two present comet reservoirs, the Oort cloud and the Kuiper Belt, in the early times of the Solar System. We address now the problem of the formation of these icy planetesimals, and the processes determining the composition of cometary matter. As underlined by Yamamoto (1991) these two steps are distinct: the chemical composition is probably fixed to a large extent early in the accretion process, well before km sized bodies are built.

Figure 8.14 summarizes the present view of Solar System formation. All the matter originates in an interstellar cloud composed of dust and gas. Close to the protosun, towards the centre of the nebula, only rocky material would condense, while volatile species remained in the gas phase. Rocky planetesimals – bodies of intermediate size, ranging between 1 m and a few km – were formed there. These rocky planetesimals represent a crucial stage in the process of planet formation; they present a great similarity with present-day asteroids. On the contrary, towards the outermost parts of the nebula, where the temperature was lower, dust and ices formed the material of icy planetesimals, similar to present-day comets.

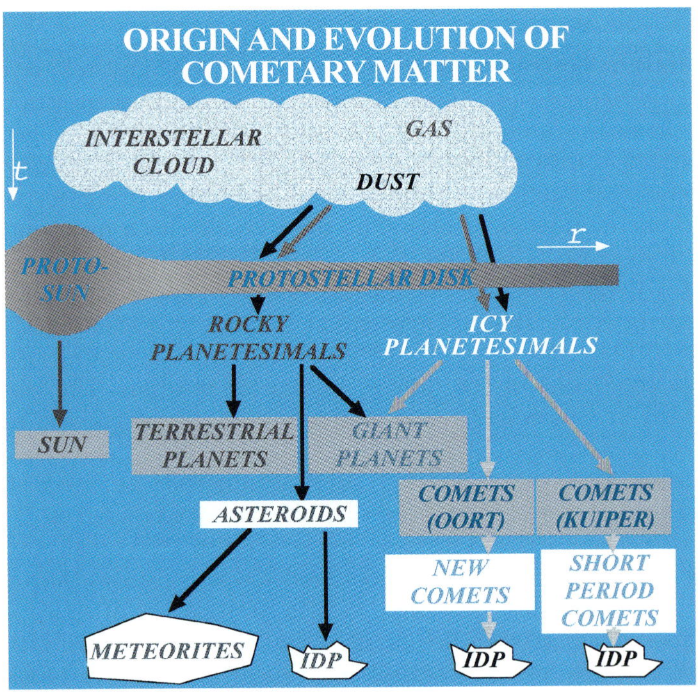

Fig. 8.14. Diagram showing the origin and the evolution of the cometary matter

These rocky and icy planetesimals intervened in very different proportions in the formation of planets. Rocky planetesimals played the major role for the formation of the Earth and other telluric planets (Mercury, Venus, Mars), while it was mostly the icy planetesimals that took part in the formation of the cores of the giant planets Jupiter, Saturn, Uranus and Neptune. Today comets are thought to be a subset of these ancient icy planetesimals, which survived until now with no or very little evolution.

This global picture, displayed in Fig. 8.14, presents in fact several simplifications. For example, there is also a small number of short-period comets that come from the Oort cloud. Moreover, the separation between rocky and icy planetesimals is not that clear, because bodies of intermediate composition and properties have been found (e.g. Phaeton, Chiron). In addition bipolar flows (see Chap. 1, Part I by T. Montmerle), which are produced in the vicinity of the protosun, may reinject a certain quantity of matter from the protosolar nebula into the collapsing protostellar envelope, allowing material highly processed in the vicinity of the Sun to be incorporated into comets at great heliocentric distances. A similar effect is obtained in the model by Shu et al. (1996) where grains are ejected from the vicinity of the protosun towards the outer parts of the disk.

8.4.2 Cometary Ices versus Interstellar Ices: the Facts

In cold dense interstellar clouds, interstellar grains are coated with ice mantles whose composition can be observed indirectly in a few regions where they evaporate: hot cores surrounding a massive star in formation inside an interstellar cloud, such as occur, for example, in the heart of the Orion Nebula, or shocked regions along bipolar flows (e.g. Kurtz et al., 2000, Bachiller et al., 2001). In the first case, the ice mantles of the grains of the collapsing envelope are thermally evaporated close to the young massive star in formation. In the second case, it is the shock wave created by the bipolar flow propagating through the interstellar medium surrounding the young star that leads to the destruction of the ice mantles and releases the molecules to the gas phase. The good correlation of molecular abundances between cometary and interstellar ices (Table 8.4 and Fig. 8.15) is one of the arguments that leads us to conclude that the matter of comets is very close to interstellar matter, suggesting similar conditions of formation, and even possibly a direct link (comets being then a frozen piece of interstellar matter).

A more straightforward comparison should arise from direct measurements of interstellar ices. Unfortunately, these measurements can only be performed in front of strong infrared sources (e.g. Ehrenfreund and Charnley 2000 and references therein). Furthermore, identification of a precise species and quantitative determination of its abundance from its IR vibrational bands are more difficult than in the case of millimetre-wave rotation spectra of gas-phase species. Table 8.4 lists the abundances in two sets of young stellar objects (YSO), respectively leading to low-mass ($< 2M_{\mathrm{Sun}}$) and high-mass stars. The former case,

Table 8.4. Cometary versus interstellar ices. Note their overall similarity. The agreement might be better in the case of ices surrounding low-mass protostars (a category to which the protosun belonged), but most of the present data are only upper limits, including NH_3 and CH_3OH. N_2, O_2, H_2 and rare gases are other important species whose ice abundances are not yet known. IS data are from Ehrenfreund and Charnley (2000); the range corresponds to 2 objects in each category (high-mass objects: W33A and N7538S; low-mass objects Elias16 and 29). Cometary data are from Table 8.1 (cometary volatiles); when no range is given, data refer to comet Hale–Bopp, except the O_2 limit, which comes from comet Halley. X-CN is the sum of various CN species (mainly the OCN^- ion in the case of IS ices)

Species	Interstellar ices high-mass YSO	Interstellar ices low-mass YSO	Cometary volatiles
H_2O	100	100	100
O_2	< 7	–	< 0.5 (*)
CO	9–16	6–25	1.7–23
CO_2	14–20	15–22	6
CH_4	2	< 1.6	0.6
C_2H_6	< 0.4	–	0.6
CH_3OH	5–22	< 4	0.9–6.2
H_2CO	1.7–7	–	0.13–1.3
HCOOH	0.4–3	–	0.09
NH_3	13–15	< 9	0.7
X-CN (∗)	1–3	< 0.4	0.08–0.25
OCS, XCS	0.05–0.3	< 0.08	0.4

which corresponds to the Sun itself, is still very incomplete. O_2 and O_3, for which only upper limits have been measured, may contain a substantial fraction of atomic O. The very volatile species H_2 and N_2, and rare gases like Ar, may be present to an unknown extent in IS ices as well as in cometary ices.

Table 8.5 gives the list of the 137 interstellar and circumstellar molecules (not including isotopic variants) detected to date (January 2004) in the gas phase. Among them are ethanol C_2H_5OH, which has been long known, and glycolaldehyde CH_2OHCHO, whose detection has been recently claimed in the ISM (Hollis et al. 2000). Glycolaldehyde belongs to the family of molecules $C_n(H_2O)_n$, which for $n > 2$ are sugars or oses. In carbonaceous chondrites (the meteorites with composition closest to comets) Cooper et al. (2001) have detected several sugar molecules. If one also takes into account the presence of H_2CO, a sugar precursor at the per cent level in comets, that sugar molecules exist in comets is a likely hypothesis.

The heaviest known interstellar molecule to date is $HC_{11}N$, which belongs to the family of cyanopolyynes (a succession of acetylene molecules linked one

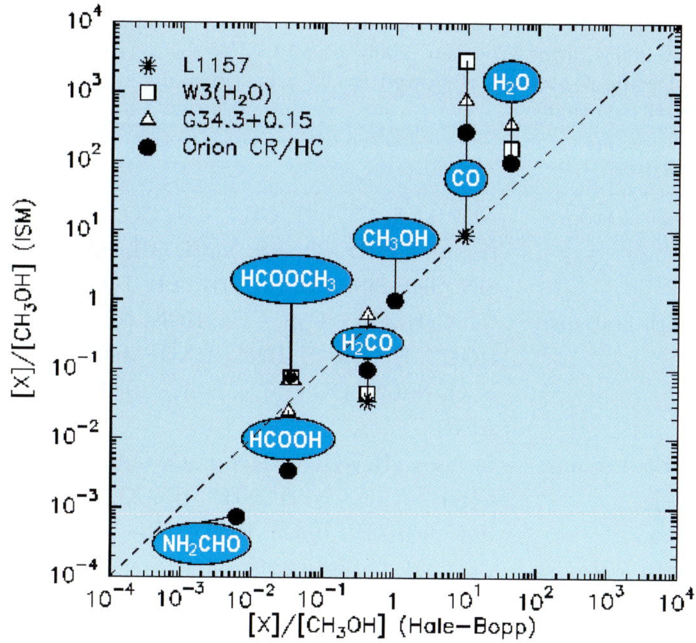

Fig. 8.15. Correlation of cometary (Hale–Bopp) and interstellar (ISM) gas-phase abundances (here for molecular species made of C, O, and H atoms; abundances are normalized to CH_3OH). The ISM regions selected for comparison are believed to release ice mantle molecules to the gas phase due to heat or shocks (Bockelée-Morvan et al., 2000)

to another to form a linear chain, and ending with a CN radical: $H(CC)_nCN$). The discovery of glycine NH_2–CH_2–COOH by Snyder et al., (e.g. 1997) in the ISM was not confirmed (Combes et al., 1996). Later searches by Ceccarelli et al., (2000) and Hollis et al., (2003) also led to upper limits. Recently, a detection has been published (Kuan et al., 2003) but it is also subject to debate. Due to the similarities between comets and the interstellar medium presented above, one can legitimately think that a number of these species exist also in comets; their detection is a challenge for new telescopes and space experiments (Sect. 8.6). In the interstellar medium grain surface chemistry also takes place. Charnley (1997) proposed a whole scheme of surface reactions to explain the formation of some complex compounds in the interstellar ices.

8.4.3 Models of Cometary Matter and Comet Nucleus Formation

According to some models, the rich IS chemistry may have been kept in comets, and thus be the ultimate source of some prebiotic compounds on Earth. This issue is especially important when discussing how common is prebiotic material

Table 8.5. List of the 137 molecules detected in the interstellar or circumstellar medium (April 2003), after Wootten (2003) and PCMI website (Boulanger and Gérin). The reality of some detections (followed by "?") is still under discussion, e.g. glycine NH_2CH_2COOH (see text)

Molecules with 2 atoms:	AlF, AlCl, C_2, CH, CH^+, CN, CO, CO^+, CP, CS, HCl, H_2, KCl, NH, NO, NS, NaCl, OH, PN, SO, SO^+, SiN, SiO, SiS, SiH, SiC, HF, SH, FeO?
Molecules with 3 atoms:	C_3, C_2H, C_2O, C_2S, CH_2, HCN, HCO, HCO^+, HCS^+, HOC^+, H_2O, H_2S, HNC, HNO, MgCN, MgNC, N_2H^+, N_2O, NaCN, OCS, SO_2, c-SiC_2, SiCN, CO_2, NH_2, $H_3{}^+$, AlNC
Molecules with 4 atoms:	C_4?, c-C_3H, l-C_3H, C_3N, C_3O, C_3S, C_2H_2, CH_3, CH_2D^+?, HCCN, $HCNH^+$, HNCO, HNCS, $HOCO^+$, H_2CO, H_2CN, H_2CS, H_3O^+, NH_3, c-SiC_3
Molecules with 5 atoms:	C_5, C_4H, C_4Si, l-C_3H_2, c-C_3H_2, C_2H_4, CH_2CN, CH_4, HC_3N, HC_2NC, HCOOH, H_2CHN, H_2C_2O H_2NCN, HNC_3, SiH_4, H_2COH^+
Molecules with 6 atoms:	C_5H, C_5O, C_5S, C_2H_4, CH_3CN, CH_3NC, CH_3OH, CH_3SH, HC_3NH^+, HC_2CHO, CH_2CHO, $HCONH_2$, l-H_2C_4, HC_4H, C_5N
Molecules with 7 atoms:	C_6H,CH_2CHCN, CH_3C_2H, HC_5N, $HCOCH_3$, NH_2CH_3 c-C_2H_4O, CH_2CHOH,
Molecules with 8 atoms:	CH_3C_3N,$HCOOCH_3$, CH_3COOH, C_7H, H_2C_6, HC_6H,CH_2OHCHO
Molecules with 9 atoms:	CH_3C_4H, CH_3CH_2CN, $(CH_3)_2O$, CH_3CH_2OH, HC_7N, C_8H
Molecules with 10 atoms:	CH_3C_5N?, $(CH_3)_2CO$, CH_2OHCH_2OH, NH_2CH_2COOH?
Molecules with 11 atoms:	HC_9N
Molecules with 12 atoms:	C_6H_6,$C_2H_5OCH_3$
Molecules with 13 atoms:	$HC_{11}N$

at the scale of our Galaxy, and the origin of chiral asymmetries (see Sect. 8.5.7 and the Chap. 3, Part II, by Cronin and Reisse). However, an important point is to realize that completely different and contradictory models are still proposed in this field, none of them being devoid of problems, although some are favoured by the present authors. A very detailed account of the pros and cons of each model can be found in Irvine et al., (2000a).

Following Yamamoto (1991) we distinguish the *formation of cometary matter* from the formation of comet nuclei: to a large extent, the size increase of the icy bodies in the protosolar nebula is decoupled from their chemical evolution.

8.4.3.1 Cometary Matter Formation

The beginning of the accretion process is the most critical step in fixing the composition of cometary matter, as the small (~ micrometre-sized) particles have the highest surface/volume ratio and maximize both surface reactions and gas-grain interactions. Furthermore, being small bodies (~ 1 – 10 km), comets did not experience substantial heating (either internal, radioactive, or external, by collisions); these processes could, however, have affected the largest KBOs, a few 100 km in diameter (see Stern (2003) for a discussion of collisions in the Kuiper Belt, McKinnon 2002 and DeSanctis et al., (2001) for models of KBO thermal evolution). The three main categories of models for the first phase differ by the degree of reprocessing of the original material.

- In the *interstellar model* of comets, cometary matter is formed very early, with no or little further processing, by sticking together *interstellar grains*, and condensing to ice volatile molecules from the original interstellar gas phase to grain mantles. This is essentially the model proposed by Greenberg (1982, 1998) and retained by Irvine et al., (1980, 1996, 2000), Yamamoto (1983, 1985, 1991), Despois (1992), Bockelee-Morvan et al., (2000). In this model ice is mostly amorphous, except in the outer layer of the nucleus heated by the Sun when the comet approaches perihelion. The strong points in favour of this model are the coexistence of oxidized (CO_2, CO, SO_2) and reduced (H_2S, CH_4, H_2S) species, as observed in IS clouds, the rough overall abundances correlation (see previous section), including complex species like HC_3N, NH_2CHO, $HCOOCH_3$, and various indications of low-temperature processes (OPR, ortho:para ratios, in H_2O and other species, deuterium enrichment, presence of very volatile species like CO and CH_4). The syntheses occurred either in the interstellar cloud or during the protosolar phase, in the outer and cold part of the protosolar nebula, and bear testimony of ISM-like chemistry (low-temperature ion–molecule and surface reactions). Indeed the DCN/HCN and HDO/H_2O ratios would mean in this hypothesis a formation temperature around 30–50 K (Meier et al., 1998), coherent with the OPR and CO volatility. Furthermore, this model is predictive to some extent, as the list of IS molecules proved to be a very good guide for the detection of new cometary species. The main problems are the presence in comets of crystalline silicates, not seen at present in the IS medium, and the deuteration level that, although high enough to imply that interstellar processes played a role, are much lower than the record values in hot cores.
- in the *pure protosolar nebula model, or 'quenching' model* (e.g. Prinn and Fegley 1989, Fegley 1993), the chemical memory has been fully or largely

lost. This model is similar in many aspects to the successful models of the inner solar nebula (terrestrial planet zone). The comet-forming material is supposed to have experienced high temperatures in a hydrogen-rich nebula in thermochemical equilibrium (temperatures above $\sim$ 500 K), which favours CO over CH_4. The concept of "kinetic inhibition" of the CO to CH_4 conversion (Lewis and Prinn 1980, Prinn and Fegley 1981) leads to quenched thermochemistry: the hot-gas composition remains in chemical equilibrium when the gas cools down, until a quenching temperature below which reactions become too slow and the composition remains unchanged. This allows CO to be the major C-bearing volatile, despite the higher thermodynamical stability of CH_4 at low T. However, species predicted as abundant by the model are not observed (NH_4HCO_3, NH_4COONH_2), the observed complex species like $HCOOCH_3$ are not predicted, and explaining the cometary volatiles' chemical composition requires mixing contributions from IS gas, protosolar disk gas, and gas from the giant planet subnebulae in the right proportion (Prinn and Fegley 1989). Thus, an increasing complexity is required to explain even the gross features (e.g. oxidized/reduced species ratio) of the observed volatile composition.

- the *partially reprocessed IS matter model* is proposed by Lunine, Gautier and coworkers (e.g. Iro et al., 2003 and refs. therein). This aims at explaining several features of the Solar System besides cometary composition, such D/H ratios and rare-gas abundances in the giant planets. The solar nebula is initially rather hot (600 K at 1 AU) and cools down with the decrease of the disk accretion rate from 10^{-5} to $10^{-8} M_{\rm Sun}$/yr. The nebula is supposed to be already isolated from the parent cloud (protostellar envelope). A key feature is that a significant fraction (depending on the heliocentric distance) of ice mantles of infalling interstellar grains vaporize through an accretion shock before reaching the nebula; the molecules then recondense as crystalline ice (as in the preceding pure protosolar nebula models), which incorporates minor species as gas clathrates or stoichiometric hydrates. The chemical composition of the gas is supposed to be unaltered, whereas the deuteration is diminished by interaction with the H_2 of the nebula. Gas composition and clathrate/hydrate stoichiometry control the final composition of cometary ices (clathrates are a class of solids in which small gas molecules like CH_4, CO_2 or noble gases occupy "cages" made up of hydrogen-bonded water molecules; other species like NH_3 form solid hydrates with well-defined composition (NH_3) $(H_2O)_n$). Another essential feature is turbulence across the disk that can transport the silicate grains processed (crystallized) in the vicinity of the Sun to the cold regions where the ice forms.

8.4.3.2 The Building of the Nuclei

In the continuous accretion process leading from 0.1-μm grains to km-sized bodies, there is a scale level, probably close to 1–10 m, after which the chemical composition of the matter is almost fixed (note, however, that very small molecules like CO could diffuse through the ice if warmed enough). At a larger size, most molecules are too deep inside to be affected by surface heating, energetic photons or particles (solar X or UV radiation, solar-wind particles and cosmic rays). Desorption processes and chemical reactions in the bulk of the ices, or on the probably large internal surface of this highly porous material, are limited, due to the very low temperature (a few tens of K to 150 K at most). Occasionally, collisions occurring during the accretion process may heat locally the cometesimals. The accretion of particles to form comets (or icy planetesimals) has been thoroughly studied by Weidenschilling (1997); his simulations show the formation of comet-size bodies from micrometre-size grains up to bodies 80 km in diameter in 2.5×10^5 yr.

8.4.4 Are Today's Comets Like Comets in the Early Solar System?

Were early comets similar to present ones? A given cometary nucleus is thought to have negligibly evolved since its formation, due to its low temperature. Exceptions may be a thin surface layer (due to particle and photon processing in the Oort cloud and the Kuiper Belt, and a few possible passages close to the Sun, see Mumma et al., 1993), and a slow volume loss of very volatile species (rare gases, CO). Occasional collisions may also have occurred in the Kuiper Belt and the Oort Cloud.

May chemistry inside the nucleus have altered the pristine composition of comets since their accretion? Navarro-Gonzalez et al., (1992) and Navarro-Gonzalez and Romero (1996) have studied theoretically the effect of ionizing radiation (cosmic rays and embedded radionuclides) on the cometary material and conclude that pristine organic compounds should not be altered except in the external layers of the comet. It is also predicted that an original enantiomeric excess of alanine would not be destroyed by radioracemisation during the decay of radionuclides. At the very most it should be attenuated, but comets could have contributed to the origin of chirality on Earth. The limitation of this work is that the kinetic data used in the calculations are for the liquid phase only, since no solid phase values are available; this may distort the result to some extent. The thermal racemization is a priori expected to be negligible at the comet low temperatures, but here also more theoretical and experimental work is required.

As stated before, the largest icy bodies (> 500 km) in the Kuiper Belt might have for a limited time reached internally at some place conditions for melting to occur. This is, of course, a cause of chemical transformation.

If comets themselves are thought to have kept their original composition, their reservoirs have undergone at least a slow evolution in number and orbital parameters (e.g. Morbidelli et al., 2000). This affects directly the possibilities of delivery to the Earth.

8.5 Delivery to the Earth

8.5.1 Shooting Stars (Meteors)

When a comet passes close to the Sun, the gases that are produced by the sublimation of the ice drag along the dust particles and release them into the interplanetary medium. Comets are thus one of the two principal sources of interplanetary dust (interplanetary dust particles, IDPs, also called "meteoroids"). The other source of IDPs is the collisions between rocky objects: asteroids or asteroid fragments. The largest cometary dust particles remain confined in the vicinity of the comet orbit and spread progressively along this orbit, forming a ring-like swarm of particles. In fact, planetary perturbations deform the orbits, which are no longer closed but transform into a tightly packed set of successive turns. Due to small differences in ejection conditions, individual particles follow different orbits, which, however, are close one to another and form a long ribbon through the Solar System. When the Earth crosses such a swarm, we experience a shower of shooting stars: during the entry into the Earth's atmosphere, particles larger than a size of about 1 mm produce, through heating, combustion and ionization of the air, a trail visible to the naked eye.

8.5.2 Overall Picture of Matter Delivery to the Earth

Several kinds of objects of the Solar System are likely to bring matter to the Earth: large-size bodies (asteroids or comets) responsible for more or less devastating impacts, small-size bodies (smaller than 10 m) called meteoroids in general, and interplanetary dust for the smallest; if a fragment of these bodies reaches the ground, it is called a meteorite (or micrometeorite). This delivered material will affect the land, the oceans, the atmosphere and the terrestrial biosphere.

In Fig. 8.16 we present an overview of the different kinds of matter delivered to the Earth, displayed according to size. Around 0.1 micrometres are particles that are probably grains coming directly from the local interstellar medium surrounding the Sun. Above 10 micrometres the range of interplanetary particles starts. The smallest ones, around 10 micrometres, are decelerated and stopped in the stratosphere, where they can be collected by airplanes: these are the "stratospheric IDPs" previously mentioned. The slightly larger particles reach the ground and are thus called micrometeorites. If one considers the delivered mass (Fig. 8.17) and not the number of particles, the distribution presents a peak towards 100 micrometres. In the size range 10–100 micrometres these particles produce in the atmosphere ion trails detectable by radar echoes; when the size exceeds about 100 micrometres (the value depending on the velocity of the particle) they appear in the visible as "shooting stars". With even larger particles, ranging from centimetre to decametre, one enters the field of the traditional meteorites. A few tens of metres is roughly

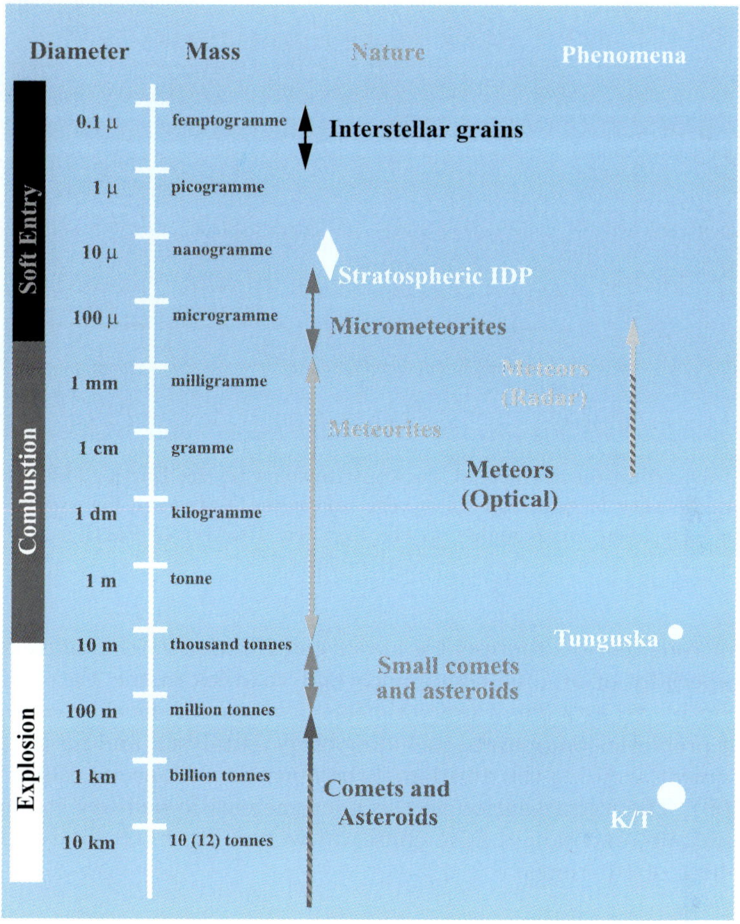

Fig. 8.16. Overview of particles delivered to Earth, displayed according to their size

the size of the bolide presumed to be responsible for the Tunguska explosion, which occurred in Siberia in June 1908, and blew down the forest over $2000\,\mathrm{km}^2$; the nature of the object is still debated, see Shuvalov and Artemieva 2002, Jopek et al., 2002 for recent discussions. At around ten kilometres diameter lies the object (comet or asteroid – e.g. Jeffers et al., 2001 – recent data favours the latter hypothesis, e.g. Kyte 1998) which would have been responsible, 65 Myr ago, for producing the K-T sedimentary layer that marks in many places the boundary of the Cretaceous and the Tertiary geological periods, and of inducing massive species extinction, most notably that of dinosaurs (e.g. Frankel 1999; but the debate is still active, e.g. Stankel 2001, Facett et al., 2001, Le Loeuff and Laurent 2001). Such an impact occurs roughly every 100 Myr.

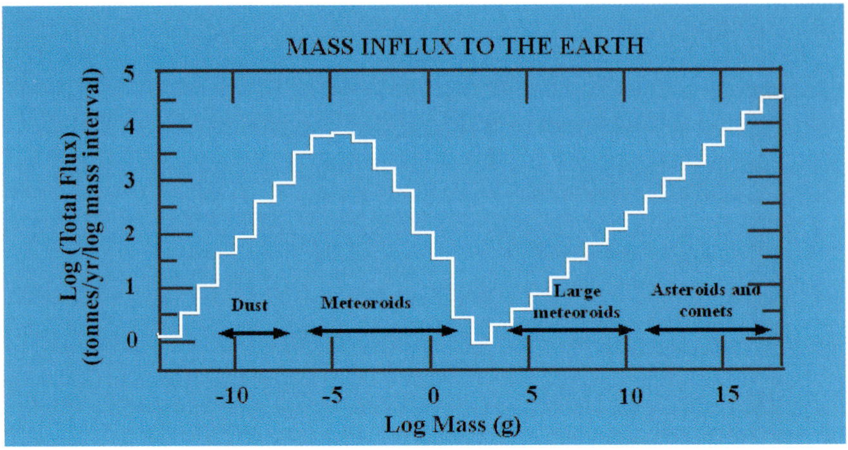

Fig. 8.17. Distribution in mass of the contributions to the Earth. The actual mean influx from the very largest objects is very uncertain. Rates are for present, and are believed to have been much higher in the first few 100 Myr of Earth's history (Steel 1996)

The micrometeorites represent a major contribution to the delivery of matter and especially of organic compounds and volatiles to the Earth; they may even be considered as a small prebiotic chemical reactor, as they contain many significant prebiotic compounds, including clays (catalyst), and insuring a natural confinement avoiding the dilution of the reactants (Maurette 1998, Maurette et al., 2000). The extrapolation of the micrometeoritic delivery in the past is highly uncertain, and ranges from conservative factors of 100 times the present rate to a high of 10^6 times.

8.5.3 Delivery of elements and Water

According to different estimations, extraterrestrial delivery to Earth over 4.5 billion years ranges from 10^{21} to 7×10^{23} kg. The lower limit of this amount that can be attributed to comets has been estimated to be 10% by Chyba et al., (1990).

- *H_2O*

It has been suggested that most of the water of the Earth oceans (1.4×10^9 km^3, roughly 10^6 times comet Halley mass) could have been imported by comet impacts during a period of heavy bombardment (e.g. Delsemme 2000). Recent dynamical models of the early Solar System (Morbidelli et al., 2000), and recent measurements of the D/H in cometary water (to be compared with D/H in the oceans (SMOW)) (Dauphas et al., 2000) both lead to a maximum of $\sim$ 10% of water of cometary origin; the latter argument assumes that the D/H ratio measured in 3 comets is typical of all comets. Petit and Morbidelli discuss further the delivery of water in their chapter. An even lower limit (1%) on cometary H_2O in ocean water has been recently presented, from indirect arguments on rare gases and metals (Dauphas and Marty 2002).

- *Carbon*

Carbon in comets is roughly 18% by mass. Chyba et al., (1990) estimated that the contribution of comets to the terrestrial carbon inventory ranges from 2×10^{19} to 10^{22} kg; while the total budget of carbon on Earth is about 9×10^{19} kg. Even if uncertainties linked to the total amount of comets impacting the Earth since its formation are not lifted, those estimations show that they brought a significant amount of carbonaceous material. The extent to which organic molecules can survive cometary impact is discussed in Sect. 8.5.5.

- *Other Elements*

The other elements required for a protocell formation are also present on comets. Phosphorus, which is involved in the synthesis of sugars, has been detected by mass spectroscopy ($m/e = 31$) in grains of comet Halley by the PUMA mass spectrometer on board Vega 1. But its abundance is very low and detection may have been affected by interference from ions such CH_2OH^+ (Kissel and Krueger 1987). However, analysis by laser probe mass spectroscopy of interplanetary dust particles (IDPs), whose cometary origin is probable, has led to the detection of PO_2 and PO_3 anions (Radicati-Di-Brozolo et al 1986). Ni and Fe have been detected in comet Ikeya-Seki and Halley.

8.5.4 Prebiotic Molecules from Comets?

Five families of compounds are considered to be the key prebiotic monomers required before starting a chemical evolution from which life would arise. These are amino acids (whose combinations lead to proteins), *purine bases* (adenine and guanine), *pyrimidines bases* (cytosine, uracil and thymine), *sugars* (whose association lead to RNA and DNA molecules), and *fatty acids* (cell walls). It is very interesting to note that, even if these compounds have not been detected in comets (maybe because of limitations in the sensitivity of telescopes, but also essentially because the synthesis of most of them requires liquid water), all but thymine can be easily produced in an early Earth environment from cometary precursors such as HCN, HC_3N, H_2CO and CO that have been firmly detected in cometary comae (Table 8.6).

Nevertheless, complex compounds of exobiological interest may also be present in comets. The only detection of adenine is a very tentative interpretation of PUMA's mass spectra of comet Halley's dust (Kissel and Krueger 1987). Such a molecule could be synthesized by HCN condensation without any

Table 8.6. Prebiotic syntheses from cometary molecules, from (Oró and Cosmovici 1997)

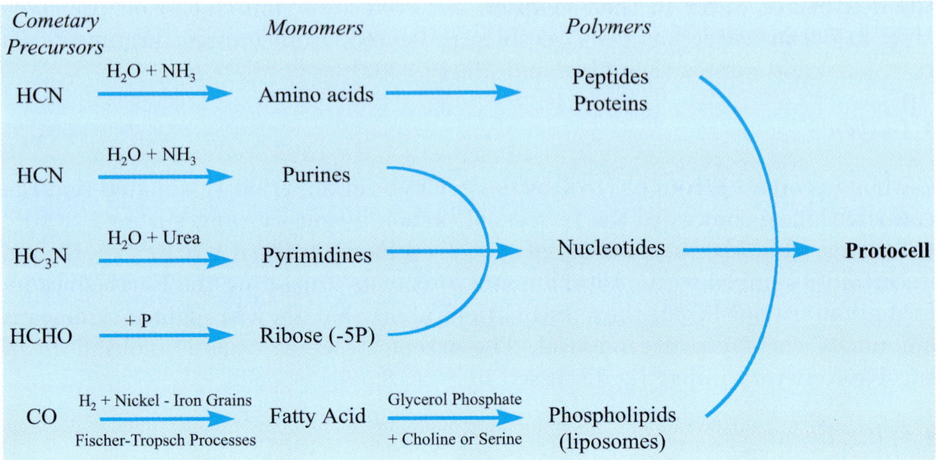

liquid water (Oro and Cosmovici 1997, Wakamutsu et al., 1966). Moreover, as already seen in this chapter, when ices relevant to the composition of interplanetary dust particles or comets are irradiated, complex molecules are formed, and among them, the simplest amino acid, glycine (Briggs et al., 1992). After an acid hydrolysis in water other amino acids, such as alanine, valine, proline and others, have also been identified (Bernstein et al., 2002; Kobayashi et al., 1995; Muñoz Caro et al., 2002).

8.5.5 Do Molecules Survive From Comets to the Earth?

Thus, comets may have imported prebiotic elements to early Earth, which, when mixed with liquid water, may have allowed the synthesis of all molecules thought to be necessary for the origin of life. To date, it is not clear as to whether the importation of such key compounds occurs at the simple stage of HCN, HC_3N, HCHO..., or more elaborate species such as amino acids, puric and pyrimidic acids and others. But before seeding the oceans, those compounds have to survive:

- if brought by small dust particles, they are subject in the interplanetary medium to the action of solar UV radiation, solar-wind particles and galactic cosmic rays; while the comet particle is slowed down and warmed up by the Earth's atmosphere, they must resist pyrolysis (destruction at very high temperature), and chemical reactions in the atmosphere. If they reach the ground, they should also survive the ensuing impact.
- alternatively, if the nucleus impacts directly the Earth, can the molecule survive the dramatic energy release?

Experiments conducted in space have shown that amino acids are quite unstable to solar UVs, but also that when somehow shielded in minerals, such as meteoritic powder, they can survive in space, and undergo no racemization (Barbier et al., 2002; Boillot et al., 2002).

The case of impacts of large bodies has been theoretically treated by Chyba et al (1990). It appears that organic compounds (even amino acids, if present) contained in cometary impactors of 100 to 200 m in size could survive a collision with Earth in a 10-bar CO_2 atmosphere thanks to an efficient aerobreaking. Comparisons with Venus and Mars lead us to think that it is the most probable composition of the primitive atmosphere. The authors note in their conclusion that "*It is intriguing that it is exactly these dense CO_2 atmospheres, where photochemical production of organic molecules should be the most difficult, in which intact cometary organics would be delivered in large amounts*". Indeed, organic syntheses are efficient in reduced atmospheres, and even in weakly reduced atmospheres such as that of Titan (N_2, CH_4), whereas they are very difficult in the oxidized environment that must have been the early Earth's atmosphere.

On the experimental scene, Blank et al (2001) conducted a series of shock experiments to assess the feasibility of the delivery of amino acids to the Earth via cometary impacts. It appears that a large fraction of the amino acids do survive impacts. It has also been shown that some chemistry occurs, which leads to the formation of peptide bonds and new compounds including amino-acid dimers.

8.5.6 Comparison of Comets with Other Likely Sources of Prebiotic Molecules

Figure 8.17 displays the total mass delivered to the Earth for each mass interval (note that these intervals are logarithmic). Two main contributions show up, which are very different in nature: while the small particles could more or less softly deliver molecules to the Earth atmosphere and surface, large-size impactors bring also large quantities of matter but of course in a very different – discontinuous and catastrophic – mode. The very high temperatures reached by gases allow the synthesis of some compounds, as well as leading to their destruction as discussed in the previous section.

In the so-called intermediate atmospheres (in which the main gases are e.g. CO_2, H_2O, N_2 instead of NH_3 and CH_4) extraterrestrial delivery plays a major role in the total budget of prebiotic molecules (Chyba and Sagan 1996, Jenniskens et al., 2000). Table 8.7 gives in kg/yr the quantity of these molecules brought by various processes in the case of reducing and neutral (or moderately oxidizing) atmospheres. The large difference between these two types of atmosphere is noticeable, in particular for the synthesis of molecules by the shock waves induced by impacts. It should be noted that if many experiments have been undertaken in the laboratory on the chemical syntheses induced by electric discharges or energetic photons, the experimental studies relevant to the entry of particles in the atmosphere are very difficult and are still in a much less elaborate stage. The direct observation of molecules deposited by the meteors in the

Table 8.7. Major sources of prebiotic organics for two candidate early Earth atmospheres. The intermediate atmosphere corresponds here to a ratio $H_2/CO_2 = 0.1$. Note the considerably larger synthesis efficiency in the case of reducing atmospheres, which are those used by Urey and Miller in their experiments. In the more probable case of an intermediate atmosphere, external delivery is a major source of organics, as 30% come from IDPs. After Chyba and Sagan (1996) for the atmospheric processes, and Ehrenfreund et al., (2002) for the hydrothermal vents and comet impact delivery estimations

Source	Reducing atmosphere (kg/yr)	Intermediate atmosphere (kg/yr)
UV photolysis	1×10^{12}	3×10^{8}
Electric discharge	3×10^{9}	3×10^{7}
IDPs	2×10^{8}	2×10^{8}
Shocks from impacts	2×10^{10}	4×10^{2}
Shocks from meteors	4×10^{9}	8×10^{1}
K/T extrapolation	2×10^{8}	2×10^{8}
Total (atmosphere)	1×10^{12}	7×10^{8}
Hydrothermal vents	1×10^{8}	1×10^{8}
Comet impact delivery	1×10^{11}	1×10^{11}

atmosphere has been attempted during the recent Leonid shower (CN: Rairden et al., 2000, HCN: Despois et al., 2000).

8.5.7 Chiral Molecules: from the Interstellar Medium to the Early Earth?

Among the most intriguing questions related to the origin of life on Earth is the origin of the chirality for some key molecules: amino acids (configuration L) and sugars (configuration D).

Even if the cometary contribution were modest in mass, it might still have played a remarkable role, allowing via cometary dust particles or impact to directly forward to the early Earth some interstellar molecules, which could have played an important part in the apparition of homochirality (e.g. Bonner et al., 1999 and references therein). Indeed, an enantiomeric excess has been measured for some nonbiological amino acids in meteorites (see Cronin and Reisse Chapt. 3, Part II). Therefore, it is conceivable that an excess was also present for the biological amino acids (for which analysis is much more difficult due to the risk of contamination by terrestrial biomolecules). Such an excess could have been transmitted to the early Earth, and constitute an initial asymmetry later amplified by prebiotic chemistry (see Commeyras et al., 2004). The possible importance of this scenario is the fact that abiotic mechanisms able to produce nonracemic mixtures are very rare.

Which physical or chemical mechanism could then have led to a non-racemic mixture in the meteorites? Bailey and collaborators highlighted conditions favourable for the buildup of chiral asymmetry in the interstellar medium, and more especially in regions of star formation (Bailey et al., 1998, Bailey 2001). They detected the presence of circularly polarized infrared radiation in vast regions, which could easily include a whole Solar System in formation. This suggests the possible existence in these same regions of circularly polarized ultraviolet radiation, which still has to be detected. This circularly polarized ultraviolet radiation could lead to a differential synthesis or destruction of one enantiomer with respect to the other. If conditions similar to those in Orion prevailed at the time of formation of our Solar System, and if these interstellar molecules could actually be preserved till their integration into the matter of the protosolar nebula, this could be a plausible and relatively natural scenario for L/D symmetry breaking between enantiomers. This scenario seems at least as probable as the assumption of a fortuitous encounter of the protosolar nebula with a neutron star (a pulsar), also suggested as a likely source of polarized light. In comets, the search for non-racemic mixtures will require in situ chemical analysis or laboratory analysis of returned samples.

8.6 Ground-based and Space Exploration of Comets: New Developments

In the Solar System, several major space missions relate to comets. ESA is preparing the ROSETTA probe that will reach comet 69P/Churyumov-Gerasimenko (the previous target, P/Wirtanen, had to be abandoned as the launch was delayed); after a ten-year cruise, it will meet the comet and follow it for almost two years, during which a lander will be deposited on the surface. On the lander, the COSAC experiment (e.g. Rosenbauer et al., 1999) will perform a chemical analysis of cometary volatiles by gas chromatography of samples taken after drilling down to 20 cm below the surface. With a detection limit of 10^{-5}–10^{-6} with respect to water, more than 100 cometary species might be detected (Szopa et al., (2003) and Crovisier (2004)). The use of chiral columns will allow the measurement of enantiomeric excess.

A US mission, STARDUST, aims to collect thousand of cometary grains and bring them back to Earth. Another probe, CONTOUR, intended to tour 3 cometary nuclei, seems unfortunately to have been definitively lost. Table 8.8 summarizes all the space efforts directed towards comets in the next few years. As for ground-based observers, they have to wait for a new bright comet, like the recent comets Hyakutake and Hale–Bopp. In the near future, large ground-based telescopes, like the Atacama Large Millimeter Array (64 antennas, 12 m in diameter, operating as an interferometer in a very good high altitude astronomical site in Chile), the Large Millimeter Telescope (50-m antenna), the Green Bank

Table 8.8. Recent, in progress and planned cometary missions (Nov. 2003)

Missions to comets	Launch	Dates	Events	Comet	Goal
DEEP SPACE 1 (NASA)	Oct. 24, 1998	Sept. 22, 2001	Flyby	P/Borrelly	Nucleus imaging, plasma
STARDUST (NASA)	Feb. 7, 1999	Jan. 2, 2004 Jan. 15, 2006	Flyby Return	P/Wild 2	Comet dust sample return
ROSETTA (ESA) (1)	Feb. 26, 2004	4/2014 11/2014 8/2015	Comet orbit Landing Perihelion	67P/Churyumov–Gerasimenko	Orbiter + lander Ice and dust in situ analysis
DEEP IMPACT (NASA)	Dec. 30, 2004	Jul. 4, 2005	Flyby + impact	P/Temple 1	500 kg projectile
CSSR (NASA) (2)	Aug. 2009	5/2017 3/2018 12/2018	Rendez-vous Comet depart Earth return	46P/Wirtanen	Comet surface sample return
Other Missions of interest	**Launch**	**Results**			**Comet-related science**
SIRTF (NASA)	Aug. 2003	2003→			Interplanetary dust disks (Exozodi)
COROT (CNES)	June 2006	2006→			Exocomets
HERSCHEL (ESA)	Feb. 2007	2007→			D/H in a variety of comets
NEW HORIZONS (NASA)	Jan. 2006	2015/2016 2017/2023	Pluto flyby KBO flyby		Pluto + Kuiper belt object imaging

(1) Rosetta Launch, foreseen 2003, has been rescheduled; previous target was 46 P/Wirtanen
(2) Project stage, not yet accepted; dates are indicative only.

Telescope (100 m antenna) will bring higher sensitivity and space resolution. Out of the Solar System, the Corot mission should make it possible to detect comets around other stars (Lecavelier des Etangs et al. 1999).

8.7 Conclusion

Comets can bring down to the Earth elements like carbon, as well as water (in an amount that remains to be determined). Moreover, they contain simple organic molecules, and it is probable that they contain more complex ones. Thus comets are likely candidates for early Earth enrichment in prebiotic molecules, either through the dust particles they release – some of which end up encountering the Earth atmosphere – or through the atmospheric chemical syntheses – and possibly also through direct delivery – taking place when cometary nuclei impact the Earth. The same impacts may also strongly affect the later development of life, either locally, or on the scale of the planet.

Many questions are central to the current debates, which we hope will find a response in coming years, and for which space missions can bring important results:

- Where and when were formed the cometary matter, and then the comets?
- How close is cometary matter to interstellar matter?
- Do comets contain amino acids and other bricks of life?
- Do these molecules present an enantiomeric excess, or are they in racemic mixture?
- What is the precise budget of the cometary contributions to the Earth (fraction of the water of the oceans; fraction of carbon; fraction of prebiotic molecules)?
  - What is the global effect of comets on life, its appearance and its development?

Out of the Solar System, is the presence of an Oort cloud or a Kuiper Belt an exceptional characteristic of our Solar System or a banal fact among main sequence stars? Given the crucial part potentially played by comets in the origin and the evolution of life on Earth, it appears fundamental to know the probability for a star with habitable planets to be also surrounded by comets. New research topics thus develop on these *exocomets* (comets orbiting around other stars). The properties of the spectra of a close star, β Pictoris – one of the rare main sequence stars around which a disc has been detected – are currently convincingly interpreted as due to a constant infall of "evaporating bodies" on this star (Lagrange et al., 2000, Karmann et al., 2001, 2003). Much is then to be expected from ambitious space experiments under development, like the Corot mission (Baglin 2003), which should make it possible to detect directly comets, as well as Earth-size planets around a wide range of other stars (Lecavelier des Etangs et al., 1999, Bordé et al., 2003).

Acknowledgement

We are very grateful for detailed reading, precious comments and many corrections on this text by W. Irvine, J. Crovisier, M. Gargaud, B. Barbier and J. Reisse.

References

A'Hearn, M. F., Millis, R. L., Schleicher, D. G., Osip, D. J., Birch, P. V. (1995). The ensemble properties of comets: Results from narrowband pghotometry of 85 comets, 1976–1992. *Icarus*, **118**, 223–270.

A'Hearn, M.F., Boehnhardt, H., Kidger, M., West, R.M. eds (1999). Proceedings of the first international conference on Comet Hale–Bopp. *Earth, Moon and Planets*, **77, 78** & **79.**

Allamandola, L. J., Sandford, S. A., Valero,G. J. (1988). Photochemical and thermal evolution of interstellar/precometary ice analogs. *Icarus*, **76**, 225–252.

Altwegg, K., Balsiger, H., Geiss, J. (1999). Composition of the Volatile Material in Halley's Coma from In Situ Measurements. *Space Science Reviews*, **90**, 3–18.

Altwegg, K., Bockelée-Morvan, D. (2003). Isotopic Abundances in Comets. *Space Science Reviews*, **106**, 139.

Arpigny, C., Jehin, E., Manfroid, J., Hutsemékers, D., Schulz, R., Stüwe, J. A., Zucconi, J., Ilyin, I. (2003). Anomalous Nitrogen Isotope Ratio in Comets, *Science*, **301**, 1522.

Bachiller, R., Pérez Gutiérrez, M., Kumar, M. S. N., Tafalla, M. (2001). Chemically active outflow L 1157, *Astronomy and Astrophysics*, **372**, 899.

Baglin, A. (2003). COROT: A minisat for pionnier science, asteroseismology and planets finding. *Advances in Space Research*, **31**, 345.

Bailey, J. (2001). Astronomical Sources of Circularly Polarized Light and the Origin of Homochirality. *Origins of Life and Evolution of the Biosphere*, **31**, 167.

Bailey, J., Chrysostomou, A., Hough, J. H., Gledhill, T. M., McCall, A., Clark, S., Menard, F., Tamura, M. (1998). Circular Polarization in Star-Formation Regions: Implications for Biomolecular Homochirality, *Science*, **281**, 672.

Ball, A. J., Gadomski, S., Banaszkiewicz, M., Spohn, T., Ahrens, T. J., Whyndham, M., Zarnecki, J. C. (2001) An instrument for in situ comet nucleus surface density profile measurement by gamma ray attenuation, *Planetary and Space Science*, **49**, 961.

Barbier, B., Henin, O., Boillot, F., Chabin, A., Chaput, D., Brack, A. (2002). Exposure of amino acids and derivatives in the Earth orbit. *Planet. Space Sci.*, **50**, 353–359.

Benkhoff, J. (1999). On the Flux of Water and Minor Volatiles from the Surface of Comet Nuclei, *Space Science Reviews*, **90**, 141.

Bernstein, M. P., Dworkin, J. P., Sandford, S. A., Cooper, G. W.,Allamandola, L. J. (2002). Racemic amino acids from the ultraviolet photolysis of interstellar ice analogues. *Nature*, **416**, 401–403.

Bernstein, M. P., Allamandola, L. J., Sandford, S. A. (1997). Complex organics in laboratory simulations of interstellar/cometary ices. *Adv. Space Res.*, **19**, 991–998.

Bernstein, M. P., Sandford, S. A., Allamandola, L. J., Chang, S. (1994). Infrared spectrum of matrix isolated Hexamethylenetetramine in Ar and H_2O at cryogenic temperatures. *J. Phys. Chem.*, **98**, 12 206–12 210.

Bernstein, M. P., Sandford, S. A., Allamandola, L. J., Chang, S., Scharberg, M. A. (1995). Organic Compounds Produced By Photolysis of Realistic Interstellar and Cometary Ice Analogs Containing Methanol. *Astrophys. J.*, **454**, 327–344.

Biver et al., (1999) Long-Term Evolution of the Outgassing of Comet Hale- Bopp from Radio Observations. *Earth, Moon and Planets*, **78**, 5–11.

Biver, N., Bockelée-Morvan, D., Crovisier, J., Colom, P., Henry, F., Moreno, R., Paubert, G., Despois, D., Lis, D. C. (2002a) Chemical Composition Diversity Among 24 Comets Observed At Radio Wavelengths. *Earth Moon and Planets*, **90(1)**, 323–333.

Biver, N., Bockelée-Morvan, D., Colom, P., Crovisier, J., Henry, F., Lellouch, E., Winnberg, A., Johansson, L. E. B., Gunnarsson, M., Rickman, H., Rantakyrö, F., Davies, J. K., Dent, W. R. F., Paubert, G., Moreno, R., Wink, J., Despois, D., Benford, D. J., Gardner, M., Lis, D. C., Mehringer, D., Phillips, T. G., Rauer, H. (2002b). The 1995–2002 Long-Term Monitoring of Comet C/1995 O1 (HALE-BOPP) at Radio Wavelength, *Earth Moon and Planets*, **90**, 5.

Blank, J. G., Miller, G. H., Ahrens, M. J., Winans, R. E. (2001). Experimental Shock Chemistry of Aqueous Amino Acid Solutions and the Cometary Delivery of Prebiotic Compounds. *Orig. Life*, **31**, 15–51.

Bockelée-Morvan, D. (1997). Cometary Volatiles: the status after comet C/1996 B2 Hyakutake. In *Molecules in Astrophysics: Probes and Processes*, IAU symposium 178, E.F. van Dishoeck ed., Kluwer, p. 219.

Bockelee-Morvan, D., Brooke, T.Y., Crovisier, J. (1995) On the origin of the 3.2 to 3.6-micrometre emission features in comets, *Icarus*, **116**, 18.

Bockelée-Morvan, D. Crovisier, J. (2002). Lessons of Comet Hale–Bopp for Coma Chemistry: *Observations and Theory. Earth Moon and Planets*, **89**, 53–71.

Bockelée-Morvan, D., Rickman, H. (1999). C/1995 O1 (Hale–Bopp): Gas Production Curves and Their Interpretation. *Earth Moon Plan.*, **79**, 55–77.

Bockelée-Morvan, D., Biver, N., Moreno, R., Colom, P., Crovisier, J., Gérard, E., Henry, F., Lism, D. C., Matthews, H., Weaver, H. A., Womack, M., Festou, M. C. (2001). Outgassing Behaviour and Composition of Comet C/1999 S4 (LINEAR) During Its Disruption. *Science*, **292**, 1339–1343.

Bockelée-Morvan, D., Gautier, D., Lis, D. C., Young, K., Keene, J., Phillips, T. G., Owen, T., Crovisier, J., Goldsmith, P. F., Bergin, E.A., Despois, D., Wootten, A. (1998). Deuterated Water in Comet C/1996 B2 (Hyakutake) and its Implications for the Origin of Comets. *Icarus*, **133**, 147–162.

Bockelée-Morvan D., Lis D. C., Wink J. E., Despois D., Crovisier J., Bachiller R., Benford D. J., Biver N., Colom P., Davies J. K., Gérard E., Germain B., Houde M., Mehringer D., Moreno R., Paubert G., Phillips T. G., Rauer H. (2000). New molecules found in comet C/1995 O1 (Hale–Bopp). Investigating the link between cometary and interstellar material., *Astronomy and Astrophysics*, **353**, 1101–1114.

Bockelée-Morvan, D., Crovisier, J., Mumma, M., Weaver, H. (2004, in press), The Composition of Cometary Volatiles, in *Comets II,* eds. M. Festou, U. Keller, H. Weaver, Univ. of Arizona Press.

Boehnhardt, H. (2002), Comet Splitting – Observations and Model Scenarios, *Earth Moon and Planets*, **89**, 91.

Boillot, F., Chabin, A., Buré, C., Venet, M., Belsky, A., Bertrand-Urbaniak, M., Delmas, A., Brack, A., Barbier, B. (2002). The Perseus Exobiology Mission on MIR: Behaviour of Amino Acids and Peptides in Earth Orbit. *Orig. Life*, **32**, 359–385.

Bonner, W. A., Mayo Greenberg, J., Rubenstein, E. (1999). The Extraterrestrial Origin of the Homochirality of Biomolecules – Rebuttal to a Critique, *Origins of Life and Evolution of the Biosphere*, **29**, 215.

Bordé, P., Rouan, D., Léger, A. (2003). Exoplanet detection capability of the COROT space mission, *Astronomy and Astrophysics*, **405**, 1137.

Boulanger, F., Gerin, M. *Interstellar and Circumstellar Molecules*, http://www.lra.ens.fr/~pcmi/table-mol-an.html.

Bouwman, J., de Koter, A., Dominik, C., Waters, L.B.F.sM. (2003). The origin of crystalline silicates in the Herbig Be star HD 100546 and in comet Hale–Bopp. *Astronomy and Astrophysics*, **401**, 577.

Briggs, R., G. Ertem, J. P. Ferris, J. M. Greenberg, P. J. McCain, C. X. Mendoza-Gomez, W. Schutte (1992). Comet Halley as an aggregate of interstellar dust and further evidence for the photochemical formation of organics in the interstellar medium. *Orig. Life*, **22**, 287–307.

Brucato, J.R., Colangeli, L., Mennella, V., Palumbo, P., Bussoletti, E. (1999). Silicates in Hale–Bopp: hints from laboratory studies. *Planetary and Space Science*, **47**, 773–779.

Brucato, J.R., Castorina, A.C., Palumbo, M.E., Satorre, M.A., Strazzulla, G. (1997). Ion Irradiation and Extended CO Emission in Cometary Comae. *Planetary and Space Science*, **45**, 835–840.

Capria, M. T. (2002). Sublimation Mechanisms of Comet Nuclei, *Earth Moon and Planets*. **89**, 161.

CBAT. *Cometary Designation System*. http://cfa-www.harvard.edu/iau/lists/CometResolution.html.

Ceccarelli, C., Loinard, L., Castets, A., Faure, A., Lefloch, B. (2000). Search for glycine in the solar type protostar IRAS 16293-2422. *Astronomy and Astrophysics*, **362**, 1122–1126.

Chamberlin, T. C., Chamberlin, R. T. (1908). *Science*, **28**, 897.

Charnley, S.B. (1997). On the nature of interstellar organic chemistry. In Astronomical and Biochemical Origins and the Search for Life in the Universe, *5th Int. Conf. On Bioastronomy, IAU Coll. 161*, Cosmovici C.B., Bowyer S. Werthimer D. eds., pp 89–96.

Charnley, S.B., Rodgers, S.D., Butner, H. M., Ehrenfreund, P. (2002a). Chemical Processes in Cometary Comae, *Earth Moon and Planets*, **90**, 349.

Charnley, S.B., Rodgers, S.D., Kuan, Y.-J., Huang, H.-C. (2002b). Biomolecules in the interstellar medium and in comets, *Advances in Space Research*, **30**, 1419.

Chen, J., Jewitt, D. (1994). On the rate at which comets split, *Icarus*, **108**, 265.

Chyba, C., Sagan, C. (1997). Comets as a Source of Prebiotic Organic Molecules for the Early Earth, in *Comets and the Origin and Evolution of Life*, eds P.J. Thomas, C.F. Chyba, C.P. McKay, Springer.

Chyba, C.F., Thomas, P.J., Brookshaw, L., Sagan, C. (1990). Cometary delivery of organic molecules to the early earth. *Science*, **249**, 249–373.

Combes F., Q-Rieu N., Wlodarczak, G. (1996). Search for interstellar glycine. *Astronomy and Astrophysics,* **308**, 618–622.

Commeyras, A., Boiteau, L., Trambouze-Vandenabeele, O., Selsis, F. (2004). *Lectures in Astrobiology – Vol. I*, Eds. M. Gargaud, B. Barbier, H. Martin, J. Reisse, Springer Verlag, [this book]

Cooper et al., (2001). Carbonaceous meteorites as a source of sugar-related organic compounds for the early Earth, *Nature*, **414**, 879–883.
Coradini, A., Capaccioni, F., Capria, M. T., de Sanctis, M. C., Espinasse, S., Orosei, R., Salomone, M., Federico, C. (1997). Transition Elements between Comets and Asteroids, *Icarus*, **129**, 337.
Cottin, H. (1999). Chimie organique de l'environnement cométaire: étude expérimentale de la contribution de la composante organique réfractaire à la phase gazeuse. Thèse, Univ. Paris 12.
Cottin, H., Gazeau, M.C., Raulin, F. (1999). Cometary organic chemistry: a review from observations, numerical and experimental simulations. *Planet. Space Sci.*, **47**, 1141–1162.
Cottin, H., Gazeau, M. C., Bénilan, Y., Raulin, F. (2001). Polyoxymethylene as parent molecule for the formaldehyde extended source in comet Halley. *The Astrophysical Journal*, **556**, 417–420.
Cottin, H., Moore, M.H., Bénilan, Y. (2003). Photodestruction of relevant interstellar molecules in ice mixtures. *The Astrophysical Journal*, **590**, 874–881.
Cottin, H., Szopa, C., Moore, M.H. (2001). Production of hexamethylenetetramine in photolyzed and irradiated interstellar cometary ice analogs. *Astrophys. J. Lett.*, **561**, L139–L142.
Cottin, H., Bachir, S., Raulin, F., Gazeau, M.C. (2002). Photodegradation of Hexamethylenetetramine by VUV and its relevance for CN and HCN extended sources in comets. *Adv. Space Res.*, **30**, 1481–1488.
Cottin, H., Bénilan, Y., Gazeau, M.-C., Raulin, F. (2004). Origin of cometary extended sources from degradation of refractory organics on grains: polyoxymethylene as formaldehyde parent molecule. *Icarus*, **167**, 397–416.
Cremonese, G., Boehnhardt, H., Crovisier, J., Rauer, H., Fitzsimmons, A., Fulle, M., Licandro, J., Pollacco, D., Tozzi, G.P., West, R.M. (1997). Neutral Sodium from Comet Hale–Bopp: A Third Type of Tail. *Astrophysical Journal Letters*, **490**, L199.
Crifo, J.F. (1988). Cometary dust sizing: comparison between optical and in-situ sampling techniques. *Particle & Particle Systems Characterization*, **5**, 38–46.
Crifo, J.F. (1991). Hydrodynamic models of the collisional coma. In *Comets in the Post-Halley Era* (eds. R. L. Newburn, M. Neugebauer, J. Rahe), pp. 937–989. Kluwer.
Crifo, J.F. (1995). A general physicochemical model of the inner coma of active comets. I. Implications of spatially distributed gas and dust production. *The Astrophysical Journal*, **445**, 470–488.
Crifo, J.F., Rodionov, A.V. (1997). The Dependence of the Circumnuclear Coma Structure on the Properties of the Nucleus. I. Comparison between a homogeneous and an Inhomogeneous Spherical Nucleus, with Application to P/Wirtanen. *Icarus*, **127**, 319.
Cronin, J., Chang, S. (1993). Organic matter in meteorites: Molecular and isotopic analysis. In *The Chemistry of Life's origin*, Greenberg et al., eds., Kluwer, 209–258.
Crovisier, J. (1999) Infrared Observations Of Volatile Molecules In Comet Hale–Bopp. *Earth, Moon, and Planets*, **79**, 125–143.
Crovisier, J. (2004, in press) The molecular complexity of comets. *Astrobiology: Future Perspectives, ISSI,* Kluwer.
Crovisier, J., Bockelée-Morvan, D., Colom, P., Despois, D., Lis, D.C. et al., (submitted 2004a). The composition of ices in comet C/1995 O1 (Hale–Bopp) from radio spectroscopy: Further results and upper limits on undetected species. *Astron. Astrophys.*

Crovisier, J., Bockelée-Morvan, D., Biver, N., Colom, P., Despois, D., Lis, D.C. (submitted 2004b). Ethylene glycol in comet C/1995 O1. *Astron. Astrophys.*

Crovisier, J. Schloerb, F.P. (1991) The Study of Comets at Radio Wavelengths. In *Comets in the Post-Halley Era*, Newburn R.L. Jr., Neugebauer M. Rahe J. eds., Kluwer, pp 149–174.

Crovisier, J., Encrenaz, T. (2000). *Comet Science: The Study of Remnants From the Birth of the Solar System*, Cambridge University Press

Crovisier, J., Leech, K., Bockelée-Morvan, D., Brooke, T.Y., Hanner, M. S., Altieri B., Keller H. U., Lellouch E. (1997). The Spectrum of Comet Hale–Bopp (C/1995 O1) Observed with the Infrared Space Observatory at 2.9 Astronomical Units from the Sun. *Science*, **275**, 1904–1907.

Dauphas, N., Robert, F., Marty, B. (2000). The Late Asteroidal and Cometary Bombardment of Earth as Recorded in Water Deuterium to Protium Ratio, *Icarus*, **148**, 508–512.

Dauphas, N., Marty, B. (2002). Inference on the nature and the mass of Earth's late veneer from noble metals and gases, *Journal of Geophysical Research (Planets)*, **107**, E12, 5129

De Bergh, C. (2004, in press). Kuiper Belt: water and organics. *Astrobiology: Future Perspectives, ISSI,* Kluwer.

Delsemme, A. H. (2000). 1999 Kuiper Prize Lecture Cometary Origin of the Biosphere. *Icarus*, **146**, 313–325.

De Sanctis, M.C., Capria, M.T., Coradini, A. (2001). Thermal Evolution and Differentiation of Edgeworth-Kuiper Belt Objects, *Astronomical Journal*, **201**, 2792.

Despois, D. (1999). Radio line observations of molecular and isotopic species in comet C/1995 O1 (Hale–Bopp). *Earth Moon and Planets*, **79**, 103–124.

Despois, D., Ricaud, P., Lautié, N., Schneider, N., Jacq T., Biver, N., Lis, D.C., Chamberlin, R.A., Phillips, T.G., Miller, M., Jenniskens, P. (2000). Search for Extraterrestrial Origin of Atomic Trace Molecules – Radio Sub-mm Observations during the Leonids. *Earth, Moon and Planets*, **82–83**, 129–140.

Despois, D. (1992). Solar System – Interstellar Medium a Chemical Memory of the Origins(rp), IAU Symp. 150: *Astrochemistry of Cosmic Phenomena*, 451.

Despois, D., Crovisier, J., Bockelée-Morvan, D., Biver, N. (2002). Comets and prebiotic chemistry: the volatile component, ESA SP-518: *Exo-Astrobiology*, 123.

Despois, D., Crovisier, J., Bockelee-Morvan, D., Colom, P. (1992). Formation of Comets: Constraints from the Abundance of Hydrogen Sulfide and Other Sulfur Species, IAU Symp. 150: *Astrochemistry of Cosmic Phenomena*, 459.

Duncan, M., Quinn, T., Tremaine, S. (1997). The formation and extent of the solar system comet cloud, *Astronomical Journal*, **94**, 1330.

Eberhardt, P., Krankowsky, D. (1995). The electron temperature in the inner coma of comet P/Halley. *Astron. Astophys.*, **295**, 795.

Edgeworth, K.E. (1943). The evolution of our planetary system, *Journal of the British Astronomical Association*, **53**, 181–186.

Ehrenfreund, P., Charnley, S.B. (2000). Organic molecules in the interstellar medium, comets and meteorites: a voyage from dark clouds to the early earth. *Annual Review of Astronomy & Astrophysics*, **38**, 427–483.

Ehrenfreund, P., Irvine, W., Becker, L., Blank, J., Brucato, J. R., Colangeli, L., Derenne, S., Despois, D., Dutrey, A., Fraaije, H., Lazcano, A., Owen, T., Robert, F.

(2002). Astrophysical and astrochemical insights into the origin of life., *Reports of Progress in Physics*, **65**, 1427.

Encrenaz, T., Knacke, R. (1991). Infrared Spectroscopy of Cometary Parent Molecules. In *Comets in the Post-Halley Era*, Newburn R.L. Jr., Neugebauer M. Rahe J. eds., Kluwer, 149–174.

Enzian, A., Klinger, J., Schwehm, G., Weissman, P. R. (1999). Temperature and Gas Production Distributions on the Surface of a Spherical Model Comet Nucleus in the Orbit of 46P/Wirtanen, *Icarus*, **138**, 74.

Fassett, J.E., Lucas, S.G., Zielinski, R.A., Budahn, J.R. (2001). Compelling New Evidence for Paleocene Dinosaurs in the Ojo Alamo Sandstone San Juan Basin, New Mexico and Colorado, USA, *Proceedings of the 32nd Symposium on Celestial Mechanics*, 3139.

Fegley, B. Jr. (1993). Chemistry of the solar nebula, in *The Chemistry of Life's Origin*, eds J.M. Greenberg, C.X. Mendoza-Gomez, V. Pirronello, NATO ASI Series C, Kluwer, **416,** 75–147.

Fernandez, J.A. (1980). On the existence of a comet belt beyond Neptune, *Monthly Notices of the Royal Astronomical Society*, **192**, 481.

Fernandez, J.A. (1997). The Formation of the Oort Cloud and the Primitive Galactic Environment, *Icarus*, **129**, 106.

Festou, M.C., Rickman, H.West R. M. (1993a). Comets. 1: Concepts and observations. *Astronomy and Astrophysics Review*, **4**, 363–447.

Festou M.C., Rickman H.West R. M. (1993b). Comets. 2: Models, evolution, origin and outlook *Astronomy and Astrophysics Review*, **5**, 37–163.

Festou, M., Keller, U., Weaver, H. (2004, in press) eds, Comets II, Univ. of Arizona Press.

Fomenkova, M.N. (1999). On the Organic Refractory Component of Cometary Dust. *Space Science Reviews*, **90**, 109–114.

Frankel, Ch. (1999). *The End of the Dinosaurs: Chicxulub Crater and Mass Extinctions* Cambridge Univ. Press.

Fray, N., Bénilan, Y., Cottin, H., Gazeau, M.-C., Crovisier, J. (submitted). The origin of the CN radical in comets: A review from observations and models. *Planetary and Space Science.*

Gargaud, M., Despois, D., Parisot, J.-P. (2001). *L'environnement de la Terre Primitive*, Presses Universitaires de Bordeaux.

Gargaud, M., Despois, D., Parisot, J.-P., Reisse, J. (2003). *Les Traces du Vivant*, Presses Universitaires de Bordeaux.

Geiss, J. (1988). Composition in Halley's comet, *Reviews in Modern Astron.*, **1**, 1–27.

Geiss, J., Altwegg, K., Balsiger, H., Graf, S. (1999). Rare Atoms, Molecules and Radicals in the Coma of P/Halley. *Space Science Reviews*, **90**, 253–268. Gerakines, P. A. M. H. Moore (2001). Carbon suboxide in astrophysical ice analogs. *Icarus*, **154**, 372–380.

Gerakines, P.A., Moore, M.H., Hudson, R.L. (2000). Carbonic acid production in H_2O:CO_2 ices. UV photolysis vs. proton bombardment. *Astron. Astophys.*, **357**, 793–800.

Gibb, E.L., Mumma, M.J., dello Russo, N., Disanti, M.A., Magee-Sauer, K., (2003) Methane in Oort cloud comets, *Icarus*, **165**, 391.

Green, D.W.E. (1999). What is improper about the term "Kuiper belt"? (or, Why name a thing after a man who didn't believe its existence?) *International Comet Quarterly*, pp 44–46, January, 1999 and http://cfa-www.harvard.edu/icq/ICQpluto2.html.

Greenberg, J.M. (1982). What are comets made of ? A model based on interstellar dust. In *Comets* (ed. L. L. Wilkening), pp. 131–163. University of Arizona Press.

Greenberg, J.M., Li, A. (1997). Silicate core-organic refractory mantle particles as interstellar dust and as aggregated in comets and stellar disks. *Adv. Space Res.*, **19**, 981–990.

Greenberg, J.M., Li, A. (1998). From interstellar dust to comets: the extended CO source in comet Halley, *Astron. Astophys.*, **332**, 374–384.

Greenberg, J.M., Li, A., Mendoza-Gómez, C. X., Schutte, W. A., Gerakines, P. A., Groot, M. d. (1995). Approaching the Interstellar Grain Organic Refractory Component. *Astrophys. J.*, **455**, L177–L180.

Greenberg, J.M. (1998). Making a comet nucleus, *Astronomy and Astrophysics*, **330**, 375.

Greenberg, J.M. (1993). Physical and Chemical composition of Comets – From interstellar space to the Earth, in *The Chemistry of Life's Origin*, eds. J.M. Greenberg, C.X. Mendoza-Gomez, V. Pirronello, NATO ASI Series C, Kluwer, **416,** pp195–208.

Greenberg, J. M. C. X. Mendoza-Gomez (1993). Interstellar dust evolution: a reservoir of prebiotic molecules. In *The Chemistry of Life's Origins* (ed. Greenberg), 1–32. Kluwer Academic.

Harker, D.E., Wooden, D.H., Woodward, C.E., Lisse, C.M. (2002). Grain Properties of Comet C/1995 O1 (Hale–Bopp). *Astrophysical Journal*, **580**, 579–597.

Hayward, T.L. Hanner, M.S. (1997). Ground-Based Thermal Infrared Observations of Comet Hale–Bopp (C/1995 O1) During 1996. *Science,* **275**, 1907–1909.

Hersant, F., Gautier, D., Huré, J.-M. (2001). A Two-dimensional Model for the Primordial Nebula Constrained by D/H Measurements in the Solar System: Implications for the Formation of Giant Planets. *Astrophysical Journal*, **554**, 391–407.

Hollis, J.M., Lovas, F.J., Jewell, P.R. (2000). Interstellar Glycolaldehyde: The First Sugar. *Astrophys. Journal Lett.*, **540**, L107–L110.

Hollis, J.M., Pedelty, J.A., Snyder, L.E., Jewell, P.R., Lovas, F.J., Palmer, P., Liu, S.-Y. (2003). A Sensitive Very Large Array Search for Small-Scale Glycine Emission toward OMC-1, *Astrophysical Journal*, **588**, 353.

Hough, J.H., Bailey, J.A., Chrysostomou, A., Gledhill, T.M., Lucas, P. W., Tamura, M. Clark, S., Yates, J., Menard, F. (2001). Circular polarisation in star-forming regions: possible implications for homochirality, *Advances in Space Research*, **27**, 313.

Hudson, R. L., Moore, M. H. (1999). Laboratory Studies of the Formation of Methanol and Other Organic Molecules by Water+Carbon Monoxide Radiolysis: Relevance to Comets, Icy Satellites, and Interstellar Ices. *Icarus*, **140**, 451–461.

Hudson, R. L., Moore, M. H. (2002). The N_3 Radical as a Discriminator between ion irradiated and UV-photolyzed astronomical ices. *Astrophys. J.*, **568**, 1095–1099.

Huebner, W. F. (1987). First polymer in space identified in comet Halley. *Science*, **237**, 628–630.

Huebner, W. F., Boice D. C., Korth A. (1989). Halley's polymeric organic molecules. *Advances in Space Research,* **9**, 29–34.

Huebner, W. F., Benkhoff, J. (1999a). From Coma Abundances to Nucleus Composition, *Space Science Reviews*, **90**, 117.

Huebner, W. F., Benkhoff, J., Capria, M. T., Coradini, A., de Sanctis, M. C., Enzian, A., Orosei, R., Prialnik, D. (1999b). Results from the Comet Nucleus Model Team at the International Space Science Institute, Bern, Switzerland, *Advances in Space Research*, **23**, 1283.

Iro, N., Gautier, D., Hersant, F., Bockelée-Morvan, D., Lunine, J. I. (2003). An interpretation of the nitrogen deficiency in comets, *Icarus*, **161**, 511.

Irvine, W. M., Schloerb F. P., Crovisier J., Fegley B., Mumma M. J. (2000a). Comets: a link between interstellar and nebular chemistry. In *Protostar and Planets IV* (eds. V. Manning, A. Boss, S. Russel), pp. 1159. University of Arizona Press.

Irvine, W. M., Senay M., Lovell A. J., Matthews H. E., McGonagle D., Meier R. (2000b). Detection of Nitrogen Sulfide in Comet Hale–Bopp. *Icarus*, **143**, 412–414.

Irvine, W., Bockelee-Morvan, D., Lis, D. C., Matthews, H. E., Biver, N., Crovisier, J., Davies, J. K., Dent, W. R. F., Gautier, D., Godfrey, P. D., Keene, J., Lowell, A. J., Owen, T. C., Phillips, T. G., Rauer, H., Schloerb, F. P., Senay, M., Young, K (1996). Spectroscopic evidence for interstellar ices in Comet Hyakutake, *Nature*, **382**, 418.

Irvine, W. M., Bergman, P., Lowe, T. B., Matthews, H., McGonagle, D., Nummelin, A., Owen, T. (2003). HCN and HNC in Comets C/2000 Wm1 (Linear) and C/2002 C1 (Ikeya-Zhang), *Origins of Life and Evolution of the Biosphere*, **33**, 609.

Irvine, W. M., Leschine, S. B., Schloerb, F. P. (1980). Thermal History, Chemical Composition and Relationship of Comets to the Origin of Life, *Nature*, **283**, 748.

Jeffers, S. V., Manley, S. P., Bailey, M. E., Asher, D. J. (2001). Near-Earth object velocity distributions and consequences for the Chicxulub impactor, Monthly Notices of the Royal *Astronomical Society*, **327**, 126.

Jenniskens, P. (2001). Meteors as a delivery vehicle for organic matter to the early Earth, in *Meteoroids 2001 Conference*, ESA SP-495, 247–254

Jenniskens, P., Baratta, G. A., Kouchi, A. Groot, M. S. D., Greenberg, J. M., Strazzulla, G. (1993). Carbon dust formation on interstellar grains. *Astronomy and Astrophysics*, **273**, 583–600.

Jessberger, E. K., Kissel, J. (1991). Chemical properties of cometary dust and a note on carbon isotopes, *ASSL* Vol. 167: IAU Colloq. 116: *Comets in the Post-Halley Era*, 1075.

Jewitt, D. C., Matthews, H. E., Owen, T., Meier. R. (1997). Measurements of $^{12}C/^{13}C$, $^{14}N/^{15}N$, and $^{32}S/^{34}S$ Ratios in Comet Hale–Bopp (C/1995 O1). *Science,* **278**, 90–93.

Jewitt, D. (2002). From Kuiper Belt object to cometary nucleus, ESA SP-500: *Asteroids, Comets, and Meteors*, ACM 2002, 11.

Jewitt, D. C., Luu, J. X. (1993). Discovery of the Candidate Kuiper Belt Object 1992 QB1. *Nature*, **362**, 730–732.

Jopek, T. J., Gonczi, R., Froeschle', C., Michel, P., Longo, G., Foschini, L. (2002). A main belt asteroid: the most probable cause of the Tunguska event, Memorie della Societa *Astronomica Italiana*, **73**, 679.

Karmann, C., Beust, H., Klinger, J. (2003). The physico-chemical history of Falling Evaporating Bodies around beta Pictoris: The sublimation of refractory material, *Astronomy and Astrophysics*, **409**, 347.

Karmann, C., Beust, H., Klinger, J. (2001). The physico-chemical history of Falling Evaporating Bodies around beta Pictoris: investigating the presence of volatiles, *Astronomy and Astrophysics*, **372**, 616.

Kasamatsu, T., Kaneko, T., Saito, T., Kobayashi, K. (1997). Formation of organic compounds in simulated interstellar media with hich energy particles. *Bulletin of the Chemical Society of Japan*, **70**, 1021–1026.

Kasting, J. F (1993). Earth's early atmosphere, *Science*, **259**, 920–926.

Kasting, J. Catling, D. (2003). Evolution of a habitable planet, *Ann. Rev. Astron. Astrophys*, **41**, 429–63.

Kissel, J. Krueger, F. R. (1987). The organic component in dust from comet Halley as mesured by the PUMA mass spectrometer on board Vega 1. *Nature*, **326**, 755–760.

Kissel, J., Krueger, F.R. et Roessler, K. (1997). Organic Chemistry in Comets From Remote and In Situ Observations. In *Comets and the Origins and Evolution of Life*, Thomas P.J., Chyba C.P. McKay C.P. eds., Springer, 69–110.

Klavetter, J. J., A'Hearn, M. F. (1994). An extended source for CN jets in Comet P/Halley. *Icarus*, **107**, 322–334.

Kobayashi, K., Kasamatsu, T., Kaneko, T., Koike, J., Oshima, T., Saito, T., Yamamoto, T., Yanagawa, H. (1995). Formation of amino acid precursors in cometary ice environments by cosmic radiation. *Adv. Space Res.*, **16**, (2)21–(2)26.

Krueger, F. R. Kissel, J. (1987). The chemical composition of the dust of comet P/Halley as measured by "Puma" on board Vega-1. *Naturwissenschaften*, **74**, 312–316.

Krueger, F. R., Korth, A., Kissel, J. (1991). The organic matter of comet Halley as inferred by joint gas phase and solid phase analyses. *Space Science Reviews*, **56**, 167–175.

Kuan, Y., Charnley, S. B., Huang, H., Tseng, W., Kisiel, Z. (2003). Interstellar Glycine, *Astrophysical Journal*, **593**, 848.

Kuiper, G.P. (1951). in *Astrophysics: A Topical Symposium*, ed. J. A. Hynek (New York: McGraw-Hill), 400.

Kurtz, S., Cesaroni, R., Churchwell, E., Hofner, P., Walmsley, C. M. (2000). Hot Molecular Cores and the Earliest Phases of High-Mass Star Formation, *Protostars and Planets* **IV**, 299.

Kyte, F.T. (1998). A meteorite from the Cretaceous/Tertiary boundary. *Nature* **396**, 237–239.

Lagrange, A.-M., Backman, D. E., Artymowicz, P. (2000). Planetary Material around Main- Sequence Stars, *Protostars and Planets* **IV**, 639.

Lamy, P., Biesecker, D. (2002), The properties of Sun-grazing coments, 34th COSPAR meeting, Houston, p 3159

Le Loeuff, J., Laurent, Y. (2001). Biodiversity of Late Maastrichtian Dinosaurs, *Proceedings of the 32nd Symposium on Celestial Mechanics*, 3126.

Lecacheux, A., Biver, N., Crovisier, J., Bockelée-Morvan, D., Baron, P., Booth, R. S., Encrenaz, P., Floren, H.-G., Frisk, U., Hjalmarson, A., Kwok, S., Mattila, K., Nordh, L., Olberg, M., Olofsson, A. O. H., Rickman, H., Sandqvist, A., von Scheele, F., Serra, G., Torchinsky, S., Volk, K., Winnberg, A. (2003). Observations of water in comets with Odin, *Astronomy and Astrophysics*, **402**, L55.

Lecavelier Des Etangs, A., Vidal-Madjar, A., Ferlet, R. (1999). Photometric stellar variation due to extra-solar comets, *Astronomy and Astrophysics*, **343**, 916.

Leonard (1930). *Leaflet Astron. Soc. Pacific*, **30**, 21–124

Levasseur-Regourd, A. C., Hadamcik, E., Renard, J. B. (1996). Evidence for two classes of comets from their polarimetric properties at large phase angles. *Astronomy and Astrophysics*, **313**, 327.

Levison, H. F., Dones, L., Duncan, M. J. (2001). The Origin of Halley-Type Comets: Probing the Inner Oort Cloud. *The Astronomical Journal*, **121**, 2253–2267.

Lewis, J. S., Prinn, R. G. (1980). Kinetic inhibition of CO and N_2 reduction in the solar nebula, *Astrophysical Journal*, **238**, 357.

Magee-Sauer, K. (2002). CSHELL Observations of Comet C/2002 C1(Ikeya-Zhang) in the 3.0-Micron Region, DPS Meeting, *BAAS*, **34**, 16.05.

Malfait, K., Waelkens, C., Waters, L. B. F. M., Vandenbussche, B., Huygen, E., de Graauw, M. S. (1998). The spectrum of the young star HD 100546 observed with the Infrared Space Observatory. *Astronomy and Astrophysics*, **332**, L25.

Maurette M. (1998). Carbonaceous Micrometeorites and the Origin of Life, *Origins Life Evolution Biosphere*, **28**, 385–412.

Maurette, M., Duprat, J., Engrand, C., Gounelle, M., Kurat, G., Matrajt, G., Toppani, A. (2000). Accretion of neon, organics, CO_2, nitrogen and water from large interplanetary dust particles on the early Earth. *Planetary and Space Science*, **48**, 1117.

McDonnell, J. A. M., Lamy, P. L., Pankiewicz, G. S. (1991). Physical properties of cometary dust. In *Comets in the Post-Halley Era*, Vol. 2, pp. 1043–1073. Kluwer.

McKinnon, W. B. (2002). On the initial thermal evolution of Kuiper Belt objects, ESA SP-500: *Asteroids, Comets, and Meteors*: ACM 2002, 29.

Meier, R., Owen, T. C., Jewitt, D. C., Matthews, H. E., Senay, M., Biver, N., Bockelee–Morvan, D., Crovisier, J., Gautier, D. (1998). Deuterium in Comet C/1995 O1 (Hale–Bopp): Detection of DCN, *Science*, **279**, 1707.

Miller S.L. (1998). The endogenous synthesis of organic compounds, in *The Molecular Origin of Life: Assembling Pieces of the Puzzle*, A. Brack ed., Cambridge, 59–85.

Mitchell, D. L., Lin, R. P., Carlson, C. W., Korth, A., Rème, H., Mendis, D. A. (1992). The origin of complex organic ions in the coma of comet Halley. *Icarus*, **98**, 125–133.

Montmerle chapter this book.

Moore, M. H., R. Khanna, B. Donn (1991). Studies of Proton Irradiated H_2O+CO_2 and H_2O+CO Ices and Analysis of Synthesized Molecules. *J. Geophys. Res.*, **96**, 17541–17545.

Morbidelli, A. Brown, M.E. (2004, in press). The Kuiper belt and the primordial evolution of the Solar System in *Comets II*, eds. M. Festou, U. Keller, H. Weaver, Univ. of Arizona Press.

Morbidelli, A., Chambers, J., Lunine, J. I., Petit, J. M., Robert, F., Valsecchi, G. B., Cyr, K. E. (2000). Source regions and time scales for the delivery of water to Earth, *Meteoritics and Planetary Science*, **35**, 1309.

Moreels, G., Clairemidi, J., Hermine, P., Brechignac, P., Rousselot, P. (1994). Detection of a polycyclic aromatic molecule in comet P/Halley. *Astronomy and Astrophysics*, **282**, 643–656.

Moreno, R. (1998). Observations millimétriques et submillimétriques des planètes géantes. Etude de Jupiter après la chute de la comète SL9. PhD Thesis, Université de Paris VI.

Mousis, O., Gautier, D., Bockelée-Morvan, D., Robert, F., Dubrulle, B., Drouart, A. (2000). Constraints on the Formation of Comets from D/H Ratios Measured in H_2O and HCN, *Icarus*,, **148**, 513.

Mumma, M. J., Weissman, P. R., Stern, S. A. (1993). Comets and the Origin of the solar system: reading the Rosetta stone. In Protostars and Planets III, eds. E.H. Levy J. I. Lunine, Univ. of Arizona Press, pp 1177–1252.

Mumma, M. J., DiSanti, M. A., Dello Russo N., Magee-Sauer, K., Gibb, E., Novak, R. (2002), The organic volatile composition of Oort cloud comets: evidence for chemical diversity in the Giant-Planets nebular region., in *Asteroids Comets Meteors 2002*, ESA SP-500, 753.

Muñoz Caro, G. M. (2003). From photoprocessing of interstellar ice to amino acids and other organics. PhD Thesis, Leiden Observatory.

Muñoz Caro, G. M., Meierhenrich, U. J., Schutte, W. A., Barbier, B., Arcones Segovia, A., Rosenbauer, H., Thiemann, W. H.-P., Brack, A., Greenberg, J. M. (2002). Amino acids from ultraviolet irradiation of interstellar ice analogues. *Nature*, **416**, 403–406.

Navarro-Gonzalez, R. A. Romero (1996). On the Survivability of an Enantiomeric Excess of Amino Acids in Comet Nuclei During the Decay of 26 A1 and Other radionuclides. *Astrophys. Space Sci.*, **236**, 49–60.

Navarro-Gonzalez, R., Ponnamperuma, C., Khanna, R. K. (1992). Computational study of radiation chemical processing in comet nuclei. *Orig. Life*, **21**, 359–374.

Newburn, R. L.,Jr., Neugebauer, M. Rahe J. eds (1991). *Comets in the Post-Halley Era*, Kluwer

Noll, K.S., Weaver, H.A., Feldman, P.D. eds. (1996). *The Collision of Comet Shoemaker-Levy 9 and Jupiter*, Cambridge University Press.

Notesco, G., Bar-Nun, A. (1996). Enrichment of CO over N_2 by their Trapping in Amorphous Ice and Implications to Comet P/Halley. *Icarus*, **122**, 118–121.

Notesco, G., Bar-Nun, A. (1997). Trapping of Methanol, Hydrogen Cyanide, and n-Hexane in Water Ice, above Its Transformation Temperature to the Crystalline Form. *Icarus*, **126**, 336–341.

Notesco, G., Laufer, D., Bar-Nun, A. (1997). The Source of the High C_2H_6/CH_4 Ratio in Comet Hyakutake. *Icarus*, **125**, 471–473.

Nuth, J. A., III, Rietmeijer, F. J. M., Hill, H. G. M. (2002). Condensation processes in astrophysical environments: The composition and structure of cometary grains. *Meteoritics and Planetary Science*, **37**, 1579–1590.

Oort, J. H. (1950). The structure of the cloud of comets surrounding the Solar System and a hypothesis concerning its origin, *Bulletin of the Astronomical Institute of the Netherlands*, **11**, 91.

Oró, J. (1961). Comets and the formation of biochemical compounds on the primitive Earth. *Nature,* **190**, 389–390.

Oró, J. (2001). Cometary molecules and Life's origin, in *First Steps in the Origin of Life in the Universe*, Chela-Flores et al., eds., Kluwer, 113–120.

Oró, J. C. B. Cosmovici (1997). Comets and Life on the primitive Earth. In *Astronomical and Biochemical Origins and the Search for Life in the Universe* (eds. C. B. Cosmovici, S. Bowyer, D. Werthimer), 97–120.

Pendleton, Y. J., Sandford, S. A., Allamandola, L. J., Tielens, A. G. G., Sellgren, K. (1994). Near-Infrared Absorption spectroscopy of interstellar hydrocarbon grains. *Astrophys. J.*, **437**, 683–696.

Petit, J.-M., Morbidelli, A. (2004). *Lectures in Astrobiology – Vol. I*, eds. M. Gargaud, B. Barbier, H. Martin, J. Reisse, Springer Verlag, [this book].

Podolak, M., Prialnik, D. (1997). ^{26}*Al and Liquid Water Environment in Comets in Comets and the Origin and Evolution of Life*, Springer.

Prialnik, D. (2002). Modeling the Comet Nucleus Interior, *Earth Moon and Planets*, **89**, 27.

Prialnik, D. (1999). Modelling Gas and Dust Release from Comet Hale–Bopp, *Earth Moon and Planets,*, **77**, 223.

Prialnik, D., Podolak, M. (1999). Changes in the Structure of Comet Nuclei Due to Radioactive Heating, *Space Science Reviews*, **90**, 169.

Prinn, R. G., Fegley, B. (1981). Kinetic inhibition of CO and N_2 reduction in circum-planetary nebulae - Implications for satellite composition, *Astrophysical Journal*, **249**, 308.

Prinn, R. G. Fegley, B. Jr. (1989). Solar Nebula Chemistry: Origin of planetary, satellite and cometary volatiles. In *Origin and Evolution of Planetary and Satellite Atmospheres*, eds S.K. Atreya, J. B. Pollack, M. Matthews (1989) Univ. Of Arizona Press, pp 78–136.

Radicati-Di-Brozolo, F., Bunch, T. E., Chang, S. (1986). Laser microprobe study of carbon in interplanetary dust particles. *Orig. Life*, **16**, 236–237.

Rairden, R. L., Jenniskens, P., Laux, C. O. (2000). Search for Organic Matter in Leonid Meteoroids, *Earth Moon and Planets*, **82**, 71.

Robert, F. (2002). Water and organic matter D/H ratios in the solar system: a record of an early irradiation of the nebula? *Planetary and Space Science*, **50**, 1227–1234.

Rocchia, R. et Robin E. (2000) L'origine extraterrestre de la crise Crétacé-Tertiaire. Pour la Science (juillet 2000) La Valse des Espèces. Dossier Hors-Série, Belin, pp 100–111.

Rodgers, S. D., Charnley, S. B. (1998). HNC and HCN in Comets. *The Astrophysical Journal Letters*, **501**, L227–L230.

Rodgers, S. D., Charnley, S. B. (2001a). Organic synthesis in the coma of Comet Hale-Bopp? *Monthly Notices of the Royal Astronomical Society*, **320,** L61–64.

Rodgers, S. D., Charnley, S. B. (2001b). On the origin of HNC in Comet Lee, *Monthly Notices of the Royal Astronomical Society*, **323**, 84–92.

Rolfe, E., Battrick, B. eds. (1987). *Proceedings of the International Symposium on the Diversity and Similarity of Comets*, ESA SP-278

Rosenbauer, H., Fuselier, S. A., Ghielmetti, A., Greenberg, J. M., Goesmann, F., Ulamec, S., Israel, G., Livi, S., MacDermott, J. A., Matsuo, T., Pillinger, C. T., Raulin, F., Roll, R., Thiemann, W. (1999). The COSAC experiment on the lander of the ROSETTA mission, *Advances in Space Research*, **23**, 333.

Sankey, J. T. (2001). Late Cretaceous Theropod Dinosaur Diversity: Latitudinal Differences in North America and Implications for the K/T Extinctions, *Proceedings of the 32nd Symposium on Celestial Mechanics*, 3148.

Shimoyama, A., Ogasawara, R. (2002). Dipeptides and Diketopiperazines in the Yamato-791198 and Murchison Carbonaceous Chondrites. *Origins of Life and the Biosphere*, **32**, 165.

Schutte, W. A., L. J. Allamandola, S. A. Sandford (1993a). Formaldehyde and organic molecule production in astrophysical ices at cryogenic temperatures. *Science*, **259**, 1143–1145.

Schutte, W. A., L. J. Allamandola, S. A. Sandford (1993b). An Experimental Study of the Organic Molecules Produced in Cometary and Interstellar Ice Analogs by Thermal Formaldehyde Reactions. *Icarus*, **104**, 118–137.

Shu, F. H., Shang, H., Lee, T. (1996). Toward an Astrophysical Theory of Chondrites, *Science*, **271**, 1545.

Shuvalov, V. V., Artemieva, N. A. (2002). Numerical modeling of Tunguska-like impacts, *Planetary and Space Science*, **50**, 181.

Snyder, L. E. (1997). The Search for Interstellar Glycine. *Origins of Life and Evolution of the Biosphere*, **27**, 115–133.

Steel, D. (1997). Cometary Impacts on the Biosphere. In *Comets and the Origins and Evolution of Life*, Thomas P.J., Chyba C.P. McKay C.P. eds, Springer, 209–242.

Stern, S. Alan (2003). The evolution of comets in the Oort cloud and Kuiper belt. *Nature*, **424**, 639.

Strazzulla, G. (1997). Ion irradiation: its relevance to the evolution of complex organics in the outer solar system. *Adv. Space Res.*, **19**, 1077–1084.

Strazzulla, G., Baratta, G. A. (1991). Laboratory study of the IR spectrum of ion-irradiated frozen benzene. *Astronomy and Astrophysics*, **241**, 310–316.

Strazzulla, G., Baratta, G. A., Johnson, R. E., Donn, B. (1991). Primordial comet mantle: irradiation production of a stable, organic crust. *Icarus*, **91**, 101–104.

Szopa, C., Sternberg, R., Raulin, F., Rosenbauer, H. (2003). What can we expect from the in situ chemical investigation of a cometary nucleus by gas chromatography: First results from laboratory studies. *Planetary and Space Science*, **51**, 863–877.

Thomas, P.J., Chyba, C.P., McKay, C.P. eds (1997). *Comets and the Origins and Evolution of Life*, Springer

Waelkens, C., Malfait K. Waters L.B.F.M. (1999) Comet Hale–Bopp, Circumstellar Dust, and the Interstellar Medium. *Earth, Moon and Planets*, **79**, 265–274.

Wakamutsu, H., Y. Yamada, T. Saito, I. Kumashiro, T. Takenishi (1966). Synthesis of Adenine by Oligomerization of Hydrogen Cyanide. *Journal of Organic Chemistry*, **31**, 2035–2036.

Walker, J. F. (1964). *Formaldehyde.* Reinhold.

Weaver, H. A., Lamy, P. L. (1999). Estimating the Size of Hale–Bopp's Nucleus, *Earth Moon and Planets*, **79**, 17.

Weidenschilling, S. J. (1997). The Origin of Comets in the Solar Nebula: A Unified Model, *Icarus*, **127**, 290.

Weiler, M., Rauer, H., Helbert, J. (2004). Optical observations of Comet 67P/ Churyumov-Gerasimenko (2004) *Astronomy and Astrophysics*, **414**, 749.

Whipple, F. L. (1950). A comet model. I. The acceleration of Comet Encke. *Astrophys. J.*, **111**, 375–394.

Whittet, D.C.B. ed., (1997). *Planetary and Interstellar Processes Relevant to the Origin of Life*, Kluwer.

Whittet, D. C. B. (1997). Is Extraterrestrial Organic Matter Relevant to the Origin of Life on Earth?, *Origins of Life and Evolution of the Biosphere*, **27**, 249.

Wolman, Y., S. L. Miller, J. Ibanez, J. Oro (1971). *Science*, **174**, 1039.

Wooden, D. H., Butner, H. M., Harker, D. E., Woodward, C. E. (2000). Mg-Rich Silicate Crystals in Comet Hale–Bopp: ISM Relics or Solar Nebula Condensates? *Icarus*, **143**, 126–137.

Wootten, H. A. (2003). http://www.cv.nrao.edu/awootten/allmols.html.

Yamamoto, T. (1991). Chemical theories on the origin of comets, ASSL Vol. 167: IAU Colloq. 116: *Comets in the Post-Halley Era*, 361.

Yamamoto, T. (1985). Formation environment of cometary nuclei in the primordial solar nebula, *Astronomy and Astrophysics*, **142**, 31–36.

Yamamoto, T., Nakagawa, N., Fukui, Y. (1983). The chemical composition and thermal history of the ice of a cometary nucleus, *Astronomy and Astrophysics*, **122**, 171.

Zahnle, K. (1998). Origins of Atmospheres, *ASP Conf. Ser. 148: Origins*, 364–391.

9 Comparative Planetology, Mars and Exobiology

Jean-Pierre Bibring

In a few decades of space exploration, our vision of the Earth and the planets has drastically evolved. What emerges is the extraordinary diversity of the Solar System objects (Table 9.1, and Plates 9.1 to 9.4), contrasting with their large original commonality: they formed at about the same date (some 4.55 ± 0.05 Gyr ago), from about the same protostellar cloud material. One of the major goals of the contemporary planetology is to account for the specificities of the planets, amongst which the fact that the Earth has hosted and preserved up to now the conditions enabling life to emerge and evolve (see for example Encrenaz et al., 2004; Weissman et al. 1999; O'Leary et al. 1999; Morrisson and Owen, 1996).

To decipher the clues of the observed present diversity, the approach is similar to the study of any dynamical evolution: one has both to search for the initial conditions of the system, and to inventory the processes operating since, under the various forces acting. Planetology is thus divided into two major classes, one dealing with the study of the "primitive" Solar System, the other with the evolution of differentiated objects, through comparative planetology.

Table 9.1. Properties of inner Solar System objects

	Mass relative to Earth	Mean density	Atmospheric pressure	Major atmospheric constituent	Mean helio-centric dist-ance (AU)
Moon	0.012	3.34	Not Applicable	Not Applicable	1
Mercury	0.056	5.43	Not Applicable	Not Applicable	0.387
Mars	0.107	3.93	600–1000 Pa	CO_2 (95.3%) N_2 (2.7%)	1.524
Venus	0.815	5.20	9×10^6 Pa	CO_2 (96.4%) N_2 (3.4%)	0.723
Earth	1	5.52	1.013×10^5 Pa	N_2 (78.9%) O_2 (20.1%)	1

1 AU = average Earth–Sun distance = 1.496×10^{11} m
Earth mass: $M_\oplus = 6 \times 10^{24}$ kg

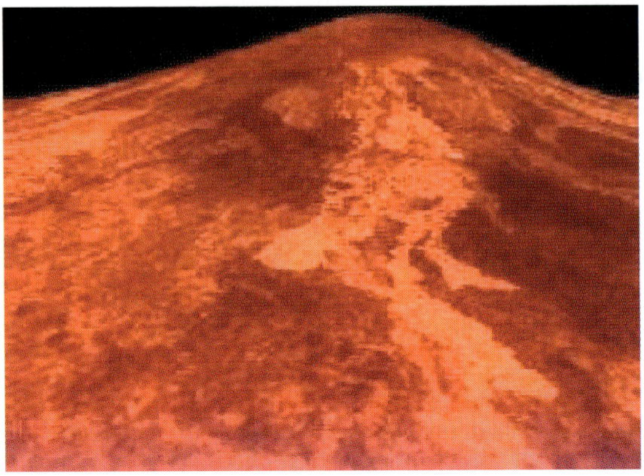

Plate 9.1. Venus lava flow in Sif Mons (from Magellan/NASA)

Plate 9.2. The blue planet (Meteosat/ESA)

9.1 The Study of the Initial Conditions of the Solar System Evolution

The processes involved in the formation of the Solar System can be addressed in several ways:

- by the numerical simulation of a collapsing interstellar cloud,
- by the observation of star formation occurring presently within the Galaxy,
- by analysing Solar System bodies and/or material having preserved, at least partially, the properties acquired at the time of their formation ("fossils"): their study has the potential to reveal both the dynamics and the composition of the early Solar System. It happens that such objects still exist, within the

families of "small bodies" (comets and asteroids, ring particles and satellites) (see Sect. 9.2). These objects, with initial sizes of a few kilometres or less, have a surface/volume ratio large enough to efficiently radiate the internal energy dissipated, so as to level out the temperature increase, thus preventing a global reprocessing of the constituent grains and molecules.

9.1.1 The Origin of Small Bodies

This existence of the various families of small objects is a result of the rapid accretion of the giant planets within the rotating protosolar nebula. These massive objects, and Jupiter particularly, induced major gravitational perturbations in both the inner and the outer Solar System, which led to the survival of a huge number of small objects.

In the inner part of the nebula, between the present orbits of Mars and Jupiter, the completion of the accretion of a planet aborted due to the jovian tidal forces; the result is the "asteroid belt", made of billions of objects of all sizes, from microscopic grains up to a few hundreds of kilometre large bodies (see Gehrels, 1979; Binzel et al., 1989; Bottke et al., 2002). The dynamical evolution of these objects is a fossil of the collisional and resonant evolution of the protoplanetary disk, thus still allowing its direct study. Through collisions and impacts, debris are ejected throughout the Solar System. Those intersecting the Earth constitute the dominant source of meteorites, which in turn are unique extraterrestrial materials sampling their asteroid parent bodies. Actually, the

Plate 9.3. The red planet, as seen from Earth orbit

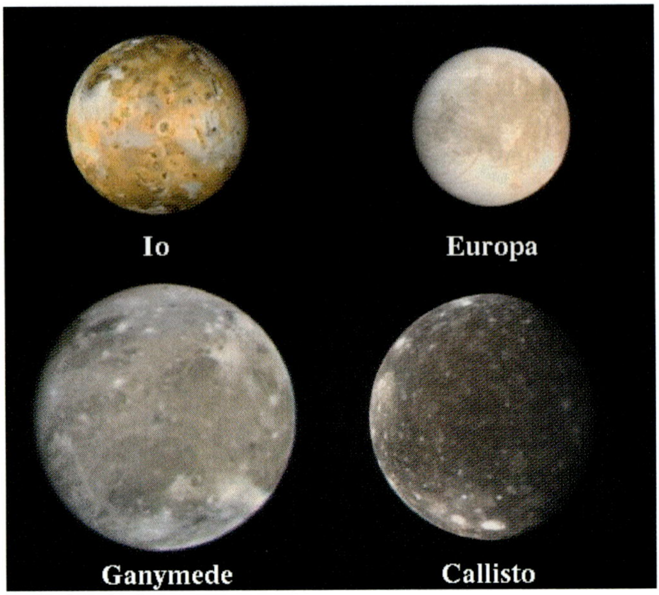

Plate 9.4. The four Galilean satellites of Jupiter (viewed at scale from images acquired by Voyager/NASA)

few thousands of meteorites so far collected on the Earth exhibit a wide variety of properties allowing them to be classified in a few families, matching the diversity of the asteroids from which they originate. Some meteorites, known as carbonaceous chondrites, still contain high abundances of volatile species, reflecting their origin from the least evolved (most primitive) objects: their study opens the possibility to directly probe the pristine material out of which inner planets formed (Norton, 2002).

In the outer part of the Solar System, the temperature was low enough to allow most molecular species, such as H_2O and CO_2, to be present as ices (only H_2 and He remain gaseous down to $< 20\,K$). These icy grains were by far more numerous than the more refractory mineral ones, so that their collisional accretion led to icy-rich objects. The rapid growth of Jupiter blocked the evolution of a number of these bodies when they reached typical sizes in the kilometre range. Moreover, this gravitational perturbation ejected a number of them at large heliocentric distances, populating vast reservoirs in the far outer Solar System, where no further thermal evolution of these bodies occurred, freezing their properties at a very primordial state. Through the subsequent gravitational perturbation by a distant planet or even a close-by star, some of these objects are reinjected into the inner Solar System, becoming comets, developing by sublimation close to the Sun, with coma and tails surrounding their nucleus (see Chap. 8, Part I by D. Despois and H. Cottin). This sublimation process releases the refractory grains, which become interplanetary, and eventually intersect the Earth. Depending on their surface/volume ratio, the smaller ones (with sizes below a few hundreds of micrometres) slow down in the atmosphere without being destroyed, reaching the stratosphere and the surface, where they can be collected as stratospheric IDPs (interplanetary dust particles) and micrometeorites, respectively. In both cases, they constitute unique extraterrestrial materials to study the composition of the protosolar nebula. The larger ones are destroyed in the atmosphere as shooting stars.

9.1.2 The Impact Rate

The impact rate of micrometeorites and meteorites on the Earth has been measured in space and on the Moon, for a very large range of object sizes. The integrated number $N(m)$ of impacts by objects with mass higher than m can be described with three power laws $N(m) = A\,m^{-\gamma}$, with values (A, γ) depending on the mass range, as shown in Table 9.2, where the values are given for the impact rate per year, on the entire Earth surface.

Micrometeorites, which essentially originate from comets, contribute dominantly to the present mass influx on the Earth, mainly through particles in the $< 10^{-6}$ g range (radius $< 100\,\mu$m typically), with a few 10^4 tons per year. Some tens of meteorites ~ 1 kg in mass fall each year on an area the size of France, while objects of 65 000 kg, similar to the one responsible for the Meteor Crater in Arizona, impact the Earth continents once every 25 000 years on average. For objects with diameters of half a kilometre, the impact rate is in the range of one per 10^8 years. Considering that the crater radius is typically 10 times that of the impacting object, the amount of material (gas and grains) ejected in the atmosphere by such an impact is large enough to efficiently filter the solar light for several months or years, so as to induce a large surface temperature drop with potentially huge climatic, and thus biological effects. Such impacts could have been responsible for some catastrophic steps in the evolution of life on Earth, as for the K/T event, some 65 millions years ago, in which dinosaurs and other giant reptiles disappeared (Chapman and Morrison, 1989; Gehrels 1994).

The impact rate discussed above can be considered stationary over billions of years, since most of the meteorites and large objects impacting the Earth, the planets and satellites come from the collisional evolution of the asteroids and other small bodies, are not likely to vary significantly with time. In contrast, this rate of impact is much smaller than that of the tremendous bombardment

Table 9.2. Parameters of the integrated number density of impacts on the Earth

Mass (in g)	$m < 10^{-7}$ g	10^{-7} g $< m < 1$ g	$m > 1$ g
A	8×10^{12}	4.5×10^7	4.5×10^7
γ	0.46	1.213	1.06

that affected all planetary surfaces over the first hundreds of millions of years following their accretion (see Sect. 9.3).

9.1.3 The Organic Content of Comets

The Comet Halley flyby missions in March 1986 revealed totally unpredicted properties of cometary nuclei in particular with respect to their carbon-rich components (Newburn et al., 1991; Altwegg et al., 1999). Actually, by being dominantly constituted of H_2O and CO_2 ices, the nucleus was expected to be fairly bright. On the contrary, it was measured with the lowest albedo ($< 5\%$) any Solar System object ever exhibited (Keller et al., 1987). The temperature of this very dark crust was estimated by its IR emission as $\sim$ 350K, much higher than the H_2O sublimation one. The IR spectral analysis also showed, in addition to the expected features dominated by those of H_2O and CO_2, a spectral structure between 3.3 and 3.4 micrometres, similar to one of the "unidentified" features observed in a number of interstellar objects, indicating the presence of complex C–H-rich species (Combes et al., 1988). In addition, the mass analysis of micrometre-sized grains present up to large distances from the nucleus showed that half of them were only constituted of C, H, O and N atoms, thus their description as "CHON grains", with no metal or refractory elements, such as Si, Al, Mg, Ca or Fe (Kissel et al., 1988).

All these analyses merge in assessing that within the cometary nucleus, a large proportion of the carbon is indeed in the form of large, refractory organic chains, the exact composition of which remaining till now essentially unknown. One of the possibilities is that the synthesis of these compounds happened during the collapse of the protosolar molecular cloud, when it reached fairly large densities ($> 10^{10}\,\mathrm{cm}^{-3}$). Comets might have sampled this very primitive material, trapped within their H_2O-rich icy matrix, and preserved their composition since. We still do not know what is the level of complexity reached by the molecular chemistry during this very early phase of cloud collapse. It might well be that this cloud, and thus the comets, contained organic molecules much more complex than those present on planetary surfaces after they formed, given the very hostile environment they faced for hundreds of millions of years (see Sect. 9.3). This is why the impact of cometary material in early oceans might have played a major role in the further evolution of terrestrial chemistry towards living organisms.

Following these pioneering discoveries of the Comet Halley flyby missions, the European Space Agency (ESA) planned a mission to sample a cometary nucleus piece of ice (CNSR, for cometary nucleus sample return) and to return it pristine to the Earth, to be thoroughly analysed in a variety of laboratories. The complexity and cost of this mission made it unaffordable within the ESA constraints. Instead, it has been decided to build a mission where the laboratory would be brought to the nucleus, to analyse in situ its constituent material, amongst a variety of other measurements. This is the Rosetta mission, scheduled to be launched in early 2004 towards the comet Churyumov-Gerasimenko.

Its operation will take place in 2014/2015. Among the expected results is the determination of the composition of the pristine organic molecules within the cometary nucleus, that might have played a key role in exobiology, as well as in terrestrial biology.

9.2 Planetary Energy Sources

Except for the nuclear fusion requiring much higher (stellar) masses to be initiated, all forces (gravitation, electromagnetism and weak interaction) act within planetary bodies, at varying levels and with specific patterns of time variation, along their history. The combination of the resulting effects produces the diversity of observed planetary properties. In the following, we summarize the major contributors to the energy sources.

Over the evolution of the planets, gravitation dissipates energy through collapse, collisions (accretion and destruction) and eventually tides. While contraction is still operating on the massive gaseous envelopes, tidal effects are acting on a number of bodies, synchronizing the orbiting and rotational movements of satellites and, as for Io, dissipating internal energy while initiating massive volcanic outflows. In some cases, the inner parts of planets, as for the Earth, is in part still heated by the accretion energy stored when they were formed.

The major electromagnetic energy source comes from the solar irradiation. The solar luminosity, $4 \times 10^{26}\,\mathrm{W}$, constitutes the dominant energy source for all inner planets. For the Earth, the solar power input amounts to some $1.8 \times 10^{17}\,\mathrm{W}$, which represents more than 1000 times the total internal power produced (see below). However, while the solar energy is key to controlling the surface and atmospheric properties and evolution, it does not contribute to the internal activity, as it is absorbed in the very shallow surface layers: the internal evolution of the planets does not depend in any essential way on their being planets, that is objects subjected to the solar radiation (and gravitational) field.

There are other ways in which the electromagnetic force acts, within planetary bodies: as an example, phase transformation, from liquid to solid (as within planetary cores) and from gas to liquid and solid, releases latent heat.

Radioactivity constitutes an important – and eventually dominant – source of energy for planetary evolution. Consider the transformation of a parent isotope P into a daughter element by alpha and/or beta decay, with a time constant τ defined by the usual radioactive law $N(P) = N_0(P)\mathrm{e}^{-t/\tau}$. A rough evaluation of the power P_u released per unit volume through this reaction is straightforward, if the number of element per unit volume $N_\mathrm{u}(P)$ is known, as well as the energy Q released per reaction: $P_\mathrm{u} = QN_\mathrm{u}(P)/\tau$.

What are the reactions playing a role? At all stages of planetary evolution, the only elements contributing to the energy production are those with radioactive time constants of order of magnitude similar to the age of the system, starting at the time of the nucleosynthesis of the relevant isotopes.

It has been recently discovered that the early Solar System material contained rather large abundances of "short-lived" radioactive species (see, for example, Kerridge and Matthews, 1988). In particular, minerals within some primitive meteorites have been shown to exhibit excesses of ^{26}Mg, proportional to their content in ^{27}Al. This excess has been attributed to the in situ decay of ^{26}Al, for which $\tau \sim 10^6$ years (Lee et al., 1976). With an initial ratio $^{26}\mathrm{Al}/^{27}\mathrm{Al}$ of $\sim 10^{-5}$, deduced from the ^{26}Mg excess, and an energy release per reaction $Q = 1.8$ MeV, one finds that the power released per unit volume in the protoplanetary bodies containing this ^{26}Al was $\sim 10^{-4}\,\mathrm{W\,m^{-3}}$. This would have been sufficient to significantly heat up the interiors of bodies tens to hundreds of km in size, enabling convection to take place. Several such "isotopic anomalies" have been identified in primitive meteorites, indicating the presence, in their protoplanetesimals parent bodies, of radioactive species with time constants of a few millions years, which constituted a major initial energy source for their early differentiation (Zinner, 1998).

Following this early phase, the only potential contributors to the production of radioactive energy are the elements with time constant of the order of the age of the Solar System: the elements with much shorter timescales, such as ^{26}Al, have been totally transformed, and therefore can no longer contribute. Those with much longer time constants are "stable" enough not to liberate energy by being transformed. The only isotopes of interest are thus the so-called long-lived radioactive nuclei: ^{238}U, ^{232}Th and ^{40}K (Table 9.3).

^{238}U and ^{232}Th are transformed into ^{208}Pb and ^{206}Pb respectively, through chains of successive α and β decays, while ^{40}K decays into ^{40}A through a single electron capture:

$$^{238}\mathrm{U} \rightarrow {}^{208}\mathrm{Pb} + 8\alpha + 6(\mathrm{e}^- + \nu)$$
$$^{232}\mathrm{Th} \rightarrow {}^{206}\mathrm{Pb} + 6\alpha + 4(\mathrm{e}^- + \nu)$$
$$^{40}\mathrm{K} + \mathrm{e}^- \rightarrow {}^{40}\mathrm{A} + \nu$$

As an example, the energy Q released by the transformation of ^{238}U into ^{208}Pb is $\sim$ 52 MeV. For a typical inner planet material, given its content in the three long-lived elements, P is of the order of a few $10^{-8}\,\mathrm{W\,m^{-3}}$. For the global Earth, the total power released is thus $\sim 4 \times 10^{13}$ W. This energy is a few thousands times smaller than the energy received from the Sun. However, as stated above, the solar energy does not contribute to the internal activity of

Table 9.3. Dominant radioactive planetary energy sources

Parent isotope	Product	Time constant τ
^{238}U	^{208}Pb	6.4×10^9 yr
^{232}Th	^{206}Pb	2×10^{10} yr
^{40}K	^{40}A	1.8×10^9 yr

the inner planets, being absorbed at its very surface: radioactivity plays the dominant role.

A very simplified model can account qualitatively for the relation between mass and level of activity. Let us consider objects of similar global composition, assumed homogeneous: the concentration of long-lived radioactive nuclei (^{238}U, ^{232}Th, ^{40}K), major contributors to the internal energy source, is then roughly identical. Therefore, each unit of volume produces a similar power. Consequently, the total power produced is proportional to the volume, thus R^3. The temperature profile at equilibrium results from the balance between this energy input, and the losses, essentially by surface radiation, which are thus proportional to R^2: with inputs varying as R^3, and losses as R^2, the thermal profile depends on R. The larger the object, the higher the gain/loss ratio: the internal temperature profile reaches higher values, leading to more intense internal activity. However, due to the intrinsic property of the energy production, which originates from the transformation of species, the energy produced decreases with time, in proportion to the decrease of the number of radioactive elements. Therefore, whatever the level of activity reached, it necessarily decreases with time, down to a level where no surface activity can be sustained. The larger the body, the higher the level of internal activity, and the longer it will last before it declines and reaches geological death. Through their size, one can then follow over time the degree of activity of the various Solar System objects: the "small bodies" can radiate efficiently the energy released in their volume, maintaining an internal temperature low enough to have avoided global metamorphism, and preserved at least partially their initial properties: they constitute "primitive bodies" with still records of the initial conditions of the Solar System formation and early evolution (see Sect. 9.1). The Moon was subjected to a short surface activity phase, dominated by mare formation, and "died" some 1.2 billion years after being formed (see Sect. 9.3). Mars, being more massive, remained active until the past hundreds of millions years: one can, in principle, "read" on its surface all steps having governed the history of inner planets, till their geological death (Sect. 9.4).

9.3 The Lunar Records of Solar System Evolution

Through lunar-orbital remote sensing, in situ investigations and intensive laboratory analyses of the returned lunar samples, the Moon has revealed key processes that took place in the Solar System along its entire history, with a special emphasis on the first billion years.

The lunar surface is dominated by the bright highland terrains, mostly constituted of anorthositic feldspar, saturated with impact craters at all scales. In contrast, dark areas, known as mare, contain pyroxene-rich basalts, with far fewer craters. Radioisotope dating has demonstrated that the highlands were formed more than 4 billion years ago, while the mare material is much younger, with crystallisation ages ranging from 3.9 down to 3.3 billion years. This dual surface composition is a precious remnant and a direct indication of the major processes that controlled the early Solar System history.

The accretion within the protosolar rotating nebula led to a vast number of solid bodies of all sizes, by far more numerous than the fairly small number of planets and satellites that we observe today. It took hundreds of millions of years to clear and clean out the Solar System from most of these objects, essentially through collisions amongst themselves, and/or towards the Sun. The few surviving bodies had their surfaces heavily cratered by this long-lasting "accretion tail". Those objects for which the further surface activity did not globally reprocess the surface have kept this record till now, testifying to the violence of this impact-driven stage. Their study indicates the rate and the level to which all Solar System objects, including those, such as the Earth, on which this record has been erased by a global geological reset, have been bombarded in their early history. The lunar highlands are typical in this respect, allowing a quantitative evaluation of the impact rate as a function of mass in the early Solar System. In particular, one observes that the more massive impacts are fewer, with some craters thousands of kilometres large, or "basins" (Plate 9.5).

While this bombardment was operating, the slow internal heating by the long-lived isotopes radioactivity built a magmatic mantle. It took more than half a billion years for the convection of this mantle to reach the bottom of the deeper, thus larger basins: their filling resulted in the mare, darker than the surrounding terrains by being constituted of Ca- and Fe-rich basaltic silicates. When they crystallised, the early bombardment had ceased, leading to far fewer impacts. The craters observed on the mare surfaces have essentially been accumulated over the past 3.3 billions of years of meteorite bombardment from the asteroid belt, enabling a precise evaluation of its rate (see Sect. 9.1.2). The large

Plate 9.5. Large impact induced lunar basins (Apollo/NASA)

contrast (by a factor larger than 1000) in the crater density between highlands and mare gives a direct indication of the intensity ratio of the bombardment that occurred during the accretion tail and subsequently, up to now, through meteoritic impacts primarily (Plate 9.6).

In addition to giving a qualitative and quantitative view of the impact history within the inner Solar System, the lunar studies have provided a number of other key indications for the formation and evolution of the Solar System. Of interest for potential exobiology implications is the understanding of the origin of the Moon (see for example Wood, 1986), the probability of the processes involved, and the potential effects of the Moon on the evolution of the Earth. The Earth–Moon system is unique in the mass ratio of the satellite relative to that of the planet. One has only recently assessed that the presence of the Moon did induce large gravitational effects on the evolution of the Earth, beyond the various and well-documented ocean tides. The obliquity of a planet, which is critical for the climatic evolution in controlling the solar irradiance, can in principle enter chaotic evolution over rather short timescales (tens of millions of years): such an evolution would then infer major climatic changes. It has been shown (Laskar et al., 1993), that in the case of the Earth, the presence of the Moon has stabilized this angle within a very short range ($\pm$ 1.3 degrees) around its present value (23.4°), maintaining the global terrestrial climatic features over sufficiently long timescales so as to favour steady chemical evolution taking place. This feature, specific to the Earth since no other planet in our Solar System has a satellite

Plate 9.6. Highland/mare contrast in crater density

with a relative mass as high as that of our Moon, could have played a role in favouring biochemical evolution.

9.4 Mars and Contemporary Comparative Planetology

Over time, Mars has always played a unique role in comparative planetology. Appearing red, the colour of blood, it was named Mars after the Latin God of War (Ares, in Greek). Later, this colour being interpreted as indicating rust, the surface of Mars was thought to be rich in ferric ions, oxidised by flooding water, thus hosting life: from a planet of death, Mars evolved towards a planet of life. The further telescopic observations tended to confirm this view. Mars was observed developing seasonal polar caps, the location of which indicating an obliquity similar to the terrestrial one: Mars was then understood as having a climate with seasons, mimicking the Earth ones. Moreover, large coloured (brownish) structures were identified on the surface of Mars, evolving with time at a seasonal rhythm: that was sufficient to support the hypothesis that Mars indeed hosted living organisms, the variation of surface colour reflecting the growth and fading of vegetation, from spring to fall.

The first space images acquired by the Mariner 4 spacecraft, in 1965, then Mariner 6 and 7, up to 1972, suddenly modified the picture, by exhibiting a geologically deserted and dead surface. However, the images revealed a number of features indicating that, in the past, Mars was an active planet (Plate 9.7).

This is where Mars draws its unique role in planetology: Mars is massive enough to have undergone an important internal activity, much higher than that of the Moon which is 9 times less massive. On the other hand, Mars is small enough to have reached its geological death, after having undergone all major steps of planetary evolution, without "global resets" that would have erased the former signatures, as happened for the Earth, which is 9 times more massive. This is why, on the Mars surface, one can still distinguish the remnants of the early cratering, as well as of the intense tectonic and volcanic activity that occurred later, in the form of spectacular structures, such as giant volcanoes, Tharsis Mons, widespread deep canyons (Valles Marineris), and fluvial networks. In summary, primarily as a result of its mass, Mars has the potential to reveal most of the steps of planetary evolution.

In the following, we describe the major global Mars properties, then the surface Martian structures, with a special emphasis on those linked to the Martian evolution with regard to potential exobiology (for general references, see in particular Kieffer et al., 1992; Jakosky, B., 1998; Kallenbach et al., 2001; Horneck and Baumstark-Khan, 2002).

9.4.1 The Global Mars Properties

Mars orbits around the Sun on a plane very close to the ecliptic one (inclination of 1.5°), along an ellipse with $a = 1.524\,\mathrm{AU}$, and a sideral period of 687 terrestrial

Plate 9.7. In the past, Mars was an active planet (Viking/NASA)

days. Its synodic period (time separating two identical configurations of the Sun–Earth–Mars system) is thus 780 terrestrial days, offering favourable "windows" to launch spacecraft to Mars about every 26 months.

The Mars orbit has a rather large eccentricity: $e = 0.0934$. The heliocentric distance thus varies by almost 20% between periapsis ($d_p = a(1-e)$) and apoapsis ($d_a = a(1+e)$), leading to close to 40% variation of the solar energy input. The periapsis is reached a few weeks before the Northern winter solstice (the periapsis occurs at a Solar longitude $L_s \sim 251°$, while the solstice is at 270°), and the apoapsis is reached at $L_s \sim 71°$, before the summer solstice ($L_s \sim 90°$).

Mars rotates around an axial tilt axis of $\sim 25.2°$ in obliquity, close to that of the Earth (23.45°); the obliquity is the angle between the planet equator and its orbital plane. In the case of Mars, the orbital plane does not strictly coincide with that of the Earth (the ecliptic), so that the angle between the Martian orbital plane and the ecliptic is 23.98°. This rotation defines the Martian day; the sidereal day lasts 24.62 h (24 h, 37 min and 22 s), while the mean solar day, or sol, taking into account the orbital period, lasts 24.66 hours (24 h, 39 min and 35 s). Consequently, the Martian year corresponds to 668.6 sols. Given these orbital properties, the 4 seasons of Mars have very different durations: 198, 184, 147, 158 terrestrial days for northern spring, summer, fall and winter, respectively.

These parameters, together with the very low surface atmospheric pressure (< 10 hPa) and a composition largely dominated by CO_2, lead to a very peculiar present Mars meteorology and climatology; a major effect is a seasonal exchange of CO_2 and, to a lesser extent of H_2O, between the atmosphere and polar caps, cycling up to 25% of the atmosphere. Thus, at latitudes varying with time, the surface is covered with frosts, the composition of which being entirely governed by the local thermodynamic equilibrium. At local summer, none of the polar caps completely sublimates: residual high albedo icy units remain, the composition and extent of which differ largely between the North and the South. The North residual cap is much larger, and dominantly constituted of H_2O, while the South one is covered by a bright, although thin, CO_2 veneer, onto a large and thick perennial H_2O ice reservoir, mixed with dust recently identified by the OMEGA instrument on board the Mars Express ESA mission.

The mass of Mars, 6.42×10^{23} kg, represents 0.107 of the mass of the Earth. With a mean radius of 3390 km, its average density is ~ 3.9, much smaller than the terrestrial one (5.52). This probably reflects a significantly lower iron content in Mars, which, together with a higher Fe concentration in its external layers, is likely to imply a much smaller iron-rich core.

The absence of a global magnetic field at present on Mars (the equatorial global magnetic field intensity is < 1 nT, to be compared with the terrestrial value of 30 000 nT) indicates that indeed, a liquid iron core, if it exists, is not capable of sustaining by its convective motion a significant dynamo effect. This was probably different in the past, since localised surface magnetic areas (up to 100 nT in intensity), remnants of a past magnetic activity, have been recently identified by the MGS mission, in the oldest, highly cratered terrains.

9.4.2 The Major Mars Surface Units

Old terrains dominate the Southern hemisphere, of mean altitude 3 to 5 km higher than the younger Northern plains. This relative age scale is indicated by the crater density, as no precise dating has been performed yet, in the absence of Mars return samples. Actually, the cratered areas resemble largely the typical lunar highlands, in having kept the record of the intense bombardment that occurred during the first hundreds of millions of years following the planetary formation (Plate 9.8). The existence of large cratered units on the surface of Mars likely indicates that there still exist Martian terrains having preserved properties acquired during the first billion years. In particular, if one looks for sites with properties dating from the times when life appeared on the Earth, these exist on Mars, in the Southern highlands.

This dichotomy between roughly the two hemispheres is a feature specific to Mars, with the level of its internal activity having induced only partial surface reprocessing. Indeed, early space missions have revealed in the Northern hemisphere a number of structures indicating an intense past activity. First of all, extinct volcanoes have been identified, which are few but very large, with altitude exceeding 20 km, and small mean slope: these "shield" volcanoes, hundreds of km in size, are the largest ever observed in the Solar System. The prominent

Plate 9.8. Mars highly cratered areas are still present in the highlands of the Southern hemisphere predominantly (Viking/NASA)

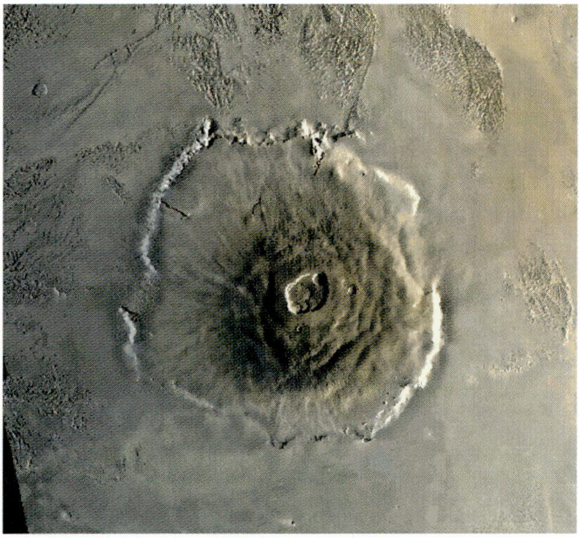

Plate 9.9. Olympus Mons, the largest of the Solar System volcano, 600 km wide (Viking/NASA)

set is located on the Tharsis Plateau, with the highest volcano, Olympus Mons, reaching an altitude of 27 km over the surrounding plains (Plate 9.9). A few impact craters can be seen in its caldera, with a number density indicating its extinction some hundreds of millions of years ago. Lava outflows from these volcanoes have covered a large fraction of the Northern plains, thus appearing much less cratered ("younger") than the Southern terrains.

The Tharsis Plateau and the associated volcanic activity constitute the prominent tectonic features on Mars. Tharsis results from a more than 5000-km wide upthrust, reaching an altitude close to 10 km. This thrust initiated a number of structures: faults, vents, grabens, canyons, etc. The giant Valles Marineris canyon network expands over 3500 km, roughly along the equator. In some areas, it is up to 10 km deep, and tens to hundreds of km wide. Stratification is also observed, originating from wind, volcanic or even possibly fluvial processing. However, no structures indicating plate tectonics have been identified: spreading ridges, subduction zones, or volcanic arcs that would result from continental collisions. The absence of plate tectonics can account for the very high altitude reached by the volcanoes. For the terrestrial intraplaques volcanism, successive outflows from same hotspots created aligned volcanoes along the lithosphere drift; on Mars, the absence of movement led to piling outflows while forming fewer volcanoes, with impressive heights.

These Mars surface structures indicate an intense past tectonic activity. This is expected for a body of that size. More surprisingly, the space remote-sensing imaging has revealed the existence of a number of surface features related to flows, indicators of a unique and intriguing past climatic evolution. These features are in the form of valleys and fluvial networks, outflow channels, gullies, all of them being dry at the present time (Plates 9.10 and 9.11). The most complex networks are located within the oldest heavily cratered southern terrains, which would indicate that they were formed during or at the end of the first billion years. Other structures associated to liquid outflows are present in both hemispheres. Some exhibit a more catastrophic origin, with a variety of blocs and debris in their beds, as if they were carried by torrential discharges (Plate 9.12). All these outflows differ from the terrestrial ones, as both their source and sink

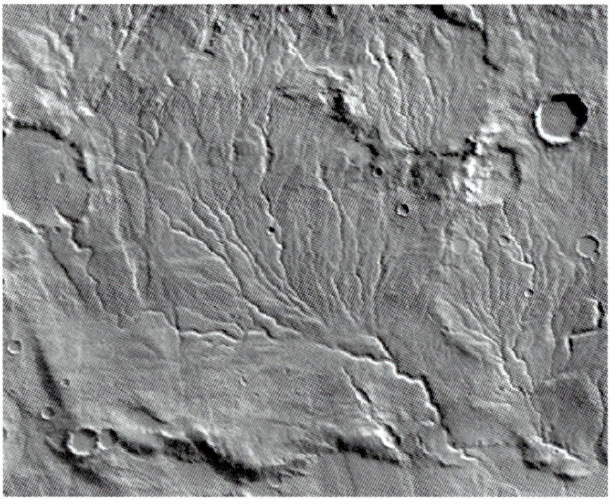

Plate 9.10. Dry valley network (Viking/NASA)

Plate 9.11. Dry valleys networks (Viking/NASA)

Plate 9.12. Outflow channel (Viking/NASA)

are not readily identified. In particular, it is still not obvious if lakes or oceans, that is permanent or at least long-lasted liquid reservoirs, ever existed at the surface of Mars. Actually, the composition of the fluid (water?) has not even been identified definitely, although if any, water is the most abundant and thermodynamically favourable molecule to be present on a planet. The structural analogy with terrestrial fluvial structures is the most convincing evidence that, in the past, water was present in abundance at the surface of Mars: Mars could then have had a climate and constituents favouring a chemical (and possibly biochemical) evolution at a period when, on the Earth, life started.

9.4.3 The Evolution of the Mars Climate

For liquid water to be perennial at the surface of Mars in the past would require the Mars climate to have drastically evolved: today, the atmospheric properties generally preclude H_2O to be stable in the liquid state.

The accurate measurements made so far indicate that CO_2, the major surface atmospheric constituent (95.3%), amounts to 6 to 10 hPa only, depending on the season: during the winter, large fractions of the CO_2 precipitates as ice, thus the seasonal polar caps extending over huge areas, and the associated pressure variations. Violent winds and storms transport sand over large distances, as evidenced by surface structures (dunes) similar to the ones commonly observed in terrestrial deserts (Plate 9.13). The minor atmospheric constituents are N_2 (2.7%) and Ar (1.6%), then CO and H_2O, which amounts to less than 0.1%. In the present thermodynamic conditions, water ice directly sublimates into vapour. Today's equilibrium is essentially between the solid and the gaseous state, through sublimation/condensation depending on the temperature.

One of the hypotheses made to account for the observed fluvial structures assumes that Mars in the past was warm and humid, which would imply that the past Martian atmosphere was dense enough for water to be present and stable as a liquid. In that case, perennial liquid reservoirs could have existed, lasting long enough to have eventually processed most of the atmospheric CO_2 into carbonates piling up on the liquid floors, and have hosted a carbon chemistry evolving in a UV-protected (liquid) environment. If that ever happened, where has this water gone today, with its potential organic species trapped in?

In support of this hypothesis is the supposed thermal evolution of Mars, given the major internal source of energy, which originates from the decay of long-lived species. Mars might have undergone, close to the end of the first billion years following its formation, much higher internal temperatures, leading to a level of activity much higher than the present one. This could have led to an intense surface tectonic and volcanic activity, providing a very efficient recycling of the atmospheric constituents. This might have maintained a high surface pressure, likely dominated by the CO_2 (possibly increasing the surface temperature to over

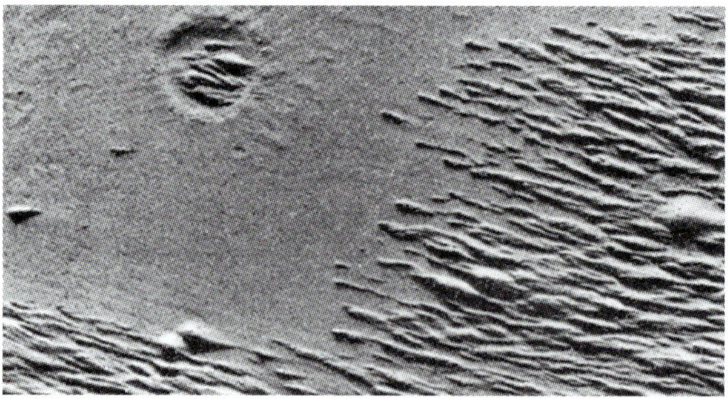

Plate 9.13. Martian desert dunes (Viking/NASA)

273 K by an efficient greenhouse effect), and H_2O abundance above that of its triple point. The fact that the present Mars atmosphere is so tenuous would then directly reflect the slow decrease with time of the radioactive energy sources. As a consequence, the trapping would progressively dominate over the recycling. The major CO_2 reservoir could then be searched in the form of carbonates, and water would reside predominantly as subsurface permafrost.

Such a scenario is based on the rough assumption that Mars, Venus and the Earth were formed with similar global content of CO_2, H_2O and N_2, scaled to the planet size (Tables 9.4 and 9.5). This seems to be verified for Venus and the Earth, at least for the CO_2, if one considers that most of the terrestrial CO_2 is in the form of carbonates, due to the existence of liquid water. A comparative atmospheric inventory for Mars, Venus and the Earth clearly indicates that, in the case of Mars, the atmospheric depletion is huge, by a factor larger than 1000.

One could then propose that Mars underwent an intense phase of internal and surface activity, developing and maintaining a dense atmosphere. This phase could have taken place as soon as the heavy bombardment dropped to a level enabling complex physicochemical processes to develop, at a time when the radioactive input reached its maximum efficiency, that is some hundreds of millions years following the planetary formation. The fluvial structures would result from these processes, preferably observed in the Southern cratered terrains having kept the record of this early period, contemporary to the formation of the first terrestrial living organisms. On Mars, a slow atmospheric decline would have followed, due to the decrease of the internal activity, governing the atmospheric recycling. Gaseous species would have been progressively trapped in solid form, leading to the present depletions.

This scenario still faces the absence of clear evidence of large volumes of permafrost and carbonates. This lack of nonambiguous identification does not

Table 9.4. Partial pressure of main atmospheric species at the surface, in bars. The relative abundance is given in % for some constituents

	Mars	Venus	Earth
CO_2	0.007 (95.3%)	87 (96.4%)	0.03
N_2	0.0004 (2.7%)	3 (3.4%)	0.8
A	0.0002	0.0004	0.009
O_2	0.00001	0.006	0.2

Table 9.5. Ratio of atmospheric mass M_a to planetary mass M_p

	Mars	Venus	Earth
Present M_a/M_p	5×10^{-8}	10^{-4}	10^{-6}
Global M_a/M_p	?	10^{-4}	10^{-4}

imply that there are not: both the sensitivity and the space resolution of the measurements performed so far might have been insufficient. With the OMEGA instrument on board the ESA/Mars Express mission, which should operate in Martian orbit starting in early 2004, followed in 2005 by the CRISM instrument on the NASA/MRO mission, the sensitivity should allow the identification of carbonates down to concentration as low as a few per cent, on scales of a few hundred metres for OMEGA, and a few tens of metres with CRISM.

As for the water, strong indications exist for underground icy-rich reservoirs. The major one comes from the structure of some impact craters: instead of being surrounded by bright radial ejecta rays, as for the Moon, made of finely impact-induced powder (thus highly reflecting through multiple scattering at grain interfaces), many of the Martian craters appear with lobate rims, as if they were build out of solidified mud (Plate 9.14). The impact would have reached a permafrost-rich subsurface material, heated and fluidised prior to being ejected. It is, however, very difficult to evaluate, from these observations, the global amount of water trapped in this form. The terrestrial oceans constitute an equivalent volume several thousands of metres deep over the entire Earth. Has Mars a water-ice reservoir of equivalent magnitude, scaled to the planet size? The MARSIS radar, on board the ESA/Mars Express mission, and then the SHARAD instrument on MRO, should provide key indications.

Subsurface water-ice reservoirs could also account for the mineralogical composition of the clay minerals: one hypothesis, which could be tested and validated

Plate 9.14. Lobate crater indicating permafrost-rich subsurface material, responsible for the mud-like ejecta (Viking/NASA)

by the on-going space missions, is that palagonite dominates. This clay, on the Earth, is in particular found resulting from the hydration that occurs when volcanism ejects magma through subsurface permafrost, as in Iceland.

9.4.4 Where Has the Nitrogen Gone?

Supposing the future missions provide evidence for solid traps of CO_2 and H_2O, a major problem would remain: in the Martian atmosphere, the nitrogen content is extremely low. N_2, the major nitrogen constituent, amounts to 2.7%, corresponding to more than 1000 times less than expected, in comparison with Venus and the Earth, if similar initial reservoirs are assumed. It is by far more complex to precipitate N_2 into minerals, than it is for CO_2. Actually, nitrates on the Earth either constitute salt formed out of nitric acid, or come from biological activity. It is very unlikely that either of these two processes, on Mars, was efficient enough to convert almost all the initial nitrogen, while on the Earth, which harboured life for almost 4 billion years, and on Venus, where acid aerosols are so abundant, most of the nitrogen has remained in the atmosphere. Moreover, the N_2/CO_2 ratios are very similar ($\sim$ 3%) on Mars and on Venus, although the concentrations are some 10^4 times higher on Venus. If indeed this similarity reflects the fact that on Mars, the initial N_2/CO_2 ratio was very close to the present one, that would require that the depletion processes of N_2 and CO_2 were both extremely efficient (99.99%) and with almost strictly the same yield (at a 10^{-4} level): it seems very unlikely that N_2 and CO_2 were removed from the atmosphere through two entirely distinct and/or uncorrelated processes. Some measurements, such as that of the Martian atmospheric $^{15}N/^{14}N$, about 1.5 times higher than the terrestrial value, suggest that Mars lost an important fraction of its atmospheric nitrogen through a mechanism leading to mass fractionation, as would the interaction with the solar wind. Would not then the CO_2 also be subjected to this escape process, making very unlikely the possibility to ever detect carbonates on the surface? It may well be that a "median" scenario has happened: a fraction only of the gaseous constituents would have left the planet, early in the planet history, leaving some to contribute to the chemical evolution of the surface.

In the hypothesis where its atmosphere remained tenuous, Mars would have been maintained cold and dry, instead of having undergone a warm and wet episode. How then to account for the observed flows, that seem to have occurred with violence in many cases? The existence of flows does not imply that they happened as stationary processes, operating at thermodynamic equilibrium. If massive volumes of subsurface water ice existed, while the atmospheric pressure was below that of the triple point, sporadic events driven by turbulent internal processes could have ejected, up to the surface, large amounts of either mud or even liquid water, rushing down slopes in digging valleys, while sublimating partially, then seeping into the plains, and rebuilding a solid icy subsurface layer. Such a cycle would exclude the existence of perennial liquid water surface reservoirs, in which CO_2 would have been dissolved, and in which a complex chemical (and possibly biochemical) activity would have taken place. In this hypothesis, the fluvial and outflow structures would have been generated by transient phenomenon, having led to major local climatic changes.

Finally, our understanding of the Mars paleoclimatic evolution still remains limited, leaving a wide diversity of scenarios still possible. It is the primary goal of the future space-exploration programs to constrain them by accurate global and local measurements. This will be summarized in Sect. 9.4.5.

9.4.5 The Martian Meteorites

In addition to space exploration, our knowledge of Mars comes from laboratory studies of meteorites supposed to originate from this planet (see Lodders, 1998, for a compilation). The vast majority of the few thousands of meteorites so far collected on the surface of the Earth originate from asteroids, the diversity of which is reflected in the variety of meteoritic classes. However, a few tens have properties that seem to indicate that they have been ejected from both the Moon and Mars. The existence of lunar samples to compare with were key to confirm the lunar origin of the presently named lunar meteorites; the Mars origin of the assumed-to-be Martian meteorites is still controversial, since no samples have been collected yet on Mars and brought back to the Earth to definitely validate the interpretation. There seems though to be compelling evidence for a Martian origin of at least some of these extraterrestrial samples.

The small family of meteorites (~ 15) assumed to originate from Mars is named SNC, since their properties are similar to that of three of these meteorites, Shergotty, Naklha and Chassigny, named after the closest location where they were found. Evidence for their Martian origin comes from their mineralogical, elemental and isotopic composition.

These meteorites are made of igneous rocks, containing inclusions that hold gases whose abundance and isotopic composition have been measured matching that of the Mars atmosphere analysed by the Viking missions.

Through the isotopic composition of specific elements, one can measure the date at which a sample crystallised for the last time, therefore dating the impact that ejected it from its parent body. According to their crystallisation ages, the SNC meteorites can be grouped into 3 broad families. The younger family is composed of basaltic rocks dominated by two pyroxene mixtures (with Shergotty ~ 180 Myr old, and others 300 and 420 Myr old); an older one is dominated by Ca-rich clinopyroxenes and olivine, with very low concentrations of Al (with Chassigny and Nakhla, ~ 1.3 Gyr old). One meteorite, collected in the Antarctic in 1984 close to the site of Allen Hills, and therefore named ALH84001, is rich in orthopyroxenes, and much older: ~ 4.5 Gyr. This meteorite was ejected from its parent body some 16 millions years ago, and fell in Antarctica 13 000 years ago.

Similar to the Martian rocks analysed in situ during the Viking and Pathfinder missions, all SNCs are richer in Fe than the terrestrial crust, probably reflecting a higher FeO content of their parent body (Mars mantle?). However, in contrast with the rocks analysed by Pathfinder, the SNCs are all poor in Al; hence, if Martian, they would derive from an Al-depleted source region. The occurrence of carbonates in SNC meteorites might suggest that these minerals were indeed crustal sinks for early CO_2. Alternatively, carbonates and other salts could have formed in restricted areas where the temperature was high enough to maintain liquid water at, or close to the surface (hydrothermal springs).

Globules of carbonates contained in ALH84001 have also been presented as possible byproducts of ancient life on Mars. However, this issue is still debated: similar carbonate structures have been produced abiologically in the lab. In ALH84001, complex organic molecules (including PAH) have been found in connection with the carbonates, as well as magnetite grains. Moreover, at the surface of some of the carbonates, elongated structures, a few hundreds of nm long, have been observed by scanning electron microscopy, and interpreted as fossils of micro-organisms (nanobacteries) of extraterrestrial (Martian) origin (Plate 9.15). These highly controversial results are still a matter of debate, against the likelihood that most, if not all, these features have been formed during the past 13 000 years the meteorite has spent on the Earth, before being collected.

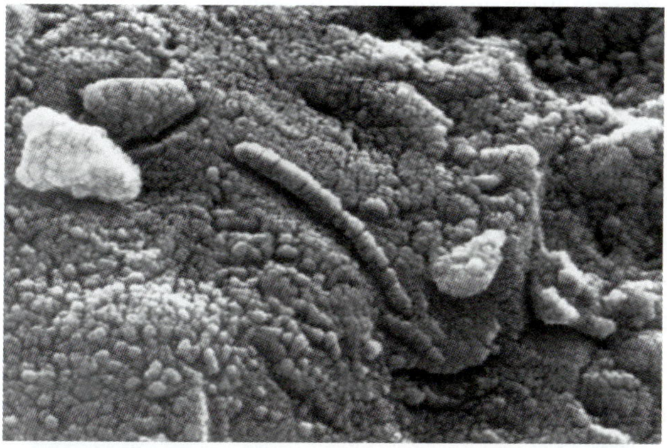

Plate 9.15. **ALH 84001 "nanobacteries"**: This high-resolution scanning electron microscope image shows an unusual tube-like structural form that is less than 1/100th the width of a human hair in size found in meteorite ALH84001 (NASA)

9.4.6 The Present and Future Mars Exploration Programs

Presently, two orbital NASA missions, MGS (Mars Global Surveyor) and Odyssey, are continuing their scientific operations.

Launched in 1996, MGS has four major instruments on board, a laser altimeter (MOLA), a thermal infrared spectrometer (TES), an imaging system (MOC)

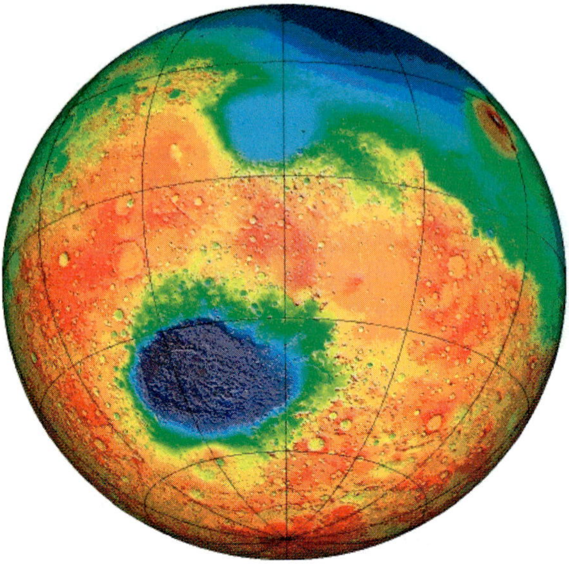

Plate 9.16. MOLA has provided the first global altimetry coverage of Mars, with false colour scale exhibiting in *dark blue* the deepest areas, in *red* the highest ones (MOLA/NASA)

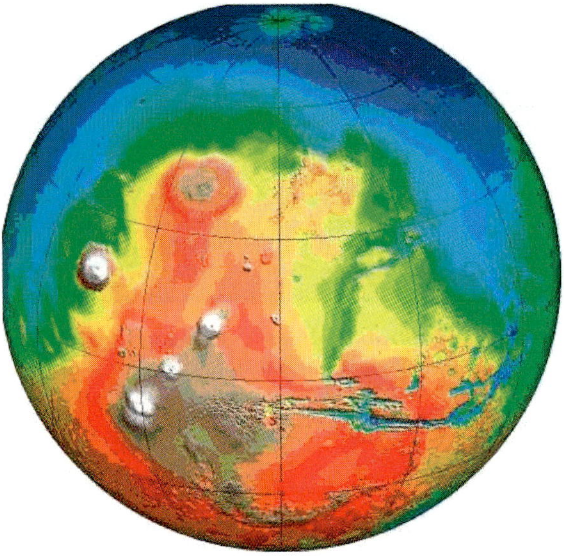

Plate 9.17. MOLA has provided the first global altimetry coverage of Mars, with false colour scale exhibiting in *dark blue* the deepest areas, in *red* the highest ones (MOLA/NASA)

and a magnetometer (MAGER). MOLA has provided the first global altimetry coverage of Mars (Plates 9.16–9.17), down to a vertical resolution of a few tens of centimetres, with a spatial sampling of a few hundreds of metres, over the entire planet. TES has provided thermal maps of the planet, as well as some mineralogical identifications. In particular, TES has identified a few areas (one is in Meridiani Planum) rich in grey, crystalline hematite (Fe_2O_3), interpreted as in-place sedimentary rock formation. They would indicate long-term stability on Mars of liquid water in the past. The MOC camera provided images down to a resolution close to 1 m of a few areas, exhibiting a huge variety of spectacular and unexpected features. As an example, MOC has identified gullies on Martian cliffs and craters walls (Plate 9.18), mostly at rather high latitudes (30° to 70°). They have been interpreted as created by liquid water having seeped onto the surface in the very recent geological past (millions of years timescales). They would result from subsurface hydrothermal activity in icy-rich permafrost. MAGER has confirmed that Mars lacks a significant global magnetic field at present (see Sect. 9.4.1). The field is dominated by superficial (crustal) sources of remanent magnetism, confined in the oldest, heavily cratered Southern highlands (Plate 9.19). This would indicate that the dynamo process operated in the very early Martian history. The tremendous outputs of the MGS investigations have led to several extensions of the mission operation phases.

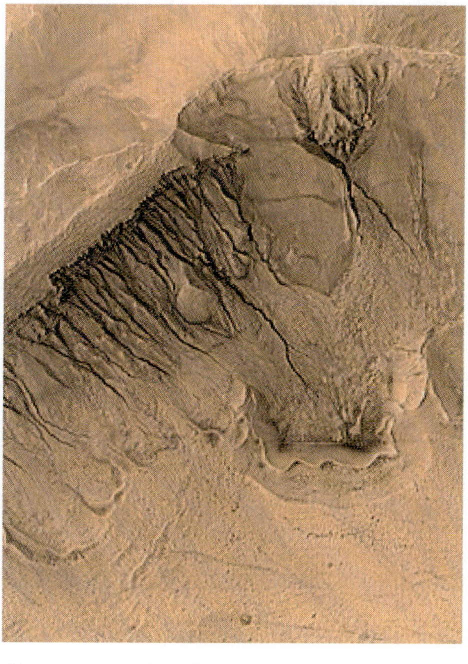

Plate 9.18. Moc gullies: These "gullies" have been discovered by the MOC/MGS camera, in a crater located at 39.0 °S, 166.1 °W (NASA/JPL/MSSS)

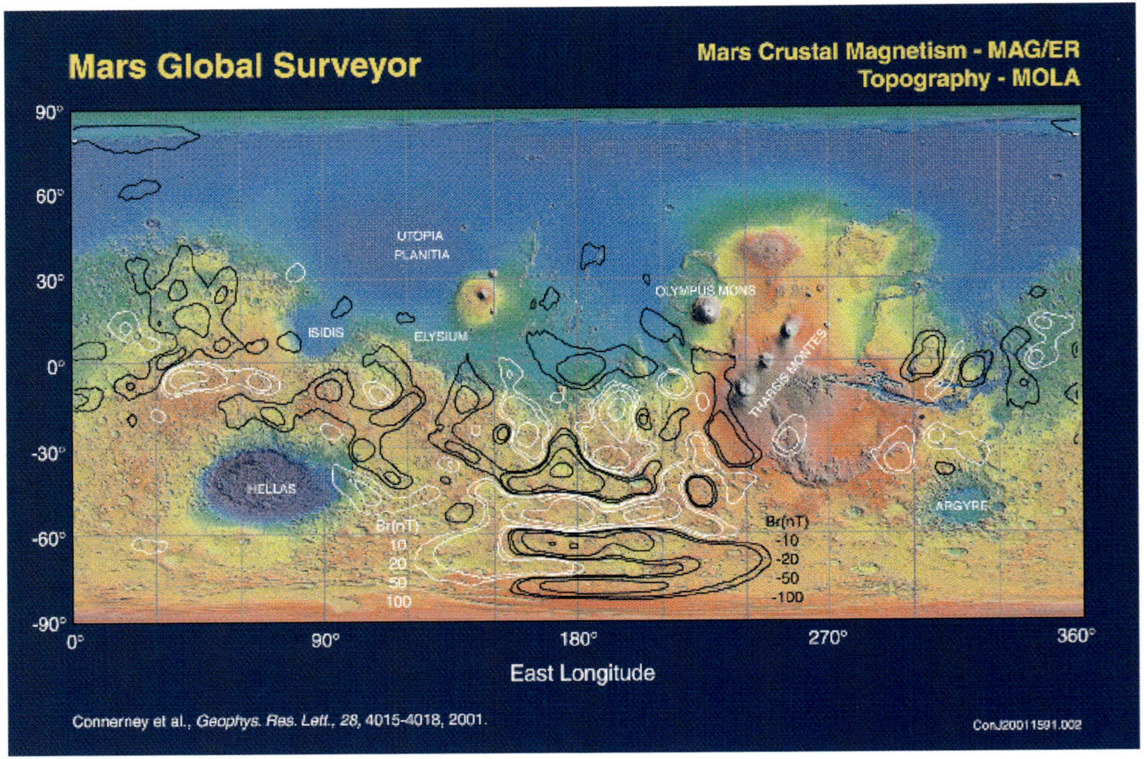

Plate 9.19. MAG/ER sources of remanent magnetism: Topographic map of Mars from the MOLA/MGS investigation with contours of constant radial magnetic field (*black*: negative, *white*: positive). Crustal magnetization appears largely confined to the ancient southern highlands. Region of extensive volcanism (e.g. Olympus Mons, Tharsis Montes) are non-magnetic as are regions surrounding the large impact basins Hellas and Argyre. Elevation relative to a reference surface (*yellow*) range up to 8 km negative (*yellow, green, light blue, dark blue*) and 8 km positive (*yellow, red, white*)

After the dramatic losses of the two missions (MCO and MPL) launched in 1998, at their arrival at Mars, the Odyssey mission, launched in early 2001 has been successfully put into orbit. Among the onboard instruments, a gamma-ray and neutron spectrometer has started to map the composition of a number of elements over the entire surface (averaged over the top metre), with special emphasis on H and on the long-lived radioactive species uranium, thorium and potassium. Massive H concentrations, likely to indicate very shallow H_2O ice, have been identified: at latitudes higher than 55°, up to the poles, the soil would contain about 50% of water ice in mass (Plate 9.20). Large zones of water-rich subsurface (with concentrations in the per cent level) have also been identified at lower latitudes, with the greatest reserves below the sands of Arabia Terra, a desert close to the equator.

The specific and highly favourable geometry of the Mars–Earth–Sun system in 2003 explains that this window has been chosen to target several missions to Mars. NASA has launched two probes with one rover each, the MERs (Mars Exploration Rovers) named Spirit and Opportunity (Plate 9.21). Spirit landed on Mars on January 4, 2004, Opportunity on January 25, 2004, in the Gusev crater and in Meridiani Planum, respectively. These sites have been selected for their potential for tracing the past hydrothermal activity on Mars. On board the MERs, both remote-sensing instrumentation, on a mast, and some in situ contact science to measure the elemental composition and some properties of samples, have been implemented. The MERs should explore hundreds of metres around their landing sites, during the subsequent few months.

For the first time, Europe with its space agency (ESA) is participating at mission level to the Mars exploration, with its Mars Express mission (Plate 9.22). Launched on June 2 from Baikonur, it got in Mars insertion December 25, 2003, for a Martian year of orbital operations. It also carried a small lander, Beagle 2,

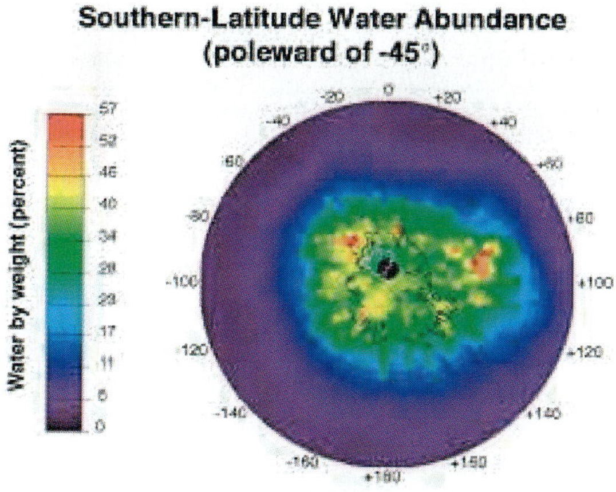

Plate 9.20. Massive H concentrations, likely to indicate very shallow H_2O ice, have been identified, by the GRS instrument on board NASA/Odyssey

Plate 9.21. MERs (Mars Exploration Rovers) named Spirit and Opportunity (artist view)

Plate 9.22. Mars Express mission, sketched as it arrived at Mars (ESA)

designed to identify records of potential past exobiological activity (Plate 9.23). Unfortunately, contact has never been received from Beagle 2 after it was released. From orbit, a global and coupled coverage of the atmosphere, surface and subsurface should be performed, during the one Martian year nominal lifetime of the mission, in 2004–2005. Mars Express should contribute to consolidating the inventory of gaseous and solid reservoirs, with an emphasis on H_2O and CO_2. As an example, the potential positive detection of carbonates would constitute strong evidence that large volumes of permanent liquid water existed, at a time indicated by their location.

Plate 9.23. Beagle 2, sketched as it might have landed (ESA)

In 2005, NASA plans to launch MRO, Mars Reconnaissance Orbiter, with high-resolution instruments to follow-on the MGS and Mars Express investigations: optical images down to a few cm in resolution would be acquired, as well as IR spectral images of areas typically ten times better resolved than with OMEGA/Mars Express, on a very limited number of sites identified as interesting/intriguing by the previous missions.

In 2007, the NASA Phoenix mission will launch a lander to the North Martian polar cap. In 2009, the NASA MSL (Mars Scientific Laboratory) will consist of a large rover, 900 kg in mass, exploring an area a few kilometres in size, with a highly sophisticated payload, to explore and quantitatively assess a potential habitat on Mars.

Carbonate-rich areas would constitute prioritised targets as future sites for in situ analyses and later collection of samples to be returned to the Earth: the availability of Mars samples to be analysed in our laboratories would have the potential for in-depth modifications of our understanding of the Mars history. Supposedly the Mars chemical evolution, within liquid water, did actually favour the building of living organisms, it cannot be excluded that one could identify microscopic fossil structures in such samples.

A more speculative step could be envisioned. Let us assume that fossil structures are discovered in Mars return samples, and subsurface water ice identified by orbital radar: thermodynamics indicate that liquid water could be stable below this ice (all scales being equally plausible, from microscopic pores to large lakes or oceans). Following the discovery of living organisms evolving in terrestrial abyssal zones, these hypothetical Martian liquid water-rich areas could constitute "niches" in which, even at present, living structures could evolve, after having remained in dormant mode for long periods. Identification of fossil structures and subsurface liquid water would then strongly suggest, as the next step in Mars exploration, deep drillings, possibly with the help of astronauts, to collect subsurface water-rich samples and look for potential extraterrestrial living organisms.

It is remarkable that forty years of space exploration, after having given to Mars the status of a planet highly inhabitable at least at its surface, severely irradiated by the solar UV field and geologically dead for millions of years, still leaves open the question of Mars hosting living structures, as extinct or possibly extant forms, in protected sites. Although they are highly unlikely, such scenarios offer such impressive perspectives for our understanding of the cosmic evolution that they deserve a major specific space programme to be initiated. Several countries have the ambition to contribute to it. This would allow our scientific communities to play a major role in what could constitute one of the most exciting scientific endeavours of the next decades.

References

Altwegg K., Ehrenfreund P., Geiss J., and Huebner W. (Edts.) (1999), Composition and Origin of Cometary Materials, Space Science Reviews **90**, 1–2, Kluwer Academic Puplishers.

Binzel R., Gehrels T., and Shapley M. (Edts.) (1989), Asteroids II, University of Arizona Press.

Bottke W.F., Cellino A., Paolicchi P., and Binzel R.P. (Edts.) (2002), Asteroids III, University of Arizona Press.

Chapman C.R. and Morrison D. (1989), Cosmic Catastrophes, Plenum Press, New York.

Combes M. et al., (1988), The 2.5–12 μm spectrum of Comet Halley from the IKS-Vega experiment, *Icarus*, **76**, 404–436.

Encrenaz T., Bibring J-P., Blanc M., Barucci M-A., Roques F., and Zarka P. (2004), The Solar System, Springer.

Gehrels T. (Ed.) (1979), Asteroids, University of Arizona Press.

Gehrels T. (Ed.) (1994), Hazards due to Comets and Asteroids, University of Arizona Press.

Horneck G. and Baumstark-Khan C. (Edts.) (2002) Astrobiology, The Quest for the Conditions of Life, Springer-Verlag.

Jakosky B. (1998), The Search for Life on Other Planets, Cambridge University Press.

Kallenbach R., Geiss J., and Hartmann W.K. (Edts.) (2001), Chronology and Evolution of Mars, *Space Sci. Rev.*, **96**, Nos 1–4, Kluwer.

Keller, H.U., Delamere W.A., Huebner W.F., Reitsema H.J., Schmidt H.U., Whipple F.L., Wilhelm K., Curdt W., Kramm J.R., Thomas N., Arpigny C., Barbieri C., Bonnet R.M., Cazes S., Coradini M., Cosmovici C.B., Hughes D.W., Jamar C., Malaise D., Schmidt K., Schmidt W.K.H., and Seige P. (1987), Comet P/Halley's nucleus and its activity. *Astron. Astrophys.*, **187**, 807–823.

Kerridge J.K. and Matthews M.S. (Eds.) (1988), Meteorites and the Early Solar System, University of Arizona Press.

Kieffer H.H., Jakosky B.M., and Snyder C.W. (1992), Mars, University of Arizona Press.

Kissel J. et al. (1988), Composition of Comet Halley dust particles from Vega observations, *Nature*, **321**, 280–282.

Laskar J., Joutel F., and Robutel P. (1993), Stabilization of the Earth's obliquity by the Moon, *Nature*, **361**, pp. 615–617.

Lee T., Papanastassiou, D.A., and Wasserburg G.J. (1976). "Demonstration of ^{26}Mg excess in Allende and evidence for ^{26}Al." *Geophys. Res. Lett.*, **3**: 109–113.

Lodders K. (1998), A survey of Shergottite, Nakhlite and Chassigny meteorites whole-rock composition, *Met. Planet. Sci.*, **33** Suppl., A183–A190.

Morrisson D. and Owen T. (1996), The Planetary System, Addison-Wesley Publishing Company.

Newburn R.L., Neugebauer M., and Rahe J. (Edts.) (1991), Comets in the Post-Halley Era, Kluwer.

Norton R. (2002), The Cambridge Encyclopedia of Meteorites, Cambridge University Press.

O'Leary B., Beatty J.K., and Chaikin A. (Edts.) (1999), The New Solar System, Cambridge University Press, Sky Publishing Corporation.

Weissman P.R., Mc Fadden L.-A., and Johnson T. V. (Edts.) (1999), Encyclopedia of the Solar System, Academic Press.

Wood J.A. (1986), Moon over Mauna Loa: A review of Hypotheses of Formation of Earth's Moon, in Origin of the Moon, Hartmann W.K., Phillips R.J. and Taylor G.J. (Eds), Lunar and Planetary Institute.

Zinner E. (1998) Stellar nucleosynthesis and the isotopic composition of presolar grains from primitive meteorites. *Ann. Rev. Earth and Planet. Sci.*, **26**, 147–188.

10 Spectroscopic Signatures of Life on Exoplanets – The Darwin and TPF Missions

Franck Selsis, Alain Léger, Marc Ollivier

The existence of other planetary systems, already discussed by ancient Greek philosophers, has been proved only very recently, first around pulsars (Wolszczan and Frail, 1992), then around solar-type stars (Mayor and Queloz, 1995). In the latter case, the planets discovered are giant planets (of the order of Jupiter's mass) and likely to be gaseous, which proved to be the case for HD 209458 b (Charbonneau et al., 2000). Up to February 1st 2004, 119 planets with masses above 0.1 Jupiter mass have been detected around 104 main sequence stars (Schneider, 2004). As the detection of a planet requires observations made over a time comparable to its period, most of them have a short period. The radial velocity method has recently been used successfully to reveal outer planets at the same orbital distance as Jupiter or Saturn.

Though some of these planets orbit in the "habitable zone" (see Appendix) of their parent star, living species are not expected on such giant planets. If this were the case, the life forms would be so different from what we know that their remote identification would be extremely unlikely. It should, however, be noted that all giant planets in our Solar System have telluric or icy satellites, so habitable satellites could orbit around some of these giant exoplanets (Williams et al., 1997).

The search for small terrestrial-type planets raises considerable scientific and philosophical interest. However, it is technically much more difficult: such planets do not sufficiently perturb the trajectory of their parent star to produce an indirect detectable feature and their brightness, 10^6 to 10^{10} times lower than the stellar one, is diluted within the diffraction pattern of the star.

Among the projects aiming at the detection of terrestrial planets, the first will search for planetary transits, observing continuously for months or years some dense fields of stars. These space missions are COROT (CNES, Rouan et al., 1998) and Kepler (NASA, Borucki et al., 1997). These missions will give us statistics on the abundance of terrestrial planets, their size, period and orbital distance.

The first projects for the direct detection of terrestrial planets are not expected before 2014 and are Darwin (ESA, Léger et al., 1996, Volonte et al., 2000) and TPF (NASA, Beichman et al., 1999). Hereafter we will refer to these missions with the common name "Darwin/TPF" to underline the strong will of both space agencies to achieve a joint project. Although other designs are currently under study, the original concept is based on "nulling interferometry". This method, combining the light from several mirrors, allows one to dim the light from the central star and to increase the contrast between the star and possible planets (Bracewell, 1978; Ollivier, 1999, see Fig. 10.1).

As soon as photons coming from the planet can be distinguished from those coming from the star, a spectral analysis is feasible within the available signal-to-noise ratio and sensitivity. The physical and chemical properties of the planets and their atmosphere can be studied. The recent detections of the upper atmosphere of a "hot Jupiter" (HD 209458 b) in absorption during a transit are the first spectral information gained on the atmosphere of an exoplanet (Charbonneau et al., 2001; Vidal Madjar et al., 2003). As life on Earth has strongly modified the planet (atmosphere, ocean, surface), can we use this fact to distinguish spectroscopically the presence of a similar ecosystem on another planet? In the particular case of Earth, O_2 is fully produced by the biosphere, less than 1 ppm coming from abiotic processes (Walker, 1977). Plants and cyanobacteria are responsible for this production by using the solar photons to extract hy-

Fig. 10.1. Darwin, the space interferometer. An artist's view of Darwin in its "free flyers" design: six telescopes of 1.5 meters receive the light from the studied planetary system and send it to a central laboratory for interferometric recombination. A eight module is used as a communication relay

drogen from water and using it to produce organic molecules from CO_2. This metabolism, called oxygenic photosynthesis, can be summarized as follows:

$$2H_2O^* + CO_2 + \text{photons} \rightarrow CH_2O + H_2O + O_2{}^*$$

(the asterisk shows that oxygen atoms forming O_2 originate from water)

Owen (1980) suggested searching for O_2 as a tracer of life. In a famous paper, Sagan et al. (1993) analyzed a spectrum of the Earth taken by the Galileo probe, searching for signatures of life. They concluded that the large amount of O_2 and the simultaneous presence of CH_4 traces are suggestive of biology. Moreover, the detection of a widespread red-absorbing pigment with no likely mineral origin supports the hypothesis of biophotosynthesis. Recently, Arnold et al. (2002) and Woolf et al. (2002) independently recorded spectra of the Earthshine on the moon (the light reflected by the Earth, then by the Moon back to the Earth) and observed the signatures of oxygen, ozone and the specific albedo of the vegetation.

Instead of searching directly for O_2, Angel et al. (1986) suggested considering O_3 and its signature in the midinfrared (9.6 µm) that produces a strong feature in the Earth's emission. In this wavelength range, the brightness contrast star/planet is 1000 times more favorable than in the visible. Léger et al. (1993) have investigated the feature of O_3 as a tracer of O_2 in planetary atmospheres and the use of O_3 as a tracer of oxygen-rich atmospheres sustained by life. This concept is at the root of Darwin, as proposed to ESA by Léger et al. (1996), and of TPF (Terrestrial Planet Finder, Beichman et al., 1999).

In the present chapter, we will merely explore the "astrobiological" possibilities of an instrument based on the original Darwin/TPF concept, able to produce a low-resolution spectrum ($\lambda/\Delta\lambda$ around 25) of the midinfrared thermal emission of a terrestrial exoplanet ($\approx 5-20$ µm). We will consider only the possible biosignatures within this spectral range. However, an alternative instrument working in the visible range is under study by NASA (see Nisenson and Papaliolios; 2001 and Labeyrie, 2002). The reader can find a complementary study of biosignatures in the visible range in a paper by DesMarais et al. (2002).

10.1 Ozone as a Biomarker

In order to validate the astrobiological strategy of a Darwin/TPF mission, it is necessary to estimate the risk of false-positive and false-negative occurrences. A false-positive case results from the detection of an abiotic feature that is wrongly attributed to some biological activity. On the other hand, a false-negative case occurs when an inhabited planet does not present any of the searched-for features, whether because the dominant metabolisms do not produce the expected bio-signature or because the biosignature is undetectable or masked by some other processes. We present here the results from various studies allowing us to test the efficiency of O_3 to trace O_2-producing ecosystems.

10.1.1 O_3 as a Tracer of O_2

Is ozone a good tracer of oxygen-rich atmospheres? Of course, this depends on the wavelength range considered. The most favorable range, in term of the contrast between the stellar and the planetary emission, is the midinfrared, around 10 micrometers, where the thermal emission of the terrestrial planets peaks. In this spectroscopic window, O_2 itself does not have vibrational transitions while O_3 exhibits a strong band centered at 9.6 µm. This band (and also H_2O bands) produces a strong feature in the Earth's infrared emission that makes our planet distinguishable from any other in the Solar System (see Figs. 10.2 and 10.3).

Ozone is produced in the atmosphere by a unique chemical reaction:

$$O + O_2 + M \rightarrow O_3 + M,$$

where M is any compound (this is a "3-body" or "association" reaction).

The oxygen atoms involved in this reaction come from the photolysis of O_2 by the solar UV radiation. On the contrary, the destruction of ozone in the Earth's atmosphere occurs through a large number of reactions, among which are some catalytic cycles involving mainly hydrogenous compounds (H, OH, HO_2), nitrogen oxides (NO_x) and chlorine compounds (ClO_x). Without these compounds, an atmosphere made of N_2 and O_2 would contain 10 times more

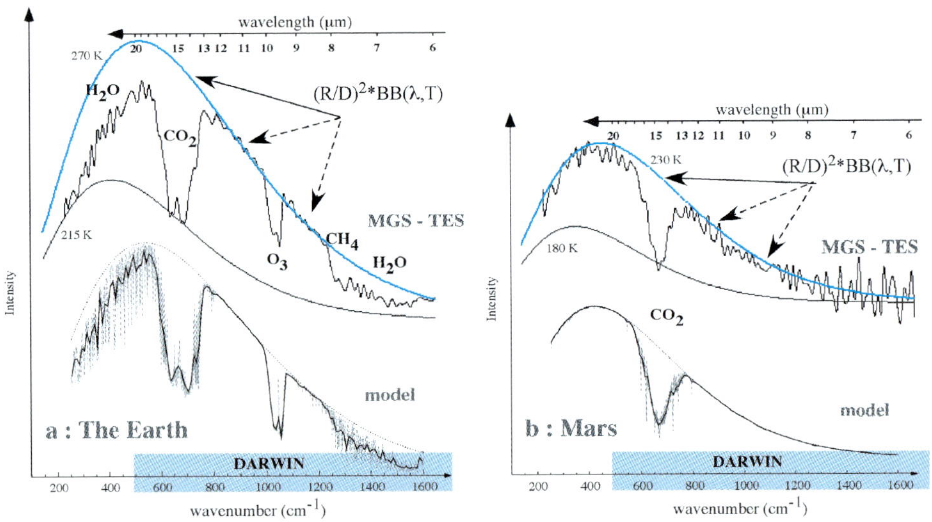

Fig. 10.2. Thermal emission of the Earth, Mars and Venus. The infrared spectra of the Earth (**a**) and Mars (**b**) were measured by the instrument TES of Mars Global Surveyor (Christensen and Pearl, 1997), and are compared to the synthetic spectra obtained with a numerical model: PHOEBE (Selsis, 2000a). The envelope of each spectrum is given by $(R/D)^2 BB(\lambda, T)$ where D is the distance to the planetary system, R the radius of the planet and BB is the Planck function

O_3, produced and destroyed only through the Chapman cycle (Brasseur and Solomon, 1984). Although, the amount of O_3 in an O_2-rich atmosphere strongly depends on the abundance of such trace compounds with various origins and it might be meaningless to "extrapolate" the chemistry of our present atmosphere to any other planet sustaining an oxygen-producing ecosystem. However, for a given atmospheric composition, the amount of O_3 is weakly sensitive to the total amount of O_2. O_3 has been described as a "logarithmic tracer" of O_2 because of this behavior: 10 times less O_2 results in 2 times less O_3 (Léger et al., 1993).

Moreover, for a given composition of the atmosphere, the amount of O_3 is very sensitive to the incoming stellar radiation. O_3 production depends on the availability of oxygen atoms and therefore on the photolysis of O_2 by the UV radiation. However, more UV radiation also produces more ozone destruction, directly by photolysis but mostly indirectly due to the increase of highly reactive radicals produced photochemically. For atmospheric compositions similar to the

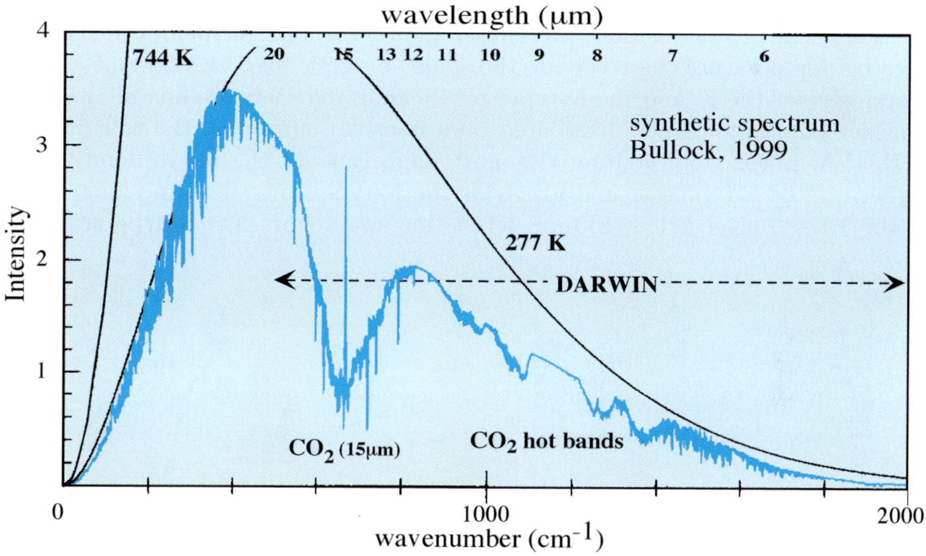

Fig. 10.3. This graph shows a synthetic spectrum of the thermal emission of Venus (energy radiated cm^{-1}). The envelope of the spectra corresponds to a black body "background emission". On Mars, this background emission comes from the surface of the planet. On Earth, part of it comes from the surface and part of it from the clouds. In the case of Venus, the background emission comes from high-altitude clouds (60 km) and from the continuous emission due to CO_2–CO_2 collisions, giving a black body temperature 400 K lower than the surface temperature. The background emission is absorbed in the atmosphere by colder atmospheric compounds: Venus and Mars spectra only exhibit CO_2 features, while the terrestrial spectrum contains also the signature of water vapor and ozone. The spectral domain presently planned for Darwin is given. The terrestrial emission at the resolution of Darwin is also shown on Fig. 10.9

Earth's, numerical simulations show a quantity of O_3 increasing with the UV flux (Selsis, 2000a). This property could play a very important role for the surface habitability by providing an ozone shield to the lethal UV radiations "tuned" to the intensity of the incoming radiation.

As shown in Fig. 10.4, the detectability of the O_3 feature, or its depth in the infrared spectrum, depends on two main parameters: the atmospheric profiles of the ozone abundance and temperature. The ability to detect O_3 with a Darwin/TPF instrument, with a low resolution (only 1 or 2 bins for the O_3 band) and a signal-to-noise ratio of about 5, requires a signature as deep as on the Earth. The complex coupling between radiation, photochemistry and temperature prevents us from generalizing the processes occurring in Earth's atmosphere and requires detailed modelling. By using a numerical code that simulates the photochemistry of a wide range of planetary atmospheres, and in order to investigate the changes in the O_3 signature, we have modelled the atmosphere of "Earth-like planets" for different spectral types of the host star. We have thus simulated a replica of our planet orbiting an F-type star (more massive and hotter than the Sun) and a K-type star (smaller and cooler than the Sun). The orbital distance was chosen in order to give the planet the Earth's effective temperature (by receiving the same energetic flux): 1.8 and 0.5 AU, respectively, for the F- and the K-type star. Scaling the energetic flux allows us to consider habitable planets irradiated by a nonsolar spectrum: the contribution of the UV range (150–400 nm, the most important for the photochemistry) is higher for the F-type star and lower for the K-type star (this is no longer true in the EUV range, below 150 nm, where low-mass stars like K-type stars are

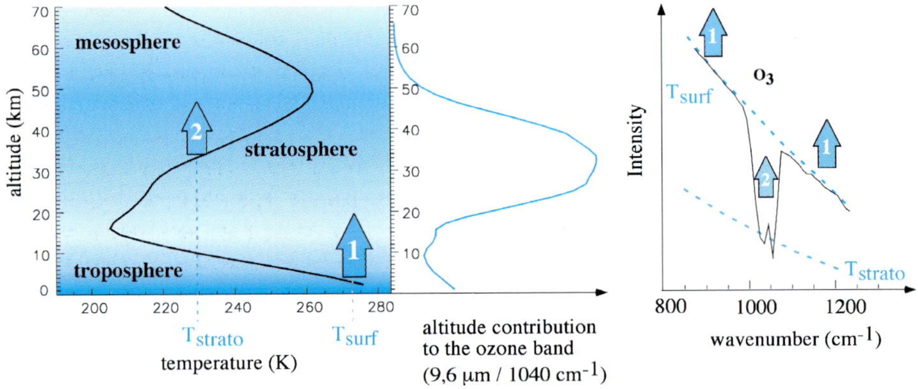

Fig. 10.4. Signature of ozone and the thermal profile. The depth of the absorption feature of O_3, and hence its detectability, depends on the atmospheric profiles of O_3 and the temperature. The envelope of the spectrum is emitted by the surface and the clouds (*arrow 1*) at the temperature of the lower atmosphere, while in the O_3 band, photons are emitted at stratospheric temperatures that are colder (*arrow 2*). The absorption is hence deeper when the temperature contrast between the surface and the middle atmosphere is high

very active). The results obtained are shown in Fig. 10.5. The planet orbiting the K star has an O_3 layer thinner than the Earth's but still exhibits a deep O_3 absorption: indeed, the low UV flux at lower altitudes than on Earth that results in a less efficient warming (because of the higher heat capacity of the dense atmospheric levels). Therefore, the ozone layer is much colder than the surface and this temperature contrast produces a strong feature in the thermal emission. The process is reversed in the case of a F-type host star. Here, the

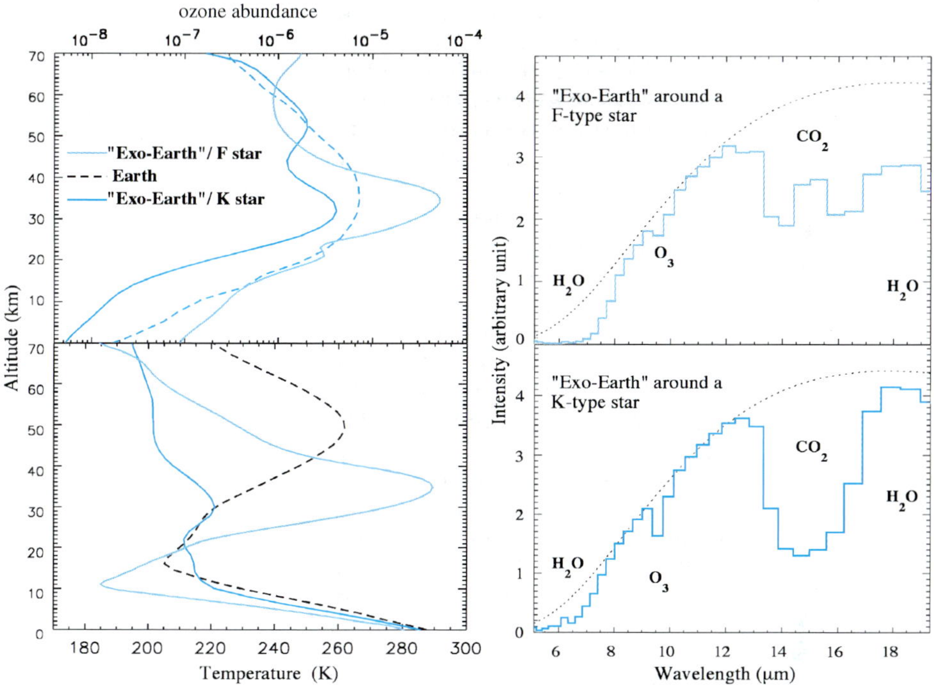

Fig. 10.5. Simulating "exo-Earths" around nonsolar-type stars. These graphs show the O_3 and thermal profiles, and the infrared spectra obtained by modeling the atmosphere of Earth-like planets around F- and K-type stars (respectively, hotter and colder than the Sun). This has to be compared to the spectrum of the Earth (Fig. 10.9) irradiated by the Sun, a G-type star. The atmosphere of these "exo-Earths" contains the same main gases (N_2, O_2, Ar, CO_2, H_2O) as the Earth's. For the "reservoir" compounds important for atmospheric chemistry (N_2O, CH_4, HCl, H_2, etc.), it is their production rate at the surface that is fixed equal to their terrestrial value. The orbital distance of the planet is chosen in order to produce a surface temperature close to Earth's. Despite a denser ozone layer, the planet around an F-type star does not exhibit a strong O_3 feature because of a low temperature contrast between the surface and the stratosphere. On the other hand, the low ozone content in the case of the K-type star is enough to produce a detectable O_3 feature because of a much higher contrast. Therefore, K stars might be more favorable for the detection of O_2-rich atmosphere through their O_3 signature than F-type stars

ozone layer is denser and warmer than the terrestrial one, exhibiting temperatures close to the surface temperature. Thus, the resulting low temperature contrast produces only a weak and barely detectable feature in the infrared spectrum. This comparison shows that the observable conditions for the search for life with the O_3 criterion may be better in the case of G- (solar) and K-type stars than in the case of F-type stars. This result is at least promising since G- and K-type stars are much more numerous than F-type stars, the latter being rare (4 closer than 10 pc) and affected by a short lifetime (less than 1 Gyr).

The cases studied differ from the Earth by only one parameter: the nature of the incoming radiation. In order to understand the sensitivity to other parameters, new simulations should be performed by changing variables such as gravity, pressure and the background composition of the atmosphere, biogenic emission of tropospheric gases such as CH_4 or N_2O, and orbital distance.

The case of M stars, the smallest, coldest and most numerous, still requires studies. They emit, sporadically, a low flux of UV but a high flux of X-rays and extreme UV (EUV). Their habitable zone, where one can expect surface temperatures compatible with liquid water, is very close to the star. If planets do form in their circumstellar zone, they should be tidally locked, always presenting the same hemisphere to their parent star. Because of this, the thermal inertia of the atmosphere has to be high enough to prevent the gases from condensing completely on the night side (Joshi et al., 1997). A CO_2 partial pressure of more than 1 bar is thus required. We discuss the further consequences of this high CO_2 pressure on the IR and visible biosignatures.

10.1.2 The Buildup of a Biogenic O_2-Rich Atmosphere[1]

The search for life through oxygen or ozone requires us to address the following double question: Assuming the existence of other biospheres, (1) does oxygenic photosynthesis have to emerge at some point in the biological/ecological evolution as a dominant metabolism and (2) should the existence of this metabolism always result in a buildup of atmospheric O_2? Again, our incomplete knowledge of the history of life on Earth does not give us the answer; however, it shows us some of the constraints on building up an O_2-rich atmosphere.

The "rise of the oxygen" on Earth occurred between -2.3 and -1.9 Gyr (see Holland, 1994, for a review of the geological records). A characteristic time for the build-up of O_2 is not known because the atmospheric level of O_2 around this period cannot be inferred with high accuracy and also because the age of the records are uncertain. What is known is that there was less than 0.002 mbar before -2.3 Gyr (Pavlov and Kasting, 2002) and that it was high enough after

[1] A recent paper (Bekker et al., 2004) reports geological records indicating a rise of atmospheric oxygen 2.32 Gyr ago. The following paragraph should be read considering this improvement in the dating.

– 1.9 Gyr to produce oxidized soils or rocks (like the red-beds), suggesting about 1% of the present O_2 atmospheric level. How fast was the increase of the O_2 partial pressure, from less than 0.002 mbar to more than 2 mbar? Faster than 400 million years is the most prudent answer.

However, the organisms producing O_2, the cyanobacteria, are believed to be older. Microfossils found in 3.5-Gyr old rocks have been described as cyanobacteria (Schopf, 1994) though this interpretation is highly debated. It is more generally admitted that cyanobacteria arose around – 2.7 Gyr (Brocks et al. 1999).

Oxygenic photosynthesis gives to an organism a considerable advantage: it allows it to grow and spread in any place where liquid water, CO_2 and light are available, and thus to populate the first few meters below the surface of the water all over the planetary surface. It would be less conceivable that organisms relying on marginal sources could change the whole planetary environment. The biogenic production of O_2 is, however, constrained by the necessity to have surface liquid water and thus a surface temperature above 273 K (at normal pressure). Indeed, the productivity of phototrophic bacteria would not be sufficient to sustain an O_2-rich atmosphere. At the beginning of the Earth's history, this constraint implied a greenhouse effect much more efficient than today due to the low luminosity of the early Sun (70% of the present one at −4 Gyr; Baraffe et al., 1998; Guinan and Ribas, 2002). Moreover, to allow a build up of O_2, the gases responsible for the greenhouse warming have to be chemically compatible with O_2: if they can be oxidized, a rise of O_2 would consume them and induce a fall in temperature to below the freezing point of water preventing further biogenic production of O_2. Presently, the surface warming is due to CO_2[1], a compound totally oxidized whose level is not affected by the presence of atmospheric oxidants. The level of CO_2 in the primitive atmosphere of the Earth is not known and only upper limits are available: < 10 mbar between −3 and −2 Gyr (Rye et al., 1995; Selsis, 2000b). Even if this upper limit, based on the absence of siderite ($FeCO_3$), is far from being robust, levels below a similar value are also derived from models of the early cycle of carbon (Sleep and Zahnle, 2001). Before the rise of O_2, it is likely that biogenic CH_4, released by methanogens (e.g. archeobacteria probably hyperthermophils), was the main greenhouse gas (Pavlov et al., 2000; Selsis, 2000b). Indeed, with this upper limit, and taking into account the evolution of the solar luminosity, CO_2 itself could not warm the surface above 273 K until about −2 Gyr. On the other hand, in the absence of atmospheric O_2, and assuming a biogenic production of CH_4 equal to the present one (it was more probably much higher), atmospheric methane would be 100 to 1000 times more abundant than today. Such high levels would produce

[1] In fact, the main greenhouse gas, in terms of radiative transfer, is water vapor. However, the abundance of H_2O depends on its saturation vapor pressure that is a function of the temperature. If a water reservoir is present at the surface, the CO_2 level (or other greenhouse gases not affected by condensation) indirectly determines the abundance of water vapor. Also, the presence of water vapor cools the surface by changing the adiabatic lapse rate of the lower atmosphere.

surface temperature well above 273 K (Pavlov et al., 2000). In a primitive environment warmed by methanogenesis, a build-up of atmospheric O_2 would fail because of the chemical destruction of methane and the surface freezing it would trigger. It is difficult to say if this climatic constraint was the main limitation on the rise of O_2, responsible for the delay between the appearance of oxygen producers and the effective rise of O_2, but the recorded age of the atmospheric change and also the global glacial period that occurred before the rise of O_2 both fit with this hypothesis (Selsis, 2002). If the need for methane was the limiting factor, then the rise of oxygen could have been much faster than the 400-Myr period. At the present rate of organic carbon burial (2 Gigatons/yr), this could be achieved in less than 1 Myr (of course, the bioproductivity was probably lower at that time).

On the other hand many other processes could have prevented the build up of O_2. First, cyanobacteria may have needed a long time to adapt to an oxic environment. Secondly, the produced oxygen may have been consumed through the oxidation of rocks, dissolved iron in the ocean, reducing volcanic emissions, and organic carbon.

The early history of the Earth's biosphere tells us that the nondetection of O_2 or O_3 is absolutely not indicative of the absence of life on an exoplanet. More generally, the absence of life cannot be inferred from the absence of a spectroscopic feature. For instance, let us consider the numerous spectroscopic studies of the planets of the Solar Systems, made with resolutions and sensitivities much higher than can possibly be achieved for a Darwin/TPF instrument: no spectral feature interpreted as a biomarker has ever been detected and nevertheless it has never been excluded that life may exist on Mars or Europa, for instance.

10.1.3 Abiotic Synthesis of O_2 and O_3 – "False-Positive" cases

The fact that, on the Earth, oxygen and indirectly ozone are byproducts of the biological activity does not mean that life is the only process able to enrich an atmosphere with these compounds. The question of the abiotic synthesis of biomarkers is crucial though very few studies have been yet dedicated to it (Rosenqvist and Chassefière, 1995; Léger et al., 1999; Selsis et al., 2002).

To qualitatively associate oxygen and life without investigating further possible abiotic sources is an error that has already been committed in the past. In the 1950s, though no precise data about the composition of the Martian atmosphere was available, it was already understood that the red color of the Martian surface was due to oxidation. Some scientists inferred from this hypothesis (true) that oxygen was the main component of the atmosphere (wrong: there is only 0.1%) and that this oxygen should have a biological origin (Spencer Jones, 1958). The presence of O_2 was interpreted as a "confirmation" of the presence of vegetation, already suggested by some seasonal color variations of the planet (now understood as seasonal dust storms). However, the 0.1% of O_2 (for a surface of 6 mbar), as well as O_3 (abundance of about 10^{-8}), has an abiotic origin: photochemistry.

The Martian atmosphere is the perfect "laboratory" to learn about the photochemical synthesis of O_2 and O_3. Indeed, if an abiotic production of O_2 does exist on Earth, it is totally masked by the biological release. The main constituent of the Martian atmosphere is carbon dioxide (CO_2: 95.3% – N_2: 2.7% – Ar: 1.6% – O_2: 0.13% – CO: 0.08% – H_2O: $\approx 0.01\%$, O_3: $\approx 10^{-6}\%$). CO_2 is photolysed by UV radiation at wavelengths below 227 nm, producing one carbon monoxide, CO and one atom of oxygen, O. The reaction between two oxygen atoms to form O_2 is much more efficient than the CO + O recombination into CO_2 (the reaction is 10^5–10^7 times faster at temperature prevailing in the Martian atmosphere: 145–270 K). This simple fact seems to be in contradiction with the stability of a CO_2 atmosphere and it took a long time to understand why CO and O_2 are not major compounds on Mars. Indeed, their accumulation should only be limited by the O_2 photolysis that destroys O_2 and "steals" part of the photons able to photolyse CO_2. Some photochemical simulations have shown (Nair et al., 1994; Selsis et al., 2002) that, in a Martian atmosphere purely made of CO_2, O_2 would reach a level of 3% (in the absence of other chemical loss of O_2 such as surface oxidation).

If O_2 is a minor component of the Martian atmosphere, it is due to the presence of another atmospheric compound playing a critical role in the photochemistry: water vapor. H_2O is only a trace gas on Mars: if condensed, the whole content of water vapor in the atmosphere would represent a 3-mm layer of water on the surface. But despite its low abundance H_2O strongly modifies the photochemistry, and above all, the processes involving O_2 and O_3.

Indeed, H_2O, like CO_2 and O_2, is photodissociated by the solar UV radiation. H_2O photolysis produces hydrogen atoms, and hydrogenated compounds, that catalyse the recombination of CO into CO_2. The following catalytic cycle is the main recombination route for CO_2 in the Martian atmosphere:

$$
\begin{array}{cc}
H^{\cdot} + O_2 + M & HO_2^{\cdot} + M \\
O + HO_2^{\cdot} & O_2 + OH^{\cdot} \\
CO + OH^{\cdot} & CO_2 + H^{\cdot} \\
\hline
\mathbf{CO + O} & \mathbf{CO_2}
\end{array}
$$

The small amount of water vapor in the atmosphere of Mars is then enough to prevent the accumulation of O_2. Moreover, the photolysis of H_2O produces, directly or indirectly, some oxidants that are much more reactive than O_2: OH, HO_2 and H_2O_2. The oxidation of the surface, and hence the loss of oxygen, is therefore much higher in their presence.

H_2O photolysis enhances the consumption of O_2 but it can also result in its production and accumulation when it is associated with a significant loss of hydrogen to space. Indeed, a fraction of the hydrogen atoms released by H_2O photolysis in the upper atmosphere have enough kinetic energy to exceed the escape velocity, inducing an "oxidation" of the atmospheric content. One can summarize this process as follows:

$$\begin{array}{cc}
4 \times (H_2O + h\nu & OH^{\cdot} + H^{\cdot}) \\
2 \times (OH^{\cdot} + OH^{\cdot} & H_2O + O) \\
O + O + M & O_2 + M \\
\multicolumn{2}{c}{\text{H escapes}} \\
\hline
\mathbf{2H_2O} & \mathbf{O_2 + (4H^{\cdot})}
\end{array}$$

On Mars, the escape of H can contribute to the observed level of O_2 (Nair et al., 1994), though it is not yet demonstrated due to uncertainties in the photochemical modelling (Selsis, 2000b).

The hydrogenous radicals produced from H_2O photolysis also very efficiently destroy ozone. Again, catalytic cycles are involved in the destruction process and the small abundances of the byproducts of water-vapor photodissociation can strongly affect the ozone content.

$$\begin{array}{lcl}
OH^{\cdot} + O & \rightarrow & H^{\cdot} + O_2 \\
H^{\cdot} + O_3 & \rightarrow & OH^{\cdot} + O_2 \\
\hline
O + O_3 & \rightarrow & 2O_2
\end{array}$$

$$\begin{array}{lcl}
OH^{\cdot} + O_3 & \rightarrow & HO_2^{\cdot} + O_2 \\
HO_2^{\cdot} + O & \rightarrow & OH^{\cdot} + O_2 \\
\hline
O + O_3 & \rightarrow & 2O_2
\end{array}$$

$$\begin{array}{lcl}
OH^{\cdot} + O_3 & \rightarrow & HO_2^{\cdot} + O_2 \\
HO_2^{\cdot} + O_3 & \rightarrow & OH^{\cdot} + 2O_2 \\
\hline
2O_3 & \rightarrow & 3O_2
\end{array}$$

The cycles above are responsible for the diurnal variation of O_3 in the upper stratosphere and mesosphere of the Earth and are the main pathways for the destruction of O_3 on Mars. This explains why the abundances of water and ozone are anticorrelated on Mars, the maximum of ozone being above the poles in winter where the atmosphere is the driest.

Mars is not a unique place in the Solar System where oxygen is produced abiotically: some icy satellites of Jupiter (Europa, Ganymede) and of Saturn (Rhea, Dione) exhibit a tenuous atmosphere ($\approx 10^{-10}$ bar) quasiexclusively made of O_2 and O_3 (Hall et al., 1995; Noll et al., 1996 and 1997). The associated column densities (density integrated over the altitude) are very low. Nevertheless, it reveals an abiotic continuous synthesis of these two molecules from the dissociation of water (ice) by UV and, mostly, charged particles trapped in giant-planet magnetospheres. The hydrogen released from water escapes from the low gravity field of the satellites, and a gas enriched in oxygen remains, partly in the atmosphere and partly in bubbles within the ice, mainly made of O_2 and O_3 (Johnson et al., 1997).

In the case of Venus, one could expect to find atmospheric oxygen. Indeed, this dry planet is believed to have been water-rich at the beginning of its history, and the hydrogen escape is generally suggested to explain the disappearance of water (Rasool and De Bergh, 1970; Kasting, 1988) should have led to considerable amounts of residual oxygen. The loss of the hydrogen content of a 100-m layer of water would result in 8 bar of pure O_2. However, molecular oxygen has not yet been observed on Venus, and its abundance in the middle atmosphere is lower than 10^{-6}. The "missing oxygen" was probably consumed through the oxidation of the surface and volcanic emissions. This illustrates the fact that, to result in an O_2 buildup on a geologically active planet, the rate of O_2 production has to exceed the oxidation rates. From this viewpoint, Venus is very different from Mars: the latter is a small planet that cooled rapidly, the surface of which is mostly inert, fully oxidized and consuming the atmospheric oxidants only at a very slow rate.

The study of the photochemistry in the planetary atmospheres of the Solar System reveals some mechanisms for the synthesis of O_2 and O_3. The production rates and the amounts of these abiotically synthesized molecules are very low: the column densities of O_2 and O_3 on Mars are respectively 6×10^{-5} and 5×10^{-4} times those of the Earth, this ratio being 10^{-10} in the case of the icy satellites. Nevertheless, in some different environments such processes could a priori lead to higher levels (as we saw in the hypothetical case of a dry Martian atmosphere) on extrasolar planets. It is therefore crucial to find some quantitative and/or qualitative criteria, accessible to remote detection, which would allow us to distinguish between biologic and abiotic origins. A first conclusion arises: a biosignature is a concept more complex than the "binary" detection of a given compound. The question "Is there oxygen (or ozone) in this atmosphere?" has to be replaced by "How much oxygen (or ozone) is there?". It is, however, not enough. The search for signs of life implies the need to gather as much information as possible in order to understand how the observed atmosphere physically and chemically works. For this, it is fundamental to take into account the instrumental characteristic. Indeed, sensitivity and resolution behave as filters, permitting us to detect given species only for amounts above a threshold: as an example, the ozone content of an "exo-Mars" or an "exo-Ganymede" could not be detected with a Darwin/TPF instrument while the ozone content of an "exo-Earth" could.

10.1.4 Numerical Simulation of O_2 and O_3 Abiotic Synthesis

In order to validate the biosignature "ozone" in the infrared, to test and refine its definition, we sought to simulate an accumulation of oxygen and ozone by abiotic processes in the atmosphere of exoplanets (Selsis et al., 2002). For this, we used a numerical code modeling the physicochemical evolution (primarily temperature and chemical composition) of a planetary atmosphere starting from an arbitrary initial state and submitted to the irradiation of a star. We sought the conditions maximizing the production of O_2 and O_3 molecules, starting from the photodissociation of CO_2 and H_2O, and we calculated in each case the infrared emission of these planets searching for the resulting spectroscopic features. We tried to produce realistic cases of "false positives", i.e. cases presenting what was hitherto regarded as a biosignature: the band of O_3 at 9.6 μm.

A mixture of CO_2 and water vapor proves to be relatively inefficient in producing O_2 because, as in the case of Mars, the hydrogenated radicals coming from the photolysis of H_2O induce a fast reformation of CO_2. However, if one considers a very dense atmosphere of CO_2 (> 1 bar) with a water reservoir on the planetary surface, the photolysis of H_2O becomes a negligible process. In this case, the radiative properties of CO_2 induce a hot lower atmosphere, whose temperature decreases quickly with altitude, up to very low values in the middle atmosphere causing the condensation of CO_2. As it goes up into the atmosphere, the water vapor also condenses when meeting this cold trap. The photons able to photolyse H_2O are also absorbed by CO_2. The altitudes where one finds the radiation that can photodissociate H_2O are then very high and only a minute quantity of water vapor reaches them. Under these conditions, CO_2 is photodissociated without a fast reformation of the molecule because of the lack of hydrogenated radicals. Thus, O_2 accumulates and O_3 forms a dense layer. However, this dense layer of O_3 absorbs the UVs at longer wavelengths than CO_2 does. As a consequence, O_3 strongly heats the middle atmosphere, modifies its thermal structure and creates a stratosphere, just like on Earth. This effect of ozone results in decreasing the efficiency of the cold trap and letting more water vapor reach high altitudes. Thus, there is a strong negative feedback limiting the accumulation of O_2 and O_3. We estimated that, in a 1-bar CO_2 wet atmosphere, without surface oxidation, and taking into account the uncertainties of modeling, the column densities of O_2 and O_3 are:

- $1.3 \times 10^{22} < \{O_2\} < 8.8 \times 10^{23}\,\mathrm{cm}^{-2}$, which is respectively 0.3% and 20% of the O_2 column densities on Earth ($4.5 \times 10^{24}\,\mathrm{cm}^{-2}$),
- $5 \times 10^{17} < \{O_3\} < 1.8 \times 10^{19}\,\mathrm{cm}^{-2}$, which is respectively 5% and 200% of that of O_3 on Earth ($9 \times 10^{18}\,\mathrm{cm}^{-2}$).

This result shows that it is theoretically possible to find an abiotic atmosphere rich in O_2 and O_3, possibly with more ozone than on Earth (see Fig. 10.6), the oxygen coming from the photolysis of CO_2 by stellar UV radiation. However, such a CO_2-rich atmosphere has a very special infrared signature. For CO_2 pressure higher than 50 mbar, characteristic absorption bands appear at 9.4 and 10.5 μm. These bands, first, characterize CO_2-rich atmospheres and, secondly, mask the signature of ozone at 9.6 μm when measurements are performed with the (low) spectral resolution of an instrument such as Darwin/TPF (see Fig. 10.7). For even higher pressures, other bands appear at 7.3 and 7.9 μm and the absorption induced by CO_2–CO_2 collisions practically erases any other spectral signature in the thermal emission.

What does occur in a CO_2 atmosphere without any water vapor? In this case, similar to the "dry Mars" considered previously, the CO_2 partial pressure, P_{CO_2}, necessary to a significant enrichment in O_2/O_3 is much weaker. Indeed, the only phenomenon that limits the conversion of CO_2 into CO and O_2 is the

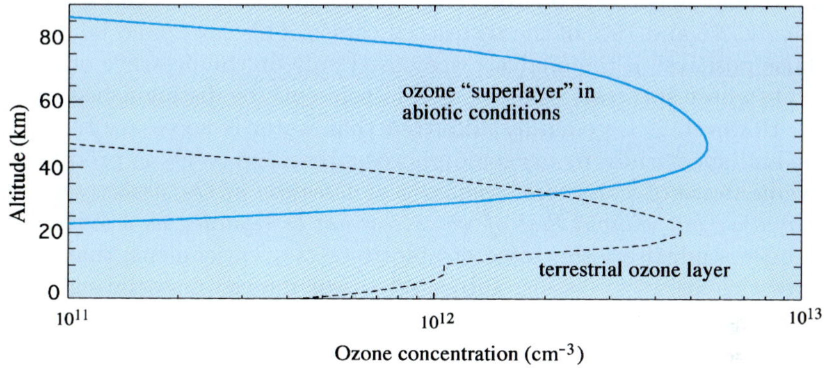

Fig. 10.6. Abiotic ozone layer. This graph shows the concentration profile of ozone photochemically produced in a 1-bar atmosphere of CO_2. One can see that this ozone layer, which is a byproduct of the presence of abiotic oxygen, is denser than the present terrestrial ozone layer. Its possible nonbiological origin could however be deduced from the CO_2 absorption features implying a high CO_2 pressure (Selsis et al., 2002)

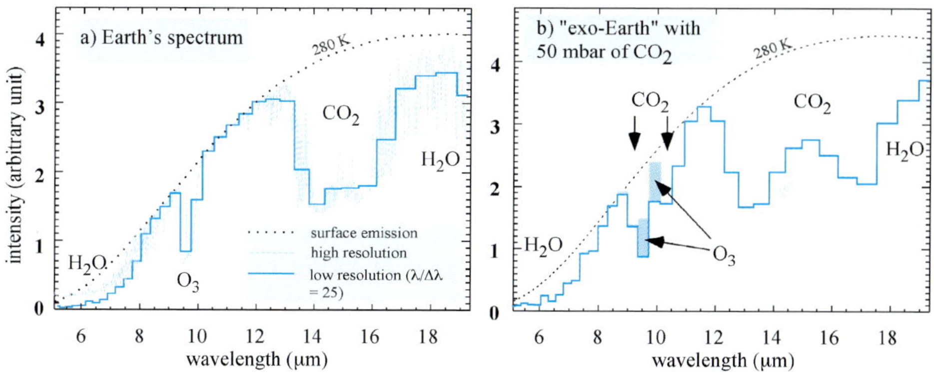

Fig. 10.7. CO_2 masking of the O_3 signature. Figure 10.9a shows the low-resolution infrared spectrum of the Earth and Fig. 10.9b the spectrum of an "exo-Earth" with the same characteristics but containing 50 mbar of CO_2 instead of 0.3 mbar (the terrestrial level). This quantity of CO_2 is necessary to warm the surface above 273 K if the planet is orbiting at 1.2 AU of a solar-type star of the age of the Sun. Due to the higher pressure of CO_2, new absorption bands appear close to the O_3 band preventing the detection of the latter at low resolution. Moreover, the presence of these masking bands indicates a high CO_2 pressure that can induce the abiotic synthesis of O_3. This means that even at a higher resolution, an O_3 signature detected simultaneously with these CO_2 bands could not be necessarily attributed to life

absorption by O_2 of the radiation able to photodissociate CO_2: the O_2 increase produces a decrease of the CO_2 dissociation rate and consequently of the O and O_2 production. Let us consider a water-free atmosphere containing 50 mbar of CO_2, knowing that for higher values of P_{CO_2}, the ozone signature is no longer detectable by an instrument with low spectral resolution. In such a case, one obtains column densities of $9 \times 10^{22}\,cm^{-2}$ for O_2 and $1.7 \times 10^{18}\,cm^{-2}$ for O_3, i.e., respectively, 2% and 19% of the terrestrial values. This case could be considered as a "false positive" if the analysis was based only on the presence of the single species O_3 whose spectral signature can, in principle, be distinguished from that of CO_2. However, it is generally admitted that water is necessary to life as we know it and specifically to oxygenic photosynthesis since O_2 is produced from the oxygen atoms of water. *Consequently, a detection of O_3, associated with the CO_2 signature but without that of water, cannot be regarded as a biosignature.*

We have studied another case of abiotic O_2/O_3 enrichment, that of an atmosphere rich in water vapor, subjected to an intense dissociating radiation and a high H atom escape. The latter can be thermal (Jeans) or nonthermal (sputtering, exothermic reactions, etc.). To induce a significant O_2 enrichment, a high photodissociation rate of water vapor is needed. It is then necessary that H_2O is the main UV absorbing species and that water vapor is present in large quantities at high altitude. This is generally not the case because of the water vapor losses by condensation, as rain or snow, when it goes into the atmosphere (cold trap). This is, for instance, what occurs during a "runaway greenhouse" scenario (Kasting, 1988): the wet atmosphere has an isothermal profile allowing the vapor to reach the high altitude where it is photodissociated.

We studied various "wet" atmospheres formed during such runaway or because of the infall of icy particles in the upper atmosphere, e.g. cometary particles or small comets. In such a situation, water vapor is present at high altitude and O_2 accumulation occurs since the H escape rate exceeds that of oxidation of the planetary surface. We tuned the surface oxidation rate in order to have an accumulation of O_2. In a 1-bar atmosphere made of N_2 and water vapor, our simulations yield an extreme upper limit of 1% for O_2. For a larger oxygen abundance, O_2 competes with H_2O for UV photodissociation and the destruction of O_2 increases while the photolysis of H_2O decreases. 1% of O_2 would be detectable through the visible/NIR lines of O_2 in the reflected spectrum of the planet. Strong H_2O lines would also be observed and at high resolution ($R > 80$) their depth and broadening may indicate a very wet atmosphere. This with the orbital distance and the stellar luminosity could warn the observer that a runaway greenhouse is occurring. Also, this abiotic mechanism for oxygen buildup cannot produce a dense IR-absorbing ozone layer. Indeed, hydrogen escape requires H_2O photolysis, which, in return, produces O_3-destroying radicals (see reactions in Sect. 10.1.3). Also, the thermal profile obtained in a runaway greenhouse scenario would not produce a deep spectroscopic signature in the middle atmosphere.

Figure 10.8 shows the deepest O_3 signature we obtained with a "wet" scenario. This most favorable case profits from not very realistic ad hoc conditions, especially because we considered the infall of pure H_2O: real cometary material would also deliver UV-absorbing species (such as CO_2) and reducing ones (such as CO or CH_4, organics and metals). If more CO_2 is added to this atmosphere the production of O_2 is strongly reduced as fewer photons can take part in water photolysis.

Wet abiotic synthesis of O_2 *would not produce* a thermal emission spectrum with the simultaneous signature of O_3, H_2O and CO_2.

The conclusions from these numerical simulations are:

- *Under abiotic conditions, an atmosphere can contain O_2 and O_3 amounts similar to, or even higher than, those current in the Earth's atmosphere.*
- *The signature of one of these molecules alone, without additional information on the composition of the atmosphere, cannot be considered as a reliable biomarker.*
- *In the 5–20 µm spectroscopic range, the triple signature of O_3–H_2O–CO_2, is a reliable biosignature that has not been reproduced so far by any abiotic photochemical modelling.*

Regarding the last conclusion, at the spectral resolution planned for Darwin/TPF instruments $\lambda/\Delta\lambda < 25$), the identification of these three species in the spectrum is a robust way to filter out a nonbiological production of O_2 and O_3. With future instruments, providing higher spectral resolution, additional

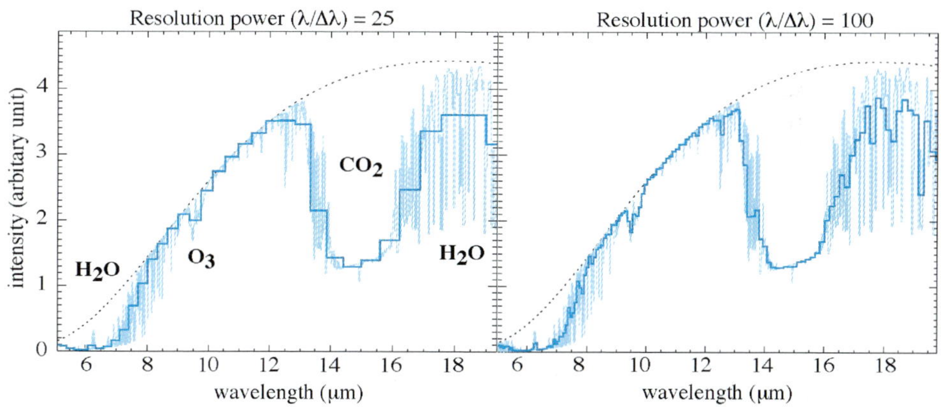

Fig. 10.8. Abiotic signature of O_3 in wet atmosphere losing H. In the case of a wet upper atmosphere, fed by external delivery of water (see text), and submitted to an intense UV/EUV irradiation, H_2O photolysis followed by the escape of H atoms may result in a buildup of O_2 and O_3. This IR spectrum was computed for the higher O_2 abundance that we obtained. Only a very thin ozone layer can form due to the radicals produced by H_2O photolysis and its signature is hardly detectable, even in this "optimized" configuration

details in the spectrum will allow the sharpening of "false-positive" eradication. For example, it will become possible to distinguish the O_3 signature from the bands that appear at high P_{CO_2}. This means that an abiotic atmosphere can present the triple signature but that the presence of the "high-pressure" band of CO_2 would make it possible to point out a possible abiotic origin of the O_3 one.

It must be noted that seeking oxygenic photosynthesis, such as is present on the Earth, by searching for the triple signature O_3–H_2O–CO_2 is no more restrictive than searching only for O_3 since H_2O and CO_2 are "ingredients" of the O_3-producing processes. Moreover, H_2O and CO_2 are expected to be abundant on habitable planets (Kasting et al., 1993).

Last but not least, it must be noted that, if the 5–20 μm range makes it possible to reject abiotic O_3 signatures even at low spectral resolution, this is probably not the case in the visible and the near-infrared range where some of the abiotic atmospheres we simulated did produce reflected light spectra with the signature of O_3, O_2, H_2O and CO_2. In addition, the CO_2 signature at $\lambda < 11\,\mu m$ is undetectable for concentrations similar to the terrestrial one (Des Marais et al., 2002).

10.1.5 False Negatives

As already mentioned, the nondetection of the O_3–H_2O–CO_2 species does not at all mean the absence of life on the observed planet. Our planet sheltered life for about 2.5 billion years without presenting such a signature. It does not even mean that there could not be a form of life based on oxygenic photosynthesis, this metabolism on Earth being older than the rise of atmospheric O_2.

On the other hand, an atmosphere rich in O_2 and O_3 may not present systematically the expected signature. Let us consider, for instance, the case of an inhabited planet similar to the Earth, but orbiting further away from its star. At its orbital distance, a CO_2 pressure higher than 50 mbar is required to ensure a mean surface temperature above 0 °C. Observed with Darwin, this "exo-Earth" would exhibit the signatures of H_2O and CO_2, the latter including in addition to the 15-μm band, additional high-pressure bands that would mask the signature of O_3 (see Fig. 10.7). A biosphere similar to the terrestrial one could not be detected under such conditions.

Observed with a "super-Darwin/TPF" and a spectrometer with a much higher resolution, the signature of O_3 could be extracted from the CO_2 bands. However, the existence of these specific CO_2 bands would imply the existence of high CO_2 pressures and therefore the possibility of a nonbiological production of O_2 and O_3. Then, one could not draw any conclusions regarding the possible existence of an ecosystem. For this reason, the thermal emission does not make it possible to search for O_2-producing ecosystems in the whole habitable zone (HZ). The value, $P_{CO_2} = 50$ mbar, roughly defines a border between the internal HZ, where a spectral signature of O_3 can be detected and the external HZ where it cannot (Fig. 10.9). In this external zone, either the greenhouse gas is CO_2, and O_3 is no longer a biomarker, or it is another gas (CH_4, NH_3, N_2O) with reducing properties incompatible with an accumulation of O_2/O_3.

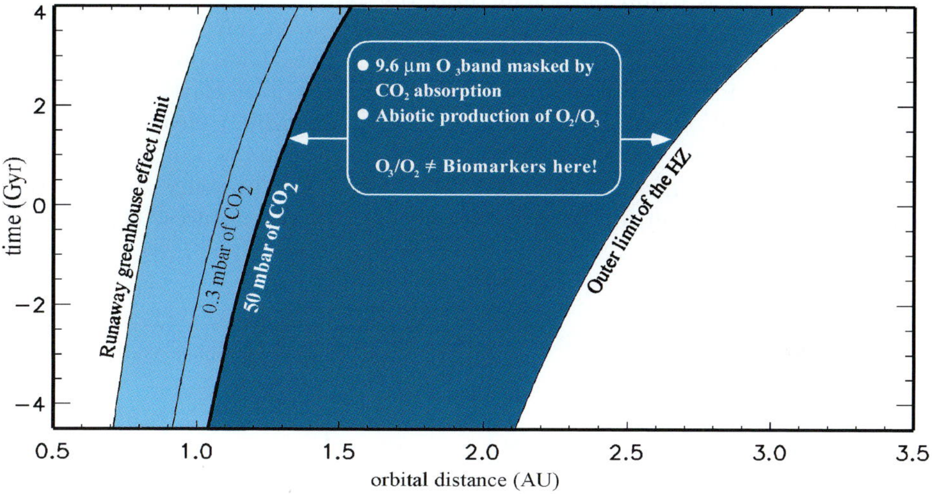

Fig. 10.9. The 9.6-μm band of O_3 as a biomarker in the habitable zone (HZ). The inner limit of the HZ comes from Kasting et al. (1993) and corresponds to a runaway greenhouse effect. The outer limit we put here is the most extended published one (from Forget and Pierrehumbert, 1997; see the Appendix). These limits change with the increase of the stellar luminosity with time. The orbital distance determines the minimum CO_2 partial pressure required to warm a planetary surface temperature above 273 K (0.3 mbar is the present atmospheric level). For CO_2 pressures above 50 mbar, two processes affect the use of O_3 as a biomarker. First, the absorption due to CO_2 masks the O_3 (Selsis et al., 2002). Therefore, this 50-mbar threshold can be used to delimit the fraction of the HZ where the 9.6μm O_3 feature can be detected (*light gray*). Secondly, at CO_2 pressures above 50 mbar, the abiotic production of O_2 can become significant and can result in ambiguous O_2/O_3 levels. Thus, in the outer part of the HZ (*dark blue*), even if a detailed visible spectrum could still reveal the presence of O_2 and O_3, however, their biologic origin would not be certain. A planet in the light blue area can also have a higher CO_2 pressure than the minimum required. This means that it is very important to get information about the CO_2 pressure to be sure that the oxygen has a biological origin. As high CO_2 pressures able to produce an absorbing O_3 layer also mask the O_3 band there is no "false-positive" risk in the IR range. On the other hand, in the reflected spectrum, the detection of O_2 or O_3 without information about the CO_2 content would remain ambiguous. In this wavelength range, getting information about the pressure using the line broadening would require a very high resolution and sensitivity and would not give the partial pressure of CO_2. The whole habitable zone of an M star would be represented in *dark blue* as the CO_2 pressure needed to prevent the atmosphere from collapsing on the dark side of the tidally locked planets is higher than 1 bar

10.1.6 The Detection of an Oxygen-Rich Atmosphere in the Reflected Spectrum

In the visible and near-infrared spectral range, the identification of an O_2-rich atmosphere faces a double problem: first the minimum resolution power required for the detection of a single O_2 line is $R = 54$, ($R \approx 70$ for the detection of 2 or 3 lines, see Fig. 10.10). Secondly, the Chappuis band of O_3, which is very wide, requires a very high sensitivity. The Earthshine spectra obtained by Woolf et al. (2002) and Arnold et al. (2002) clearly show how difficult the detection of this feature on extrasolar planets would be. Observations in the near-infrared are needed to detect the presence of CO_2 that is a crucial information for both habitability and biosignatures. High resolu-

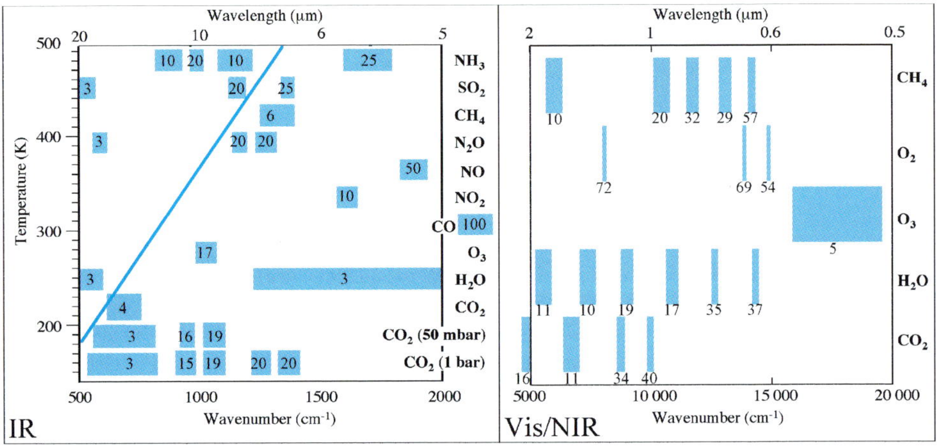

Fig. 10.10. Potential spectroscopic signatures in the midinfrared [5–20 µm] and in the visible-near-infrared [0.5–2 µm] ranges. The resolution power ($R = \lambda/\Delta\lambda$) required for detection is associated to each molecular feature (see text). *Left* (midinfrared, from Selsis, 2000a): All these compounds are present (at least as trace gases) in the present Earth's atmosphere though only CO_2, H_2O and O_3 are effectively detectable at low resolution ($\lambda/\Delta\lambda < 25$). The detection of the other species would imply levels higher than their terrestrial values (see Table 10.1). The level required for detection and the consequent width of the bands has been estimated from synthetic spectra computed for various abundances, such as those given in Fig. 10.2. One notes in the case of CO_2 that, even at low resolution, some quantitative information can be inferred. We also give the relation between the brightness temperature and λ_{max}, the wavelength at which Planck's law $BB(T, \lambda)$ is maximal when expressed as the number of photons emitted per time unit and frequency (or wavelength) unit. *Right* (visible/near-infrared, from Des Marais et al., 2002, and Ollivier, 2004): here, the minimum resolution power is for present atmospheric levels (absorption lines can be broader for higher abundance). One can see that near-infrared is required for CO_2 detection. The Chappuis band of O_3 is a very broad but faint feature that requires a high signal-to-noise ratio

tion ($R > 100$) would be required to retrieve information about the pressure but would not give the CO_2 partial pressure, which is the important parameter to rule out abiotic photochemistry as the origin of O_2. O_2 and O_3 detections are thus ambiguous biomarkers, at low resolution, in the reflected spectrum.

A planet experiencing a runaway greenhouse effect may concentrate O_2 in its atmosphere but not O_3 because of the radicals produced by H_2O photodis-

Table 10.1. The signatures and detectability of biological compounds in the 5–20 µm band. Abundances necessary to obtain a detection with an instrument such as Darwin/TPF (estimated for a 1-bar atmosphere) and abundances in the Earth's current atmosphere (low refers to the lower atmosphere, < 15 km, and mid to the middle or upper one, > 15 km). The threshold for detectability is estimated using synthetic spectra such as presented in Fig. 10.11 (ppm: 10^{-6}, ppb: 10^{-9}). The resolution required for detection is shown in Fig. 10.10

Component	Band location (µm)	Minimal detectable abundance	Terrestrial abundance
CH_4	7.5	10 ppm	2 ppm
N_2O	7.8–8.5–17	1–10 ppm	0.3 ppm
NO	5.4	1 ppm	< 1 ppb (low) 10 ppb (mid)
NO_2	6.2	10–100 ppb	1 ppb (low) 0.1 ppb (mid)
NH_3	9.6–11	1–10 ppm	0.01 ppb

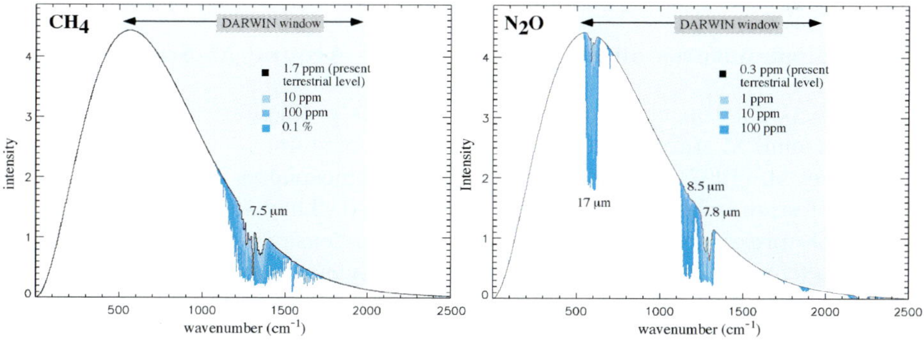

Fig. 10.11. Infrared signature of methane (CH_4) and nitrous oxide (N_2O). These synthetic spectra have been computed for a "terrestrial" atmosphere containing various abundances of CH_4 and N_2O. Only the individual signature of each compound is taken into account. In fact, their detectability depends also on the presence of other absorbing species that can overlap and mask their features (Selsis, 2000a)

sociation (except during the period starting when all the water is lost and ending when all the O_2 is consumed). It was estimated by Kasting (1988) that the runaway phase, if it occurred on Venus, should have lasted for only a small fraction of the planet's history, less than 100 Myr, for a water reservoir comparable to the terrestrial one. However, theories about planetary formation suggest the existence of water-rich terrestrial planets (Léger et al., 2004; Raymond et al., 2004) having an inexhaustible reservoir, able to feed a runaway greenhouse process during a large fraction of the planet's lifetime. Therefore, an abiotic signature of O_2 can be detected in the reflected spectrum of a planet even for low CO_2 levels. A way to filter out such false-positive cases would be to validate only the planets with both O_2 and O_3 signature, but this strategy faces an observational problem: The ozone absorption is a broad but very faint feature (compared to the O_2 lines) that requires a very high sensitivity and could possibly be masked by the Rayleigh scattering (Woolf et al., 2002).

The simultaneous detections of O_2 and CH_4, requiring moderate resolution, were suggested as a biosignature by Lovelock (1980) and Sagan et al. (1993) and remain the most reliable and accessible biomarkers in the visible/NIR range.

The spectral signature of chlorophyll (the "red-edge" seen at 700–750 nm and due to the absorption band of the pigment) has been suggested as a marker for vegetation on exoplanets (Arnold et al., 2002; Woolf et al., 2002). These authors have performed observations of the Moon's Earthshine as an analog to the observation of an extrasolar Earth-like planet. To show up in the data, the red-edge of chlorophyll requires the use of a model for the atmospheric transmission because it is mixed with the 760-nm line of O_2. Such a procedure could obviously not be done in the case of an unknown exoplanet. Also, pigments with other spectral properties exist and thus the terrestrial vegetation signatures may not be universal.

10.1.7 Biosignatures on Habitable Planets Around M-Stars?

These effects due to the CO_2 pressure lead to the problematic case of habitable planets around M stars, the most common type of stars.

Joshi et al. (1997) have modeled planetary atmospheres inside the habitable zone of M stars, where planets are tidally locked. They showed that the minimum CO_2 pressure required to prevent the condensation and collapse of the atmosphere on the night side is 1–1.5 bar. This means that Fig. 10.9 would have no equivalent for M stars: Indeed a CO_2 pressure higher than 1–1.5 bar is required for habitability over the entire habitable zone and not only for its outer part. This has two implications: the first is that the 9.6 μm band of O_3 will not show up. The second is that the efficient abiotic photochemical production of O_2 and O_3 can occur in the whole HZ so any spectroscopic evidence of these molecules (for instance in the visible range) would not necessarily be indicative of the presence of life.

10.2 Others Biomarkers

Oxygenic photosynthesis is only one among many possible metabolisms invented by life on Earth. Other sources of carbon are used (CO, CH_4, organic molecules) as well as many sources of hydrogen or "electron donors" (H_2S, S, H_2, CH_4, organic molecules, NH_3, NO_2, Fe^{2+}, Mn^{2+}, SO_4^{2-}, to quote only a few of them). The output of these various metabolisms are present in our atmosphere as trace materials, with abundances of about, or lower than, a ppm (part per million). Table 10.1 gives a list of the terrestrial atmospheric compounds that are produced by biological activity and that are potentially detectable in the thermal emission of a planet.

One can see from Table 10.1, that none of these compounds could be detected for the current terrestrial abundances with Darwin/TPF. These abundances could have been higher in the past and this was probably the case for CH_4. The methanogenic archaea that produce CH_4 by consuming CO_2 and H_2 are very primitive organisms and appeared prior to the oxygen producers. Before the buildup of atmospheric oxygen, which occurred around -2 Gyr, the photochemical lifetime of CH_4 in the atmosphere was much longer. Nowadays, the total production of CH_4 by the biosphere is 2×10^{14} g/yr (GEIA, 2002) which induces in the atmosphere a CH_4 abundance of 2×10^{-6} (2 ppm). In an atmosphere without oxygen (where O_2 would be replaced, for instance, by N_2, CO_2 or Ar) the same production would lead to an abundance between 100 and 1000 times larger and CH_4 would then be a major greenhouse gas, easily detectable in the planetary thermal emission (Fig. 10.11). In addition, methane is currently produced by anaerobic organisms that are confined in "marginal" environments (marsh, sediments, animal digestive systems, underwater hydrothermal sources; GEIA, 2002). Before the rise of atmospheric O_2 on the Earth, these methanogens were spread over a much wider biotope and the total CH_4 production was higher than today. The recent discovery of anaerobic methanotrophic organisms (using CH_4 as a source of carbon, Hinrichs et al., 1999) indicates that the climate before oxygenation was controlled by the combined action of methanogenic and the methanotrophic species.

The Earth could have exhibited a strong biological CH_4 signature during an important period of its existence (Schindler and Kasting, 2000). CH_4 could thus be quite an interesting biomarker, complementary to O_3, which could be characteristic of primitive worlds and/or worlds that require a strong greenhouse effect that cannot be provided by CO_2. It is generally accepted that only a very small fraction of the CH_4 emitted into our present atmosphere comes from non-biological sources. This "geothermal" fraction, lower than 0.04% of the total production, is released by underwater hydrothermal systems. The abiotic formation process is believed to rely on the oxidation of iron by water that releases H_2 that, in the presence of CO_2 and under specific temperature and pressure conditions, gives CH_4 (Holm and Andersson, 1998). On the other hand, some authors suggest that this abiotic CH_4 flow could have been more important in the past or could exist on other planets (Pavlov et al., 2000). Consequently, it is necessary to carry out the same type of study for CH_4 as for ozone in order to know whether the detection of a CH_4-rich terrestrial planetary atmosphere does imply the presence of an ecosystem.

Methane is also an abundant compound in the external and cold part of our Solar System: It is a major component of Titan's atmosphere and is part of the composition of giant gas planets. According to models (Prinn, 1993), the carbon present in the protosolar nebula, before planet formation, was mainly in the form of CH_4 in the external region and in CO in its inner region. The limit beyond which CH_4 condenses, under the pressures and temperatures in this nebula, is at about 10 AU from the Sun, which explains why one finds CH_4 ice on Pluto, Triton, and Kuiper Belt objects. Cometary ices typically contain 0.5–1% of CH_4 (Crovisier, 1994) and a large fraction of their mass is made of organic matter. Comet impacts on a terrestrial planet could thus be a significant source of CH_4 (a part of which would be a byproduct of the organic matter after the impact). 1% of the mass of a Hale–Bopp-sized comet converted into methane after an impact would result in an atmospheric level of CH_4 equal to the present terrestrial one. This confirms that, in the absence of more thorough study, it is not possible to consider CH_4 as a reliable biomarker.

Among the other components that have a direct or secondary biological origin, one finds several nitrogen compounds, mainly the dinitrogen oxide, N_2O, for which no significant abiotic source is known. N_2O is produced by bacteria (known as *nitrifying* and *denitrifying*) in the ground and oceans, at a rate of 10^{13} g/yr (GEIA, 2002). Its content in the atmosphere, although small (0.3 ppm), plays a major part in photochemistry. It is indeed the main source of nitrogen oxides (NO and NO_2) that are the main destroyers of ozone in the middle and lower atmosphere. It is an optically very active gas in the infrared and, regarding the greenhouse effect, it is even more effective than CH_4, for a given mass. In a terrestrial thermal emission spectrum, the signature of N_2O at 7.8 μm is detectable but a quantity at least 10 times larger would be needed to detect N_2O on an exoplanet with Darwin/TPF. It is especially difficult to state whether such quantities are realistic and if they could have occurred on the Earth, in the past, and particular before its oxygenation.

Atmospheric N_2O is oxidized and photolysed in the upper atmosphere, where it produces NO and NO_2 that are themselves midinfrared-active compounds. However, the minimum quantities for the detection of these two gases are much higher than their current terrestrial abundance and it is considered that they can also be produced by abiotic sources related to lightning (especially in volcanic plumes, Navarro-Gonzalez, 1998) and impacts (Prinn and Fegley, 1987) at a rate larger that coming indirectly from the biosphere.

Another nitrogen compound is present at a trace level in our atmosphere: ammonia, produced by nitrogen-fixing bacteria. NH_3 has an extremely short lifetime because of our current oxidizing environment but also because of its photolysis by UV. This explains its very low abundance (0.01 ppb) that is 2×10^5 times lower than that of CH_4, in spite of a production rate only 5 times weaker

(7.5×10^{13} g/yr; GEIA, 2002). As for CH_4, there are abiotic sources of NH_3 in the outer Solar System. The necessary concentration for the detection of NH_3 (1–10 ppm) is so high that it would imply a large and continuous production source, which would be difficult to assign to an abiotic process. It is also not obvious that any ecosystem would be able to produce such quantities of NH_3.

10.3 Temperature and Radius

If the surface temperature of a planet does not directly reveal the presence of life, it can be a particularly useful data when combined with the composition of a possible atmosphere and to the orbital parameters of the planet. In the case of the Earth, it is striking to note that the surface temperature always remained within the reasonable "limits" for life (Crowley, 1983) whereas the luminosity of the Sun has increased by approximately 40% since its formation (Baraffe et al., 1998; Guinan and Ribas, 2002). The coldest times in the Earth's history are the "snowball Earth events" characterized by an ice cover down to the equator. At the beginning of these events, the runaway ice-albedo feedback makes the global mean temperature drop to −50 °C for a few tens of thousands of years (Hoffman and Schrag, 2002). This temperature drop is followed by a period of a few million years during which the mean temperature is around −10 °C. Snowball events occurred near the beginning (2.45–2.22 Gyr ago) and at the end (0.73–0.58 Gyr ago) of the Proterozoic. The most ancient low-latitude glaciations may be a consequence of the first release of biological oxygen (Selsis, 2002). A description of the Neoproterozoic glacial events has recently been published by Ridgwell et al. (2003).

The hottest periods (for which we have no reliable geological records) may have occurred before the Proterozoic when the atmosphere was anoxic. Indeed, in the absence of an oxidizing atmosphere, the levels of atmospheric methane (produced by methanogens) could have been 100–1000 times higher than today, resulting in a strong greenhouse warming. It is difficult to estimate the surface temperature reached because the photolysis of methane also leads to the formation of Titan-like hazes that increase the albedo. However, mean surface temperatures above 30 °C are possible after the appearance of methanogenesis and before the rise of oxygen.

At the same time, the atmosphere underwent deep transformations. It is probable that the influence of life on the atmosphere and the volatile compounds produced by the geophysical cycles very strongly force this temperature within limits favorable to life. Is the propagation and the evolution of terrestrial life a consequence of favorable and random climatic conditions? Or do we have to consider the conditions on the Earth as the consequence of a "homeostasis" linked to the biological activity and its consequences for the geophysics processes?

If it is difficult to answer this question today, the history of our planet seems to show that life played an important role in maintaining favorable, if not optimal, conditions. In particular, the biological genesis of methane probably counterbalanced the weak solar luminosity before the rise of atmospheric O_2 (Pavlov et al., 2000; Selsis, 2002).

Simultaneously with the information on the chemical composition of the atmosphere, it would be very interesting to gather information on the temperature in order to accumulate signs of life and signs of habitability. It is clear that ozone in the atmosphere of a planet whose surface is heated to 300 °C can only be considered as being of biological origin with some difficulty.

What can we know, remotely, about the temperature of an exoplanet? This question is far from being easy. In theory, spectroscopy gives access to information on the temperature. The thermal profiles of the planets in the Solar System could thus be deduced from the structure of their spectra (obtained over a wide spectral range: UV, visible, infrared, radio). This, however, requires a spectral resolution and a sensitivity that are well beyond the performance of an instrument of the Darwin/TPF type.

The question must thus be adapted to the observational means that will initially be available.

When a planet is detected, and observed for a sufficiently long time, it is possible to compute its orbital parameters (period, semimajor axis, eccentricity, inclination of the system compared to line of sight). It is then possible to deduce the quantity of energy received per unit of area, starting from the luminosity of its star and the semimajor axis of the planetary orbit. This quantity has to be compared with the $1370\,\mathrm{W\,m^{-2}}$ received by the Earth. As shown in Fig. 10.12 this gives us very little information on the temperature. We can then measure the quantity of energy coming from the planet in the spectral range of the instrument. It can be, for example, the energy emitted in the midinfrared (5–20 μm spectral range planned for Darwin) or reflected in the visible-near IR as proposed for alternative concepts of Darwin and TPF (Nisenson and Papaliolios, 2001; Labeyrie 2002).

With a dense atmosphere, the temperature (and thus the thermal emission) has a much more uniform distribution over the planetary surface than without one. In particular, the day–night contrasts are considerably smoothed. Figure 10.13 shows the Earth as seen in the visible and thermal IR ranges. In the case of a Mercury- or Moon-like planet, both images would be extremely similar. Indeed, because the temperatures are much higher during daytime (Mercury: $T_{\mathrm{SS}} = 700\,\mathrm{K}$, $T_{\mathrm{night}} = 150\,\mathrm{K}$; Moon: $T_{\mathrm{SS}} = 380\,\mathrm{K}$, $T_{\mathrm{night}} = 150\,\mathrm{K}$; SS: subsolar point) and, as the integrated thermal emission varies as T^4, the contribution of the night side to the global IR emission is negligible. On the contrary, planets like the Earth (or Venus as an extreme case) with a high thermal inertia provided by an atmosphere (and by the hydrosphere if there is one) have much smaller day–night differences. Therefore, the thermal flux received by the observer would not vary as the visual phase ϕ. This is illustrated in Fig. 10.14. To produce this graph, we considered a simplified inner solar system (circular orbits and obliquities = 0°) seen with an inclination of 45° (the average inclination for randomly

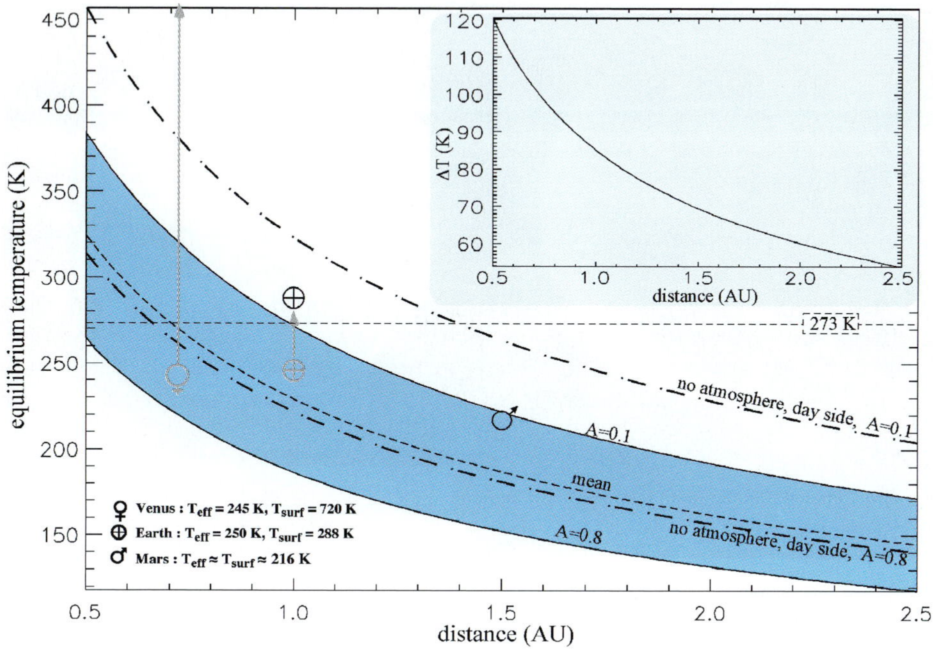

Fig. 10.12. Equilibrium temperature and albedo. The orbital distance of a planet and its albedo (the fraction of the incoming light reflected to space) determine an equilibrium temperature $T_{\rm eq}$, corresponding to the temperature of a black body that re-emits all it absorbs. The *solid curves* give $T_{\rm eq}$ for two extreme albedos observed in the Solar Sytem (0.1 and 0.8) and delimit a blue area of possible values of $T_{\rm eq}$ (the *dashed line* shows the average of them). The *dotted-dashed lines* represent the day-side temperature when the illuminated side is the only one that radiates ("no atmosphere" case). $T_{\rm eq}$ for Venus, Earth and Mars is indicated with a *light gray symbol*. The *black symbols* show the measured surface temperature on these planets (for Venus this is too high to appear). The surface of the Earth, and especially of Venus, are heated by the greenhouse effect of the atmosphere: these two planets have very different albedos but a similar effective temperature; however, their surface temperature differs by more than 400 K. The *graph* in the *upper-right part* shows the uncertainty on $T_{\rm eq}$ (i.e. the difference between the two *solid curves*). One can see that the orbital distance by itself is poorly indicative of the surface temperature of a planet

oriented systems). The day and night emissions have been averaged to compute a dayside and a nightside mean value. For the Earth, this was done with MODTRAN (Berk et al., 1989) taking into account the cloudiness, the cloud types and the area covered by land and sea. For Mars, we used the Martian Climate Database (Lewis et al., 1999) that provides the total thermal emission into space at the top of the atmosphere. On Venus, the day and night sides emit roughly the same amount of energy. Mercury and the Moon are assumed to be Lambertian spheres, which is a good approximation according to Lawson et al. (2000).

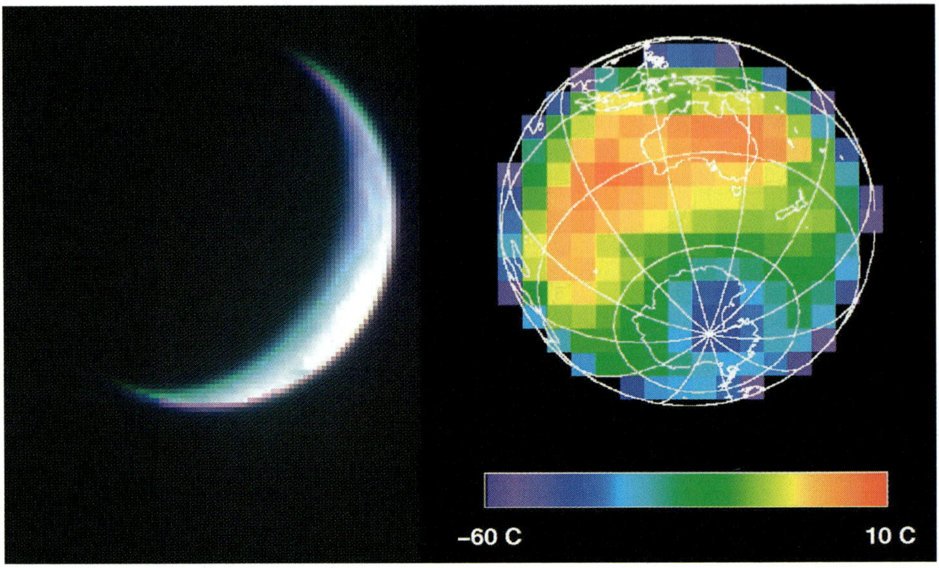

Fig. 10.13. These two images were acquired at the same time by Mars Odyssey using a visible camera (*left*) and the thermal emission imaging system (THEMIS, Murray et al., 2003). These images illustrate what could be seen (without spatial resolution) by the two alternative projects for Darwin/TPF: an optical coronographic single telescope observing the stellar light reflected by the planet and an IR interferometer observing the thermal proper emission of the planet. While the reflected light comes only from the illuminated fraction of the planet the thermal emission comes from the whole planetary disk with low spatial variations. A planet without an atmosphere like Mercury or the Moon would produce a similar image in both spectral ranges. Image credit: NASA/JPL/Arizona State University

One can see that the variation of the thermal flux for planets having a dense atmosphere does not follow the phase variation. On a Mars-like planet, the day/night contrast is still strong but the thin CO_2 atmosphere moderates the variations of the thermal flux. On an Earth-like planet, the differences between the day and night are very small, due to the thermal inertia of the atmosphere, the ocean and the clouds. Note that if the Earth–Moon system is not spatially resolved, (Darwin/TPF will not resolve planet–satellite systems) the global variations of the IR flux are enhanced by the Moon. As an extreme case, the IR brightness of a Venus-like planet remains constant, whatever the geometry of the observation. The curves plotted on this graph do not show the realistic variations of the IR flux from the planets of the Solar System because the seasonal changes due to the obliquity (or the eccentricity) are not included. However, these seasonal changes are not correlated with the phase and additional seasonal modulations would not prevent the detection of the presence of an atmosphere and may give some information about the obliquity of the planet.

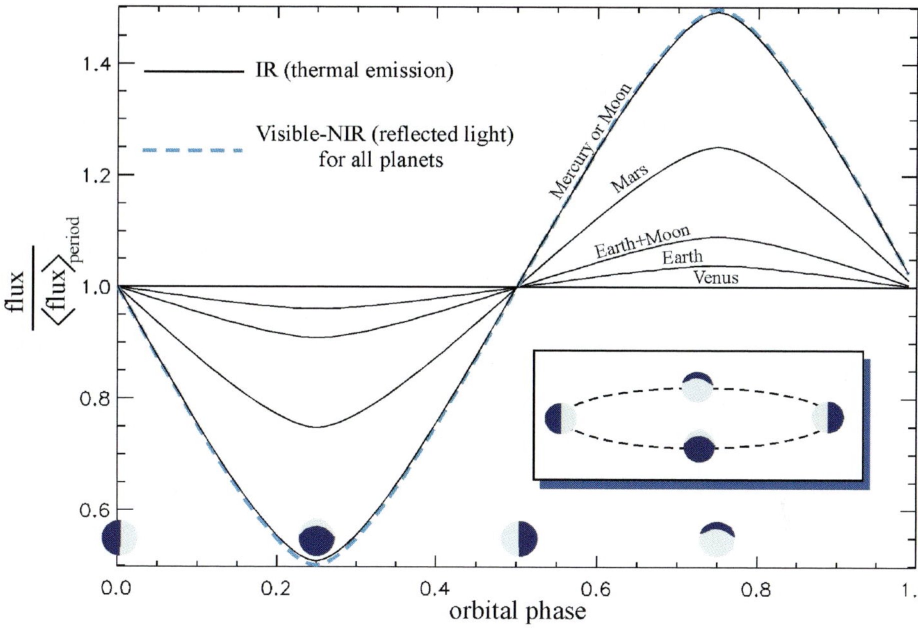

Fig. 10.14. Variation of the IR flux over an orbital period for different kinds of planets. The reflected light received from a planet is directly proportional to the illuminated fraction of the planet (visual phase: ϕ, indicated by a dashed line). In return, the IR emission only follows the variation of ϕ for atmosphereless planets. In this simple model, taking only into account the thermal emission averaged over the day and night hemispheres, the planets have null obliquities and circular orbits seen with an inclination of 45°. With this inclination, the reflected flux is 3 times higher at its maximum than at its minimum $\phi_{\max}/\phi_{\min} = 0.75/0.25$. Real variations of the observed thermal flux would be modulated by seasonal changes that are not correlated with the visual phase

Therefore, by providing the orbital and the integrated thermal emission along the orbit (no spectral resolution is required here), the follow-up of a detected planet would reveal the presence of an atmosphere, provided that the inclination is not too close to 0°. It would also show the stability of the temperature, which is an important parameter for habitability. One question still remains: could a planet without atmosphere rotate fast enough too smooth its day–night contrast and thus too mimic the presence of an atmosphere? At least not in the habitable zone of very low mass stars (K and M) where the tidal braking of the rotation is strong. Anyway, the variation of the IR brightness would surely reveal the absence of a dense atmosphere and allow us to reject nonhabitable worlds.

With a low spectral resolution ($\lambda/\Delta\lambda > 10$), as shown in Fig. 10.2, the mean brightness temperature and the radius of the planet can be obtained by fitting the envelope of the thermal emission by a Planck function. The ability to retrieve the temperature relies on the existence and identification of spectral windows probing the same atmospheric levels or the surface (indicated by dashed arrows on the graph). Such identification faces some difficulties. For instance, a strong absorption by water, smoothed by a low resolution, may produce a Planck-like spectral shape and thus in a significant error in the temperature and the radius. There are however, atmospheric windows that can be used in most of the cases, especially between 8 and 11 μm as seen in Fig. 10.10.

The average brightness temperature of the Earth is a correct estimate of the average surface temperature, slightly lowered by about 10 K due to the contribution from the top of the clouds. On Venus and Mars, the brightness temperature gives, respectively, a good estimate of the atmospheric temperature at 60 km and of the surface.

As said previously, the brightness temperature can differ from the surface temperature. Let us compare the infrared spectra of Venus, Earth and Mars, three telluric planets having a dense atmosphere (Figs. 10.2 and 10.3). Does the black-body temperature deduced from the emission spectra give a good approximation to the surface temperature? The answer is "yes" in the case of Mars where the emitting surface is the planetary surface. However, it is difficult to define an average Martian temperature because the temperature varies with time and latitude. In the Earth's case, the total planetary spectrum is a mixture of surface and clouds emission, the latter occurring at lower temperature. The temperature given by the envelope of the spectrum is thus slightly lower than the average surface temperature. This temperature also depends on the observation geometry, as well as seasons. It can thus differ from the total average temperature of the Earth ($\sim$ 288 K). In the extreme case of Venus, the spectrum envelope gives a temperature of 277 K, to be compared to a surface temperature of 740 K. Observed in the midinfrared Venus could be considered as a habitable planet, with a mean temperature just above the triple point of water. The reason for this discrepancy comes from the fact that the Venusian atmosphere is completely opaque below 60 km because of the permanent cloud cover. However, a Venusian atmosphere without any cloud would still be opaque below an altitude similar to that of the clouds because of the absorption continuum, induced at high pressure, by CO_2–CO_2 collisions. With low-resolution spectral observations, it is difficult to know, without ambiguity, if the lower atmosphere contributes or not to the spectrum, and thus if the temperature reflects the surface conditions.

It should, however, be noted that the presence of a cloud layer, or more generally of an opaque atmosphere, does not prevent a deduction of the planet diameter starting from its thermal IR spectrum.

The accuracy of the radius determination will depend on the quality of the fit (and thus on the sensitivity and resolution of the spectrum), the precision of the Sun–star distance (known with a precision of better than 2% for G and K stars within 20 pc of the Sun) and also the distribution of brightness temperatures over the planetary surface. Concerning the latter, we can consider a Lambertian sphere (like the Moon) as a "worst case" exhibiting high temperature contrasts. In such a case, the inferred radius would be 10% smaller than the real one. However, in this case, the variation of the thermal flux with the phase (see Fig. 10.14) would reveal the absence of atmosphere and the estimate could be readjusted. When the brightness temperature is stable along the orbit, the estimated radius is reliable to better than 10%.

On the other hand, the reflected flux only allows one to infer the product $A \times R^2$ where A is the albedo and R the planetary radius. The first generation of optical instruments will be very far from the angular resolution required to directly measure an exoplanet radius. Such a measurement can nowadays be performed only if the planet passes in front of its parent star (this case is called "transit"), by an accurate photometric technique. This is probably the main weakness of the characterization in the visible range. Indeed, the knowledge of the planetary radius is crucial for the general understanding of the physical and chemical processes occurring on the planet (tectonics, hydrogen loss to space).

10.4 Other Exo-/Astrobiological Aspects of the Darwin/TPF Missions

As exciting as the search for life signatures can be, it remains an uncertain objective, which is neither the single nor the first motivation for a mission such as Darwin/TPF. These observatories will be wonderful tools for the study of telluric planets, their atmosphere, their formation and their evolution. To do planetary science with Darwin/TPF is also essential for astrobiology because this should bring many elements improving our knowledge of terrestrial planets and possible sites for life. We should in particular get clues to answer the following question: are our planetary system and our planet common objects in the Universe or on the contrary extremely marginal?

First, the existence itself of telluric planets in the internal part of a planetary system, in the famous "habitable zone" of stars, is considered by the models of planetary formation but cannot be confirmed with our current observation tools. The first major element will thus be the statistical information on the distribution of these small planets. On this point, other space observatories should give us information before Darwin; first COROT, then the Kepler mission, based on planetary transits observation should have the sensitivity to detect Earth-sized planets. Contrary to these other missions, Darwin/TPF will give us this information for nearby systems ($<$ 10parsecs, approximately) for which we have access to complementary data concerning the presence or not of giant planets, right now detectable by the radial velocity method. The influence of giant planets on the habitability of internal planets is not yet well known but probably important, regarding the formation of internal telluric planets itself, the origin of water and the volatile compounds on these planets, the bombardment of these planets by asteroids or comets (Lunine, 2001).

For each system studied, the observed properties of internal planets (such as their distribution, mass, orbit, the presence or absence of an atmosphere, its chemical composition), confronted with the properties of giant planets of the system, will give us precious clues about the origin, the formation and the evolution of the planetary systems, the atmospheres and perhaps life.

Such an instrument would also inform us considerably about the concept of "habitable zone": are these zones filled with planets as in the Solar System? Do these planets have atmospheres? And if yes what are their composition? Is CO_2 the only greenhouse gas that proves to provide habitability (any efficient greenhouse gas produces detectable features in the thermal-emission spectrum)? How do the characteristics of these planets vary with the spectral type and the metallicity of their star, the distribution of giant planets or the age of the system?

The prospect of being able to observe very young planetary systems (less than 0.5–1 Gyr) is very interesting. The atmospheres of the Solar System planets have evolved and differ from their primitive composition. The nature of the prebiotic environment on Earth is probably one of the keys to understanding the origins of life. However, it is possible that only the observation of young exoplanets could bring part of the answer.

The χ^1 Ori star, located at 8.7 pc from the Earth is a G1V star, comparable, or at least very similar, to the Sun (Guinan and Ribas, 2002); its mass is $0.99\,M_{\mathrm{Sun}}$ and its age is estimated at 300 Myr. At an equivalent age, the Earth already had oceans (Wilde et al., 2001) and an atmosphere, of unknown composition. Exposed to an intense bombardment, life could not settle there even if, perhaps, it could already appear. Does χ^1 Ori have planets? If yes, what are they made of? An instrument such as Darwin/TPF could maybe highlight these points.

10.5 Conclusion

In this chapter, we considered the information given by low-resolution ($\lambda/\Delta\lambda \sim 25$) midinfrared spectral range (5–20 μm) spectra on the nature and the atmospheric composition of extrasolar planets, and particularly, telluric ones. Assuming earth-like biotic or at least comparable conditions, we reviewed potential biosignatures and detailed how they can be searched for. In particular, we showed that ozone, often considered as a biomarker, can, if detected alone, lead to false detection of life. A more robust criterion of life can be found in the simultaneous detection of ozone, water, and carbon dioxide.

More generally, we showed that the thermal IR spectral range is particularly well suited for exoplanetology and the study of the exoplanets atmospheres. In addition to spectral features, the thermal IR spectral range can lead to important physical parameters such as the size and an estimation of the temperature of the planet. The development of space observatories aiming at detecting and analyzing extrasolar planets should lead to major breakthroughs in that scientific domain in a near future.

Appendix: Elements Concerning the Habitable Zones

The habitable zone (HZ) around a star is defined as the region in which water can exist in the liquid state, in a permanent way, on the planet surface. Kasting et al. (1993, 1997) estimated the limits of this zone for various types of stars. The inner limit of the area is determined by the capacity of the planet to keep its water reservoir. Beyond a threshold of luminosity, a runaway greenhouse effect starts: water is vaporized, transported towards the upper atmosphere, photodissociated and hydrogen escapes thermally. Kasting estimated this limit to be at 0.84 AU for the current Sun. Venus, which is at 0.7 AU, could have lost its water reservoir at the time of such a phase.

The outer limit depends on the composition of the atmosphere and on the greenhouse gases it contains. Kasting, as well as other authors (see the review paper by Forget, 1998), considered the case of atmospheres of CO_2. In this case, dense atmospheres may lead to heat the surface above 0°C but beyond several bars, the increase in the atmospheric albedo decreases the efficiency of the greenhouse effect. This outer limit was considered by Kasting at 1.67 AU[2] and was extended to 2.5 AU taking into account the radiative effects of the CO_2 clouds (Forget and Pierrehumbert, 1997). Mars (1.5 AU) is thus inside the HZ and certain authors considered the "terraforming" of this planet (McKay et al. 1991).

Given that the luminosity of a star increases during its life, the inner and outer limits gradually move away from the star, as shown in Fig. 10.9. The continuous habitability zone (CHZ) is defined as the region that remains habitable as long as the star stays on the main sequence. Venus was probably within the solar habitable zone in the past but is no longer included in the CHZ.

The argument for a definition of the habitable zone based on CO_2 is that the amount of this greenhouse gas must be self-regulated by the CO_2 cycle (Walker et al., 1981). If the CO_2 level is too low, water freezes and CO_2 cannot precipitate in the form of carbonate any longer, whereas it is always produced by volcanoes; its abundance thus increases until the frozen surface becomes liquid again. If there is too much CO_2, the high surface temperatures makes the atmosphere very wet, leading to strong precipitation, strong surface weathering and an increase of the CO_2 precipitation rate into carbonate. However, the self-regulation of CO_2 and climate is currently extensively discussed (Sleep and Zahnle, 2001, Franck et al., 2000): it is not at all certain that terrestrial planets can maintain very

[2] It is important to mention the difficulties such estimations assume, regarding the modelling of complex atmospheric processes: in a dense CO_2 atmosphere, CO_2 condenses, leading to clouds, which heating or cooling effect is still discussed. In addition, and as mentioned in this section, such an atmosphere produces an ozone layer that can prevent CO_2 condensation. Finally, surface-temperature calculations, based on the CO_2 greenhouse effect are based on high CO_2 pressure and on the absorption properties of CO_2–CO_2 collisions. This last point is not yet well known.

high CO_2 levels in the atmosphere, in the presence of liquid water. It is possible, for instance, that the habitability of the terrestrial surface was due, in the past, to high atmospheric methane levels and not only to CO_2.

The presence of a planet in the HZ does not mean that this planet is habitable. Mars, for example, would be habitable with an atmosphere of about 1 bar of CO_2, i.e. 170 times denser than the current Martian atmosphere. It does mean, on the other hand, that a planet orbiting outside the HZ would not be habitable. But the concept of a "habitable zone" should a priori not only be based on the properties of CO_2 because, as mentioned before, other atmospheric compounds can play the role of greenhouse gases (CH_4, N_2O, NO_x, some sulfur compounds, NH_3). Some of them are more efficient than CO_2, and could push back the outer boundary of the HZ. However, since CO_2 is the only one among these gases that is "compatible" with an oxidising atmosphere, this definition of the HZ can be restricted to the biologically O_2-enriched atmospheres.

The organic chemistry on which life such as we know it is based uses the volatile elements C, H, O and N. It is interesting to note that if the telluric planets were made up only of material formed at the level of their orbit, they would be probably stripped of organic matter, water, and maybe atmosphere. If the internal telluric planets do not receive a veneer of volatile-rich material (asteroids, comets or interplanetary dust) nor originate from a mixture with the planetary embryos formed further in protostellar nebula, the HZ as previously defined loses its meaning.

The habitability we talked about is in fact a surface habitability, which depends on the solar luminosity and the presence of an atmosphere. The current interrogations about the possible presence of life in an ocean under the cold surface of Europa or in the Martian subsoil show that the concept of habitable zone must be widened. Indeed, energy sources other than the radiation can maintain a liquid-water reservoir. On Europa, this energy is provided by the tidal forces due to the eccentricity of the orbit around Jupiter. One can thus define a habitable zone around giant planets (Reynolds et al., 1987).

In fact, the concept of habitable zone widened a lot after the discovery of ecosystems in environments qualified as "extreme". Lithotrophic Bacteria, not requiring luminous energy, can live in rocks, at great depth, provided that the rock is sufficiently porous to let water infiltrate. In this case, the geothermal energy makes the rock habitable. The Earth, ejected out of the Solar System, and thus out of its habitable zone, would certainly keep a biosphere in its crust. However, there is no idea, at the present time, of the conditions (perhaps more restrictive) necessary to the emergence of life. A habitable planet must, of course, satisfy these unknown conditions to become an inhabited world.

References

Specialised

Angel J. R. P., Cheng A. Y. S., and Woolf N. J. (1986). A space telescope for infrared spectroscopy of Earth-like planets. *Nature*, **322**, 341

Arnold L., Gillet S., Lardière O., Riaud P., and Schneider J. (2002). A test for the search for life on extrasolar planets – Looking for the terrestrial vegetation in the Earthshine spectrum. *A&A*, **392**, 231–237

Baraffe I., Chabrier G., Allard F., and Hauschildt P. H. (1998). Evolutionary models for solar metallicity low-mass stars: mass-magnitude relationships and color-magnitude diagrams. *A&A*, **337**, 403–412

Beichman C. A., Woolf N. J., and Lindensmith C. A. (eds.) (1999). *The Terrestrial Planet Finder (TPF): a NASA Origins Program to Search for Habitable Planets*, JPL Publications, Pasadena, Ca

Bekker A., Holland H. D., Wang P.-L., Rumble III, D., Stein H. J., Hannah J.L., Coetzee L.L., and Beukes N.J. (2004). Dating the rise of atmospheric oxygen, *Nature*, **427**, 117–120

Berk A, Bernstein L.S., and Robertson D.C. (1989). MODTRAN: A MODerate resolution for LOWTRAN 7, *Tech. Rep.*, *Air Force Geophysics Laboratory*, Bedford, Mass

Borucki W.J., Koch D.G., Dunham E.W., and Jenkins J.M. (1997). The Kepler mission: A mission to determine the frequency of inner planets near the habitable zone of a wide range of stars, in *ASP Conf. Ser. 119: Planets Beyond the Solar System and the Next Generation of Space Missions*, p. 153

Bracewell R.N. (1978). Detecting nonsolar planets by spinning infrared interferometer. *Nature*, **274**, 780

Brasseur G. and Solomon S. (1984). *Aeronomy of the Middle Atmosphere: Chemistry and Physics of the Stratosphere and Mesosphere*, Dordrecht, D. Reidel Publishing Co.

Brocks J.J., Logan G.A., Buick R., and Summons R.E. (1999). Archean molecular fossils and the early rise of Eukaryotes. *Science*, **285**, 1033–1036

Charbonneau D., Brown T.M., Latham D.W. and Mayor M. (2000). Detection of Planetary Transits Across a Sun-like Star. *ApJ. Lett.*, **529**, L45

Charbonneau D., Brown T.M., Noyes R.W., and Gilliland R.L. (2001). Detection of an extrasolar planet atmosphere. *Ap. J.*, **568**, 377

Christensen P.R. and Pearl J.C. (1997). Initial data from the Mars Global Surveyor thermal emission spectrometer experiments: Observations of the Earth. *J. Geophys. Res.*, **102**, 10875

Crovisier J. (1994). Molecular abundances in comets, in *IAU Symp. 160: Asteroids, Comets, Meteors*, **160**, 313

Crowley T.J. (1983). The geologic record of climatic change. *Rev. Geophys. Space Phys.*, **21**, 828–877

Des Marais D.J., Harwit M.O., Jucks K.W., Kasting J.F., Lin D.N.C., Lunine J.I., Schneider J., Seager S., Traub W.A., and Woolf N.J. (2002). Remote sensing of planetary properties and biosignatures on extrasolar terrestrial planets. *Astrobiology*, **2**, 153

Forget F. and Pierrehumbert R.T. (1997). Warming early Mars with carbon dioxide clouds that scatter infrared radiation. *Science*, **278**, 1273

Forget F. (1998). Habitable zone around stars. *Earth, Moon and Planets*, **81**, 59–72

Franck S., Block A., von Bloh W., Bounama C., Schellnhuber H., and Svirezhev Y. (2000). Habitable zone for Earth-like planets in the solar system. *Plan. Space. Sci.*, **48**, 1099–1105

GEIA (2002). Global Emissions Inventory Activity (GEIA) – http://weather.engin.umich.edu/geia/

Guinan E. and Ribas I. (2002). Our changing Sun: The role of solar nuclear evolution and magnetic activity on Earth's atmosphere and climate, in *ASP Conf. Series*, **269**, 85–106

Hall D.T., Strobel D.F., Feldman P.D., McGrath M.A., and Weaver H.A. (1995). Detection of an oxygen atmosphere on Jupiter's moon Europa. *Nature*, **373**, 677

Hinrichs K.-U., Hayes J.M., Sylva S.P., Brewert P.G., and Delong E.F. (1999). Methane-consuming archeabacteria in marine sediments. *Nature*, **398**, 802–805

Hoffman P.F. and Schrag, D.P. (2002) The snowball Earth hypothesis: testing the limits of global change. *Terra Nova*, **14**, 129–155

Holland H.D. (1994). Early Proterozoic atmospheric change, in *Early life on Earth, Nobel Symposium No.* **84**, 237–244

Holm N.G. and Andersson E.M. (1998). In *The Molecular Origins of Life*, p. 86–99, Cambridge Univ. Press

Johnson R.E. and Jesser W.A. (1997). O_2/O_3 microatmospheres in the surface of Ganymede. *ApJ. Lett.*, **480**, L79

Joshi M.M., Haberle R.M., and Reynolds R.T. (1997). Simulations of the atmospheres of synchronously rotating terrestrial planets orbiting M dwarfs: Conditions for atmospheric collapse and the implications for habitability. *Icarus*, **129**, 450–465

Kasting J.F. (1988). Runaway and moist greenhouse atmospheres and the evolution of Earth and Venus. *Icarus*, **74**, 472

Kasting J.F., Whitmire D.P., and Reynolds R.T. (1993). Habitable zones around main sequence, *Icarus*, **101**, 108

Kasting, J.F. (1997). Habitable zones around low mass stars and the search for extraterrestrial life. *Origins of Life and Evolution of the Biosphere*, **27**, 291–307

Labeyrie A. (2002). Hypertelescopes and exo-Earths coronography in *Proceedings of the 36th ESLAB Symposium: Earth-like Planets and Moons*, 245–249, ESA Publications

Lawson S.L., Jakosky B.M., Park H., and Mellon M.T. (2000). Brightness temperatures of the lunar surface: Calibration and global analysis of the Clementine long-wave infrared camera data, *J. Geophys. Res.*, **105**, 4273

Léger A., Mariotti J.M., Mennesson B., Ollivier M., Puget J.L., Rouan D., and Schneider J. (1996). Could we search for primitive life on extrasolar planets in the near future? The DARWIN project. *Icarus*, **123**, 249

Léger A., Ollivier M., Altwegg K., and Woolf N.J. (1999). Is the presence of H_2O and O_3 in an exoplanet a reliable signature of a biological activity? *A&A*, **341**, 304

Léger A., Pirre M., and Marceau F.J. (1993). Search for primitive life on a distant planet: relevance of O_2 and O_3 detections. *A&A*, **277**, 309

Léger A., Selsis F., Sotin C., Guillot T., Despois D., Lammer H., Ollivier M., Brachet F., Labèque A., Valette C. (2004). A new family of planets? "ocean-planets". *Icarus*, in press

Lewis S.R., Collins M., Read P.L., Forget F., Hourdin F., Fournier R., Hourdin C., Talagrand O., and Huot J. (1999). A climate database for Mars, *J. Geophys. Res.*, **104**, 24177

Lovelock J.E., (1980). The Recognition of Alien Biospheres, *Cosmic Search Summer*, 2, 2
Lunine J.I., (2001). The occurrence of Jovian planets and the habitability of planetary systems, *Proc. Natl. Acad. Sci. USA*, **98**, 809–814
MacKay C.P., Toon O.B and Kasting J.F., (1991). Making Mars habitable. *Nature*, **352**, 489
Mayor M. and Queloz D. (1995). A Jupiter-mass companion to a solar-type star. *Nature*, **378**, 355
Murray K.C., Christensen P.R., Mehall G.L., Gorelick N.S., Harris J.C., Bender K.C., and Cherednik L.L. (2003) In *Lunar and Planetary Institute Conference Abstracts*, 1363
Nair H., Allen M., Anbar A.D., Yung Y.L., and Clancy R.T. (1994). A photochemical model of the Martian atmosphere. *Icarus*, **111**, 124–150
Navarro-Gonzalez R., Molina M.J., and Molina L.T. (1998). Nitrogen fixation by volcanic lightning in the early Earth. *Geophys. Res. Lett.*, **25**, 3123
Nisenson P. and Papaliolios C. (2001). Detection of Earth-like planets using apodized telescopes. *ApJ. Lett.*, **548**, L201
Noll K.S., Johnson R.E., Lane A.L., Domingue D.L., and Weaver H.A. (1996). Detection of ozone on Ganymede. *Science*, **273**, 341
Noll K.S., Roush T.L., Cruikshank D.P., Johnson R.E., and Pendleton Y. J. (1997). Detection of ozone on Saturn's satellites Rhea and Dione. *Nature*, **388**, 45
Ollivier M. (1999). Contribution to the search for exoplanets. Nulling interferometry for the DARWIN mission, PhD Thesis, Université de Paris XI
Ollivier M. (2004). in Proceedings of IAP colloquium "Extrasolar Planets: Today and Tomorrow", *ASP Conf Series* (in press)
Owen T. (1980). The search for early forms of life in other planetary systems: future possibilities afforded by techniques in *Strategies for the Search for Life in the Universe*, M. Papagiannis (ed.), Reidel
Pavlov A.A., Kasting J.F. (2002). Mass-independent fractionation of sulfur isotopes in Archean sediments: Strong evidences for an anoxic Archean atmosphere, *Astrobiology*, **2–1**, 27–41
Pavlov A.A., Kasting J.F., Brown L.L., Rages K.A., and Freedman R. (2000). Greenhouse warming by CH_4 in the atmosphere of early Earth. *J. Geophys. Res.*, **105**, 11981
Prinn R.G. and Fegley B. (1987). Bolide impacts, acid rain, and biospheric traumas at the Cretaceous-Tertiary boundary. *Earth and Planetary Science Letters*, **83**, 1–4
Prinn R.G. (1993). Chemistry and evolution of gaseous circumstellar disks, in *Protostars and Planets III*, p. 1005–1028, Univ. Arizona Press
Rasool S.I. and De Bergh C. (1970). The runaway greenhouse and accumulation of CO_2 in the Venus atmosphere. *Nature*, **226**, 1037
Raymond S.N., Quinn T.R., Lunine J.I. (submitted to *Icarus*). Making other Earths: Dynamical simulations of terrestrial planet formation and water delivery
Reynolds R.T., MacKay C.P., and Kasting J.F. (1987). Europa, tidally heated oceans, and habitable zones around giant planets. *Adv. Space Res.*, **7**, 125
Ridgwell A.J., Kennedy M.J., and Caldeira K. (2003) Carbonate deposition, climate stability, and Neoproterozoic ice ages. *Science*, **302**, 859–862
Rosenqvist J. and Chassefiere E. (1995). Inorganic chemistry of O_2 in a dense primitive atmosphere. *Planet. Space Sci.*, **43**, 3

Rouan D., Baglin A., Copet E., Schneider J., Barge P., Deleuil M., Vuillemin A., and Léger A. (1998). The exosolar planets program of the COROT satellite. *Earth, Moon, Planets*, **81**, 79

Rye R., Kuo P.H., and Holland H. (1995). Atmospheric carbon dioxide concentrations before 2.2 billions years ago. *Nature*, **378**, 603

Sagan C., Thompson W.R., Carlson R., Gurnett D., and Hord C. (1993). A search for life on Earth from the Galileo spacecraft. *Nature*, **365**, 715

Schindler T.L. and Kasting J.F. (2000). Synthetic spectra of simulated terrestrial atmospheres containing possible biomarker gases, *Icarus*, **145**, 262

Schneider J. (2003). *The extrasolar planets encyclopedia*, http://www.obs.pm.fr/planets

Schopf J.W. (1994). The oldest known records of life: Early Archean stromatolites, microfossils, and organic matter, in *Early life on Earth, Nobel Symposium No. 84*, 193

Selsis F. (2000a). Darwin and the atmospheres of terrestrial planets in *DARWIN and Astronomy – The Infrared Space Interferometer*, ESA SP-451, p. 133–140, ESA Publications

Selsis F. (2000b). *Modeling the physical and chemical evolution of planetary atmospheres. Application to early Earth's atmosphere and terrestrial exoplanets*, PhD Thesis, Université de Bordeaux I

Selsis F. (2002). Occurrence and detectability of O_2-rich atmosphere in circumstellar "habitable zones", in *ASP Conf. Series*, **269**, 273–281

Selsis F., Despois D., and Parisot J.-P. (2002). Signature of life on extrasolar planets: Can Darwin produce false positive detection? *A&A*, **388**, 985–1003

Sleep N.H. and Zahnle K. (2001). Carbon dioxide cycling and implications for climate on ancient Earth. *J. Geophys. Res.*, **106**, 1373

Spencer Jones H. (1959). *Life on Other Worlds*, Hodder and Stoughton, London

Vidal-Madjar A., Lecavelier des Etangs A., Désert J.-M., Ballester G. E., Ferlet R., Hébrart G. and Mayor M. (2003). An extended upper atmosphere around the extrasolar planet HD209458b. *Nature*, **422**, 143–146

Volonte S., Laurance R., Whitcomb G., Karlsson A., Fridlund M., Ollivier M., Gondoin P., Guideroni B., Granato G.L., Amils R., and Smith M. (2000). *Darwin: the Infrared Space Interferometer*, Technical report, ESA

Walker J.C.G. (1977). *Evolution of the Atmosphere*, New York: Macmillan

Walker J.C.G., Hays P.B., and Kasting J.F. (1981). A negative feedback mechanism for the long-term stabilization of Earth's surface temperature. *J. Geophys. Res.*, **86**, 9776

Wilde S.A., Valley J.W., Peck W.H., and Graham C.M. (2001). Evidence from detrital zircons for the existence of continental crust and oceans on the Earth 4.4 Gyr ago. *Nature*, **409**, 175

Williams D.M., Kasting J.F., and Wade R.A. (1997). Habitable moons around extrasolar giant planets. *Nature*, **385**, 234–236

Wolszczan A. and Frail D.A. (1992). A planetary system around the millisecond pulsar PSR1257+12. *Nature*, **355**, 145

Woolf N.J., Smith P.S., Traub W.A., and Jucks K.W. (2002). The spectrum of Earthshine. A Pale Blue Dot observed from the ground. *Ap. J.*, **574**, 430–433

Bibliography

Jakosky B. (1998) *Search for Life on Other Planets*, Cambridge University Press

Kump L.R., Kasting J.F., and Crane R.G. (2003). *The Earth System*, 2nd Edition, Prentice may

Lunine J.I. (1998). *Earth: Evolution of a Habitable World*, Cambridge University Press

Ward P. and Brownlee D. (2000, new edition: 2003) *Rare Earth: Why Complex Life is Uncommon in the Universe.* Copernicus Books

Part II

From Prebiotic Chemistry to the Origin of Life on Earth

Second Part Introduction

Chemical evolution, which started very early in our Universe, is not only a preliminary but also an essential step to any biological evolution. All the discussions concerning a hypothetical RNA world evolving towards a DNA world is based on the assumption that deoxyribose derives from ribose or, at least, that the replacement of ribose by deoxyribose was an important step in biological evolution. Many other examples of the interdependence between chemical evolution and biological evolution are known.

Unfortunately, we have no molecular fossils or relics of the prebiotic world and our knowledge about molecular evolution during the prebiotic period is based on models and scenarios. These scenarios themselves are based on hypotheses concerning the physico-chemical conditions present at the surface of the young Earth. They are also supported by experimental simulations and by indirect observations. Indeed, interstellar clouds that can be observed today are probably not very different from the protosolar nebula and chondritic and micrometeoritic matter that we can study in our laboratories today are probably similar to the chondritic or cometary matter that fell on the young Earth in large quantities during the prebiotic period. LUCA (our last universal common ancestor) was most probably a unicellular entity with a membrane, a metabolism, a reproduction capability and therefore a genetic code. LUCA itself was most probably not the most primitive form of life on Earth. Chemists and biologists must introduce constraints into the scenarios they suggest for the transition steps from nonlife to life. These constraints are based on what we know about the chemical elements and their reactivity, about the physicochemical laws, and very importantly, about the conditions prevailing on the young Earth when the transition occurred. The study of extreme biotopes on Earth, such as the ices above the Vostock Lake or the vicinity of oceanic black smokers, yields information about the diversity of life and also about the eventual universal requirements for life.

Together with astronomers, planetologists and geologists, biologists and chemists involved in exobiological studies participate in the elaboration of these models. As already mentioned, the chemist has no relics, no molecular fossils of the prebiotic period but he/she is the only one who can perform experimental simulations to help a model progress. The biochemist and the biologist are the only ones who work with living species, they know how living species have evolved and they are ready to take the risk to extrapolate towards the past: their approach could be described as a top-down approach while the chemists have a bottom-up approach. The hope (or better the dream) is that, one day, chemists and biologists could reach a similar conclusion about what LUCA looked like!

For the chemist, as for all "specialists" involved in exobiology, the most important contribution to the field can only come from their capacity to interact efficiently with other specialists. Exobiology is probably the best example of an interdisciplinary science: all natural sciences but also mathematics and informatics participate to its development. The next ensemble of chapters is a clear example of the work of chemists, biochemists and biologists, working in different fields but searching to contribute to the understanding of what remains an important problem: the emergence of life from nonlife.

1 A Rational Approach to the Origin of Life: From Amphiphilic Molecules to Protocells. Some Plausible Solutions, and Some Real Problems

Guy Ourisson, Yoichi Nakatani

Synopsis Self-organisation of amphiphiles in water into closed vesicles leads automatically to self-complexification into "protocells". However, some real problems are usually not even mentioned in the various theories of the origin of Life. The present discussion is a follow up of our initial publications (Ourisson and Nakatani, 1994, 1999; see Maddox, 1994)

1.1 From Amphiphilic Molecules to "Protocells" by Understandable Processes. Self-Organisation and Self-Complexification

Our thesis will be that the known general universal laws of the physical world can go a long way towards showing how simple molecular systems can spontaneously become self-organised by understandable processes. This is not an original statement. However, we shall also show that self-organisation automatically carries with it the seeds of self-complexification. Without requiring new "steps", it leads to novel properties, all singularly necessary for life to accrue at some later stage. We believe that this is an original set of far-reaching assumptions, even though it does not (yet?) lead to the definitive solution. It will serve us finally to delineate some problems that have seldom or never been raised, and that would require, sooner or later, careful study and eventually a solution to help us understand how life begun on Earth. These "hidden problems" include:

- The need to have had somewhere a concentrated brew, the "primordial soup" of Oparin (Oparin, 1968)
- The need to have had an efficient synthesis of long chain lipids
- The need to have had in the same brew high enough concentrations of phosphoric acid derivatives
- The need to surmount the unexpected and inherent difficulty of phosphorylation with phosphoric anhydrides
- The need to have had sufficient concentrations of all the "bricks of life", however they are produced
- And the need to understand the formation of the cytoskeleton of microtubules

1.2 Water and Self-Organisation of Amphiphiles

1.2.1 The Structure of Liquid Water

The unique structure of water, dominated by an irregular network of hydrogen bonds, is the key to many of its unusual properties (Fig. 1.1). Dissolution of a substance carrying many hydrophilic groups, for instance a sugar, is possible because it leads to the replacement of many homologous hydrogen bonds (water to water) by many heterologous ones (water to OH). Around each O atom, the H atoms of the same molecule, and those of neighbouring ones, are arranged statistically along tetrahedral directions. The energy required is small, and the foreign solute is well accepted (i.e. it is soluble; Fig. 1.1). In some cases (e.g., that of the sugar trehalose), the new network of hydrogen bonds is so well accommodated that the solute in fact better organises the molecular network of water into quasitetrahedral arrangements (see Tanford 1978, Blokzijl 1993, Lemienx 1996). Rather similar molecular properties characterise other solvents, such as ethanediol (Larsen 1984) or formamide (Auvray, 1991); these solvents appear, however, to have no significance for life as we know it. (For an up-to-date discussion of the structure of water, see for instance Westhof 1993).

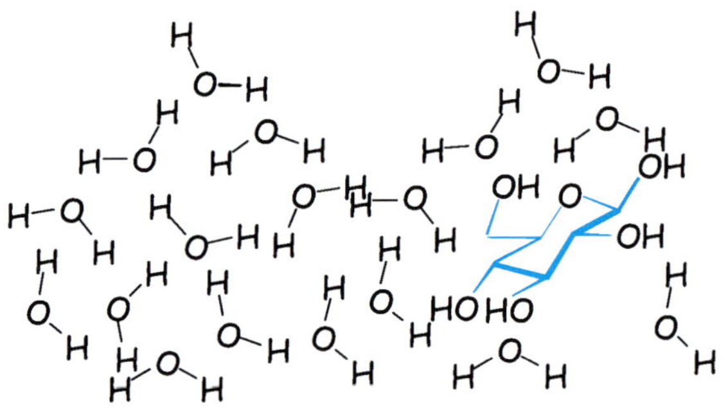

Fig. 1.1. Schematic structure of water and of water containing a molecule of glucose. The hydroxyl groups of the substrate can engage in hydrogen bonding and only little distort the network of H bonds

1.2.2 The Structure of Amphiphile-water Mixtures

By contrast, a molecule like a long-chain alcohol, despite its terminal hydroxyl, is not soluble, because its insertion into the network of hydrogen bonds requires breaking many of them without compensation. It is akin to boring a hole into this compact arrangement, and costs too much energy (Fig. 1.2). Rather than being dissolved in the water phase, the molecules of such an "amphiphile" reach for the surface, and are dissolved only through their polar head-groups (see Wang, 1994; Ron, 2000). At the same time, their long chains are rejected from the water

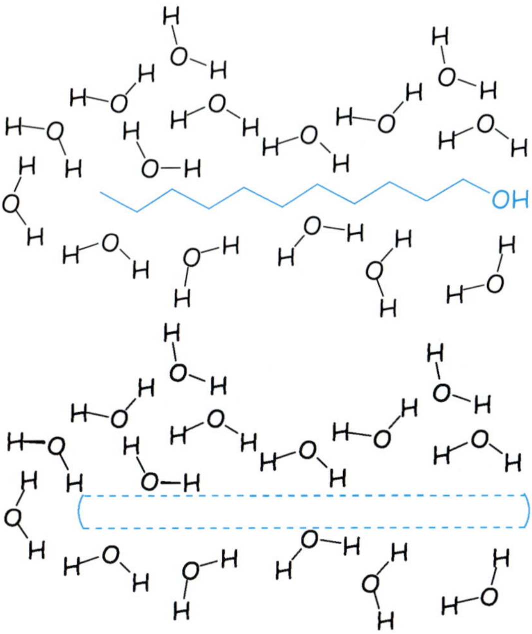

Fig. 1.2. Dissolving a long-chain alcohol in water would require "boring a hole" in the system of H-bonds

to its surface where they self-organise into a monolayer of hydrocarbon chains, more or less compact depending on the side-pressure (Fig. 1.3) (for a detailed review of monolayers, see Kuzmenko et al., 2001; for an analysis of the relevance of monolayers for chiral amplification, a topic not covered in the present chapter but obviously relevant to the origin of life, see Weissbuch et al., 2003).

Amphiphilic molecules of the proper shape (depending on the relative sizes of the polar head-group and the side-chain (Israelachvili et al., 1980), can self-organise in water. It is then possible to obtain a variety of arrangements where the polar heads are grouped together to maximise their contact with the surrounding water, while the lipidic side-chains are segregated. Depending on the particular shape and concentration of the amphiphile, one can then obtain *micelles* (in which the hydrophobic chains are segregated and the hydrophilic heads coat the surface), or larger objects, *vesicles* (syn.: single-walled *liposomes*). These are the systems of particular interest to us. In these, the lipidic chains, all approximately parallel, are ordered into a double layer coating the surface of the vesicle, and separate the inside water from the outside water (Fig. 1.4). They are formed spontaneously by hydration of a lipid film (sonication gives very small vesicles). Vesicles can be of micrometric size, and therefore can be observed by modern optical microscopy; they are then called "giant vesicles". They can be filtered through polycarbonate filters of homogeneous bore size, and have then a narrow diameter distribution, which can be evaluated by light diffusion.

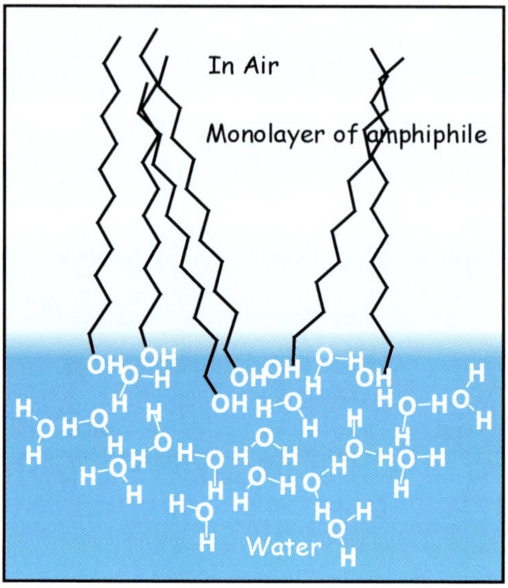

Fig. 1.3. Molecules of a long-chain alcohol self-organise into a monolayer on the surface of water

In these spontaneously formed systems, three distinct phases can be recognised: the water outside, the water inside, and the inside of the bilayer, which is lipidic. This leads, for instance, to the selective extraction into the lipidic membrane of any lipophilic molecule present, for instance a lipophilic dye such as Nile Red. It also leads to the phase separation of racemic and nonracemic amphiphilic α-amino acid derivatives at the air/water interface, a phenomenon that may be important to explain the homochirality of the present peptide amino-acids (Weissbuch et al., 2003).

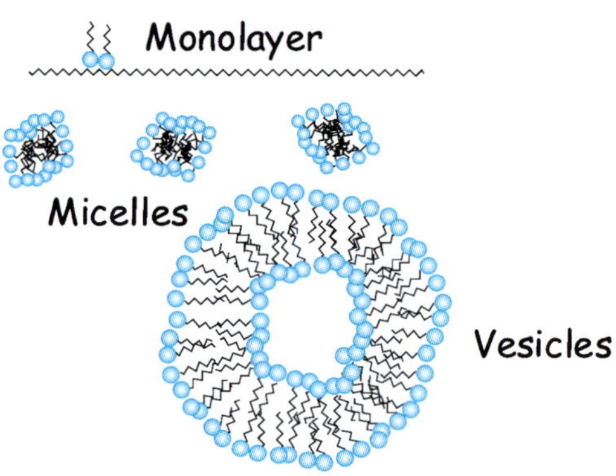

Fig. 1.4. Amphiphiles at the surface of and inside water

1.3 Properties Ensuing from the Self-organisation of Amphiphiles

1.3.1 Extraction and Orientation

The inside of the bilayer is not only lipophilic, but also highly anisotropic: it is a fibrous medium, and any nonisotropic lipophilic molecule will therefore be extracted and oriented. This is obvious, but it is also experimentally proved in at least two cases: that of cholesterol, which has been shown to be oriented parallel to the lipidic chains (with its OH group at the surface) (Nakatani et al., 1996), and that of the hydrocarbon β-carotene, which lies, depending on the lipid used, parallel to the surface, probably in the midlayer space, or embedded in the lipids (Nordén et al. 1977) (Fig. 1.5).

A very interesting discussion of the possibilities offered by this selective extraction is given in Deamer et al. (1994).

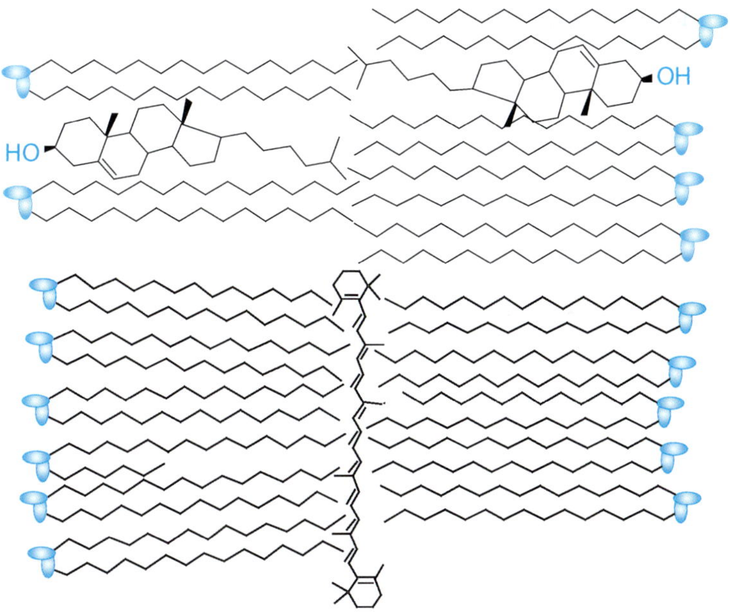

Fig. 1.5. Cholesterol is inserted in lipidic membranes perpendicularly to the surface, whereas β-carotene is oriented parallel to the surface

1.3.2 Increased Concentration and Condensation

Another property induced by the self-organisation of amphiphiles into vesicles is that the selective extraction of lipophilic substances into the double-layer will lead to increased concentrations, and that this opens up the possibility of selective reactions within this double layer. This has been demonstrated in three cases, by Fukuda in Urawa (Fukuda et al. 1981), by Ringsdorf in Mainz (Folda et al., 1982) and by Luisi and Walde in Zurich (Bachmann et al., 1991, Walde et al., 1994; Blocher et al., 1999), with somewhat hydrophobic of amino-acid derivatives that, in the presence of vesicles, underwent nonenzymatic condensation into dipeptides or small peptides (Fig. 1.6).

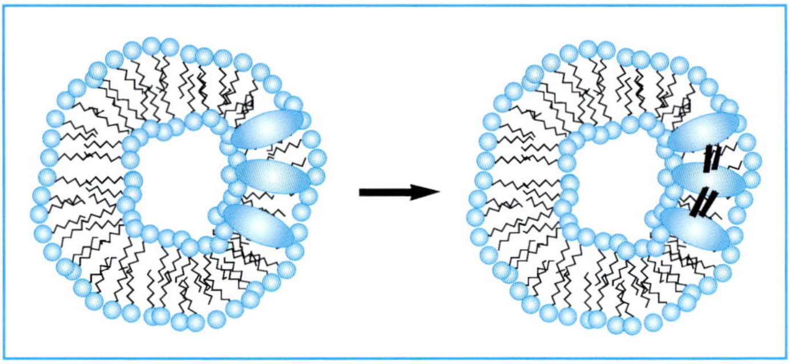

Fig. 1.6. Hydrophobic derivatives of amino-acids are selectively extracted into the double layer of vesicles, where they are spontaneously condensed into derivatives of polypeptides

1.3.3 Vectorial Properties

Self-organisation also leads to vectorial properties: the outer lipids, on a convex surface, are intrinsically different from the inner ones, which are on a concave surface. This had been demonstrated by biochemical arguments already in 1972, in the case of erythrocytes (Bretscher, 1972). It also leads to consequences measurable by NMR, as had been first shown by Chrzeszczyk et al. (1977). We have shown it quite directly in a ^{31}P NMR study of mixed vesicles composed of two different phospholipids, giving each two peaks of different intensities (Fig. 1.7); the ratio of these intensities corresponds to the ratio of the number of molecules in the outside and the inside layers of the vesicle wall, as deduced from the average size of the vesicles (evaluated by light diffusion) (Lee et al., 2002); the same doubling of peaks cam also be observed on vesicles formed of a single phospholipid, at the proper pH. If the NMR signals of the inside and the outside head-groups differ, all their properties must also be sensitive to their location: for instance, they must have different pKa values (Swairjo et al., 1994). Water inside the vesicle must be different from water outside because the outside layer is convex, the inside one concave (and a little more crowded); this is the basis for these "vectorial properties". These differences H must be more marked for small vesicles than for larger ones, but might well be carried over from small ones when they grow. It would be quite interesting to be able to demonstrate whether a measurable pH difference is spontaneously established, and also whether a lipophilic polypeptide is inserted in a predictable orientation (amino-end out?). We have not yet set out to do this.

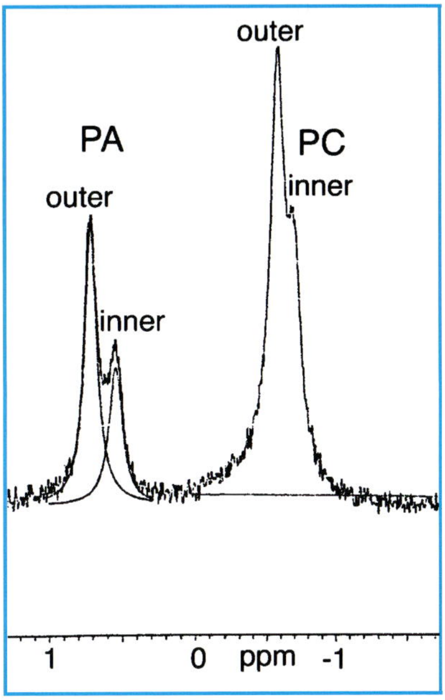

Fig. 1.7. ^{31}P NMR spectrum of vesicles of two phospholipids. Each P-containing head-group gives rise to a double peak; the more intense doublet is due to the major PC in the mixture, the smaller one to the phytanyl phosphate (here, PA). Each is a doublet, the minor peak being due to the inner head-groups, the major one to the outer ones

1.3.4 Coating the Vesicles

Another novel property arising from self-organisation is the possibility to coat the vesicle with a carbohydrate wrapping, analogous to the bacterial cell wall. This was first demonstrated by Sunamoto with the large fungal polysaccharide pullulane, by labelling it with a fluorescent tag making it visible by optical microscopy, and by linking it to cholesterol, as an anchor, into the outside layer (we shall return below to this concept of anchors) (Ueda et al., 1998) (Fig. 1.8).

We have also shown with Akiyoshi that the same polysaccharide, labelled not with cholesterol but with the highly branched phytol, acts in the same way to cover the outside of vesicles made of geranylgeranyl phosphate (Ghosh et al., 2000).

It still remains to check whether these "coatings" add stability to the vesicles.

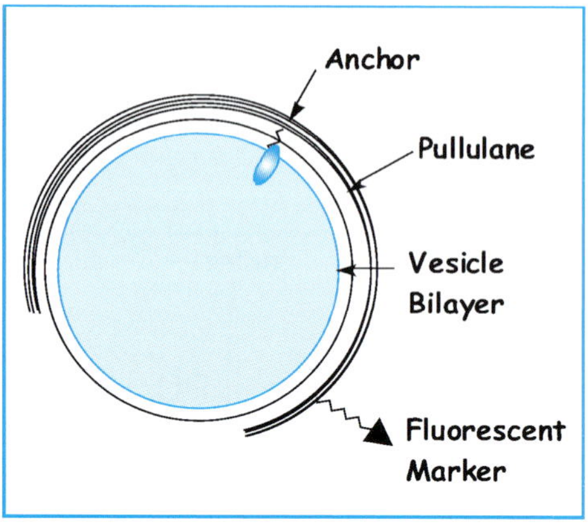

Fig. 1.8. Schematic representation of a double-layer vesicle coated with pullulane, linked to a membrane anchor and labelled with a fluorescent tag

1.3.5 Vesicles and Nucleic Acids; Vesicles as Protocells

Several authors have shown that it is possible to entrap nucleic acid molecules into vesicles, which then act as tight microcontainers. The first success in this field was obtained by enclosing Simian virus 40 (5 kb) into small vesicles of phospholipids ($D = 400$ nm) (Fraley et al., 1980). Next, success was achieved with small (1–20 kb) DNAs, like salmon sperm DNA, enclosed into egg lecithin vesicles (3 ± 5 µm) by dehydration/rehydration (Deamer, Barchfeld, 1982) and with plasmid DNA (3.3 kb) by using the same method, or the reverse-phase method, or the freeze-thaw method (Monnard et al., 1997). Yoshikawa's group in Kyoto has more recently shown that, in the presence of M_g^{2+} ions, one can introduce into vesicles, by several techniques, giant DNA (10^2–10^5 kb) or RNA molecules, large enough to be observed by optical microscopy. The same can be done with their compacted histone complexes. One can use forceful insertion of an individual nucleic acid molecule, under the microscope, pushing it through the membrane by holding it with a laser tweezer, in much the same way as one carries out in vitro fecundation (Nomura et al., 2001). The nucleic acid molecules can be observed undergoing Brownian movement inside the vesicle, "bouncing" on the inner wall. They are then transcribed into m-RNA molecules that are efficiently protected from the action of nucleases introduced in the surrounding medium: the membranes form effective barriers (Tsumoto et al. 2001; Fischer et al., 2002). What is very remarkable is that nucleic acid molecules efficiently find their way into vesicles during the spontaneous formation of these vesicles when a lipid film is left in the presence of a suitable buffer containing the nucleic acid molecules during hydration–dehydration cycles (Deamer, Barchfeld, 1982, Tsumoto et al., 2001). It looks as though they are preferentially entrapped into these newly formed vesicles, as a surprisingly large proportion of them (typically 35–50%) do contain nucleic acid molecules.

Jay and Gilbert showed that this entrapment was facilitated by the presence of a basic protein, lysozyme (Jay, Gilbert, 1987), and the clay montmorillonite facilitates the conversion of fatty acid micelles into vesicles, often encapsulating clay particles which may be of significance for the formation of protocells. (Hanczyc, 2003).

Other very important biochemical reactions have been successfully carried out inside vesicles: the polymerase chain reaction (Oberholzer et al., 1995-1), and the enzymatic RNA replication in self-reproducing vesicles (a non-natural model obtained from oleic acid/sodium oleate) (Oberholzer et al., 1995-2), see also: Walde, Ishikawa, 2001.

Even more spectacular has been the synthesis of proteins inside vesicles: Oberholzer et al., (1999) used non-native RNA (Poly-U) as a template for the synthesis of the non-natural polypeptide polyPhe, and Yu et al., (2001) described the synthesis of a functional protein in giant vesicles (but did not show reliably whether the synthesis was occurring inside or outside the vesicles); finally the synthesis, more efficient inside vesicles than outside, of the green fluorescent protein (GFP) by a plasmid encoding the GFP gene, trapped inside the vesicle, has been described by Nomura et al., (2003).

To sum up, phospholipid vesicles can be considered as model protocells. That they are still very far from being cell models is obvious. That it would still be very worthwhile to study further their spontaneous complexification is, in our view, just as obvious.

This provides additional support, as well as more precise structural detail, for the seminal hypothesis of Deamer and of Morowitz, regarding the role of vesicles as likely precursors of early cells (Deamer, 1986, Morowitz et al., 1988). For more general considerations on this topic, and novel hypotheses, see for instance Szostak et al., (2001).

We shall now turn to a review of a series of questions often hidden when discussing the hypotheses about the origin of life.

1.4 The Nature of the Primitive Amphiphiles

1.4.1 The Modernity of N-Acyl Lipids

So far, we have mentioned "amphiphiles" without really discussing their nature. The best known amphiphiles, those of Bacteria and of Eucaryotes, are quite complex molecules, requiring for their biosynthesis three distinct pathways: the one producing long n-acyl chains, saturated or not, by a series of condensation of C_2 units and reductions, and those required for the C_3 unit of glycerol and for the C_2 unit of choline (Fig. 1.9). The other types of "standard" amphiphiles carry head-groups just as complex or more complex; none can be considered as possibly produced by "prebiotic" mechanisms. Usually, proponents of an early compartmentation into vesicles refer, in an attempt to "explain" a prebiotic formation of the long n-acyl chains, to a remarkable

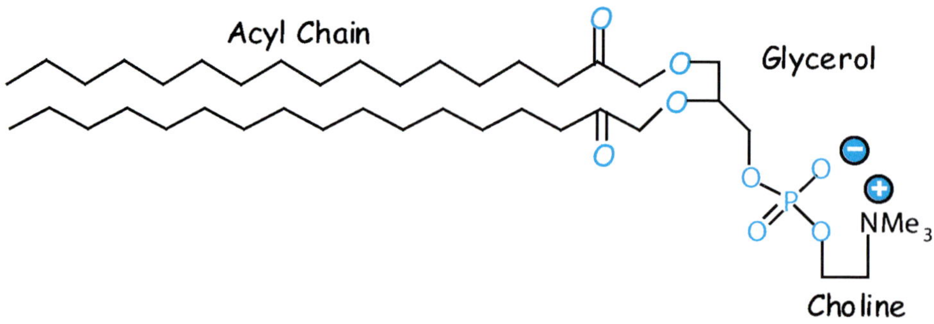

Fig. 1.9. A typical eucaryotic phospholipid: dipalmitoylphosphatidylcholine (DPPC)

process: the Fischer–Tropsch (F–T) reaction, which indeed converts C_1 or C_2 units into longer chains; however, to our knowledge, even the most complex variants of the F–T reaction do not lead to any significant amount of long-chain (C_{14-20}) terminally functionalised n-acids or alcohols. We think that any "explanation" by recourse to this reaction is rather a formal invocation. We also consider as doubtful the contention that meteorites could have been the source of these lipids, as they contain amphiphilic molecules; they do, but in very small amounts, and their complex mixtures containing many aromatic acids are certainly not able to form membranes (Deamer et al., 1994, pp. 112–114).

1.4.2 The Archæal Lipids and Their Synthesis

In recent years, another family of membrane lipids has been discovered in the "new" phylum of Archæa: polyprenyl lipids, usually in C_{20} or, by dimerisation, in C_{40}. These would be just as difficult to obtain under prebiotic conditions; but they display some characteristics much more compatible with a primitive origin: their polyprenyl chains are biosynthesised by C_5 increments (not C_2 as in the biosynthesis of n-acyl chains), which reduces, of course, the number of steps required; these steps are simple alkylations of double bonds (not complex condensations and reductions), and the chemistry involved is "elementary" and has been duplicated (from C_5 to C_{10}) by simple heating of suitable precursors with a montmorillonite clay (Desaubry et al. 2003) (Fig. 1.10).

These steps are, of course, repetitive, and could lead in principle, with the same substrates, from C_{10} to C_{15}, then to C_{20}. Selectivity could be ensured as soon as there is a phase separation at a certain length of the lipophilic chain, with the proper head-group. The complex head-groups present in modern archæal phospholipids call for an independent evolution. We have postulated that the simplest possible polar head, a phosphoric acid salt, might already have the necessary characteristics. We have therefore synthesised a series of phosphate esters containing one or two polyprenyl chains, and have demonstrated that

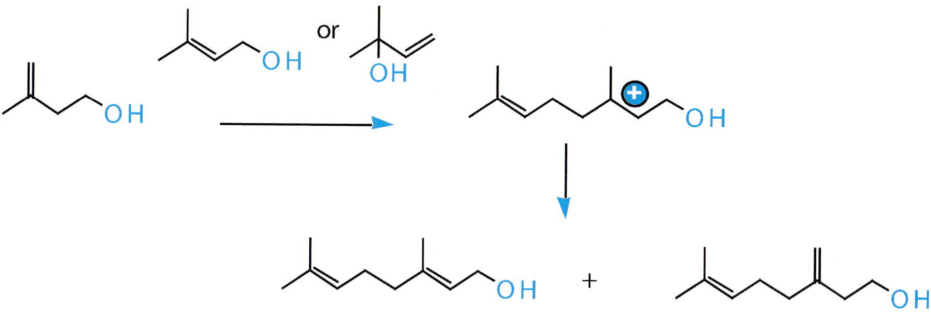

Fig. 1.10. Synthesis of C_{10} terpenic chains from C_5 units

indeed single- or double-chain polyprenyl phosphates do form vesicles, provided their chain(s) contain in toto at least 15 carbon atoms (Birault et al., 1996, Pozzi et al., 1996).

One of the most easily accessible of these "primitive" phosphates is the C_{20} phytyl phosphate (Fig. 1.11), as phytol is a cheap commodity material. We have used this phytyl phosphate extensively, as well as the very similar geranylgeranyl phosphate (Fig. 1.12) to study various properties of polyprenyl phosphate membranes, and retain, as a hypothesis, a scenario involving these phosphates as primitive membrane constituents. It is then possible to postulate a series of hypothetical successive steps, rendering their structures progressively more complex. A complete "genealogical tree" of polyterpenes has thereby be constructed (Ourisson, 1986); however, as it is entirely hypothetical, and relates more to the evolution than to the origin of living organisms, we shall not develop that point here but only present this genealogical tree without comments (Fig. 1.13).

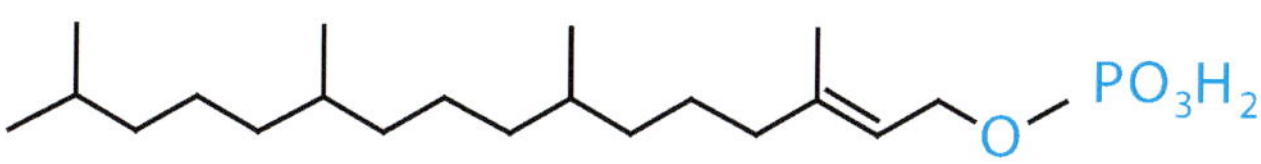

Fig. 1.11. Phytyl phosphate

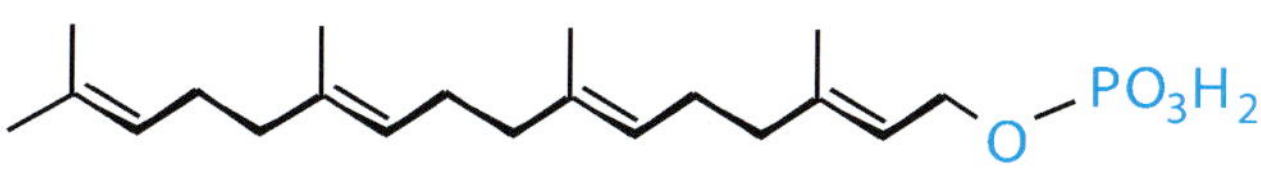

Fig. 1.12. Geranylgeranyl phosphate

Polyprenyl phosphates could be the precursors of *all* polyprenyl derivatives found in memberbranes (Sect. 1.5.4).

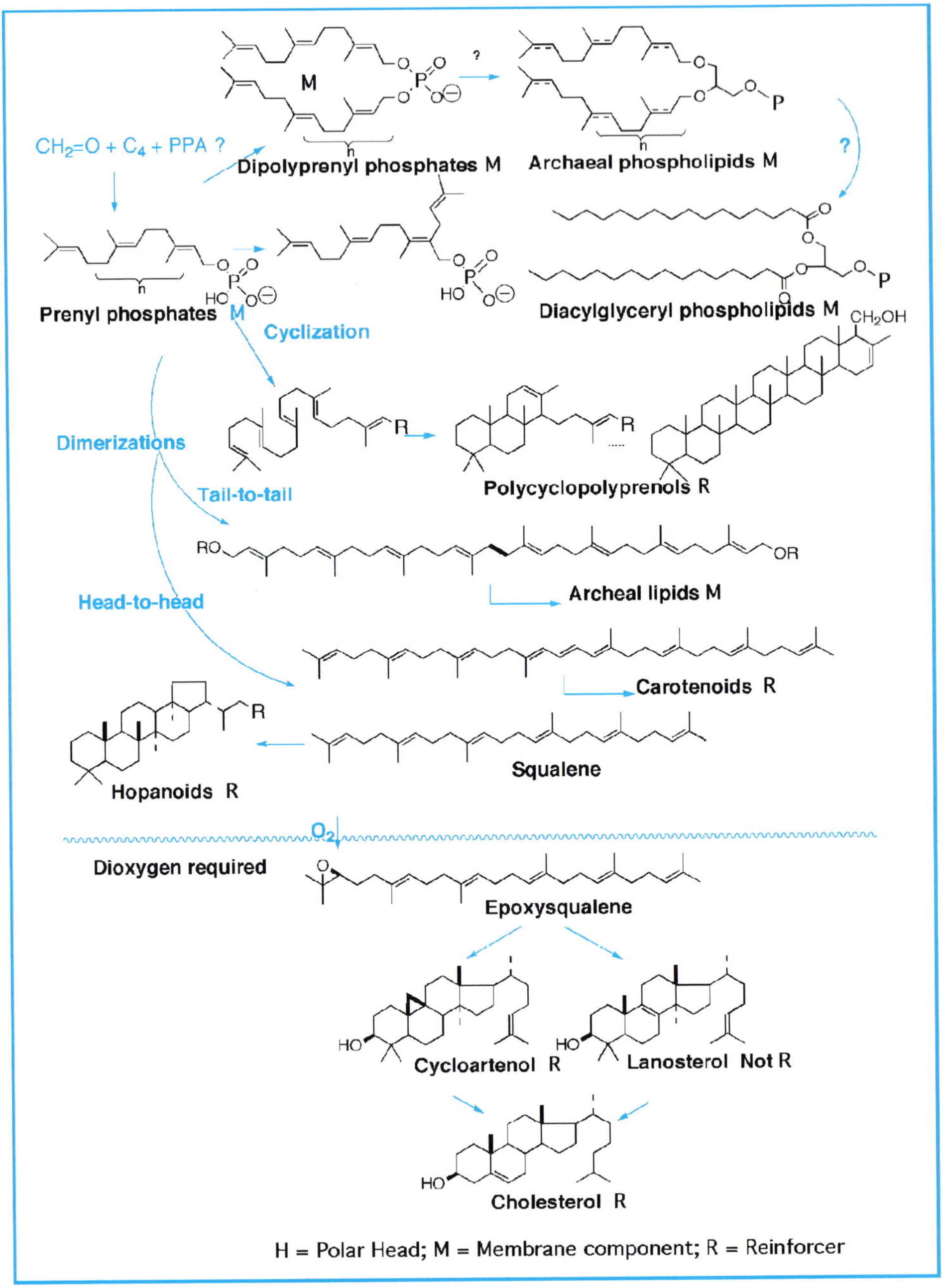

Fig. 1.13. Hypothetical evolution of membrane-related terpenoids (adapted from Ourisson, 1986)

1.4.3 The Terpenoids as Universal Metabolites

We must, however, note that polyterpenes are not "secondary metabolites" as usually considered: they are universal, necessary constituents of all living organisms, and must therefore be promoted to the dignity of essential metabolites. They occur as:

- Constituents of the archæal lipids
- Ubiquitous anchors binding proteins to cell membranes by farnesyl or geranylgeranyl chains
- Ubiquitous constituents of electron transporters across membranes (ubiquinones)
- Ubiquitous membrane constituents involved in the glucosylation of proteins ($C_{\approx 100}$ dolichyl phosphates)
- Light sensors: retinal of the light-driven proton pump of *Halobacterium*, bacteriorhodopsin, and of the light-sensing protein of our eyes, rhodopsin
- Constituents of the photosynthetic systems (the phytyl group of chlorophyll)
- Reinforcers of phospholipidic membranes (cholesterol or equivalents such as phytosterols in Eucaryotes, hopanoids in Bacteria and α,ω-dihydroxylated carotenoids in Bacteria (*St. aureus* and *Halobacterium*)
- Hormones and pheromones of invertebrates and vertebrates
- Vertebrate hormones (vitamin A, calciferol)
- Thousands of pheromones, attractants, repellents, defense substances, toxic alkaloids, etc., in all groups of living organisms.

Thus, a terpenic origin of membrane lipids is a quite tenable hypothesis. However, we shall now list a series of problems that are seldom or never discussed in the present context, and for which we have no solution to propose.

1.5 Some Remaining Problems

1.5.1 The Problem of the Synthesis of Ingredients

From the brief description we have given of the formation of vesicles in water, it is clear that a minimal concentration must be attained locally before any self-organisation sets in (this is the critical micellar concentration, the cmc). Two separate questions arise:

- Where did the necessary ingredients come from?
- How did they reach high enough local concentrations?

The first question has found, in principle, a satisfactory solution in the classical Miller experiment, now 50 years old (Miller, 1953): sparks in a suitable atmosphere (methane, ammonia, dihydrogen, water) lead to the formation of a series of amino-acids and nucleic acids bases; these "bricks of life" are remarkably easy to obtain. With M. Devienne, we have ourselves obtained very similar results, including not only the synthesis of proteogenic amino-acids, but also that of the four nucleic bases and of many other complex substances; these were formed by high-energy bombardment of graphite targets with molecular beams of the desired atoms, simulating plausible interstellar events (Devienne et al., 1998, 2002). Similar results have also been obtained by UV irradiation of ice containing the necessary ingredients (Bernstein et al., 2002; Muñoz Caro et al., 2002). In short, it looks as though any input of enough energy into a set of molecules containing the necessary atoms, followed by quenching, gives rise to a bevy of "bricks of life".

1.5.2 The Problem of Local Concentration

This does not begin to address the question of local concentration. When Oparin spoke of a "primitive brew" (Oparin, 1968), he had in mind a continuous process of formation of the necessary constituents and of accumulation in the primitive oceans, because there was, of course, no biodegradation in the absence of living organisms. This hypothesis may be compatible with Miller's process (Miller, 1953) (despite all the criticisms raised against his choice of composition of a supposedly "primitive", highly reducing atmosphere, already postulated by Oparin). It is, however, hardly compatible with one of the other two mentioned above: interstellar dust or dirty ice crystals may be the site of synthesis of the "bricks of life", but these can hardly have fallen on the primitive Earth in large enough amounts. Anyhow, given the huge volume of the oceans, obtaining large enough concentrations globally appears to be a critical process. We do not see how it could be helped by Darwin's poetic hypothesis of the shallow lagoon by the side of the sea, periodically replenished by waves with dilute ingredients, and constantly concentrated by evaporation. Unless one simultaneously accepts J. Monod's contention that "*L'Univers n'était pas gros de la Vie*" (Monod, 1970), which implies that Life started not because some general conditions were right, but because some contingent coincidence of circumstances made it possible.

1.5.3 The Problem of the Prevalence of Phosphates

Another problem comes from the prevalence of phosphates in all the head-groups of membrane lipids (some sulfate head-groups do exist, but very seldom). In fact, the eminent role of phosphates in many aspects of biochemistry has been emphasised and assigned to the specific properties of this group (Westheimer, 1987).

We should in this respect be conscious that this ubiquitous role of phosphates is in sharp contrast with the low abundance of phosphorus in the universe, and in particular in the Earth's crust, where one evaluates its abundance at only 0.12%. It is well known that, in agriculture, the availability of P is often the growth-limiting factor. Most of the phosphate on Earth is anyway present as calcium phosphate rock, itself derived from the accumulation of fossil debris and therefore not providing an answer to the question of the prebiotic availability of phosphate. Phosphoric and polyphosphoric acids have been reported from volcanic sources (Yamagata et al., 1991) but again these are such localised and limited occurrences that they do not begin to answer the question of the origin of phosphates in prebiotic systems, for which we have no answer.

1.5.4 The Problem of Phosphorylation by Phosphoric Anhydrides

In the preceding paragraph, we have pointed out that phosphates are rather rare on Earth. Furthermore, some aspects of their reactivity are also problematic. On paper, it appears easy to form phosphates from alcohols. Heating phosphoric acid spontaneously gives polyphosphoric acids of various complexities. These polyphosphoric acids are anhydrides, and phosphorylation of an alcohol by any one of them should be easy; however, this is not the case. For instance, a careful study of the phosphorylation of several alcohols with cyclotriphosphoric acid (Fig. 1.14) has shown that this does proceed, but not well, and only at pH 12 (Baba et al., 1990); it requires therefore that the alcohol be present partly as an alkoxide, even though the cyclotriphosphoric acid is then to a large extent anionised.

Not only the phosphorylation of alcohols, but also the simple hydrolysis of these polyphosphates is not an efficient process; the cyclotriphosphates used in domestic detergents, in foods, in tooth pastes, "should" be promptly hydrolysed in water, and precipitated as insoluble calcium phosphate; this is not the case, and complex processes are required to "dephosphate" used waters. The various phosphoric anhydrides involved in biological-energy conversion, in particular adenosinetriphosphate, are also remarkably stable: they require ATP-ase to be hydrolysed. The fact is that all these phosphoric acid anhydrides have, at neutral pH, ionised hydroxyl groups, and are therefore anionic and more difficult to attack by a negatively charged species. Whatever the reason, the fact is that it is not easy to phosphorylate an alcohol! For further discussions on this point, see Baltcheffsky and Baltcheffsky, 1992.

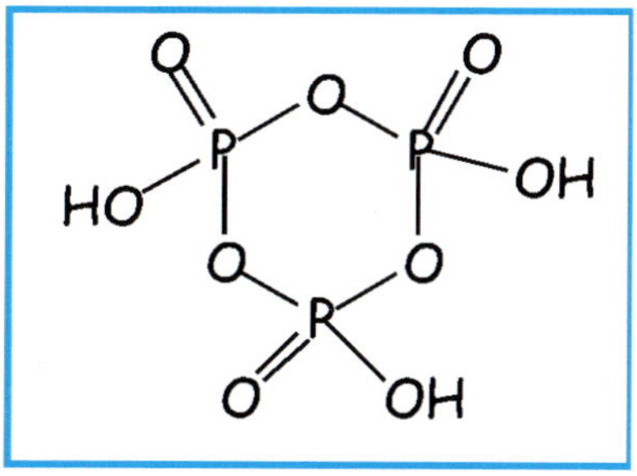

Fig. 1.14. Cyclotriphosphoric acid

1.5.5 The Problem of the C_5 Unit

The isopentenyl, C_5 units postulated to condense into the C_{15+} phosphates involved in membrane formation could be formed by a variety of processes (Fig. 1.15). Biochemically, they are formed either by the classical Lynen–Bloch–Cornforth mevalonic pathway, or by the nonmevalonic pathway more recently discovered by Rohmer (Rohmer et al., 1993, 1996). Both are much too complex to have close nonbiological analogues. However, several reactions could give rise to the branched isopentenyl unit. One is the Prins condensation of formaldehyde and isobutene, to an acetal of isopentanediol (Arundale and Mikeska, 1952; Brace, 1955) or, in acetic acid, to the acetate of isopentenol (Blomquist and Verdol, 1955). Another would be a Paterno–Buchi photochemical condensation of formaldehyde and isobutene, or of acetone and ethylene, to give an acetal of isobutanediol (L. Desaubry, personal communnication). However, none of these appears to be even remotely compatible with prebiotic conditions. We prefer to consider that the abiotic origin of the C_5 unit is still an unsolved problem.

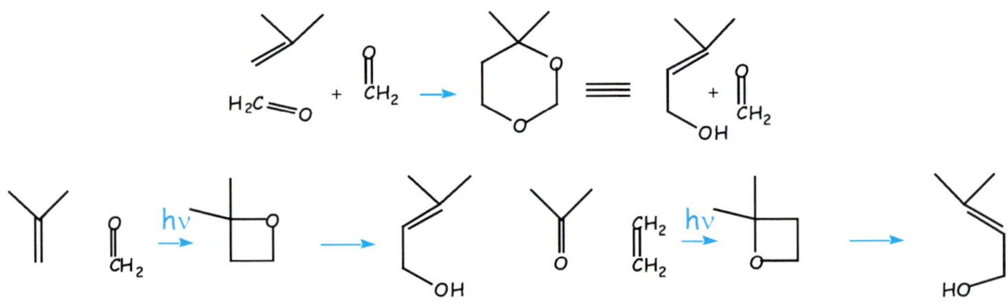

Fig. 1.15. Three possible routes to isopentenol

1.5.6 The Problem of the Cytoskeleton

We consider, of course, that our hypothesis of the early involvement of polyprenyl phosphates to form protocells warrants consideration. We must, however, admit that we cannot make, at the moment, any hypothesis to explain, even in hypothetical terms, the emergence of the cytoskeleton: microtubules in particular. Some such hypotheses are absolutely needed to proceed further.

References

Arundale E., Mikeska L.A., (1952) The olefin-aldehyde condensation, The Prins reaction. *Chem. Rev.*, **51**: 505–555

Auvray X., Perche T., Anthore R., Petipas C., Rico I., Lattes A., (1992) Structure of lyotropic phases formed by sodium dodecyl sulfate in polar solvents, *Langmuir*, **7**: 2385–2393, and many papers by I. Rico and A. Lattes

Bachmann P.A., Walde P., Luisi P.L., Lang J., (1991) *J. Am. Chem. Soc.*, **113**: 8204

Baltcheffsky M., Baltcheffsky H., (1992) Inorganic pyrophosphate and inorganic pyrophosphatases, in Ernster L. (Ed.) *Molecular Mechanisms in Bioenergetics*, New Comprehensive Biochemistry, Elsevier, Amsterdam, 331–348

Bernstein M.P., Dworkin J.P., Sandford S.A., Cooper G.W., Allamandol L.J., (2002) Racemic amino acids from the ultra-violet photolysis of interstellar ice analogues, *Nature*, **416**: 401–403

Birault V., Pozzi G., Plobeck N., Eifler St., Schmutz M., Palanché T., Raya J., Brisson A., Nakatani Y., Ourisson G., (1996) Di(polyprenyl) phosphates as models for primitive membrane constituents: synthesis and phase properties. *Chem. Eur. J.*, **2**: 789–799

Blocher M., Liu D., Walde P., Luisi P.L., (1999) Liposome-assisted selective polycondensation of α-amino acids and peptides, *Macromolecules*, **32**: 7332–7334

W. Blokzijl, J.B.F.N. Engberts (1993). Hydrophobic effects. Opinions and Facts, *Angew. Chem. Int. Ed. Engl.* **32**: 1545–1579

Blomquist A.T., Verdol J.A., (1955) The thermal isobutylene-formaldehyde condensation, *J. Am. Chem. Soc.*, **77**: 78–80

Bretscher M., (1972) Phosphatidyl-ethanolamine: Differential labelling in intact cells and cell ghosts of human erythrocytes by a membrane-impermeable reagent. *J. Mol. Biol.*, **71**, 523–528. – Asymmetrical lipid bilayer structure for biological membranes. *Nature New Biology*, **236**: 11–12

Chrzeszczyk A., Wishnia A., Springer, Jr C.S. (1977) The intrinsic structural asymmetry of highly curved phospholipid bilayer membranes, *Biochim. Biophys. Actan.*, **470**: 161–169

Brace N.O., (1955) The uncatalyzed thermal addition of formaldehyde to olefins, *J. Am. Chem. Soc.*, **77**: 4566–4668

Deamer D.W., (1986) Role of amphiphilic compounds in the evolution of membrane structure on the early Earth, *Origins of Life and Evolution of the Biosphere*, **17**: 3–25

Deamer D.W., Barchfeld G.L., (1982) Encapsulation of macromolecules by lipid vesicles under simulated prebiotic conditions. *J. Molec. Evol.*, **18**: 203–206

Deamer D.W., Harang Mahon E., Bosco G., (1994) in *Early Life on Earth*, Nobel Symposium No 84, Columbia Univesity Press, New York, Self-assembly and function of primitive membrane structures, 107–123

Desaubry L., Nakatani Y., Ourisson G., (2003) Toward higher polyprenols under "prebiotic" conditions. *Tetrahedron Lett.*, **44**: 6959–6961

Devienne F.M., Barnabé C., Couderc M., Ourisson G., (1998) Synthesis of biological compounds in quasi-interstellar conditions. C.*R. Acad. Sci. Paris*, **1, Série IIc**: 435–439

Devienne F.M., Barnabé C., Ourisson G., (2002) Synthesis of further biological compounds in interstellar-like conditions. *C.R. Chimie*, **5**: 651–653

Fischer A., Franco A, Oberholzer T., (2002) Giant vesicles as microreactors for enzymatic mRNA synthesis. *Chem. Bio. Chem.*, **3**: 409–417

Folda T., Gros L., Ringsdorf H., (1983) Formation of oriented polypeptides and polyamides in monolayers and liposomes. *Macromol. Chem. Rapid Commun.*, **3**: 167–174

Fraley R., Subramani S., Berg P., Papahadjopoulos D., (1980) Introduction of liposome-encapsulated SV40 DNA into cells. *J. Biol. Chem.*, **255**: 10431–10435

Fukuda K., Shibasaki Y., Nakahara H., (1981) Polycondensation of long-chain esters of α-amino acids in monolayers at air/water interface and in multilayers on solid surface, *J. Macromol. Sci.-Chem.*, **A(15)**: 999–1014

Ghosh S., Lee S.J., Ito K., Akiyoshi K, Sunamoto J., Nakatani Y., Ourisson G., (2000) Molecular recognition on giant vesicles: coating of phytyl phosphate vesicles with a polysaccharide bearing phytyl chains, *Chem. Commun.*, 267–268

Hanczyk M.M., Fujikawa S.M., Szostak J.W., (2003) Experimental models of primitive cellular compartments: encapsulation, growth, and division, *Science*, **302**: 618–622

Israelachvili J.N., Marcelja S., Horn R.G., (1980) Physical principles of membrane organization, *Quart. Rev. Biophys.*, **13**: 121–200

Jay D.G., Gilbert W. (1987) Basic protein enhances the incorporation of DNA into lipid vesicles: model for the formation of primordial cells *Proc. Natl. Acad. Sci. USA*, **84**: 1978–1980

Kuzmenko I., Rapaport H., Kjaer K., Als-Nielsen J., Weissbuch I., Lahav M., Leiserowitz L., (2001) Design and characterization of crystalline thin film architectures at the air-liquid interface: simplicity to complexity, *Chem. Rev.*, **101**: 1659–1696

Lee S., Desaubry L., Nakatani, Y., Ourisson G., (2002) Vectorial properties of small vesicles. *C.R. Chimie*, **5**: 331–335

R.U. Lemieux, (1996) How water provides the impetus for molecular recongnition in aqueous solution, *Acc. Chem. Res.* **29**: 375–380

Maddox J., (1994) Origin of the first cell membranes? *Nature*, **371**: 101

Miller S.J., (1953) A production of amino-acids under possible primitive Earth conditions. *Science*, **117**: 528–529

Monnard P.A., Oberholzer T., Luisi P.L., (1997) Entrapment of nucleic acids in liposomes. *Biochim. Biophys. Acta*, **1329**: 39–50

Monod J., (1970) *Le Hasard et la Nécessité.* Le Seuil, Paris

Morowitz H.J., Heinz., D., Deamer D.W., (1988) The chemical logic of a minimum protocell. *Origins Life Evol. Biosphere*, **18**: 281–287

Muñoz Caro G.M., Meierhenrich U.J., Schutte W.A., Barbier R., Arcones Segovia A., Rosenbauer H., Thiemann W.H.-P., Brack A., Greenberg J.M., (2002) Amino-acids from ultra-violet irradiation of interstellar ice analogues. *Nature*, **416**: 403–406

Nakatani Y., Yamamoto M., Diyizou Y, Warnock W., Dollé V., Hahn W., Milon A., Ourisson G., (1996) Studies on the totography of biomembranes: regioselective photolabelling in vesicles with the tandem use of cholesterol and a photoactivable transmembrane phospholipidic probe. *Chem. Eur. J.*, **2**: 129–138

Nomura S.M., Yoshikawa Y., Yoshikawa K., Dannenmuller O., Chasserot-Golaz S., Ourisson G., Nakatani Y., (2001) Towards proto: cells: "primitive" lipid vesicles encapsulating giant DNA and its histone complex. *Chem. Bio. Chem.* **2**: 457–459

Nomura S.M., Tsumoto K., Hamada T., Akiyoshi K., Nakatani Y., Yoshikawa K., (2003) Gene expression within cell-sized lipid vesicles. *Chem. Bio. Chem.*, in press

Nordén B., Lindblom G.,Joná I., (1977) Linear spectroscopy as a tool for studying molecular orientation in model membrane systems. *J. Phys. Chem.* **81**: 2086–2093

Oberholzer T., Albrizio, M., Luisi P.L., (1995-1) Polymerase chain reaction in liposomes *Curr. Biol.*, **2**, 677–682

Oberholzer T., Nierhaus K.H., Luisi P.L., (1999) Protein expression in liposomes *Biochem. Biophys. Res. Commun.*, **261**: 238–241

Oberholzer T., Wick R., Luisi P.L., Biebricher C.K., (1995-2) Enzymatic RNA replication in self-reproducing vesicles: an approach to a minimal cell. *Biochem., Biophys. Res. Commun.*, **207**: 250–257

Oparin A.I., (1968) *Genesis and Evolutionary Development of Life*, Academic Press, New York N.Y.

Ourisson G., (1986) Vom Erdöl zur Evolution der Biomembranen (Heinrich-Wieland Lecture) *Nachr. Chem., Tech. Lab.*, **34**: 8–14

Ourisson G., Nakatani Y., (1994) The terpenoid theory of the origin of cellular life: the evolution of terpenoids to cholesterol. *Chem. Biol.*, **1**: 11–23

Ourisson G., Nakatani Y., (1999) Origins of cellular life: molecular foundations and new approaches. *Tetrahedron*, **55**: 3183–3190

Pozzi G., Birault V., Werner B., Dannenmuller O., Nakatani Y., Ourisson G., Terakawa S., (1996) Single-chain polyprenyl phosphates form "primitive" membranes. *Angew. Chem. Int. Ed. Engl.*, **35**: 177–179

Rohmer M., Knani M., Simonin P., Sutter B., Sahm H., (1993) A novel pathway for the early steps leading to isopentenyl diphosphate. *Biochem. J.*, **295**: 517–524

Rohmer M., Seemann M., Horbach S., Bringer-Meyer S., Sahm H., (1996) Glyceraldehyde 3-phosphate and pyruvate as precursors of isoprenic units as an alternative non-mevalonic pathway for terpenoid biosynthesis. *J. Am. Chem. Soc.*, **118**: 2564–2566

Ron E., Huang, J.Y., Popovitz-Biro R., Kjaer K., Bouwman W.G., Howes P.B., Als-Nielsen J., Ron shen Y., Lahav M., Leiserowitz L., (2000) Absolute orientation of molecules of amphiphilic alcohols in crystalline monolayers at the air-water interface. *J. Phys. Chem.*, **104**: 6843–6850

Szostak J.W., Bartel D.P., Luisi P.L., (2001) Synthesizing life. *Nature*, **409**: 387–390

Swairjo M.A., Seaton B.A., Roberts M.F., (1994) Effect of vesicle composition and curvature on the dissociation of phosphatidic acid in small unilamellar vesicles – a ^{31}P-NMR study. *Biochim. Biophys. Acta*, **191**: 354–361

Tanford C., (1978) The hydrophobic effect and the organization of living matter. *Science*, **200**: 1012–1018

Tsumoto K., Nomura S.M., Nakatani Y., Yoshikawa K., (2000) Giant liposome as a biochemical reactor: transcription of DNA and transportation by laser tweezers. *Langmuir*, **17**: 7225–7228.

Ueda T., Lee S.L., Nakatani Y., Ourisson G., Sunamoto J., (1998) Coating of POPC giant liposomes with hydroxylated polysaccharide. *Chem. Lett.*, 417–418

Walde P., Ichikawa S., (2001) Enzymes Inside Lipid Vesicles. Preparation, reactivity and applications. *Biomol. Eng.*, **18**: 143–177

Walde P., Wick R., Fresta M., Mangone A., Luisi P.L., (1994) Autopoietic self-reproduction of fatty acid vesicles. *J. Am. Chem. Soc.*, **116**: 11649–11654

Wang J.-L., Leveiller F., Jacquemain D., Kjaer K., Als-Nielsen J., Lahav M., Leiserowitz L., (1994) Two-dimensional structures of crystalline self-aggregates of amphiphilic alcohols at the air-water interface as studied by grazing incidence synchrotron X-ray diffraction and lattice energy calculations. *J. Am. Chem. Soc.*, **116**: 1192–1204

Weissbuch, I., Zepik, H., Bolbach, G., Shavit, E., Tang, M., Jensen, T.R., Kjaer, K., Leiserowitz L., Lahav, M., (2003) Homochiral oligopeptides by chiral amplification within two-dimensional crystalline self-assemblies at the air-water interface; relevance to biomolecular handedness. *Chem. Eur. J.*, **9**: 1782–1794

Weissbuch I., Rubinstein I., Weygand M.J., Kjaer K., Leiserowitz L., Lahav M., (2003) Crystalline phase separation of racemic and nonracemic zwitterionic α-amino acid amphiphiles in a phospholipidic environment at the air/water interface: a grazing-incidence X-ray diffraction study. *Helv. Chim. Acta*, **86**: 3867–3874

Westheimer F.H., (1987) Why Nature chose phosphates? *Science*, **235**: 1173–1178

Westhof E., (ed.) *Water and Biological Molecules.* MacMillan Press, London 1993, Chap. 1, Savage H.F.J.: "Water structure", Chap. 2 Ludemann, H.D., "Thermodynamic and dynamic properties of water"

Yamagata Y., Watanabe H., Saitoh M., Namba T., (1991) Volcanic production of polyphosphate under primitive Earth conditions. *Nature*, **353**: 516–519

Yu W., Sato K., Wakabayashi M., Nakaishi T., Ko-Mitamura E.P., Shima Y., Urabe I., Yomo T., (2001) Synthesis of functional protein in liposomes *J. Biosc. Bioeng.*, **92**: 590–593

2 Prebiotic Chemistry: Laboratory Experiments and Planetary Observation

François Raulin, Patrice Coll, Rafael Navarro-González

The majority of the current theories on the origins of life are based on the concept of chemical evolution, e.g., a process in which life arises from nonliving matter in accordance with the ordinary laws of physics and chemistry. In this framework it is important to note that the primitive terrestrial environment differed substantially from the present Earth in that there was no free oxygen available in the atmosphere and there was a large abundance of organic compounds that were transformed chemically, by a process referred to as prebiotic chemistry into the first self-replicating systems. This concept was originally introduced by Oparin and Haldane in the 1920s (Oparin, 1924; Haldane, 1929); according to these authors, the evolution of the primordial atmosphere led to the formation of key organic compounds that were reactive intermediates. Upon dissolution in the primitive oceans, these compounds were transformed into amino acids and other molecules relevant for life, constituting the so-called "primordial soup" from which the first heterotrophic cells emerged. This first experimental study in support of the process of chemical evolution was carried out almost five decades later by Miller (1953) whose work came to consolidate this theory.

Later it was realized that the primitive terrestrial atmosphere was not as reducing as originally thought by Oparin and Miller, and under such conditions the synthesis of key organic molecules was not so favorable. Therefore, it was necessary to consider the extraterrestrial inputs of organic material (from comets, meteorites and micrometeorites) to Earth (Oro, 1961) and the synthesis of organic material in the vicinity of hydrothermal sources, from the reduction of carbon dioxide in the presence of hydrogen sulfide and iron sulfide (Baross, Hoffman, 1985). However, the atmospheric synthesis of organic material has always been of great interest, particularly in the exploration of the Solar System such as in the atmosphere of Titan, the largest satellite of Saturn. Even though the environmental conditions of Titan are quite different from our planet, especially in terms of the temperature, there are several analogies to the early terrestrial environment. The study of the chemistry and physical chemistry of these planets offers us an exceptional means of testing the theories and models of the terrestrial primitive environment and prebiotic chemistry that took place approximately 4 billion years ago since most of the geophysical and geochemical evidence of that period was wiped out away.

In this chapter, we discuss three complementary exobiological topics related to the studies of the origin of life and prebiotic chemistry. The first part describes the experimental approach based on the development of what is frequently referred to as laboratory simulations. The first laboratory simulation of the early Earth is not only relevant from a historical point of view but also from exobiology as this type of simulation is routinely used today to model the chemical processes occurring in planetary atmospheres. The second part gives an overview of the elementary steps of prebiotic chemistry: an organic chemistry in liquid water and plausible conditions of the primitive terrestrial environment that likely led to the formation of complex organic compounds that are of biological interest. The last section shows that many of the data obtained in laboratory simulations (as described in Sects. 2.1 and 2.2) are applicable to environments different from the primitive Earth. Combining such laboratory results with astronomical observations and theoretical models, one can then derive relevant information for understanding these astronomical environments, as well as enabling us to test our conclusions in situ.

2.1 Simulation Experiments and Photochemical Models

2.1.1 An Historical View of Miller's Experiment and the Development of a New Field: Prebiotic Chemistry

One of the pioneering experiments specifically designed to test the theory of chemical evolution is the famous celebrated experiment of Miller (also known as the Miller–Urey Experiment; Miller, 1953). Miller studied a model of the primitive Earth's atmosphere in a glass reactor by subjecting it to spark discharges in the presence of liquid water. He initially used a mixture of methane, ammonia, hydrogen, and water vapor, which represented, according to Urey, a model of the early atmosphere of the Earth. These gases are thermochemically stable at ambient temperature and it was thought by Urey in the early 1950s that the Earth accreted at low temperature from an interstellar gas-dust cloud. After several days of sparking, Miller demonstrated the formation of some key organic compounds such as formaldehyde (HCHO) and hydrogen cyanide (HCN) that later were shown to be precursors in the synthesis of several biological and non-biological amino acids. This unique experiment was a pioneering study in the field and was considered as the initiator of a new field of research called prebiotic chemistry, i.e. chemistry of carbonaceous compounds in the absence of life but under conditions relevant to the primitive terrestrial environment. Miller's experiment was conducted with a gas mixture at relatively high pressure (approximately 1 bar), simulating the effect of lightning flashes of thunderstorms in the low terrestrial troposphere.

Lightning activity was certainly not the most abundant energy source in the Earth's atmosphere, compared to the ultraviolet radiation from the Sun. The choice of Miller was most probably guided by experimental constraints. The system of irradiation by electronic impact used by Miller was relatively easy to implement in the laboratory and it allowed the transformation of methane and ammonia (but also of molecular nitrogen if this last molecule is used as a source of nitrogen atoms instead of ammonia). A photochemical system is more complicated to put into practice, and introduces strong constraints in terms of energy levels that can be excited. Indeed, most traditional UV lamps (such as low pressure mercury lamps providing a quasimonochromatic emission to 185 nm), do not allow the photodissociation of methane. However, one can use UV lamps emitting at lower wavelengths (lower than approximately 150 nm) to photodissociate methane. An alternative is to introduce a new compound (such as hydrogen sulfide, H_2S) into the initial gas mixture that can be photodissociated with higher wavelengths and whose products of photodissociation are sufficiently energetic to cause the dissociation of methane.

Since the first publication of Miller's experiment in 1953, this type of experiment has been performed more than several hundred times, by Miller and others, with some variations. The most common modifications are the nature of the initial gas mixture (less reducing or neutral) and the type of energy used to drive the chemical reactions. These studies have shown that electron impact from lightning and solar UV radiation, which are the two most important energy sources in planetary atmospheres, play a complementary role in the synthesis of a large suite of organic compounds.

This experimental approach relies on the use of analytical techniques of chemical analysis, in particular molecular characterization. Their nature and their performances have evolved considerably since the 1950s. During the early 1970s, these experiments became increasingly more complex due to the availability of newly developed analytical techniques with greater sensitivity, allowing in particular the separation and quantification of enantiomers. Thus the experiments carried out by Sagan and his team (Sagan, Khare, 1971) made it possible to show for the first time that amino acids formed during these experiments and later chemically derivatized for their separation by gas chromatography (GC) are produced in racemic form, as one might expect, and thus do not come from an unspecified biological contamination, during the irradiation itself or incorporated during the analytical stage. These results were corroborated later with the advance of new and more powerful analytical techniques (more sensitive and to higher resolution), either by GC or by high-performance liquid chromatography (HPLC). For instance, the work of Kobayashi and his team on the chiral analysis of organic materials, including amino acids (Kobayashi et al., 1989), demonstrated that some amino acids detected at ultralow level in some experiments may actually originate by contamination rather than by synthesis during the preprocessing of the sample, e.g., during hydrolysis.

2.1.2 An Overview of Experimental and Theoretical Data

During these experiments, one could obtain a very broad mixture of compounds of biological interest: a hundred amino acids, including the major proteinie amino acids, more than a dozen purine and pyrimidine bases, including those used in DNA, and many other organics. One can classify them in two categories. The first one corresponds to simple, volatile compounds of low molecular mass such as HCN and other nitriles, HCHO and other carbonyl compounds, carboxylic acids, thiols, amines etc. The second category contains more complex refractory organic compounds of the macromolecular type, whose structures are still not well defined, and are often called "tholins" (see below). Many of these compounds, which may be simple or complex, always lead to compounds of biological interest upon treatment in an aqueous solution.

The experimental evidence indicates that carbon, in the form of methane (CH_4), carbon monoxide (CO) or even carbon dioxide (CO_2), in the presence of various mixing ratios of hydrogen (H_2), can lead to gas phase organic synthesis. Mixtures that mainly contain an oxidized form of carbon (CO_2), in the absence of hydrogen, are not able to yield organic compounds regardless of the type of energy used. Experimental work carried out with intermediate gas mixtures, formed with various mixing ratios of CO_2 and CH_4, leads to the same conclusions: only atmospheres corresponding to a mixing ratio $CH_4/CO_2 > 1$ results in notable outputs of organic compounds (Miller, Schlesinger, 1984; Hattori et al., 1984). This conclusion is also confirmed by photochemical models that were developed in the 1980s (Table 2.1). The first photochemical model of the evolution of the Earth's primitive atmosphere, was published by Pinto et al. (1980), and considered a gas mixture that was mainly composed of carbon dioxide and molecular nitrogen (dinitrogen) with low mixing ratios of molecular hydrogen (dihydrogen) (about 0.01) and carbon monoxide (0.2 ppm). The model included

Table 2.1. Production of formaldehyde in the terrestrial primitive atmosphere according to photochemical models. These models are based on a chemical composition $N_2(0.8)$–H_2O (0.012) (mole fractions), on a total pressure close to the current pressure. Various percentages (expressed in mole fractions) of CO_2 (PAL = present atmospheric level = current mole fraction = 2.4×10^{-4}), CO and H_2 are added. A UV flux= 1 means an intensity equal to the present intesity. T-Tauri means a higher intensity (X1000) considered as similar as the UV-flux of the young Sun during its T-Tauri phase

	Pinto et al. (1980)	Canuto et al. (1983)	Kasting et al. (1984)
No. of reactions	39	47	134
CO_2	1 PAL	1 and 100 PAL	1–1000 PAL
CO	2×10^{-7}	$\sim 10^{-6}$	$\sim 3 \times 10^{-4}$
H_2	8×10^{-3}	17×10^{-6} and 10^{-3}	10^{-5}–10^{-3}
UV flux	1	1 and T-Tauri phase	1
HCHO yield (mole.cm^{-2}.s^{-1})	3×10^{8}	6×10^{3} to 10^{10}	3×10^{8} to 9×10^{8}

approximately 40 chemical reactions, and took into account the possible formation of formaldehyde in the atmosphere with its subsequent transport into the hydrosphere by rain processes. The model predicted that solar UV photons led to an extremely inefficient production of formaldehyde, unable to sustain a moderate concentration of formaldehyde in the oceans that would be required for prebiotic chemistry (see below). More elaborate models (Table 2.1), taking into account the influence of CO_2 pressure and that of solar flux were published in subsequent years (Canuto et al., 1983; Kasting et al., 1984), but were unable to increase the production of formaldehyde. These models, however, do not include the chemistry of nitrogen since molecular nitrogen is photochemically stable in the atmosphere.

2.1.3 New Scenario for Prebiotic Chemistry

It seems relatively well established today that the terrestrial primitive atmosphere was mainly composed of CO_2 and not CH_4, and would have never contained a significant mixing ratio of H_2. Although the experimental data presented above concern mainly the organic synthesis in the gas phase and do not take into account any possible catalytic effects, in particular due to the presence of solid phases (dust or volcanic ash) suitable for inducing heterogeneous processes, the main conclusion drawn from experimental and theoretical studies is that the atmosphere has very likely not played a key role in the formation of organic compounds in the primitive Earth (Navarro-González et al., 1998; 2001).

An alternative mechanism for the production of organics in the early Earth comes from deep hydrothermal vents. These hydrothermal sources are rich in carbon dioxide, hydrogen sulfide and iron sulfide. In the presence of liquid water at high pressures and temperatures, thermodynamical models and laboratory studies (Hennet et al., 1992; Holm, 1992; Marshall, 1994; Brack, 1998; Holm, Andersson, 1998) indicate that carbon dioxide can be reduced into biologically relevant organic compounds. These conditions can also allow the elongation of oligopeptides (Imai et al., 1999). The idea of prebiotic synthesis under the extreme conditions of deep hydrothermal vents is also supported by Günther Wächtershäuser (Wächtershäuser, 1990) who thinks that the first prebiotic systems were capable of replication from an inorganic matrix, namely pyrite, which was quite abundant in the primitive environment. The iron sulfide constituting the pyrite would have the ability to reduce the atmospheric carbon dioxide in the presence of hydrogen sulfide into organic compounds in a process referred to as prebiotic chemiosynthesis. All the requirements of this assumption, in particular the ingredients, are present in the vicinity of the hydrothermal sources.

Finally, if the organic synthesis in the early terrestrial atmosphere were an unefficient process , the origin of the prebiotic molecules could have been essentially exogeneous i.e. deposits of molecules coming from space as minor constituents of meteorities, comets and other extraterrestrial objects falling on the young Earth (Briggs, 1961). A notable fraction of the meteorites that reach the surface of our planet indeed contains many complex organic compounds that could take part in the process of chemical evolution. However, as we will show in the third part of this chapter, the synthesis of organic compounds in the atmosphere is of great interest for exobiology, because they could intervene in a significant way in many extraterrestrial planetary atmospheres.

2.2 Elementary Prebiotic Chemistry in Aqueous Solution

The laboratory simulations of the early Earth are also important to understand the mechanisms by which biologically relevant molecules are made from the raw materials. These chemical pathways are generally referred to as prebiotic chemistry (Miller, Orgel, 1974; Raulin, 1990; Brack, 1998). Important in this regard are the roles of a limited number of key organic molecules such as nitriles (mainly HCN), and aldehydes (mainly formaldehyde). It is the chemical transformations of these compounds in aqueous solution that lead to the building blocks of life.

Today, the knowledge of the prebiotic synthesis of the main building blocks of life (the twenty amino acids of proteins and the five bases of the nucleic acids) is satisfactory. One can assemble prebiotically the amino acids into chains of polypeptides that make up proteins (see, for example, Brack, 1998; Zubay, 2000; and references therein). One can also obtain microstructures, with a membrane similar to those of living cells (see Chap. 1, Part II by Ourisson and Nakatami). On the other hand, neither the prebiotic synthesis of nucleosides (ribose and deoxyribose) nor that of nucleotides and polynucleotides have yet been conclusively demonstrated. The suggestion of a RNA world (see chapter by Maurel) implies that nucleic acids were the first living molecules on the planet; however, the extreme difficulty of their prebiotic synthesis seems to contradict this possibility. The prebiotic synthesis of proteins or at least of polypeptides is more or less much more accessible although there is still a lot of work to realize in this field where new pathways can be unraveled.

2.2.1 Prebiotic Chemistry of HCN: Strecker Reaction or Oligomerization (see Box 2.1)

The main mechanism of formation of the amino acids in Miller's experiments is the reaction referred as "Strecker raction" (see Box 2.1, well known in organic chemistry for more than one century) (Miller, Orgel, 1974): reaction of HCN with ammonia and an aldehyde in aqueous medium leading to an aminonitrile (step 1), followed by hydrolysis of this aminonitrile into an amino-amide (step 2) and finally into the amino acid (step 3).

Indeed, in Miller's experiment, amino acids are not obtained directly, but are released from a precursor after hydrolysis. However, it has not yet been demonstrated in Miller-like simulation experiments that the precursors to the

Box 2.1. Basic prebiotic chemistry

Prebiotic chemistry is organic chemistry in aqueous solution, under plausible conditions for the primitive terrestrial environment, leading to compounds of biological interest. Elementary prebiotic chemistry uses simple and reactive organic compounds, such HCN, HCHO, HC_3N or their oligomers.

Prebiotic synthesis of amino acids by Strecker reaction

Basic ingredients: aldehyde (RCHO), ammonia (NH_3), hydrogen cyanide (HCN)

* Reaction of HCN with ammonia in aqueous medium leading to an aminonitrile:

$$R-CHO + HCN + NH_3 \quad \rightarrow \quad R-CH(NH_2)-CN + H_2O \tag{2.1}$$

* followed by the hydrolysis of this aminonitrile into an amino-amide then an amino acid:

$$R-CH(NH_2)-CN + H_2O \rightarrow \quad R-CH(NH_2)-CONH_2 \tag{2.2}$$

$$R-CH(NH_2)-CONH_2 + H_2O \rightarrow R-CH(NH_2)-COOH + NH_3 \tag{2.3}$$

Case of the sulfurated amino acids: example of methionine ($R = CH_3$–S–$(CH_2)_2$)

* step (1) above uses a sulphurated aldehyde obtained prebiotically by reaction between methanethiol and acrolein:

$$CH_3-HS + CH_2 = CH-CHO \rightarrow CH_3-S-(CH_2)_2-CHO \tag{2.4}$$

Prebiotic chemistry of HCN

$$HCN + CN^- \overset{+H^+}{\rightarrow} \underset{\text{Dimer}}{H-C(=NH)-CN} \overset{+HCN}{\rightarrow}$$

$$\underset{\text{Trimer}}{NC-CH(NH_2)-CN} \overset{+HCN}{\rightarrow} \underset{\text{tetramer}}{NC-C(NH_2) = C(NH_2)-CN} \overset{+nHCN}{\rightarrow}$$

$$\text{oligomers} \overset{+H_2O}{\rightarrow} \text{amino acids purines and pyrimidines}$$

Prebiotic chemistry of HC_3N

$$HCC-CN + NCO^-(\text{cyanate}) \rightarrow \text{cytosine} + \text{uracil}$$

$$HCC-CN + H_2O \overset{\text{urea}}{\rightarrow} OHC-CH_2-CN \rightarrow \text{cytosine} + \text{uracil}$$

Box 2.1. Basic prebiotic chemistry (continued)

Prebiotic chemistry of HCHO

$$HCHO + HCHO \overset{HCHO}{\rightarrow} CHO-CH_2OH \overset{HCHO}{\rightarrow}$$

$$CHO-CHOH-CH_2OH \overset{HCHO}{\rightarrow} C_4$$

$$C_4 + HCHO \rightarrow C_5 \quad \text{etc...}$$

amino acids are the corresponding aminonitriles. Moreover, such a mechanism implies the (improbable) formation of a broad range of aldehydes required to yield the corresponding amino acids.

Another mechanism involves the polymerization of HCN. This idea was suggested by Matthews (see, for example, Matthews, Moser, 1967; Matthews, 1979; Refs. therein), which caused much controversy in all the conferences on the origin of life and prebiotic chemistry. The idea is tempting: it rests on the assumption that the polymerization of HCN leads to a polymer of well-defined structure, including a side-group imine. This group is suitable for forming, under conditions of soft hydrolysis, a peptide bond, leading thus directly to a polypeptide, without requiring the formation of amino acids. The latter can also be formed starting from polymers of HCN during more severe hydrolysis conditions. This mechanism has the enormous advantage of removing the difficult stage of condensation of the amino acids in the prebiotic synthesis of polypeptides. However, this assumption was constantly criticized and largely disputed, in particular by Ferris (see, for example, Ferris, 1979). This last author claimed that the proposed polymeric structure of HCN has not been supported by analytical means and, furthermore, that its formation depends on the prior formation of a hypothetical HCN dimer that has an improbable regular structure. Ferris, on the other hand, supports the role of the formation of HCN oligomers, with structures still poorly known but most probably much more irregular (Ferris, Hagan, 1984 and Refs. therein). The formation of these oligomers is dependent on the polymerization of the HCN tetramer, diaminomaleonitrile (DAMN, not to be confused with damned, although sometimes this type of compound makes one think of diabolical properties). In any case, without knowing their chemical structure, it is shown that hydrolysis of the HCN oligomers/polymers yields a large variety of amino acids (Ferris, Hagan, 1984 and Refs. therein).

Finally, the last mechanism that should be mentioned calls upon even more complex compounds generically referred to as "tholins" (see below) for the formation of amino acids, but this assumption is not in contradiction with that of Ferris: HCN oligomers can also be regarded as made of tholins.

In any case, it is clearly established that HCN in aqueous solution is easily polymerized into oligomers or "polymers" (actually macromolecules built with nonrepetitive structures) that release a complex suite of organic compounds, including biological amino acids, purines (such as adenine) and pyrimidines upon hydrolysis. The mechanism of HCN polymerization in aqueous solution was studied thoroughly in the 1960s, within the framework of prebiotic chemistry and is now well established (Toupance et al., 1970; Ferris, Hagan, 1984). The first stage is the nucleophilic attack of cyanide (CN^-) into hydrogen cyanide (HCN). This nucleophilic attack can only occur under a narrow range of pH for the reaction to proceed rapidly. Indeed, it was shown (Toupance et al., 1970) that the rate limiting step is precisely the dimerisation occurring at a pH close to the pKa of the acid/base couple: HCN/CN^- that is to say, pH 8 to 9, and at temperatures ranging between 0 and 60 °C.

The chemistry of cyanoacetylene, HC_3N, can lead in aqueous solution to the prebiotic synthesis of pyrimidines. However, the direct involvement of this compound does not seem very probable, taking into account its strong reactivity, in particular with ammonia and water. Indeed, this compound is hydrolyzed very easily into cyanoacetaldehyde (Ferris et al., 1968). In fact, it is cyanoacetaldehyde that has an important role in prebiotic chemistry, in spite of pH constraints, taking into account its easy dimerization (Raulin, Toupance, 1975). In particular, the reaction between cyanoacetaldehyde and guanidine provides a relatively simple and efficient pathway for the prebiotic synthesis of cytosine and uracil (Ferris et al., 1974).

2.2.2 Prebiotic Chemistry of HCHO, Formose Reaction

Methanal (commonly called formaldehyde or formol) also undergoes a polymerization reaction in aqueous solution – a reaction known as formose reaction, which is quite relevant in prebiotic chemistry since it leads to the formation of sugars, and among them ribose. This process of polymerization is to some extent similar to the HCN polymerization. It starts by the attack of a monomer molecule, playing the role of nucleophilic reagent, on the carbon of a second monomer molecule, leading, by nucleophilic addition, to a dimer. This dimer can itself undergo the attack of the monomer to give a trimer, etc. One can thus obtain in aqueous solution sugars with $2, 3, \ldots, n$ carbon atoms (C_2, C_3, ..., C_n), including pentoses (C_5 such as ribose) and hexoses (C_6 such as glucose).

However, for this reaction to proceed it is required to have high concentrations of formaldehyde and very alkaline solutions. These are unlikely conditions for the terrestrial primitive environment. Moreover, this reaction leads to an extremely complex mixture of sugars. Living systems only use one of these sugars in their genetic material: ribose (in the case of RNA). Therefore one of the biggest puzzles in the field of prebiotic chemistry is the problem of selection. In the absence of living systems, what was the natural process that led 1) to the selection of ribose among hundreds of sugars potentially available in the environment of the primitive Earth and 2) to its specific incorporation in polynucleotides? If the prebiotic synthesis of the amino acids, the building blocks of proteins, and polypeptides is well established, this is not the case for nucleotides, the building blocks of nucleic acids and polynucleotides (Shapiro, 1988).

2.2.3 Prebiotic Chemistry of Tholins

In all the experiments simulating the evolution of a gas mixture under an energy-source effect there is always the formation of complex organic material of two types. The first type corresponds to simple volatile organic compounds able to participate in prebiotic reactions, such as HCN, HCHO, HC_3N, etc. (see Box 2.1). The second type of products is much less well defined; they are macromolecular products with a bidimensional structure (mainly made of polyaromatic groups) or tridimensional structures, branched or not but still very poorly characterized.

In order to refer generically to this type of material, Sagan and Khare (1979) proposed the word "tholins", derived from the Greek word "tholos", which means muddy. The name "tholins" is assigned to refractory macromolecular organic materials, often obtained in a viscous form, which are systematically formed in simulation experiments. Their composition and chemical structure (which is still very poorly known) depends on parameters such as initial gas composition, energy source, temperature and pressure, etc., among others. However, regardless of their mode of formation, they release biological amino acids upon acid hydrolysis (Khare et al., 1986). Therefore they could be the precursors of these biological compounds in aqueous solution. Many studies were developed during recent years on the chemical and physical properties of "tholins", mainly in the case of Titan, but this topic will be discussed in a next section.

2.3 Application of These Laboratory Experimental Data to Space Studies

The study of the different planetary objects in the Solar System in particular with the possibility of space exploration, enables us today to test the preceding results, which were obtained in the laboratory through simulations (Table 2.2) or those coming from modeling. Thus comparative planetology is an essential approach for exobiology. It obviously makes it possible to seek on other objects for traces of life, biological activity or prebiotic chemistry. But it also makes it possible to observe and study in a real planetary environment, the processes that could have played a key role in the terrestrial chemical evolution process towards the origins of life, and that were until now accessible through numerical modeling or laboratory simulation experiments.

2.3.1 Telluric Planets

Of the three telluric planets in the Solar System that contain an atmosphere, the Earth atmosphere seems for the moment the only one to contain organic compounds in detectable amounts (Table 2.2).

Indeed the systematic exploration of Venus, thanks to the many Soviet missions, and the American Pioneer probes, allowed, among other things, the detailed chemical analysis of its atmosphere. With the exception of COS, which can be regarded as border-line cases, no organic compound was detected in the atmosphere of Venus. The same is true of the exploration of Mars: no organic compound was detected in the Martian atmosphere nor in the first centimeters of the Martian ground. The lack of detection of organic compounds in the atmospheres of Venus and Mars confirms forecasts made starting from laboratory data described in the first part (Table 2.2). Indeed, these atmospheres consist mainly of CO_2 (more than 90%) and N_2 (a few %). Their chemistry is currently dominated by the photochemistry of CO_2 and H_2O, which, in the absence

Table 2.2. Comparison of planetary observations and laboratory synthesis of organics in gas phase (simulation experiments under electron impact)

Gas mixture	Detected products (in particular organics) during the experiment	*Relevant planetary atmosphere* associated with this mixture and detected organics
$CH_4 + NH_3 + H_2O$ $(+H_2)$	RH (saturated & unsaturated), HCN, RCN (mainly saturated) HCHO, other aldehydes, ketones, alcohols, RCO_2H	*Giant planets* hydrocarbons in C_2 & C_3, C_6H_6 (*Jupiter*) HCN *(Neptune)*
	Tholins → amino acids, N-heterocycles	Chromophores of the Great Red Spot? (*Jupiter*)
$CH_4 + N_2 + H_2O$	RH (saturated and unsaturated), HCN, RCN (saturated and unsaturated) including HC_3N and C_2N_2 HCHO, other aldehydes, ketones, alcohols, RCO_2H	(Without important fraction of H_2O) Titan & Triton hydrocarbons in C_2–C_4, C_6H_6, nitriles in C_2–C_4 (*Titan*)
	Tholins → amino acids, N-heterocycles	Tholins in aerosols? (*Titan*) Tholins on the surface? (*Triton*)
$CO + NH_3 + H_2O$	HCN Tholins → amino acids	
$CO_2 + N_2 + H_2O +$ CO/H_2	RH (mainly saturated) HCN, other RCN HCHO, other aldehydes, ketones Tholins → amino acids	*Primitive Earth*?
$CO_2 + N_2 + H_2O$	No organic compounds	*Primitive Earth*? Mars, Venus: no organics
$CO_2 + N_2 + H_2O +$ O_2	No organic compounds	*Current Earth:* many organics, including HCN, HCHO, C_2N_2

of significant quantities of CO cannot lead to notable concentration of organic compounds, such as formaldehyde.

Only the terrestrial atmosphere even if strongly oxidizing contains organic compounds: they are in disequilibrium and it is a sign of life on our planet. However, it cannot be excluded that Mars may have had life at some time in the past.

2.3.2 Giant Planets and Their Satellites

All the giant planets have atmospheres with notable concentrations of methane, which induces an organic chemistry (Table 2.2), possibly coupled with the chemistry of ammonia (Jupiter), phosphine (Saturn) or even dinitrogen (Neptune), although the latter is for the moment only a presumption.

Indeed in all these planets, many organic compounds have already been detected (Table 2.2). Leaving aside the case of Neptune's atmosphere where HCN was clearly detected (Rosenquist et al., 1992; Marten et al., 1993; Lellouch et al., 1994), the list is limited for the moment to hydrocarbons, whose formation is easily explained by the photochemistry of methane.

It is fair to say that although the photochemistry of methane has been extensively studied both theoretically and experimentally, there are still significant gaps in our understanding of its chemistry; this is case, for instance, with some primary processes that lead to significant uncertainties in photochemical models (Dobrijevic and Parisot, 1998; Smith, 1999). This is illustrated in Table 2.3, by the great variability in the data reported in the literature for Lyman alpha photons (Yung, deMore, 1999; Mordaunt et al., 1993; Smith and Raulin, 1999). It is thus essential to develop experimental techniques to determine the photochemical parameters of this molecule and of this photo-products (C_2 and C_3 hydrocarbons) as well as the main branching ratios, and their variation with wavelength, temperature, pressure as well as the absorption coefficients among others.

The chemistry of methane is particularly active in Titan's atmosphere, Saturn's largest satellite. With a diameter of more than 5100 km, this satellite is

Table 2.3. Methane photodissociation products by Lyman alpha photons 121.6 nm: variability of the literature data

	Branching ratio			
	Pre-1993 data, Yung and deMore, 1999 (and refs included)	Mordaunt et al., 1993 Diagram 1	Mordaunt et al., 1993 Diagram 2	Smith and Raulin 1999
$CH_3 + H$	0	0.49	0.51	0.41
$^1CH_2 + H_2$	0.41	0	0.24	0.53
$^3CH_2 + H + H$	0.51	0	0.25	0
$CH + H_2 + H$	0.08	0.51	0	0.06

in size, the second largest moon in the Solar System. Titan is the only satellite with a dense atmosphere, composed mainly of dinitrogen and methane, with small quantities of dihydrogen. This atmosphere is nearly five times denser than the Earth's, with a surface temperature of 90–100 K and a surface pressure of 1.5 bar. The atmosphere of Titan is very rich in organic compounds. They are present not only in the gas phase but also very likely in the aerosols of which thick layers mask the surface of the satellite.

As in the Earth's atmosphere, with water vapor and carbon dioxide on the one hand and clouds on the other hand, Titan's atmosphere also contains greenhouse gases, (condensed CH_4, equivalent to terrestrial H_2O; noncondensable H_2, equivalent to terrestrial CO_2), and antigreenhouse compounds (aerosols). The thermal profile of the lower atmosphere of Titan is very similar – although the temperatures are much lower there – to that of the Earth, with a troposphere (90–70 K), a tropopause (70 K) and a stratosphere (70–175 K). Moreover, the models of the surface of Titan suggest that it is covered – at least partially – with lakes or seas of methane and ethane. Also, the organic chemistry seems to be present in the three components of what one can call, always by analogy with our planet, the "geofluids" of Titan: air (gas atmosphere), aerosols (solid atmosphere) and surface (oceans) (Fig. 2.1 and Fig. 2.2).

In Titan's atmosphere, the chemistry of methane is coupled with that of dinitrogen, resulting in the formation of many nitrogenated organic compounds in both gas and particulate phases: hydrocarbons, nitriles and tholins. There are complex chemical and physical couplings among the three components of the geofluid of Titan that control the formation and the evolution of the organic matter.

Indeed on Titan there are couplings between the photochemistry of the atmosphere – producing minor constituents and submicrometer particles formed by photooligomerization – and the surface chemistry induced by the cosmic rays.

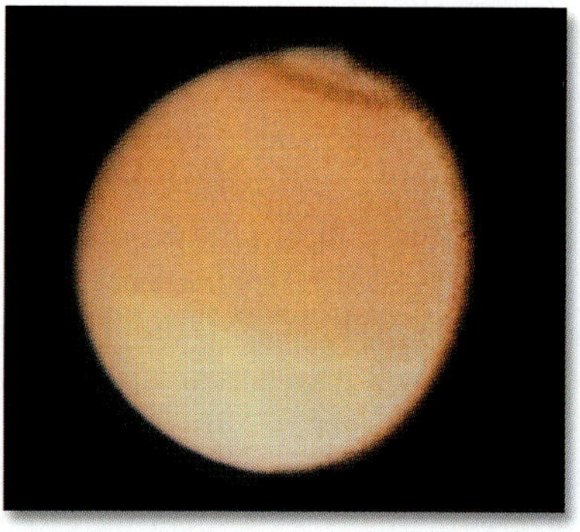

Fig. 2.1. A picture of Titan taken from 2.3 million km by Voyager 2 in 1981. It shows the presence of haze layers covering the whole disk of the satellite. South hemisphere appears brighter than the north and a dark ring is observed near the north pole (credit: NASA)

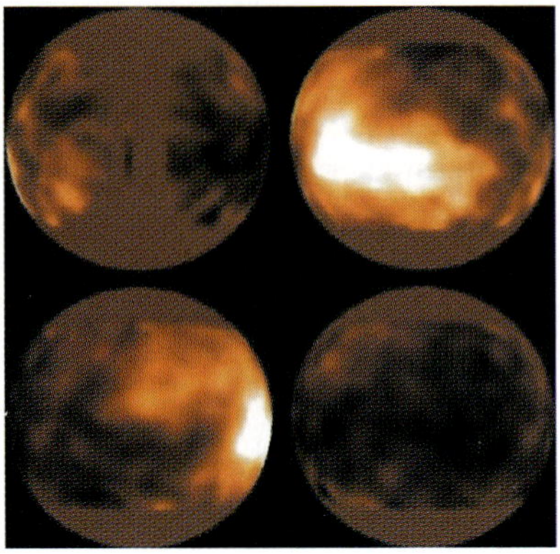

Fig. 2.2. Images of Titan's surface from the Hubble Space Telescope, using near infrared spectral windows. These images obtained by LPL scientists (Univ. Arizona) show bright and dark features over the surface (from the NASA Gallery: http://www.jpl.nasa.gov/cassini/)

Couplings are also involved between the microphysical processes of evolution of aerosols in the upper atmosphere towards the surface, the processes of molecular and turbulent diffusions, and those of condensation in the atmosphere and solubilization in the liquid phase of the lakes or seas which may be present on the Titan surface. Taking into account its many analogies with the Earth (Clarke, Ferris, 1997), the absence of liquid water on its surface, Titan is particularly relevant to understand the process of chemical evolution under anhydrous conditions (Raulin, Owen, 2002). It is interesting to note that all the organic compounds that were detected in the atmosphere of Titan (Table 2.4), were also produced in simulation experiments, including dicyanoacetylene, C_4N_2, an unstable compound at ambient temperature, whose detection in the simulation experiments has been reported (Coll et al., 1999b).

It should be noted that all of these compounds are hydrocarbons and H-, C-, N-containing compounds, mainly organics. The latter are essentialy nitriles, as was also envisaged in laboratory studies (Thompson et al., 1991; De Vanssay et al., 1995; Coll et al., 1998; 1999b; and refs. therein).

Many other organics, including again hydrocarbons and nitriles, are also formed during these experiments and it seems very probable that they are also present in the atmosphere of Titan. This statement is strongly supported by the very recent detection of benzene in Titan's atmosphere through ISO observations (Coustenis et al., 2003; and refs. therein), the presence of which was expected from the results of simulation experiments. Also, the detection of water varpour, in the atmosphere (Coustenis et al., 1998) although at very low concentration, together with CO and CO_2, allows us to consider the possible presence of oxygenated organic compounds, such as methanal and methanol in the atmosphere of Titan, as assumed in the photochemical models of this atmosphere (Yung et al., 1984; Toublanc et al., 1995; Lara et al., 1996). Recent experiments on the modeling of the atmosphere of Titan including traces of CO, show that the main O-containing organic product is oxirane (Coll et al., 2003).

These oxygenated organic compounds that have been identified in the simulation experiments have so far not been detected in the atmosphere of Titan. Their non-detection is probably due to their low concentration and to the observation techniques.

These simulation experiments also lead to the formation of solid, refractory products: macromolecular compounds containing H, C, and N atoms. These "tholins" may be similar to the solid particles of unknown composition that make up fog layers in Titan's upper atmosphere. Tholins have been studied in

Table 2.4. Chemical composition of Titan's stratosphere, and comparison with simulation experiments data (adapted, according to Gautier and Raulin, 1997)

Compounds	Stratosphere mixing ratio (E = Equ.; N = North Pole)	Production in simulation experiments*
• **Main constituents**		
Nitrogen N_2	0.90–0.99	
Methane CH_4	0.017–0.045	
Argon Ar	0–0.06?	
Hydrogen H_2	0.0006–0.0014	
• **Hydrocarbons**		
Ethane C_2H_6	1.3×10^{-5} E	Maj.
Acetylene C_2H_2	2.2×10^{-6} E	Maj.
Propane C_3H_8	7.0×10^{-7} E	++
Ethylene C_2H_4	9.0×10^{-8} E	++
Propyne C_3H_4	1.7×10^{-8} N	+
Diacetylene C_4H_2	2.2×10^{-8} N	+**
Benzene C_6H_6	few 10^{-9}****	+
• **N-Organics**		
Hydrogen cyanide HCN	6.0×10^{-7} N	Maj.
Cyanoacetylene HC_3N	7.0×10^{-8} N	++
Cyanogen C_2N_2	4.5×10^{-9} N	+
Acetonitrile CH_3CN	few 10^{-9}	++
Dicyanoacetylene C_4N_2	Solid phase N	+
• **O-Compounds**		
Carbon monoxide CO	2.0×10^{-5}	
Carbon dioxide CO_2	1.4×10^{-8} E	
Water H_2O	few 10^{-9}***	

* average relative abundance, Maj. = major compound >> ++ >> + ** Coll et al., 1999b; *** Coustenis et al., 1998; **** Coustenis et al., 2003

detail by Sagan and his team (Khare et al., 1981; 1984; 1986; McDonald et al., 1994; and refs. included), their elemental composition determined (McKay, 1996; Coll et al., 1998; 1999a; and refs. therein), and information on the structure was obtained from analysis by pyrolysis coupled to gas chromatography and mass spectrometry (Ehrenfreund, 1995, Coll et al., 1995; 1997, 1999a), and by gel-filtration chromatography of the water soluble fraction (McDonald et al., 1994). The results indicate an elemental composition with a very variable C/N ratio depending on the experimental conditions (Table 2.5). This ratio varies from 1 to 11, and seems to indicate a complex structure including the CN, NH_2, aliphatic and aromatic groups, with an average molecular mass of about 500 to 1000 daltons. Some studies were even carried out to determine the potential assimilation of these materials by micro-organisms, in order to know if they could be used as potential food for hypothetical "Titanic" micro-organisms (Stocker et al., 1990).

Future observational investigations of Titan, which will be carried out within the framework of the Cassini–Huygens mission, either by remote-sensing measurements thanks to the instruments of the Cassini orbiter – which orbits Saturn since July 2004 – or *in situ* measurements from the Huygens probe (Lebreton, 1997; Matson et al., 2002; Lebreton and Matson, 2002), should allow the detection of many compounds identified in the laboratory but not yet observed in Titan's atmosphere (see Box 2.2 and Fig. 2.3).

One of the interests of the experimental simulations is precisely to provide information – at least qualitative – on the nature of the compounds likely to be present in a planetary atmosphere, including those not yet detected. The results of these experiments can thus be used as a guide for the preparation of observational campaigns – by remote sensing or in situ measurements – and for data interpretation.

At LISA a spectroscopic data base in UV and IR has been developed for more than 15 years, for organic compounds of planetological interest (Raulin et al., 1998; Bénilan et al., 2000). The already collected data base relates to hydrocarbons and nitriles as well as to less stable compounds (at ambient temperature),

Table 2.5. C/N ratios in laboratory Titan's tholins according experiments parameters

Experiment	Temperature (K)	Pressure (mbar)	Contamination by oxygen at the time of the sampling?	C/N
Sagan et al., 1984	300	~ 1–0.1	Contamination	1.9
Coll et al., 1995	100	~ 760	No contamination	11
McKay, 1996	300	~ 760	Contamination	5.5
Coll et al., 1997	300	~ 0.1	No contamination	1.7
Coll et al., 1999a	100	~ 0.1	No contamination	2.8

Box 2.2. The Cassini–Huygens Mission

The Cassini–Huygens mission (Matson et al., 2002), developed jointly by the American (NASA) and European (ESA) space agencies includes the sending of an orbiter (Cassini) around Saturn and of a probe in the atmosphere of Titan. Launch took place in October 1997. After two flybys of Venus, then of the Earth in August 1999, and Jupiter in December 2000, Cassini arrived in the Saturn system in July 2004, and the Huygens probe entered into the atmosphere of Titan on January 14th, 2005. The nominal duration of the Cassini part of the mission is 4 years. The artificial satellite of Saturn, Cassini, contains twelve scientific instruments. Most of them, in particular the IR spectroscopy measurements (CIRS, VIMS) and UV (UVIS), are essential for exobiology. It is the same for most of the six instruments of the Huygens probe, in particular the GC-MS and ACP experiments. The mission also includes 9 interdisciplinary scientists (IDS), who will carry out scientific investigations based on the integrated use of the future Cassini–Huygens data. It is necessary to note the very strong European participation in this mission. (http://www.jpl.nasa.gov/cassini; http://sci.esa.int/huygens)

The Cassini Orbiter

– **Scientific instrument**	**Acronym**	**Principal Investigator**
Optical remote sensing measurements		
– Infrared spectrometer	CIRS	V. Kunde, USA
– Imagery	ISS	C. Porco, USA
– Ultraviolet spectrograph	UVIS	L. Esposito, USA
– Visible/near infrared spectrometer	VIMS	R.H. Brown, USA
Fields of particles and waves		
– Spectrometer of plasma	CAPS	D. Young, USA
– Analyzer of cosmic dusts	CDA	E. Grün, Germany
– Mass spectrometry of ions & neutrals	INMS	J.H. Waite, USA
– Magnetometer	MAG	D. Southwood, GB
– Imagery of the magnetosphere	MIMI	S. Krimigis, USA
– Measurement of radio waves & plasma	RPWS	D. Gurnett, USA
Microwave remote sensing measurements		
– Radar	Radar	C. Elachi, USA
– Radio Measurements	RSS	A.J. Kliore, USA hack

Box 2.2. The Cassini–Huygens Mission(continued)

– **Interdisciplinary program**	**IDS**
– Magnetosphere and plasmas	M. Blanc, France
– Dust and rings	J.N. Cuzzi, USA
– Magnetosphere and plasmas	T.I. Gombosi, USA
– Atmospheres	T. Owen, USA
– Satellites and asteroids	L.A. Soderblom, USA
– Aeronomy and interaction with solar wind	D.F. Strobel, USA

The Huygens Probe

– **Scientific instrument**	**Acronym**	**Principal Investigator**
– Gas phase chromatograph coupled to a mass spectrometer	GC-MS	H. Niemann, USA
– Collector and pyrolyser of aerosols	ACP	G. Israel, France
– Measurement of the atmospheric structure	HASI	M. Fulchignoni, Italy
– Descent imagery and spectral radiometry	DISR	M. Tomasko, USA
– Doppler measurement of the winds	DWE	M. Bird, Germany
– Measuring Instrument of surface	SSP	J.C. Zarnecki, GB

Interdisciplinary program	**IDS**
– Aeronomy	D. Gautier, France
– Interactions between atmosphere and surface	J. Lunine, USA
– Chemistry and exobiology	F. Raulin, France

such as polynes and cyanopolyynes. The data set includes frequency and intensity of the bands or the lines, including low temperatures data, which are of particular interest for the atmosphere of Titan.

Experimental simulations also allow the production of laboratory analogues of these atmospheric aerosols. The availability of analogues is also crucial for the preparation of the instruments denoted to in situ analysis and for the interpretation of observed data. They allow testing and calibration of the instrument before and after launching such as the GC-MS and ACP experiments to be performed by the Huygens probe. But they are also essential to determine the physical and chemical parameters of the atmospheric aerosols whose knowledge is essential

Fig. 2.3. Artist view of the Huygens probe entering Titan's atmosphere after release from the Cassini orbiter (from the NASA Gallery: http://www.jpl.nasa.gov/cassini/)

to data processing. The optical properties measured on tholins obtained during laboratory experiments are of major importance to interpret observations made by satellites (Khar, 1984). New determinations of these parameters have recently been made (Ramirez et al., 2002), using Titan tholins obtained under more realistic conditions (in term of gas-mix design, total pressure and temperature) and with an experimental protocol avoiding contamination by the atmosphere of the laboratory.

2.4 Conclusions

Since the first experiment by Miller, in 1953, models of the primitive terrestrial environment and more particularly of its atmosphere have dramatically changed. According to the models currently available and accepted by the scientific community, it seems probable today that the primitive atmosphere of the Earth was never sufficiently reducing to allow atmospheric organic syntheses at least in homogeneous phase, to play a major role in the origin of organics on the prebiotic Earth.

Elemental prebiotic chemistry, based on chemical transformation in aqueous solution of simple organic molecules, such as HCN or HCHO or their oligomers or co-oligomers (equivalent then to what others called tholins) remains however of interest. These basic ingredients could have been formed at the ocean floor, via heterogeneous reactions at the liquid solid interface in submarine vents, or have been imported on the Earth by meteorites, micrometeorites or comets.

Nevertheless, the atmospheric organic syntheses, are still of great importance, because they can occur in many extraterrestrial environments, in particular in the atmosphere of giant planets and especially in Titan's environment. Indeed, chemical processes in Titan atmosphere are the source of many organics, including some of the simple molecules which are key compounds in prebiotic chemistry.

With the development of space exploration, simulation experiments, coupled with theoretical modeling and in situ observations remain an important tool for the understanding of the physical chemistry of planetary environments. The study of atmospheric organic chemistry in extraterrestrial objects is an essential approach to try to understand the primitive terrestrial environment and the organic processes which triggered the emergence of life.

References

Baross J.A., Hoffman, S.E. (1985). Submarine hydrothermal vents and associated gradient environments as sites for the origin and evolution of life. *Origins Life* **15**, 327–345.

Bénilan Y., Smith N.R., Jolly A., Raulin F. (2000). The long wavelength range temperature variations of the mid-UV acetylene absorption coefficient, *Planet. Space Sci.,* **48**, 463–471.

Brack A. ed. (1998). *The Molecular Origins of Life: Assembling Parts of the Puzzle*, Cambridge University Press, Cambridge.

Briggs M.H. (1961). Organic constituents of meteorites. *Nature*, **191(4794)**, 1137–1140.

Canuto V.M., Levine J.S., Augustsson T.R., Imhoff C.L., Giampapa M.S. (1983). The young sun and the atmosphere and photochemistry of the early Earth, *Nature,* **305**, 281–286.

Clarke D.W., Ferris J.P. (1997). Chemical evolution on Titan: comparisons to the prebiotic Earth, *Origins of Life and Evol. Biosphere,* **27** 225–248.

Coll P., Coscia D., Gazeau M.-C., de Vanssay E., Guillemin J.-C., Raulin F. (1995). Organic chemistry in Titan's atmosphere: new data from laboratory simulations at low temperature, *Adv. Space Res.,* **16** (2), 93–104.

Coll P., Coscia D., Gazeau M.-C., Raulin F. (1997). New planetary atmosphere simulations: application to the organic aerosols of Titan, *Adv. Space Res.,* **19** (7), 1113–1119.

Coll P., Coscia D., Gazeau M.-C., Raulin F. (1998). Review and latest results of laboratory investigation of Titan's aerosols, *Origins of Life and Evol. Biosph.,* **28**, 195–213.

Coll P., Coscia D., Smith N.R., Gazeau M.-C., Ramirez S.I., Cernogora G., Israel G., Raulin F. (1999a). Experimental laboratory simulation of Titan's atmosphere: aerosols and gas phase, *Planet. Space Sci.,* **47** (10,11), 1331–1340.

Coll P., Guillemin J.-C., Gazeau M.-C., Raulin F. (1999b). Report and implications of the first observation of C_4N_2in laboratory simulations of Titan's atmosphere, *Planet. Space Sci.,* **47** (12), 1433–1440.

Coll P., Bernard J.-M., Navarro-González R., Raulin F. (2003). Oxirane: An exotic oxygenated organic compound in Titan? *Astrophys. J.*, **589**, 700–703.

Coustenis A., Salama A., Lellouch E., Encrenaz Th., Bjoraker G.L., Samuelson R.E, De Graauw Th., Feuchtgruber H., Kessler M.F. (1998). Evidence for water vapor in Titan' s atmosphere from ISO/SWS data, *Astron. Astrophys.,* **336,** L85–L89.

Coustenis A., Salama A., Schulz B., Ott S., Lellouch E., Encrenaz Th., Gautier D., Feuchtgruber H. (2003). Titan's atmosphere from ISO mid-infrared spectroscopy, *Icarus,* **161**, 383–403.

De Vanssay E., Gazeau M.-C., Guillemin J.-C., Raulin F. (1995). Experimental simulation of Titan's organic chemistry at low temperature, *Planetary Space Sci.,* **43**, 25–31.

Dobrijevic M., Parisot J.-P. (1998). Effect of chemical kinetics uncertainties on hydrocarbon production in the stratosphere of Neptune, *Planet. Space Sci.,* **46** (5), 491–505.

Ehrenfreund P., Boon J.P., Commander J., Sagan C., Thompson W.R., Khare B.N. (1995). Analytical pyrolysis experiments of Titan aerosol analogues in preparation for the Cassini–Huygens mission, *Adv. Space Res.,* **15** (3), 335–342.

Ferris J.P. (1979). HCN did not condense to give heteropolypeptides on the primitive earth, *Science,* **203**, 1135.

Ferris J.P., Hagan W.J. (1984). HCN and chemical evolution: the possible role of cyano compounds in prebiotic synthesis, *Tetrahedron,* **40** (7), 1093–1120.

Ferris J.P., Sanchez R.A., Orgel L.E. (1968). Studies in prebiotic synthesis III Synthesis of pyrimidine from cyanoacetaldehyde, *J. Mol. Evol.,* **3**, 301–309.

Ferris J.P., Zameck O.S., Altbuch A.M., Freiman H. (1974). Chemical evolution XVIII Synthesis of pyrimidines from guanidine and cyanoacetylene and cyanate, *J Mol. Biol.,* **33**, 693–704.

Gautier D., Raulin F. (1997). Chemical Composition of Titan's atmosphere, *Special ESA Publication,* ***SP*** **1177**, 359–364.

Haldane J.B.S. (1929). The Origin of Life, *The Rationalist Annual,* **148**, 3–11.

Hattori Y., Kinjo M., Ishigami M., Nagano K. (1984). Formation of amino-acids from CH_4-rich or CO_2-rich model atmosphere, *Origins of Life,* **14** (1–4), 145–150.

Hennet R.J.-C., Holm N.G., Engel M.H. (1992). Abiotic synthesis of amino acids under hydrothermal conditions and the origin of life: a perpetual phenomenon? *Naturwissenschaften,* **79**, 361–365.

Holm N.G., Guest Editor (1992). Marine hydrothermal systems and the origin of life, *Origins of Life and Evol. Biosph.,* **22** (1–4), 1–191.

Holm N.G., Andersson E.M. (1998). Hydrothermal systems, in *The Molecular Origins of Life: Assembling Pieces of the Puzzle,* ed. A. Brack, p. 86–99, Cambridge University Press, Cambridge.

Imai E-I., Honda H., Hatori K., Brack A., Matsuno K. (1999). Elongation of oligopeptides in a simulated submarine hydrothermal system. *Science,* **283**, 831–833.

Kasting J.F., Pollack J.B., Crisp D. (1984). Effects of high CO_2 levels on surface temperature and atmospheric oxidation state of the early Earth, *J. Atmosph. Chem.,* **1**, 403–428; and refs therein.

Khare, B.N., Sagan C., Zumberge J.E., Sklarew D.S., Nagy B. (1981). Organic solids produced by electrical discharges in reducing atmospheres: tholin molecular analysis, *Icarus,* **48**, 290–297.

Khare, B.N., Sagan C., Arakawa E.T., Suits F., Callicott T.A., Williams M.W. (1984). Optical constant of organic tholins produced in a simulated Titanian atmosphere: from software X-rays to microwave frequencies, *Icarus,* **60**, 127–137.

Khare, B.N., Sagan C., Ogino H., Nagy B., Er C., Schram K.H., Arakawa E.T. (1986). Amino acids derived from Titan tholins, *Icarus,* **68**, 176–184.

Kobayashi K., Oshima T., Yanagawa H. (1989). Abiotic synthesis of amino acids by proton irradiation of a mixture of carbon monoxide, nitrogen and water, *Chem. Letters,* **1989** (9), 1527–1535.

Lara L.M., Lellouch E., Lopez-Moreno J.J., Rodrigo R. (1996). Vertical distribution of Titan's atmospheric neutral constituents *J. Geophys. Res.,* **101** (E10), 23261–23283.

Lebreton J.P., European Space Agency (1997). *Huygens: Science, Payload and Mission,* ESA SP-1177.

Lebreton J.-P., Matson D.L. (2002). The Huygens probe: science, payload and mission overview, *Space Science Rev.*, **104** (1–4), 59–100.

Lellouch E., Romani P.N., Rosenqvist J. (1994). The vertical distribution and origin of HCN in Neptune's atmosphere, *Icarus,* **108**, 112–136.

Marshall W.L. (1994). Hydrothermal synthesis of amino acids, *Geochim. Cosmochim. Acta,* **58** (9), 2099–2106.

Marten A., Gautier D., Owen T., Sanders D.B., Matthews H.E., Owen T.C., Atreya S.K., Tilanus R.P.J., Deane J.R. (1993). First observation of CO and HCN on Neptune and Uranus at millimeter wavelength and their implications for atmospheric chemistry, *Astrophys. J.,* **406**, 285–297.

Matson D.L., Spilker L.J., Lebreton J.-P. (2002). The Cassini/Huygens mission to the Saturnian system, *Space Science Rev.*, **104** (1–4), 1–58.

Matthews C.N. (1979). Reply to Ferris J.P. (1979), *Science,* **203** 1136.

Matthews C.N., Moser R.E. (1967). Peptide synthesis from hydrogen cyanide and water, *Nature,* **215**, 1230–1234.

McDonald G.D., Thompson W.R., Heinrich M., Khare B.N., Sagan C. (1994). Chemical investigation of Titan and Triton tholins, *Icarus,* **108**, 137–145.

McKay C.P. (1996). Elemental composition, solubility, and optical properties of Titan's organic haze, *Planet. Space Sci.,* **44** (8), 741–747.

Miller S.L. (1953). A production of amino-acids under possible primitive earth conditions, *Science,* **117**, 528–529.

Miller S.L., Orgel L. (1974). *The Origins of Life on the Earth,* Prentice Hall, N. Jersey.

Miller S.L., Schlesinger G. (1984). Carbon and energy yields in prebiotic syntheses using atmospheres containing CH_4, CO and CO_2, *Origins of Life,* **14** (1–4), 83–90, and refs. included.

Mordaunt D.H., Lambert I.R., Morley G.P., Ashfold M.N.R., Dixon R.N., Western C.M., Schnieder L., Welge K.H. (1993), Primary product channels in the photodissociation of methane at 121.6 nm, *J. Chem. Phys.,* **98** (3), 2054–2065.

Navarro-González R., Molina M.J., Molina L.T. (1998). Nitrogen Fixation by Volcanic Lightning in the Early Earth. *Geophys. Res. Lett.*, **25**, 3123–3126.

Navarro-González R., McKay C.P., Nna Mvondo D. (2001). A possible nitrogen crisis for archaean life due to reduced nitrogen fixation by lightning, *Nature* **412**, 61–64.

Oparin, A.I. (1924). *Proiskhozhdenie Zhizni,* Izd. Moskovshii. Rabochii, Moscow.

Oro J. (1961). Comets and the formation of biochemical compunds on the primitive earth. *Nature* **190**, 389–390.

Pinto J.P., Gladstone G.R., Yung Y.L. (1980). Photochemical production of formaldehyde in Earth's primitive atmosphere, *Science,* **210**, 183–185.

Ramírez S.I., Coll P., Da Silva A., Navarro-González R., Lafait J., Raulin F. (2002). Complex Refractive index of Titan's aerosol analogues in the 200–900 nm domain, *Icarus,* **156** (2), 515–530.

Raulin F. (1990). Prebiotic syntheses of biologically interesting monomers in aqueous solutions: facts and constraints, *J. British Interplanet. Soc.,* **43**, 39–45.

Raulin F., Coll P., Bénilan Y., Coscia D., Gazeau M.-C., Khlifi M., Bruston P. (1998). Titan's atmosphere: new data of exobiological importance, in *Planetary Systems: The Long View* eds. L.M. Celnikier & J Trân Thanh Vân, p. 435–441, Editions Frontières, Gif/Yvette, France.

Raulin F., Owen T. (2002). Organic chemistry and exobiology on Titan, *Space Science Rev.*, **104** (1–4), 377–394.

Raulin F., Toupance G. (1975). Etude cinétique de l'évolution du cyanoacétaldéhyde en solution aqueuse, *Bull. Soc. Chim.*, 1975 (1–2), 188–195.
Rosenqvist J., Lellouch E., Romani P.N., Paubert G., Encrenaz Th. (1992). Millimeter wave observations of Saturn, Uranus and Neptune: CO and HCN on Neptune, *Astrophys. J.*, **392**, L99–L102.
Sagan C., Khare B.N. (1971). Long wavelength UV photoproduction of amino acids on the primitive earth, *Science,* **173**, 417–420.
Sagan C., Khare B.N. (1979). Tholins: Organic chemistry of interstellar grains and gas, *Nature,* **277**, 102–107.
Sagan C., Khare B.N., Lewis J. (1984). Organic matter in the solar system, in *Saturn* University of Arizona Press, Tucson., pp. 788–807.
Shapiro R. (1988). Prebiotic ribose synthesis: a critical analysis, *Origins Life Evol. Biosphere,* **18**, 71–85.
Smith, N.R. (1999). Sensibilité des modèles théoriques de l'atmosphère de Titan aux incertitudes sur la photochimie des hydrocarbures simples, These de Doctorat, Université Paris 12.
Smith N.R., Raulin F. (1999). Modeling of methane photolysis in the reducing atmospheres of the outer solar system, *J. Geophys. Res.*, **104** (E1), 1873–1877.
Stocker C., Boston P.J., Mancinelli R.L., Segal W.D., Khare B.N., Sagan C. (1990). Microbial metabolism of Tholins, *Icarus,* **85**, 241–256.
Thompson W., Todd H., Schwartz J., Khare B.N., Sagan C. (1991). Plasma discharge in $N_2 + CH_4$ at low pressures: experimental results and applications to Titan, *Icarus,* **90**, 57–73.
Toublanc D., Parisot J.-P., Brillet J., Gautier D., Raulin F., McKay C.P. (1995). Photochemical modeling of Titan's atmosphere, *Icarus,* **13**, 2–26.
Toupance G., Sebban G., Buvet R. (1970). Etape initiale de la polymérisation de l'acide cyanhydrique et synthèses prébiologiques, *J. Chim. Phys.*, **67** (10), 1870–1874.
Wachtershäuser G. (1990). The case for the chemoautotrophic origin of life in an iron-sulfur world. *Origins Life Evol. Biosphere* **20**, 173–176.
Yung Y.L., Allen M., Pinto J.P. (1984). Photochemistry of the atmosphere of Titan: comparison between model and observations, *Astrophys. J. Suppl. Ser.,* **55**, 465–506.
Yung Y.L., DeMore W.B. (1999). *Photochemistry of Planetary Atmospheres*, Oxford Univ. Press, Oxford.
Zubay G. (2000). *Origins of Life on the Earth and in the Cosmos,* Academic Press, San Diego.

3 Chirality and the Origin of Homochirality

John Cronin, Jacques Reisse

3.1 Chirality: Basic Concepts

> "I call any geometrical figure, or group of points, chiral, and say it has chirality, if its image in a plane mirror, ideally realized, cannot be brought to coincide with itself".

This definition of chirality was given by Kelvin in May 1893, at the time of a conference of the Oxford University Junior Scientific Club (Kelvin, 1904). The definition was illustrated by the following comment:

> "Two equal and similar right hands are homochirally similar. Equal and similar right and left hands are heterochirally similar or allochirally similar (but heterochirally is better). These are also called "enantiomorphs", after a usage introduced, I believe, by German writers".

The hand as metaphor is deeply rooted in the history of stereochemistry. (The term chiral derives from the Greek *cheir*, hand.) As Mislow (1996) has pointed out, Kelvin's geometric definition of chirality is equivalent to that given many years later by Vladimir Prelog in his 1975 Nobel Prize lecture:

> "an object is chiral if it cannot be brought into congruence with its mirror image by translation or rotation."

This is now the commonly accepted definition of chirality.

Every object must be either chiral or achiral and, if it is chiral, it can exist a priori in two non-superposable mirror image, i.e., enantiomorphic/enantiomeric forms. Symmetry group theory provides a mathematical criterion by which it can be unequivocally determined whether an object is chiral or achiral. The criterion is expressed as follows: an object is achiral if, and only if, it possesses an improper axis of rotation of the order n. The symmetry operation associated with this element of symmetry corresponds to a rotation of $360°/n$ around the axis, accompanied by reflection in a plane perpendicular to the axis (see Fig. 3.1).

Here S_1 corresponds to a plane of symmetry (σ) and S_2 corresponds to a center of inversion (i). On the other hand, S_4 does not have an equivalent notation. Occasionally one still finds in the chemical literature the formal criterion stated incorrectly by referring only to the absence of a center and plane of symmetry.

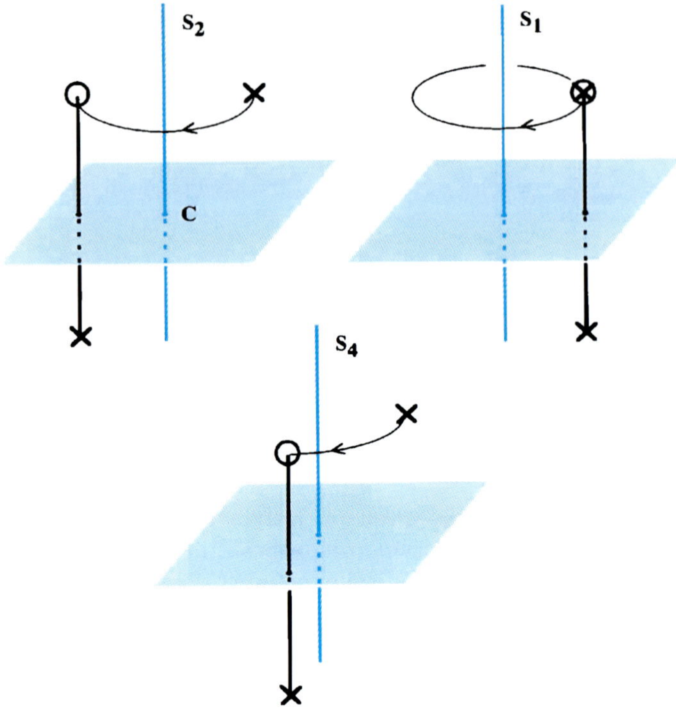

Fig. 3.1. The operations S_1, S_2, and S_4 performed on a material point, X, i.e., rotation about an angle $360°/n$ ($n = 1$ for S_1, $n = 2$ for S_2 and $n = 4$ for S_4) accompanied by reflection in a plane perpendicular to the axis

Nevertheless, it is true that molecular chirality is most commonly due to the absence of a center and (or) planes of symmetry.

A tetrahedron whose four vertices are rendered different, for example by numbering as in Fig. 3.2, is a chiral object of particular interest in chemistry. Indeed, if the center of the tetrahedron is occupied by an atom, for example a carbon atom, and the vertices correspond to different atoms or different groups of atoms bonded to the central atom, the central atom is said to be *asymmetric*. The exchange of two atoms or groups generates the enantiomer, i.e., the nonsuperposable mirror image structure. This is the significant property of the asymmetric tetrahedron from the chirality point of view. The chirality of the molecules found in living things is due to the fact that they contain one, or sometimes many, four-bonded carbon atoms that are asymmetric in this way.

Geometrical chirality can be discussed in spaces of any dimension. The chemist is generally interested in three-dimensional space (E^3) although work of the Mislow group at Princeton has also been directed to quantification of the geometrical chirality of two-dimensional objects such as triangles (Auf Der Heyde et al., 1991; Buda and Mislow, 1991; Buda et al., 1992; Mislow, 1997).

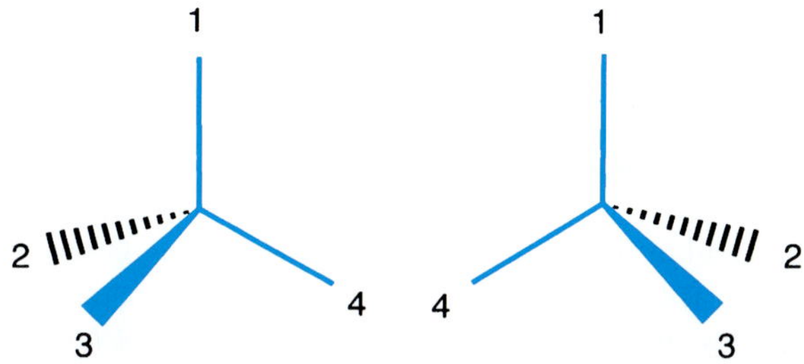

Fig. 3.2. An asymmetric tetrahedron: a chiral object of particular importance in chemistry

Very recently, the problem of the most chiral triangle has been considered by Rassat and Fowler (2003) using a very elegant approach based on quantum mechanics (calculation of eigenfunctions of the Schrödinger equation for the particle confined to an equilateral triangular box).

Two-dimensional space is also of interest because one must often represent a three-dimensional object as a two-dimensional picture, for example, on a page (see Fig. 3.3). Certain rules for handling these representations are imposed by the fact that two enantiomeric objects in two-dimensional space can become congruent (superposable) after passage through three-dimensional space.

Thus it is necessary to be careful in the use of the two-dimensional Fisher projection. For example, if one wishes to determine whether two Fisher projections correspond to the same molecule or, instead, represent two enantiomers, one must test their congruence by translations and rotations only within the plane in which they are represented and not move one or the other through the third dimension.

When the object is a molecule, the Kelvin–Prelog definition remains valid even though molecular objects have some unique properties relating to their dynamics (Mislow, 1997).

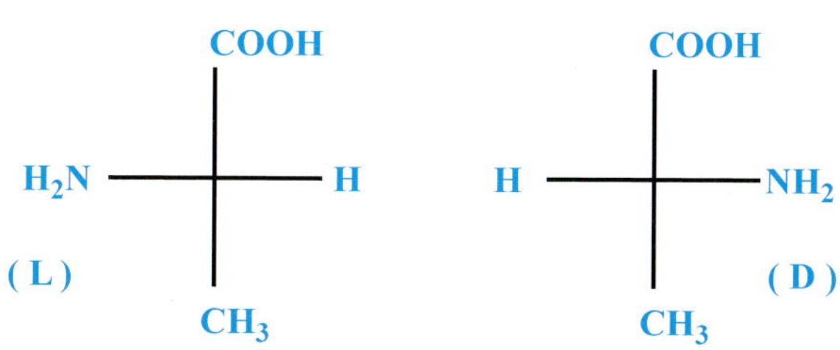

Fig. 3.3. Fisher projection of the two enantiomers of alanine. By convention, the two atoms or groups of atoms (here H and NH_2) at the ends of the *horizontal line* (*bonds*) are viewed as being in front of the plane of the page and the two groups of atoms at the ends of the *vertical line* (here COOH and CH_3) are viewed as being behind the plane of the page. The asymmetric carbon atom is not explicitly shown but is understood to be located at the intersection of the lines

For example, the methane molecule (CH_4) is commonly described as a tetrahedral structure. However, it is sometimes necessary to consider the variations from this ideal structure that can result from motions of the C–H bonds. In reality, the regular tetrahedron is a model that describes the average structure of a great number of methane molecules at any moment (or the average structure of a particular methane molecule taken over time). However, if one were able to photograph, using an infinitely short exposure time, a collection of methane molecules, one would see that only a very small number of them are regular tetrahedrons. The majority of the molecules would be seen to have atomic configurations that are chiral. Of course, one would see among them, in equal numbers, the enantiomeric configurations as well.

An equimolar mixture of enantiomers (molecules that differ only because they are of opposite chiralities) is called a racemic mixture. The term "racemic" was first used by Gay-Lussac to designate an organic acid that lacked optical activity but was of identical composition to optically active tartaric acid (Racemic comes from the Latin, *racemus*, bunch of grapes). Pasteur later showed racemic acid to be composed of equal amounts of the two enantiomers of tartaric acid (see Sect. 3.3) and the term racemic subsequently came to be used as a descriptive term for any such equimolar mixture of enantiomers. This definition is subject to some qualification on the basis of statistical variation, a subject that will be discussed further in Sects. 3.7 and 3.10.

It is important to note that some molecules do not contain asymmetric atoms and owe their chirality to an overall structure that is helical and thereby asymmetric. Although the homochiral molecules isolated from living things owe their chirality primarily to the presence of one or several asymmetric carbon atoms, some biopolymers (proteins, polysaccharides, and DNA) that have asymmetric carbon atoms in their monomeric units (amino acids, sugars, and deoxynucleotides, respectively) also adopt helical conformations. Interestingly enough, at the macromolecular level a hypothetical right helix and a hypothetical left helix made of the same chiral subunits would not be enantiomers but diastereoisomers.

The chirality we have been concerned with up to this point is observed in objects for which the formal interconversion of one enantiomeric form to the other can be carried out by an inversion of space (denoted P), compared to a fixed arbitrary origin. The effect of this inversion is identical to a reflection in a plane mirror. This geometrical chirality does not depend on the overall motion of the object; however, there is another type of chirality associated with movement. An example is a half-cone rotating around its axis. In this case, the inversion of time (operation T), which consists of replacing "t" by "$-t$" and reversing the direction of the movement, transforms the dynamic chiral object into its image. On this basis, Barron (1982) introduced a distinction between "true" chirality and "false" chirality. "False" chirality is associated with objects, like the half-cone example, for which the object-image interconversion can be carried out, not only by inversion of space, but also by inversion of time (followed eventually by rotation in space). Within the framework of this chapter we shall deal only with "true" chirality; however, we must point out that true chirality can be associated with movement (and thus time), for example, in the case of circularly polarized radiation. Indeed, using a wave description, the electric field vector (just like the magnetic field vector) of circularly polarized radiation describes a helix. This helix is palindromic, i.e., the horizontally oriented helix is identical whether traversed left to right or right to left. Thus the inversion of time (which corresponds to the inversion of the direction of the propagation of the light) does not modify the sign of the helicity. On the other hand, the inversion of space transforms right-circularly polarized radiation into left-circularly polarized radiation, i.e., radiation opposite in the sign of its helicity. Thus, one can conclude that circularly polarized radiation possesses true chirality.

In this context, linearly polarized radiation can be described, figuratively, as "racemic" radiation since it is composed of both right-circularly polarized and left-circularly polarized radiation. When such racemic radiation traverses a chiral medium, one of the circularly polarized components is propagated less rapidly than the other and the plane of the polarized light beam is rotated; this phenomenon is called optical rotation (or optical rotatory dispersion when the optical rotation is measured as a function of wavelength). In addition, if the medium is chromophoric, one of the two circularly polarized components is absorbed more than the other, which results in the phenomenon of circular dichroism. The individual enantiomers of chiral compounds differ only with respect to the signs associated with these optical phenomena and have otherwise identical physical and chemical properties except, of course if they interact or react with other homochiral molecules (see Sect. 3.2).

Before ending this section, it may be useful to describe the ways in which enantiomers are specified. In the past, it was usual to designate them with (+) and (−) symbols or with the letters *d* (dextrorotatory) and *l* (levorotatory) preceeding the name of the chiral compound. The former was simply the sign of the optical rotation measured with light of the wavelength of the sodium D-line emission. According to a suggestion of Fisher, the D and L nomenclature was based on the relationship between the asymmetric centers of the enantiomers in question and that of a chiral reference compound, the glyceraldehyde of (+) optical rotation (at the D-line wavelength of Na), which Fisher designated D-glyceraldehyde. This assignment depends on the ability to chemically transform the enantiomer into glyceraldehyde. For example, the amino acid serine can be converted to glyceraldehyde by a series of reactions that does not affect the configuration of the groups attached to its asymmetric carbon atom. The product was found to be (−)glyceraldehyde, the levorotatory enantiomer, and thus natural serine is designated L-serine (see Fig. 3.4). The D/L nomenclature is well established for the sugars and amino acids and their derivatives and will be used in this chapter. In Fig. 3.3 the enantiomers of alanine are designated in this way.

In 1950, it was shown by X-ray diffraction carried out by the Bijvoet group that the configuration arbitrarily assigned by Fisher to D-glyceraldehyde was correct (there was a 50% chance!) and since then the D/L nomenclature also

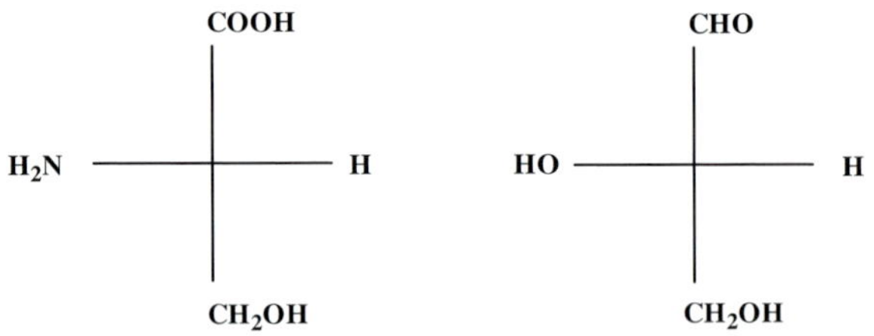

Fig. 3.4. Fisher projection of L-serine and L-glyceraldehyde

correctly denotes absolute configuration. Today, the configurations of chiral molecules are most commonly described by the Cahn–Ingold–Prelog (CIP) system recommended by the IUPAC (International Union of Pure and Applied Chemistry). This nomenclature, which is independent of any measurement of optical rotation, absolutely defines the configuration of every asymmetric center (stereocenter) in a molecule and designates each one as S or R (sinister or rectus).

The reader interested in a concise introduction to stereochemistry and molecular chirality is referred to Mislow (1965). One will find an exhaustive treatment of these subjects in Eliel and Wilen (1994).

3.2 Reactivity of Chiral Molecules

The behavior of chiral molecules when they are not alone is an important consideration. Chemists and biochemists are frequently interested in molecules when they interact (or react) with other identical or different molecules and models are necessary that allow the properties of such sets of interacting molecules to be described. A molecule that is achiral when it is alone can become chiral when it interacts with a chiral molecule. An extreme case is the "chiralisation" of xenon atoms trapped in a chiral molecular cage (Bartik et al., 2000, 2001).

Neglecting for the moment the very small differences in energy due to violation of parity (discussed in Sect. 3.8), two enantiomers have the same internal energy and are characterized by identical reactivity if the second reactant is achiral. What is true for a covalent bond making/breaking chemical reactions is also true for weaker intermolecular interactions if the interaction is with an achiral material. However, for the same reason that a right foot slides easily into a right shoe and with some difficulty into a left shoe, the reaction/interaction will occur at different rates for enantiomers if the molecule with which it reacts/interacts is itself an enantiomer. Enantiomers D1 and L1 will react with reactant D2 with different rates. The interacting pairs (D1–D2) and L1–D2), whether they are dissociable complexes, reaction transition states, or stable molecules, are called *diastereomers*, meaning that they are stereoisomers (but not enantiomers) and therefore have different internal energies and different properties. The symbols

1 and 2 are not necessarily meant to indicate different molecules: they may be the same or they may be different.

The work of Blount and Idelson (1956) presents an interesting example. They observed that the polymerization of a glutamic acid derivative occurs 20 times faster starting with molecules of the same chirality (only D–D or L–L reactions) rather than with a racemic mixture (D–L and L–D reactions can occur). Matsuura et al. (1965) found that the partial polymerization of a reactive derivative of alanine having a small enantiomeric excess led to an increased enantiomeric excess in the polypeptide products, i.e., they showed that these differences in reactivity could be used to enhance the enantiomeric excess in the product. This is an example of a kinetic resolution (Kagan et al., 1988). The term "resolution" in this context means a process allowing the separation of enantiomers. The possible importance of such processes to the origin of homochirality will be examined in Sect. 3.10. Molecular-recognition processes, which are of enormous importance in biology, are frequently highly chirospecific when both the ligand and the binding molecule are chiral. They represent examples of diastereomeric interactions in which one is generally unfavorable to the point of insignificance. For example, the transport of D-glucose through the membrane of the red cell is facilitated by a chiral transport molecule, a protein.

L-glucose is essentially not recognized by this protein and thus its only mode of penetration into the red blood cell is by unfacilitated diffusion, a much slower process (Rawn, 1989).

3.3 Pasteur and the Discovery of Molecular Chirality

The crystallization of chiral compounds provides many interesting examples of the interaction of homochiral or heterochiral molecules. In fact, it was Pasteur's careful studies of crystals of sodium-ammonium paratartrate that led to the discovery of molecular chirality. Tartaric acid obtained from wine was known to have optical rotation, whereas paratartaric acid lacked optical activity but seemed identical to tartaric acid in all other respects.

In Pasteur's notebooks (Valléry-Radot, 1968), one finds this description of his separation of the enantiomorphic crystals he obtained by crystallization of the sodium-ammonium salt of paratartaric acid.

> "The happy idea came to me to just orient my crystals in a plane perpendicular to the observer, and then I saw that in this confused mass of crystals of paratartrate there were two kinds of them with respect to the distribution of the asymmetric facets. In the one case, the facet of asymmetry close to me was inclined on my line, relative to the plane of orientation of which I spoke, while the others, the asymmetric facet was inclined to my left. In other words, the paratartrate presented itself as formed of two kinds of crystals, one asymmetric to the right, the other one asymmetric to the left. A new idea, very naturally, soon came

> to me. The crystals, asymmetric to the right, that I could manually separate from the others, were absolutely identical in form to those of right tartrate".

Pasteur had made a simple discovery with profound implications. He had found that paratartaric acid was a racemic mixture of D- and L-tartaric acids (see Fig. 3.5) but, more importantly, he had shown for the first time that a single organic compound can exist in two forms that differ in the sign of optical rotation and, underlying this, in their molecular asymmetry. The exact nature of this asymmetry was unknown at the time although, according to Mislow (1996), Pasteur explicitly considered that it might be due to an asymmetric tetrahedral arrangement of atoms. It was left to van't Hoff and Le Bel to fully develop the idea that tetrahedral carbon atoms bounded to four different proups are the basis for the chirality of organic molecules such as the tartaric acids and their salts.

The structures of D- and L-tartaric acid are shown in Fig. 3.5. It can be seen that tartaric acid has two asymmetric carbon atoms and, because of this, its stereoisomerism is a little more complicated than we have seen thus far. In the D- and L-tartaric acids, the two asymmetric carbon atoms are of the same configuration (two "right hands" in one case, two "left hands" in the other); however, there exists another form, known as mesotartaric acid, in which the two asymmetric carbon atoms have opposite configurations (one "left hand" and one "right hand"). Consequently mesotartaric acid is optically inactive as a result of internal compensation, that is, the chirality associated with one asymmet-

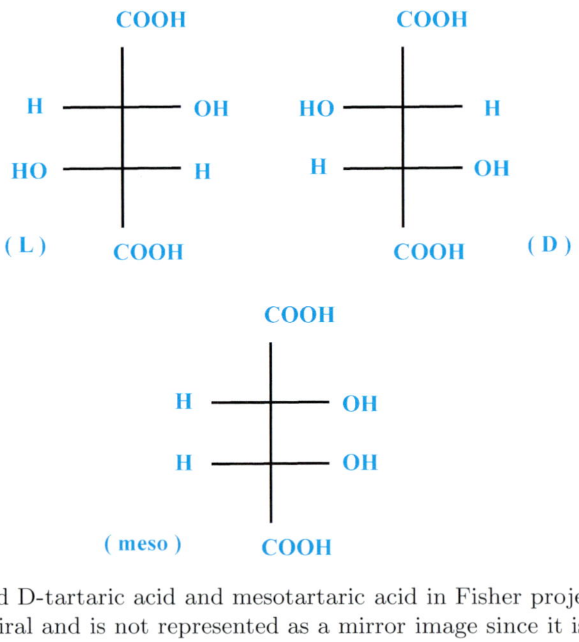

Fig. 3.5. L- and D-tartaric acid and mesotartaric acid in Fisher projection (mesotartaric acid is achiral and is not represented as a mirror image since it is identical)

ric carbon atom is exactly compensated intramolecularly by the chirality of the other asymmetric carbon atom. One might think of mesotartaric acid as being analogous to a right hand and a left hand coupled palm against palm. In contrast to the internal compensation observed in these meso compounds, one can consider a racemic mixture, such as paratartaric acid, as optically inactive due to "external" compensation since the compensation results from the presence of opposed chiralities in different molecules present in equal numbers.

Readers who are interested in the respective roles of Pasteur, van't Hoff, and Le Bel in establishing the foundations of stereochemistry will read with interest the special review volume of Tetrahedron published in 1974 on the occasion of the 100th anniversary of the publication of the work of van't Hoff and Le Bel. The preface to this volume (Robinson, 1974) and an article by Mason (2002) make apparent the pioneering role that Pasteur played in 1860. An article by Lardner et al. (1967) also furnishes information on historical aspects of the chemistry of tetrahedral carbon.

3.4 Crystals and Crystallization

Pasteur's discovery resulted from a combination of exceptional scientific ability and a stroke of good luck. Indeed, the crystallization of a racemic mixture is not always accompanied by enantiomer segregation into enantiomorphic crystals, a process that is unfavorable from the entropic point of view. Generally, molecules of opposite chiralities crystallize together to give a single type of crystal, i.e., a crystal racemate that does not allow separation of the right- and left-handed molecules. Also, the temperature of crystallization is a sensitive factor in determining the type of crystals obtained and this is particularly true in the case of the crystals studied by Pasteur.

Mason (1982), Jacques et al. (1981), Collet (1980), and Collet (1990) thoroughly discuss the question of crystallization of mixtures of enantiomers, the phase diagrams of these mixtures, and the reasons why the crystallization of a racemic mixture more frequently gives crystals of the racemate rather than a conglomerate of crystals composed individually of only one enantiomer. According to Collet (1990), only approximately 10% of racemic mixtures crystallize in the form of conglomerates.

While considering the crystallization of chiral molecules, it is important to note that some achiral molecules can form chiral crystals, which can exist as enantiomorphs. A well-known example is quartz, in which the macroscopic chirality of the crystals arises from the helical arrangement of the achiral SiO_2 units of which they are composed (see Fig. 3.6). On a global basis quartz is racemic, a conclusion based on examination of 27,053 quartz crystals (Frondel, 1978).

There are several examples (see Bonner, 1996) of achiral substances that, on crystallization, give enantiomorphic conglomerates. For example, sodium chlorate ($NaClO_3$), when crystallized from supersaturated solutions without agitation, always gives a conglomerate in which there are equal numbers of right and

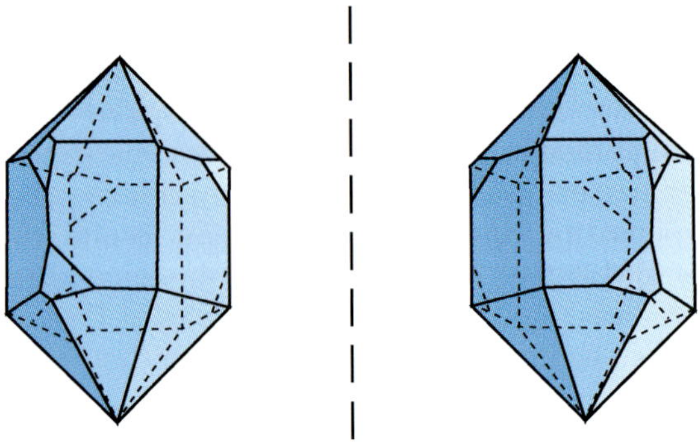

Fig. 3.6. Enantiomorphic crystals of quartz

left crystals. On the other hand, when crystallization is brought about with agitation, sometimes right crystals predominate and sometimes left crystals (Kondepudi et al., 1999). In this case, each experiment gives a spontaneous symmetry breaking, a spectacular phenomenon in which an achiral solution yields a crystalline solid in which the crystals are homochiral! The interpretation of these observations is straightforward: primary nucleation leads to formation of the first chiral crystal and this crystal is then broken by agitation. The fragments, dispersed in the liquid phase, serve as crystallization nuclei and induce secondary crystallization leading, by a cascade effect, to a symmetry breaking (McBride and Carter, 1991). As in the case of quartz, sodium chlorate has been the subject of many studies aimed at showing statistically that the number of right crystals is equal to the number of left crystals. Such crystal counts might at first seem trivial but, as we will see in Sect. 3.8, this is not the case. They aim at eventually obtaining evidence, at the macroscopic level, of an important physical phenomenon called violation of parity.

There are also substances that, in the molten state or in solution, are known to exist as a rapid equilibrium between enantiomers (that is, there is rapid interconversion or racemization) and that, upon crystallization, give homochiral crystals (Havinga, 1954; Kondepudi et al., 1999). In this case there is again a cascade effect in which the separation of the first crystals provides nuclei for further crystallization, effectively displacing the equilibrium toward the first enantiomer to crystallize. These examples of symmetry breaking illustrate the amplification of an initially very slight enantiomeric excess represented here by the first crystal formed. In Sect. 3.10 we shall return to the general problem of the amplification of enantiomeric excesses.

3.5 Homochirality and Life

The importance of Pasteur's discovery to biology was immediately apparent to him. At the time, it was already well known that optical rotation was a property of natural liquids such as turpentine, oil of lemon, and oil of laurel, and therefore must be a property of the molecules, per se. Optical rotation was also known to be a property of solutions of sugars, camphor, tartaric acid and, as we have seen, tartrates. Pasteur's work gave a deeper meaning to these observations by making it clear that molecules, as well as crystals, could exist as asymmetric pairs. Because the direction of optical rotation correlated with a particular asymmetry, and because optical rotation had been observed in so many biological materials, it became clear that the use of molecules of a particular asymmetry was a fundamental property of life. Then the question of the origin of homochirality could be asked for the first time, i.e., how did life choose between the two enantiomeric possibilities with which it was almost always presented? Against the advice of his mentors, the young Pasteur set about attempting to answer this question.

It is ironic that in the second half of the 19th century molecular chirality came to provide one of the last bastions for vitalism, which Pasteur's later research effectively discredited (see Mason, 1982). According to the advocates of this theory, the matter of living things is qualitatively different from that of inanimate matter and one of the differences is that the chiral molecules extracted from living things are always homochiral, whereas, when the chemist synthesizes these same compounds in the laboratory, a mixture of enantiomers in equal amounts, that is, a racemic mixture, is always obtained.

Although Pasteur was unable to answer the fundamental question of the origin of homochirality, he did clearly perceive why chemists were unable to prepare homochiral molecules. At a conference in 1883, he declared:

> "Indeed whenever the chemist in his laboratory combines elements or products born of the elements, he brings into play only nonasymmetric forces. For this reason the syntheses which he carries out never show asymmetry.... I would try asymmetric combinations of elements.... I would make them react under the influence of magnets, solenoids, elliptically polarized light – finally, under the influence of everything which I could imagine to exert asymmetric actions." (Vallery-Radot, 1968)

This statement is similar to, but earlier than, the more general one of Curie, which is now called the symmetry principle:

> "When certain causes produce certain effects, the elements of symmetry of the causes must be found in the produced effects. When certain effects reveal a certain asymmetry, this asymmetry must be found in the causes which gave them birth" (Curie, 1894).

Pasteur made additional comments that are quite interesting in an origin of life context

> "If the immediate principles of life are asymmetric, it is because, in their development, they are governed by asymmetric cosmic forces; therein, in my opinion, is a link between life on the surface of the Earth

and the cosmos, i.e., the entirety of forces spread throughout the universe".

It is remarkable that more than a century ago Pasteur, because of his need to discover "asymmetric forces" that could bring about the symmetry breaking necessary for the homochirality of biomolecules, came to view life on Earth in a cosmic context. In fact, most of the questions that arise today in connection with the origin of the homochirality of life have roots in the work of the 19th century scientists, Pasteur, van't Hoff, and Le Bel.

3.6 The Why and When of Homochirality

Chiral molecules dominate organic chemistry. For example, if we consider the acyclic saturated aliphatic α-amino acids, we see that glycine, the only two-carbon member of the series, is achiral; however, alanine, the C_3 member of the series, is chiral; one of the two C_4 isomers is chiral; all of the three C_5 isomers (including valine) are chiral; and six of the seven C_6 isomers (including leucine and isoleucine) are chiral. In general, the more atoms there are in molecules, the more isomers there are and the larger the number of isomers that are chiral. Clearly, chirality must be a fact of life unless life is somehow constrained to use only the very simplest molecules.

Given that organic chemistry is dominated by chiral molecules, why is biochemistry so rigorously homochiral? The answer to this question lies in understanding the importance of biopolymer structure. Life is based essentially on polymers of two fundamental types, proteins and nucleic acids (DNA and various RNAs). Both of these biopolymers result from the polymerization of chiral monomers and the biological function of both is strongly dependent on their existence in precisely defined three-dimensional structures (conformations). The right-handed double-helical structure of DNA and the unique globular structures of enzymes provide vivid examples of these functionally essential three-dimensional structures.

In the case of proteins, the monomeric units are 19 L-amino acids (to which we could add two very rare L-amino-acids units) plus the achiral amino acid, glycine. In the biosynthesis of a particular protein these amino acids are polymerized to give a chain (polypeptide) having a unique length and amino-acid sequence determined by a particular gene. The resulting polypeptide folds, spontaneously in many cases, to assume the conformation necessary for its biological function. If the amino acids were incorporated in the correct sequence but could vary randomly with respect to their chirality, an enormous number of diastereomeric polypeptides would result (2^{100} for a 100-amino-acid polypeptide). Only a very small fraction of these might be expected to fold correctly to give the exact three-dimensional structure (conformation) required for proper function. Therefore, specification of a particular amino acid as well as its chirality is essential at each step in the biosynthesis of a polypeptide chain in order to insure that all of the products assume the correct three-dimensional structure. The homochirality of amino acids simply precludes chiral variation by providing amino acids of only one configuration (L) for protein synthesis. Of course, this homochirality is now reinforced by the strict chiral specificity displayed by the enzymes of contemporary organisms. The great majority of these enzymes are themselves proteins.

A priori, proteins could have been composed of D-amino acids. If this were the case, the chiralities of all the other constituents of cells would also be of the opposite chirality. This is illustrated by observations made on the enzyme HIV-1 protease when chemically synthesized from D-amino acids. This synthetic enzyme exhibits catalytic activity identical to that of the natural enzyme except that it is specific for the enantiomer of the natural substrate (Milton et al., 1992).

The question has been raised whether molecular homochirality is an absolute requirement for life and many believe that it is. For example, Bonner (1996) finds heterochiral life to be inconceivable. Nevertheless, if, as we have argued above, it is biosynthetic reproducibility of protein structure that life requires, in principle, this does not demand amino-acid homochirality. One can imagine an alternative life form that is both similar and different from terrestrial life: similar in having a triplet nucleotide genetic code, but different in that it specifies both the D and L enantiomers of the 19 chiral amino acids. Its polypeptides could then contain, in precisely ordered sequence, both D and L enantiomers of the chiral amino acids. Each polypeptide would fold and assume a unique three-dimensional structure and would have an active site and display a particular activity just as a homochiral polypeptide does. Such a requirement for both enantiomers of amino acids would, of course, lead to some metabolic duplication of effort. If both D- and L-amino acids were required, it would be necessary, inter alia, to have dual sets of t-RNAs and amino acyl t-RNA synthetases, proteases specific for D–X and L–X peptide bonds, a chirally nonspecific peptidyl transferase, and racemases in order to ensure a balanced supply of both sets of enantiomers.

In the case of the nucleic acids, random heterochirality of the deoxyribonucleotide and ribonucleotide monomeric units of DNA and RNA, respectively, would be just as disastrous for the function of the nucleic acids as is random amino-acid heterochirality for proteins. It would make the regular double-helical structure of DNA impossible as well as the replication and transcription of DNA, during which an extensive regular double-stranded structure is formed transiently (Gol'danski and Kuz'min, 1988). However, in this case, specified heterochirality as suggested for the amino acids of proteins would be equally disastrous because it would also preclude formation of the unique higher-order structure, the antiparallel double helix, that is required for function. Consequently, homochirality is a necessity for a ribose/deoxyribose-nucleotide-based genetic system.

On the other hand, achirality is a possibility for nucleic-acid-like molecules. Achiral nucleic-acid analogues have been synthesized, for example, the peptide nucleic acid (PNA) of Nielsen (1993; 1996). This nucleic-acid analogue, which is completely devoid of asymmetric centers, has a linear array of bases, i.e., the essential feature necessary for encoding and translating information, and can take on the double-helical structure necessary for function (Miller, 1997). If such an informational macromolecule were to provide the genetic memory necessary to reproduce protein amino-acid sequence and, if it specified the use of both D- and L-amino acids, a PNA-protein biochemistry is conceivable in which overall homochirality would not be a necessity. If terrestrial life began on such an achiral-heterochiral basis, the strict homochirality that characterizes life now must have been an outcome of very early evolution and been achieved before the appearance of the Last Universal Common Ancestor (LUCA).

Thus, it is possible to imagine life that is at least partially heterochiral if, but only if, the chiralitry of each amino-acid subunit were fully specified. But how likely is the origin of life on such a basis? If life originated as a PNA-world, catalytic molecules, the analogues of ribozymes, would have been necessary and these would have been asymmetric in their folded forms. Their active sites would have been asymmetric and those that catalyzed key steps, e.g., transamination, in the formation of amino acids would have had a strong tendency to be enantioselective. Therefore, amino-acid heterochirality would have been unlikely. On the other hand, if proteins were necessary for the origin of life, it is difficult to see how the necessary reproducibility of three-dimensional structure could have been achieved without homochirality first having been achieved abiotically. Consequently, we conclude that, although partially heterochiral life is conceivable, its origin and/or evolution on this basis seems unlikely.

The alternative to an evolutionary origin of homochirality is that it was achieved prior to the origin of life. In this case, complex molecules could have been used from the beginning since the problem of chiral variations in monomer sequences that would complicate the production of reproducible biopolymers would have been solved. In the following sections we shall consider some plausible prebiotic processes that might have (a) created a symmetry breaking in chiral molecules and (b) brought about amplification of the possibly small initial enantiomeric excesses that might have resulted.

3.7 Origin of Homochirality and Spontaneous Symmetry Breaking

Many hypotheses have been put forward to explain how homochirality, or at least a significant imbalance in the amounts of enantiomers, might have appeared on the prebiotic Earth as a result of a spontaneous symmetry breaking. For example, a symmetry breaking can occur by the spontaneous separation of a racemic mixture into its constituent enantiomers during crystallization if there is the formation of a conglomerate or a unique L or D crystal (see Sect. 3.4). However, it should be noted that the permissive conditions are tightly constrained and would have a low probability of being met on the primitive Earth. For example, the crystallization of a racemic mixture as a conglomerate followed by separation of the enantiomorphic crystals requires both a high concentration (saturation) as well as the correct temperature for crystallization as a conglomerate and not a racemate. Also, the initial mixture of enantiomers must be relatively pure because the formation of crystals requires that the compound not be a minor constituent of a complex mixture. Finally, some process must be available by which the crystals, once formed, can be spontaneously separated, i.e., an abiotic equivalent of Pasteur's human intervention.

Another possibility is presented by enantiomers in rapid equilibrium in the liquid state that can crystallize as a conglomerate. Here, again, total conversion of the mixture to one enantiomorphic crystal type can be achieved through secondary crystallization caused by strong agitation. Very demanding conditions, even more than in the previous case, must be met in order to observe this behavior. Asakura et al. (2002) continued the work of Kondepudi et al. (1999) in this area. The temperature of crystallization has a narrow range and the excesses of one homochiral crystalline form in comparison with the other varies by an order of magnitude! We concur with the opinion expressed by Bonner in his 1994 review that spontaneous resolution under racemizing conditions, of which there are numerous laboratory examples, probably does not constitute a mechanism of importance for the natural appearance of homochiralty.

Another spontaneous symmetry-breaking scenario is the selective adsorption of one enantiomer from a racemic mixture at the surface of a chiral crystal. As noted in Sect. 3.4, quartz crystals are asymmetric as a consequence of the right- or left-handed helical arrangement of the SiO_2 units and d-quartz and l-quartz selectively adsorb the enantiomers of amino acids. This enantioselective behavior was observed as early as 1935, a role for it in the origin of homochirality suggested by 1938, and the observations more recently confirmed (Bonner, 1996). In verifying this work, Bonner found that the phenomenon required carefully controlled anhydrous conditions, conditions quite unlikely to have been met on the primitive Earth. Furthermore, a survey of terrestrial quartz crystals has shown them to be racemic and, consequently, incapable of giving rise to an overall enantiomeric excess, that is, beyond what might be produced by a single crystal. This limitation to an extremely small locale is a general criticism that can be raised of the hypotheses for the origin of homochirality based on enantioselective adsorption by minerals/crystals. Nevertheless, such hypotheses continue to be suggested as the cause for homochirality on the primitive Earth.

The locality limitation is particularly severe when the adsorptive enantiomorphic surfaces are on the same crystal. For example, enantioselective adsorption has been observed with monoclinic, centrosymmetric and therefore achiral glycine crystals (Weissbuch et al., 1984). Although glycine is achiral, the crystals present enantiotopic faces, that is, their two-dimensional faces are nonsuperposable mirror images of each other. If such crystals float on the surface of a solution containing a racemic mixture of an amino acid and, if only one of the two enantiotopic faces is immersed in the solution, that face can act as a specific adsorbant and enantioselective adsorption approaching 100% can be obtained. More recently, Hazen et al. (2001) have carried out enantioselective adsorption experiments using one of the asymmetric faces of single calcite crystals and gone on to develop a rather comprehensive hypotesis for the origin of self-replicating homochiral peptides.

Clay minerals have been of interest as possible prebiotic catalysts for many years. Many are achiral and thus incapable of playing a role in the origin of homochirality. Others, for example kaolin, are, in principle, chiral, although it has not been possible as yet to demonstrate the chirality of single crystals. Many years ago it appeared that the chirality of kaolin might have been the basis for observations of enantiospecific adsorption and polymerization of amino acids (Jackson, 1971); however, in a careful subsequent study neither effect could be reproduced (Bonner and Flores, 1975). The possibility of such a role for kaolin was later resurrected in a theoretical study that assumed the existence of asymmetric crystals as a result of parity violation (see Sect. 3.8) and showed that enantiospecific amino-acid adsorption is possible (Julg, 1989; Julg et al., 1989).

Following Mills (1932), Dunitz (1996) and Siegel (1998) have suggested that homochirality could have arisen from statistical fluctuations from the equimolar condition that defines the racemic state. As these authors point out, it is incorrect to consider a racemic mixture as consisting of exactly the same number of enantiomeric molecules. If the synthesis of an enantiomeric mixture from achiral starting materials is repeated many times, the products obtained will have different enantiomeric compositions. A plot of frequency vs. enantiomeric composition will give a bell-curve centered on the exactly equimolar composition and with a width proportional to the square root of the variance. The statistics that apply to this situation are the same as those that apply to tossing a coin, that is, an independent event with a binary outcome having a probability $p = 0.5$ for each outcome repeated n times and under the same conditions (Droesbeke, 2001). In this case one obtains a binomial distribution of outcomes with variance proportional to n. If n is equal to 10^{24}, the distribution width will therefore be 10^{12}. According to these authors, a spontaneous symmetry breaking due to statistical fluctuation, followed by its amplification, can be the basis for the origin of homochirality. Mills (1932) explicitly states

> "we might account on the basis of the laws of probability for the existence of an initial minute bias towards one optical system or the other; and this would then, if the principles which I have endeavoured to explain are justified, eventually lead to the complete optical activity of the molecularly dissymmetric components of all living matter."

Furthermore, he stresses the importance of such fluctuations for microscopic prebiotic systems containing, for example, 10^6–10^8 molecules in which the statistical enantiomeric excess, although small in absolute value, can be large relative to the total number of molecules.

In dealing with evolution, Monod (1970) famously employed the dichotomy of chance versus necessity, distinctions that are useful in discussing the origin of homochirality. In the cases described above it is clear that, whatever the cause of the symmetry breaking, the probabilities of formation of both enantiomers are exactly the same and whichever one becomes predominant is a matter of chance. In the section that follows, we will consider an aspect of atomic physics that necessarily gives one particular enantiomer a slight advantage, that is, to use coin tossing as a metaphor, a situation in which the two faces of the coin do not have equal probabilities.

3.8 Origin of Homochirality and Parity Violation

If the homochirality of terrestrial life was a "necessity" and not a matter of "chance" selection, one must ask in what way the biological amino acids of the L-series and monosaccharides of the D-series could have been advantaged. An answer has been suggested by the recognition of a basic asymmetry of matter, the violation of parity that was previously alluded to in regard to the enumeration of quartz crystal enantiomorphs and studies of the handedness of sodium chlorate crystals. This parity violation constitutes a symmetry breaking at the level of the basic laws of atomic physics (Zee, 1999).

Until the 1950s, it was believed that quantum-mechanical operators, wave functions, and the observables that derive from them, preserved parity (P) and were thus independent of the operations of symmetry, that is, inversion with respect to a point or reflection with respect to a plane. In 1956, two young physicists, Tsung-Dao Lee and Chen Ning Yang, suggested that this conservation of parity applied only to quantum systems in which the controlling forces are the intranuclear strong interactions and the electromagnetic interactions. On the other hand, according to Lee and Yang, systems controlled by the weak intranuclear interactions were likely to undergo transformations that did not preserve parity.

Nonconservation of parity was soon experimentally demonstrated in the β-decay of ^{60}Co by Wu and collaborators in 1957. In their experiments, carried out at a temperature close to 0 K and in a static magnetic field in order to create a privileged direction of space, the dissociation of a neutron to a proton, an electron, and an antineutrino was observed. The fact that the intensity of the emission of electrons was not the same parallel and antiparallel with respect to the orientation of the magnetic field constituted proof that the angular momentum of the electrons was preferentially directed antiparallel with respect to the moment associated with the linear movement, i.e., that the electron spin preferentially describes a left-handed helix.

In 1974 Vester reported the first attempts to experimentally induce chirality through the interaction of an achiral reactant with β-radiation from various radioactive sources. He carried out numerous experiments but obtained only false positive and artifactual results. This fruitless work used β-radiation but also the associated circularly polarized light (bremsstrahlung). Bonner (1996) describes later work, including some by his own group. He carried out β-irradiations (exceeding ten years in some cases) using β-radiation from various elements (^{32}P, ^{14}C) and also bremsstrahlung in the hope of obtaining an enantioselective degradation of racemic amino acids but was unable to observe a significant effect.

In addition to the chirality of β-radiation, there is another consequence of the violation of parity that originates in the so-called electroweak interactions that result from unification of the intranuclear weak interactions and electromagnetic interactions. The resulting z force interacts with both the nucleus and electrons and, because of its parity-violating character, distinguishes between left and right. Yamagata (1966) pointed out that because of this coupling, enantiomers cannot have the same energy and that this fact could have relevance to the origin of biological homochirality. The energy difference between enantiomers is called the parity-violating energy difference (PVED). It results from the CPT theorem that states that physics is invariant if, simultaneously, one inverts space (operation P), replaces the elementary particles by their antiparticles (operation C for "loads conjugation") and inverts time, that is, the direction of movement (operation T). To explain the existence of the PVED between enantiomeric molecules, it suffices to take only CP into consideration since in this case time does not play a role.

The apparently equal amounts of enantiomers present in racemic mixtures (i.e., equilibrium mixtures) indicate that K_{eq} must be extremely close to 1.0 and the PVED therefore exceedingly small. To date, it has not been possible to experimentally measure its value; however, there are experimental approaches, in principle, and it may be possible to accomplish this in the future (Quack, 2002). In the meantime, it has been necessary to rely on theoretical methods for determination of PVED. Yamagata's early attempt at such a calculation overestimated its value by several orders of magnitude but, over the last 20 years, increasingly sophisticated ab initio quantum-mechanical calculations have been applied to the problem (reviewed by Quack, 2002). These calculations have generally given PVED values of about $10^{-14}\,\text{J}\,\text{mol}^{-1}$ corresponding to enantiomeric excesses of about $10^{-15}\%$, although more recent calculations have given larger values for PVED by about one to two orders of magnitude (Quack, 2002).

In addition to the magnitude of the PVED, the question of whether it favors the L- or D-enantiomer of the amino acids and monosaccharides has been of great interest with respect to the origin of terrestrial homochirality. Recent calculations for the amino acids, alanine, valine, serine, and aspartate, and the simplest of the monosaccharides, glyceraldehyde, have shown the PVED to have the "correct" sign, i.e., to favor the biological enantiomers (Zanasi et al., 1999; Mac Dermott, 2002). However, such results are controversial, as their outcome is dependent on both the conformation assumed for the molecule in question as well as the particular calculation method that is used (Buschmann et al., 2000). Bonner

(2000), after a careful recent review of the literature in this area, is unconvinced of a causal link between parity violation and biological homochirality.

3.9 Origin of Homochirality and Photochemistry

As noted in Sect. 3.5, Pasteur included elliptically polarized light among the "asymmetric forces" that he suggested might be capable of inducing enantioselectivity. Nevertheless, it was necessary to wait almost 70 years for an experimental demonstration of this phenomenon. In 1929 Kuhn and Braun showed that partial photolysis of a racemic mixture by irradiation with UV circularly polarized light (UVCPL) gave rise to an enantiomeric excess in the residue. In 1974, Balavoine et al. provided a theoretical treatment of this process that allows calculation of the enantiomeric excess obtained as a function of the degree of photolysis and the differences in the extinction coefficients of the enantiomers for UVCPL. Both Norden (1977) and Flores et al. (1977) demonstrated that aminoacids are subject to enantioselective photolysis by UVCPL. Flores et al. obtained enantiomeric excesses of 1.98% and 2.50% in originally racemic leucine solutions when photolysed to the extent of 59% and 75%, respectively. In this case, the difference between the extinction coefficients relative to their average value (g factor) is about 2%. Recently, Nishino et al. (2001) studied the reaction mechanism and found that acidic conditions are required and that glycine is one of the products of the reaction. Bonner and Bean (2000) have shown that enantioselective photolysis of racemic leucine can also be observed with elliptically polarized light. A role for UVCPL in the origin of the homochirality of terrestrial amino acids has been discussed by Norden (1977; 1978), Rubenstein and Bonner (1983) and by Bonner and Rubenstein (1987). Greenberg (1996) showed that partial UVCPL photolysis of a racemic mixture of the amino acid tryptophan in his laboratory model of low-temperature interstellar-grain chemistry gave an enantiomeric excess in the residue.

Circularly polarized light can also be used to effectuate enantioselective syntheses starting with an achiral reactant. In this case, CPL is the only chiral element. The first example of this was published by Kagan et al. in 1971 and concerned the synthesis of helicene (see also articles by Buchardt (1974), Mason (1982), Rau (1983), and Inoue (1992) which furnish other examples of enantioselective photosyntheses). In Sect. 3.11, we shall discuss the possible role of UV circularly or elliptically polarized light in the formation of the enantiomeric excesses found in the amino acids extracted from certain meteorites.

The results of these experiments are consistent with the Curie principle in that circularly or elliptically polarized light has true chirality and, therefore, a photolysis or a photosynthesis induced by such light must lead to nonracemic products. A recent paper by Rikken and Raupbach (2000; see also the commentary of Barron (2000)) shows that nonpolarized laser light can also lead to enantioselectivity during a photochemical reaction if the irradiated system is placed in a magnetic field and the field is not orthogonal to the propagation direction of the light. Indeed, the combination of a magnetic field B and a wave vector $\boldsymbol{k}$ (along the propagation direction of the light) is endowed with true chirality in so far as $\boldsymbol{k}$ and B are not orthogonal. Reflection in a mirror inverts the propagation direction of the light and therefore the orientation of $\boldsymbol{k}$ but leaves B unchanged since it is an axial vector. The theory of the dielectric constant of an isotropic environment, with or without an external magnetic field, allows quantification of the response of the material environment to the electric field of the electromagnetic radiation. The theoretical expression that follows from this treatment (Jorissen and Cerf, 2002) shows the dielectric constant as an expansion into a series. If one assumes the response of the environment to be linear, one notes that there are four terms that depend on the static magnetic field B in which the sample is placed. These four terms correspond, respectively, to the magnetic rotatory dispersion, the magnetic circular dichroism, the magnetochiral dispersion and the *magnetochiral dichroism*. The results obtained by Rikken and Raupbach (2000) point to the same phenomenon as the one that is responsible for magnetochiral dichroism. There appears a factor $g{\cdot}B$ that measures the difference between the extinction coefficients of the unpolarized light by the two enantiomers if these are placed in a field B. If B is zero (absence of external field), $g{\cdot}B$ is therefore also zero and enantioselective photochemistry is not observed. According to Wagnière and Meier (1983), g, in the case of the magnetochiral effect, is several orders of magnitude less than the g characterizing the different response of the two enantiomers to circularly polarized light. This means that to obtain significant enantioselectivity, B must be increased, typically by a few Tesla. (For comparison, the Earth's magnetic field at the surface of the planet is about 10^{-4} Tesla.) Could there exist, at the heart of a protosolar cloud, environments where such large magnetic fields prevailed and where, at the same time, photochemical syntheses produced molecules of biological interest? This remains an open question: the surroundings of a neutron star could be a favorable place in the sense that an enormous magnetic field exists there as well as synchrotron emission of electromagnetic radiation. This light can also be circularly polarized (see Sect. 3.11) but in this case CPL is not necessary for enantioselective syntheses.

Although UVCPL and magnetochiral photochemistry are usually considered to be extrasolar effects, it is necessary to consider the possibility of photosyntheses or enantioselective photolyses on the primitive Earth. In fact, the solar light incident on the Earth is circularly polarized as a result of scattering from atmospheric aerosols (Wolstencroft, 1985; Jorissen and Cerf, 2002). Nevertheless, this effect is canceled, on average, on a daily basis and it is necessary to imagine special topographies such that photosynthesis cannot occur during a part of the day or to invoke light-intensity differences between the morning and evening. The weak circular polarization of the scattered light along with the very constrained conditions that must be satisfied suggests that enantioselective photosynthesis in the atmosphere of the early Earth would not be a very likely occurrence.

3.10 Amplification of Enantiomeric Excesses

3.10.1 Introduction

Thus far we have dealt with several ways in which the symmetry, that is, the perfect enantiomeric equivalence, of racemic compounds can be broken. These symmetry breaking can be nearly quantitative, as in the case of crystallization, or can be minuscule, for example, as predicted for the PVED or allowed by natural statistical deviation from the equivalence of the racemic state. In this section we shall consider mechanisms by which enantiomeric excesses might be amplified, even to homochirality. Bonner's reviews (1991; 1996) critically discuss many of these mechanisms.

3.10.2 Kinetic Resolution

Kinetic resolution (Kagan and Fiaud, 1988), which takes advantage of the different reaction rates of diastereomeric transition states, is particularly pertinent in the case of amino-acid polymerization, a reaction that is commonly assumed to have occurred under prebiotic conditions. As noted previously (Sect. 3.2), transition states for the coupling of chiral amino acids can be diastereomeric (e.g., L L vs. L D) and, as a result, they have different energies and their reactions occur at different rates. Consequently, a homochiral amino-acid pair might react more rapidly than a heterochiral pair. Likewise, the transition states for the next addition can differ similarly (e.g., L L–L vs. D L–L) and their polymerization rates will differ (end-group diasteroisomerism effect).

When about eight amino acids have been joined in this way, the resulting peptide can begin to assume secondary structure, for example, a helical twist. Helixes are themselves chiral and their handedness can be dependent on the chirality of the constituent amino acids. The chirality of such helical coils can reinforce the effect of end-group diasteroisomerism in promoting the preferential addition of a particular amino acid enantiomer (helix effect). The unexpectedly large increase (20×) in the rate of polymerization for derivatives of D- or L-benzyl glutamate compared with that of racemic DL-benzyl glutamate suggested that the helicity of the product might be playing a significant role (Blout and Idelson, 1956). This led Wald (1957) to suggest that the α-helical structure of polypeptides might favor the further addition of amino acids of the same chirality as those that determined the handedness of the helix and that this could have been of significance in the origin of the homochirality of biological amino acids. Later, Spach (1974) demonstrated that the enantioselective effect observed in the polymerization of the benzyl glutamates arises mainly from the helical structure of the product. However, a positive helix effect is not observed with all amino acids. Blair and Bonner (1980) found substantial increases in enantiomeric excess in the polypeptide product when a derivative of leucine was polymerized but enantiomeric decreases with the corresponding valine derivative. Brack and Spach (1981) have pointed out that formation of another type of secondary structure assumed by some polypeptides, that is, the β-pleated sheet, can also give rise to enantiomeric enrichments.

A third effect, the greater stability to hydrolysis of the helical polymer in comparison with the corresponding random coil (stability effect), can further enhance the enantiomeric excesses achieved by polymerization. An example of the amplification that can be achieved by amino-acid polymerization combined with partial hydrolysis has been given by Blair et al. (1981). They found that partial polymerization of a leucine derivative having a 31% L-enantiomeric excess gave polypeptide products with an enantiomeric excess of 45% and that after partial hydrolysis of this material the enantiomeric excess in the surviving residue was 55%. Blair et al. (1981) suggested that repeated wet–dry cycles on the prebiotic Earth might have led to partial polymerization of amino acids followed by partial hydrolysis of the polypeptide products and thus similarly enhanced any initially small enantiomeric excesses.

The work of Eschenmoser et al. (Bolli, 1997) gives an example of what the authors describe as a deracemization process that possibly could occur during polymerization. Activated tetramers of ribonucleotides containing ribose as a pyranose ring were involved in further polymerizations. 8-mers (octamers), 12-mers and even 16-mers are obtained. A strong chiroselectivity was observed and the presence of tetramers containing L-ribose instead of D-ribose had no major influence on the polymerization rates. If we consider one mole of a 50-mer based on 4 different but homochiral ribonucleotides it is impossible to form all the possible sequences because 4^{50} is much larger than the Avogadro number. Following Eschenmoser et al. the final mixture of polymers must necessarily be nonracemic leading therefore to a deracemization process without any chiral external influence. Siegel (1998) also has pointed out the practical impossibility of generating all possible enantiomers in a mixture of polymers with a high (or even moderate) degree of polymerization by starting from racemic mixtures of 20 different monomers.

Recently, Zepik et al. (2002) have given an interesting example of amplification of an enantiomeric excess by polymerization in the two-dimensional environment provided by a water/air interface. The activated monomers of amino acids (lysine and glutamic acid) were functionalized with long-chain hydrocarbons to make them amphiphilic. As a result, they accumulate at the water/air interface where they associate in crystalline two-dimensional aggregates. When the starting mixture was enantiomerically unbalanced, they observed the formation of a racemate along with an enantiomorphic phase that polymerizes, yielding homochiral oligopeptides. In this case intermolecular packing at the interface provides structural order analogous to the helix effect described in the preceding paragraph for homogeneous polymerization. A mechanism for separation of the enantiomeric and racemic phases was not suggested by the authors. Although this phenomenon could be of prebiotic interest with respect to hypothetical chiral amphiphilic compounds it should be noted that amino acids and monosaccharides are quite water soluble and, unless chemically modified, do not accumulate at a water/air interface.

Takats et al. (2003) have shown that clustering of serine molecules is a efficient process for chirality amplification as well as for chirality tranfer to other amino acids or sugars. Although this is not polymerization in the strict sence, it does provide an example of an amino acid based suparmolecular system with possible relevance to prebiotic chemitry.

3.10.3 Chiral Catalysis

Chiral catalysis allows the transfer of chirality from a catalyst to the reaction product and, in many cases, impressive enantiomeric excesses are achieved in the product when the catalyst is enantiomerically pure. Sometimes very significant nonlinear effects are also observed in such catalytic reactions, that is, when the catalyst is not enantiomerically pure, a greater or lesser enantiomeric excess than that of the catalyst may appear in the product (Girard and Kagan, 1998). In the former case, this amounts to an amplification of chirality.

The work of Feringa et al. (1995) describes a physical system in which a small amount of an enantiomer stimulates a large effect. Here an achiral mesophase (achiral liquid crystals), when seeded with a very small quantity of a chiral additive, gives a transition phase leading to a chiral environment at the macroscopic level (transition between an achiral nematic phase and a cholesteric chiral phase). The chiral additive in this case is a photoisomerizable dopant that allows, according to the wavelength of irradiation, the induction of a cholesteric phase of either right or left helicity (chiroptical switching). Of course, the presence of achiral or chiral mesophases on the primitive Earth is an area of sheer speculation.

3.10.4 Asymmetric Autocatalysis: Theoretical Models

Asymmetric autocatalysis can be considered to be a special case of chiral catalysis in which a chiral reaction product acts as the catalyst. Franck (1953) proposed an early theoretical model for such a process, that is, for the amplification of an enantiomeric excess by means of a chemical reaction in which the product is a chiral compound and one enantiomer (for example, the L-enantiomer) catalyzes its own formation while inhibiting the formation of the other enantiomer (the D-enantiomer). The crystallization from solution of an enantiomeric mixture in rapid equilibrium is an example of such a process. In this case, a very slight initial excess of one enantiomer due to a random fluctuation in the racemic mixture can trigger a crystallization cascade that carries the system far from equilibrium, i.e., toward a large enantiomeric excess of the randomly favored enantiomer (see Sect. 3.4).

Following Franck, Decker (1974) systematically studied four different chemical systems all characterized by the presence of autocatalytic feedback. He showed that such systems show bifurcation phenomena that manifest themselves by the sudden amplification of a very small initial enantiomeric excess. Decker explicitly placed this work in the context of the Brussels school of the thermodynamics of open systems far from equilibrium. Kondepudi, based on previous work by Prigogine and Nicolis, has published several papers (Kondepudi and Nelson, 1984; 1985; Kondepudi, 1996; Prigogine and Kondepudi, 1999) devoted to an in-depth study of a system composed of five interdependent chemical reactions likely to show bifurcation phenomena. The system, also a modification of the autocatalytic model of Franck, is as follows:

$$
\begin{aligned}
&\mathrm{A} + \mathrm{B} \rightarrow \mathrm{X(L)}\\
&\mathrm{A} + \mathrm{B} \rightarrow \mathrm{X(D)}\\
&\mathrm{X(L)} + \mathrm{A} + \mathrm{B} \rightarrow 2\mathrm{X(L)}\\
&\mathrm{X(D)} + \mathrm{A} + \mathrm{B} \rightarrow 2\mathrm{X(D)}\\
&\mathrm{X(L)} + \mathrm{X(D)} \rightarrow \mathrm{Y}
\end{aligned}
$$

The scheme has two autocatalytic steps and an irreversible mutual destruction step, features that appear to be essential in systems that show symmetry breaking. The system is open, that is, A and B, the chiral reactants, are maintained at constant concentrations throughout the simulation experiment and the product Y is eliminated, thereby rendering the reverse reaction impossible and making the evolution of the system irreversible.

Kondepudi et al. are interested in the magnitude of X(L)–X(D), the enantiomeric concentration difference, in relation to the concentration of the perfect racemic mixture (X(L)–X(D) = 0). They carried out several simulations corresponding to different values of the molar concentrations product $\mathrm{A} \cdot \mathrm{B}$ and observe that there exists a critical value of product $\mathrm{A} \cdot \mathrm{B}$ beyond which the symmetrical solution X(L) = X(D) becomes unstable. This means that a random statistical fluctuation can provoke bifurcation of the system toward a state in which X(L) is no longer equal to X(D) and that corresponds to a spontaneous symmetry breaking. However, if at the start there is a chiral perturbation of the system, for example by the presence of a very small excess of X(D) or of X(L), a bifurcation will be observed when the critical value is attained leading to an amplification of the initial enantiomeric excess (Fig 3.7).

According to Kondepudi et al., this mechanism will allow amplification of initial excesses as small as those that are predicted by the violation of parity (PVED) (see Sect. 3.8) if the accumulation of A and of B can continue in a very large volume (typically a lake of some km^2) for long time periods (typically tens of thousands of years), in this way reaching the critical value of the concentration product $\mathrm{A} \cdot \mathrm{B}$. Kondepudi and Nelson (1984) note that in their simulations they cannot account for the effect of chance chiral impurities in the environment that would affect the kinetics of the system. They conclude that

> "When one finds a real chemical system that breaks chiral symmetry, in order to preserve the effects of weak-neutral current interactions, the

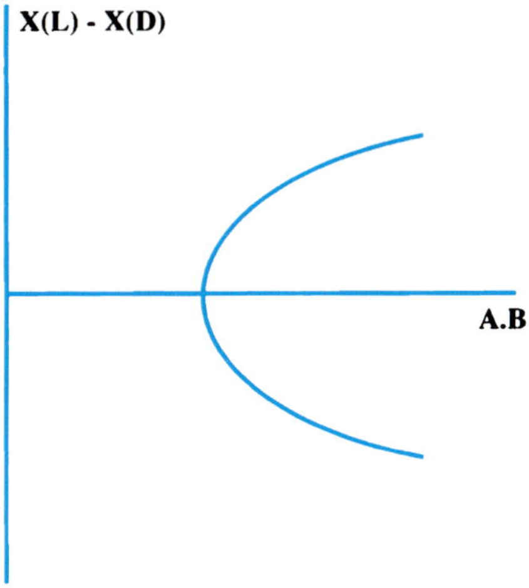

Fig. 3.7. Example of bifurcation predicted by the theoretical model of Kondepudi. A and B are the molar concentrations of the achiral reactants A and B while X(D) and X(L) are the molar concentrations of the chiral products X(L) and X(D) (see text for discussion)

system must be prepared with sufficient purity that the selection occurs due to this interaction and not due to the presence of a chiral impurity."

These condition and constraints raise serious questions as to the significance of such a theoretical model in the context of prebiotic chemistry. However, as will be seen in what follows, the general point is quite significant in evaluating experimental systems of this type.

3.10.5 Asymmetric Autocatalysis: Experimental Data

Recently, asymmetric autocatalysis has been experimentally shown by the group of Soai to give spectacular amplifications (Soai et al., 1995; 2000; Shibata et al., 1998). Soai and his coworkers have studied in detail the addition of an achiral organozinc derivative to an achiral aromatic aldehyde (Soai reaction). The reaction ultimately yields a mixture of enantiomeric alcohols. These alcohols then act as catalysts of their own synthesis. If the reaction is carried out in the presence of a chiral initiator such as leucine with 2% L-enantiomeric excess, the mixture of alcohols obtained at the end of the reaction shows an enantiomeric excess of 21% in one of the enantiomers. If the initiator is enriched in D-leucine, at the end of the reaction the mixture of alcohols is enriched in the other enantiomer and the enantiomeric excess is 23%. The interpretation of these results is as follows. The presence of the chiral initiator leads to very small enantiomeric excess in the first alcohols formed. These then act as the asymmetric catalyst. Since the alcohol is catalyzing its own synthesis an autocatalytic system is operating that amplifies the enantiomeric excess. Soai et al. have observed that an initiator with an enantiomeric excess of only 0.1% can give a mixture of alcohols with an enantiomeric excess greater than 80%! The authors point out that the enantiomeric excess in the leucine used as the initiator is similar to those that can be obtained by enantioselective photodecomposition of a racemic mixture of leucine using UVCPL (Sect. 3.9). They conclude that autocatalysis in the presence of a chiral initiator could have played a role in the appearance of terrestrial homochirality. Furthermore, these researchers (Soai, 1999) have observed that homochiral quartz crystals are also able to play the role of initiator and, starting with the same achiral reactants, give enantiomeric excesses of greater than 80% in the product alcohol. Depending on whether the quartz crystals are l or d, a preponderance of one or the other alcohol is observed.

Subsequently, Singleton and Vo (2002) attempted to use the Soai reaction to test whether the statistically allowed enantiomeric excesses that occur in "racemic" mixtures might be sufficient to initiate the formation of products with a substantial enantiomeric excess. They were unable to demonstrate this for an unexpected reason: the chiral impurities present in even purified solvents are capable of initiating the amplification process in this autocatalytic system and effectively overwhelm any weaker effects. Thus, as Kondepudi and Nelson pointed out (see above), the ubiquity of homochiral molecules in terrestrial environments poses a severe experimental problem for those who would hope to attribute chiral effects to minute enantiomeric excesses such as the statistically allowed variations of racemic mixtures or those that might arise from PVEDs.

Very recently, Soai et al. (2003) appear to have accomplished the amplification of enantiomeric excesses having only a statistical origin. They carried out the synthesis of the chiral pyrimidyl alcohol from only the achiral reagents (the pyrimidyl aldehyde and diisopropyl zinc) in pure solvents and obtained enantiomeric excesses ranging from 29–91% with each enantiomer predominating in approximately half of the cases (18*R*, 19*S*). As Mislow (2003) points out in an interesting commentary on this work, this is an example of absolute asymmetric synthesis, i.e., the formation of enantiomerically enriched products without the influence of any pre-existing chirality. Apparently, the random statistical fluctuations in the chirality of the initially formed catalytic zinc complex are amplified and determine the asymmetric outcome of the overall reaction. To be consistent with Curie's symmetry principle "this [product] asymmetry must be found in the causes that gave them birth" and in this case no causal asymmetric influence appears to be available other than random statistical fluctuation.

These results are of obvious importance in the context of the origin of homochirality problem although, as Avalos et al. (2000) have pointed out, the particular reaction studied by the Soai group does not readily proceed under prebiotic conditions. Organometallic chemistry requires rather restrictive nonaqueous conditions but it is hoped that in the near future similar autocatalytic reactions will be discovered that are more compatible with the constraints imposed by prebiotic chemistry. If such are found, it will be possible to agree with

Siegel (2002) when he states that *"in the prebiotic world examples of dominant molecular handedness were already likely to be abundant."*

3.10.6 On the Possibility to Amplifying Enantiomeric Excesses due to Parity Violation

Kinetic resolution and other related processes associated particularly with phase changes like crystallization can lead to amplification only when starting with mixtures having initial enantiomeric excesses that are very much greater (many orders of magnitude) than those that might be provided by the PVED. (see Sect. 3.8.) Consequently, amplification of these extremely small enantiomeric excesses is a subject that has received considerable attention and is treated separately here although, as discussed below, it has much in common with the problem of amplifying statistical excesses in racemic mixtures.

Yamagata (1966), who recognized very early the possible implications of parity violation for the origin of homochirality, also speculated as to how such small enantiomeric excesses might be amplified. He proposed an "Accumulation Principle," that is, the amplification of a very small enantiomeric bias when it is repeatedly multiplied by itself, for example, when it affects each step in a linear reaction sequence composed of a very large number of discrete steps. According to Yamagata, this principle would be applicable to either polymerization or crystallization phenomena. However, the application to polymerization considers only the possibility of formation of two completely homochiral polymers (enantiomers) and fails to consider the inevitable formation of numerous heterochiral polymers (diastereomers). Likewise, the proposal of irreversible consecutive steps is an oversimplification of crystallization. An excess of the l enantiomorph in natural quartz has been cited as possible evidence for the influence of PVED on crystallization (Tranter, 1985); however, as noted in Sect. 3.4, more recent evaluation of the data has failed to confirm a significant enantiomorphic excess (Frondel, 1978). Bonner (1999) has recently evaluated the Accumulation Principle and found it to be incapable of producing amplification by either polymerization or crystallization (see also Avalos, 2000).

Work by Szabo-Nagy et al. (1999) is purported to be the first experimental demonstration of the PVED. The authors prepared a racemic mixture of the *tris*(1,2-ethanediamine) complexes of Co^{3+} and Ir^{3+} by mixing the enantiomers to zero circular dichroism (CD). They then crystallized numerous samples and measured the CD of their solutions. A distribution of CD values was obtained, which, in the case of the metallocomplexes, was shifted from zero. On this basis, the authors claim that undetectably small enantiomeric excesses due to the PVED were amplified by crystallization. The observation of an effect only with the metallocomplexes and not with the racemic mixture of tartaric acids was attributed to the higher atomic number of the atom at the asymmetric center and the resulting larger PVED. The results are controversial (Avalos, 2000). It is not clear that the amplification is attributable to the PVED rather than deviations from the exactly racemic state and/or the effect of chiral impurities on the crystallization process. Circular dichroism is not sufficiently sensitive to detect small enantiomeric exess, even ones much larger than expected taking into account parity violation or statistical fluctuations.

Salam (1993) has proposed another mechanism by which the minuscule enantiomeric excesses due to PVEDs might be amplified. Salam, who received the Nobel Prize for his work on unification of the weak interactions and electromagnetic interactions, envisages an interconversion of the less stable enantiomer to the more stable one in the crystalline phase below some critical temperature. This cooperative effect would occur by a second-order phase transition and is similar to the transition observed when an electrical conducting material becomes a supraconductor. Figureau et al. (1995) attempted to observe a configurational change in racemic cystine over the temperature range 0.6–77K but failed to obtain a positive result. More recently, Wang et al. (2000) published results that are presented as a partial experimental confirmation of the Salam hypothesis. These authors claim to have observed a phase transition at 270 K by measuring the difference between the specific heats of D- and L-alanine in the solid crystalline phase. Measurements of magnetic susceptibility and by Raman spectroscopy also appear to show unique behavior at this temperature. Wang et al. also found that the specific heat of D-alanine is greater than that of L-alanine. These results are interesting but also surprising, notably because of the high transition temperature reported. One hopes that further studies along these lines will be carried out in order to confirm these results and exclude the possibility of artifacts.

The results of Soai et al. (2003) on the amplification of the enantiomeric excesses that occur in racemic materials as a result of random statistical variation require a rethinking of the possible role of enantiomeric excesses due to the PVED. In the Soai reaction initially 0.1 mMole of the chiral zinc complex was formed and it can easily be calculated that the statistically allowed enantiomeric excess in even this relatively large amount of material would be of the order of 10^{-8}%. Since random enantiomeric excesses of this magnitude will always exist and can, as Soai and his coworkers have demonstrated, be amplified in an autocatalytic reaction to as much as 90%, the presence of much smaller enantiomeric excesses of the order 10^{-15} to 10^{-13}% attributable to the PVED would have no significance.

In summary, it is apparent from this brief review that amplification processes are not lacking and many have been subjected to thorough laboratory testing, in part because of the importance of obtaining pure enantiomers for the pharmaceutical industry (Triggle, 1997). It has also become clear that experimental studies concerning the amplification of extremely small enantiomeric excesses can be subject to misinterpretation due to the presence of chiral contaminants. Finally, it should also be stressed that in the context of research on the origin of the homochiralty of life, amplification mechanisms that require pure crystals, enantiomer solutions near saturation, anhydrous conditions, or reactors on the scale of a lake that are free of chiral impurities for time periods measured in centuries or millenia, are unrealistic.

3.11 Exogenous Origin of Homochirality

The exogenous delivery of organic matter to the primitive Earth is now widely acknowledged to have been an important aspect of its prebiotic chemistry (Chyba and Sagan, 1992). There is little doubt that organic matter of meteoritic, micrometeoritic, and cometary origin accumulated in the hydrosphere of the primitive Earth (for reviews see the chapters by D. Despois, by F. Robert and by A. Morbidelli and D. Benest in Gargaud et al., 2001). Therefore it is understandable that the question of chiral organic matter in meteorites, particularly in carbonaceous chondrites has interested researchers for several decades (see reviews of Mullie and Reisse, 1987; Reisse and Mullie, 1993; Cronin and Pizzarello, 2000). In this regard, it is important to recognize that contamination of meteoritic matter by terrestrial compounds is almost unavoidable and that this can easily give artifacts. The work carried out by Meischein et al. (1966) on optical activity was pioneering in this respect. A meticulous analysis of an extract of the Homestead meteorite led these authors to the conclusion that the optical activity observed in meteoritic extracts by earlier workers was a result of contamination. However, more recently some of the amino acids detected in the Murchison and Murray meteorites have been found to be nonracemic (Cronin and Pizzarello 1997; Engel and Macko, 1997; Pizzarello and Cronin, 1998; 2000; Pizzarello et al., 2003), a discovery of considerable interest with respect to the origin of homochirality. It should be noted that the method of analysis used is no longer measurement of optical rotation or rotatory dispersion but rather gas chromatography (GC) on chiral supports coupled to mass spectrometry (MS) or isotope ratio mass spectrometry (IRMS), more sensitive and specific methods than those that were available in 1966.

There remains a great risk of contamination in the analysis of those meteoritic amino acids that are also present in the biosphere and this cannot be neglected regardless of the analytical methods used. In order to reduce this risk, Cronin and Pizzarello (1997) studied α-methyl amino acids that are either unknown or rarely found in the biosphere. These authors initially reported significant enantiomeric excesses (1 to 9% depending on the amino acid) in favor of the L-enantiomers and more extensive later analyses of isovaline have shown even larger enantiomeric excesses (Pizzarello et al., 2003).

On the basis of these results, it seems very probable that certain of the amino acids present on the primitive Earth had enantiomeric excesses and that these excesses were large enough to be susceptible to amplification by the processes discussed previously, notably by polymerization. It is necessary to add that the α-methyl amino acids, which are relatively abundant in carbonaceous chondrites, are particularly good candidates for amplification by polymerization.

As discussed in Sect. 3.10, the formation of secondary structures of the alpha-helical or beta-pleated-sheet type enhances amplification by polymerization and it has been shown (Altman et al., 1988; Formaggio et al., 1995) that the α-methyl amino acids give polyp eptides with very stable secondary structure of a helical type.

The origin of the enantiomeric excesses found in meteoritic amino acids is an important question. These amino acids have general characteristics that clearly indicate their production by abiotic processes (Kvenvolden et al., 1970; Cronin and Chang, 1993). They are nearly racemic, structurally diverse, with all possible isomeric forms represented, and are found in amounts that decrease exponentially within homologous series. Furthermore, they are isotopically distinct from their terrestrial counterparts, being substantially enriched in the heavier stable isotopes of C, H, and N (Epstein et al., 1987; Engel et al., 1990; Pizzarello et al., 1991; 2003; Engel and Macko, 1997). The isotopic enrichment, particularly in deuterium, is suggestive of chemistry at very low temperature, e.g., the ion–molecule reactions that occur in cold interstellar clouds (Wannier, 1980; Penzias, 1980). Thus, it is possible that meteoritic amino acids are of presolar origin (Epstein et al., 1987), although it is has been suggested that appropriate low-temperature conditions may have also existed within the solar nebula (Aikawa and Herbst, 2001).

As discussed previously (Sect. 3.9) UV circularly polarized light (UVCPL), when absorbed by a racemic organic compound, leads to photolysis with preferential survival of increasing amounts of one enantiomer as the photolysis proceeds. According to Rubenstein et al. (1983) and Bonner and Rubenstein (1987), a neutron star can produce UVCPL as synchrotron radiation resulting from the circulation of an electron plasma in the equatorial plane of the star. In addition, circular polarization has been observed within interstellar clouds and attributed to Mie scattering of light by aligned interstellar grains (Bailey et al., 1998). Bonner and Rubenstein go on to suggest that the UVCPL generated in these ways might affect the organic matter of interstellar clouds or the organic matter at or near the surface of a primitive planet thus giving rise to enantiomeric excesses in its chiral components. This hypothesis was adopted as a possible explanation for the asymmetry observed in meteoritic amino acids (Cronin and Pizzarello, 1997; Engel and Macko, 1997). Greenberg (1996) determined that it is probable that interstellar grains were, at some point in their existence, subjected to circularly or elliptically polarized radiation emitted by a neutron star and therefore concluded that a "*significant fraction of Solar Systems started off with significant enantiomeric excesses.*" The possibility of obtaining UVCPL from a neutron star has been questioned by Roberts (1984), and Mason (1997) has expressed doubt, based on the Kuhn–Condon effect, whether the broad band UVCPL emitted by a neutron star could achieve preferential photolysis. The latter objection has been effectively countered by Bonner et al. (1999). Mie scattering is not subject to these objections and a strong case can be made for its greater significance as a source of interstellar UVCPL (Bailey, 2001). Cerf and Jorissen (2000) and

Jorissen and Cerf (2002) have also discussed the pros and cons of a UVCPL scenario for the origin of enantiomeric excesses.

It is also necessary to consider quantitative aspects of the UVCPL photolysis hypothesis in evaluating its relevance to the origin of the enantiomeric excesses observed in meteorites. L-enantiomeric excesses up to 9% were initially observed in the isovaline from the Murchison meteorite and an excess as high as 15% has been measured for this amino acid more recently (Pizzarello et al., 2003). Although not impossible, enantiomeric excesses of this magnitude are difficult to achieve by UVCPL photolysis. The extinction coefficients of the enantiomers of aliphatic amino acids for UVCPL are not very different ($g \cong 0.02$), both enantiomers are subject to photolysis, and substantial enantiomeric excesses are achieved only as the reaction approaches completion. For example, it can be seen from the theoretical treatment of Balavoine et al. (1974) that a practical limit for *ee* of about 9% is reached for a fixed population of aliphatic amino acid molecules subjected to irradiation with UVCPL. (For $g \cong 0.02$, $ee = 9.2\%$ at 99.99% decomposition.) It seems likely that in any natural setting UVCPL irradiation would occur under conditions inferior to the optimized conditions used in laboratory studies and, consequently, that enantiomeric excesses values smaller than maximal, perhaps much smaller, would be observed if asymmetric photolysis of a racemate were the operative mechanism. From this perspective, some of the L-excesses measured for the meteoritic amino acids seem quite large, which raises doubt about UVCPL as the cause if the process is conceived of in terms of the laboratory model, i.e., a single exposure of a fixed population of molecules. In principle, UVCPL can effect an enantiomeric excess in an amino acid not only by the asymmetric photolysis of its racemate but also by synthesis; however, such photosyntheses are also limited to enantiomeric excess values governed by the g-value, i.e., approximately 2%.

However, two possibilities envisioned by Balavoine et al. (1974) seem worth considering in this context. If a mechanism existed whereby the residual fraction of photolyzed amino acids could accumulate from a large volume and experience a second exposure to UVCPL, this time starting with the small enantiomeric excess achieved previously, a further enhancement in the excess could be attained, although again at the expense of the total amount of amino acid surviving. Such a process of accumulation and re-exposure is conceivable as an interstellar cloud collapses in the process of nebula formation. The second possibility is the secondary formation of amino acids by an asymmetric catalyst. In this case one must imagine the formation of an asymmetric catalyst by the preferential photolysis of a racemic compound and this enantiomerically enriched catalyst then acting to promote the formation of amino acids with substantial enantiomeric excesses. As described in Sect. 3.10.5, in some autocatalytic reactions even small enantiomeric excesses in a chiral initiator can have dramatic nonlinear effects on the chirality of the reaction products.

Magnetochiral photochemistry (Rikken and Raupbach, 2000), which was discussed in Sect. 3.9, is possibly another explanation for the enantiomeric excesses observed in meteoritic amino acids. In this case, photochemistry with unpolarized light propagated parallel to a magnetic field would have been the asymmetric agent that influenced the organic chemistry of the interstellar medium in a region that contributed to meteoritic matter. The enantiomeric excesses achieved would have depended on the magnetic field strength. Presumably, magnetic fields of widely varying field strength exist near stars embedded in interstellar clouds.

3.12 Hypothesis and Summary

As we have seen, life on Earth is inextricably linked to homochirality and the origin of the latter remains a key unanswered question within the larger origin of life problem. In the foregoing sections we have attempted to provide the reader with a basic understanding of chirality as it pertains to organic compounds and then to review some of the more significant ideas that have been put forward with respect to the origin of the homochirality of life. In this section we describe a scenario based on our own evaluation of the many mechanisms that have been suggested. We assume that homochirality was achieved by prebiotic chemistry, i.e., by purely physical/chemical processes operating before the origin of life. We include within "prebiotic chemistry", chemistry that took place in the presolar cloud, as well as chemistry that might have unfolded from it within the solar nebula and on the early Earth. This may encompass reactions of molecules that eventually, in the course of evolution, completely or largely disappeared, for example, the α-methyl amino acids. In evaluating the suggested mechanisms we have considered whether they seem to have a reasonable chance of having operated within the constraints set by the conditions, to the extent that we can know them, that obtained on the prebiotic Earth. We have also been guided by Occam's Razor, the dictum that the best hypothesis is the one that minimizes the number of required assumptions.

Hypothesis: Terrestrial homochirality is a consequence of the postaccretional provision of organic compounds, particularly nonracemic amino acids, by the fall of meteorites, micrometeorites, and possibly comets on the primitive Earth. The enantiomeric excesses that initially characterized the amino acids were amplified at the time of pre-biotic polymerizations alternating with partial hydrolyses. The resulting polypeptides (not necessarily constituted of the contemporary protein amino acids) were necessarily asymmetric and some were endowed with catalytic properties. These polypeptides (protoenzymes) influenced the chirality of the products formed in their catalytic reactions by either the production of a specific enantiomer from achiral reactants or by the selection of a particular enantiomer in the case of chiral reactants.

This hypothesis is based on the following observations and experimental results described in more detail previously:

- Enantiomeric excesses ranging up to 15% are observed in α-methyl-α-amino acids isolated from carbonaceous chondrites impacting the Earth now. Such amino acids were very likely present and probably more abundant on the primitive Earth when meteoritic, micrometeoritic and cometary infalls were more intense than they are today.

- Partial polymerization of amino acids coupled with partial hydrolysis is an effective amplification mechanism of small enantiomeric excesses. Polymerization reactions are a necessary feature of prebiotic chemistry.
- The α-methyl amino acids form unusually stable helical structures and are thus particularly subject to amplification of their initial enantiomeric excesses by polymerization. Furthermore, they do not readily racemize and the gains achieved in their enantiomeric purity are not susceptible to loss as in the case of their α-hydrogenated counterparts.
- Random polypeptides are sometimes endowed with catalytic activity (Fox and Krampitz, 1964).
- Folded helical polypeptide conformations are inherently asymmetric and the binding of chiral substrates will necessarily be chirospecific and lead to pervasive chiral specificity and eventual homochirality.

The symmetry breaking postulated in this hypothesis is presumed to be due to irradiation of the organic matter of the presolar cloud by UVCPL produced by either a stellar source or by scattering. In this respect, the hypothesis is a more specific version of the one suggested by Pasteur more than a century ago, according to which terrestrial homochirality could be the consequence of *"des actions dissymétriques dont nous pressentons l'existence enveloppante et cosmique"* (Valéry-Radot, 1968). The production of enantiomeric excesses in amino acids by UVCPL is well documented by laboratory studies carried out even under low-temperature interstellar-grain conditions. In this case, the terrestrial preference for L-amino acids rather than D-amino acids would have been a matter of chance and life on extrasolar planets could have the same or the opposite chirality with equal probability as nothing requires that circularly polarized light have the same helicity in all protostellar clouds. The magnetochiral effect is also an attractive possibility; however, it has been subject to laboratory verification only with respect to the displacement of an equilibrium mixture and has not, as yet, been shown to affect stable enantiomers, i.e., enantiomers like those of amino acids that are not ordinarily in rapid equilibrium.

The hypothesis proposes the prebiotic amplification of the small initial enantiomeric excesses by polymerization. Insofar as polymerization leads to even weakly catalytic polypeptides, it also has the potential for providing, at the culmination of the prebiotic period, not only homochiral amino acids, but also a more extensive organic milieu with not only chiral specificity but also specificity with respect to the compounds represented. This latter specificity would arise as a consequence of the statistical impossibility of the formation of all polypeptide sequences, i.e., all catalytic activities, as well as the improbability of the formation of their enantiomeric forms. Although this hypothesis provides an explanation for the origin of both chiral and molecular specificity, it should be noted that it is silent with respect to the vexing problem of how certain catalytic polypeptides, formed at first by chance, could be reproduced, that is, in regard to the essential event that marked the transition from a prebiotic to a biotic world.

The possibility that the small enantiomeric excesses in the amino acids present on the early Earth acted as chiral initiators in Soai-type autocatalytic reactions is an attractive idea. In this way, large enantiomeric excesses could conceivably have been rapidly achieved. We expect that further research in this area will lead to the demonstration of similar effects in prebiotically more realistic aqueous phase reactions but, unless/until this is achieved, Occam's Razor causes us to favor the well-established polymerization–hydrolysis amplification mechanism.

Insofar as this hypothesis postulates "polypeptides first" with respect to the establishment of homochirality as well as early catalytic activities, it is incompatible with an exclusively RNA world, although it would not be incompatible with an RNA world in which, once homochirality was established, polypeptides played a subsidiary role. One might imagine that polypeptides in which α-methyl amino acids were important had, for reasons unknown, a limited future and that the rise of the more familiar nucleic acid-protein world awaited the exclusive use of the more easily (bio)synthesized α-hydrogen amino acids for protein synthesis. As a corollary to this hypothesis, one might postulate that an early catalytic polypeptide had the ability to polymerize formaldehyde to D-ribose and thus solved one of the two chirality problems that must be dealt with on the way to an RNA-world.

With regard to other possible symmetry-breaking mechanisms, we would note that hypotheses invoking specific adsorption on natural mineral surfaces to break symmetry suffer from the fact that, insofar as is known, the enantiomorphic forms of minerals occur equally; thus, unless the origin of homochirality was spatially confined to a very small area/volume, any enantiomeric excess achieved would be countered by a nearby equal and opposite excess. As discussed previously, symmetry breaking by crystallization effects requires a concentration and degree of purity that seem unrealistic under prebiotic conditions.

The violation of parity is difficult to dismiss because of the intuition that such fundamental asymmetry might have far-reaching consequences. However, enantiomeric excesses in chiral molecules deriving from parity violation have not been demonstrated. Furthermore, it is not possible to see how such very small excesses could have emerged from the "noise" of random statistical fluctuation to a degree that would allow their amplification. The possibility that this "noise" itself, i.e., the statistically allowed excursions away from the perfect racemic state, provided the symmetry breaking for the origin of homochirality is an interesting possibility made more plausible by the demonstration of their apparent amplification in the "Soai reaction." If a similar reaction can be shown to occur under aqueous, prebiotic conditions a statistical symmetry breaking must be seriously considered for the origin of homochirality.

3.13 Homochirality Analyses in the Solar System and Beyond

Exobiologists have taken an interest in homochirality as a signature of life and attention has been given to the analysis of chirality (or, more precisely, enantiomeric excesses) beyond the terrestrial environment by the SETH (search for extraterrestrial homochirality) project. SETH could be accomplished remotely on various Solar System bodies by landers equipped with detectors of enantiomeric excesses such as miniaturized polarimeters (Mac Dermott, 1996) or chromatographs with chiral support materials (Meierhenrich et al., 1999). The Rosetta mission to the Churyumov-Gerasimenko comet which was launched in February 2004 and that will reach the comet in November 2014 involves a SETH. SETH also encompasses the broader goal of finding the remains of life and could, for example, contribute to the validation of chemical fossils on other bodies of the Solar System such as Mars. However, diagenesis is accompanied by partial and ultimately complete racemization of chiral molecules and this poses a problem for the use of chirality as a signature of past life. This record is written with ink that inevitably fades with the passage of time. It inevitably fades in terrestrial sediments and must in all sediments (Bada and Miller, 1987; Bada and McDonald, 1995). SETH assumes the necessity of homochirality for life and as long as we know only one form of life it is not possible to know with certainty whether homochirality is an absolute requirement for life although it seems highly probable that it is. The enantiomeric excesses found in meteoritic amino acids add another complication as they are of abiotic origin and widely distributed. It seems clear that drawing the correct inferences from extraterrestrial chirality measurements will require their interpretation within a broad chemical, morphological, and geological context. It is also necessary to remember that, if extant or extinct life on Mars were discovered to have the same molecular homochirality as that of terrestrial life, the meaning of this discovery would not be unambiguous. Transfers of matter between Mars and the Earth are relatively frequent on a time scale of the age of these planets and such transfers may allow the insemination of one planet by the other (Mileikowsky, 2000).

The study of homochirality outside the Solar System, *SEXSOH* (search for extrasolar homochirality) will probably be a goal of future missions but at present, it is too early to predict when (and even how) observations might be carried out.

Dedication

This chapter is dedicated to Kurt Mislow in recognition of his 80th birthday and with appreciation for his many experimental and theoretical contributions to the modern understanding of chirality and stereochemistry in general.

Bibliography

General

Avalos M., Babiano R., Cintas P., Jimenez J.L., Palacios J.C. (2000). From parity to chirality: Chemical implications revisited. *Tetrahedron Asymmetry*, **11**, 2845–2874.

Bouchiat M.A., Pottier L. (1984). An atomic preference between left and right. *Scientific American*, 76–86.

Brack A. (July, 1998), L'asymétrie du vivant in "Pour la Science – Dossier Hors Série," Paris (France) "Les symétries de la nature" (this special issue contains many interesting articles for those who wish to integrate chirality into the general problem of symmetry and the breaking of symmetry in biology, chemistry, physics, and cosmology.)

Cline B.C. (editor) (1996). *Physical Origin of Homochirality on Earth*, American Institute of Physics, Woodbury, New York (USA).

Feringa B.L., van Velden A. (1999). Absolute asymmetry synthesis: the origin, control and amplification of chirality. *Angew. Chem. Int. Ed.*, **38**, 3418–3438.

Jacques J. (1992). *La Molécule et son Double*. Hachette, Paris.

Mason S.F. (1982). *Molecular Optical Activity and the Chiral Discriminations*. Cambridge University Press, Cambridge (UK).

Mislow, K. (199). *Molecular Chirality*. Chapter 1 in "Topics in stereochemistry," vol 22, S.E. Denmark, ed. J. Wiley and Sons, New York (USA).

Nordén B. (1978). The Asymmetry of Life. *J. Mol. Evol.*, **11**, 313–332.

W.J. Lough and I. Wainer (Eds.) (2002). *Chirality in Natural and Applied Science*. Blackwell Science Ltd-CRC Press, USA and Canada.

References

Aikawa Y., Herbst, E. (2001). Two-dimensional distribution and column densities of gaseous molecules in protoplanetary disks. II. Deuterated species and UV-shielding by ambient clouds. *Astron. Astrophys.*, **372**, 1107–1117.

Altman E., Altman K.H., Nebel K., Mutter M. (1988). Conformational studies on host-guest peptides containing chiral alpha-methyl-alpha-amino acids. *Int. J. Pept. Protein Res.*, **32**, 344–351.

Asakura S., Soga T., Uchida T., Osanai S., Kondepudi D.K. (2002). Probability distribution of enantiomeric excess in unstirred and stirred crystallization of binaphthyl melt. *Chirality*, **14**, 85–89.

Auf Der Heyde T.P.E., Buda A.B., Mislow K. (1991). Desymmetrization and degree of chirality. *J. of Mathematical Chemistry*, **6**, 255–265.

Avalos M., Babiano R., Cintas P., Jimenez J.L., Palacios J.C. (2000). From parity to chirality: Chemical implications revisited. *Tetrahedron: Asymmetry*, **11**, 2854–2874.

Bada J.L., McDonald G.D. (1995). Amino acid racemization on Mars: Implications for the preservation of biomolecules from an extinct Martian biota. *Icarus*, **114**, 139–143.

Bada J.L., Miller S.L. (1987). Racemization and the origin of optimalised active organic compounds in living organisms. *Biosystems*, **20**, 21–26.

Bailey J., Chrysostomou A., Hough J.H., Gledhill T.M., McCall A., Clark S., Ménard F., Tamura M. (1998). Circular polarization in star formation regions: Implications for biomolecular homochirality. *Science*, **281**, 672–674.

Bailey J. (2001). Circularly polarized light and the origin of homochirality. *Origins Life Evol. Biosphere*, **31**, 167–183.

Balavoine G., Moradapour A., Kagan H.B. (1974). Preparation of chiral compounds with high optical purity with circularly polarized light, a model for the prebiotic generation of optical activity. *J. Am. Chem. Soc.*, **96**, 5152–5158.

Barron L.D. (1982). *Molecular Light Scattering and Optical Activity*, Cambridge University Press, Cambridge.

Barron L.D. (2000). Chirality, Magnetism and Light. *Nature*, **405**, 895–896.

Barron L.D. (2002). Chirality at the sub-molecular level: true and false chirality, p. 53–84 in Chirality in *Natural and Applied Science*, Eds. W.J. Lough and I.W. Wainer, Blackwell Science Ltd. CRC Press, USA and Canada.

Bartik K., El Haouaj M., Luhmer M., Collet A., Reisse J. (2000). Can monoatomic xenon become chiral? *Chem. Phys.Chem.*, **4**, 221–324.

Bartik K., Luhmer M., Collet A., Reisse J. (2001). Molecular polarization and molecular chiralization: The first example of a chiralized xenon atom. *Chirality*, **13**, 2–6.

Blair N.E., Bonner W.A. (1980). Experiments on the amplification of optical activity. *Origins of Life*, **10**, 255–263.

Blair N.E., Dirbas F.M., Bonner W.A. (1981). Stereoselective hydrolysis of leucine oligomers. *Tetrahedron*, **37**, 27–29.

Blout E.R, Idelson M. (1956). Polypeptides VI. Poly-alpha-L-glutamic acid. Preparation and helix-coil conversions. *J. Am. Chem. Soc.*, **78**, 497–498.

Bolli M, Micura R., Eschenmoser A. (1997). Pyranosyl-RNA/ Chiroselective self-assembly of base sequences by ligative oligomerization of tetranucleotide-2′, 3′-cyclophosphates (with a commentary concerning the origin of biomolecular homochirality), *Chemistry and Biology*, **4**, 309–320.

Bonner W.A., Flores J.J. (1975). Experiments on the origin of optical activity. *Origins of Life*, **6**, 187–194.

Bonner W.A., Rubenstein E. (1987). Supernovae, neutron stars and biomolecular chirality. *Biosystems* **20**, 99–111.

Bonner W.A (1991). The origin and amplification of biomolecular chirality. *Orig. Life Evol. Biosphere*, **21**, 59–11.

Bonner W.A. (1994). Enantioselective autocatalysis - spontaneous resolution and the prebiotic generation of chirality. *Orig. Life Evol. Biosphere*, **24**, 63–78.

Bonner W.A. (1996). The quest for chiralitry in *Physical Origin of Homochirality on Earth*, ed. D.B. Cline, p. 17–49, American Institute of Physics, Proc. 379, Woodbury, New York.

Bonner W.A., Rubenstein E., Brown G.S. (1999). Extraterrestrial handedness: A reply. *Orig. Life Evol. Biosphere*, **30**, 329–332.

Bonner W.A. (1999). Chirality amplification – the accumulation principle revisited. *Orig. Life Evol. Biosphere*, **29**, 615–623.

Bonner W.A., Bean B.D. (2000). Asymmetric photolysis with elliptically polarized light. *Orig. Life Evol. Biosphere*, **30**, 513–517.

Bonner W.A. (2000). Parity violation and the evolution of biomolecular homochirality. *Chirality*, **12**, 114–126.

Brack A., Spach G. (1981). Enantiomer enrichment in early peptides. *Origins of Life*, **11**, 135–142.

Buda A.B., Mislow K. (1991). On geometric measure of chirality. *J. of Molecular Structure (Theochem)*, **232**, 1–12.

Buda A.B, Auf der Heyde T., Mislow K. (1992). On quantifying chirality. *Angew. Chem. Int. Ed. Engl.*, **31**, 989–1007.

Buchardt O. (1974). Photochemistry with circularly polarized light. *Angew. Chem. Int. Ed. Engl.*, **13**, 179–185.

Buschmann H., Thede R., Heller D. (2000). New developments in the origins of the homochirality of biologically relevant molecules. *Angew. Chem. Int. Ed.*, **39**, 4033–4036.

Cerf C., Jorissen A. (2000). Is amino-acid homochirality due to asymmetric photolysis in space? *Space Science Reviews*, **92**, 603–612.

Chyba C.F., Sagan, C. (1992). Endogeneous production, exogeneous delivery and impact-shock synthesis of organic molecules: An inventory for the origins of life. *Nature*, **355**, 125–132.

Collet A., Brienne M-J., Jacques J. (1980). Optical Resolution by Direct Crystallization of Enantiomer Mixtures. *Chem. Rev.*, **80**, 215–230.

Collet A. (1990). The Homochiral versus heterochiral packing dilemma in *Problems and Wonders of Chiral Molecules*, ed. M. Simonyi, Akademia Kiado, Budapest.

Cronin J.R., Chang S. (1993). Organic matter in meteorites: Molecular and isotopic analyses of the Murchison meteorite in *The Chemistry of Life's Origins*, eds. J.M. Greenberg, et al., p 209–258, Kluwer Acad. Pub., Netherlands.

Cronin J.R., Pizzarello S. (1997). Enantiomeric exesses in meteoritic amino acids. *Science*, **275**, 951–955.

Cronin J.R., Pizzarello S. (2000). Chirality of meteoritic organic matter: A brief review. in *Perspectives in Amino Acid and Protein Geochemistry*, eds. Goodfriend G. et al., p 15–22, Oxford University Press, Oxford, New York.

Curie P. (1894). Sur la symétrie dans les phénomènes physiques, symétrie d'un champ électrique et magnétique. *J. Chim. Phys.*, 3éme série, t.III, 393–402.

Decker P. (1974). The origin of stochastic information (noise) in bioids: Open systems which can exist in several steady states. *J. Mol. Evol.*, **4**, 49–65.

Droesbeke J.J. (2001). in *Eléments de Statistique*, 4ème édition, p 206–207, Editions de l'Université de Bruxelles).

Dunitz J.D. (1996). Symmetry arguments in chemistry. *Proc. Natl. Acad. Sci. USA*, **93**, 14260–14266.

Engel M.H., Macko S.A., Silfer J.A. (1990). Carbon isotope composition of individual amino acids in the Murchison meteorite. *Nature*, **348**, 47–49.

Engel M.H., Macko S.A. (1997). Isotopic evidence for extraterrestrial non-racemic amino acids in the Murchison meteorite. *Nature*, **389**, 265–268.

Eliel E.M., Wilen S.H.(1994). *Stereochemistry of Organic Compounds*. J. Wiley and Sons, New York (USA).

Epstein S., Krishnamurthy R.V., Cronin J.R., Pizzarello S., Yuen G.U. (1987). Unusual stable isotope ratios in amino acids and carboxylic extracts from the Murchison meteorite. *Nature*, **326**, 477–479.

Feringa B.L., Huck N.P., van Doren H.K. (1995). Chiroptical switching between liquid crystalline phases. *J. Am. Chem. Soc.*, **117**, 9929–9930.

Figereau A., Duval E., Boukenter A. (1995). Can biological homochirality result from a phase transition? *Orig. Life Evol. Biosphere*, **25**, 211–217.

Flores J.J., Bonner W.A., Massey G.A. (1977). Asymmetric photolysis of (R,S)-leucine with circularly polarized light. *J. Am. Chem. Soc.*, **99**, 3622–3625.

Formaggio F., Crisma M., Bonora G.M., Pantano M., Valle G., Toniolo C., Aubry A., Bayeul D., Kamphuis J. (1995). (*R*)-Isovaline homopeptides adopt the left-handed helical structure. *Peptide Research*, **8**, 6–14.

Fox S.W., Krampitz G. (1964). Catalytic decomposition of glucose in aqueous solution by thermal proteoids. *Nature*, **203**, 1362–1364.

Franck F.C. (1953). On spontaneous asymmetric synthesis. *Biochim. Biophys. Acta*, **11**, 459–463.

Frondel C. (1978). Characters of quartz fibers. *Am. Mineral.*, **63**, 17–27.

Gargaud M, Despois D, Parisot J-P. (Eds.), (2001). *L'environnement de la Terre Primitive*, Presses Universitaires de Bordeaux, France.

Girard C., Kagan H.B. (1998). Nonlinear effects in asymmetric synthesis and stereoselective reactions: Ten years of investigations. *Angew. Chem. Int. Ed.*, **37**, 2922–2959.

Gol'danski V.I., Kuz'min V.V. (1988). Spontaneous mirror symmetry breaking in nature and origin of life. *Z. Phys. Chem.*, **269**, 216–274.

Greenberg J.M. (1996). Chirality in interstellar dust and in comets: Life from dead stars, in *Physical Origin of Homochirality on Earth*, Ed.D.B. Cline, p. 185–186, American Institute of Physics; Proc. 379, Woodbury, New York.

Havinga E. (1954). Spontaneous formation of optically active substances. *Biochem. Biophys. Acta*, **38**, 171–174.

Hazen R.M., Filley R.F, Goodfriend G.A. (2001). Selective adsorption of L- and D-amino acids on calcite: Implications for biochemical homochirality. *Proc. Nat. Acad. Sci. USA*, **98**, 5487–5490.

Inoue Y. (1992). Asymmetric photochemical reactions in solution. *Chem. Rev.*, **92**, 741–770.

Jackson T.A. (1971). Preferential polymerization and adsorption of L-optical isomers of amino acids relative to D-optical isomers on kaolinite templates. *Chem. Geol.* **7**, 295–306.

Jacques J., Collet A., Wilen S.H. (1981). *Enantiomers, Racemates and Resolution*, J. Wiley and Sons, New York.

Jorissen A., Cerf C. (2002). Photoreactions as the Origin of Biomolecular Homochirality: A critical review. *Origins Life Evol. Biosphere*, **32**, 129–142.

Julg A., Favier A., Ozias, Y. (1989). A theoretical study of the difference in the behavior of L- and D-alanine toward the two inverse forms of kaolinite. *Struc. Chem.*, **1**, 137–141.

Julg A. (1989). Origin of the L-homochirality of amino-acids in the proteins of living organisms in Molecules in Physics, *Chemistry and Biology*, vol. IV, Ed. J. Maruani, p. 33–52, Kluwer Academic Pub., Netherlands.

Kagan H., Moradpour A., Nicoud J.F., Balavoine G., Martin R.H., Cosyn J.P. (1971). Photochemistry with circularly polarised light. Asymmetric synthesis of octa- and nonahelicene. *Tetrahedron Lett.*, **22**, 2479–2482.

Kagan H.P., Fiaud J.C. (1988). *Kinetic Resolution in Topics in Stereochemistry*, vol 18. eds. Eliel, E.L. and Willen, S.H., p. 249–330, John Wiley, New York.

Kelvin W.T. (1904). *Baltimore Lectures on Molecular Dynamics and the Wave Theory of Light*, p. 602–642, C.J. Clay, London.

Kondepudi D.K., Nelson G.W. (1984). Chiral symmetry-breaking and its sensitivity in non-equilibrium chemical systems. *Physica*, **125A**, 465–496.

Kondepudi D.K., Nelson G.W. (1985). Weak neutral currents and the origin of biomolecular chirality. *Nature*, **314**, 438–441.

Kondepudi D.K. (1996). Selection of handedness in prebiotic chemical processes, in *Physical Origin of Homochirality in Life*, Ed. Cline D.B., p. 65–72, American Institute of Physics, AIP Conference 379, Woodbury, New York.

Kondepudi D.K., Laudadio J., Asakura K. (1999). Chiral symmetry breaking in stirred crystallization of 1,1′-binaphthyl melt. *J. Am. Chem. Soc.*, **121**, 1448–1451.

Kuhn W., Braun E. (1929). Photochemische erzeugung optisch aktiver stoffe. *Naturwiss.*, **17**, 227–228.

Kvenvolden K., Lawless J., Pering K., Peterson E., Flores J., Ponnamperuma C., Kaplan I.R., Moore C. (1970). Evidence for extraterrestrial aminoacids and hydrocarbons in the Murchison meteorite. *Nature*, **228**, 923–926.

Larder D.F. (1967). Historical aspects of the tetrahedron chemistry. *J. Chem. Ed.*, **44**, 661–666.

Lee T.D., Yang C.N. (1956). Question of parity conservation in weak interactions. *Phys. Rev.*, **104**, 254–258.

Mc Bride J.M., Carter R.L. (1991) Spontaneous resolution by stirred crystallization. *Angew. Chem. Int. Ed.*, Engl. **30**, 293–295.

Mac Dermott A.J. (1996). The weak force and SETH: The search for extra-terrestrial homochirality, in *Physical Origin of Homochirality on Earth*, Ed.D.B. Cline, p. 241–254, American Institute of Physics, Proc. 379, Woodbury, New York.

Mac Dermott A.J. (2002). The origin of biomolecular chirality, p. 23–52 in *Chirality in Natural and Applied Science*, Eds. W.J. Lough and I.W. Wainer, Blackwell Science Ltd. CRC Press, USA and Canada.

Mason S.F. (1982). *Molecular Optical Activity and the Chiral Discriminations*. Cambridge University Press, Cambridge.

Mason S.F. (1997). Extraterrestrial handedness. *Nature*, **389**, 804.

Mason S.F. (2002). Pasteur on molecular handedness – and the sequel, p. 1–19 in *Chirality in Natural and Applied Science*, Eds. W.J. Lough and I.W. Wainer, Blackwell Science Ltd. CRC Press, USA and Canada.

Matsuura K., Inoue S., Tsuruta T. (1965) Asymmetric selection in the copolymerization of N-carboxy-L- and D-alanine. *Makromol. Chem.*, **85**, 284–290.

Meierhenrich U., Thiemann W.H-P., Rosenbauer, H. (1999). Molecular parity violation via comets. *Chirality*, **11**, 575–582.

Meinschein W.G., Frondel C., Laur P., Mislow K. (1966). Meteorites: Optical activity in organic matter. *Science*, **154**, 377–380.

Mileikovwsky C., Cucinotta F., Wilson F., Gladman B., Hornek G., Lindegren, L., Melosh J., Rickman H., Valtonen M., Zheng, J.Q. (2000). Natural transfer of viable microbes in space, Part 1: From Mars to Earth and from Earth to Mars. *Icarus*, **145**, 391–427.

Miller S.L. (1997). Peptide nucleic acids and prebiotic chemistry. *Nature Struct. Biol.*, **4**, 167–169.

Mills W.H. (1932). Some aspects of stereochemistry. *Chem. and Ind.*, 750–759.

Milton R.C. deL, Milton S.F.C., Kent, S.B.H. (1992). Total chemical synthesis of a D-enzyme. The enantiomers of HIV-1 protease show demonstration of remplaçable chiral-substrate-specificity. *Science*, **256**, 1445–1448.

Mislow K. (1965). *Introduction to Stereochemistry*, W.A. Benjamin, New York.

Mislow K. (1996). A commentary on the topological chirality and achirality of molecules. *Croat. Chim. Acta*, **69**, 485–511.

Mislow K. (1997). Fuzzy restrictions and inherent uncertainties in chirality studies, in *Fuzzy Logic in Chemistry*, ed. D.H. Rouvray, p. 65–88, Academic Press, San Diego (USA).

Monod J. (1970) *Le Hasard et la Nécessité; Essai sur la Philosophie Naturelle de la Biologie Moderne*, Editions du Seuil, Paris.

Mullie F., Reisse J. (1987). Organic matter in carbonaceous chondrites. *Topics in Current Chemistry (Spinger Verlag)*, **139,** 83–117.

Nielsen P.E. (1993). Peptide-Nucleic Acid (PNA)–A model structure for the primordial genetic code. *Origins Life Evol. Biosphere*, **23**, 323–327.

Nielsen P.E. (1996). Peptide Nucleic Acid (PNA). Implications for the origin of the genetic material and the homochirality of life, pages in Physical Origin of Homochirality in Life, *AIP Conference Proceedings* 379. Ed. D.B. Cline, p. 55–61,Woodbury, New York.

Nishino H., Kosaka A, Hembury G.A., Shitomi H., Onuki H., Inoue I. (2001). Mechanism of pH-dependent photolysis of aliphatic aminoacids and enantiomeric enrichment of racemic leucine by circularly polarized light. *Org. Letters*, **3**, 921–924.

Nordén B. (1977). Was photoresolution of amino acids the origin of optical activity in life. Nature, **266**, 567–568.

Nordén B. (1978). The Asymmetry of Life. *J. Mol. Evol* **11**, 313–332.

Penzias A.A. (1980). Nuclear processing and isotopes in the Galaxy. *Science*, **208**, 663–669.

Pizzarello S., Krishnamurty R.V., Epstein S., Cronin J.R. (1991). Isotopic analyses of amino acids from the Murchison meteorite. *Geochim. Cosmochim. Acta*, **55**, 905–910.

Pizzarello S., Cronin J.R. (1998). Alanine enantiomers in the Murchison meteorite. *Nature*, **394**, 236.

Pizzarello S., Cronin J.R. (2000). Non-racemic amino acids in the Murchison and Murray meteorites. *Geochim. Cosmochim. Acta*, **64**, 329–338.

Pizzarello S., Zolensky M., Turk K.A. (2003). Nonracemic isovaline in the Murchison meteorite: Chiral distribution and mineral association. *Geochim. Cosmochim. Acta*, **67**, 1589–1595.

Quack M. (2002). How important is parity violation for molecular and biomolecular chirality? *Angew. Chem. Int. Ed.*, **41**, 4618–4630.

Prigogine I., Kondepudi D.K. (1999). *Thermodynamique. Des Moteurs Thermiques aux Structures Dissipatives*, ed. Odile Jacob, Paris.

Rassat A, Fowler P.V. (2003). Any scalene triangle is the most chiral triangle. *Helvetic Chimica Acta*, **86**, 1728–1740.

Rau H. (1983). Asymmetric Photochemistry in Solution. *Chem. Rev.*, **83**, 355–547.

Rawn J.D. (1989). *Biochemistry*, Neil Patterson Pub. Carolina Biological Supply Company, Burlington, North Carolina.

Rikken G.L., Raupach E. (2000). Enantioselective Magnetochiral Photochemistry. *Nature*, **405**, 932–935.

Reisse J., Mullie F. (1993). On the origins of organic matter in carbonaceous chondrites. *Pure and Applied Chemistry*, **65**, 1281–1292.

Reisse J. (2001). in *L'environnement de la Terre Primitive*, eds. Gargaud M., Despois D., Parisot J.P., p. 323–342, Presses Universitaires de Bordeaux, Bordeaux, France.

Roberts J.A. (1984). Supernovae and life. *Nature*, **308**, 318.

Robinson R. (1974). Preface of the Van't Hoff-Le Bel Centenary Volume. *Tetrahedron*, **30**, 1477–1486.

Rubenstein E., Bonner W.A., Noyes H.P., Brown G.S. (1983). Supernovae and life. *Nature*, **306**, 118–120.

Salam A. (1993). The Origin of Chirality, the Role of Phase Transition and their Induction in Amino acids in *Chem. Evol. and Origin of Life*, eds. Ponnamperuma C, Chela-Flores J., Deepak Pub., Hampton, Virginia, USA.

Shibata T., Yamamoto J., Matsumoto N., Yonekubo S., Osanai S., Soai K. (1998). Amplification of a Slight Enantiomeric Imbalance in Molecules Based on Asymmetric Autocatalysis: The First Correlation between High Enantiomeric Enrichment in a Chiral Molecule and Circularly Polarized Light. *J. Am. Chem. Soc.*, **120**, 12157–12158.

Siegel J.S. (1998). Homochiral imperative of molecular evolution. *Chirality*, **10**, 24–27.

Siegel J.S. (2002). Shattered Mirrors. *Nature*, **419**, 346–347.

Singleton D.A., Vo L.K. (2002). Enantioselective synthesis without discrete optically active additives. *J. Am. Chem. Soc.*, **124**, 10010–10011.

Soai K., Shibata T., Morioka H., Choji K. (1995). Asymmetric autocatalysis and amplification of enantiomeric excess of a chiral molecule. *Nature*, **378**, 767–768.

Soai K., Osanai S., Kadowaki K., Yonebuko S., Shibata T., Sato I. (1999). d- and l-Quartz-promoted highly enantioselective synthesis of a chiral organic compound. *J. Am. Chem. Soc.*, **121**, 1235–1236.

Soai K., Shibata T., Sato I. (2000). Enantioselective automultiplication of chiral molecules by asymmetric autocatalysis. *Acc. Chem. Res.*, **33**, 382–390.

Spach P.G. (1974). Polymérizsation des énantiomères d'un acide α-aminé. Stéréosélection and amplification de l'asymétrie. *Chimia*, **28**, 500–503.

Szabo-Nagy A., Keszthelyi L. (1999). Demonstration of the Parity-Violating Energy Difference between Enantiomers. *Proc. Natl. Acad. Sci. USA*, **96**, 4225–4255.

Takats Z, Nanita S.C., Cooks R.G. (2003). Serine octamer reactions: indicators of prebiotic relevance. *Angew. Chem. Int. Ed.*, **42**, 3521–3523.

Tranter G.E. (1985). Parity-violating energy differences of chiral minerals and the origin of biomolecular homochirality. *Nature*, **318**, 172–173.

Triggle D.J. (1997). Stereoselectivity of drug action. *Drug Discovery Today*, **2**, 138–147.

Valéry-Radot P. (1968). *Pages Illustres de Pasteur.* Hachette (Paris).

Vester F. (1974). The (hi)story of the induction of molecular asymmetry by the intrinsic asymmetry in β-decay. *J. Mol. Evol.*, **4**, 1–13.

Wagnière G., Meier A. (1983). Difference in the absorption coefficient of arbitrarily polarized light in a magnetic field. *Experientia*, **39**, 1090–1091.

Wald G. (1957). The origin of optical activity. *Ann. N.Y. Acad. Sci.*, **69**, 353–358.

Wang W., Yi F., Ni Y., Jin Z., Tang Y. (2000). Parity violation of electroweak force in phase transition of single crystals of D- and L- alanine and valine. *J. of Biological Physics*, **26**, 51–65.

Wannier P.G.A. (1980). Nuclear abundances and evolution of the interstellar medium. *Ann. Rev. Astron. Astrophys.*, **18**, 399–437.

Weissbuch I., Addadi L., Berkovitch-Yellin Z., Gatti E., Lahav M., Leiserowitz L. (1984). Spontaneous generation and amplification of optical activity in alpha amino-acids by enantioselective occlusion in centrosymmetric crystals of glycine. *Nature*, **310**, 161–164.

Wolstencroft R.D. (1985). Astronomical sources of circularly polarized light and their role in determining chirality on Earth. in IAU Symp. 112. *The Search for Extraterrestrial Life*, p. 171–175, D. Reidel, Dordrecht.

Wu C.S., Ambler E., Hayward R.W., Hoppes D.D., Hudson R.P. (1957). Experimental test of parity conservation in beta-decay. *Phys. Rev.*, **105**, 1413–1415

Yamagata Y. (1966). A hypothesis for the asymmetric appearance of biomolecules on Earth. *J. Theor. Biol.*, **11**, 495–498.

Zanasi R., Lazeretti P., Ligabue A. and Soncini A. (1999) in *Advances in BioChirality* eds. Palyi, G. et al., ch. 29, pp 377–385, Elsevier, Amsterdam.

Zee A. (1999). *Fearful Symmetry: The Search for Beauty in Modern Physics*, p 224, Princeton Science Library.

Zepik H., Shavit E., Tang M., Jensen T.R., Kjaer K., Bolbach G., Leiserowitz L., Weissbuch I., Lahav M. (2002). Amplification of oligopeptides in two-dimensional crystalline self-assemblies on water. *Science*, **295**, 1266–1269.

4 Peptide Emergence, Evolution and Selection on the Primitive Earth

I. Convergent Formation of *N*-Carbamoyl Amino Acids Rather than Free α-Amino Acids?

Auguste Commeyras, Laurent Boiteau, Odile Vandenabeele-Trambouze, Franck Selsis

Abstract After summarising current knowledge about the origins of primitive Earth organic matter, we focus our attention solely on α-amino acids and their derivatives. We then analyze the mechanism for the formation of these compounds, under both extraterrestrial and primitive-Earth conditions, and show that a "multicomponent system" consisting of prebiotic molecules (hydrogen cyanide, several carbonyl compounds, ammonia, alkyl amines, carbonic anhydride, sodium bicarbonate, borate, cyanic acid) may have been the precursors of these essential compounds. We show that this multicomponent system leads reversibly to several intermediate nitriles, which irreversibly evolve, first to α-amino acids and *N*-carbamoyl amino acids via selective catalytic processes, and then to *N*-carbamoyl amino acids alone.

4.1 Introduction

The Earth is 4.6 billion years old (4.6 Gyr). Two million years after it was formed, it had attained 80% of its present-day mass, the continents and oceans were already present, the plate-tectonic process had already begun (Martin 1986, 1987; Maas and McCulloch 1991; Carlson 1996; Vervoort et al. 1996; Mojzsis et al. 2001; Wilde et al. 2001), and there was also an atmosphere, though admittedly little is known about it (Selsis 2000). The last 20 per cent of the Earth's mass (1.2×10^{21} tons) was added by meteorites, micrometeorites and comets between -4.4 and -4.1 Gyr (Maurette 2001). The emergence of life may possibly have taken place in this period (acceptedly before -3.5 Gyr); as a consequence its substrate, namely organic matter, was obviously of both exogenous and endogenous origins. Qualitatively and quantitatively assessing each of these two different origins is a question of current importance that we examine, limiting ourselves to amino acids.

Investigation of the chemical mechanisms involved in the formation of both exogenous and endogenous amino acids should eventually provide information on the prebiotic environment. Such research needs first to acquire knowledge on meteoritic amino acids and related compounds (Sects. 4.2 and 4.3) and on the chemical reactions responsible for their synthesis (Sects. 4.4 and 4.5), in order to find out their potential prebiotic relevance (Sect. 4.6) as well as physical constraints on parent bodies (exogenous synthesis).

We conclude by studying a series of reactions that could have led to the formation of peptides, under prebiotic conditions.

4.2 Organic Molecules on the Primitive Earth

An exhaustive inventory of the origins of all the organic molecules occurring on the primitive Earth 4 Gyr ago was carried out by Chyba and Sagan (1992). They distinguished three probable sources:

- Exogenous contribution from meteorites, micrometeorites, interplanetary dust particles (IDP), comets, etc.
- Endogenous synthesis associated with impact of such bodies
- Endogenous production associated with other available energy sources (solar UV, lightning, radioactivity, etc.)

The total contribution of these three sources would have been heavily dependent on the composition of the primitive atmosphere. It would, in fact, have ranged from $\approx 10^{11}\,\mathrm{kg\,y^{-1}}$ for a reductive atmosphere to $\approx 10^{8}\,\mathrm{kg\,y^{-1}}$ for a slightly reductive atmosphere (atmospheric $[H_2]/[CO_2]$ ratio ≈ 0.1) or less for even lower hydrogen levels. The presence of methane (CH_4) in the primitive atmosphere would have oriented endogenous syntheses towards a considerable production of nitriles, especially the simplest of these, hydrocyanic acid HCN, which amount could have reached ca. $10^{10}\,\mathrm{kg\,y^{-1}}$. Assuming the volume of the primitive oceans to have been identical to that of the present day, considering only the soluble part of the organic matter ($\approx 40\%$ of the total determined by Chyba and Sagan (1992), see Cronin and Chang 1993), and supposing this organic matter to be rather stable under its supposed conditions, the primitive oceans could have contained (at a steady state) between ca. $0.4 \times 10^{-3}\,\mathrm{g\,L^{-1}}$ of total organic matter in the case of a neutral atmosphere and $0.4\,\mathrm{g\,L^{-1}}$ in the case of a reductive atmosphere. Given the imprecision of current knowledge about the composition of the primitive atmosphere, three different eventualities can be considered:

- A highly reductive atmosphere in which syntheses due to impacts would have predominated over endogenous production due to electrical discharges and solar UV
- A slightly reductive atmosphere in which, by contrast, endogenous syntheses (solar UV, lightning) would have predominated, augmented by the contribution of organic matter from IDPs
- A very slightly reductive, or nonreductive atmosphere in which the contribution of organic matter from IDPs would have predominated

We shall now examine the qualitative aspects linked to these eventualities.

4.3 Exogenous Amino Acids and Related Compounds

Analysis of the organic matter contained in meteorites, especially carbonaceous chondrites, regularly collected on Earth's surface (Cronin and Chang 1993; Maurette 1998), has provided an approach to extraterrestrial organic chemistry (the study of dense clouds is another possible approach, doubtless more direct, but restricted mostly to volatile materials). This organic matter could correspond to that formed in the early Solar System, or even in the pre-Solar System (Cronin and Pizzarello 1983; Cronin and Chang 1993; Cooper and Cronin 1995; Cooper et al. 2001). It is mainly (> 60%) macromolecular, insoluble and difficult to characterise. The soluble part (< 40%) is a complex mixture of molecular organic compounds in which the following have been characterized: acids (carboxylic, dicarboxylic, hydroxyl, sulfonic, phosphonic); amines; amides; nitrogenous heterocycles including purines and pyrimidines; alcohols; sugars; carbonyl compounds; aliphatic and aromatic hydrocarbons; amino acids and hydantoins. Structures of compounds dealt with in this chapter are shown in Fig. 4.1.

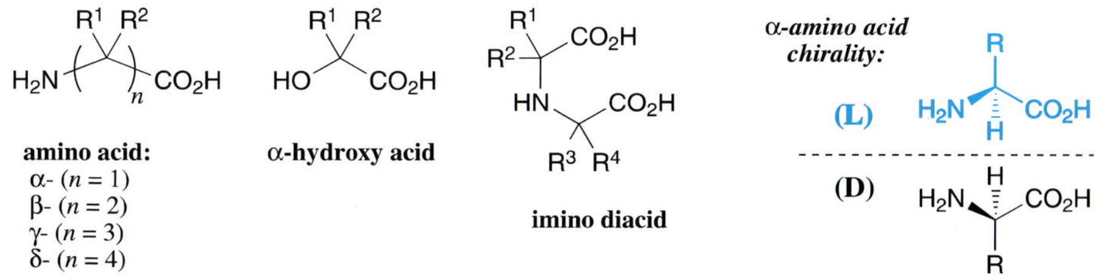

Fig. 4.1. General structure of amino acids and related compounds. All proteinic α-amino acids have the L configuration (*blue*)

4.3.1 Exhaustive Survey of Exogenous Amino Acids

More than 70 different amino acids have been characterized, principally in the Murchison and Murray meteorites (both being carbonaceous chondrites) (Tables 4.1–4.4), representing a total abundance of 75 µg per g of meteorite. Over a set of 7 other carbonaceous chondrites, the average mass of extracted amino acids was only 0.6 µg per g. It should, nevertheless, be noted that the mass of organic molecules present in chondrites decreases as a function of the time the materials have spent on the Earth (Cronin and Pizzarello 1983), thus probably leading to an underassessment of the amount of exogenous organic matter. Thus, even if the value obtained for the Murchison meteorite is representative of chondrites and if this value is doubled to take into account the existence of amino acid precursors in the total carbonaceous matter (Cronin and Pizzarello 1983; Cooper and Cronin 1995), the abundance of meteoritic amino acids remains quantitatively low, even considering that chondritic micrometeorites could have

Table 4.1. The 20 protein α-amino acids: names, 3-letter and 1-letter codes, and numbers of C atoms in their alkyl chains. *blue*: α-amino acids detected in carbonaceous chondrites

Name	3-lett code	1-lett. code	# C atoms	Name	3-lett code	1-lett. code	# C atoms
Alanine	Ala	A	3	Leucine	Leu	L	6
Arginine	Arg	R	5+1	Lysine	Lys	K	6
Aspartic Acid	Asp	D	4	Methionine	Met	M	4+1
Asparagine	Asn	N	4	Phenyl-alanine	Phe	F	3+6
Cysteine	Cys	C	3	Proline	Pro	P	5
Glutamic Acid	Glu	E	5	Serine*	Ser	S	3
Glutamine	Gln	Q	5	Threonine	Thr	T	4
Glycine	Gly	G	2	Tryptophane	Trp	W	3+8
Histidine	His	H	3+3	Tyrosine	Tyr	Y	3+6
Isoleucine	Ile	I	6	Valine	Val	V	5

* though very often detected, Ser is widely considered as a terrestrial contaminant.

delivered larger amounts of organic matter than meteorites themselves. This low rate of exogenous carbon delivery may nonetheless have been of great *qualitative* importance (see Sect. 4.3.4) for prebiotic chemistry processes on the primitive Earth.

4.3.2 Formation Mechanisms of Exogenous Amino Acids

Out of more than 70 identified extraterrestrial amino acids (Tables 4.2–4.4), 46 (including 8 diacids) have their amine function in the α-position, 13 in the β-position, 10 in the γ-position and 1 in the δ-position. The formation rate and/or the stability of α-isomers therefore appears favoured in extraterrestrial environments.

4.3.2.1 α-Amino Acids

The formation mechanism of α-amino acids has been the subject of a great deal of attention. In the historic experiment of Miller (1953) α-amino acids and α-hydroxy acids were associated with imino diacids $HN(CR_2CO_2H)_2$, these three compound classes altogether being considered as *the* signature of Urey–Miller experiments. Miller (1957) explained the formation of the α-amino acids α-hydroxy acids by the Strecker reaction (4.1), without being able to suggest a mechanism

Table 4.2. Names and structures of α-amino acids detected in carbonaceous chondrites

2–5 carbon atoms		6,7 carbon atoms		7 carbon atoms (contd)	
glycine	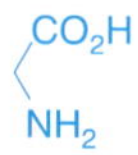	leucine	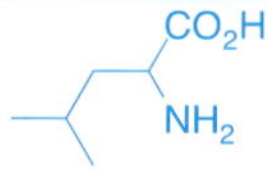	_2-amino-4-methyl caproic acid*_	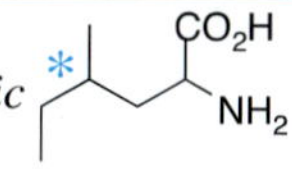
alanine	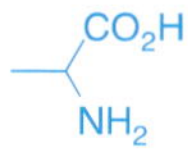	_isoleucine*_	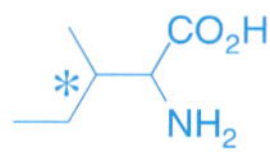	2-amino-4,4-dimethyl valeric acid	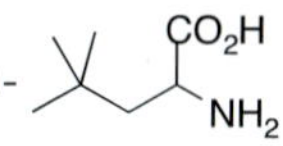
sarcosine	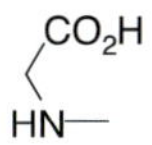	norleucine	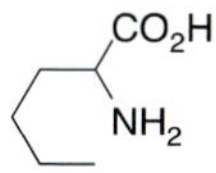	2-amino-2-ethyl-valeric acid	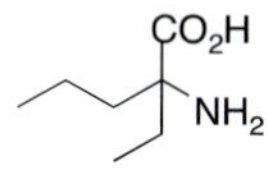
2-amino-butyric acid	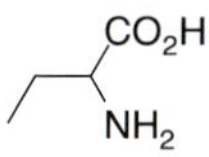	**2-methyl-norvaline**	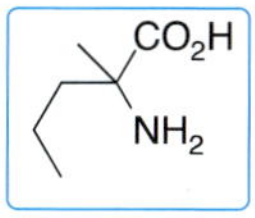	2-amino-3-ethylvaleric acid	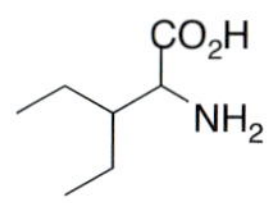
2-amino isobutyric acid	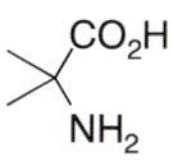	**2-amino-2,3-dimethyl butyric acid**	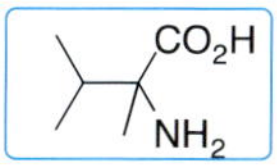	2-amino-5-methyl caproic acid	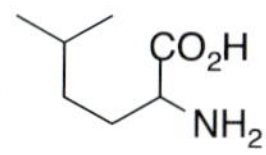
N-methyl-alanine	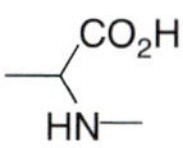	2-amino-2-ethylbutyric-acid	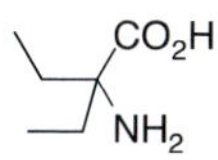	_2-amino-3-methyl caproic acid*_	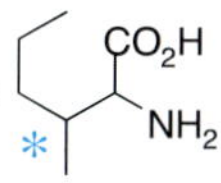
N-ethyl-glycine	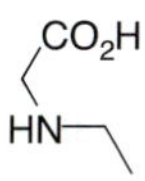	pseudoleucine	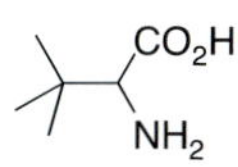	_2-amino3,4-dimethyl-valeric acid*_	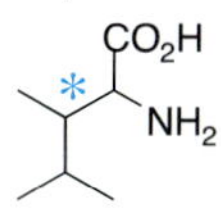
valine	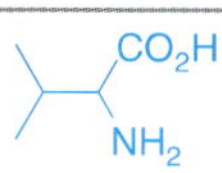	cycloleucine	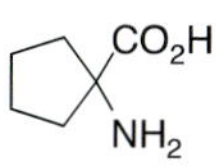	2-amino-2,3,3-trimethyl butyric acid	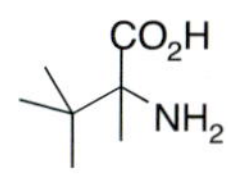
proline	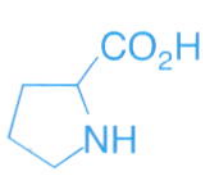	2-amino-heptanoic acid	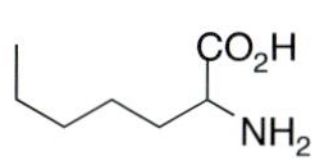	2-amino-2-ethyl-3-methylbutyric acid	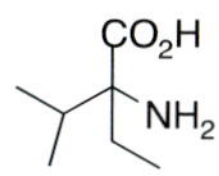
isovaline	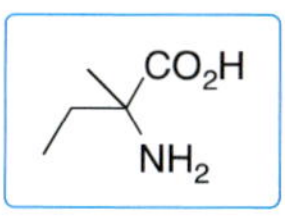	**_2-amino-2,3-dimethyl valeric acid*_**	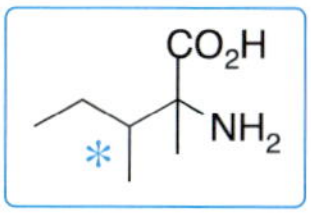	2-amino-3,3-dimethyl valeric acid	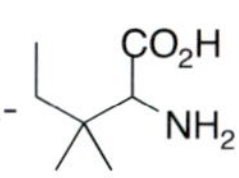
norvaline	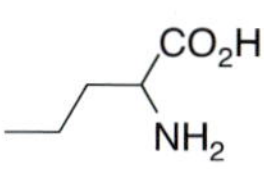	2-amino-2,4-dimethyl valeric acid	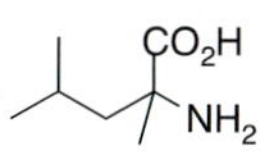	**2-amino-2-methyl caproic acid**	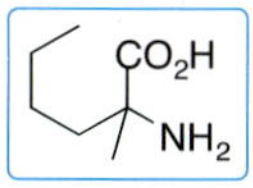

In _blue_: protein amino acids (only 8; Ser although very often detected, is widely considered as a terrestrial contaminant). * _(names in italic)_: diastereomer pairs (counted as two compounds; most amino acids are enantiomer pairs and counted as a single compound). _Boxed (names in bold)_: amino acids detected as non racemic in meteorites (1–9% _ee_ of L configuration; the methyl group in place of the α-hydrogen makes them resistant to racemization).

Table 4.3. Names and structures of α-amino diacids detected in carbonaceous chondrites

Succinic type		**Glutaric type**		**Others**	
aspartic acid	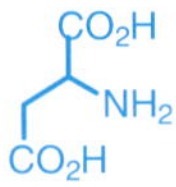	glutamic acid	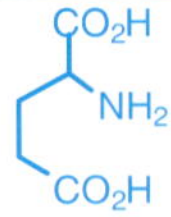	2-amino pimelic acid	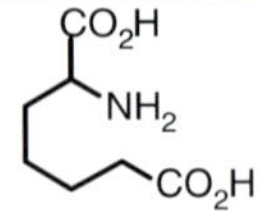
2-amino-2-methyl succinic acid	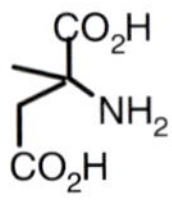	2-amino-2-methyl glutaric acid	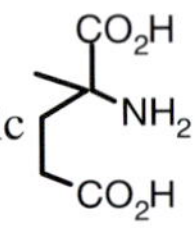	2-amino adipic acid	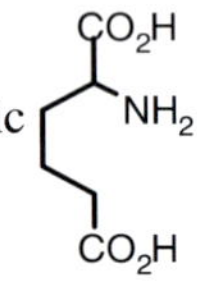
2-amino-3-methyl succinic acid*	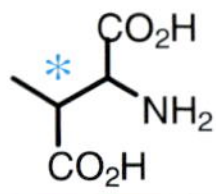				

(see notes on Table 4.2)

Table 4.4. Names and structures of non-α-amino diacids detected in carbonaceous chondrites

β-amino acids		**β- (cont.), γ-amino acids**		**γ- (cont.), δ-amino acids**	
β-alanine	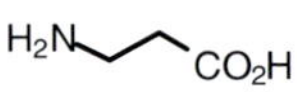	_3-amino-2-ethyl butyric acid*_	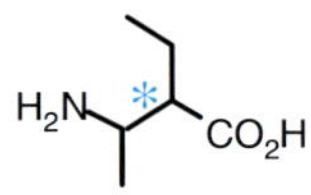	_4-amino-2-methyl valeric acid*_	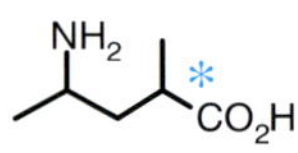
3-amino-butyric acid	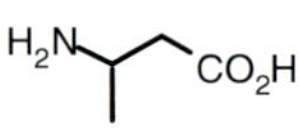	3-amino-3-methyl valeric acid	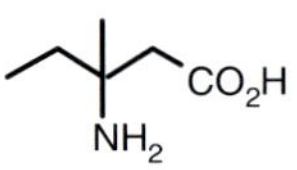	_uncertain structure (either of these 2):_	
3-amino-2-methyl-propionic acid	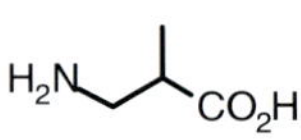	3-amino-2,3-dimethyl butyric acid	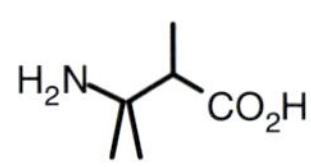	4-amino-caproic acid	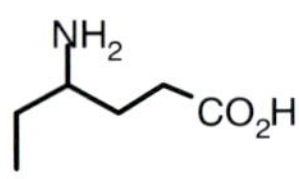
3-amino-2-methyl butyric acid	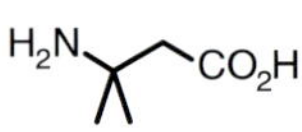	4-amino-butyric acid	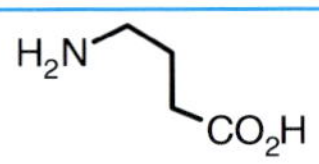	4-amino-3-ethyl butyric acid	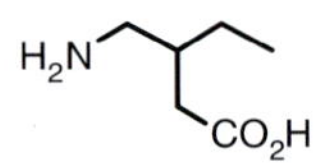
3-amino-3-methylbutyric acid*	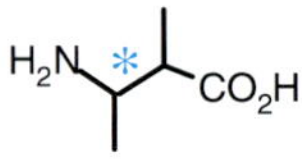	4-amino-2-methylbutyric acid	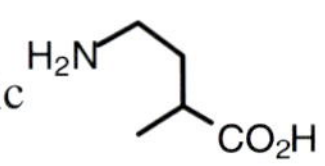	_uncertain structure (either of these 2):_	
3-amino-pivalic acid	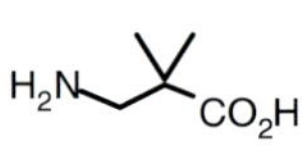	4-amino-3-methylbutyric acid	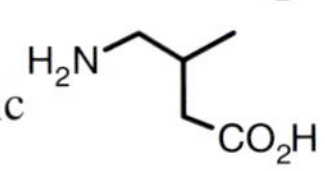	4-amino-4-methylvaleric acid	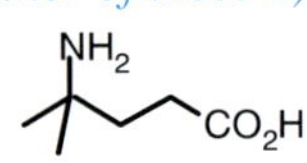
3-amino-valeric acid	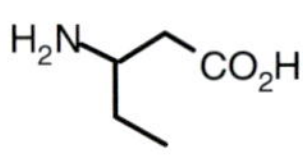	4-amino-valeric acid	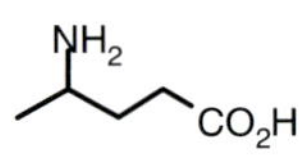	4-amino-3,3-dimethyl butyric acid	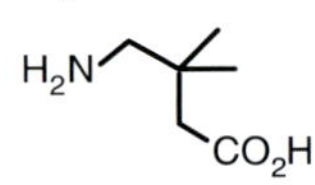
3-amino-2-ethyl-propionic acid	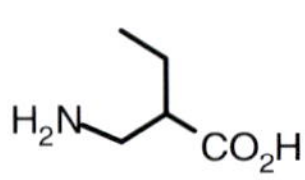	_4-amino-3-methylvaleric acid*_	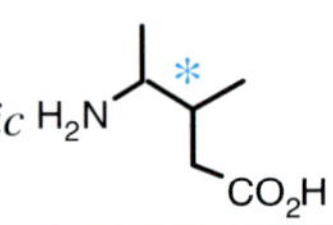	5-amino valeric acid	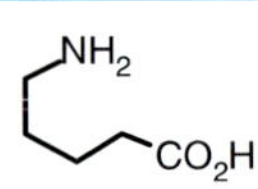

(see notes on Table 4.2)

for the imino-diacid formation. He nevertheless intuitively considered these three classes altogether to be also the signature of the Strecker reaction.

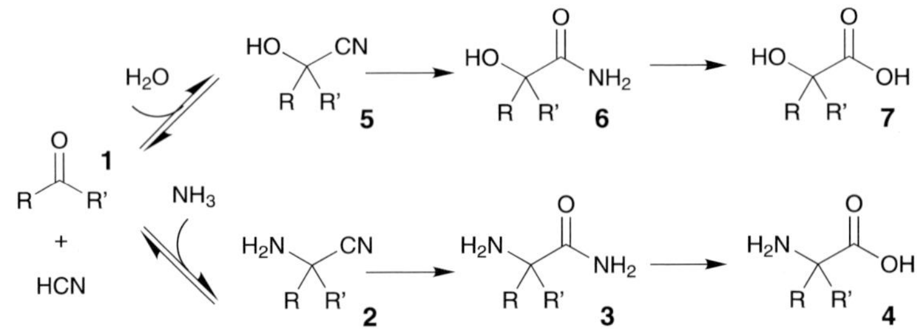

(4.1)

The early (and still accepted) hypothesis for meteoritic α-amino acid synthesis considered the Strecker reaction as the most plausible although only two (α-hydroxy- and α-amino acids) out of its three-component signature had been identified in meteorites (Peltzer et al. 1984; Cronin and Chang 1993; Botta et al. 2002). The recent discovery of imino diacids in meteorites (Pizzarello and Cooper 2001) completed this signature and confirmed Miller's hypothesis of the Strecker reaction for meteoritic α-amino acid synthesis. A better understanding of chemical mechanisms (see Sect. 4.5.2) explains the formation of imino diacids through Strecker processes.

If meteoritic α-amino acids actually originated through Strecker reactions, then their precursor carbonyl compounds are easily deduced by applying retrosynthesis to the set of meteoritic amino acids (Tables 4.2 and 4.2). This provides the 35 carbonyl derivatives listed in Table 4.5, where all possible isomers are present. The higher their molecular mass, the lower their abundance (relatively to that of α-amino acids).

The presence of all the possible isomers of these carbonyl derivatives, some of which being still present in meteorites (Jungclaus et al. 1976), clearly shows that they have previously been synthesized by nonselective processes, possibly photochemical. In the long term, this whole set of carbonyl derivatives is to be taken into consideration in a global kinetic model to explain the presence of exogenous and endogenous α-amino acids on the primitive Earth.

4.3.2.2 Non α-Amino Acids

The proposed mechanism (Cronin and Chang 1993) (4.2) for the formation of γ-amino acids involves the following initial molecules: α, β unsaturated aldehyde **8** variously substituted, hydrocyanic acid and ammonia. After addition of two HCN molecules, cyanohydrins (α-hydroxy nitriles) **9** are obtained. After hydration and hydrolysis (Strecker reaction), α-amino nitriles **10** lead to **11** and **12** (when R^1, R^2, and R^3 are hydrogens, **12** is glutamic acid). The following reaction sequence: cyclization of **12**, decarboxylation of **13** and hydrolysis of **14** produces γ-amino acids **15**, which might therefore appear as degradation products of α-amino 1,3-diacids **12**. If so, this would increase the count of exogenous α-amino acids from 46 to about 54, out of 70.

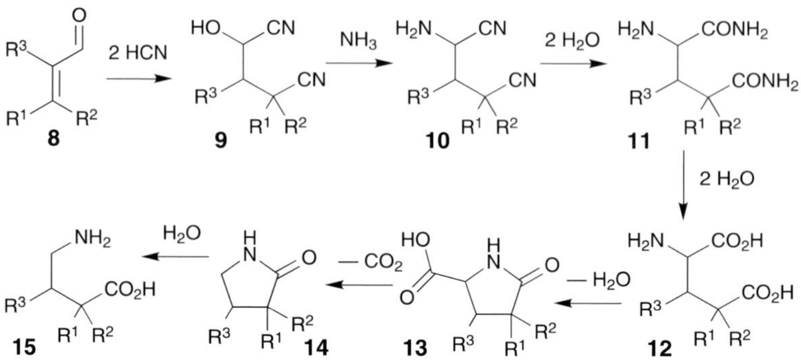

(4.2)

β-amino acids could have been formed according to (4.3) by addition of ammonia on acrylonitrile **16** ($R^1 = R^2 = R^3 = H$) or its substituted derivatives ($R^1, R^2, R^3 \neq H$), followed by hydrolysis of the nitrile function (Cronin and Chang 1993).

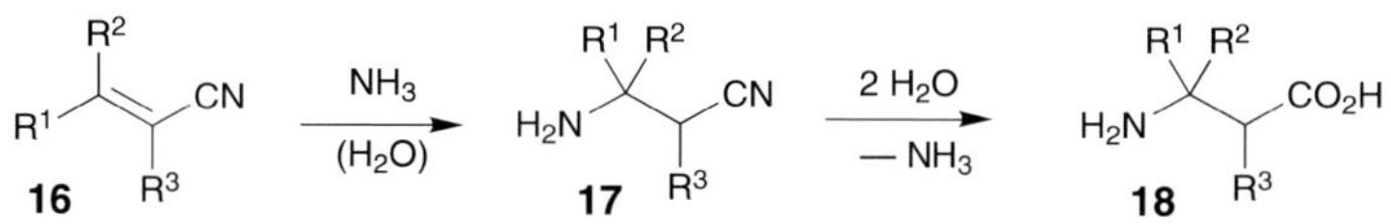

(4.3)

The proposed mechanism (Cronin and Chang 1993) (4.4) for the formation of the unique δ-amino acid **21** detected in the meteorites, is similar to that in (4.3), but starting with cyanobutadiene **19** as the initial substrate. This process involves, together with the above addition reaction, a reduction step on **20**.

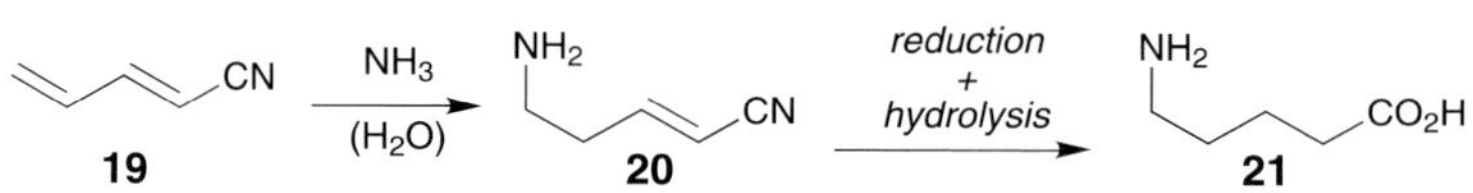

(4.4)

The formation of proline **25** can be explained either by the mechanism described by (4.5) (detection of pyrrolidine (saturated **22**) in interstellar clouds is consistent with this explanation) (Cronin and Chang 1993), or by cyclization of $H_2N–CH_2–CH_2CH_2CHO$.[1]

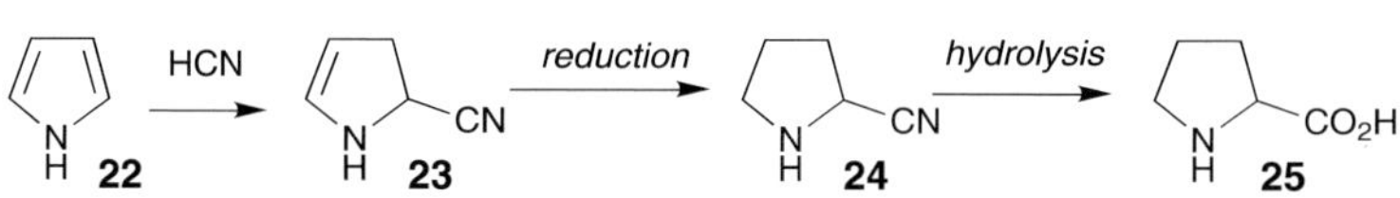

(4.5)

[1] Another possible pathway involves cyanohydrins $H_2N–(CH_2)_n–CH(CN)–OH$, the "Strecker precursors" of ornithine ($n = 3$) and lysine ($n = 4$): if present, their intramolecular substitution would lead to proline and pipecolic acid, respectively (both compounds being identified in Murchinson and Murray meteorites, but neither lysine nor ornithine).

Table 4.5. The 35 carbonyl derivatives precursor of meteoritic α-amino acids and α-amino diacids. In *blue:* compounds bearing a nitrile group

#C	Aldehydes	Ketones
1		
2	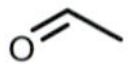	
3		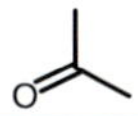
4	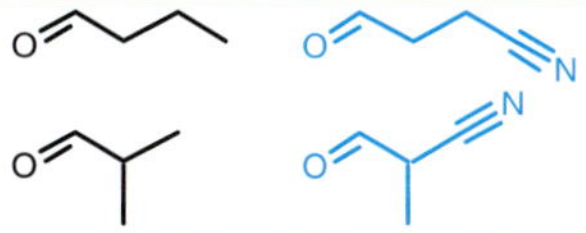	
5	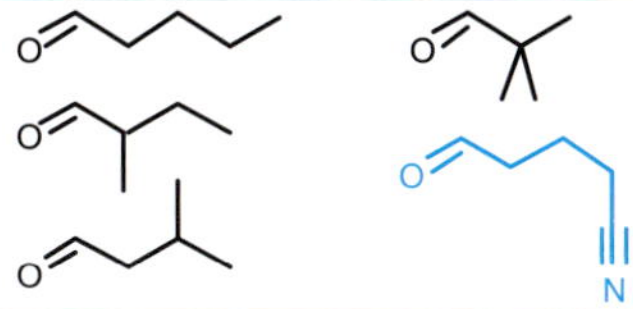	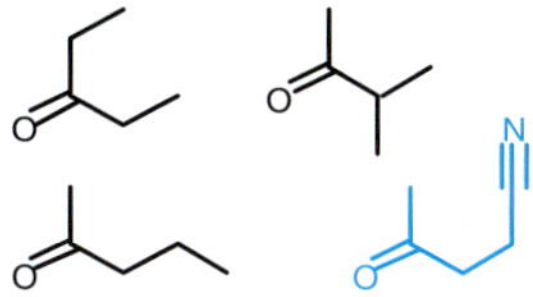
6	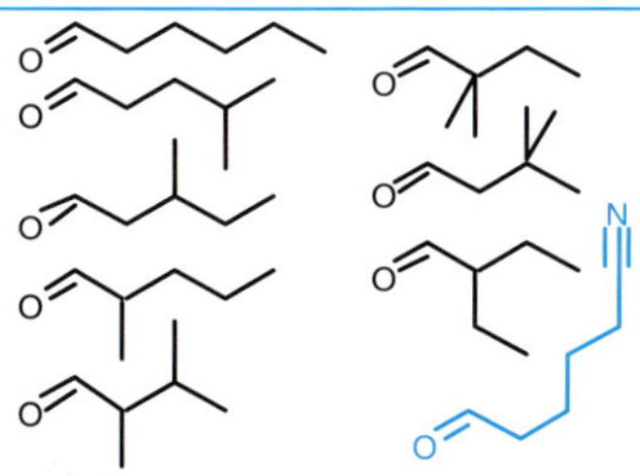	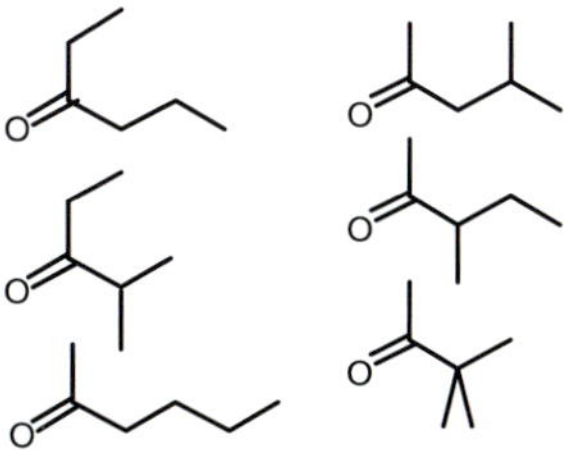

4.3.3 Other Meteoritic Compounds Closely Related to Amino Acids

Cooper and Cronin (1995) and subsequently Shimoyama and Ogasawara (2002) identified the compounds shown in Fig. 4.2 in the Murchison and Yamoto-791198 meteorites.

Lactames **13** and **14** are intermediates in γ-amino acid formation from α-amino acids (4.2). Compounds **26** to **30** are amide or imide derivatives. Compounds **31** are hydantoins (imidazolidine-2,4-diones) which can result from two pathways (Fig. 4.3):

- Either the reaction of cyanic acid on α-amino acids (formed through the Strecker reaction), followed by cyclization (Cooper and Cronin 1995), in which case hydantoins both substituted and unsubstituted on nitrogen 1 can be formed.
- Or the Bücherer–Bergs reaction (Taillades et al. 1998) via the cyclization of isocyanates **35** (see Sect. 4.5.2). In this case only 1-unsubstituted hydantoins are accessible.

Actually only 1-unsubstituted hydantoins have been identified in meteorites, which rather supports the hypothesis of a Bücherer–Bergs pathway.

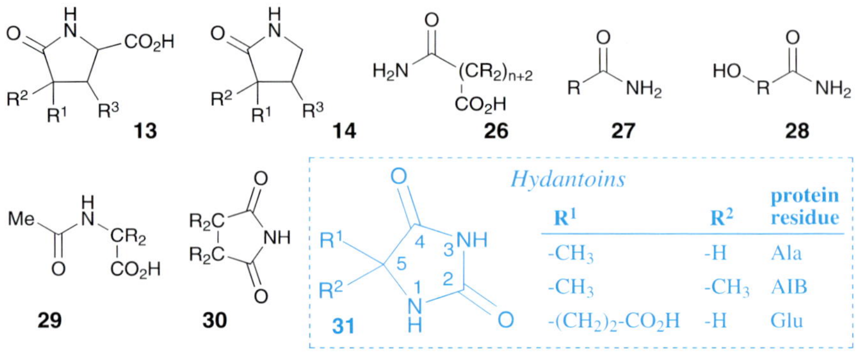

Fig. 4.2. Amides, imides and hydantoins discovered in the Murchison and Yamoto-791198 meteorites

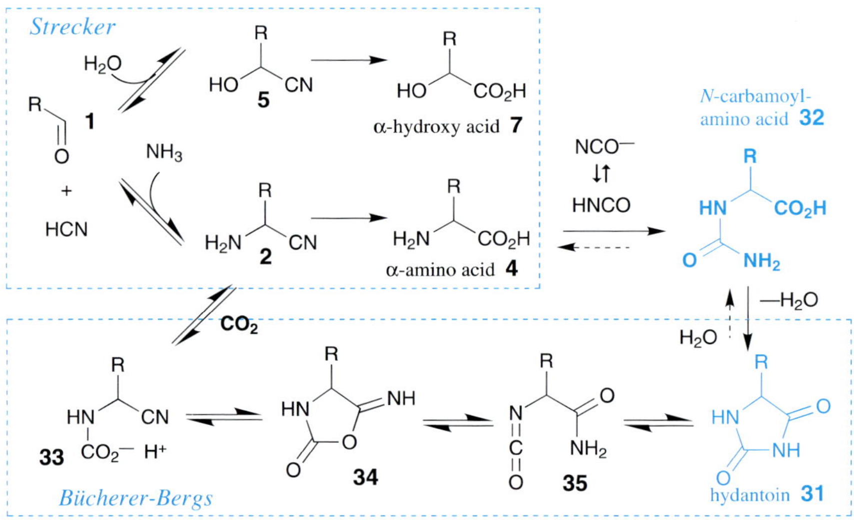

Fig. 4.3. Pathways of the Strecker and Bücherer–Bergs reactions and *N*-carbamoylation of α-amino acid

4.3.4 Non-Racemic Exogenous α-Amino Acids

(see also Cronin and Reisse Chap. 3, Part II)

Six extraterrestrial α-amino acids (framed structures in Tables 4.2–4.4) extracted from the Murchison and Murray meteorites were found as nonracemic (Cronin and Pizzarello 1997; Pizzarello and Cronin 2000). Their enantiomeric excesses (L) are between 1.0 and 9.2%. The presence of a CH_3 group in the α position instead of an H, argues for their nonbiological origin, which is confirmed by their D/H and $^{13}C/^{12}C$ isotopic ratios. This enantiomeric excess could be the result, not of chemoselectivity, but rather of enantioselective photodegradation of the initial racemic mixture in the presolar cloud. This photodegradation could have been caused by circularly polarized UV light. Circularly polarized light, although infrared, has been observed in the Orion cloud, meanwhile the origin of such radiation is the focus of numerous questions (Jorissen and Cerf 2002). This hypothesis of partial photodegradation of racemic mixtures is nonetheless consistent with laboratory results obtained as early as 1977 (Flores et al. 1977). The mechanism of this reaction was recently developed further (Nishino et al. 2001).

The existence of these nonracemic compounds opens up a wide research area in the field of the homochiral synthesis of peptides. They could have induced a symmetry breaking during the production of peptides by a "molecular engine" described in the next chapter of this book.

4.3.5 Exogenous Peptides

It is only recently that the first exogenous peptides were looked for and found in the Murchison and Yamoto-791198 meteorites. The authors of these works (Shimoyama and Ogasawara 2002) showed, however, that the only peptides present are glycylglycine (Gly-Gly) and the corresponding diketopiperazine. The abundance of these compounds, (of the order of ca. $20\,\mathrm{pmol\,g^{-1}}$), and, above all, the absence of both Ala-Gly and Gly-Ala, despite the presence of Ala and Gly in meteorites, led Shimoyama to conclude that these compounds were probably formed by classical dehydration processes, rather than via the copolymerization of NCA, which should have produced sequential macromolecules (see our next chapter in this book). The potential exogenous imports thus appear to be of minor importance.

4.3.6 Conclusion

α-Amino acids are therefore major compounds of the meteorite amino acid pool, both in structural diversity (54/70) and abundance, including 8 of the 20 protein amino acids[2]. Their formation through Strecker reactions is consistent with the presence in meteorites of the three compound classes characteristic of this kind of reaction: α-hydroxy- and α-amino acids, imino diacids. The – yet unexplained – formation of imino diacids is detailed in Sect. 4.5 devoted to chemical studies. Since β- and δ-amino acids are rather marginal, we shall not go into detail of their synthesis in that section, focusing rather on Strecker and Bücherer–Bergs processes.

[2] 10 if we include asparagine and glutamine (ω-amide forms of aspartic and glutamic acids, respectively). Though never detected in meteorites, these amino acids were very probably formed, but would have been readily and completely hydrolysed (into aspartic and glutamic acids respectively) during either meteorite ageing or extraction process.

4.4 Endogenous Organic Matter

The endogenous production of organic matter is unknown due to the absence of any fossil record. It must have greatly depended on the environment, especially on how reducing or oxidizing this environment was.

Knowledge of the different sources of available energy and their respective contributions to the endogenous production of organic matter, has evolved over time. While solar UV has always been considered to have played an important quantitative role (Chyba and Sagan 1992) the origin of organic nitrogen chemistry nonetheless still needs to be explained, given the short wavelengths ($\lambda < 100$nm) required for the photolysis of N_2 (Selsis et al. 1996; Selsis 2000). In this area, the (somewhat underestimated) importance of lightning and meteoritic impacts was recently re-evaluated (Navarro-Gonzalez et al. 1998, 2001), together with its consequences on the production of nitric oxide, NO. This aspect will be presented in Commeyras et al., Chap. 5, Part II, for which it is of high importance: a part of so-produced NO, remained in the atmosphere, might have played a fundamental role in the further synthesis of peptides.

Meanwhile, the other part of this NO, dissolved in the primitive ocean, could have been reduced to $NH_3$by Fe^{2+}in the ocean, even in the presence of high levels of cyanide (Summers and Chang 1993; Summers and Lerner 1998). The ammonia thus produced could have made a major contribution to primitive organic synthesis and especially to that of α-amino acids (see Sect. 4.6.2). In addition to its role as a major source for early-Earth nitrogen chemistry, NO may also have served as an energy source for carbon–oxygen chemistry, which, however, could also be sustained by solar UV.

4.4.1 Endogenous α-Amino Acids

It was the historic experiment of Miller (Miller 1953) that showed that α-amino acids could have been formed on the primitive Earth from a reductive atmosphere. In the absence of reliable information on the composition of this atmosphere, the experiment was repeated on numerous occasions (Miller 1998) under different conditions, intended to simulate many of the different possible compositions $\{CH_4, NH_3, H_2, H_2O\}$, $\{CH_4, NH_3, H_2O\}$, $\{CH_4, H_2, N_2, NH_3, H_2O\}$, $\{CH_4, H_2, N_2, H_2O\}$, $\{CO, H_2, N_2, H_2O\}$, $\{CO_2, H_2, N_2, H_2O\}$, under various H_2/CH_4, H_2/CO or H_2/CO_2 ratios. The formation of amino acids was observed in all cases, even unfavourable ones (CO_2, H_2, N_2, H_2O), provided the H_2/CO_2 ratio was greater than 2, thus involving large amounts of hydrogen in the atmosphere[3]. If we complement these atmospheric models by the sulfydryl function (–SH), phenylacetylene and indole (obtained by flash pyrolysis simulating the electrical discharges of gaseous CH_4, NH_3 mixtures, see Miller 1998), we obtain

[3] Recent observations show that the gases issuing from black smokers (undersea hot springs) contain considerable hydrogen amounts (Brack, lecture in Montpellier, 2002).

17 protein α-amino acids, plus a collection of nonprotein amino acids analogous to those discovered in meteorites, together with some imino diacids.

As early as 1955, Stanley Miller showed that carbonyl derivatives, hydrocyanic acid and ammonia were the raw materials that produced these α-amino acids. He proposed the Strecker reaction as the formation mechanism of these compounds (Miller 1957, 1998). This proposition was used once again to explain the formation of some of the exogenous α-amino acids (see Sect. 4.3.2). It is remarkable to see how Miller's proposition remains of current importance.

In addition to these synthetic pathways, others have been explored. We can cite:

- The one starting from HCN (Oró 1961) leading to low percentages of glycine and traces of alanine and aspartic acid.
- The one starting from 2-amino acrylonitrile $H_2C{=}C(NH_2)CN$ (Ksander et al. 1987), which leads, in the absence of oxygen and water, to numerous organic compounds such as glutamic acid, and a precursor of the corrinoid cycle of vitamin B12 obtained with a very high yield (50%).

This last synthesis could have been prebiotic. The ease of synthesis of an essential building block of vitamin B12 interested Eschenmoser et al. (Ksander et al. 1987), who wondered whether the complexity of vitamin B12 was only apparent, and whether the selectivities of such syntheses starting from simple reagents might not be more common in the field of prebiotic chemistry than we had imagined. The authors suggested that research efforts should be developed in that direction.

We show in the next section of this chapter that the synthesis of *N*-carbamoyl amino acids could be a second example of selective prebiotic synthesis. To present these results, we first summarize the knowledge acquired through studying a multicomponent reaction system made up of the following molecules $\{H_2O, HCN, R^1R^2CO, NH_3, R^3NH_2, R^3R^4NH, CO_2, B(OH)_4{}^-\}$. Secondly, we use this knowledge to try to understand the formation dynamics of α-amino acids and *N*-carbamoyl amino acids on the primitive Earth.

4.5 Formation Mechanisms of α-Amino Acids and *N*-Carbamoyl Amino Acids Via Strecker and Bücherer–Bergs reactions

The aim of this section is to present the reaction mechanisms of Strecker and Bücherer–Bergs reactions, involving 4 initial components $\{H_2O, R^1R^2CO, HCN, NH_3\}$ and 5 initial components $\{H_2O, R^1R^2CO, HCN, NH_3, CO_2\}$, respectively.

We shall also examine extensions of these multicomponent systems, including borate[4] $B(OH)_4{}^-$ (see Sect. 4.5.2), primary or secondary amines (R^3NH_2 or R^3R^4NH, respectively, see Sect. 4.5.3), the fate of both Strecker and Bücherer–Bergs reactions being different for each instance. Considering then the global system made up of all 8 above components, the combination of Strecker and Bücherer–Bergs reactions together with those of borate[4] will lead to a set of α-amino acids, N-carbamoyl amino acids, α-hydroxy acids and traces of imino diacids. The great structural diversity of several reactants: R^1R^2CO, R^3NH_2, R^3R^4NH (see Tables 4.1–4.5), will logically result in highly complex mixtures.

If such complexity is examined through combinatorial chemistry techniques only, we can get information on the set of final products only, losing the dynamic and mechanistic aspects such as equilibria and kinetics. This latter information, which is nonetheless indispensable for the understanding and modelling of dynamic processes, needs to be sought by means of more detailed, less global experimental approaches. That is the price to be paid for the quantitative study of chemical-evolution processes.

Our essential results are summarized in Commeyras et al., 2003. On this basis, and for the sake of clarity, the behaviour of the above 8-component system can be schematically presented as a first group of reversible steps producing a wide variety of chemical intermediates (Fig. 4.4), followed by a second group of irreversible steps resulting into a limited number of more stable species

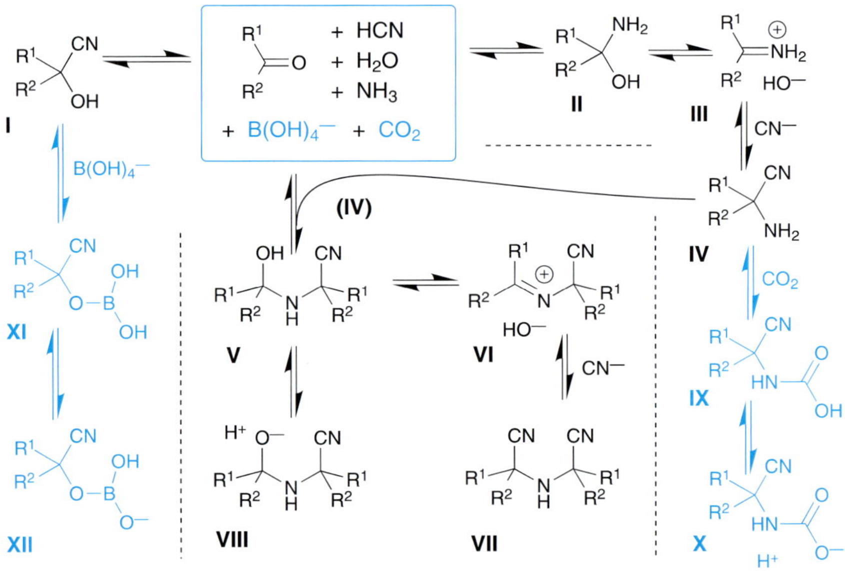

Fig. 4.4. Reversible reactions and products (at equilibrium) of the 6-component system $\{H_2O, HCN, R^1R^2CO, NH_3, CO_2, B(OH)_4{}^-\}$

[4] Boron is a non-negligible component of the Earth's crust (0.03%), with borates involved in the composition of glass (13%). Therefore it may be worthwhile analyzing the potential participation of borates in organic molecules formation under both prebiotic and laboratory conditions, because of their catalytic role in some reactions.

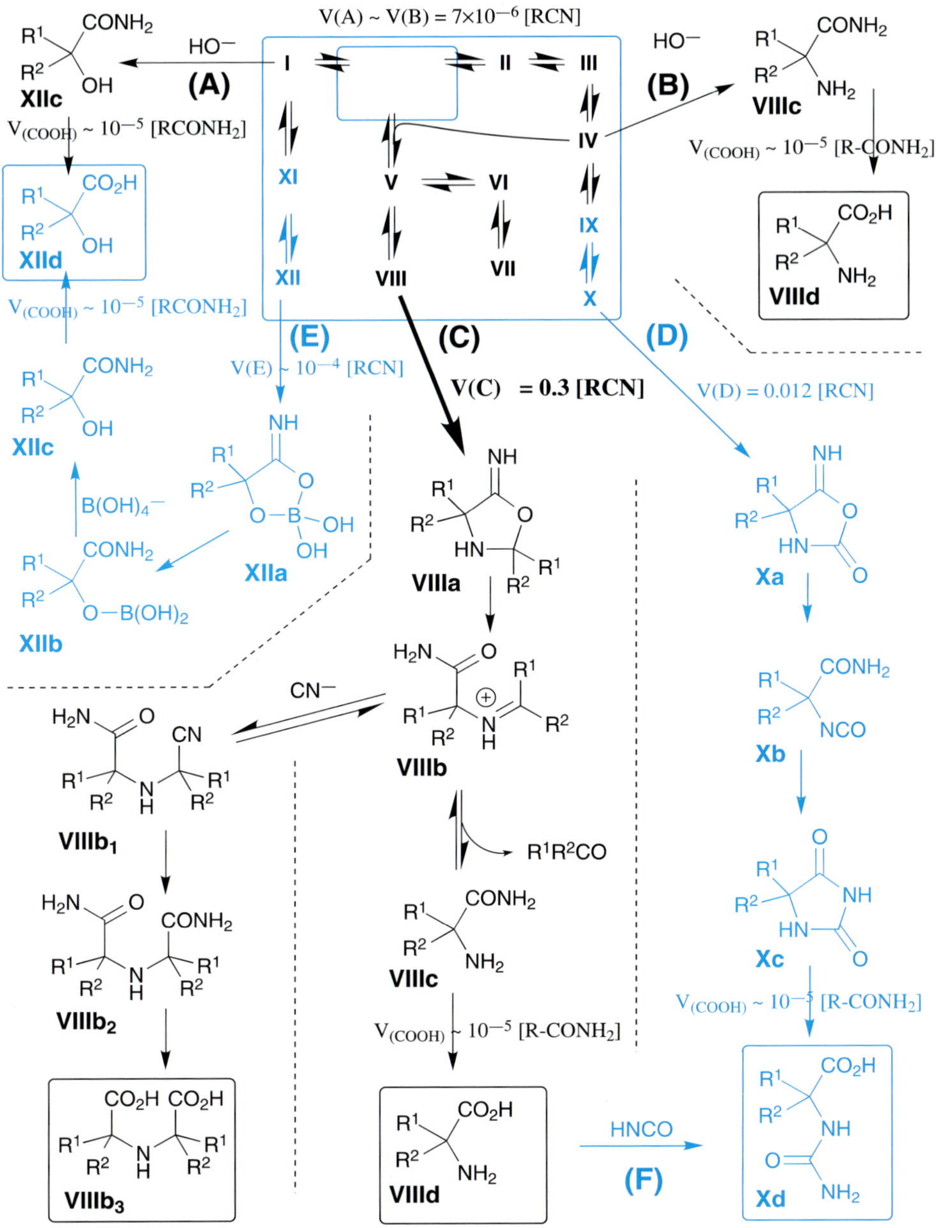

Fig. 4.5. The set of (competing) irreversible reactions in the reaction network from the 6-component system {H_2O, HCN, R^1R^2CO, NH_3, CO_2, $B(OH)_4^-$}, showing pathways (A)–(E)

(Fig. 4.5).[5] The special case of formaldehyde in the presence of high amounts of ammonia will not be examined however, since we considered it as too far from reasonable prebiotic conditions (see Sect. 4.6.2).

4.5.1 The Set of Reversible Reactions

We shall first examine the set of reversible reactions of the initial 6-component system $\{H_2O, HCN, R^1R^2CO, NH_3, CO_2, B(OH)_4{}^-\}$ (additional constraints introduced by primary and secondary amines R^3NH_2 and R^3R^4NH will be examined later). This set of reactions[6] (Fig. 4.4) is the starting point of Strecker and Bücherer–Bergs reactions that involve further, irreversible steps.

The reactions able to take place in this initial system (box on top of Fig. 4.4) mostly depend on the reactivity of the carbonyl derivatives R^1R^2CO. They can react:

- with water to give hydrates (not shown)
- with hydrocyanic acid to give cyanohydrins **I** (fast)[7]
- with ammonia to give α-amino alcohols **II**

Amino alcohols **II** (slowly) lose an HO^- anion to produce iminium cations **III**, which then react with the cyanide anion to give α-amino nitriles **IV**, the amine function of which reacts with carbonyl derivatives to produce α-amino alcohols **V**. **V** are subject to two competitive reactions: either they lose an HO^-, yielding iminium cations **VI**, which further react with cyanide to produce α-amino dinitriles **VII**, or **V** lose a proton to give alcoholate ions **VIII**.

CO_2 specifically reacts with α-amino nitriles **IV** to give carbamic acids **IX** that ionize into carbamates **X**. Boric acid $B(OH)_3$ (in equilibrium with borate $B(OH)_4{}^-$) specifically reacts with the cyanohydrins **I** to produce boric esters **XI**, which ionize into borates **XII**.

Under certain conditions (experimental but probably not prebiotic), the equilibria corresponding to these reversible reactions can be reached, thus constituting a reservoir of various nitrile families (cyanohydrins, amino nitriles, amino dinitriles, etc.). Their relative abundance depends on reactant structures (the nature of R^1, R^2), initial stoichiometry and concentration, and on the experimental conditions, especially pH, temperature and CO_2 pressure (Commeyras et al. 2003).

While such information is easily accessible experimentally, the main concern is then to determine whether this product distribution at equilibrium is representative or not of the final product distribution when further irreversible steps are

[5] This separation into two sets is made for the sake of clarity. Under prebiotic conditions, the equilibrium corresponding to the first set is very unlikely to be reached due to both reactant (e.g. ammonia) fluxes and competitive reactions from the second set.

[6] Equilibria of both hydration of carbonyl derivatives R^1R^2CO and carbonation of ammonia NH_3 (which are also part of this set) have been omitted to make Fig. 4.4 simpler.

[7] Molecular structures involved in Figs. 4.4– 4.7 are labelled with bold Roman figures.

involved. In other words whether the final product distribution is under thermodynamic or kinetic control. This requires determination of kinetic data on both reversible and irreversible steps.

4.5.2 The Set of Irreversible Reactions

In these reaction media, irreversibility appears when nitriles (–CN) are transformed into amides ($-CONH_2$). Five different, competitive pathways (A–E, Fig. 4.5) can bring about this transformation[8]. Their starting points are the species **I**, **IV**, **VIII**, **X**, **XII** respectively, as shown in Fig. 4.4.

The first two pathways (A) and (B) are well known as the *Strecker reaction*[9] (Strecker 1850; Miller 1957; see Sect. 4.3.2). They correspond to the intermolecular attack of the hydroxide anion (HO^-) on the nitrile group of the cyanohydrins **I** and α-amino nitriles **IV**. They can explain the formation of both α-hydroxy amides **XIIc** and α-amino amides **VIIIc**, but not the formation of imino diacids **VIIIb$_3$**.

The other three reaction pathways (C), (D), (E) evidenced by us, result in the hydration of the nitrile group of **I** and **IV** in a more efficient way than both pathways (A) and (B). In addition, the reaction pathway (C) can explain the formation of imino diacids **VIIIb$_3$**.

The starting point of the pathway (C) is the alcoholate anions **VIII**. These anions react *intra*molecularly with their nitrile group, to give the α-amino amides **VIIIc** via two intermediates **VIIIa** and **VIIIb**. This pathway (C) leads to the same products as (B), but approximately 10^5 times faster, through catalysis by carbonyl derivatives.

Concerning imino diacids **VIIIb$_3$**, Garrel (personal communication) recently showed that they originate from **VIIIb**. When the initial reagent concentrations are such as $[R^1R^2CO] > [CN^-] \gg [NH_3]$, the addition of cyanide to intermediates **VIIIb** occurs, that forms new nitriles **VIIIb$_1$** as minor products, which then evolve into **VIIIb$_3$**. Therefore we consider that this reaction pathway (C) and its side-products must be completely included into the Strecker reaction, which should not be limited to reaction pathways (A) and (B).

The starting point of the pathway (D) is the carbamate anions **X**, which react *intra*molecularly with their nitrile group to give the hydantoins **Xc** via two intermediates **Xa** and **Xb**. This reaction is almost as fast as (C), and is known as the Bücherer–Bergs reaction.

---

[8] An additional pathway to amino acids and involving hydrogen peroxide, was not taken into consideration because poorly compatible with prebiotic condition (Rossi et al. 1996).

[9] The name "Strecker reaction" was initially given to the synthesis of α-amino acids by *acidic hydrolysis* (quenching of equilibrated aqueous alkaline) mixtures of carbonyl compounds, hydrocyanic acid and ammonia, referring to his original experiments. Later this name covered any experimental conditions using reaction pathways (A)+(B)+(C).

The starting point of the pathway (E) is the borate anions **XII**, which react *intra*molecularly with their nitrile group to give the α-hydroxy amides **XIIc**, regenerating, en route, the boric acid, which thus behaves as a catalyst in this reaction.

Kinetic studies enabled us to determine the kinetic laws and constants (4.6–4.10) of these reactions (Taillades et al. 1998; Commeyras et al. 2003), the following equations being valid in the pH 5–9 range, with values given at 25°C:

$$v_{(\mathrm{A})} = k_{(\mathrm{A})}[\mathrm{HO}^-][\mathbf{I}] \quad k_{(\mathrm{A})} = 3.8 \times 10^{-4}\,\mathrm{L\,mol^{-1}\,s^{-1}} \tag{4.6}$$

$$v_{(\mathrm{B})} = k_{(\mathrm{B})}[\mathrm{HO}^-][\mathbf{IV}] \quad k_{(\mathrm{B})} = 1.4 \times 10^{-3}\,\mathrm{L\,mol^{-1}\,s^{-1}} \tag{4.7}$$

$$v_{(\mathrm{C})} = k_{(\mathrm{C})}[\mathrm{HO}^-][\mathrm{R^1R^2CO}][\mathbf{IV}] \quad k_{(\mathrm{C})} = 83\,\mathrm{L^2\,mol^{-2}\,s^{-1}} \tag{4.8}$$

$$v_{(\mathrm{D})} = \frac{k_{(\mathrm{D})}}{1 + \frac{K_\mathrm{w}}{K_1}\,[\mathrm{HO}^-]}\,[\mathrm{CO_2}]^t\,[\mathbf{IV}] \quad k_{(\mathrm{D})} = 1.2 \times 10^{-4}\,\mathrm{L\,mol^{-1}\,s^{-1}} \tag{4.9}$$

$$v_{(\mathrm{E})} = k_{(\mathrm{E})}[\mathrm{B(OH)_3}]^t\,[\mathrm{HO}^-]\,[\mathbf{I}] \quad k_{(\mathrm{E})} = 180\,\mathrm{L^2\,mol^{-2}\,s^{-1}}, \tag{4.10}$$

with K_w and K_1 in (4.9) being the dissociation constants of water and of carbonic acid respectively (first dissociation), $[\mathrm{CO_2}]^t$ in (4.9) and $[\mathrm{B(OH)_3}]^t$ in (4.10) being the total concentration (acid + base forms) of carbonate + CO_2 and borate, respectively. It is noteworthy that above pH 7.5 the rate law (4.9) is independent of the pH. Further studies will be necessary to determine the temperature dependence of these rate constants.

In a first approach, however, at any pH between 5 and 9, and provided the concentration of either carbonyl compounds, carbonate or borate are above the millimolar range, the reaction pathways (A) and (B) (the only ones dealt with in previous prebiotic studies, see Sect. 4.3.2) should play a minor role comparatively to much faster pathways (C), (D), (E) using the same starting nitriles **I**, **IV**, **VIII**. The latter pathways would thus be responsible for the formation of α-amino amides **VIIIc**, hydantoins **Xc**, and α-hydroxyamides **XIIc**. Nevertheless, such a conclusion must be adapted to prebiotic conditions, where the concentrations of the various nitrile families **I**, **IV**, **VIII** should be far from identical.

Studying the subsequent evolution of these α-amino amides **VIIIc**, hydantoins **Xc**, and α-hydroxyamides **XIIc** into α-amino acids **VIIId**, *N*-carbamoyl amino acids **Xd** and α-hydroxy acids **XIId**, respectively, showed that these three hydrolytic reactions have low but equivalent rates ($v_{(\mathrm{CO_2H})} \approx 10^{-5}[\mathrm{R-CONH_2}]$, see Commeyras et al. 2003). From this we can deduce that these reactions are incapable of modifying the selectivities induced by reactions (C), (D) and (E).

If the multicomponent system under consideration (Fig. 4.4) is completed with cyanate (the actual reacting species being cyanic acid HNCO), the system is driven to evolve towards *N*-carbamoyl amino acids (CAA) **Xd** through pathway (F). Kinetic studies of this reaction (Taillades et al. 2001) showed that the operating conditions required for carrying out this reaction are fully compatible with those of the other reactions described in Fig. 4.5. In addition, cyanic acid also reacts with amino amides **VIIIc** to give *N*-carbamoylamino *amides*.[10] We observed (unpublished results) that the nitrosation of such compounds by gaseous NO_x (see next chapter of this book) quantitatively yields hydantoins, the hydrolysis of which produces *N*-carbamoylamino *acids*.

Globally, we can thus conclude that:

The "multicomponent" system represented in Figs. 4.4 and 4.5 is kinetically controlled by irreversible reactions proceeding through reaction pathways (C), (D), (E).

This "multicomponent" system convergently evolves towards two families of compounds: α-hydroxy acids (minor) and *N*-carbamoyl amino acids (major).

4.5.3 Fate of Primary and Secondary Amines (R^3NH_2, R^3R^4NH)

Primary amines R^3NH_2 follow the same reaction pathways as ammonia, with some limitations. Indeed while they can take pathways (B) and (C) to form α-NR^3-amino acids **VIIId**, they are blocked on pathway (D) by the impossibility of forming isocyanates **Xb**, so that hydantoins **Xc** substituted on nitrogen-1 are inaccessible. A roundabout way to obtain such substituted hydantoins is to react cyanate with α-amino acids-NR^3 according to pathway (F), then to dehydrate the NR^3-carbamoylamino acids **Xd** thus obtained. However, no such product has been identified in meteorites, which leads to the conclusion that these cyanates only played a minor role in the parent bodies, doubtlessly for kinetic reasons (see Sect. 4.7). This obviously does not exclude their having played a role on the primitive Earth. In the parent bodies, hydantoins could have been formed via the Bücherer–Bergs reaction.

The reactivity of secondary amines (R^3R^4NH) is even more severely limited. They can only lead to α-NR^3R^4-aminonitriles **IV** and then α-NR^3R^4-amino acids **VIIId** through pathway (B).

Taking these limitations into consideration provides interesting information in the analysis of prebiotic scenarios (see Sect. 4.6).

4.5.4 Conclusion

On the basis of the above chemical data, it is possible to explain the formation of imino diacids through Strecker processes, which therefore include the three pathways A, B, C allowing the initial multicomponent mixture {H_2O,

[10] Cyanic acid reacts with amines in general; its reaction with ammonia produces urea (this simple reaction is not included in Fig. 4.5).

HCN, R^1R^2CO, R^3R^4NH, CO_2, $B(OH)_4^-\}$ to be driven out of equilibrium. The Bücherer–Bergs reaction is a fourth pathway D allowing this equilibrium to be broken, and leading to CAA. The kinetics of pathways C, D, E that predominate over A and B determine the relative hydroxy acid/amino acid/CAA formation ratio. The presence of cyanic acid may have convergently driven AA evolution towards CAA, which would then become the most abundant products of this chemical process.

In the next section we examine possible prebiotic implications of such chemistry.

4.6 Prebiotic Formation of α-Amino Amides and Hydantoins Through Strecker and Bücherer–Bergs Reactions

Dynamic study of complex systems generally provides information about the preferential directions taken by the system under the pressure of a given environment. In the case of chemical evolution, these studies enable us to imagine how the molecular complexification operated, and what were the most probable mechanisms for this complexification.

4.6.1 Formation of Exogenous α-Amino Amides and Hydantoins

4.6.1.1 Environment and Reagents

Both Strecker and Bücherer–Bergs reactions require the presence of (rather liquid) water, which puts constraints on the places where they might have occurred. Exogenous organic matter delivered by meteorites is considered to have been formed in parent bodies, the history of which indicates that they could have met temperature ranges of ca. 263–298 K (Clayton and Mayeda 1984), with the possible presence of (interstitial or permafrost) liquid water.[11]

The analysis of meteorite extracts led Peltzer et al. (1984) to propose the Strecker reaction to explain the formation of α-amino acids in parent bodies. The presence of α-hydroxy acids and (recently discovered) of imino diacids (Pizzarello and Cooper, 2001) strongly confirm this hypothesis.

Besides, investigation by Lerner (1997) on the Strecker reaction in aqueous solution simulating the putative parent-body composition ($[NH_3] = 2\,mM$, $[CN] = 5\,mM$, [aldehydes] $= 7\,mM$, [ketones] $= 0.75\,mM$) at various temperatures (263 and 295 K), shows the formation of the 3 components of the Miller–Strecker signature (α-hydroxy- and α-amino acids, imino diacids) after a 4-month reaction. When the reaction was performed at 295 K, only the compounds

[11] The synthesis of amino acids and related compounds in interstellar molecular clouds is also probable; however, the preservation of such organics from presolar origin during accretion of the (hot) protoplanetary disc is very doubtful.

derivating from aldehydes were observed, whereas at 263 K derivatives from both aldehydes and ketones were observed.[12] The formation of these imino diacids appeared not to be influenced by the presence of meteoritic minerals.

From these data we can reasonably consider that meteoritic amino acids were synthesised in parent bodies through the Strecker reaction. We can add that the temperature was probably lower than 263 K (considering the abundance of α-alkyl, α-amino acids in meteorites), and that the ammonia concentration was quite low compared to that of other reagents such as HCN or carbonyl compounds.

We must, however, take care on requirements put on the prebiotic environments to fit the Strecker reaction with meteorite amino acids, since their analysis through extraction processes gives merely an estimation of meteorite content that can easily be modified by the extraction process itself.

4.6.1.2 Suitability of Reactions to Prebiotic Conditions

We have seen that reaction pathway (C) is catalyzed by carbonyl derivatives. When the catalyst used is formaldehyde, the activation energy of this reaction is zero (Pascal et al. 1980; Commeyras et al. 2003). Such low activation energy values are generally the consequence of favoured intramolecular reactions, as it is the case of the hydration of nitriles by alcoholate anions **VIII** in pathway (C). Noteworthy that the rate constant for this reaction ($k_{H_2CO} = 3.4 \times 10^7\,L^2\,mol^{-2}\,min^{-1}$) is of the same magnitude order as enzymatic reaction rate constants. Current research into the emergence of the catalytic activities of enzymes makes reference to such reactions (Pascal 2003).

A similar cyclization of carbamates **X** occurs in pathway (D), for which the (yet unknown) activation energy is probably quite low, thus allowing the reaction to occur at low temperatures. However, this step is probably not the rate-determining one for hydantoin **Xc** formation, which might rather be the opening of **Xa** with formation of isocyanate **Xb** (requiring base catalysis). Further investigation is necessary to confirm this hypothesis.

Conversely to (D) and similarly to (C), the (fast) cyclisation of **XII** in pathway (E) is probably the rate-determining step in borate-catalysed formation of α-hydroxyamides **XIIc**. However, there is no information about the concentration of borate in prebiotic media.

The above examples of intramolecular catalysis are certainly not unique in prebiotic chemistry; they could justify continuing research in this direction, especially since they have already led to the development of energy-economical, waste-free industrial processes (Commeyras et al. 1976; Taillades et al. 1986).

[12] We explain this by the higher thermolability of tertiary α-amino nitriles (ketone derivatives) compared to that of secondary α-amino nitriles (aldehyde derivatives), the former thus being unable to react with carbonyl compounds (pathway C on Fig. 4.5) at higher temperature to form the α-alkyl amino amides and acids (Commeyras et al. 2003).

4.6.2 Endogenous Formation of α-Amino Amides and Hydantoins

To try to explain the formation of endogenous α-amino acids it is obviously desirable to have information about the nature and concentrations of the species present on the primitive Earth, and about the pH and temperature of the oceans. This knowledge is still imprecise and can thus only lead to hypotheses.

4.6.2.1 Availability of Starting Compounds

The evaluation range for the partial pressure of CO_2 is very wide, going from 10^{-4} bar (Pinto et al. 1980) to more likely estimates, ranging between 0.1 and 10 bar (Owen et al. 1979; Kasting and Ackerman 1986; Mojzsis et al. 1999).

While the formation of carbonyl derivatives and cyanide does not seem to be problematic, evaluation of their relative and especially absolute concentrations remains highly uncertain. Values of $2 \times 10^{-2}\,\mathrm{mol\,L^{-1}}$ for these different constituents (Pinto et al. 1980) have been proposed. To our knowledge, these estimations have not been updated.

Concerning ammonia, its origins and concentration have been the subject of numerous investigations. The initial proposition, that it was formed by reduction of nitrogen by atmospheric hydrogen ($N_2 + 3\ H_2 \rightarrow 2\ NH_3$) (Bada and Miller 1968), has been contested. Two other reactions have been proposed. The first is based on hydrolysis of cyanic acid ($HNCO + H_2O \rightarrow NH_3 + CO_2$), itself formed by electrical discharges in a hypothetical primitive atmosphere made up of N_2, CO_2 and H_2, (Yamagata and Mohri 1982). The other involves reduction of NO by Fe^{2+}, which was common in the primitive ocean (Summers and Chang 1993, Summers 1999). The formation of NO in considerable quantities (Prinn and Fegley 1987; Navarro-Gonzalez et al. 1998, 2001) could thus have led to regular and sufficient production of NH_3, a requirement for the production of amino acids. If, however, we take into account the fact that NH_3 readily photolyzes, it is likely that the ammonia concentration in the primitive atmosphere was still very low. Summers (1999) estimated the (pH dependent) steady-state concentration of ammonia (both NH_3 and $NH_4{}^+$) in primitive Earth oceans to have remained within the $2 \times 10^{-7}\,\mathrm{mol\,L^{-1}}$ (pH 8.5) to $3 \times 10^{-5}\,\mathrm{mol\,L^{-1}}$ (pH 5) limits. These very low concentrations led these authors to doubt the efficiency of the Strecker process, although afterwards, being unable to find an alternative, they accepted it as probable.

These facts must therefore be taken into consideration when examining the endogenous synthesis of α-amino acids.

4.6.2.2 Thermodynamic Aspects

From a thermodynamic point of view, taking into account these low reactant concentrations and the equilibrium constants K_{cya} and K_{an} associated to the reactions forming cyanohydrins **I** and α-amino nitriles **IV** respectively, it is easy to estimate their relative concentration. In the case of formaldehyde, $K_{cya} = 4.76 \times 10^5\,L\,mol^{-1}$ and $K_{an} = 4 \times 10^7\,L^2\,mol^{-2}$, therefore at equilibrium the concentration ratio of **I** to **IV** would be of the order of 1000/1. Nevertheless, the situation is different due to the fact that the relative concentrations of these two products is determined by their formation rates (kinetic control) and not by their relative stability (thermodynamic control).

4.6.2.3 Kinetic Aspects

From a kinetic point of view, cyanohydrins are formed within a few minutes ($t_{1/2} = 1$ min at any pH) whereas α-amino nitriles are formed much more slowly ($t_{1/2} = 6$ h, maximum rate at pH 9).[13] As soon as formed, α-amino nitriles **IV** react, partly decomposing back to initial reactants (with a pH-dependent rate, maximum above pH 5–6) and partly hydrating into α-amino amides **VIIIc** via the irreversible reaction (C). We have shown that this hydration reaction (C) is 10 times faster than the decomposition reaction (Pascal et al. 1978; Commeyras et al. 2003). The consequence (Fig. 4.6) is that the formation equilibria of the α-amino nitriles **IV** should never be attained under prebiotic conditions, because these compounds **IV** are immediately and irreversibly transformed into α-amino amides. We can add that the Bücherer–Bergs reaction (D), leading to the hydantoins **Xc**, should have amplified this process of withdrawing α-amino nitriles from the reversible reaction system.

Therefore even for ammonia concentrations of $10^{-6}\,mol\,L^{-1}$, equilibrium shifts towards amino amides and hydantoins may have allowed the major formation of these compounds to the detriment of hydroxy amides. Then the summary of thermodynamic and kinetic data seems in agreement with: (1) a similar abundance of amino acids compared to hydroxy acids in meteorites, and (2) the fact that compounds originating from irreversible reactions of α-amino nitriles α-amino amides, hydantoins) might have formed more abundantly on the primitive Earth in presence of carbonyl derivatives and ammonia (in lower concentration), if the pH of primitive oceans was around 5–6.

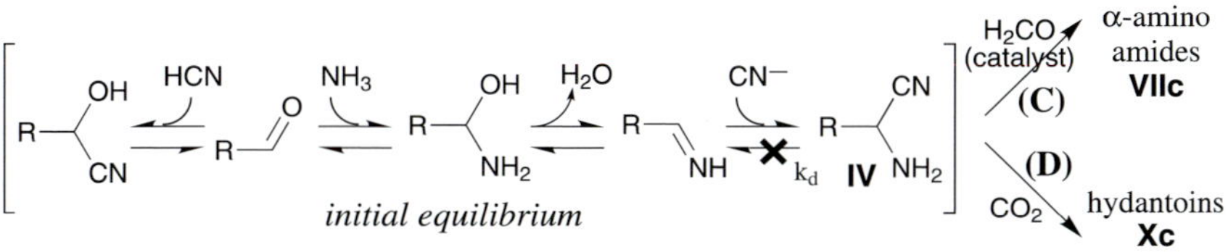

Fig. 4.6. Rapid and irreversible reaction of α-amino nitriles through pathways (C) and (D) does not enable the formation equilibrium of α-amino nitriles to be established

[13] Cyanohydrins (α-hydroxy nitriles) **I** are stable in aqueous solution, unlike carbonyl derivatives or hydrocyanic acid. We can thus state that they protected both carbonyl derivatives and hydrocyanic acid against degradation, enabling these compounds to wait for the gradual arrival of ammonia.

In summary, it is reasonable to conclude that the formation rates of α-amino amides and hydantoins on the primitive Earth, via the Strecker and Bücherer–Bergs reactions, could have been controlled by the formation rate of ammonia.

4.7 Convergent Evolution Towards *N*-Carbamoyl Amino Acids under Prebiotic Conditions

On the primitive Earth, even when controlled by the ammonia formation rate, a "multicomponent" system represented by Figs. 4.4 and 4.5 could only have spontaneously evolved (Fig. 4.7) into the three families of compounds analyzed above (see Sect. 4.5.2), i.e. a minority of α-hydroxy amides **XIIc** (omitted in Fig. 4.7), and a majority of α-amino amides **VIIIc** and hydantoins **Xc**. The subsequent evolution of these compounds into α-hydroxy acids **XIId**, α-amino acids **VIIId** and *N*-carbamoyl amino acids **Xd** with equivalent reaction rates ($3.6 \times 10^{-5}\,h^{-1}$ and $2.2 \times 10^{-5}\,h^{-1}$ measured at pH 8 and 25 °C, see Commeyras et al. 2003) cannot have influenced the selectivity generated by the fast steps (C) (D) (E) investigated above.

Concerning the *N*-carbamoylation reaction, it most certainly played an important prebiotic role[14], but it can only have been involved in the slow step

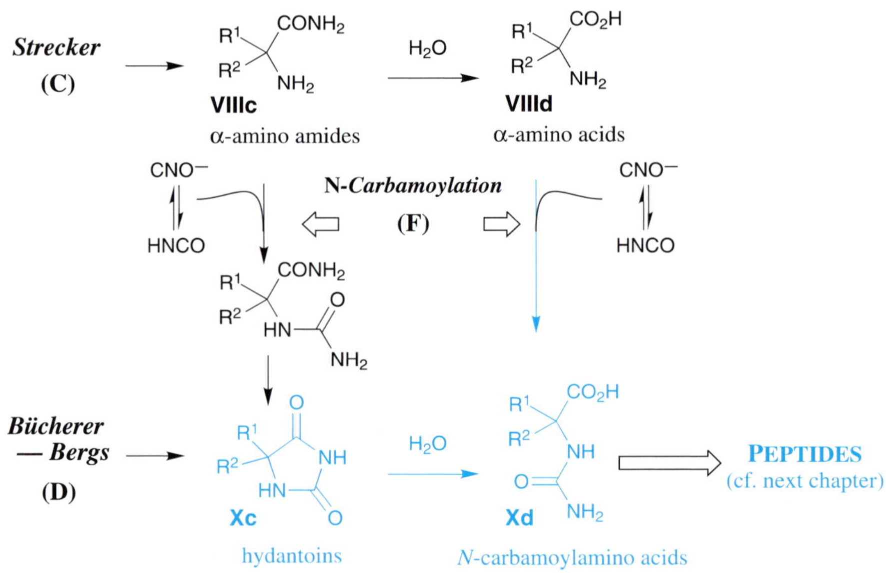

Fig. 4.7. Convergent evolution of reaction pathways (C), (D) and (F) towards *N*-carbamoyl amino acids and thence to peptides

[14] The estimated pH of the primitive ocean (5–7) (Mojzsis et al. 1999) is favorable to carbamoylation reactions. Cyanic acid HNCO is formed in model gas mixtures of

($k_{(F)} \approx 1.7 \times 10^{-3}\,\mathrm{L\,mol^{-1}\,s^{-1}}$ at 50°C and pH 6.5) (Taillades et al. 2001; Commeyras et al. 2003), considerably later than the preceding fast steps which controlled selectivity.[15]

It is therefore reasonable to consider cyanate (in equilibrium with cyanic acid) in the primitive ocean at pH 5–7, to have slowly transformed α-amino amides **VIIIc** and α-amino acids **VIIId** into N-carbamoyl amino acids **Xd** by two different pathways: either via N-carbamoyl amino amides and hydantoins, or directly from α-amino acids (Fig. 4.7).

The case of meteorites, where amino acids would be more abundant than both hydantoins and (yet unobserved) N-carbamoyl amino acids, appears contradictory with a convergent N-carbamoyl amino acid synthesis. Although the measured ratio may be biased from the original meteorite composition (by e.g. impact or extraction processes), this is probably rather due to quite different conditions having occurred in parent bodies comparatively to the primitive Earth, especially concerning CO_2 and cyanic acid abundance. Further analytic and chemical investigations should be able to provide useful information on these parent-body conditions.

In summary, on the primitive Earth, the set of reaction processes Fig. 4.7 could have converged not towards free α-amino acids **VIIId**, but rather towards N-carbamoyl amino acids **Xd**. In the next chapter we shall examine how these N-carbamoyl amino acids could have led to peptides.

4.8 Conclusions

In this chapter, we have shown that the multicomponent system $\{H_2O, HCN, R^1R^2CO, NH_3, R^3NH_2, R^3R^4NH, CO_2, B(OH)_4{}^-\}$ could have been the origin of the α-amino acids and derivatives discovered in meteorites and formed on the primitive Earth.

The formation mechanisms of these derivatives have been exhaustively examined, and the Strecker and Bücherer–Bergs reactions can be said to have played a major, even unique, role in the formation of these compounds. It was these reactions that led the above multi-component system, made up of essential prebiotic molecules, to evolve quickly, majoritarily and selectively into α-amino acids and N-carbamoyl amino acids.

Under prebiotic conditions, the formation rates of these compounds could have been controlled by the ammonia formation rate.

The involvement of cyanic acid, an essential prebiotic molecule, in addition to the above reactions, in a subsequent slower step, could have led to N-carbamoyl amino acids alone.

the primitive atmosphere (Yamagata and Mohri 1982) and is detected as its trimer (cyanuric acid) in meteorites (Hayatsu et al. 1975).

[15] This additional example confirms (were it still necessary) the absolute necessity of studying the dynamics of potentially prebiotic systems in order to understand their chemical evolution.

The knowledge acquired about these reactions enables us to envisage kinetic modelling of the whole system, the only possible way to understand the respective contributions of all the forces involved.

Acknowledgement

For their fruitful collaboration to this work, we are indebted to our colleagues from Montpellier: Hélène Collet, Jacques Taillades, Laurence Garrel, Robert Pascal, Alain Rousset, Jean-Christophe Rossi, Hervé Cottet, Jean-Philippe Biron, Olivier Lagrille, Raphaël Plasson, and Grégoire Danger, as well as to Hervé Martin from the laboratory Magmas & Volcans at the University of Clermont-Ferrand (France), and to Michel Dobrijevic from the Observatoire Aquitain des Sciences de l'Univers (Bordeaux, France).

We are also grateful to:

The Centre National de la Recherche Scientifique (CNRS), especially its Chemical Science Department and the Institut National des Sciences de l'Univers (INSU); the Exobiology Research Group of the Centre National d'Etudes Spatiales (CNES); and the European Community through the COST D27 action (prebiotic chemistry), for their support.

The companies Aventis Animal Nutrition (France) and Degussa (Germany) for financial supports.

Professor Dr. Jacques Reisse from the Université Libre de Bruxelles (Belgium), and Dr. Muriel Gargaud from the Observatoire Aquitain des Sciences de l'Univers (Bordeaux, France) for many helpful discussions.

References

Bada J, Miller SL (1968). Ammonium ion concentration in the primitive ocean. *Science* **159**: 423–425.

Botta O, Glavin DP, Kminek G, Bada J (2002). Relative amino acid concentrations as a signature for parent body processes of carbonaceous chondrites. *Orig. Life Evol. Biosphere* **32**: 143–163.

Carlson RW (1996). Where has all the old crust gone? *Nature* **379**: 581–582.

Chyba CF, Sagan C (1992). Endogenous production, exogenous delivery and impact-shock synthesis of organic molecules: an inventory for the origins of life. *Nature* **355**: 125–132.

Clayton RN, Mayeda TK (1984). The oxygen isotope record in Murchison and other carbonaceous chondrites. *Earth Planet. Sci. Lett.* **67**: 151–161.

Commeyras A, Taillades J, Mion L, Pascal R, Lasperas M, Rousset A (1976). Procédé d'hydrolyse catalytique chimique d'alpha-aminonitrile ou de leurs sels. French Patent nr 76 365 20.

Commeyras A, Taillades J, Collet H, Boiteau L, Pascal R, Vandenabeele-Trambouze O, Pascal R, Rousset A, Garrel L, Rossi JC, Cottet H, Biron JP, Lagrille O, Plasson R, Souaid E, Selsis F, Dobrijevic M (2003). Approche dynamique de la synthèse des peptides et de leurs précurseurs sur la Terre primitive. In: *Les Traces du Vivant.*

Gargaud M, Despois D, Parisot JP (eds.), Presses Universitaires de Bordeaux, (Chap. 5) pp 115–162 (see especially the appendix).
Cooper GW, Cronin JR (1995). Linear and cyclic aliphatic carboxamides of the Murchison meteorite: hydrolyzable derivatives of amino acids and other carboxylic acids. *Geochim. Cosmochim. Acta* **59**: 1003–1015.
Cooper GW, Kimmich N, Beslisle W, Sarinana J, Brabham K, Garrel L (2001). Carbonaceous meteorites as a source of sugar-related organic compounds for the early Earth. *Nature* **414**: 879–883.
Cronin JR, Pizzarello S (1983). Amino acids in meteorite. *Adv. Space Res.* **3**: 5–18.
Cronin JR, Chang S (1993). Organic Matter In Meteorites: Molecular and Isotopic Analyses of the Murchison Meteorite. In: *The Chemistry of Life's Origins,* Greenberg JM et al. (eds.), Kluwer Academic Publishers, Dordrecht, The Netherlands, pp 209–258.
Cronin JR, Pizzarello S (1997). Enantiomeric excesses in meteoritic amino acids. *Science* **14**, 275(5302): 951–955.
Flores JJ, Bonner WA, Massey GA (1977). Asymmetric photolysis of (RS)-leucine with circularly polarized ultraviolet light. *J. Am. Chem. Soc.* **99**: 3622–3625.
Hayatsu R, Studier MH, Moore LP, Anders E (1975). Purines and trieazines in the Murchison meteorite. *Geochim. Cosmochim. Acta* **39**: 471–488.
Jorissen A, Cerf C (2002). Asymmetric photoreactions as the origin of biomolecular homochirality: a critical review. *Orig. Life Evol. Biosphere* **32**: 129–142.
Jungclaus GA, Yuen GU, Moore CB (1976). Evidence for the presence of low molecular weight alcools and carbonyl compounds in the Murchison météorite. *Meteoritics* **11**: 231–237.
Kasting JF, Ackerman TP (1986). Climatic consequences of very high carbon dioxide levels in the Earth's early atmosphere. *Science* **234**: 1383–1985.
Ksander G, Bold G, Lattmann R, Lehmann C, Früh T, Xiang Y, Inomata K, Buser H, Schreiber J, Zass E, Eschenmoser A (1987). Chemie der α-aminonitrile. *Helv. Chim. Acta* **70**: 1115–1172.
Lerner NR (1997). Influence of Allende minerals on deuterium retention of products of the Strecker synthesis. *Geochim. Cosmochim. Acta* **61**: 4885–4893.
Maas R, McCulloch MT (1991). The provenance of Archean clastic metasediments in the Narryer Gneiss complex, western Australia: trace elements geochemistry, Nd isotopes, and U-Pb ages for detrital zircons. *Geochim. Cosmochim. Acta* **55**: 1915–1932.
Martin H (1986). Effect of steeper Archean geothermal gradient on geochemistry of subduction-zone magmas. *Geology* **14**: 753–756.
Martin H (1987). Petrogenesis of Archaean trondhjemites, tonalites and granodiorites from eastern Finland: major and trace element geochemistry. *J. Petrology* **28**: 921–953.
Maurette M (1995). Were micrometeorites a source of prebiotic molecules on the early Earth? *Adv. Space Res.* **15**: 113–126.
Maurette M (1998). Micrometeorites on the early Earth. In: *The Molecular Origin of Life: Assembling Pieces of the Puzzle.* Brack A (ed.), Cambridge University Press, Cambridge, UK, pp 147–186.
Maurette M (2001). La matière extraterrestre primitive et les mystères de nos origines. In: *L'environnement de la Terre primitive.* Gargaud M, Despois D, Parisot JP (eds.), Presses Universitaires de Bordeaux, Bordeaux, Fr, pp 99–127.

Miller SL (1953). Production of aminoacids under possible primitive Earth conditions. *Science* **117**: 528–529.

Miller SL (1957). The mecanism of synthesis of amino acids by electric discharges. *Biochim. Biophys. Acta* **23**: 480–489.

Miller SL (1998). The endogenous synthesis of organic compounds. In: *The Molecular Origins of Life*. Brack A (ed.). Cambridge University Press, Cambridge, UK, pp 59–85.

Mojzsis SJ, Krishnamurthy R, Arrhenius G (1999). Before RNA and after: Geophysical and geochemical constraints on molecular evolution. In: *The RNA Word (second edition)*. Gesteland RF, Cech TR, Athins JF (eds.), Cold Spring Harbor University Press, NY, USA, pp 1–47.

Mojzsis SJ, Harrison MT, Pidgeon RT (2001). Oxygen-isotope evidence from ancient zircons for liquid water at the Earth's surface 4,300 Myr ago. *Nature* **409**: 178–181.

Nagata Y (1999). D-amino acids in nature. In: *Advances in BioChirality*. Pályi G, Zucchi C, Caglioti L (eds.), Elsevier, Amsterdam , The Netherlands, pp 271–283.

Navarro-Gonzalez R, Molina MJ, Molina LT (1998). Nitrogen fixation by volcanic lightning in the early Earth. *Geophys. Res. Lett.* **25**: 3123–3126.

Navarro-Gonzalez R, McKay CP, Nna Mvondo D (2001). A possible nitrogen crisis for Archaean life due to reduced nitrogen fixation by lightning. *Nature* **412**: 61–64.

Nishino H, Kosaka A, Hembury G, Shitomi H, Onuky H, Inoue Y (2001). Mechanism of pH-Dependent Photolysis of Aliphatic Amino Acids and Enantiomeric Enrichment of Racemic Leucine by Circularly Polarized Light. *Organic Lett.* **3**: 921–924.

Oró J (1961). Amino-acid synthesis from hydrogen cyanide under possible primitive Earth conditions. *Nature* **190**: 389–390.

Owen T, Cess RD, Ramanathan V (1979). Enhanced carbon dioxide greenhouse to compensate for reduced solar luminosity on early Earth. *Nature* **277**: 640–642.

Pascal R (2003). Catalysis by Induced Intramolecularity: What Can Be Learned by Mimicking Enzymes with Carbonyl Compounds that Covalently Bind Substrates? *Eur. J. Org. Chem*, pp 1813–1824.

Pascal R, Taillades J, Commeyras A (1978). Systèmes de Strecker et Apparentés. X. Décomposition et hydratation en milieu aqueux basique des α-aminonitriles secondaires. Processus d'hydratation autocatalytique et catalyse par l'acétone. *Tetrahedron* **34**: 2275–2281.

Pascal R, Taillades J, Commeyras A (1980). Systèmes de Strecker et Apparentés. XII. Catalyse par les aldéhydes de l'hydratation intramoléculaire des α-aminonitriles. *Tetrahedron* **36**: 2999–3008.

Peltzer ET, Bada JL, Schlesinger G, Miller SL (1984). The chemical conditions on the parent body of the Murchison meteorite: some conclusions based on amino, hydroxy, and dicarboxylic acids. *Adv. Space Res.* **4**: 69–74.

Pinto JP, Gladstone GR, Yung YL (1980). Photochemical production of formaldehyde in Earth's primitive atmosphere. *Science* **210**: 183–185.

Pizzarello S, Cronin JR (2000). Non-racemic amino acids in the Murray and Murchison meteorites. *Geochim. Cosmochim. Acta* **64**: 329–338.

Pizzarello S, Cooper GW (2001). Molecular and chiral analyses of some protein amino acid derivatives in the Murchison and Murray meteorite. *Meteorit. Planet. Sci.* **36**: 897–909.

Prinn RG, Fegley B (1987). Bolide impacts, acid rain, and biospheric traumas at the Cretaceous-Tertiary boundary. *Earth Planet. Sci. Lett.* **83**: 1–4.

Rossi JC, Garrel L, Taillades J, Commeyras A (1996). Hydrolyse et oxydation d'α-aminonitriles en présence de solution aqueuse basique de H_2O_2. *CR Acad. Sci. Paris* **322**: 767–773.

Selsis F (2000). *Modèle d'évolution physico-chimique des atmosphères de planètes telluriques. Application à l'atmosphère primitive terrestre et aux planètes extrasolaires.* PhD thesis, Université Bordeaux 1 (France).

Selsis F, Parisot JP, Dobrijevic M, Toublanc D (1996). *Photochemical modeling of the primitive atmosphere of telluric planets.* 11th International Conference on the Origin of Life, Orléans (France).

Shimoyama A, Ogasawara R (2002). Dipeptides and Diketopiperazines in the Yamato-791198 and Murchison Carbonaceous Chondrites. *Orig. Life Evol. Biosphere* **32**: 165–179.

Strecker A (1850). *Liebigs Ann. Chem.* **75**: 27–51.

Summers DP (1999). Sources and sinks for ammonia and nitrite on the early Earth and the reaction of nitrite with ammonia. *Origins of Life and Evolution of the Biosphere* **29**: 33–46.

Summers DP, Chang S (1993). Prebiotic ammonia from reduction of nitrite by iron(II) on the early Earth. *Nature* **365**: 630–633.

Summers DP, Lerner NR (1998). Ammonia from iron (II) reduction of nitrite and the Strecker synthesis: do iron (II) and cyanide interfere with each other? *Origin of Life Evolution of the Biosphere* **28**: 1–11.

Taillades J, Beuzelin I, Garrel L, Tabacik V, Commeyras A (1998). N-carbamoyl-α-aminoacids rather than free α-aminoacids formation in the primitive hydrosphere: a novel proposal for the emergence of prebiotic peptides. *Orig. Life Evol. Biosphere* **28**: 61–77.

Taillades J, Brugidou J, Pascal R, Sola R, Mion L, Commeyras A (1986). Nouvelles voies de synthèse d'acides α-aminés. *L'Actualité Chimique*, pp 13–20.

Taillades J, Boiteau L, Beuzelin I, Lagrille O, Biron JP, Vayaboury W, Vandenabeele-Trambouze O, Giani O, Commeyras A (2001). A pH-dependent cyanate reactivity model: application to preparative N-carbamoylation of amino acids. *Perkin Trans. 2*, pp 1247–1253.

Vervoort JD, Patchett PJ, Gehrels GE, Nutman AJ (1996). Constraints on early Earth differentiation from hafnium and neodymium isotopes. *Nature* **379**: 624–627.

Wilde SA, Valley JW, Peck WH, Graham CM (2001). Evidence from detrital zircons for the existence of continental crust and oceans on the Earth 4.4 Ga ago. *Nature* **409**: 175–178.

Yamagata Y, Mohri T (1982). Formation of cyanate and carbamyl phosphate by electric discharges of model primitive gas. *Orig. Life* **12**: 41–44.

5 Peptide Emergence, Evolution and Selection on the Primitive Earth
II. The *Primary Pump* Scenario

Auguste Commeyras, Laurent Boiteau, Odile Vandenabeele-Trambouze, Franck Selsis

Abstract We propose a dynamic scenario for the emergence and evolution of peptides on the primitive Earth, through a molecular engine (the primary pump), which works at ambient temperature and continuously generates, elongates and complexifies sequential peptides. This new scenario is based on a cyclic chemical reaction sequence that could have taken place on tidal beaches; it requires a buffered ocean, emerged land and a nitrosating atmosphere. We show that the primitive Earth during the Hadean may have satisfied all of these requirements.

This scenario is not necessarily what actually happened, but it represents a global approach of peptide prebiotic synthesis, and most of its parts are accessible to experiment. As it develops, it may open up a gateway to the emergence of homochirality and the catalytic activities of peptides.

5.1 From N-carbarmoyl Amino Acids (CAA) to Peptides

5.1.1 Introduction

The hypothesis of the prebiotic RNA world (Gilbert 1986; Gesteland et al. 1999) is hampered by the infeasibility of producing these macromolecules without the aid of enzymatic catalysts or modern organic synthetic tools (Joyce and Orgel 1999). In the absence of any satisfactory scenario to explain the origin of peptides, the coemergence of RNA and peptides as suggested by Kauffman (1993) or De Duve (1998), remains itself a fragile hypothesis. The prebiotic emergence of peptides meets two major obstacles: the activation of amino acids, then the possibility of forming long peptide chains under aqueous conditions, namely the only reasonable prebiotic solvent until now. Three different approaches have been developed in this direction.

The first approach focuses on peptide synthesis under extraterrestrial conditions (e.g. interstellar ice grains), through the analysis of meteoritic content as the only available record. This approach remains disappointing: only the Gly-Gly dipeptide has actually been detected in the Murchison and Yamoto-791198 meteorites (Shimoyama and Ogasawara 2002). Experimental models of interstellar ice do not show evidence of peptide formation.[1]

[1] In our opinion, the formation of peptides in an extraterrestrial ice analogue claimed by Munoz Caro et al. (2002) results from a hasty interpretation of predictable results.

The second consists in studying the behaviour of amino acids at high temperature. At 100–180 °C racemic compounds are obtained with insufficient evidence for the presence of peptide bonds (Rohlfing 1976; Fox and Dose 1977). Wet/dry cycles at 25–100 °C, in the presence or absence of clays or silicates, lead to dipeptides and traces of racemic trimers to pentamers (Lahav et al. 1978; Rode et al. 1997). At 100 °C in the presence of NiS/FeS and thiols to simulate volcanic zones or undersea vents, only dipeptides and traces of racemic tripeptides are detected (Huber and Wächtershäuser 1998). This approach is therefore highly contested (Bengston and Edstrom 1999).

The third approach investigates activated α-amino acids (esters, thioesters, N-carboxy anhydrides) or activating agents (carbonyl diimidazoles or carbodiimides) with amino acids. It is more encouraging since it effectively allows for oligocondensation under smooth, nonracemising conditions, whether in homogenous aqueous solution, on the surface of rocks, (Paecht-Horowitz and Eirich 1988; Orgel 1989, 1998; Bohler et al. 1996; Weber 1998), or in double-layer membranes (Luisi et al. 2000); however, it has not yet led to the expected *sequential* peptides[2] (Luisi 2000). For this approach to be credible we need to understand how activated molecules, which are highly reactive and sensitive to hydrolysis, could have been continuously formed on the primitive Earth. The relevance to prebiotic chemistry of activating agents such as carbonyl diimidazole or dicyclohexyl carbodiimide, or of amino acid N-carboxy anhydrides (NCA) remains contested (Liu and Orgel 1998; Blair and Bonner 1980, 1981).

Therefore the question of the prebiotic formation of long peptides remains open. We shall address this question by investigating a cyclic reaction scheme occurring in an open system out of equilibrium. The dynamics of such system (which is an important feature of living organisms) is very likely to result in self-organisation, as highlighted by, e.g. Nicolis and Prigogine (1977). It has been the object of numerous, mostly theoretical investigations in many fields – including the origins of life. Meanwhile, very few experimental studies have been published on the behaviour of such open chemical systems, despite their fundamental interest. Investigations (including experimental) on the prebiotic emergence and (self-) organisation of complex chemical reaction schemes, including their dynamical aspects, seems therefore important.

In the previous chapter we showed that the conditions that existed on the primitive Earth could have selectively led to N-carbamoyl amino acids (CAA), which have, however, been considered for long as an evolutionary cul-de-sac under prebiotic conditions. In this chapter we shall examine a possible activation pathway from CAA to sequential peptides through amino-acid N-carboxy anhy-

---

In such experiments amino acids identified after acidic hydrolysis (6 N HCl, 120 °C for 6 h) probably do not actually originate from the hydrolysis of peptides, but rather of α-amino nitriles, namely the Strecker precursors of observed α-amino acids.

[2] In this chapter, sequential peptides are referred to as peptides made of different amino-acid residues, (of optionally known sequence), oppositely to homopeptides in which all residues are identical.

drides (NCA), based on a dynamic process, that might also have contributed to the emergence of homochirality and of catalytic activity. Our attention will be focused on the relevance of the required conditions with respect to the primitive Earth environment.

5.1.2 The Primary Pump

Our approach of the prebiotic emergence of peptides is based on a cycle of chemical reactions forming the basis of a molecular engine, called by us the primary pump and shown in Fig. 5.1 (Commeyras et al. 2001, 2002).

- **First cycle.**

The first step **(1)** is the formation of *N*-carbamoyl amino acids (CAA) either by reaction of free amino acids with cyanate, or in other ways (see previous chapter). The second step **(2)** is the concentration of these CAA by water evaporation. This could have taken place on the shores of Hadean continents, at

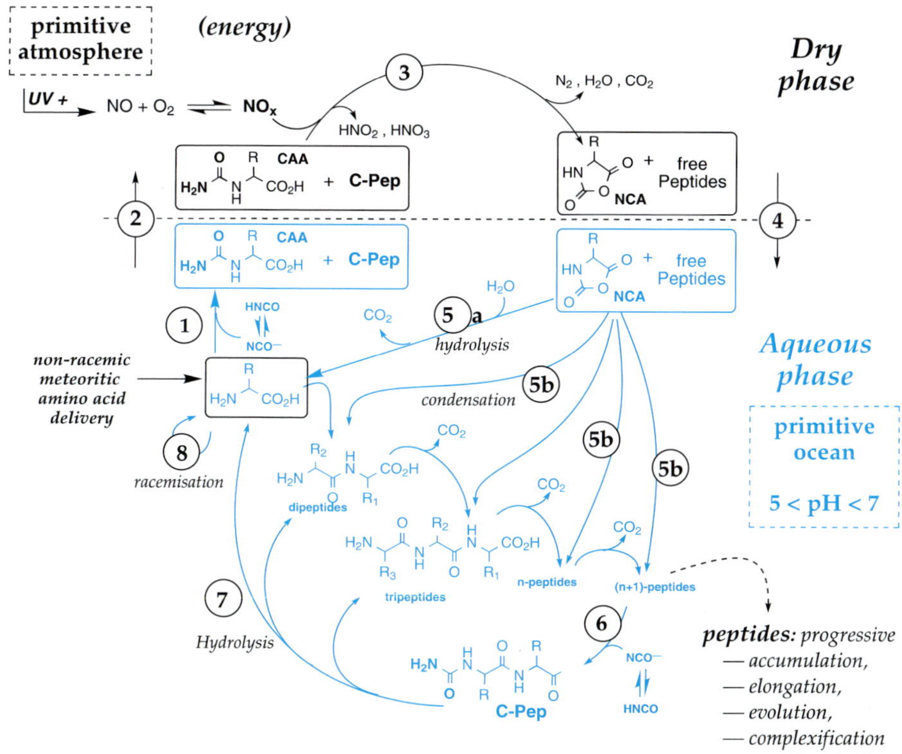

Fig. 5.1. The primary pump: a molecular engine leading to the production of evolutionary sequential peptides, starting from CAA and from NCO^-, NO, O_2, HCO_3Na, H_2O

low tide. In the dry phase, NO_x[3] from the atmosphere (see Sect. 5.2.3 and previous chapter) react with the CAA (nitrosation), leading to NCA through the cyclisation of unstable nitroso urea intermediates, step **(3)**. This reaction is quantitative at ambient temperature, producing stoichiometric amounts of nitrogen, water and nitrous/nitric acid together with NCA. In such acidic medium NCA degradation is relatively slow; within a few hours they undergo hydrolysis ($t_{1/2} \approx 1\,\mathrm{h}$) into α-amino acids, thus being recycled. At step **(4)** the rising tide quickly puts NCA into contact with ocean water (this is also conceivable for tidal pools or lagoons), the pH of which (5–7) allows peptide formation from NCA in competition with their hydrolysis, step **(5)**. This step is complete within a few minutes. Afterwards, the so-produced peptides can react with cyanate to give *N*-carbamoyl peptides, step **(6)**: this reaction occurs as easily as amino-acid *N*-carbamoylation. Over longer time in the hydrosphere, peptides and *N*-carbamoyl peptides undergo slow hydrolysis, step **(7)**: this leads to shorter peptides, eventually down to α-amino acids. Meanwhile both undergo *N*-carbamoylation.

- **Next cycles.**

The reaction of NCA with *N*-carbamoyl peptides is impossible, thus theoretically preventing peptide elongation in subsequent drying/wetting cycles. However, when *N*-carbamoyl peptides arrive on the shore and are dried, they are immediately unprotected by reaction with NO_x (Collet et al. 1999; Lagrille et al. 2002), step **(3)**. This reaction occurs under the same conditions as NCA formation from CAA. Therefore at the next high tide, unprotected peptides are elongated by reaction with newly formed NCA.

Such successive elongation/partial hydrolysis cycles may have allowed peptide chemical evolution. The intrinsic rate of peptide formation is higher than that of peptide hydrolysis, which should be favourable to peptide accumulation; however, actual rates are highly dependent on the reactant fluxes and, moreover, on the dry/wet alternation period.

When considering peptides, it is difficult not to associate homochirality with emergence, as hypothesized by George Wald (1957). It is well known that the extraterrestrial input of α-amino acids was not totally racemic, see for instance Pizzarello and Cronin (2000). If these nonracemic α-amino acids played a role in the synthesis of peptides, then their (low) enantiomeric excesses may have been amplified. Such amplification phenomena have been demonstrated by Blair and Bonner (1980), however, in organic solvents. If in addition, amino acids or CAA racemise faster than peptide-chain residues[4], then peptide prebiotic synthesis

[3] In this chapter NO_x is referred to as the result of mixing NO and O_2 in any ratio, which can equivalently be obtained in other ways, e.g. mixing NO and NO_2.

[4] In aqueous solution it is likely that free amino acids epimerise slower than peptide residues. However, the latter may epimerise slower than some amino-acid derivatives such as CAA (recent, unpublished results of our group are consistent with this possibility), or even slower than free amino acids/peptide end-residues if epimerisation catalysts are involved.

may have progressed towards homochirality. A great deal of work still needs to be done to test this hypothesis.

5.2 Environmental Requirements

Each step of the so-called primary pump has been studied carefully and can be considered as well known. Nevertheless, for the primary pump to produce peptides, all steps must be well-controlled. Under prebiotic conditions, is it conceivable that, spontaneously, the correct sequence of steps took place and gave peptides? What are the minimal requirements to fulfill such a goal? Without any doubt, oceans and emerged land are required and they must be in contact with an atmosphere containing nitrogen oxides. The time window for this scenario lies between -4.4 Gyr and -3.8 Gyr. The following discussion will consider the probability of these environmental requirements, with special emphasis on the pH of the aqueous phases, which is critical in our scenario.

5.2.1 Primitive Earth

The oldest terrestrial rocks recognized so far are the Acasta gneisses in Canada, dated at -4.030 ± 0.003 Gyr (Bowring and Williams 1999), which consist of a tonalite, trondhjemite and granodiorite suite (TTG) that is typical of the primitive Archaean continental crust. The oldest known terrestrial materials are zircon crystals, extracted very recently from Jack Hill metaquartzites (Australia) and dated at -4.404 ± 0.008 Gyr (Wilde et al. 2001). Their chemical characteristics demonstrate that they crystallized in TTG-like magmas, thus establishing the existence of a stable continental crust as early as 4.4 Gyr ago. In addition, hafnium and neodymium isotopic studies in both old zircons and early Archaean TTG (McCulloch and Bennet 1993; Wilde et al. 2001) show that, when these materials were generated, significant volumes of continental crust had already been extracted from the terrestrial mantle.[5] Based on these isotopic data, McCulloch and Bennet (1993) calculated that between 8 and 15% of the volume of the present-day continental crust had already been extracted prior to -4.0 Gyr.

Petrogenetic studies performed on the Archaean TTG continental crust have shown that it was generated by partial melting of hydrous basalts (Martin 1986, 1987). Available water lowers the temperature of basalt solidus by several hundred degrees, and is therefore absolutely necessary for melting basalt and generating TTG. Basalts, generated at great depth by anhydrous melting of mantle peridotite, contain only very small amounts of water. Today they incorporate water through alteration by seawater, for instance in midocean-ridge systems. As an essential consequence, the existence of a Hadean continental crust implies

[5] Isotopic deviations $\varepsilon_{\mathrm{Hft}}$ (for Hf) and $\varepsilon_{\mathrm{Ndt}}$ (for Nd) were both found to be positive: $\varepsilon_{\mathrm{Hft}}$ (resp., $\varepsilon_{\mathrm{Ndt}}$) represents the difference between the isotopic ratio $^{176}\mathrm{Hf}/^{177}\mathrm{Hf}$ (resp., $^{143}\mathrm{Nd}/^{144}\mathrm{Nd}$) of a rock and that of chondrites, taken as the reference.

that huge volumes of liquid water must have been available on the surface of the primitive Earth. Mojzsis et al. (2001) and Wilde et al. (2001) have analyzed the oxygen isotopic composition of the 4.4- and 4.3-Gyr zircons: the $\delta^{18}O$ (5.4–15.0‰) clearly indicates that the source of the magma in which these zircons crystallized had already strongly reacted with a hydrosphere. This demonstrates that oceans existed as early as 4.4 Gyr ago on the Earth's surface; this conclusion is also supported by the fact that the Isua gneisses in Greenland (dated at 3.865 Gyr) consist of sediments accumulated in an aqueous environment.

The existence of Hadean beaches acted upon by ocean tides (the presence of the Moon at that time is now widely accepted) is therefore shown to be a consistent hypothesis.

5.2.2 Primitive Atmosphere

The abundance of atmospheric nitrogen oxides (NO_x) is a limiting factor for the efficiency of nitrosation and thus for the production of peptides through the primary pump. Unfortunately, the nature (pressure, temperature, composition) of the prebiotic atmosphere remains obscure in the absence of geological archives, preventing us from addressing quantitatively this point. Still, we can discuss the availability of NO_x considering the few constraints on the primitive terrestrial environment.

The formation of NO_x is considered to occur through two successive steps: (1) formation of nitric oxide NO through activation of N_2; (2) formation of higher NO_x species by reaction of NO with atomic or molecular oxygen. The major oxygen source is photodissociation of CO_2, with a mixing ratio highly dependent on CO_2 pressure (Rosenqvist and Chassefière 1995; Selsis et al. 2002). In a 1-bar CO_2 atmosphere, O_2 pressures up to 20 mbar can be maintained at a steady state in this way, though this would depend upon there being no O_2 consumption at the Earth's surface. Such an uptake would almost certainly have occurred, through the oxidation of rocks and reducing volcanic gases (Selsis et al. 2002), the amount of which is unknown, however. While it has been suggested that the early upper mantle was more reducing than it is today (Pavlov et al. 2000), its present oxidation state could have been reached as early as 4.3 Gyr ago (Tolstikhin and Marty 1998). We performed computer simulation of atmosphere models under photochemical activation, using the computer code PHOEBE (photochemistry for exobiology and exoplanets, Selsis et al. 2002), starting from $N_2 + CO_2$ background atmospheres including water vapour at thermodynamic equilibrium. For initial mixtures of 0.8 bar of N_2 plus 0.2 bar (A) or 3.2 bar (B) of CO_2, O_2 mixing ratios of 10^{-4} and 10^{-3} were found for configurations A and B, respectively, if no O_2 uptake is taken in account. Supposing this O_2 uptake to be at the same level as today, the calculated mixing ratio drops to 10^{-6} and 5×10^{-5} for configurations A and B, respectively. An example of such simulation (configuration A) taking in account other NO sources (see infra) is shown in Fig. 5.2.

Concerning NO formation, the partial pressure of N_2 is thought to have reached a level close to its present value (0.8 bar) earlier than 4.3 Gyr (Tolstikhin

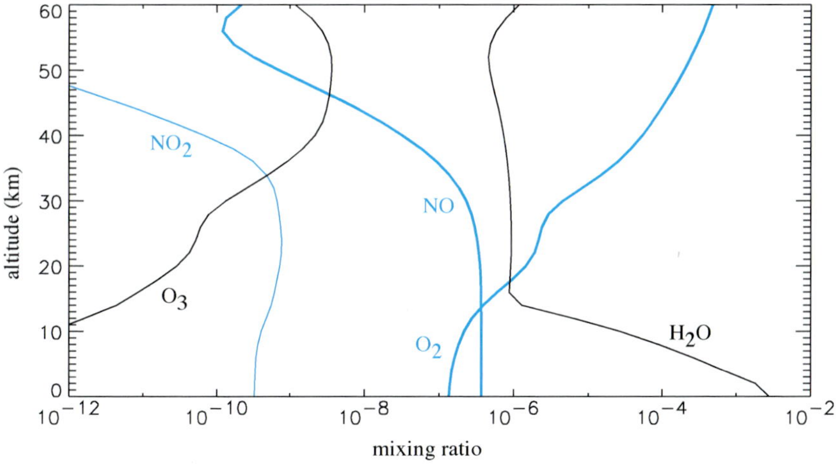

Fig. 5.2. Simulation of the primitive Earth atmosphere composition with the numerical code PHOEBE. The 1-bar background atmosphere of N_2 (80%) + CO_2 (20%) includes NO production by volcanic lightning (Navarro-Gonzalez et al. 1998)

and Marty 1998). A high CO_2 partial pressure (p_{CO_2}) is usually assumed, although a recent study (Sleep and Zahnle 2001) claims that the formation of carbonates was so efficient on the Early Earth that CO_2-dominated atmospheres have occurred only for geologically brief periods following giant impacts. In our case, we assume p_{CO_2} to be at least high enough to insure a mean surface temperature above 0 °C. Depending on the period considered (from 4.3 to 3.6 Gyr ago), this implies a minimum p_{CO_2} being in the range 100–400 mbar (Caldeira and Kasting 1992). If p_{CO_2} was lower, two cases can be considered:

1. No other greenhouse gas warmed the surface and the Earth was globally frozen: surface liquid water was then restricted to the vicinity of active volcanic zones, or at the surface of ice near the equator and during the warmest day hours. The efficiency of the primary pump in such conditions has not been addressed yet.
2. The surface is warmed above 0 °C by CH_4 produced in hydrothermal systems (Pavlov et al. 2000): In that situation the presence of abundant reduced species in the atmosphere considerably lowers the photochemical lifetime of the NO_x and inhibits the nitrosation.

Thus, a prebiotic atmosphere for which nitrosation cannot be a priori ruled out contains at least 100 mbar of CO_2 but no abiotic methane. This picture is compatible with our knowledge of the primitive Earth as recently reviewed by Kasting and Catling (2003).

In the abiotic early atmosphere, NO_x have to be produced from N_2 (and not from the biogenic N_2O as it is now). The N–N bond energy is particularly high and consequently, short wavelengths are required for the photolysis of N_2 ($\lambda <$

100 nm). Under such conditions the photochemistry initiated by solar UV does not produce significant levels of NO_x (Selsis 2000). Such a conclusion is valid for the present-day situation, but what about the Hadean era? Indeed, observations of young Sun-like stars (Guinan and Ribas 2000) leads to the conclusion that the soft X-ray and hard UV (XUV: 1–100 nm) emission of the young Sun was very probably much higher than its present value. The luminosity in the XUV range (L_X) varies as $L_X(t)/L_{X,Sun} = 6.16 \times t^{-1.19}$ where t is the age of the star in Gyr, and $L_{X,Sun}$ the present solar XUV luminosity. The photochemistry initiated by this intense XUV flux on the early Earth still has to be explored but it surely led to a more efficient production of NO_x at high and middle altitudes. However, even under such irradiation conditions and in the presence of an atmosphere without reduced species such as CH_4 and H_2, the surface mixing ratio of cumulated NO_x species do not reach 10^{-9}.

Another, more efficient way to produce NO_x consists in shocking up to more than 1500 K a gas mixture containing N_2 and oxygen-bearing species (O_2, CO_2, H_2O) and cooling it as rapidly as possible. In this case, the oxygen molecules, or atoms generated at high temperature, react with N_2 producing NO. Such chemistry occurs during lightning, meteors and impacts.

The efficiency of these processes in our present O_2-rich atmosphere is higher than in an anoxic atmosphere where the reaction $N + O_2 \rightarrow NO + O$ does not occur. In the absence of O_2, the efficiency depends on the ratio CO_2/N_2. When the ratio $[N_2]/[CO_2]$ is above 5, the formation of NO (through the reaction $N + CO_2 \rightarrow NO + CO$) becomes inefficient because of the fast recombination of N atoms into N_2 through the three-body reaction $2N + N_2 \rightarrow 2N_2$. But when the CO_2 and N_2 abundances are comparable, the energy yield of the NO production is only 10 times lower than in an O_2-rich atmosphere (Navarro-Gonzalez et al. 2001).

Navarro-Gonzalez et al. (1998, 2001) studied both experimentally and theoretically the production of NO by lightning in CO_2-rich atmospheres. Considering the early volcanic activity and the number of lightning strikes associated to volcanic plumes, these authors estimated the continuous production of NO on Early Earth and found 3×10^9–3×10^{10} mol y^{-1} for an atmosphere with $[CO_2]/[N_2] > 0.3$. As an upper limit, if all of these radicals could be used by the primary pump, this rate would allow a NCA production of 10^{-9} mol cm^{-2} s^{-1}. Of course, a fraction of the produced NO is reduced by volcanic gases, or reacts to form HNO_3 that is rained out through acid rains. This continuous production occurring in the lower atmosphere could maintain an average abundance of about 10^{-6}, 100 times higher than in the present-day atmosphere.

In the case of meteors, few quantitative studies have been done (see Jenniskens et al. 2000, for the meteor-induced chemistry in a CO_2-rich atmosphere). NO is produced in the plasma of the meteor trails in which air temperature is briefly raised to about 5000 K: NO and NO_2 lines have thus been observed in Leonid meteor spectra (see, for instance, Carbary et al. 2003). The NO_x produced by these small-sized bodies is generated in the upper atmosphere, above

50 km and, even in the absence of detailed modelling, this source does not seem to contribute to the delivery of NO_x to the lower atmosphere.

More studies are available for larger impacts because of the environmental consequences they may induce. Prinn and Fegley (1987) modelled the K/T impact[6] of a 5×10^{14}-kg asteroid, or a 1.25×10^{16}-kg comet. They suggested that the explosion could have produced from 5×10^{14} to 1.2×10^{17} moles of NO, explaining by the same way the acidic content of the worldwide K/T geological layer. However, Zahnle et al. (1988) showed in a short note that Prinn and Fegley overestimated these numbers by considering an instantaneous cooling. Later, Turco et al. (1982) estimated that the Tungunska bolid (probably a comet) that disrupted in the atmosphere in 1908 generated about 3×10^{12} moles of NO between 10 and 50 km of altitude. In an early, CO_2-dominated atmosphere, this would produce in a single event 3×10^{11} moles of NO, the same amount as the whole production of NO by volcanic lightning. Again, most of the NO is generated in the middle atmosphere and rained out in the form of nitric acid. Further modeling would be required to know how much NO reaches the surface. Indeed, this source may be very efficient and nearly continuous as Tunguska-like events may have occurred 0.1–1 times a year on the prebiotic Earth. In a review paper addressing the environmental consequence of large impacts, Toon et al. (1994) inferred a power law giving the mass production of NO as a function of the impact energy.

We summarized these results in Fig. 5.3, using Navarro-Gonzalez et al. (2001) for scaling the result to a CO_2-dominated atmosphere. We applied the standard evolution of the impact rate: an exponential decrease with a timescale of 144 Myr (Chyba 1990) until –3.5 Gyr ago. The impact flux 3.5 Gyr ago, equal to the present-day one, is taken from Morrison et al. (1994). We assume that the impactors are asteroids with a typical density of $3\,g\,cm^{-3}$, a mass distribution that is constant in time, and an average impact velocity of $20\,km\,s^{-1}$.

For impact energies above 10^{21} J (equivalent to an impactor diameter of 3 km), the induced NO is well mixed down to the surface. For smaller bodies, 40 m–3 km, the abundance of NO peaks up to 10^{-3} in the middle atmosphere and diffuses to the surface where it can reach 10^{-6} as an upper limit. Impactors with diameter below 40 m do not deliver efficiently NO to the surface. For the largest impacts (similar to those that formed the main Lunar Basins around 3.9 Gyr ago) the entire atmosphere is heated to $T > 1500$ K and may contain 0.1% NO, (10^{19} moles). Because such hot conditions obviously frustrate organic chemistry, we did not consider impacts with energy above 10^{25} J (equivalent to a 50-km impactor) as a source for "useful NO". Converted in mol yr^{-1}, the production of NO by impactors in the 3–50 km range is quite low ($< 1.3 \times 10^7\,mol\,yr^{-1}$, see Fig. 5.3), but every 10^4–10^5 years, the abundance of NO reaches 10^{-3} and

[6] Of a ca. 10-km diameter object, having occurred 65 Myr ago at the Cretaceous/Tertiary era (K/T) boundary, and plausibly responsible for the extinction of the dinosaurs. Such impact energy is estimated between 10^{22} and 10^{24} J (equivalent to 0.25 to 25 billion Hiroshima bombs).

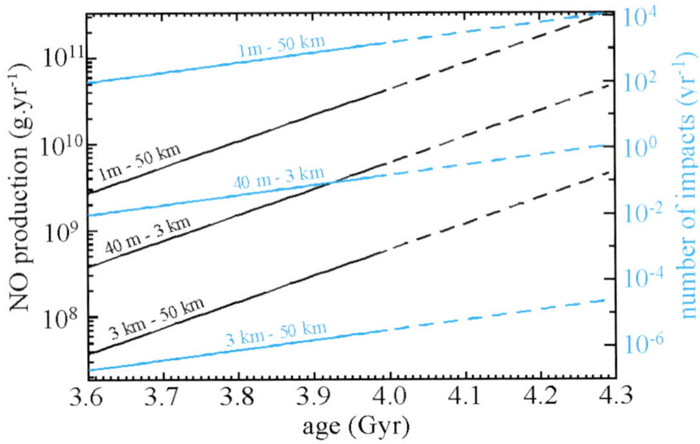

Fig. 5.3. Nitric-oxide production by impacts. *Black lines*: impact frequency as a function of age, for three diameter ranges of impactors. *Blue lines*: corresponding production of nitric oxide. *Dashed lines*: extrapolation beyond 4.0 Gyr (age of oldest available Lunar cratering records)

decreases with a typical timescale of 1 yr (much shorter than the typical interval between two impacts of this size).

In summary, the continuous source of NO on the early Earth was due to volcanic lightning. In the most favourable conditions (a negligible abiotic production of methane and a moderate release of reduced volcanic species) it could have maintained an average abundance of NO around 10^{-6}, this level being higher or lower, by at least one order of magnitude, depending on the proximity (in space and time) to volcanic plumes. Violent and stochastic impacts resulted in higher abundance (up to 10^{-3}) decreasing slowly for months or years. From the present knowledge, we can thus conclude that, if the primary pump did contribute to the production of peptides on the prebiotic Earth, it may have been fed with NO by single and violent events (impacts or volcanic eruptions) rather than continuously. Complementary experimentation and modelling will be necessary anyway in the future to substantiate this very hypothetical scenario.

5.2.3 About the pH of Primitive Oceans

5.2.3.1 Requirements from the Primary Pump

The pH of the aqueous phase (the primitive ocean) is a critical parameter of our scenario, especially for steps **(1)** and **(5)**. Concerning step **(1)**, the *N*-carbamoylation of amino acids occurs at any pH comprised between 5 and 12, with a maximum rate at pH 7–8. CAA are stable over the whole pH 5–12 range. Concerning step **(5)**, the polycondensation of NCA in water has been studied from the middle 1950s. (Bartlett and Dittmer 1957; Bartlett and Jones 1957). Below pH 4, only NCA hydrolysis occurs. Above pH 4, there is a competition between NCA hydrolysis and NCA condensation on amino acids or peptides, with increasing condensation yields while increasing the pH. In the pH range 4–6.5 this competition creates the most favourable conditions for an efficient kinetic selectivity in peptide formation, thus allowing the chemical evolution of the growing peptide pool. This is important especially for the selection of homochiral peptides.

5.2.3.2 Environmental Data.

With respect to these requirements, it is important to have an idea of the pH of the primitive ocean. Due to the presence of large amounts of CO_2 in the primitive Earth atmosphere, (Kasting 1993; Zahnle 1998), the early ocean was probably initially quite acidic (its pH being possibly as low as 3.5). The pH should then have gradually increased to its current value (8.2) through progressive extraction of alkaline materials (Na^+, K^+, Ca^{2+}, etc.) from reducing rocks.

Therefore the primary pump process could have started as soon as the pH of the ocean reached 4, and could have worked until it reached ca. 6.5. This pH window is estimated to have taken place during the Hadean. Moreover, the pH of the ocean may have remained around 6.3 (pK_A of carbonic acid) for a long time due to the buffer effect. More precise knowledge of this primitive-ocean pH chronology is necessary since it would allow the corresponding time windows to be refined.

5.2.3.3 pH Change During Water Evaporation (Step 2).

A possible drawback in the activation of CAA can be the acidification of the medium during evaporation: CAA are well known to convert into hydantoins under acidic conditions (usually pH < 2.5), what we did not observe in our experiments simulating prebiotic conditions. In fact the evaporation of an aqueous solution of pH 6.5 containing (bi)carbonate salts as the major components does not result into its acidification since carbonic acid will evaporate as well as CO_2 during the concentration process (in fact we even observed an increase in pH; ultimately an alkaline salt of the CAA is obtained). Therefore the possibility of hydantoin formation during step **(2)** appears very limited. Conversely, a part of the NO_x from the atmosphere will be necessary to neutralise alkali before the nitrosation takes place.

5.3 Investigation of the Primary Pump

In the above section we have seen that α-amino acids N-carboxy anhydrides (NCA), may have been synthesized in a fairly continuous way on the primitive

Earth. This lends credence to the argument for peptide synthesis through activated α-amino acids (see introduction) as it argues in favour of the prebiotic relevance of NCA, which are also the major monomers for the preparation of synthetic polypeptides from laboratory to industrial scale.

At the same time our approach acquires a dynamic dimension. It gives prebiotic significance to Blair and Bonner's results concerning the amplification of low enantiomeric excesses during polymerization of NCA, even though their results were obtained in nonaqueous solvents (Blair and Bonner 1980, 1981). It provides arguments for carrying out the necessary research to evaluate the potentialities of the primary pump. We investigated (see next Section) the different steps of the primary pump, first separately, focusing on steps **(1)**, **(3)** and **(5)**, then in a more integrated way.

5.3.1 Step-By-Step Experimental Investigation

5.3.1.1 Step (1): *N*-carbamoylation

Cyanate (NCO^-) irreversibly reacts with α-amino acids to produce CAA acids. Study of this reaction shows that it can occur in aqueous media over a very wide pH range: 5 to 13 (Taillades et al. 2001), with the rate law (5.1) involving the hydrolysis of cyanate (k_0), the catalysis by carbonates (k_1), the formation of urea (k_2) and the N-carbamoylation (k_3) of amino acids (AA) (t = total: acid + base forms).

$$v = \mathrm{d}[\mathrm{NCO}]^t/\mathrm{d}t = [\mathrm{NCO}]^t \times (k_0 + k_1 \times [\mathrm{CO_3}]^t + k_2 \times [\mathrm{NH_3}]^t + k_3 \times [\mathrm{AA}]^t). \quad (5.1)$$

At 50 °C and pH 6.5, the N-carbamoylation rate constants k_3 were measured: $1.7 \times 10^{-3}\,\mathrm{L\,mol^{-1}\,s^{-1}}$ for glycine, $1.3 \times 10^{-3}\,\mathrm{L\,mol^{-1}\,s^{-1}}$ for valine, and $3.5 \times 10^{-3}\,\mathrm{L\,mol^{-1}\,s^{-1}}$ for threonine. This reaction is possible only above pH 5, with almost quantitative yields at pH 7 for cyanate to α-amino acid ratios above 1.5.

In addition to its interest for the quantification of a future computer kinetic model, this reaction might allow refinement of the moment the primary pump start-up, namely when the pH of primitive oceans reached 5 (see Sect. 5.2.3).

5.3.1.2 Step (3): CAA Nitrosation

The gaseous mixture O_2/NO reacts on CAA even in the absence of water, as shown in Fig. 5.4. (Collet et al. 1996; Taillades et al. 1999). Photo (a) shows crystals of anhydrous N-carbamoyl valine. In the presence of an excess of anhydrous gaseous mixture O_2 + NO N-carbamoyl valine is quantitatively (according to NMR analysis) converted into valine-NCA at 20 °C within half an hour (photo c) (Collet et al. 1996). The middle photo (b) shows the surface of the crystal during the reaction. The bubbles that are forming are of nitrogen. The liquid consists of the water and nitrous acid produced by the reaction (step **(3)**). Even

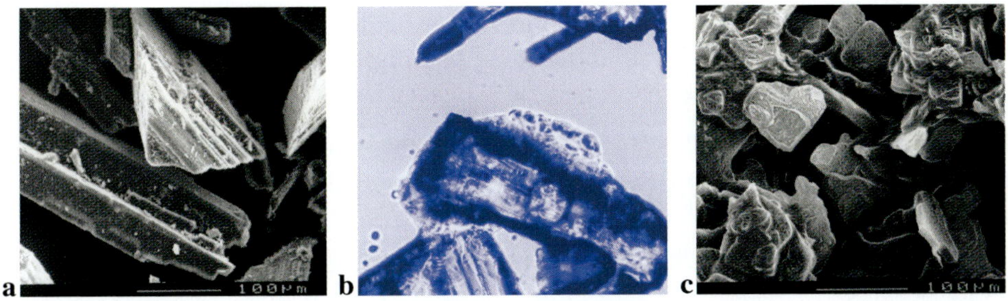

Fig. 5.4. (From left to right) (**a**) crystals of *N*-carbamoyl valine (SEM); (**b**) action of $NO + O_2$ (optical microscopy, false colours) (**c**): valine-NCA (SEM). Approximately same magnification for **a**–**c**

under initially anhydrous conditions, the solid CAA is completely converted into solid NCA (Taillades et al. 1999).

The reaction efficiency depends on the O_2/NO ratio. In the absence of O_2 there is no reaction. When the O_2/NO ratio is equal to, or greater than, 1 the reactivity is below its maximum, which is observed for an O_2/NO ratio of 0.25 (this ratio probably remained quite low during the Hadean). The NO_x gaseous mixture actually contains the species NO, O_2, N_2O_4 and N_2O_3 in equilibrium, the ratio of which depends on the O_2/NO mixing ratio. We evaluated the rate constants of CAA nitrosation in the solid-gas phase by N_2O_4, N_2O_3 (nitrosating agents present in the gaseous NO_x mixture) and also by nitrous acid HNO_2 (a side-product of both CAA nitrosation and N_2O_3/N_2O_4 hydrolysis). The values obtained at 25°C ($k_{N_2O_4} = 1.7\,s^{-1}, k_{N_2O_3} = 10^8\,s^{-1}$ and $k_{HNO_2} = 0.1\,s^{-1}$, respectively) show the very high reactivity of N_2O_3 (Lagrille 2001): even traces of this species in the primitive atmosphere might have been sufficient for activating CAA into NCA.

Further studies will be necessary to evaluate the working limits of the primary pump, in terms of, e.g. minimum NO_x/CAA ratio, or NO_x concentration in the atmosphere.

5.3.1.3 Step (5): Insights in Peptide Synthesis

According to George Wald's suggestions (Wald 1957) the α-helical secondary structure of a growing homochiral polypeptide chain might ensure the selection of α-amino acids of the same chirality as those already present in the chain during its lengthening, thus preserving and increasing the homochirality of the growing chain.

These suggestions were proved by Idelson and Blout (1958) and Lundberg and Doty (1957), by studying the polymerization (in dioxane) of the NCA. Stereoselectivity was negligible at the simple peptide level and much higher for the α-helix at higher degrees of polymerization. In the same paper Wald said "If one could grow such polymers in a reversible system in which synthesis was partly balanced by hydrolysis, the opportunity for selection would be greatly improved." Obviously, Wald was underlining the importance of water. Our results giving prebiotic status to NCA led us to carry out these studies again using water as solvent.

Oligocondensation of NCA in water: stereoselectivity.

NCA in aqueous solutions undergo two competitive reactions during step **(5)**, hydrolysis and
condensation (aminolysis), with different kinetic laws:

$$v_{\mathrm{hyd}} = k_{\mathrm{hyd}} \times [\mathrm{HO^-}] \times [\mathrm{NCA}], \tag{5.2}$$

and

$$v_{\mathrm{cond}} = k_{\mathrm{cond}} \times [\mathrm{R{-}NH_2}] \times [\mathrm{NCA}], \tag{5.3}$$

where R–NH_2 is the amine form of amino acid or peptide.

We reconsidered the work carried out by Bartlett (Bartlett and Dittmer 1957; Bartlett and Jones 1957), using capillary electrophoresis. With this technique, we directly measured the rate of NCA hydrolysis and of peptide formation (Plasson et al. 2002). Partial results for valine-NCA give $k_{\mathrm{hyd}} = 4.40 \times 10^{-4}\,\mathrm{s}^{-1}$ for hydrolysis, and $k_{\mathrm{LL}} = 1.65 \times 10^{-2}\,\mathrm{L\,mol^{-1}\,s^{-1}}$, and $k_{\mathrm{LD}} = 1.14 \times 10^{-2}\,\mathrm{L\,mol^{-1}\,s^{-1}}$ for the formation of Val_2 from L-Val-NCA reacting on L-Val and D-Val, respectively.

So, in water, the hydrolysis rate of the NCA is low compared to their condensation rate. The condensation rate constant k_{LL} is much greater than k_{LD}, and for the formation of dipeptide, the $k_{\mathrm{LL}}/k_{\mathrm{LD}}$ ratio is already much greater than 1. This ratio must be higher for the synthesis of larger peptides (octapeptides and above).

Fast Measurement of Peptide pK_A.

In water, the (pH-dependent) formation rate of peptides depends on their pK_A, which are, however, not very well known. In our group, Cottet et al. (personal communication, to be published) have developed a fast method for measuring the pK_A of oligopeptides using capillary electrophoresis (CE), with no need to isolate them from the mixture. Figure 5.5 illustrates the pk_A measurement of oligoglycines.

From this information, we deduce that, in aqueous solution, longer peptides (lower pK_A) react faster with NCA than the shorter ones (higher pK_A), assuming their intrinsic rate constants of coupling with NCA to be identical (Fig. 5.6). This last point needs to be confirmed.

On the basis of these experimental data, we are running computer simulation to investigate possible enantiomeric amplification of the peptide pool by the primary pump.

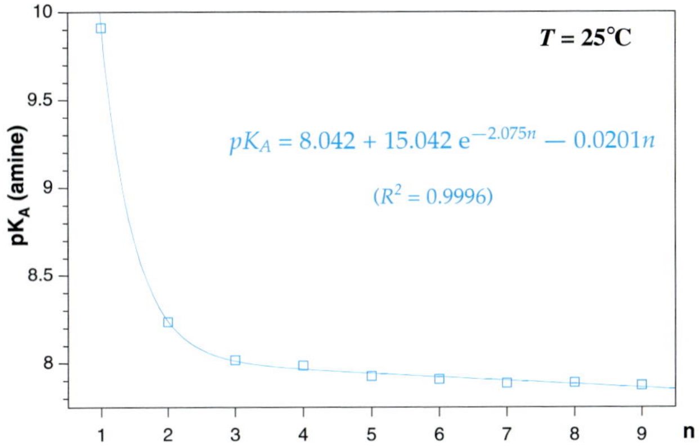

Fig. 5.5. pK_A of oligoglycines Gly_n (1–9 residues) as a function of n (Cottet, to be published)

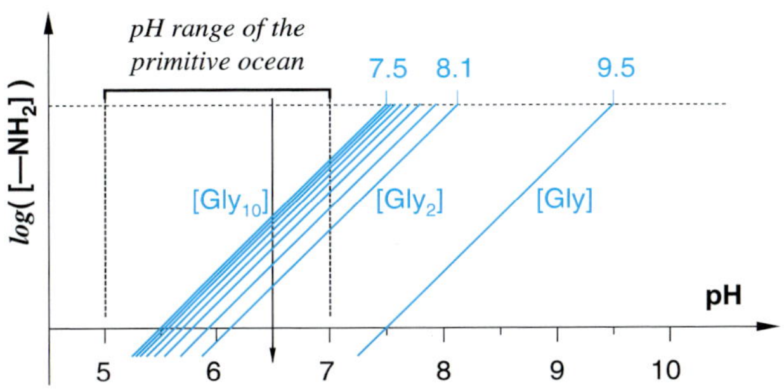

Fig. 5.6. In aqueous solution, for equal concentration of peptides, the concentration of the nonprotonated amino group of the longer peptides (lower pK_A) is higher than the concentration of the nonprotonated amino group of shorter peptides

5.3.2 Integrated Experimental Approach: Chemoselectivity

This physicochemical approach needs to be checked by synthesis techniques because, in water, other parameters are involved in the lengthening of peptide chains, and these parameters need to be understood.

We investigated the chemoselectivity of peptide synthesis under the conditions of the primary pump, by experimentally cycling its essential steps **(1–5)** several times ($NaHCO_3$ buffer 100 mM, pH 6.7) containing amino acids or their *N*-carbamoyl derivatives (Commeyras et al. 2002). Results using two amino acids valine (V) and glutamic acid (E) are shown in Fig. 5.7. Many peptides are formed but not all of them. Figure 5.8 shows the distribution of peptides formed in these experiments and shows also a very effective chemoselectivity during the elongation process.

With four amino acids (glycine, alanine, valine, glutamic acid) as starting materials, MALDI-TOF analysis shows that the length of the peptides increases with the cycle number, but insoluble peptides are formed. The insoluble fraction (separated after 7 cycles) is made up of peptides whose molecular weigh is distributed around 800 daltons, whereas the mass of the soluble compounds is higher than 2000, thus evidencing the effect of additional selectivity factors, physical rather than chemical.

The stimulating aspect of this point is that a stress (here possibly due to hydrophobicity of peptide chains) separates the insoluble peptides. The only

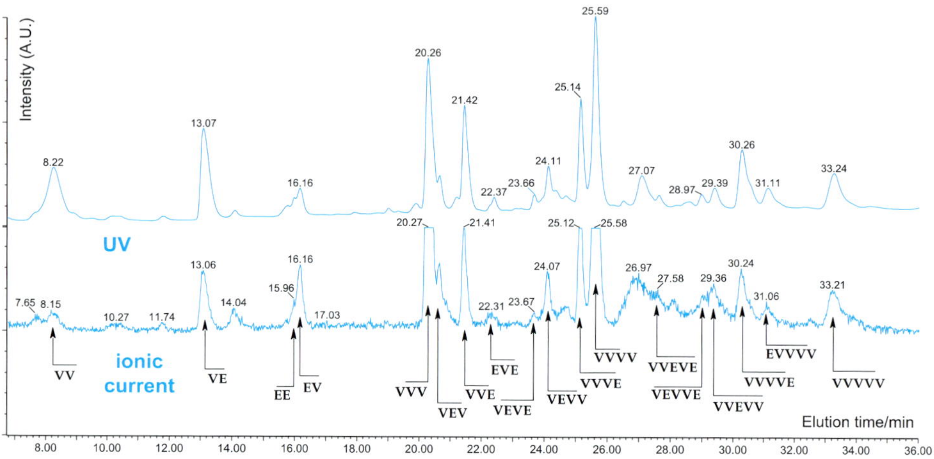

Fig. 5.7. Peptide elongation and selection after 5 cycles under primary pump conditions, running with 2 amino acids valine (V) and glutamic acid (E). Analysis by HPLC/MS (ESI/Q-TOF)

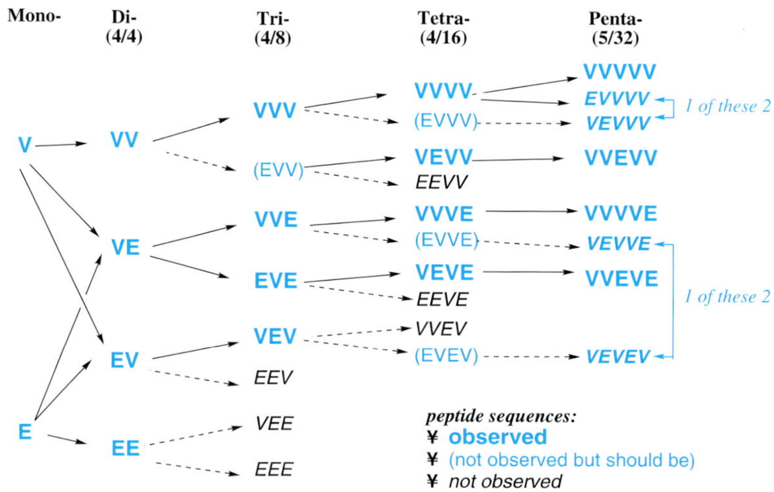

Fig. 5.8. Observed chemoselectivity during the formation of peptides under primary pump conditions, from 2 amino acids (V, E) as starting materials

peptides capable of evolving in the short term are the soluble ones, while the others – inappropriately adapted – are "frozen" as formed, and probably slowly hydrolyzed.

In conclusion we can consider that in aqueous solution, the above-described selectivities could have considerably decreased the number of possible peptides formed ($2n^x$ with $n > 20$ and $x > 100$), and have thus made this synthesis realistic. Detailed analysis of these various selectivities will require additional research.

5.4 Energy

Chemical evolution involves the interconversion of energy under its various forms, from *physical*: heat, electromagnetic (radioactivity, UV light, etc.), mechanic (tides, impacts, etc.) to *chemical*, through the formation of activated molecules, which could prefigure the increasing complexity of metabolic intermediates in the living world.

The primary-pump scenario is one example of this first step of converting energy from physical to chemical, in a permanent, continuous way: primary energy sources (heat, UV light, etc.) promote the formation of small activated molecules such as cyanic acid and NO_x, then followed by anhydrides such as NCA, responsible for the polymerisation of amino acids. As an energy carrier, NCA may also have been involved in other processes. Furthermore, as chemical structures increased in complexity, other activated/activating species should have emerged. Why not therefore consider NCA as possible ancestors (another could be thioesters, see De Duve 1998) of adenosine triphosphate (ATP), the "modern", somewhat ultimate energy carrier in today's biological world?

The kinetic model of the *primary pump* shown in Fig. 5.1 could, in a future study, be broadened to include how it was fed with matter and energy as shown in Fig. 5.9.

The energy aspect, with photochemical reactions leading to the essential prebiotic molecules, for which there are already rich sources of information and methods, could be developed through the utilization of, dedicated software such as PHOEBE (photo-chemistry for exobiology and exoplanets: see Selsis et al. 2002).

Concerning the feeding of matter, the selective synthesis process of CAA is described in the previous chapter. We showed that about 50 carbonyl derivatives, combined with ammonia, methylamine, ethylamine, hydrocyanic acid, and carbonic anhydride, are sufficient to model the formation of all meteoritic α-amino acids and terrestrial peptides. The mechanisms of the reactions involved in these syntheses are known, along with most of their kinetic constants (Commeyras et al. 2003).

Although complex, only kinetic modelling of the whole system can provide understanding or insight into the evolution of such systems.

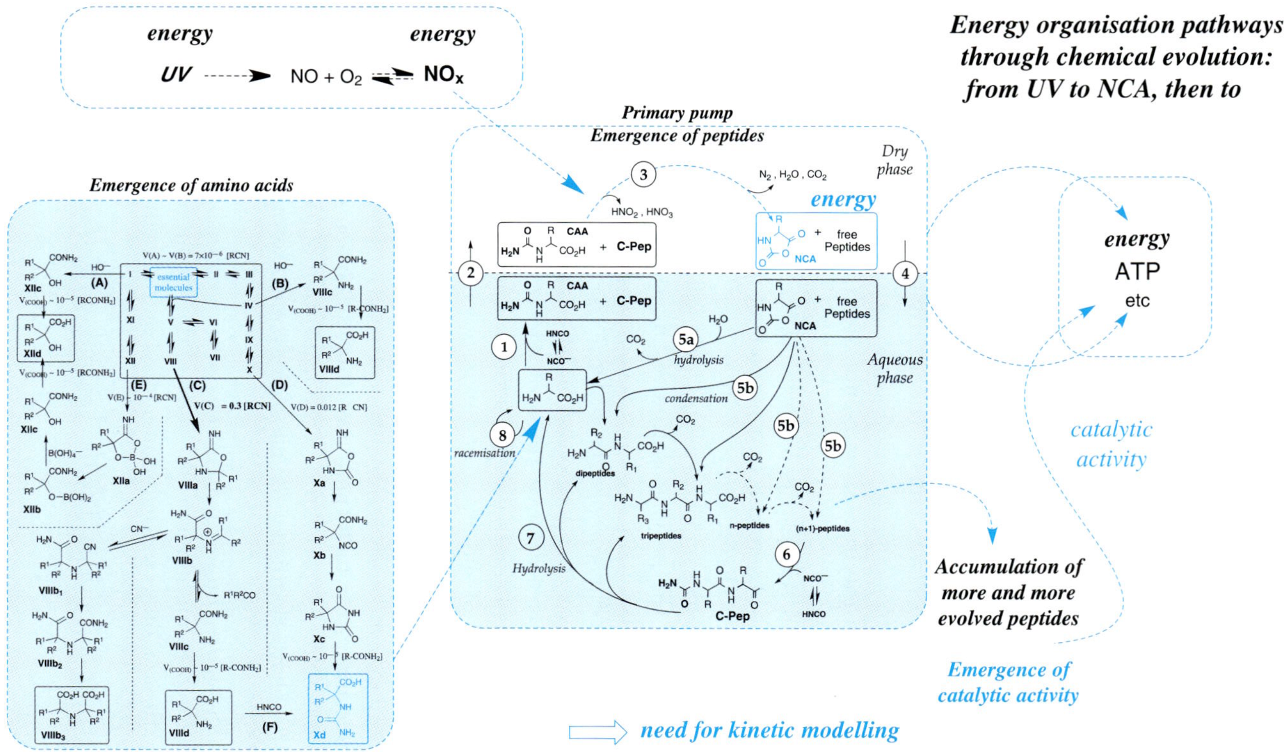

Fig. 5.9. Global view of the primary pump with matter and energy feeding

5.5 Conclusions and Perspectives

Our primary-pump scenario appears to be the first that could have been capable of supplying *sequential* peptides under rather realistic prebiotic conditions. Maintained out of thermodynamic equilibrium by continuous/sufficient inputs of matter (amino acids) and energy (through activated molecules such as cyanate and NO_x), this dissipative system has the ability to recycle its reagents and to increase the complexity of the reaction products (peptides). The minimum requirements for the prebiotic relevance of the model, namely the presence of oceans, emerged land and nitrogen oxides in the atmosphere, seem to have been fulfilled on the primitive Earth and the primary pump could work as soon as the pH of the oceans rose to 4–5. Our results underline the importance of such a buffering effect and suggest the most suitable locations for this process were beaches, rather than closed lakes or lagoons without any contact with the ocean. The emerging (eventually homochiral) peptides should progressively have begun to act as catalysts (Pascal 2003), which may have helped the emergence of self-replicating systems. In addition, the role of NO in the metabolism of currently living organisms, being the remnant of a past, much more important contribution (e.g. as a source of either organic nitrogen or energy), raises questions about a possible key role of this compound in prebiotic chemistry.

Although several questions remain unanswered, such as the effective abundance of NO_x in the atmosphere, or of cyanate in the hydrosphere in regard to the requirements of the primary pump, this model presents the unique advantage to allow experimental testing. This scenario should work as well with NCA produced by other prebiotic pathways. To our opinion, NCA should be considered as a major prebiotic intermediate.

Acknowledgement

For their fruitful collaboration to this work, we are indebted to our colleagues from Montpellier: Hélène Collet, Jacques Taillades, Laurence Garrel, Robert Pascal, Jean-Christophe Rossi, Hervé Cottet, Jean-Philippe Biron, Olivier Lagrille, Raphaël Plasson, and Grégoire Danger, as well as to Hervé Martin from the laboratory Magmas & Volcans at the University of Clermont-Ferrand (France), and to Michel Dobrijevic from the Observatoire Aquitain des Sciences de l'Univers (Bordeaux, France).

We are also grateful to:

- The Centre National de la Recherche Scientifique (CNRS), especially its Chemical Science Department and the Institut National des Sciences de l'Univers (INSU); the Exobiology Research Group of the Centre National d'Etudes Spatiales (CNES); and the European Community through the COST D27 action (prebiotic chemistry), for their support.
- The companies Aventis Animal Nutrition (France) and Degussa (Germany) for financial supports.

– Prof. Dr. Jacques Reisse from the Université Libre de Bruxelles (Belgium), and Dr. Muriel Gargaud from the Observatoire Aquitain des Sciences de l'Univers (Bordeaux, France) for many helpful discussions.

References

Bartlett PD, Dittmer DC (1957). A kinetic study of the Leuchs anhydrides in aqueous solution. II. *J. Am. Chem. Soc.* **79**: 2159–2161.

Bartlett PD, Jones RH (1957). A kinetic study of the Leuchs anhydrides in aqueous solution. I. *J. Am. Chem. Soc.* **79**: 2153–2159.

Bengston M, Edstrom ED (1999). A new method for testing models of prebiotic peptide assembly. In: *Advances in BioChirality*, Pályi G, Zucchi C and Caglioti L (eds), Elsevier. pp 115–123.

Blair NE, Bonner WA (1980). Experiments on the amplification of optical activity. *Origin of Life and Evolution of the Biosphere* **10**: 255–263.

Blair NE, Bonner WA (1981). A model for the enantiomeric enrichment of polypeptides on the primitive Earth. *Origin of Life* **11**: 331–335.

Bohler C, Hill AR, Orgel LE (1996). Catalysis of the oligomerization of o-phosphoserine aspartic acid, or glutamic acid by cationic micelles. *Origins of Life and Evolution of the Biosphere* **26**: 2–5.

Bowring SA, Williams IS (1999). Priscoan (7,00–4,03 Ga) orthogneisses from northwestern Canada. *Contribution to Mineralogy and Petrology* **134**: 3–16.

Caldeira K, Kasting JF (1992). Susceptibility of the early Earth to irreversible glaciation caused by carbon dioxide clouds. *Nature* **359**: 226–228.

Carbary JF, Morrison D, Romick GJ, Yee JH (2003). Leonid meteor spectrum from 110 to 860 nm. *Icarus* **161**: 223–234.

Chyba CF (1990). Impact delivery and erosion of planetary oceans in the early inner solar system. *Nature* **343**: 129–133.

Collet H, Boiteau L, Taillades J, Commeyras A (1999). Solid phase decarbamoylation of *N*-carbamoylpeptides and monoalkylureas using gaseous NO_x: a new simple deprotection reaction with minimum waste. *Tetrahedron Letters* **40**: 3355–3358.

Collet H, Bied C, Mion L, Taillades J, Commeyras A (1996). A new simple and quantitative synthesis of α-Aminoacid-*N*-Carboxyanhydrides. *Tetrahedron Letters* **37**: 9043–9046.

Commeyras A, Collet H, Boiteau L, Taillades J, Vandenabeele-Trambouze O, Cottet H, Biron J-P, Plasson R, Mion L, Lagrille O, Martin H, Selsis F, Dobrijevic M (2002). Prebiotic Synthesis of Sequential Peptides on the Hadean Beach by a Molecular Engine Working with Nitrogen Oxides as Energy Sources. *Polymer International* **51**: 661–665.

Commeyras A, Taillades J, Collet H, Mion L, Boiteau L, Trambouze-Vandenabeele O, Cottet H, Biron JP, Schué F, Giani O, Lagrille O, Plasson R, Vayaboury W, Martin H, Selsis F, Dobrijevic M, Geffard M (2001). La Terre, matrice de la Vie: émergence avant-gardiste des peptides sur les plages de l'Hadéen. In: *L'environnement de la Terre Primitive*, Gargaud M, Despois D, Parisot JP (eds.), Presses Universitaires de Bordeaux, pp 361–380.

Commeyras A, Taillades J, Collet H, Boiteau L, Pascal R, Vandenabeele-Trambouze O, Pascal R, Rousset A, Garrel L, Rossi JC, Cottet H, Biron JP, Lagrille O, Plasson

R, Souaid E, Selsis F, Dobrijevic M (2003). Approche dynamique de la synthèse des peptides et de leurs précurseurs sur la Terre primitive. In: *Les Traces du Vivant*, Gargaud M, Despois D, Parisot JP (eds.). Presses Universitaires de Bordeaux, Bordeaux, France, pp 115–162.

De Duve C (1998). Clues from present-day biology: the thioester world. In: *The Molecular Origin of Life*, Brack A (ed.), Cambridge University Press. Cambridge, UK, pp 219–236.

Fox SW, Dose H (1977). *Molecular Evolution*, New York, Academic Press. New York, (USA).

Gesteland RF, Cech TR, Athins JF (1999). *The RNA Word.* 2nd edition, Cold Spring Harbor Laboratory Press. New York (USA).

Gilbert W (1986). The RNA World. *Nature* **319**: 618.

Guinan E, Ribas I (2002). Our changing Sun: the role of solar nuclear evolution and magnetic activity on Earth's atmosphere and climate. *ASP Conference Series*, **269**: 85–106.

Huber C, Wächtershäuser G (1998). Peptides by activation of amino acids with CO on (Ni,Fe)S surfaces: implications for the origin of life. *Science* **281**: 670–672.

Idelson M, Blout ER (1958). Polypeptides. XVIII. A kinetic study of the polymerisation of amino acid *N*-carboxyanhydrides initiated by strong bases. *J. Am. Chem. Soc.* **79**: 2387–2393.

Jenniskens P, Wilson MA, Packan D, Laux CO, Krüger CH, Boyd ID, Popova OP, Fonda M (2000). Meteors: A delivery mechanism of organic matter to the early Earth. *Earth, Moon and Planets* **82/83**: 57–70.

Joyce GF, Orgel LE (1999). Prospects for understanding the origin of the RNA world. In: *The RNA World.* 2nd edn, Cold Spring Harbor Laboratory Press (New York), pp 49–77.

Kasting JF (1993). Earth's early atmosphere. *Science* **259**: 920–926.

Kasting JK, Catling D (2003). Evolution of a habitable planet. *Annual Review of Astronomy and Astrophysics* **41**: 429–463.

Kauffman SA (1993). *The Origin of Order. Self-organisation and Selection in Evolution*, 1st edn. Oxford University Press (see especially Chap. 7).

Lagrille O (2001). *Nitrosation de N-Carbamoylaminoacides Solides par le mélange gazeux NO/O_2. Synthèse de N-Carboxyanhydrides (Anhydrides de Leuchs).* PhD thesis, Université de Montpellier 2 (France).

Lagrille O, Taillades J, Boiteau L, Commeyras A (2002). *N*-Carbamoyl Derivatives and their nitrosation by gazeous NO_x–A new promissing tool in stepwise peptide synthesis. *Eur. J. Org. Chem*: 1026–1032.

Lahav N, Wite DH, Chang S (1978). Peptide formation in the prebiotic Era: thermal condensation of glycine in fluctuating clay environments. *Science* **201**: 67–69.

Liu R, Orgel LE (1998). Polymerization of β-amino acids in aqueous solution. *Orig. Life Evol. Biosphere* **28**: 47–60.

Luisi PL (2000). L'assemblage des macromolécules. *La Recherche* **336**: 25–29.

Luisi PL, Walde P, Blocher M, Liu D (2000). Research on the origin of Life: membrane-assisted polycondensations of amino acids and peptides. *Chimia* **54**: 52–53.

Lundberg RD, Doty P (1957). Polypeptides. XVII. A Study of the Kinetics of the Primary Amine-initiated Polymerisation of *N*-carboxy-anhydrides with special reference to configurational and stereochemical effects. *J. Am. Chem. Soc.* **78**: 3961–3972.

Martin H (1986). Effect of steeper Archean geothermal gradient on geochemistry of subduction-zone magmas. *Geology* **14**: 753–756.

Martin H (1987). Petrogenesis of Archaean trondhjemites, tonalites and granodiorites from eastern Finland: major and trace element geochemistry. *Journal of Petrology* **28**: 921–953.

McCulloch MT, Bennet VC (1993). Evolution of the early Earth: constraints from ^{143}Nd-^{142}Nd isotopic systematics. *Lithos* **30**: 237–255.

Mojzsis SJ, Krishnamurthy R, Arrhenius G (1999). Before RNA and after: Geophysical and geochemical constraints on molecular evolution. In: *The RNA Word* 2nd edn. Gesteland RF, Cech TR and Athins JF (eds.), Cold Spring Harbor Laboratory Press, pp 1–47.

Mojzsis SJ, Harrison MT, Pidgeon RT (2001). Oxygen-isotope evidence from ancient zircons for liquid water at the Earth's surface 4,300 Myr ago. *Nature* **409**: 178–181.

Morrison D, Chapman CR, Slovic P (1994). The impact hazard. In: *Hazards due to Comets and Asteroids*, Gehrels T, Matthews MS and Schumann A (eds.). Univ. Arizona Press, Tucson, USA, pp 59–92.

Munoz Caro GM, Meierhenrich UJ, Schutte WA, Barbier B, Arcone Segovia A, Rosenbauer H, Thiemann WHP, Brack A, Greenberg GM (2002). Amino acids from ultraviolet irradiation of interstellar ice analogues. *Nature* **416**: 403–406.

Navarro-Gonzalez R, Molina MJ, Molina LT (1998). Nitrogen fixation by volcanic lightning in the early Earth. *Geophys. Res. Lett.* **25**: 3123–3126.

Navarro-Gonzalez R, McKay CP, Nna Mvondo D (2001). A possible nitrogen crisis for Archaean life due to reduced nitrogen fixation by lightning. *Nature* **412**: 61–64.

Nicolis G, Prigogine I (1977). *Self-organization in Nonequilibrium Systems.* Wiley & Sons. New York (USA).

Orgel LE (1989). The origin of polynucleotide-directed protein synthesis. *J. Mol. Evol.* **29**: 465–474.

Orgel LE (1998). Polymerization on the rocks: Theoretical introduction. *Origins of Life and Evolution of the Biosphere* **28**: 227–237.

Paecht-Horowitz M, Eirich FR (1988). *Origins of Life and Evolution of the Biosphere* **18**: 359.

Pascal R (2003). Catalysis by induced intramolecularity: what can be learned by mimicking enzymes with carbonyl compounds that covalently bind substrates? *Eur. J. Org. Chem.*: 1813–1824.

Pavlov AA, Kasting JF, Brown LL, Rages KA, Freedman R (2000). Greenhouse warming by CH_4 in the atmosphere of early Earth. *J. Geophys. Res.* **105**: 11981–11990.

Pizzarello S, Cronin JR (2000). Non-racemic amino acids in the Murray and Murchison meteorites. *Géochimica Cosmochemica Acta* **64**: 329–338.

Plasson R, Biron JP, Cottet H, Commeyras A, Taillades J (2002). Kinetic study of α-aminoacids *N*-carboxyanhydrides polymérisation in aqueous solution using capillary electrophoresis. *J. Chrom. A* **952**: 239–248.

Plasson R. (2003). *Origine moléculaire de la vie: étude de la polymérisation de aminoacides N-carboxyanhydrides dans des conditions prébiotiques, par électrophorèse capillaire.* PhD thesis, Université de Montpellier 2 (France).

Prinn RG, Fegley B (1987). Bolide impacts, acid rain, and biospheric traumas at the Cretaceous-Tertiary boundary. *Earth and Planetary Science Letters* **83**: 1–4.

Rode BM, Eder AH, Yongyai Y (1997). Amino acid sequence preferences of the salt-induced peptide formation reaction in comparison to archaic cell protein composition. *Inorganica Chemica Acta* **254**:309–314.

Rohlfing DL (1976). Thermal polyamino acids: synthesis at less than 100 °C. *Science* **193**: 68–69.

Rosenqvist J, Chassefière E (1995). Inorganic chemistry of O_2 in a dense primitive atmosphere. *Planet. Space Sci.* **43**: 3–10.

Selsis F (2000). *Modèle d'évolution physico-chimique des atmosphères de planètes telluriques. Application à l'atmosphère primitive terrestre et aux planètes extrasolaires.* PhD thesis, Université Bordeaux 1, France.

Selsis F, Despois D, Parisot J-P (2002). Signature of life on exoplanets: Can Darwin produce false positive detections? *Astronomy and Astrophysics* **388**: 985–1003.

Shimoyama A, Ogasawara R (2002). Dipeptides and Diketopiperazines in the Yamato-791198 and Murchison Carbonaceous Chondrites. *Orig. Life Evol. Biosphere* **32**: 165–179.

Sleep NH, Zahnle K (2001). Carbon dioxide cycling and implications for climate on ancient Earth. *Journal of Geophysical Research* **106**: 1373–1399.

Taillades J, Collet H, Garrel L, Beuzelin I, Boiteau L, Choukroun H, Commeyras A (1999). *N*-Carbamoylaminoacid solid-gas nitrosation by NO/NO_x : a new route to oligopeptides via α-aminoacid *N*-carboxyanhydride. Prebiotic implications. *Journal of Molecular Evolution* **48**: 638–645.

Taillades J, Boiteau L, Beuzelin I, Lagrille O, Biron JP, Vayaboury W, Vandenabeele-Trambouze O, Giani O, Commeyras A (2001). A pH-dependent cyanate reactivity model: application to preparative *N*-carbamoylation of amino acids. *Perkin Trans. 2:* 1247–1253.

Tolstikhin IN, Marty B (1998). The evolution of terrestrial volatiles: a view from helium, neon, argon and nitrogen isotope modelling. *Chem. Geol.* **147**: 27.

Toon OB, Zahnle K, Turco RP, Covey C (1994). *Environmental perturbations caused by asteroid impacts, in Hazards due to comets and asteroids.* Univ. Arizona Press, Tucson, USA, pp 791–826.

Turco RP, Toon OB, Park C, Whitten RC, Pollack JB, Noerdlinger P (1982). An analysis of the physical, chemical, optical, and historical consequence of the 1908 Tungunska meteor fall. *Icarus* **50**: 1–52.

Wald G (1957). The origin of Optical Activity. *Ann. NY Acad. Sci.* **69**: 352–368.

Weber AL (1998). Prebiotic aminoacid thioester synthesis: Thiol-dependent amino acid synthesis from formose substrate (formaldehyde and glycolaldehyde) and ammonia. *Origins of Life and Evolution of the Biosphere* **28**: 259–270.

Wilde SA, Valley JW, Peck WH, Graham CM (2001). Evidence from detrital zircons for the existence of continental crust and oceans on the Earth 4.4 Ga ago. *Nature* **409**: 175–178.

Zahnle K (1998). Origins of atmospheres. *ASP Conf Series 148: Origins*: 364.

Zahnle K, Kasting JF, Sleep N (1988). Impact production of NO and reduced species. In: LPI Contribution 673 (abstracts of the *Topical conference on global catastrophes in Earth history: interdisciplinary conference on impacts, volcanism, and mass mortality*, Snowbird, Utah, Oct. 20–23, 1988), Lunar & Planetary Institute, Houston, pp 223–224.

6 The RNA World: Hypotheses, Facts and Experimental Results

Marie-Christine Maurel, Anne-Lise Haenni

A biochemical world that would have existed before the contemporary DNA-RNA-protein world, and named in 1986 "The RNA World" by Walter Gilbert (Gilbert, 1986), such a world had already been proposed during the preceding decades by Carl Woese, Francis Crick and Leslie Orgel (Woese, 1965; Crick, 1968; Orgel, 1968).

By demonstrating the remarkable diversity of the RNA molecule, molecular biology proved these predictions. RNA, present in all living cells, performs structural and metabolic functions many of which were unsuspected only a few years ago. A truly modern "RNA world" exists in each cell; it contains RNAs in various forms, short and long fragments, single and double-stranded, endowed with multiple roles (informational, catalytic, that can serve as templates, guides, defense, etc.), certain molecules even being capable of carrying out several of these functions.

Are the sources of this RNA world to be found in the bygone living world?

6.1 The Modern RNA World

6.1.1 Where in the Living Cell is RNA Found?

Synthesized (transcribed) in the nucleus, mature messenger RNAs (mRNAs), transfer RNAs (tRNAs) and ribosomal RNAs (rRNAs) are exported as single strands to the cytoplasm of the cell after various maturation steps. A ribonucleic acid (RNA) is formed by linking nucleotides[1], themselves composed of heterocyclic bases associated with a sugar, β-D-ribofuranose, and a phosphate molecule (phosphoric acid). The four main nucleotides contain the heterocyclic purine (adenine and guanine) or pyrimidine (cytosine and uracil) bases[2]. However, RNAs, in particular rRNAs and tRNAs contain a very large diversity of modified nucleotides, since more that a hundred modified nucleotides[3] have now been identified in these two classes of molecules (Grosjean and Benne, 1998).

---

[1] To yield a polyribonucleotide

[2] Adenine, A; guanine, G; cytosine, C; uracil, U

[3] Post-transcriptional modifications

RNAs are usually single stranded[4]. Nevertheless, these strands can base pair locally or over long stretches (intramolecular pairing). Finally, from a structural point of view, they contain a reactive hydroxyl group in the 2' position of ribose (a group that is absent in DNA). The stacking forces and pairing of bases produce "stems and helices"; defined structures bring together the helices and the regions separating them, into "motifs".

RNA helices: Through the action of the stacking forces, the skeleton of the single strand by itself tends to take the shape of a simple, right-handed and irregular helix. However, the important conformation is the double helix composed of two strands of RNA or of RNA/DNA (hybrids formed transiently during transcription) or that occurs when two distantly located complementary segments of the same RNA base pair.

The motifs identified are bulges, elbows, or loops.

Hairpins are other important structural motifs related to certain functions of RNAs. They can lead to interactions with special sequences, such as the GNRA loops[5], seven-base-long loops, etc. Large RNAs possess independent domains formed by the arrangement of a certain number of motifs. An RNA molecule can adopt several reversible conformations, depending on the presence of ions, specific surfaces or bound ligands. RNAs possess a repertory of structures reminiscent of proteins (motifs or domains) allowing them to express certain functions such as catalysis. Finally, non-Watson–Crick base pairs[6] are frequently encountered in RNAs (G-U pairs are common) and modified bases are involved, and by their strong steric hindrance with the bases, the 2' OH groups of the ribose moieties tend to prevent folding in the B helical conformation[7].

6.1.1.1 The Three Large Classes of RNA

- Messenger RNAs (mRNAs of 400 to 6000 nucleotides) are the copy of DNA genes[8]. The RNA transcripts are considerably modified in the nucleus during maturation, and during transcription of DNA into RNA, short hybrids of the A conformation appear. Their life is short in prokaryotes (a few minutes to a few dozen minutes) and can be of several hours in higher eukaryotes; mRNAs correspond to only a few per cent of the total cellular RNAs. The step-by-step decoding of the mRNA by the ribosome known as translation is regulated by specific proteins, and in some cases also by hairpin motifs and/or by pseudoknots.

[4] Paired *two-stranded* RNAs are exceptions found in a few rare viruses

[5] N is any nucleotide, R is a purine nucleotide

[6] See glossary. Watson–Crick pairings are the standard pairs (A-U and G-C)

[7] The bends they impose to the plane of the bases – of about 20° – on the axis results in a structure resembling the A conformation (also designated RNA 11 to stress the 11 base pairs per turn). The A form of RNA double helices is characterized by 11 base pairs per helical turn (instead of 10 for the B form), and by bending of the base pairs by 16°/helical axis (instead of 20° for DNA A)

[8] A gene is a fragment of DNA whose information is expressed via the genetic code

Pseudoknots result from base pairing between nucleotides within a loop and complementary nucleotides outside of the loop.

• Transfer RNAs (tRNAs) are small molecules whose maximum length is about 100 nucleotides. They are strongly conserved and are involved in the central metabolism of all types of cells. Their main function is to ensure the interaction between the codon presented by the mRNA and the specific amino acid (corresponding to this codon) and contained in the anticodon of the aminoacyl-tRNA. tRNAs possess two extremely specific sites: the first is the sequence CCA located at the 3' OH of the molecule; the second site is located in a loop that contains the anticodon. The cloverleaf-shaped secondary structure (Fig. 6.1) possesses several motifs. tRNAs also serve as primers during replication of certain viruses and are involved in the activity of telomerases. Synthesized as pre-tRNAs they undergo a maturation step during which RNAse P cleaves off a short fragment from the 5' end of the RNA (Guerrier-Takada et al., 1983). As already mentioned, tRNAs contain a large number of modified bases that are probably the most visible "relics" of an ancient RNA world (Cermakian and Cedergren, 1998).

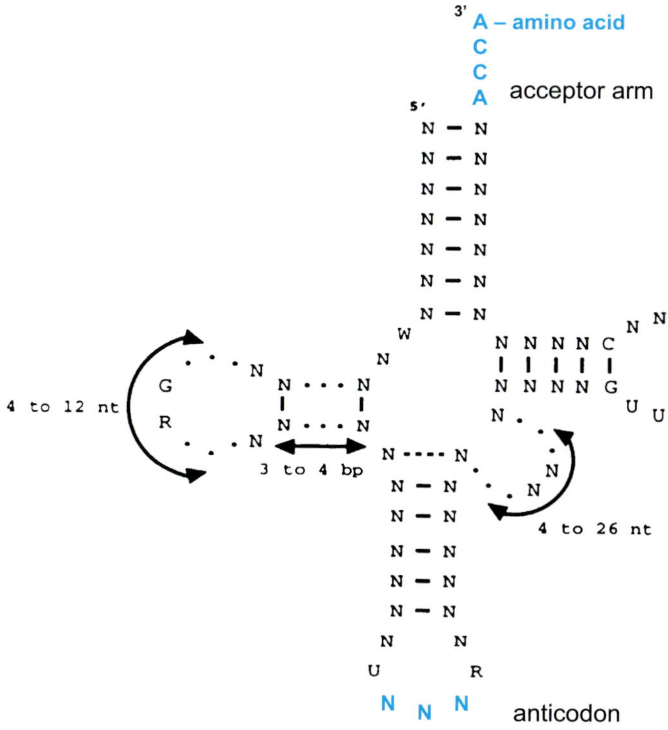

Fig. 6.1. Secondary cloverleaf structure of a tRNA. *Arrows* indicate number of nucleotides in the loop, stem and bulge

• The size of the ribosomal RNAs (rRNAs) is variable, from 120 to 4718 nucleotides. rRNAs are located in the ribosome, the site of protein synthesis. In addition to about fifty proteins, the prokaryotic ribosome contains three rRNAs and the eukaryotic ribosome four rRNAs. The rRNAs are methylated (sometimes in the 2'OH position of the ribose, protecting the polymer from hydrolysis). Their typical secondary structure is remarkably well conserved (Fig. 6.2).

They possess complex global tertiary conformations that compact the molecule into different domains, and it has now been clearly demonstrated that the rRNA catalyzes the formation of the peptide bond during protein biosynthesis (Ban et al., 2000; Nissen et al., 2000; Ysupov et al., 2001).

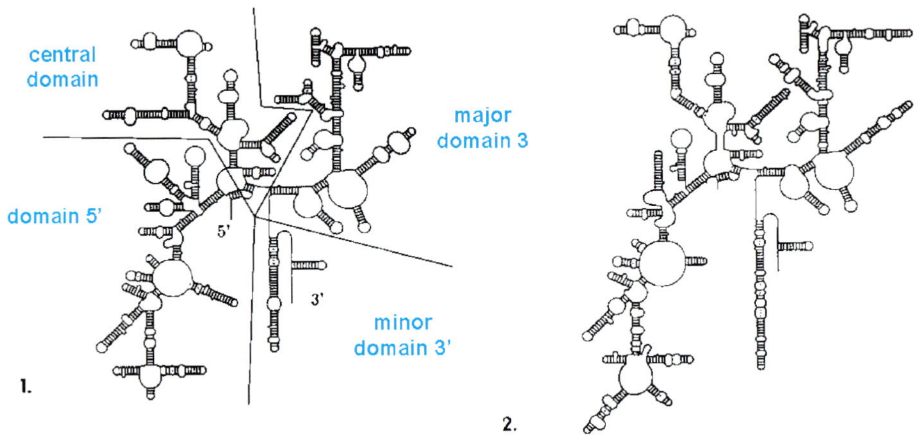

Fig. 6.2. Typical secondary structure 1) 16S rRNA of the bacterium *Escherichia coli*, 2) 18S rRNA of the yeast *Saccharomyces cerevisiae*

6.1.1.2 Noncoding RNAs (ncRNAs)

In addition to rRNAs, tRNAs and mRNAs a variety of RNA molecules have been discovered that possess very diverse functions in the living cell (Maurel, 1992; Meli et al., 2001; Zamore, 2002; Grosshans and Slack, 2002; Westhof, 2002). Before involvement of the ribosome, the RNA transcripts must undergo maturation steps. In eukaryotes, these post-transcriptional modification steps require the participation of small RNAs, the snoRNAs (small nucleolar RNAs) that together with proteins, form the snoRNP (small nucleolar ribonucleoprotein particles). Over 150 snoRNPs have been described in eukaryotes (in different lineages). They form a snoRNP complex, the snorposome, that participates in RNA maturation. The origin of the modification systems is still unknown. One of the various hypotheses put forward suggests that the snoRNAs of the RNA world would have been involved in the assembly of the protoribosomes, and more generally in the scaffolding of ribozymes (Terns and Terns, 2002).

Moreover, large snRNPs (small nuclear ribonucleoprotein particles) responsible for intron excision from pre-mRNAs have been identified. Each snRNP is composed of snRNA and about a dozen snRNP proteins. Two classes of such spliceosomes cleave different introns, whereas excision and ligation of the exons is achieved by the same biochemical mechanism (Tarn and Steitz, 1997). Spliceosomes are restricted to eukaryotes, even though containing introns bacteria have been reported.

The telomerase is an enzyme that uses a small RNA as primer during replication to elongate the linear DNA located at the end of eukaryotic chromosomes (Maizels et al., 1999).

Vault RNAs are ribonucleoprotein particles located in the cytoplasm of eukaryotes (Kong et al., 2000). They are associated with the nuclear "pore complex"; their function has not been clearly defined, but their structure suggests that they may be involved in cell transport or in the assembly of macromolecules. The history of the evolution of vault RNAs remains unknown, but these RNAs could have participated in primitive compartmentation.

Finally, an RNA-protein complex, the SrpRNA (signal recognition particle RNA) is highly conserved in the three kingdoms (Wild et al., 2002). It is involved in translation, and during secretion of proteins from the plasma membrane or from the endoplasmic reticulum.

About 15 years ago, the existence of a correcting mechanism, «editing», was demonstrated (Lamond, 1988). This co- or post-transcriptional mechanism modifies the sequence of the mRNA by the insertion or deletion of nucleotides, or by the modification of bases. Up to 55% modifications can take place with respect to the gene (in this case it is designated "cryptogene"). The sites where editing takes place are determined by the structure of the RNA, or by guide-RNAs (Stuart and Panigrahi, 2002). In kinetoplastid protozoa, guide RNAs are required to edit mitochondrial pre-mRNAs by inserting or deleting uridylate residues in precise sites (Kable et al., 1997).

Finally, the tmRNA (transfer-messenger RNA) is a stable cytoplasmic RNA found in eubacteria. tmRNAs contain a $tRNA^{Ala}$-like structure (with pairing between the 5' and 3' ends) and an internal reading frame that codes for a short peptide (peptide tag) (Fig. 6.3). It is thus at variance with the strict definition of snRNAs, since it encompasses a short reading frame. It performs a new type of recently discovered translation, known as trans-translation, during which a peptide is synthesized starting from two distinct mRNAs. tmRNA acts as tRNA and as mRNA to "help" ribosomes that are blocked on a trunctated mRNA lacking a termination codon. tmRNA participates by adding alanine to the growing peptide chain. Thus, tmRNA plays a dual role: as $tRNA^{Ala}$ it can be aminoacylated by the corresponding alanyl-tRNA synthetase, and as mRNA its open reading frame can be translated by the ribosome (Withey and Friedman, 2002; Valle et al., 2003). Could tmRNA be a bacterial adaptation, or could it have been lost by the archae and the eukaryae?

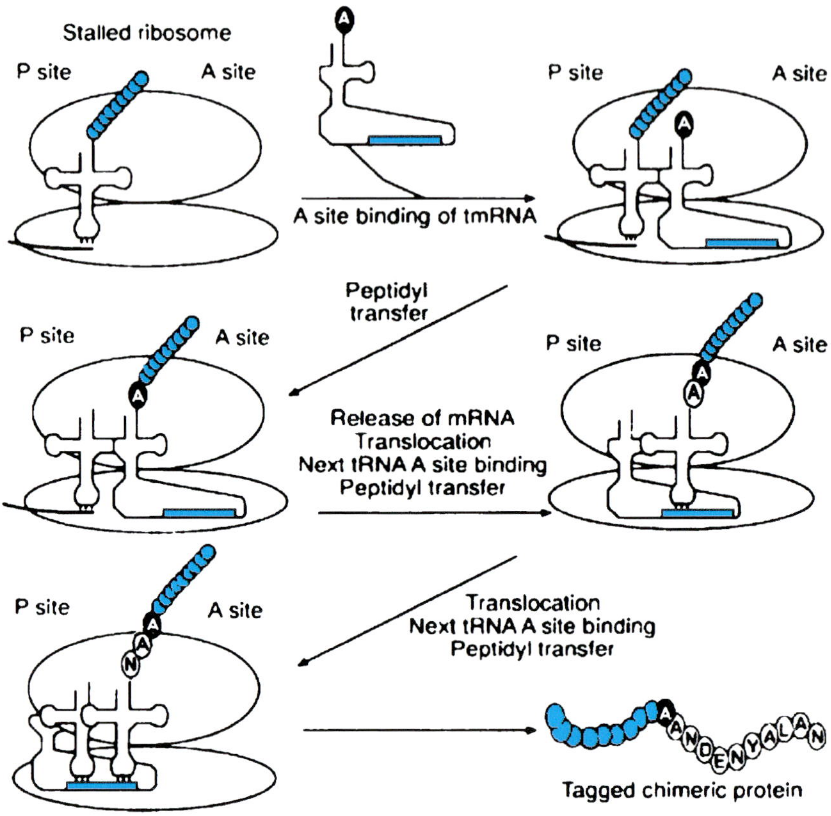

Fig. 6.3. How tmRNA functions

A eukaryotic system distantly related to tmRNA has recently been described (Barends et al., 2003) in the single-stranded Turnip yellow mosaic virus (TYMV) RNA. The 3' end of the viral genome harbors a tRNA-like structure that is indispensable for the virus viability and can be valylated. During protein biosynthesis programmed with valylated TYMV RNA, the valine residue is N-terminally incorporated into the viral polyprotein, thereby introducing a novel mechanism of initiating protein synthesis (Fig. 6.4). Here again, the viral RNA would be bifunctional, serving both as tRNA and as mRNA.

It will be interesting to determine whether other viral RNAs whose 3' end bears an aminoacylatable tRNA-like structure (Fechter et al., 2001) can also donate their amino acid for mRNA translation.

Viroids are subviral plant pathogens responsible for economically important diseases. They are small (246–401 nucleotides), single-stranded closed circular RNA molecules characterized by a highly compact secondary structure. They are devoid of coding capacity and replicate autonomously in the plant host. Two families of viroids have been characterized, the Pospiviroidae (type-member:

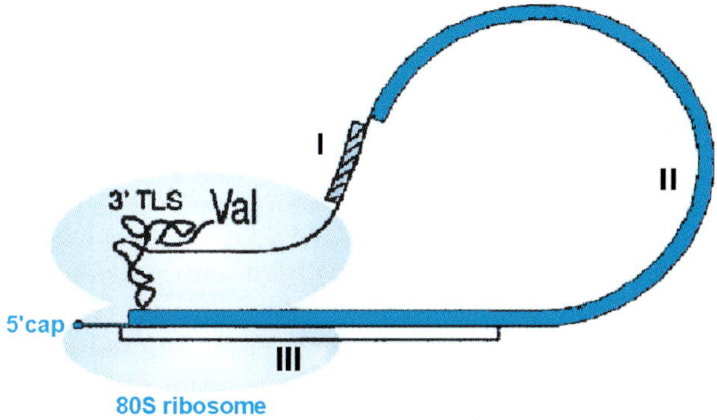

Fig. 6.4. Model of the tRNA-like structure-mediated internal initiation mechanism of TYMV RNA for polyprotein translation. I: Coat protein gene; II: Polyprotein gene; III: Movement protein gene; Adapted from Barends et al., 2003

Potato spindle tuber viroid, PSTVd) that replicates in the nucleus, and the Avsunviroidae (type-member: Avocado sun blotch viroid, ASBVd) that replicates in chloroplasts and possesses conserved hammerhead structures in the viroid and in the complementary RNA orientation. It has been suggested that the presence of hammerhead structures could reflect the early appearance of viroids in the course of evolution; they could correspond to "living fossils" of the primitive RNA world (Diener, 2001).

The few ncRNAs described here are probably only the tip of a huge iceberg (Bachellerie et al., 2002) since most of the transcriptional output of superior eukaryotes is nonprotein coding (97% for human). These ncRNAs could constitute a real RNA world in complex organisms (Eddy, 2001; Mattick, 2003). Their study may open new perspectives about the importance of RNA in primitive life. Certain RNAs that are presently being investigated are those involved in RNA interference (RNAi): the RNAs responsible for RNAi are the small interfering RNAs that target and cleave mRNAs (Nykanen et al., 2001). *Micro-RNAs*, another class of small RNAs, are involved in translation regulation (Grosshans and Slack, 2002). In eukaryotes, guide snoRNAs participate in selecting the sites on rRNAs that undergo modifications such as Ψ formation or 2'-O-methylation (Lafontaine and Tollervey, 1998).

6.2 An RNA World at the Origin of Life?

The scenario of evolution postulates that an ancestral molecular world existed originally that was common to all the present forms of life; the functional properties of nucleic acids and proteins as we see them today would have been produced by molecules of ribonucleic acids (Joyce, 1989; Orgel, 1989; Benner et al., 1989, 1993; Joyce and Orgel, 1999; Gesteland et al., 1999; Bartel and Unrau, 1999; McGinness et al., 2002; Joyce, 2002).

6.2.1 Facts

As we have seen, RNAs occupy a pivotal role in the cell metabolism of all living organisms and several biochemical observations resulting from the study of contemporary metabolism should be stressed. For instance, throughout its life cycle, the cell produces deoxyribonucleotides required for the synthesis of DNA that derive from ribonucleotides of the RNA. Thymine, a DNA specific base is obtained by transformation (methylation) of uracil a RNA specific base, and RNAs serve as obligatory primers during DNA synthesis (Fig. 6.5). Finally, the demonstration that RNAs act as catalysts is an additional argument in favour of the presence of RNAs before DNA during evolution.

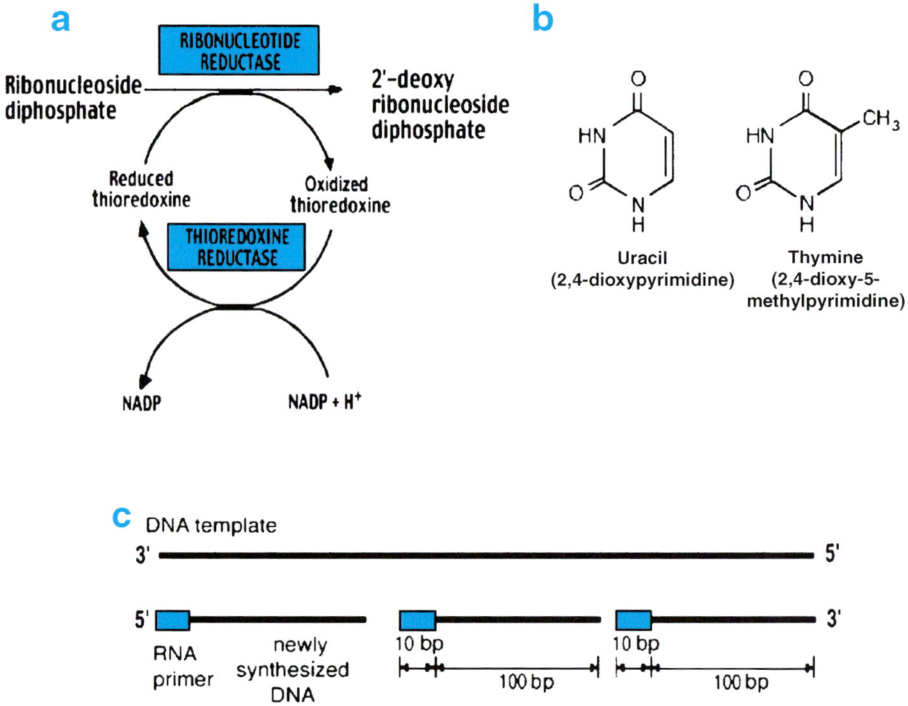

Fig. 6.5. Facts in favour of an RNA world. (**a**) Synthesis of deoxyribonucleotide; (**b**) Structure of uracil and thymine; (**c**) DNA synthesis primed by RNA

6.2.2 Hypotheses

DNA replication triggered by ribonucleotide primers can be considered as a modified transcription process during which polymerisation of RNA is "replaced" by that of DNA. In addition, DNA a double-stranded molecule lacking a hydroxyl group in 2' of the desoxyfuranose, appears more stable than RNA. Therefore it seems highly likely that RNA arose before DNA during biochemical evolution, and for this reason DNA is sometimes considered as modified RNA better suited for the conservation of genetic information. This genetic privilege would constitute a logical step in an evolutionary process during which other molecules could have preceded RNA and transmitted genetic information.

The idea of an "RNA" world rests primarily on three fundamental hypotheses, developed by Joyce and Orgel (1999):

- during a certain period in evolution, genetic continuity was assured by RNA replication,
- replication was based on Watson–Crick type base pairing,
- genetically coded proteins were not involved in catalysis.

6.2.3 But What do We Know about Primitive Replication?

Synthesis of a strand complementary of the template was studied extremely thoroughly in vitro in the group of Orgel (Inoue and Orgel, 1983; Joyce and Orgel, 1986; Orgel, 1992). During this directed synthesis, the mononucleotides (activated under the form of 5′-phosphorimidazolides) are positioned according to the Watson and Crick pairing rules along a preformed polypyrimidine template. Since these monomers are activated, they can bind together to form the complementary strand (Fig. 6.6). Orgel and his coworkers showed that starting from activated monomers, it is possible in certain conditions to copy a large number of oligonucleotide sequences containing one or two different nucleotides in the absence of enzyme (Hill et al., 1993).

Ferris and his coworkers spent some 15 years studying the assembly of RNA oligomers on the surface of montmorillonite (clay of Montmorillon in the Vienne region in France) (Ferris, 1987; Ferris and Ertem, 1992). The monomers used, nucleoside 5′-phosphorimidazolides, were probably not prebiotic molecules. Nevertheless, experimental results demonstrated that minerals that serve as adsorbing surfaces and as catalysts (Paecht-Horowitz et al., 1970; Ferris et al., 1996), can lead to accumulation of long oligonucleotides, as soon as activated monomers are available. One can thus envisage that activated mononucleotides assembled into oligomers on the montmorillonite surface or on an equivalent mineral surface. The longest strands serving as templates, direct synthesis of a complementary strand starting from monomers or short oligomers, and double-stranded RNA molecules accumulate. Finally, a double RNA helix of which one strand is endowed with RNA polymerase activity, would dissociate to copy the complementary strand to produce a second polymerase that would copy the first to produce a second complementary strand, and so forth. The RNA world would thus have emerged from a mixture of activated nucleotides. However, a mixture of activated nucleotides would need to have been available! In addition, this nucleotide chemistry is restricted in another way, since a copy of the template can be started only if the nucleotides are homochirals (Joyce et al., 1987).

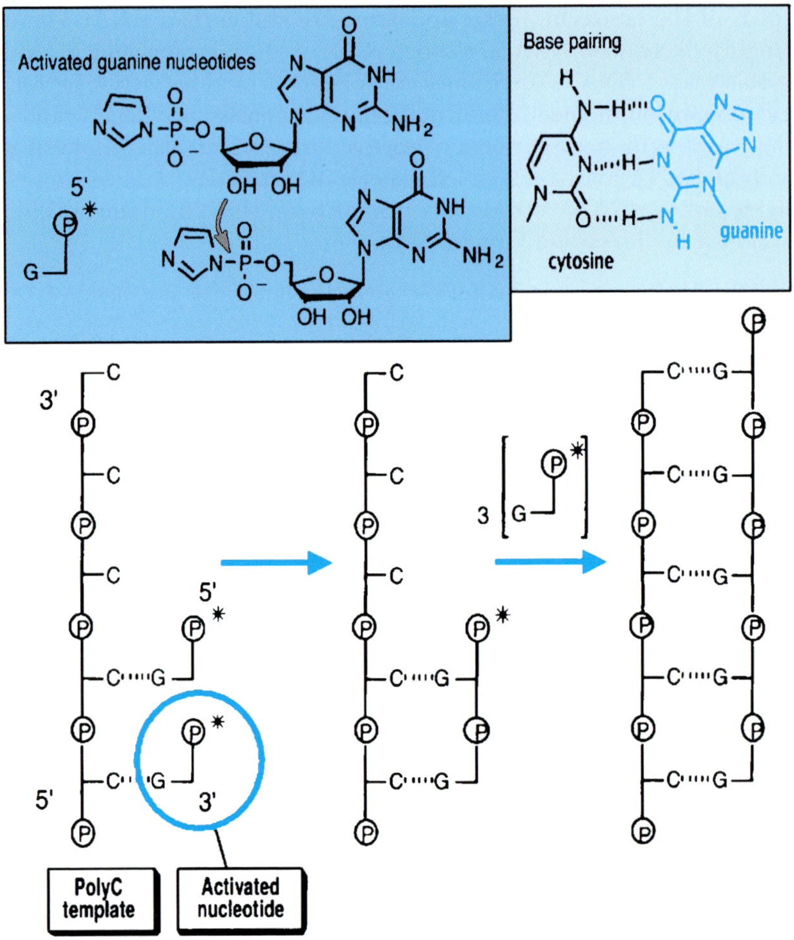

Fig. 6.6. Template-directed RNA synthesis

Finally, when either the first replicative molecule, the template or one of its elements (nucleotides) is to be synthesized from the original building blocks in particular the sugars that are constituents of nucleotides, a number of difficulties are encountered (Sutherland and Whitfield, 1997). Synthesis of the sugars from formaldehyde produces a complex mixture in which ribose is in low amounts. On the other hand, production of a nucleoside from a base and a sugar leads to numerous isomers, and no synthesis of pyrimidine nucleosides has so far been achieved in prebiotic conditions. Finally, phosphorylation of nucleosides also tends to produce complex mixtures (Ferris, 1987). Onset of nucleic acid replication is almost inconceivable if one does not envisage a simpler mechanism for the prebiotic synthesis of nucleotides. Eschenmoser succeeded in producing 2,4-diphosphate ribose during a potentially prebiotic reaction between glycol aldehyde[9] monophosphate and formaldehyde (Eschenmoser, 1999). It is thus possible that direct prebiotic nucleotide synthesis can occur by an alternative chemical pathway. Nevertheless, it is more likely that a certain organized form of chemistry preceded the RNA world, hence the notion of "genetic take-over". Since the ribose-phosphate skeleton is theoretically not indispensable for the transfer of genetic information, it is logical to propose that a simpler replication system would have appeared before the RNA molecule.

6.3 A Pre-RNA World

6.3.1 Evolutive Usurpation

During the evolutionary process, a first genetic inorganic material, would have been replaced by organic material. The hypothesis of a precursor of nucleic acid[10] (Cairns-Smith, 1966, 1982) is a relatively ancient idea, but it is only within the last few years that research has been oriented towards the study of molecules simpler than present-day RNAs, yet capable of autoreplication. Models with predictably retroactive activities can thus be tested experimentally.

6.3.2 Alternative Genetic Systems

In the peptide nucleic acids (PNA) of Nielsen and coworkers, the ribofuranose-phosphate skeleton is replaced by a polyamidic skeleton on which purine and pyrimidine bases are grafted (Fig. 6.7). PNAs form very stable double helices with an RNA or a complementary DNA (Egholm et al., 1993) and can serve as template for the synthesis of RNA, or vice versa (Schmidt et al., 1997). PNA-DNA chimeras containing two types of monomers have been produced on DNA or PNA templates (Koppitz et al., 1998). The information can be transferred from PNAs (achiral monomers) to RNA during directed synthesis; the double-helical molecule with a single complementary RNA strand is stable. Transition from a "PNA world" to an "RNA world" is hence possible. Nevertheless, the formation of oligomers from PNA monomers seems particularly difficult in prebiotic conditions.

Eschenmoser (1994) explored the properties of nucleic acid analogues in which ribofuranose is replaced by one of its isomers, ribopyranose (Furanose, 5–membered ring; pyranose, 6–membered ring). p-RNAs (pyranosyl RNAs) (Fig. 6.7) form more stable double helices (with Watson–Crick pairings) than RNA with ribofuranose. In addition, the double helices of p-RNA wind and unwind more easily than those formed with standard nucleic acids, and this should facilitate their separation during replication. p-RNAs could therefore constitute good candidates as precursor genetic systems, but a p-RNA strand cannot pair

[9] Recently shown to exist in interstellar clouds and comets (Cooper et al., 2001)
[10] This is the idea of genetic take-over developed by Cairns-Smith in the 1960s

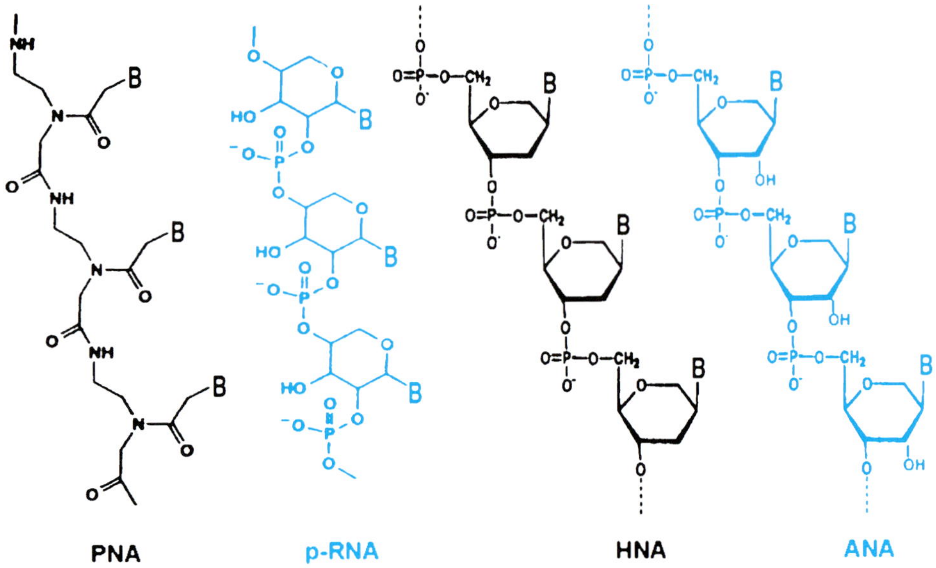

Fig. 6.7. Alternative genetic systems (B = base)

with an RNA of complementary sequence, and this makes it difficult to imagine a transition from p-RNA to RNA.

The group of Eschenmoser recently replaced the ribose moiety by a four-carbon sugar, threose, whose prebiotic synthesis seems easier. The resulting oligonucleotides designated TNAs, (3' → 2')-α-L-threose nucleic acid (Fig. 6.8), can form a double helix with RNA (Schöning et al., 2000). TNA is capable of antiparallel, Watson–Crick pairing with complementary DNA, RNA and TNA oligonucleotides. Furthermore, Szostak and his collaborators have recently found that certain DNA polymerases can copy limited stretches of a TNA template, despite significant differences in the sugar-phosphate backbone, (Chaput and Szostak, 2003).

Finally, hexitol nucleic acids (HNA) (Fig. 6.7), whose skeleton is composed of 1, 5-anhydrohexitol (six-membered cyclic hexitol) and their isomers altritol nucleic acids (ANA), form stable duplexes with complementary oligonucleotides, and are very efficient templates since they favour assembly of a complementary strand during directed synthesis (Kozlov et al., 1999a, 1999b, 2000). The shape of the duplexes formed is reminiscent of that of DNA in the A form. Double-helical DNA is mainly in the B form[11], whereas the double helices of RNA in the DNA-RNA hybrids adopt the A form[12]. Kozlov et al. (1999c) have demonstrated that

[11] d-ribose in the 2'-endo form. The characteristics of these forms are indicated above in the text

[12] In this case, the sugar is in the 3'-endo conformation. In the A form of RNA double helices, there are 11 base pairs per helical turn (instead of 10 for the B form); the inclination of the base pairs is 16°/helical axis (20° for DNA A)

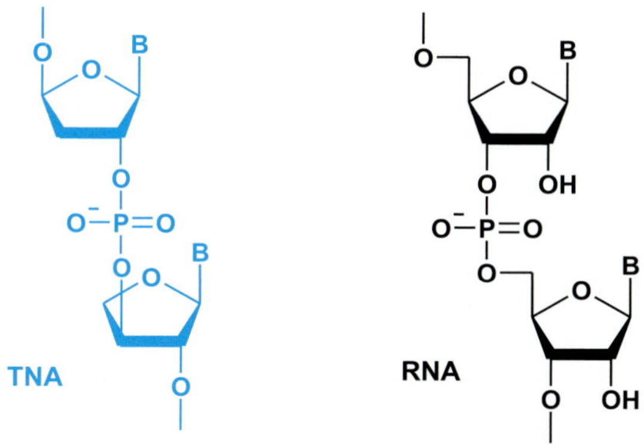

Fig. 6.8. Structure of TNA and RNA

the more the template is in the A form, the better the efficiency of directed synthesis. Based on these studies one can imagine an entire series of templates that would supply the "good" structural preorganization. Furthermore, these same authors have shown that RNA partially preorganized in the A form, is a more efficient matrix than single-stranded DNA. Finally, whatever the precursor skeleton adapted to the formation of stable duplexes may have been, the bond at the mineral surface could have imposed the necessary geometrical constraints: yet this still remains to be experimentally demonstrated.

This leads us to two major conclusions, namely that on the one hand a transition may have occurred between two different systems without loss of information, and that on the other hand the HNA and ANA nucleic acids are very efficient templates. Even if it is difficult to imagine prebiotic synthesis of these molecules, they are good model systems that show the importance of a necessary structural preorganization for directed synthesis by a template.

From the point of view of evolution, the studies described previously demonstrate that other molecules capable of transmitting hereditary information may have preceded our present day nucleic acids. This is what Cairns-Smith coined the "take-over" (Cairns-Smith, 1982), the evolutionary encroachment or genetic take-over, or to some extent what François Jacob (1970) calls genetic tinkering, in other words, making new material from the old. This also sheds light on the precision with which the various elements or processes progressively adjusted themselves, thanks to successive trials and errors.

6.4 Optimizing the Functional Capacities of Ribonucleic Acids

6.4.1 Coenzymes and Modified Nucleosides

The nucleotides that by post-transcriptional modification can today acquire the majority of functional groups present in amino acids, possess a great potential diversity that is expressed at the level of ribonucleotide coenzymes (several coenzymes derive from AMP), and of the modified bases of tRNAs (Fig. 6.9). The role of cofactors at all steps of the metabolism and their distribution within the three kingdoms suggest that a great variety of nucleotides was present in the ancestor common to all forms of life.

Several authors have underscored the possible presence of coenzymes before the appearance of the translation machinery (White, 1976). Proteins would have appeared only at a later stage, coenzymes and ribozymes being fossil traces of past catalysts. Indeed, in the living cell, only a minority of enzymes function without coenzyme; they are mostly hydrolases, and apart from this group, 70% of the enzymes require a coenzyme. If metal coenzymes involved in catalysis are considered, the number of enzymes that depend on coenzymes increases further. Present-day coenzymes, indispensable cofactors for many proteins, would be living fossils of primitive metabolism catalysts.

Most coenzymes are nucleotides (NAD, NADP, FAD, coenzyme A, ATP, etc.) or contain heterocyclic nitrogen bases that can originate from nucleotides (thiamine pyrophosphate, tetrahydrofolate, pyridoxal phosphate, etc.).

Coenzymes would be vestiges of catalytic nucleic enzymes that preceded ribosomal protein synthesis, and tRNAs can be viewed as large coenzymes participating in the transfer of amino acids. It is even possible to consider that catalytic groups that were part of nucleic enzymes were incorporated in specific amino acids rather than being "retained" as coenzymes. This could be the case of imidazole, the functional group of histidine, whose present synthesis in the cell is triggered by a nucleotide.

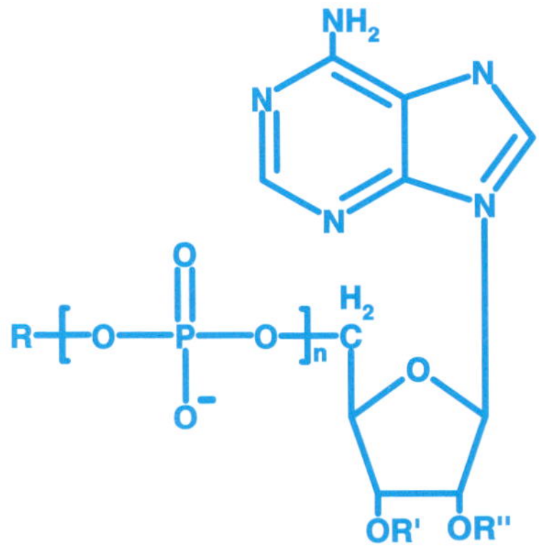

Coenzyme	R	R'	R''	n
Activated methionine	methionine	H	H	0
Amino acid adenylate	amino acid	H	H	1
Activated sulfate	SO_3^{2-}	H	PO_3^{2-}	1
Cyclic 3'-5' AMP	H	H	PO_3^{2-}	1
NAD		H	H	2
NADP		PO_3^{2-}	H	2
FAD		H	H	2
CoA-SH		H	H	2

Fig. 6.9. List of coenzymes derived from AMP

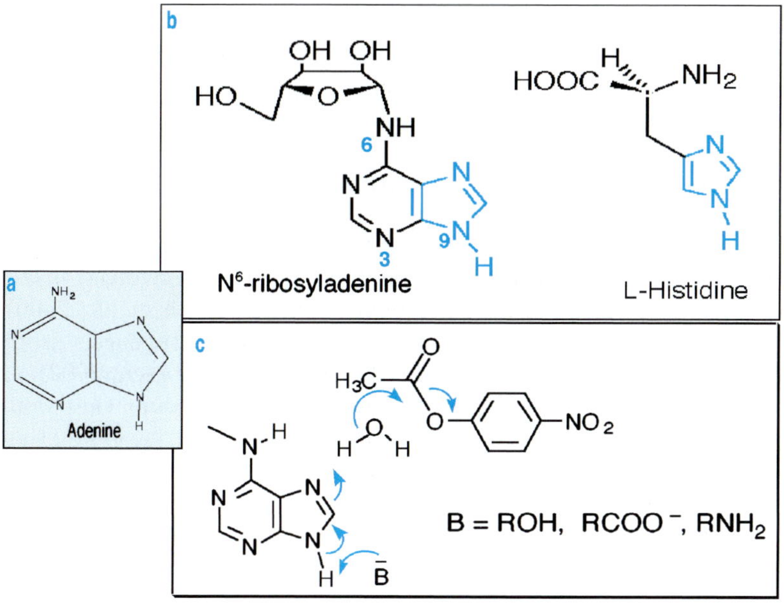

Fig. 6.10. (**a**) Adenine; (**b**) Comparison of modified adenosine and histidine; (**c**) Catalytic activity of adenine residue

The modified nucleosides present today in RNAs result from post-transcriptional modifications. Nevertheless, modified nucleosides could have been present in the primitive world and their distribution would have become established in the RNAs of the three living kingdoms (Cermakian and Cedergren, 1998).

Our working hypothesis is based on the demonstration of esterase activity in a nucleoside analogue N^6-ribosyladenine (Fuller et al., 1972; Maurel and Ninio, 1987). This activity, which is due to the presence of an imidazole group that is free and available for catalysis, is comparable to that of histidine placed in the same conditions (Fig. 6.10). We have studied the kinetic behaviour of this type of catalyst (Ricard et al., 1996) and have shown that the catalytic effect increases greatly when the catalytic element, pseudohistidine, is placed in a favourable environment within a macromolecule (Décout et al., 1995). Moreover, primitive nucleotides were not necessarily restricted to the standard nucleotides encountered today, and because of their replicative and catalytic properties, the N6 and N3 substituted derivatives of purines could have constituted essential links between the nucleic acid world and the protein world.

6.4.2 The Case of Adenine

Purine nucleotides, and in particular those containing adenine, participate in a large variety of cellular biochemical processes (Maurel and Décout, 1999).

Their best-known function is that of monomeric precursors of RNAs and DNAs. Nevertheless, derivatives of adenine are universal cofactors. They serve in biological systems as a source of energy (ATP), allosteric regulators of enzymatic activity and regulation signals (cyclic AMP). They are also found as acceptors during oxidative phosphorylation (ADP), as components of coenzymes (such as in FAD, NAD, NADP, coenzyme A), as transfer agents of methyl groups and of S-adenosylmethionine, as possible precursors of polyprenoids in C5 (adenosylhopane) (Neunlist et al., 1987), and – last but not least – adenine 2451 conserved within the large rRNA in the three kingdoms, would be involved in acid-base catalysis during the formation of the peptide bond (Muth et al., 2000). However, this role of adenine has been refuted based on mutagenesis studies and phylogenetic comparisons (Muth et al. 2001; Green and Lorsch, 2002).

On the other hand, biosynthesis of an amino acid, histidine, that would have appeared late in evolution, begins with 5-phosphoribosyl-1-phosphate that forms N'-(5-phosphoribosyl)-ATP by condensation with ATP. This reaction is akin to the initial reaction of purine biosynthesis. Finally, the ease with which purine bases are formed in prebiotic conditions[13] (Orò, 1960) suggests that these bases were probably essential components of an early genetic system. The first genetic system was probably capable of forming base pairs of the Watson–Crick type, Hoogsteen and other atypical associations, by hydrogen bonds as they still appear today in RNA. It probably contained a different skeleton from that of RNA, and no doubt also modified bases, thereby adding chemical functions, but also hydrophobic groups, and functions such as amine, thiol, imidazole, etc. Wächterhäuser (1988) also suggested novel pairings of the purine–purine type.

Originally, the principle probably rested on forced cooperation of genetic and functional components, rather than on selection by individual competition. It may have first entailed testing and improvements (learning by trial and error) of the informational content of the genes, i.e. linking the genotype (sequence) to the phenotype (shape). One can consider that in such a system the unforeseen was faced, so that the living organism would need to adapt favourably and rapidly.

6.4.3 Mimicking Darwinian Evolution

Most of the "rational" biochemical approaches consist of deducing the active sequence of a nucleic acid or protein from a primary sequence, or in synthesizing a defined compound by modelling and structural analysis. However, "real life", that of our ancestors as also that of our cells, does not proceed in this manner. The hunter-gatherers of prehistory survived only thanks to their extraordinary capacity to recognize objects. In addition, survival of a population in a new environment is often linked to the appearance of a few variants to which random mutations conferred the power to adapt and exploit the new situation to their advantage. Combinatorial methods, by modelling these observations[14], have now

[13] Purines have also been found in the Murchison meteorite

[14] And by giving access to many related molecules that can be sorted

become the alternative to the rational concept. Selection in vitro requires no information concerning the sequence of the molecules, and replaces the pre-established adjustments between the molecule and its target. What is needed, is to mimic the processes of evolution at the molecular level.

Indeed, it has been known since the experiments of Spiegelman (1971) and his colleagues (Kramer et al., 1974) that populations of different molecules capable of reproducing themselves in a hereditary manner, can evolve and adapt to an appropriate environment. Spiegelman, the inventor of non-natural selection indeed demonstrated in the 1960s, that RNA populations can evolve when they replicate with the help of an enzyme, the replicase of the bacteriophage $Q\beta$. A population of macromolecules can thus comply with the prerequisites of Darwinian theory, and must find a form adapted to recognition of the target in a sufficiently rich population. Coexistence in the same entity of shape and sequence, can favour the emergence of favourable candidates by means of a selection step (linked to the shape) and an amplification step (linked to the sequence) at the end of this molecular evolutionary process. A selection of this type could have occurred during early molecular evolution, some 3.5 billion years ago.

The original polymers more or less related to RNA and formed in the primitive world must have randomly contained the A, U, G and C bases. There are over one million possible sequences for a decanucleotide composed of 10 monomers A, U, G, C, and over 10^{12} sequences for a polynucleotide of 20 monomers[15]. Nature does not appear to have exploited all the possible combinations before having reached the remarkable functional unity of the living world, and given the immense number of possibilities it is also useless to try to explore experimentally, one by one, all the potentially functional sequences.

The SELEX method (systematic evolution of ligands by exponential enrichment) (Tuerk and Gold, 1990) is an efficient, quasi automatic method based on repeated cycles of reproductive selection of those individuals that are best adapted to a given function. Established in the 1990s, this method makes it possible to obtain new structures, aptamers, selected through their aptitude to recognize other molecules (Ellington and Szostak, 1990). Aptamers are capable to recognize targets as small as metal ions, or as large as cells. They can interact with a great variety of molecules that are important for primitive metabolism, like amino acids, porphyrines, nucleotide factors, coenzymes, small peptides and short oligonucleotides (Illangasekare and Yarus, 1997; Jadhav and Yarus, 2002; Joyce, 2002; McGinness et al., 2002; Reader and Joyce, 2002).

At the molecular level, the Darwinian behaviour requires that a method of selection (RNA-aptamers), of amplification of selected species, and of mutations (introduction of variants in the population by means of mutations) be established. Through several cycles of selection, amplification and mutations,

[15] For a nucleic acid of 200 nucleotides, 10^{120} different sequences are theoretically possible, and for a small protein containing 200 amino acids, 10^{280} arrangements are possible! Which also applies to the protein world (phage display and combinatorial synthesis of peptides)

populations of molecules are "pushed" to evolve towards novel properties. The molecules presenting the best "aptitudes" are selected and a new generation will thereby come out. Evolutionary processes performed experimentally thus make it possible for molecules to emerge that have not yet been produced by Nature, or allow the re-emergence of precursor molecules that have strongly diverged or naturally disappeared.

In practice, how does one proceed? A "library" of oligonucleotides is a conformations population containing at least one particular conformation able to regonize the molecular target we are interested in (Fig. 6.11). The protocol is composed of five steps: the creation of double-stranded DNA carrying the random "box"[16] flanked by regions required for amplification; transcription of this DNA into single-stranded RNA; selection; production of a DNA population by reverse transcription and PCR of the sequences retained during the selection step, then cloning and sequencing of the strands obtained after a certain number of selection and amplification cycles.

From a vast combination of nucleic acids, one can isolate aptamers that possess catalytic properties (RNA ligation, cleavage or synthesis of a peptide bond, transfer of an aminoacyl group, etc.). The first nucleic acids could possess independent domains, separated by flexible segments, creating reversible conformational motifs, dependent on ions and bound ligands. Thus, a peptide that is 10 amino-acids long can recognize fine structural differences within a micro-RNA helix (discrimination can be made between two closely placed microhelices). Just as protein and antibodies, RNA molecules can present hollows, cavities, or slits that make these specific molecular recognitions possible. RNAs must "behave as proteins". Whatever the chronology and the order of appearance of the various

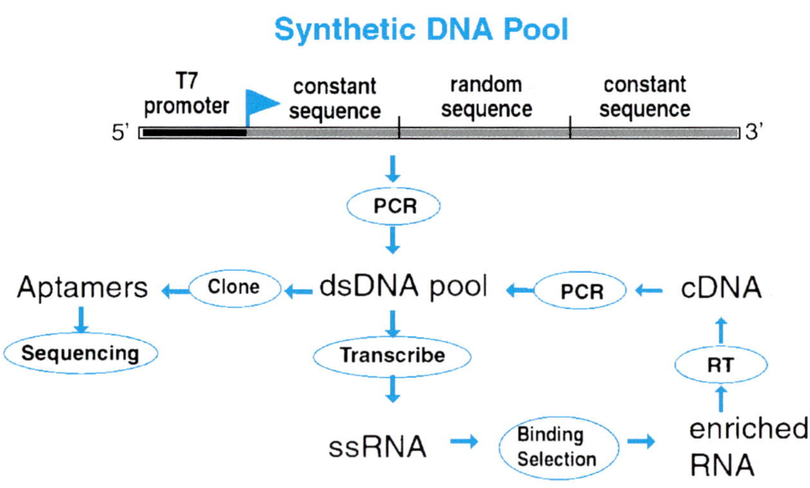

Fig. 6.11. The SELEX method (adapted from Wilson and Szostak, 1999)

[16] Region of defined length, for instance, of some 50 randomly aligned nucleotides

classes of molecules, the importance lies in the shape, the scaffolding and the architecture that have allowed functional associations.

Starting from a heterogeneous population of RNAs with 10^{15} variants (a population of 10^{15} different molecules) we have selected 5 populations of RNAs capable of specifically recognizing adenine after about ten generations (Meli et al., 2002). When cloned, sequenced and modelled, the best one among the individuals of these populations, has a shape reminiscent of a claw capable of grasping adenine. Is it the exact copy of a primitive ribo-organism that feeds on prebiotic adenine in prebiotic conditions? Functional and structural studies presently under way will highlight other activities, other conformations, etc.

Following this line of investigation we have selected two adenine-dependent ribozymes capable of triggering reversible cleavage reactions (Fig. 6.12). One of them is also active with imidazole alone. This result leads to very important perspectives (Meli et al., 2003).

A considerable amount of research has been focused on the selection of ribozymes in vitro. Recently, it was demonstrated that a ribozyme is capable of continuous evolution, adding successively up to 3 nucleotides to the initial molecule (McGuinnes, 2002). It is also possible to construct a ribozyme with only two different nucleotides, 2,6-diaminopurine and uracil (Reader and Joyce, 2002). Finally, Bartel and coworkers have selected a ribozyme-polymerase, capable of self-amplification (Johnston et al., 2001).

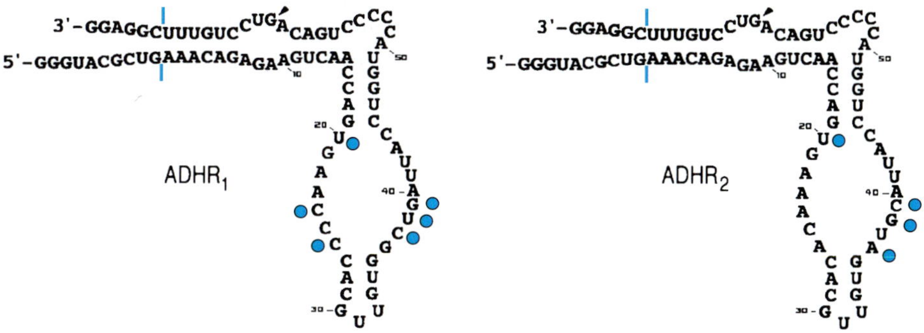

Fig. 6.12. Adenine-dependent hairpin ribozymes (ADHR). *Arrowheads*: cleavage sites; *Grey dots*: degenerated (mutated) sites; *Vertical bars*: separation between the primer binding region and the random sequence

6.4.4 Other Perspectives

Very little is known to date about the behaviour of macromolecules in "extreme" environments. How do structures behave? What are the major modifications observed? What are the conditions of structural and functional stability? How are the dynamics of the macromolecules and their interactions affected? What are the possibilities of conserving biological macromolecules in very ancient soils or in meteorites? Can we find traces of these macromolecules as molecular biosignatures, and if so in what form (Maurel and Zaccaï, 2001; Tehei et al., 2002)?

The selection of thermohalophilic aptamers, RNAs resistant to high temperatures (80 °C) in the presence of salt (halites that are 30 million years old), undertaken in our laboratory, will perhaps allow us to answer some of these questions, that are fundamental for the search of past traces of life, and of life on other planets.

6.5 Conclusion

The RNA world thus contains innumerable perspectives. The combination of methods available today are the best adapted to explore the vast combinations of nucleic acids but also of peptides. Will they make it possible to reconstitute the first steps of the living world? Attractive simulations may emerge, opening new evolutionary paths that have not been envisaged or that Nature has not yet explored.

The RNA world, at whatever step we place it in the history of the living world, must be considered as a step in the history of life, an important step in the evolution of the contemporary cellular world. Because of its strong explanatory power, it also constitutes an important opening in the scientific study of the origin of life. Even if this concept does not explain how life appeared, it nevertheless promises a great number of experimental breakthroughs.

Acknowledgement

Figure 6.4 is reprinted from *Cell*, 2003,112, Barends S., Bink H.H.J., van den Worm S.H.E., Pleij C.W.A., Kraal B. Entrapping ribosomes for viral translation: tRNA mimicry as a molecular trojan horse. Copyright 2003, with permission from Elsevier. We also thank Dr. G.F. Joyce for his constructive comments on the manuscript.

References

Bachellerie J.P., Cavaillé J., Hüttenhofer A. (2002). The expanding snoRNA world. *Biochimie*, **84**, 775–790.

Ban N., Nissen P., Hansen J., Moore P.B., Steitz, T.A. (2000). The complete atomic structure of the large ribosomal subunit at 2.4 Å resolution. *Science*, **289**, 905–920.

Barends S., Bink H.H.J., van den Worm S.H.E., Pleij C.W.A., Kraal B. (2003). Entrapping ribosomes for viral translation: tRNA mimicry as a molecular trojan horse. *Cell*, **112**, 123–129.

Bartel D.P., Unrau P.J. (1999). Constructing an RNA world. *Trends Biochem. Sci.*, **24**, 9–13.

Benner S.A., Ellington A.D., Tauer A. (1989). Modern metabolism as a palimpsest of the RNA world. *Proc. Natl. Acad. Sci. USA*, **86**, 7054–7058.

Benner S.A., Cohen M.A., Gonnet G.H., Berkowitz D.B., Johnson K.P. (1993). Reading the palimpsest: contemporary biochemical data and the RNA World, in *The RNA World*, eds. Gesteland R.F., Atkins J.F., p. 27–70, Cold Spring Harbor Laboratory Press: Cold Spring Harbor NY.

Cairns-Smith A.G. (1966). The origin of life and the nature of the primitive gene. *J. Theor. Biol.*, **10**, 53–88.

Cairns-Smith A.G. (1982). *Genetic Takeover and the Mineral Origins of Life*, Cambridge University Press, Cambridge.

Cermakian N., Cedergren, R. (1998). Modified nucleosides always were: an evolutionary model, in *Modification and Editing of RNA*, ed. Grosjean H., Benne R., p. 535–541, ASM Press, Washington, D.C.

Chaput J.C., Szostak J.W. (2003). TNA synthesis by DNA polymerase. *J. Am. Chem. Soc.*, **125**, 9274–9275.

Cooper G., Kimmich N., Belisle W., Sarinana J., Brabham K., Garrel L. (2001). Sugar-related organic compounds in carbonaceous meteorites. *Nature*, **414**, 879–883.

Crick F.H. (1968). The origin of the genetic code. *J. Mol. Biol.*, **38**, 367–379.

Décout J-L., Vergne J., Maurel M-C. (1995). Synthesis and catalytic activity of adenine containing polyamines. *Macromol. Chem. Phys.*, **196**, 2615–2624.

Diener TO. (2001). The viroid: biological oddity or evolutionary fossil? *Adv. Virus. Res.* **57**, 137–184.

Eddy S.R. (2001). Non-coding RNA genes and the modern RNA world. *Nature Reviews Genetic*, **2**, 919–929.

Egholm M., Buchardt O., Christensen L., Behrens C., Freier S.M., Driver D.A., Berg R.H., Kim S.K., Norden B., Nielsen P.E. (1993). PNA hybridizes to complementary oligonucleotides obeying the Watson–Crick hydrogen-bonding rules. *Nature*, **365**, 566–568.

Ellington A.D., Szostak J.W. (1990). In vitro selection of RNA molecules that bind specific ligands. *Nature*, **346**, 818–822.

Eschenmoser A. (1994). Chemistry of potentially prebiological natural products. *Origins Life Evol. Biosphere*, **24**, 389–423.

Eschenmoser A. (1999). Chemical etiology of nucleic acid structure. *Science*, **284**, 2118–2124.

Fechter P., Rudonger-Thirion J., Florentz C., Giegé R. (2001). Novel features in the tRNA-like world of plant viral RNAs. *Cell. Mol. Life. Sci.* **58**, 1547–1561.

Ferris J.P. (1987). Prebiotic synthesis: problems and challenges *Cold Spring Harbor Symp. Quant. Biol.*, **LII**, 29–39.

Ferris J.P., Ertem G. (1992). Oligomerization of ribonucleotides on montmorillonite: reaction of the 5′ phosphorimidazolide of adenosine. *Science*, **257**, 1387–1389.

Ferris J.P., Hill A.R., Liu R., Orgel L.E. (1996). Synthesis of long prebiotic oligomers on mineral surfaces. *Nature*, **381**, 59–61.

Fuller W.D., Sanchez R.A., Orgel L.E. (1972). Studies in prebiotic synthesis. *J. Mol. Biol.*, **67**, 25–33.

Gesteland R.F., Cech T.R., Atkins J.F. (eds.) (1999). *The RNA World*, second edition, Cold Spring Harbor Laboratory Press, Cold Spring Harbor, NY.

Gilbert W. (1986). The RNA world. *Nature*, **319**, 618.

Green R., Lorsch J.R. (2002). The path to perdition is paved with protons. *Cell*, **110**, 665–668.

Grosjean H., Benne R. (eds.) (1998). *Modification and Editing of RNA*, ASM Press, Washington, D.C.

Grosshans H., Slack F.J. (2002). Micro-RNAs: small is plentiful. *J. Cell Biol.*, **156**, 17–21.

Guerrier-Takada C., Gardiner K., Marsh T., Pace N., Altman S. (1983). The RNA moiety of ribonuclease P is the catalytic subunit of the enzyme. *Cell*, **35**, 849–857.

Hill A.R., Orgel L.E., Wu T. (1993). The limits of template-directed synthesis with nucleoside-5′-phosphoro (2-methyl) imidazolides. *Orig. Life Evol. Biosphere*, **23**, 285–290.

Illangasekare M., Yarus M. (1997). Small-molecule substrate interactions with a self aminoacylating ribozyme. *J. Mol. Evol.*, **54**, 298–311.

Inoue T., Orgel L.E. (1983). A non-enzymatic RNA polymerase model. *Science*, **219**, 859–862.

Jacob F. (1970). *La Logique du Vivant*, Gallimard, Paris.

Jadahv V.R., Yarus M. (2002). Coenzymes as coribozymes. *Biochimie*, **84**, 877–888.

Johnston W.K., Unrau P.J., Lawrence M.S., Glasner M.E., Bartel D.P. (2001). RNA-catalyzed RNA polymerization: Accurate and general RNA-templated primer extension. *Science*, **292**, 1319–1325.

Joyce G.F., Orgel L.E. (1986). Non-enzymic template-directed synthesis on RNA random copolymers: poly (C,G) templates. *J. Mol. Biol.*, **188**, 433–441.

Joyce G.F., Schwartz A.W., Miller S.L., Orgel L.E. (1987). The case for an ancestral genetic system involving simple analogues of the nucleotides. *Proc. Natl. Acad. Sci. USA*, **84**, 4398–4402.

Joyce G.F. (1989). RNA evolution and the origins of life. *Nature*, **338**, 217–224.

Joyce G.F., Orgel L.E. (1999). Prospects for understanding the origin of the RNA world, in *The RNA World*, eds. Gesteland R.F., Cech T.R., Atkins J.F. p. 49–77, Cold Spring Harbor Laboratory Press: Cold Spring Harbor, NY.

Joyce G.F. (2002). The antiquity of RNA-based evolution. *Nature*, **418**, 214–221.

Kable M.L., Heidmann S., Stuart K.D. (1997). RNA editing: getting U into RNA. *Trends Biochem. Sci.*, **22**, 162–166.

Kong L.B., Siva A.C., Kickhoefer V.A., Rome L.H., Stewart P.L. (2000). RNA location and modeling of a WD40 repeat domain within the vault. *RNA*, **6**, 890–900.

Koppitz M., Nielsen P.E., Orgel L.E. (1998). Formation of oligonucleotide-PNA-chimeras by template-directed ligation. *J. Am. Chem. Soc.*, **120**, 4563–4569.

Kozlov I., Politis P.K., Pitsch S., Herdewijn P., Orgel L.E. (1999a). A highly enantioselective hexitol nucleic acid template for nonenzymatic oligoguanylate synthesis. *J. Am. Chem. Soc.*, **121**, 1108–1109.

Kozlov I., De Bouvere B., Van Aerschot A. Herdewijn P., Orgel L.E. (1999b). Efficient transfer of information from hexitol nucleic acids to RNA during nonenzymatic oligomerization. *J. Am. Chem. Soc.*, **121**, 5856–5859.

Kozlov I., Politis P.K., Van Aerschot A. Busson R., Herdewijn P., Orgel L.E. (1999c). Nonenzymatic synthesis of RNA and DNA oligomers on hexitol nucleic acid templates: the importance of A structure. *J. Am. Chem. Soc.*, **121**, 2653–2656.

Kozlov I., Zielinski M., Allart B., Kerremans L., Van Aerschot A, Busson R., Herdewijn P., Orgel L.E. (2000). Nonenzymatic template-directed reactions on altritol oligomers, preorganized analogues of oligonucleotides. *Chem. Eur. J.*, **6**, 151–155.

Kramer F.R., Mills D.R., Cole P.E., Nishihara T., Spiegelman S. (1974). *J. Mol. Biol.*, **89**, 719–736.

Lafontaine D.L., Tollervey D. (1998). Birth of the snoRNPs: the evolution of the modification-guide snoRNAs. *Trends Biochem. Sci.*, **83**, 383–388.

Lamond A.I. (1988). RNA editing and the mysterious undecovered genes of trypanosomatid mitochondria. *Trends Biochem. Sci.*, **13**, 283–284.

Maizels N., Weiner A.M., Yue D., Shi P.Y. (1999). New evidence for the genomic tag hypothesis: archaeal CCA-adding enzymes and DNA substrates. *Biol. Bull.*, **196**, 331–333.

Mattick J.S. (2003). Challenging the dogma: the hidden layer of non-protein-coding RNAs in complex organisms. *BioEssays*, **25**, 930–939.

Maurel M.-C., Ninio J. (1987). Catalysis by a prebiotic nucleotide analog of histidine. *Biochimie*, **69**, 551–553.

Maurel M.-C. (1992). RNA in evolution. *J. Evol. Biol.*, **2**, 173–188.

Maurel M.-C., Décout J.-L. (1999). Origins of life: molecular foundations and new approaches. *Tetrahedron*, **55**, 3141–3182.

Maurel M.-C., Zaccaï G. (2001). Why Biologists should support the exploration of Mars. *BioEssays*, **23**, 977–978.

McGinness K.E., Wright M.C., Joyce G.F. (2002). Continuous in vitro evolution of a ribozyme that catalyzes three successive nucleotidyl addition reactions. *Chem. Biol.*, **9**, 585–596.

Meli M., Albert-Fournier B., Maurel M.-C. (2001). Recent findings in the modern RNA world. *Int. Microbiol.* **4**, 5–11.

Meli M., Vergne J., Décout J.-L., Maurel M.-C. (2002). Adenine-aptamer complexes. A bipartite RNA site which binds the adenine nucleic base. *J. Biol. Chem.*, **277**, 2104–2111.

Meli M., Vergne J., Maurel M.-C. (2003). *In vitro* selection of adenine-dependent hairpin ribozymes. *J. Biol. Chem.*, **278**, 9835–9842.

Muth G.W., Ortoleva-Donnelly L., Strobel S.A. (2000). A single adenosine with a neutral pKa in the ribosomal peptidyl transferase center. *Science*, **289**, 947–950.

Muth G.W., Chen L., Kosek A.B., Strobel S.A. (2001). PH-dependent conformational flexibility within the ribosomal peptidyl transferase center. *RNA*, **7**, 1403–1415.

Neunlist S., Bisseret P., Rohmer M. (1987). The hopanoids of the purple non-sulfur bacteria *Rhodopseudomonas palustris* and *Rhodopseudomonas acidophila* and the absolute configuration of bacteriohopanetetrol. *Eur. J. Biochem.*, **87**, 245–252.

Nissen P., Hansen J., Ban N., Moore P.B., Steitz T.A. (2000). The structural basis of ribosome activity in peptide bond synthesis. *Science*, **289**, 920–930.

Nykanen A., Haley B., Zamore P.D. (2001). ATP requirements and small interfering RNA structure in the RNA interference pathway. *Cell*, **107**, 309–321.

Orgel L.E. (1968). Evolution of the genetic apparatus. *J. Mol. Biol.*, **38**, 381–393.

Orgel L.E. (1989). Was RNA the first genetic polymer? in *Evolutionary Tinkering in Gene Expression* eds. M. Grunberg-Manago et al., p. 215–224, Plenum Press, London.

Orgel L.E. (1992). Molecular replication. *Nature*, **358**, 203–209.

Orò J. (1960). *Biochem. Biophys. Res. Comm.*, **2**, 407–412.

Paecht-Horowitz M., Berger J., Katchalsky A. (1970). Prebiotic synthesis of polypeptides by heterogeneous polycondensation of amino acid adenylates. *Nature*, **7**, 847–850.

Reader J.S., Joyce G.F. (2002). A ribozyme composed of only two different nucleotides. *Nature*, **420**, 841–844.

Ricard J., Vergne J., Décout J.-L., Maurel M.-C. (1996). The origin of kinetic cooperativity in prebiotic catalysts. *J. Mol. Evol.*, **43**, 315–325.

Schmidt J.G., Nielsen P.E., Orgel L.E. (1997). Information transfer from peptide nucleic acid RNA by template-directed syntheses. *Nucl. Acids Res.*, **25**, 4797–4802.

Schöning K.-U., Scholz P., Guntha S., Wu X., Krishnamurthy R., Eschenmoser A. (2000). Chemical etiology of nucleic acid structure: the α-threo-furanosyl-(3'$\rightarrow$2') oligonucleotide system. *Science*, **290**, 1347–1351.

Spiegelman S. (1971). An in vitro analysis of a replicating molecule. *Quaterly Review Biophys.*, **4**, 213–253.

Stuart K., Panigrahi A.K. (2002). RNA editing: complexity and complications. *Mol. Microbiol.*, **45**, 591–596.

Sutherland J.D., Whitfield J.N. (1997). Prebiotic chemistry: a bioorganic perspective. *Tetrahedron*, **53**, 11493–11527.

Tarn W.Y., Steitz J.A. (1997). Pre-mRNA splicing: the discovery of a new spliceosome doubles the challenge. *Trends Biochem. Sci.*, **22**, 132–137.

Tehei M., Franzetti B., Maurel M.-C., Vergne J., Hountondji C., Zaccaï G. (2002). Salt and the search for traces of life. *Extremophiles*, **6**, 427–430.

Terns M.P., Terns R.M. (2002). Small nucleolar RNAs: versatile transacting molecules of ancient evolutionary origin. *Gene Expr.*, **10**, 17–39.

Tuerk, C., Gold, L. (1990). Systematic evolution of ligands by exponential enrichment: RNA ligands to bacteriophage T4 DNA polymerase. *Science*, **249**, 505–510.

Valle M., Gillet R., Kaur S., Henne A., Ramakrishnan V., Frank J. (2003). Visualizing tmRNA entry into a stalled ribosome. *Science*, **300**, 127–130.

Wächtershäuser G. (1988). An all-purine precursor of nucleic acids. *Proc. Natl. Acad. Sci. USA*, **85**, 1134–1135.

Westhof E. (2002). Foreword. *Biochimie*, **84**, 687–689.

White H.B. (1976). Coenzymes as fossils of an earlier metabolic state. *J. Mol. Evol.*, **7**, 101–104.

Wild K., Weichenrieder O., Strub K., Sinning I., Cusack S. (2002). Towards the structure of the mammalian signal recognition particle. *Curr. Opinion Struct. Biol.*, **12**, 72–81.

Wilson D.S., Szostak J.W. (1999). In vitro selection of functional nucleic acids. *Annu. Rev. Biochem.*, **68**, 611–647.

Withey J.H., Friedman D.I. (2002). The biological roles of trans-translation. *Curr. Opinion Microbiol.*, **5**, 154–159.

Woese C.R. (1965). On the evolution of the genetic code. *Proc. Natl. Acad. Sci. USA*, **54**, 1546–1552.

Yusupov M.M., Yusupova G.Z., Baucom A., Lieberman K., Earnest T.N., Cate J.H., Noller H.F. (2001). Crystal Structure of the Ribosome at 5.5 Å resolution. *Science*, **292**, 883–896.

Zamore P.D. (2002). Ancient Pathways Programmed By Small RNAs. *Science*, **296**, 1265–1269.

7 Looking for the Most 'Primitive' Life Forms: Pitfalls and Progresses

Simonetta Gribaldo, Patrick Forterre

Two complementary approaches can be applied to try to understand how life appeared on our planet. The first, illustrated by Miller's seminal experiment, consists in trying to reproduce – either theoretically or/and practically – the conditions of primitive Earth, and to figure out which possible scenarios might have led from such conditions to the apparition of life as we know it today. This is the type of approach frequently adopted by chemists and physicists. The second approach, favoured by biologists, consists in looking at current organisms for hints of their most remote history. In this chapter we will mostly discuss this second approach.

Clearly, no barrier exists between researchers who look at the past to understand present time, and those who do the opposite. In fact, in order to reach a coherent view of the transition from unanimated matter to life, the scenarios suggested by both approaches must eventually converge. It is sensible to keep in mind this dichotomy of approaches, in order to be able to reach, in the end, their complementation. In fact, such a dichotomy raises a fundamental question that is rarely taken into consideration, i.e. at what time, over the history of life evolution on our planet, should the Darwinian notion of natural selection be called into play? In other words, when did chemistry give way to biology? The answer to this question is crucial to estimate the probability of apparition of life on other planets, in a starting environment compatible with life development.

In this chapter we will mostly focus on the quest, in the present-day biosphere, of "relic" cells that would have conserved the features harboured by the last common ancestor of all extant life forms, i.e. the LUCA (Last Universal Common Ancestor; Forterre, 1996a). Over the last several years, a few authors have suggested that life on our planet might have arisen at high temperatures and that LUCA was itself a hyperthermophile (i.e., thriving at temperatures close to the boiling point of water). These two hypotheses (a hot origin of life and a hot-loving LUCA) have become very popular amongst exobiologists. We will thus dedicate a large part of this chapter to discuss this issue. Moreover, given the current revolution in biology due to the sequencing of complete genomes from a growing number of diverse organisms, we will also focus on the most recent advances of comparative genomics, and whether they can disclose any detail on the nature of LUCA.

7.1 Simpler Doesn't Necessarily Mean Older!

One of the most traditional approaches of biologists interested in life origin issues is to search for the simplest extant cellular life forms, with the assumption that their features may date back to the first cells that appeared on our planet. This would make it easier to trace back the steps that drove to these first cells. In such a quest, simplicity and small size are often considered as ancestral characters (see, for example, the interest given to the supposedly primitive martian "nanobacteria"). However, this type of reasoning is risky and has already resulted in erroneous judgements; in fact, what is simple is not necessarily primitive, but can be the product of secondary simplification. For example, mycoplasma have been historically considered as candidates for primitive cells, since they are the smallest extant Bacteria. However, we now know that they are indeed ancient – and "complex" – Bacteria that have lost their cell wall and most of their biosynthesis abilities, due to adaptation to a parasitic lifestyle (Fig. 7.1). Nonetheless, mycoplasma still represent precious model organisms to both experimental and theoretical research on minimal genomes. Indeed, the smallest mycoplasma genome harbours less than 500 genes for a total of 500 000 pairs of nucleotides (0.5 mega bases, or MB), while the size of classical bacterial

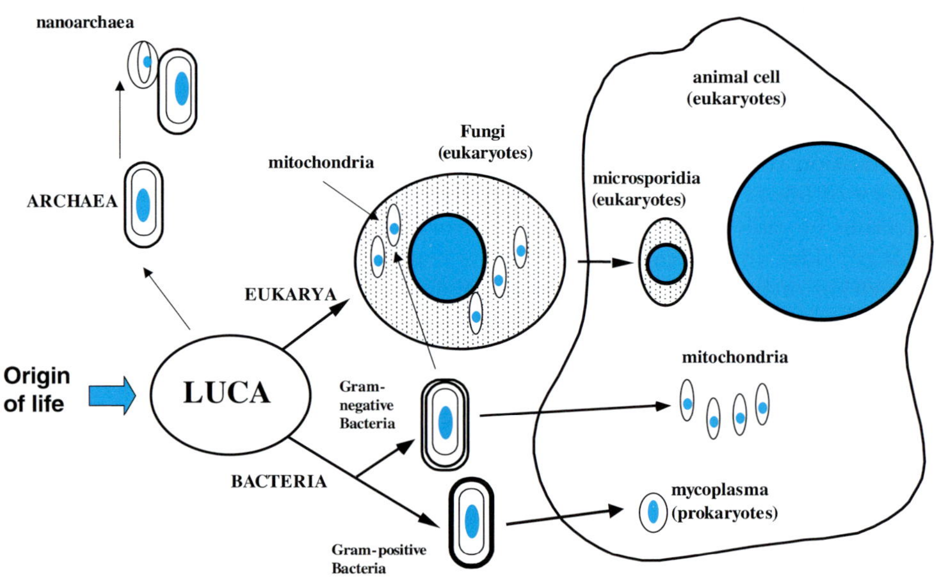

Fig. 7.1. Evolutionary mechanisms based on simplification and size reduction have played a major role in life's history. This sketch shows the origin of mycoplasma from Gram-positive Bacteria, the origin of mitochondria from Gram-negative Bacteria (note the loss of cell wall in both cases), and the origin of microsporidia from Fungi following mitochondrial loss. In all these cases, a reduction of the size of their genomes (*blue*) has occurred

genomes ranges from 1.5 to 10 MB. Recently, the interest in minimal genomes was renewed by the description of a nanosized hyperthermophilic archaeon from a submarine hot vent (Huber, 2002). This is the smallest archaeal genome – 0.5 MB – identified to date. Here again we are most likely in front of the product of a secondary simplification since this species leads a parasitic life onto the surface of a specific "normal-sized" archaeal host (Fig. 7.1). The same conclusions are also valid for eukaryal minimal genomes. In fact, it was recently shown that microsporidia, unicellular amitochondriate eukaryal parasites of animal cells, are not primitive Eukarya, as early molecular phylogenetic analyses seemed to indicate, but secondarily simplified fungi (Fig. 7.1) (Vivares, 2000). The smallest known microsporidial genome, sequenced at Génoscope, is of 2.9 MB, similarly to a medium-sized bacterium.

In conclusion, although the first cells were certainly very simple, the simplest extant cellular forms are instead their extremely evolved and specialized descendants; this hampers any attempt to reconstruct the past by starting from the present, and all the more any extrapolation of our knowledge on the present-day terrestrial biosphere to possible extraterrestrial ones.

7.2 Hyperthermophiles are not Primitives, but are Remnants from Thermophilic Organisms

7.2.1 Hyperthermophiles and the Hypothesis of a Hot Origin of Life

The hypothesis that life appeared in extremely hot environments (submarine hydrothermal vents) or directly within the Earth's mantle, either within or nearby oceanic ridges (interfaces between cold-fluid infiltrations and gas produced at their contact with the magma) is appealing to most exobiologists. Indeed, under such a hypothesis, the numbers of either solar or extra-solar planets that may have hosted life would rise enormously, given that initial atmosphere conditions and distance from a star no longer come into play. Moreover, if life arose in such conditions (i.e., far from Earth's surface) its first evolution steps would have been protected from the cataclysmic meteoritic bombardment that took place at the same time as planet formation. Consequently, hyperthermophiles (hot-loving micro-organisms thriving at temperatures from 80° to 113°C Forterre, 1999b) now replace mycoplasma as candidates for "primitive cells" in the majority of generic publications on life origins and textbooks (Madigan, 2000). A number of authors, such as Carl Woese and Norman Pace in the USA, and Carl Stetter and Gunther Wachtershäuser in Europe, have indeed supported the hypothesis of a hot origin of life up to LUCA (Pace 1991). According to these authors, LUCA would have appeared and lived on a still-heated Earth, and extant hyperthermophiles would be its direct descendants (see Ciba Foundation publications, 1996; and Wiegel, 1998). Indeed, terrestrial and submarine volcanic biotopes where microbiologists search for hyperthermophiles are environments traditionally associated to primitive Earth in the collective imaginary.

7.2.2 Hyperthermophiles are Complex Prokaryotes

The evidence that all extant hyperthermophiles are prokaryotes (cells without a nucleus) is a priori in agreement with a hyperthermophilic LUCA, since prokaryotes are generally considered as more representative of primitive cells with respect to Eukarya (we will get back on this important issue later). Indeed, most hyperthermophiles are Archaea (Fig. 7.2) – a definition evoking presumed ancient origins and an archaic nature.

However, molecular analyses have failed to highlight any primitive character in hyperthermophiles, but rather the opposite (Forterre, 1996b). For example, life at high temperatures needs specific lipids able to keep the cytoplasmic membrane impermeable to protons; these are giant lipids made up of two "classic" phospholipids covalently linked (tail-to-tail) by a sophisticated enzymatic system (see Sect. 7.2.3) (Alberts, 2000). Moreover, ribosomal and transfer RNAs from hyperthermophiles harbour modifications that increase their stability at high temperatures (Noon, 1998). Such modifications (frequently the addition of methyl groups) allow a larger range of hydrophobic interactions involved in RNA structuring that may protect RNA molecules from thermal degradation (see Sect. 7.2.4).

Giant lipids and RNA modifications do exist also in mesophilic (thriving at temperatures between 20–40°C and psycrophilic (cold-loving) organisms (Shouten, 2000). Instead, an exclusivity of hyperthermophiles is represented by reverse gyrase, an enzyme that introduces positive supercoiling in DNA molecules (Fig. 7.3) (Forterre, 1995a). In supercoiled DNA helices, the axis of the Watson and Crick double helix folds on itself in the same sense as the double helix (which is conventionally assumed to be positive). This positive supercoiling increases the number of topological links between the two strands of the DNA molecule, making more difficult the local opening-up of the helix, and consequently preventing the risk of denaturation that might occur at temperatures of 80°C and higher. Although the essential role of reverse gyrase at high temperature remains to be rigorously tested, this is strongly suggested by the systematic presence of this enzyme in hyperthermophiles. Indeed, comparative genomic analyses have shown that reverse gyrase is the only protein

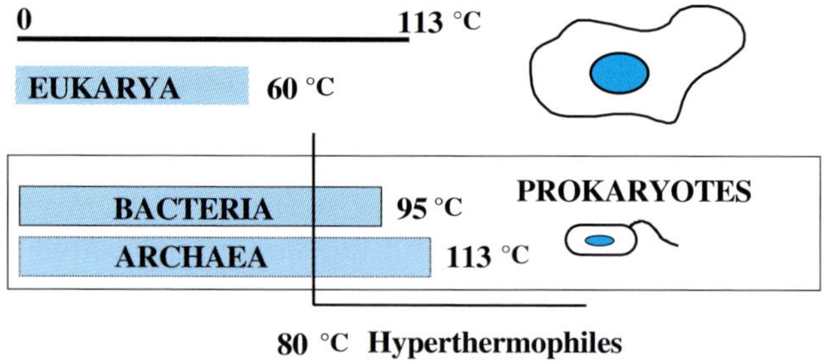

Fig. 7.2. Upper limits of life on Earth for the three major groups of cellular organisms

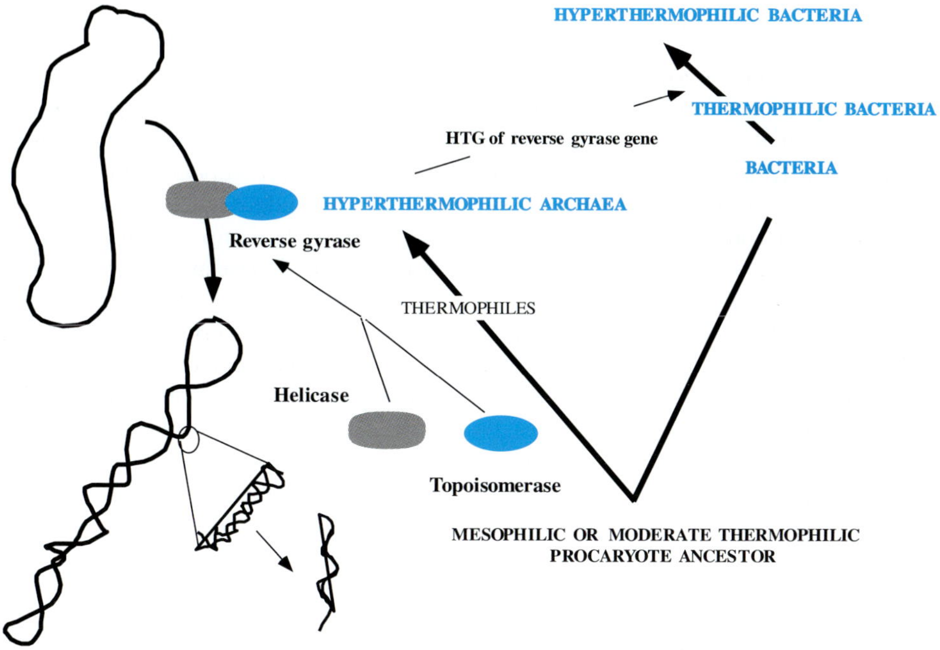

Fig. 7.3. A character specific to hyperthermophiles: the reverse gyrase. This enzyme, made up by the fusion of two 'classic' proteins, a helicase and a topoisomerase, transforms relaxed DNA into positive supercoiled DNA by the use of ATP energy. This enzyme might help to prevent denaturation of the double-helix structure at high temperature

displaying such a distribution (Forterre, 2002a), with one exception (see below). Molecular analyses have shown that reverse gyrase is a relatively modern enzyme, i.e., it appeared quite recently in evolution, by the fusion of the genes coding for two different proteins (a DNA topoisomerase and a DNA helicase) (Fig. 7.3), each belonging to a protein family already specialised at the time of the fusion (Confalonieri, 1993). Thus, if reverse gyrase is crucial to the functioning of DNA at temperatures higher than 80 °C, hyperthermophiles would have had a late emergence, subsequent to the "invention" of this enzyme (Forterre, 1995a). Reverse gyrase likely appeared in a thermophilic organism thriving between 70 and 80 °C, and whose proteins were thus already adapted to high temperatures. This is consistent with the recent evidence of a gene coding for reverse gyrase in the genome of *Thermoanaerobacter tencongensis*, a thermophilic bacterium. A late emergence of hyperthermophiles would exclude a direct link between extant hot-loving organisms and a hot origin of life. However, the possibility that life arose in warm environments (50–80 °C) remains a possible scenario.

7.2.3 Origin of Hyperthermophily

As far as we know, the apparition of reverse gyrase in all bacterial and archaeal hyperthermophiles seems to have been at the origin of adaptation to hot environments. Molecular phylogenies have allowed retracing the evolution of this protein (Forterre, 2000b). Reverse gyrase most likely appeared in Archaea and was later transferred to some Bacteria by independent lateral gene transfers from different archaeal sources (Fig. 7.3). Indeed, comparative genomics has shown that such transfers are very frequent both between and within the two prokaryotic groups, and that they boost adaptation to new biotopes. In hyperthermophilic Bacteria such as *Thermotoga maritima* (upper growth temperature 96 °C) and *Aquifex aeolicus* (upper growth temperature 95 °C) a large fraction of genes are indeed much more similar to their archaeal counterparts rather than to their bacterial ones (Aravind, 1998; Nelson, 1999), indicating that they were likely acquired from Archaea by horizontal transfer. This evidence, together with the fact that reverse gyrase in hyperthermophilic Bacteria is likely of archaeal origin, suggests that Archaea might have been the first organisms to overcome the 80 °C barrier and that they transferred this ability to Bacteria. Nonetheless, Bacteria did not succeed in adapting to temperatures as high as those of Archaea (maximal growing temperature 113 °C) (Fig. 7.2).

Indeed, all extant Archaea might have a hyperthermophilic origin. This hypothesis is suggested by a few lines of evidence. For instance, the unique structure of archaeal glycerolipids indicates that they possibly arose as an adaptation for survival at high temperatures. Archaeal glygerolipids are glycerol and phytanol di- or tetraethers (highly hydrophobic long isoprenoid chains). In hyperthermophilic Archaea such tetraethers form a monolayer at the level of the cytoplasm membrane (Fig. 7.4). Tetraether lipids are particularly adapted to extremely hot environments, since ether links are very stable and the formation of monolayers prevents high-temperature-driven proton leakage through the membrane, a key feature to maintain a ATP-producing proton gradients. Tetraether lipids are also found in some mesophilic and psycrophilic Archaea, suggesting that they can maintain membrane fluidity over a large temperature spectrum. This type of lipids might thus have appeared in mesophiles and permitted adaptation to hyperthermophily. Alternatively, they might be an ancestral character dating back to a hyperthermophilic archaeal ancestor. This latter hypothesis, i.e. a hyperthermophilic origin of Archaea, is supported by rRNA-based phylogenies, where hyperthermophilic lineages emerge as the most basal offshoots (reviewed in Forterre, 2002b). A recent archaeal phylogeny based on ribosomal proteins also showed a similar basal placement of hyperthermophilic species (Matte-Taillez, 2002). By contrast, hyperthermophily in Bacteria appears to be a secondary adaptation. Bacterial glycerolipids, such as eukaryal ones, are generally glycerol and fatty acid esters. However, those of hyperthermophilic Bacteria tend to mimic the archaeal ones. Some of them are based on ether links that can form monolayer structures

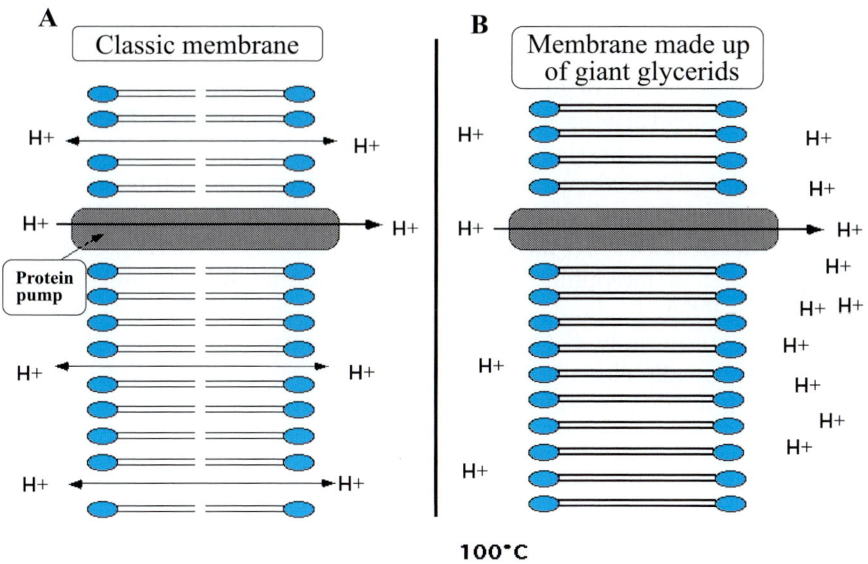

Fig. 7.4. Proton permeability of cell membranes at high temperature (100 °C). Lipid polar heads are drawn in *blue*. The protein pump creates a proton gradient that allows the production of cellular energy. At 100 °C this gradient cannot be created since bilayer classic membranes become permeable to protons at this temperature (**A**). In hyperthermophiles, the nonpermeability of the membrane to protons at high temperature is granted by a monolayer glycerid structure and/or by higher hydrophobicity of lipid tails (**B**)

(Schouten, 2000), while some others are based on tetraesters, as suggested by the identification in *Thermotoga maritima* of long fatty acid chains harbouring an acid function at each end ("diabolic acids"). Despite such archaeal-like features, hyperthermophilic bacterial glycerolipids remain strongly bacterial in essence. In fact, they do not harbour any isoprenic chain, and the absolute configuration of the asymetric carbon atom of the glycerolipids are different in mesophilic Bacteria and Eukarya, on the one hand, in Archea, one the other hand. Significantly, lipids from hyperthermophilic Bacteria are different between lineages. This strongly suggests an independent adaptation to hyperthermophily in different bacterial lineages, consistently with the fact that they likely recruited their reverse gyrases from different archaeal sources (Forterre, 2000a).

7.2.4 LUCA was Probably not a Hyperthermophile

Adaptation to temperatures higher than 80 °C is thus a character that likely appeared in Archaea and was punctually transferred to some Bacteria. If the root of the life tree lies in the bacterial branch, this implies a nonhyperthermophilic LUCA. Indeed, this is consistent with a simulation analysis that reconstructed the hypothetical ancestral ribosomal RNA of LUCA and obtained a molecule whose content of guanine and cytosine (G-C content) is characteristic of extant organisms thriving at temperatures lower than 70 °C (Galtier, 1999). Ribosomal RNAs of hyperthermophiles are in fact very rich in G-C, a feature that stabilises their numerous minihelixes since three hydrogen bonds exist between G-C base pairs and only two between A-U base pairs (Fig. 7.5). Since the reconstructed ancestral ribosomal RNA is not very G-C rich, LUCA may have rather been a mesophilic or moderately thermophilic organism. As opposed to their RNA, it should be noted that the genomic DNA of hyperthermophiles is not especially G-C rich. This is not surprising, given that cellular DNA is "topologically closed", i.e. its two strands cannot rotate freely one over the other; such a DNA is stable up to a temperature of at least 107 °C (Marguet, 1994).

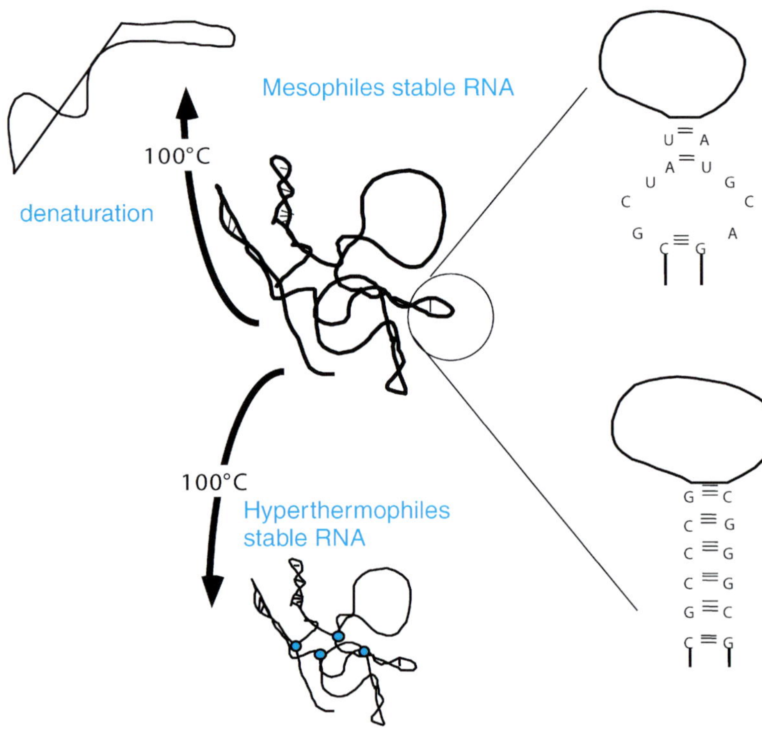

Fig. 7.5. Stable RNA protection at high temperature. Stable RNA molecules (ribosomal RNA, transfer RNA, etc.) are composed of a single polynucleotide chain folded on itself to form a number of 'hairpins' corresponding to stem-loop structures. In hyperthermophiles, these hairpins are stabilized by longer stems and a higher content of guanine (G) and cytosine (C). The overall globular structure can be stabilised by the addition of proteins and/or chemical groups at specific positions (*blue circles*)

7.2.5 Temperature and the RNA World

The fact that LUCA was not a hyperthermophile does not necessarily exclude the possibility of a hot origin of life, since a very long period of evolution occurred between the apparition of life and that of LUCA (Forterre, 1995a; Lazcano, 1996). Life might thus have appeared at very high temperatures, gone through a cold evolution period, and diversified at the whole range of temperatures from 0 to 110 °C! It is likely that the step corresponding to a "RNA world" occurred at relatively low temperatures (Joyce, 1988; Forterre, 1995b). Indeed, RNA is a very fragile molecule due to the reactive oxygen at position 2′ of ribose (absent in DNA, deoxyribonucleic acid) that can cause the breaking of nucleotide bonds (Fig. 7.6). This reaction, faster at high temperature, as are all chemical reactions, leads to rapid degradation of RNA by hydrolysis at temperatures close to 100 °C. Moreover, this thermodegradation is highly stimulated *in vitro* by magnesium chloride, a cofactor necessary to the activity of ribozymes (RNA enzymes), which themselves were essential components of metabolism in a RNA world.

However, a hot RNA world cannot yet be totally excluded. Actually, it has been shown that RNA thermodegradation can be inhibited by strong concentrations of monovalent ions (> 100 mM potassium chloride) (Hetcke, 1999; Tehei, 2002). Thus, a hot RNA world might have existed in combination with monovalent salts and/or unknown RNA-stabilizing molecules. Given its thermosensi-

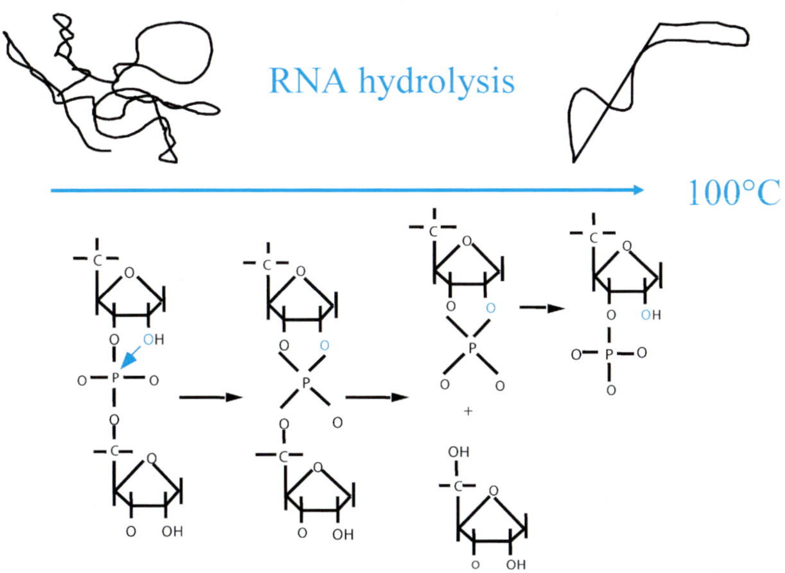

Fig. 7.6. RNA hydrolysis at high temperature. The link between two ribose molecules can be cleaved at the level of the phosphate by the ribose oxygen at position 2′ (*blue*)

bility, this scenario implies also the absence of magnesium chloride, and consequently the presence of ribozymes depending only on monovalent ions. It would be interesting to test such dramatic constraints by in vitro RNA selection tests.

In extant hyperthermophiles, stable RNAs (transfer RNA, ribosomal RNA) are protected against thermodegradation by methylation of a number of ribose 2′ oxygens and by association with proteins (Fig. 7.5). Messenger RNA is already *per se* highly unstable in prokaryotes (half-life of about one minute). mRNA is, in fact, used by ribosomes for protein synthesis (translation). RNA instability at high temperatures might perhaps explain why most thermophilic Eukarya (such as some fungi) are unable to grow beyond 60°C (Forterre, 1995b) (Fig. 7.1). In fact, in Eukarya, messenger RNA has to be stable enough to survive over its transfer from the nucleus (transcription site) to the cytoplasm (translation site). It was even proposed that selection pressure for adaptation to thermophily might have been the driving force for the appearance of the prokaryotic phenotype (e.g., rapid macromolecular turnover; elimination of the nuclear membrane to allow coupling between transcription and translation) – the "thermoreduction hypothesis" (Forterre, 1995b).

7.3 Comparative Genomics: a Novel Approach to Retrace Our Most Distant Past

7.3.1 Simple or Complex LUCA? A RNA or a DNA Genome?

While the study of mycoplasma or hyperthermophiles does not provide many insights into the nature of LUCA, comparative genomics seems to be a much more promising approach (Koonin, 2000). At the begining of 2004, 169 genomes (131 bacterial, 17 archaeal, and 21 eukaryal) have been completely sequenced and are publicly available in genomic data banks (http://wit.integratedgenomics.com/GOLD/). Their comparison has permitted the identification of approximately a hundred genes that are universally distributed in all these genomes and thus were likely present in LUCA. The majority of the proteins encoded by these genes are involved in either protein or RNA synthesis (indeed mostly ribosomal proteins) (they can be easily viewed on the COG website at http://www.ncbi.nlm.nih.gov/COG/).

In general, these proteins are much closer between Eukarya and Archaea than between Bacteria and either of the other two domains (Olsen, 1997). Moreover, a certain number of translation and transcription proteins are shared only between Eukarya and Archaea. Finally – and remarkably – all bacterial proteins involved in DNA replication are very different from their analogues in the other two domains; conversely, they are very similar between Archaea and Eukarya (Mushegian, 1996; Olsen, 1997; Edgell, 1997; Myllykallio, 2000).

These observations have been interpreted by invoking two opposite scenarios. In the first, LUCA would have been much simpler than any extant organism, i.e. it would have been a progenote, an organism still belonging to an RNA world (or to the very dawn of a DNA world, prior to the apparition of DNA replication) (Fig. 7.7A) (Mushegian, 1996; Leipe, 1999). This scenario is consistent with the rooting of the universal tree of life between Bacteria on one side and Archaea/Eukarya on the other (Woese, 2000). This rooting was proposed

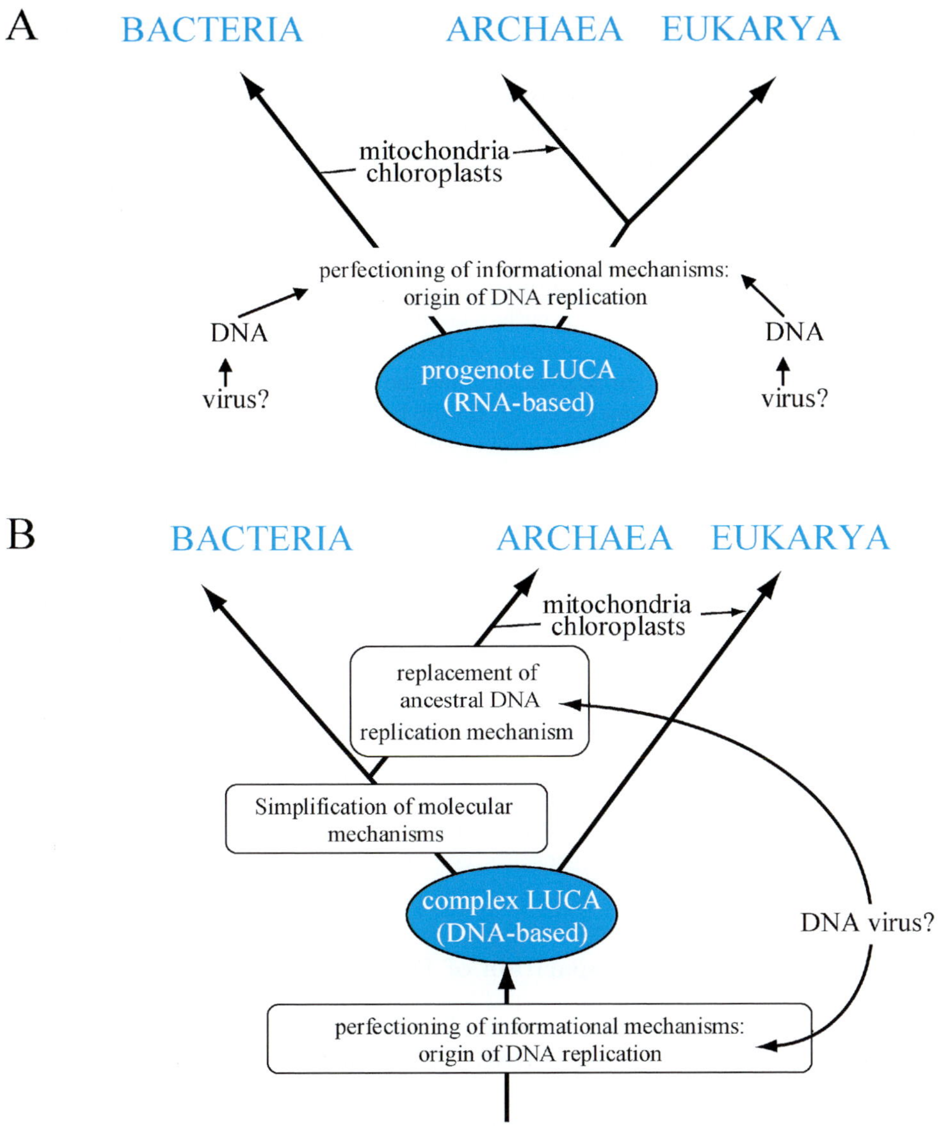

Fig. 7.7. Two alternative hypotheses to explain evidence from comparative genomics: a simple (**A**) or complex (**B**) Last Universal Common Ancestor (LUCA)

about 15 years ago, on the basis of phylogenetic analyses of paralogue protein couples (proteins coded by duplicated genes) present in all three domains of life. Since these couples are the products of a gene duplication that occurred prior to the separation of the three lines of descent, i.e. in LUCA, the tree built on the first member of the duplicated couple provides the outgroup to the tree built on the second one, and vice versa, thus producing two rooted universal phylogenies (Iwabe, 1989; Gogarten, 1989). However, such analyses should be treated with caution, due to the methodological problems affecting the reconstruction of such ancient relationships (Gribaldo, 2002).

In opposition to the idea of a primitive LUCA, a few authors have put forward the hypothesis of a complex ancestor, whose molecular biology was closer to that of extant Eukarya rather than that of extant prokaryotes (Fig. 7.7B) (Forterre, 1999c; Penny, 1999). This hypothesis rests on the presence in Eukarya of a number of RNA-based mechanisms that may possibly be relics of an RNA world. This is particularly the case of a eukaryal mechanism producing functional messenger RNAs by the cutting off of noncoding regions (exons) and the joining of coding ones (introns). This mechanism, called «splicing», is performed by a molecular complex (the spliceosome) composed of RNA and proteins. No counterpart of the spliceosome exists in prokaryotes. Since such a structure is similar to that of the spliceosome, the ribosome itself might be an ancient ribozyme. In this alternative scenario, the root of the universal tree of life would rather lie between Eukarya and Archaea/Bacteria, rendering prokaryotes monophyletic and indicating all characters common to Eukarya and Archaea as ancestral. In this case, the functional differences between bacterial informational proteins and their archaeal/eukaryal counterparts would be explained by a dramatic acceleration of evolution of these proteins in Bacteria, or even by their replacement by proteins of distinct origins (particularly with regards to DNA replication enzymes), incluting viral ones (Forterre, 1999a). Indeed, the genomes of many DNA viruses code for replication proteins that are very different from their cellular analogues as well as from other viruses, suggesting that viral and cellular DNA replication proteins diverged a long time ago.

For the time being, no evidence allows favouring either a complex or a simple LUCA. The answer might be provided by a more accurate analysis of the ribosomal machinery, since it presents very well preserved features in all three domains of life.

7.3.2 A Key Step: the Apparition of DNA

Another important issue is represented by the period comprised between the apparition of proteins and that of DNA. This period might be called the "RNA world's second age", to distinguish it from the first age, dominated by ribozymes. This age witnessed the progressive assemblage of the complex systems of protein synthesis and of the genetic code in LUCA, together with the apparition of RNA viruses, composed of proteins and RNA. Large metabolic pathways of synthesis and degradation might also date back to this period, together with the apparition of bioenergetic mechanisms and cellular membranes, as we know them today.

DNA might have appeared in two steps towards the end of this period. The first step would have been the apparition of ribonucleotide reductase (RNR), an enzyme that produces deoxyribonucleotides (substrates for DNA polymerases) starting from ribonucleotides (substrates for RNA polymerases) (Fig. 7.8). Ancestral RNRs were already most likely proteins, since the mechanism of ribose reduction involves highly active radical groups that would have not been stable in an RNA molecule (Freeland, 1999). The second step would have been the appari-

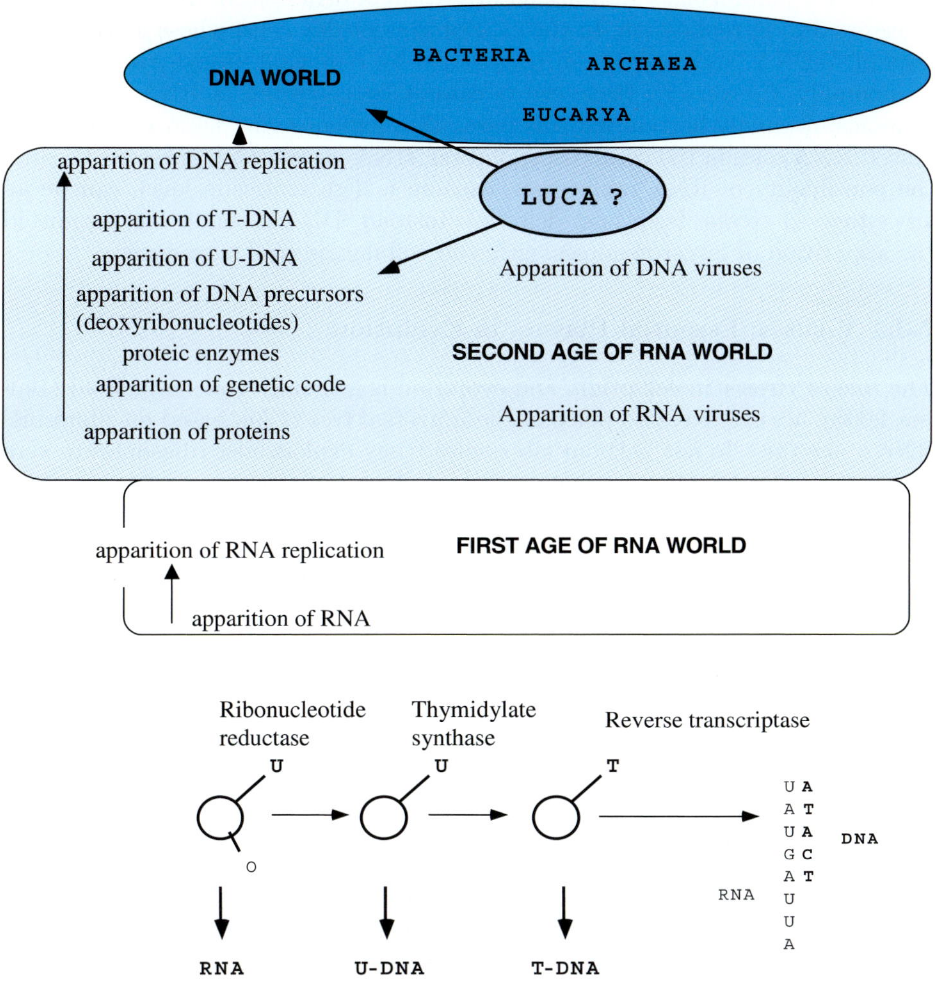

Fig. 7.8. The first steps of life evolution on Earth based on molecular biology data: the transformation of a ribonucleotide into a deoxyribonucleotide and the appearance of DNA from RNA

tion of enzymatic proteins necessary to deoxyribonucleotide polymerisation (reverse transcriptase and DNA polymerase), and the formation of deoxythymidine monophosphate (dTMP) starting from uracyl monophosphate (dUMP), leading to the replacement of uracyl by thymine in DNA. The fact that this transition occurred at the level of deoxynucleotides (base+desoxyribose+phosphate) shows that the apparition of U-DNA (uracyl-containing DNA) must have predated that of present-time T-DNA (thymine-containing DNA) (Fig. 7.8).

DNA is more stable than RNA due to the absence of oxygen at position 2′ of ribose. Moreover, it can be more reliably replicated than RNA due to the replacement of uracyl by thymidine. Indeed, DNA and RNA cytosines tend to transform spontaneously into uracyl in vivo by a deamination reaction, leading to errors during replication. In the case of present-day cells, whose genome consists of T-DNA, such errors can be avoided by removing uracyl, recognized as an anomaly. Conversely, these errors cannot be corrected in RNA or U-DNA genomes, normally harbouring this base. This explains why DNA cells have replaced RNA ones in the course of evolution. RNA viruses have survived because the non-fidelity of RNA replication, causing a high mutation level, can be an advantage to escape from host defences. Instead, DNA stability has permitted the apparition of larger genomes, either in cellular or viral organisms.

7.3.3 Viruses: Essential Players in Evolution

The role of viruses in cell origin and evolution is generally underestimated (Balter, 2000). Viruses have no place in the universal tree of life based on ribosomal RNA since they do not harbour ribosomes (they exploit host ribosomes to synthesise their proteins). Nonetheless, it is well known that viruses have played, and still play, a crucial role in gene transfer, so frequent in prokaryotes. Viruses have also played an important role in the apparition of new proteins, thanks to their well-developed ability to recombine and to their particularly high mutation rate. It might also be argued whether viruses have indeed "invented" DNA (Forterre, 2002c). In fact, some viruses do modify their DNA to make it resistant to nucleases produced by the cells they infect, and DNA can be indeed considered as a modified RNA. As mentioned above, the advantages of DNA over RNA (higher stability and higher replication fidelity) perfectly explain why DNA cell populations outcompeted RNA ones, although they cannot account for the initial selective advantage of the first DNA organism. The apparition of DNA in a virus would have provided an immediate selective advantage, giving a satisfactory answer, in terms of selective pressure, to the apparition of this molecule. If viruses invented DNA and DNA-replicating proteins, these proteins might have had the possibility to diverge very early over the course of evolution. This would explain why viral-replication proteins are often very divergent from virus to virus, and very different from their cellular homologues. In such a scenario, different cellular lineages may have acquired very different DNA replication proteins from different viral sources. This would indeed explain an important evolutionary puzzle, i.e., the fact that, despite being very similar in terms of overall mechanisms and enzymatic steps, several essential proteins involved in identical steps of cellular DNA replication are evolutionary unrelated in Archaea and Bacteria. Moreover, the rising number of complete genomes in public databases is revealing the existence of a number of other evolutionary unrelated (or very distantly related) protein families performing essential and common steps in DNA replication and metabolism in the three domains of life, Archaea, Bacteria, and Eukarya. This is, for instance, the case of topoisomerases, essential enzymes involved in the handling of DNA topological constraints that arise in a number of basic cellular processes (i.e., replication, transcription, recombination, chromatin remodelling). In fact, two types of DNA topoisomerases, TopoIV in Bacteria and TopoVI in Archaea/Eukarya, perform essentially the same function, despite being evolutionary unrelated. This peculiar distribution is consistent with a possible viral origin of these two enzymes (Gadelle, 2003). Another stunning example is that of tymidylate synthase (ThyA), a key enzyme that catalyzes the transformation of deoxyuracyl monophosphate (dUMP) into deoxythymidine monophosphate (dTMP), one of the four fundamental components of DNA. Recent investigations have evidenced that a number of bacterial species harbour an evolutionary unrelated analogue (ThyX), which catalyzes the same reaction (Myllykallio, 2002). Here again, the presence of numerous ThyX and ThyA viral homologues leaves open the possibility of a viral origin for these enzymes. The possibility that an enzyme involved into a fundamental step of DNA metabolism may be of viral origin is consistent with the hypothesis that T-DNA might have first appeared in viruses (Forterre, 2002c).

Returning to the two scenarios for the origin of LUCA discussed above, the introduction of viral DNA into cells might have occurred by two different ways (Fig. 7.8). The first would have been a double independent transfer of DNA and its replication machinery from a viral source to an RNA-based LUCA, once in the line leading to Bacteria, and another in that leading to Archaea and Eukarya. This scenario, although implying a bacterial rooting for the tree of life, is nonetheless consistent with an RNA-based LUCA, either a simple prokaryotic-like organism, or a fairly complex protoeukaryal one, containing perhaps RNA genes with introns and exons. In the latter case, present-day Bacteria would be the product of a radical secondary simplification (Forterre, 1995b). The second scenario would posit that a transfer of DNA and its machinery occurred from a viral source prior to LUCA, leading to a DNA-based LUCA. Another transfer would have followed either in the bacterial branch, or in the branch common to Archaea and Eukarya, leading to a radical replacement of the ancestral replication machinery by an analogue and nonhomologue system (Forterre, 1999a).

7.3.4 The Origin of the Nucleus: a Further Puzzle

Evidently, a number of unknown variables are involved in retracing the history of life on our planet. One of them is the origin of the eukaryal cell nucleus. We now know that mitochondria and chloroplasts are remnants from Bacteria that engaged endosymbioses with ancestral eukaryal cells (Fig. 7.1). By analogy, a few authors have suggested that the nucleus also originated from the endosymbiosis of an archaeon within a bacterial host, in order to explain the archaeal nature of eukaryal proteins involved in the expression and replication of the genomic material (reviewed in Lopez-Garcia, 1999). In this model, a total disappearance of the ancestral archaeal membrane has to be invoked. Indeed, the actual nuclear membrane corresponds to an expansion of the endoplasmic reticulum (the net of vesicles present in all eukaryal cells) and does not contain any archaeal lipid. The big concern with such a model is the origin of nuclear pores, whose structure is utterly complex and has no counterpart in prokaryotes. Moreover, it is difficult to understand why the evolution of the nucleus would have occurred by a complexification of the archaeal endosymbiont, when endosymbionts normally go through a reductive evolution of most of their genome (simplification). Finally, this model does not explain the disappearance of the whole genome of the bacterial host, together with all its informational proteins. The recent completion of the first eukaryotic genomes has triggered comparative genomics approaches aimed at testing alternative hypotheses on the origin of the eukaryotic cell. The fact that eukaryotic proteins functionally related to nuclear functions share higher similarity with their archaeal homologues, while those related to metabolisms and cytoplasmic activities are more similar to their bacterial counterparts has been used as a strong support for an archaeal endosymbiotic origin of the nucleus in a bacterial host (Horiike, 2001). However, the reliability of such an approach has been seriously questioned (Poole, 2001; Rotte, 2001). More recently, searches for sequence homologues across complete genomes from members of the three domains allowed the identification of 347 proteins that are unique to Eukarya (Hartman, 2002). These proteins were thus assumed to represent genuine 'eukaryal signatures' that would have evolved in a protoeukaryotic anucleate ancestor that was named "chronocyte" (Hartman, 2002). The cronocyte would have thus been a complex cell already provided with many important novelties, including a cytoskeleton apparatus. The nucleus would have originated by the engulfment of an archaeal symbiont. The authors invoke an RNA-based host in order to explain the origin of the nuclear envelope due to the need to physically isolate the RNA genome from the archaeal symbiont DNA genome. The complement of eukaryal proteins with bacterial origin would thus have been introduced into the eukaryal genome subsequently, by LTG from mitochondria to the nucleus.

However, these theories on the origin of the nucleus from an archaeal symbiosis (either in a bacterial or a protoeukaryal host) are at odds with the evidence that a substantial number of eukaryal nuclear proteins are not of archaeal origin (e.g., eukaryotic type II DNA topoisomerases or some replicative DNA polymerases). In some cases, these proteins have homologues in viral genomes instead (Gadelle, 2003; Villarreal, 2000). This has led some authors to propose that viruses may have played an important role in the assembling of eukaryotic-specific molecular processes, up to the very origin of the nucleus ("viral eukaryogenesis"; Bell, 2001; Takemura, 2001). Indeed, some stunning affinities exist between the process of nuclear envelope dissolution during eukaryal cellular division, and the cycle of some DNA viruses such as Pox viruses. Moreover, nuclear pores are similar to those that some RNA viruses utilise to excrete their viral messenger RNA into the cellular cytoplasm of the host. It will be interesting to test this possibility by including viral sequences into genomic approaches and attempting to dissect the impact (and timing) of a viral contribution to the eukaryotic genome.

7.4 Conclusions and Perspectives

7.4.1 More Data are Needed

In conclusion, the quest for primitive cells in the present-day biosphere has led, up to date, only to wrong conclusions. It seems that no cells preceding LUCA have survived, and that all extant cells are equally evolved. In particular, the study of hyperthermophiles has not unveiled any obvious primitive life forms. However, it has largely contributed to the debate on a hot origin of life and to the clarification of the steps leading to hyperthermophily. Conversely, the identification of three domains in the cellular world and the systematic comparison of these three domains at the genomic and molecular level is extremely instructive. Molecular mechanisms common to all three domains allow to retrace a part of life history prior to the apparition of LUCA, and to answer a few precise questions on both the nature of LUCA and the mechanisms that have led to present-day cellular types.

The sequencing of an ever-growing number of genomes should provide an answer to all these questions. It would be particularly interesting to analyse the genomes of a number of protists (unicellular eukaryotes), and of novel archaeal and bacterial lineages. In most cases, the existence of these lineages is known only from ribosomal RNA sequences amplified randomly by PCR (polymerase chain reaction) from either natural or artificial biotope samples. One of the major future challenges of biology will be indeed to cultivate these organisms in order to be able to perform physiological and genomic analyses. It would be equally crucial to sequence a number of viral genomes in order to test the hypotheses discussed in this chapter on the role of viruses during the first steps of the DNA world. For example, some viruses harbour genomes larger than those of some Bacteria (La Scola, 2003). The sequencing of such genomes might reveal many surprises. Viruses represent the majority of the biosphere (each cellular species being the target of many different viruses) and we know only a tiny fraction of this diversity.

7.4.2 To not Forget Darwin!

In this chapter, we implicitly assumed that all evolutionary steps prior to LUCA and involving RNA, proteins (via RNA), and DNA (via proteins) occurred in a cellular world. We reckoned it essential since, to us, natural selection acts on individuals, either cellular or viral, competing one with each other (the core of the hypothesis on a viral origin of DNA). Nonetheless, this notion is not evident to all researchers involved in the study of life origins and the first steps of evolution. To some of them, evolution could have occurred in a noncellular medium up to the emergence of a RNA world, viruses, or even LUCA! Is it really possible to imagine evolution by natural selection in similar conditions? Is this not a return to pre-Darwininan concepts? Similarly, it is very fashionable these days to discuss of a primitive and nondifferentiated cellular world where continuous gene transfers prevented any speciation (Woese, 2000). Again, is this scenario compatible with a natural selection process?

It would thus be sensible to bring Darwin back in our discussions on life origin, particularly for researchers dealing with the domain of evolution but not having a specific preparation on the subject (the great majority indeed). This remark is valid not only for non-biologists, but also for molecular biologists and geneticists (Kupic, 2000).

References

General

Ciba Foundation, *Evolution of Hydrothermal Ecosystems on Earth (and Mars?)* (1996). J. Wiley and Son, Chichester

Madigan M.T., Martinko J.M, Parker J., *Brock Biology of Microorganisms* (2000) 9th edition, Prentice Hall, USA

Wiegel J., Adams W.W., *Thermophiles, the Keys to Molecular Evolution and the Origin of Life?* (1998) Taylor and Francis, London

Specialised

Alberts, S.V., Van de Vossenberg, J.L., Driessen, A.J., Konings, W.N. (2000) Adaptation of archaeal cell membrane to heat stress. *Front. Biosci.* **5**, D813–820

Aravind, L., Tatusov, R.L, Wolf, Y.L., Walker, D.R., Koonin, E.V. (1998) Evidence for massive gene exchange between archaeal and Bacterial hyperthermophiles. *Trends Genet.* **15**, 298–299

Bell, P.J.L. (2001) Viral eukaryogenesis: was the ancestor of the nucleus a complex DNA virus? *J. Mol. Evol.* **53**, 251–256

Ciba Fondation, *Evolution of Hydrothermal Ecosystems on Earth (and Mars?)* 1996, J. Wiley and Son, Chichester

Confalonieri, F., Elie, C., Nadal, M., Bouthier de la Tour, C., Fortere, P., Duguet, M. (1993) Reverse gyrase: a helicase-like domain and a type I topoisomerase in the same polypeptide. *Proc. Natl. Acad. Sci. USA* **90**, 4753–4757

Edgell, D.R., Doolittle, W.F. (1997) Archaea and the origin(s) of DNA replication proteins. *Cell* **89**, 995–998

Forterre, P., Confalonieri, F., Charbonnier, F.,Duguet, M. (1995a) Speculations on the origin of life and thermophily *Origin of Life and Evolution of the Biosphere* **25**, 235–249

Forterre, P. (1995b) Thermoreduction, a hypothesis for the origin of prokaryotes, *C. R. Acad. Sci. Paris* **318**, 415–422

Forterre, P. (1996a) A la recherche de LUCA. Compte rendu du colloque de la fondation des Treilles, http://www-archbac.u-psud.fr/Meetings/LesTreilles/Treilles_frm.html

Forterre, P. (1996b) A hot topic: the origin of hyperthermophiles. *Cell* **85**, 789–792

Forterre, P. (1999a) Displacement of cellular proteins by functional analogues from plasmids or viruses could explain puzzling phylogenies of many DNA informational proteins. *Mol. Microbiol.* **33**, 457–465

Forterre, P. (1999b) Les hyperthermophiles sont-ils nos ancêtres? *La Recherche* **317**, 36–43

Forterre, P., Philippe, H. (1999c) Where is the root of the universal tree of life? *Bioessays.* **10**, 871–879

Forterre, P., Bouthier de la Tour, C., Philippe, H., Duguet, M. (2000b) Reverse gyrase from hyperthermophiles, probable transfer of a thermoadaptation trait from Archaea to Bacteria. *Trends Genet.* **16**, 152–154

Forterre, P. (2002a) A hot story from complarative genomics: reverse gyrase is the only hyperthermophile-specific protein. *Trends Genet.* **18**, 236–238

Forterre, P., Brochier, C., Philippe, H. (2002b) Evolution of the Archaea. *Theor. Popul. Biol.* **61**, 409–422

Forterre, P. (2002c) The origin of DNA genomes and DNA replication proteins. *Curr. Opin. Microbiol.* **5**, 525–32

Freeland, S.J., Knight, R.D., Landweber, L.F. (1999) Do proteins predate DNA? *Science* **286**, 690–691

Gadelle, D., Filée, J., Buhler, C., Forterre P. (2003) Phylogenomics of type II DNA topoisomerases. *Bioessays* **25**, 232–242

Galtier, N., Tourasse, N., Gouy, M. (1999) A non hyperthermophilic common ancestor to extant life forms. *Science* **283**, 220–221

Gribaldo, S., Philippe, H. (2002) Ancient phylogenetic relationships. *Theor. Popul. Biol.* **61**, 391–408

Gogarten, J.P., Kibak, H., Dittrich, P., Taiz, L., Bowman, E.J., Bowman, B.J., Manolson, M.F., Poole, R.J., Date, T., Oshima, T. (1989) Evolution of the vacuolar H^+-ATPase: implications for the origin of Eukarya. *Proc. Natl. Acad. Sci. USA* **86**, 6661–6665

Hartman, H. and Fedorov, A. (2002) The origin of the eukaryotic cell: a genomic investigation. *Proc. Natl. Acad. Sci. USA* **99**, 1420–1425

Hethke, C., Bergerat, A., Hausner, W., Forterre, P., Thomm, M. (1999) Cell-free transcription at 95 degrees: thermostability of transcriptional components and DNA topology requirements of *Pyrococcus* transcription *Genetics* **152**, 1325–1333

Horiike, T., Hamada, K., Kanaya, S., Shinozawa, T. (2001) Origin of eukaryotic cell nuclei by symbiosis of Archaea in Bacteria is revealed by homology-hit analysis. *Nat. Cell. Biol.* **3(2)**, 210–214

Huber, H., Hohn, M.J., Rachel, R., Fuchs, T., Wimmer, V.C., Stetter, K.O. (2002) A new phylum of Archaea represented by a nanosized hyperthermophilic symbiont. *Nature* **417**, 63–67

Iwabe, N., Kuma, K.-I., Hasegawa, M., Osawa, S., Miyata, T. (1989) Evolutionary relationship of Archaea, euBacteria and Eukarya inferred from phylogenetic trees of duplicated genes. *Proc. Natl. Acad. Sci.* **86**, 9355–9359

Joyce, G. (1988) Hydrothermal vents too hot? *Nature* **334**, 564–565

Koonin, E.V., Aravind, L., Kondrashov, A.S. (2000) The impact of comparative genomics on our understanding of evolution, *Cell* **101**, 573–6

Kupic, J.P., Sonigo, P. (2000), *Ni Dieu ni Maître.* Editions Odile Jacob, Paris

La Scola, B., Audic, S., Robert, C., Jungang, L., de Lamballerie, X., Drancourt, M., Birtles, R., Claverie, J.M., Raoult, D. (2003) A giant virus in amoebae. *Science* **299**, 2033

Lazcano, A., Miller, S.L. (1996) The origin and early evolution of life: prebiotic chemistry, the pre-RNA world, and time. *Cell* **14**, 793–798

Leipe, D.D., Aravind, L., Koonin, E.V. (1999) Did DNA replication evolve twice independently? *Nucleic Acids Res.* **27**, 3389–3401

Lopez-Garcia, P., Moreira, D. (1999) Metabolic symbiosis at the origin of Eukarya, *Trends Biochem. Sci.* **24**, 88–93

Matte-Tailliez, O., Brochier, C., Forterre, P., Philippe, H. (2002) Archaeal phylogeny based on ribosomal proteins. *Mol. Biol. Evol.* **19**, 631–639

Marguet, E., Forterre, P. (1994) DNA stability at temperatures typical for hyperthermophiles. *Nucl. Acid Res.* **22**, 1681–1686

Myllykallio, H., Lipowski, G., Leduc, D., Filée, J., Forterre, P., Liebl, U. (2002) An alternative flavin-dependent mechanism for thymidylate synthesis. *Science* **297**, 105–107

Nelson, K.E. et al. (1999) Evidence for lateral gene transfer between Archaea and bacteria from genome sequence of Thermotoga maritima. *Nature* **399**, 323–329.

Noon, K.R., Bruenger, E., McCloskey, J.A. (1998) Post-transcriptional modifications in 16S and 23S rRNAs of the archaeal hyperthermophile *Sulfolobus solfataricus. J. Bacteriol.* **80**, 2883–2888

Madigan, M.T., Martinko, J.M., Parker, J. (2000*) Brock Biology of Micro-organisms*, 9th edition, Prentice Hall, USA

Mushegian, A., Koonin, E.V. (1996) A minimal gene set for cellular life derived by comparison of complete Bacterial genomes. *Proc. Natl. Acad. Sci.* **93**, 10 268–10 273

Myllykallio, H., Lopez, P., Lopez-Garcia, P., Hailig, R., Saurin, W., Zivanovic, Y., Philippe, H., Forterre, P. (2000) *Science* **288**, 2212–2215

Nelson, K.E., et al. (1999) Evidence for lateral gene transfer between archaea and Bacteria from genome sequence of *Thermotoga maritima. Nature* **399**, 323–329.

Olsen, G., Woese, C.R. (1997) Archaeal genomics: an overview. *Cell* **89**, 991–994

Pace, N. (1991) Origin of life-facing up the physical setting. *Cell* **65**, 531–533

Penny, D., Poole, A.M. (1999) the nature of the universal ancestor. *Curr. Opin. Genet. Devel.* **9**, 672–677

Poole, A., Penny, D. (2001) Does endo-symbiosis explain the origin of the nucleus? *Nat. Cell. Biol.* **3(8)**, E173–174

Rotte, C., Martin, W. (2001) Does endo-symbiosis explain the origin of the nucleus? *Nat. Cell. Biol.* **3(8)**, E173–174

Schouten, S., Hopmans, E.C., Pancost, R.D., Damsté, J.S.S. (2000) Widespread occurrence of structurally diverse tetraether membrane lipids: evidence for the ubiquitous presence of low-temperature relatives of hyperthermophiles. *Proc. Natl. Acad. Sci. USA* **97**, 14 421–14 426

Takemura, M. (2001) Poxviruses and the origin of the eukaryotic nucleus. *J. Mol. Evol.* **52**, 419–425

Tehei, M., Franzetti, B., Maurel, M.C., Vergne, J., Hountondji, C., Zaccai, G. (2002) The search for traces of life: the protective effect of salt on biological macromolecules. *Extremophiles* **5**, 427–430

Villarreal, L.P., DeFilippis, V.R. (2000) A hypothesis for DNA viruses as the origin of eukaryotic replication proteins. *J. Virol.* **74(15)**, 7079–7084

Vivares, C.P., Méténier, G. (2000) Towards the minimal eukaryal parasitic genome. *Curr. Opin. Microbiol.* **5**, 463–467

Wiegel, J., Adams, W.W. (1998) *Thermophiles, the keys to molecular evolution and the origin of life?* Taylor and Francis eds. 339 p London

Woese, C.R. (2000) Interpreting the universal phylogenetic tree. *Proc. Natl. Acad. Sci. USA* **97**, 8392–8396

8 The Universal Tree of Life: From Simple to Complex or From Complex to Simple

Henner Brinkmann, Hervé Philippe

The idea of Zuckerkandl and Pauling put forward in 1965 (Zuckerkandl and Pauling 1965), that the macromolecules, which are the genetic inheritance of cells, could conserve traces of the evolutionary history of the extant organisms, profoundly changed our view of the universal tree of life. In the late 1970s Carl Woese promoted the use of the ribosomal RNA as a molecular marker, which allowed the definite confirmation of the endosymbiotic origin of mitochondria and plastids, and to define the major lines of microbial evolution. The prime discovery that Prokaryotes are composed of two very distinct groups, Eubacteria and Archaebacteria, was completely unexpected and triggered a great number of new research fields.

In this chapter we will first explain the principles of molecular phylogenetic reconstruction methods. Then, we will present the predominant view of the universal tree of life (often called the tree of Woese) and the main conclusions about the ancient evolution, which were deduced from it in the early 1990s. The end of the 1990s was dominated by something that is best described as the crisis of molecular phylogenetic reconstructions, because two major problems appeared almost simultaneously: (1) the proof of serious tree reconstruction artefacts and (2) a much higher than expected frequency of lateral gene transfers/duplications, which implied that the phylogeny of a given gene does not necessarily reflect the evolutionary history of the organism. However, a clear progress in the methods of molecular tree reconstruction together with a high number of sequences mostly produced by genome-sequencing projects did permit several important errors present in the original tree to be corrected. These corrections have naturally consequences for the hypotheses about the ancestral/early evolution (e.g., the origin of Eukaryotes) and also to a certain degree for the origin of life.

8.1 Principles of Tree-Reconstruction Methods

Phylogenetic inference consists of using what one can observe from organisms (e.g., either DNA sequences of the extant organisms or morphology for both extant and fossil) in order to infer their relationships. The first step is to find comparable characters, a task that had taken morphologists several centuries to obtain a rigorous approach. In 1848 Richard Owen defined the concept of homology, by applying the principle of connectivity of Geoffroy Saint-Hilaire:

two structures (or organs using the terminology of that epoch) in two different species are homologous (and therefore comparable) irrespective of their forms, if they are connected in the same way to identical structures. The same principle is used for DNA sequences. In fact, during the alignment step, one tries to maximise the number of identical nucleotides between two (or n) sequences (equivalent to the organs of Owen) for finding the homologous nucleotides (differing between the two or n species compared, but linked in the same way to identical nucleotide positions). The alignment step represents a problem that can be easily formalized and it is the object of an intense algorithmic research (Hickson et al. 2000). Theoretically, this objectivity may look like progress compared to the phylogeny based on morphology (the famous art of the taxonomists), but in reality the molecular phylogenies remain quite subjective. The current approach consists, in fact, in using an automatic alignment program, manually refining the result and then choosing the nonambiguously aligned regions (also manually). Even if certain programs allow formalizing the last step, e.g., (Castresana 2000), there is still a large theoretical and algorithmic progress to be realized in order to render the sequence alignment step efficient and objective.

We will assume in the following that we possess several perfectly aligned sequences and that we want to reconstruct their phylogenetic relationships. Figure 8.1 shows a tree, based on the evolution of eleven amino-acid positions from four species. Starting from the aligned sequences of the four extant species (Fig. 8.1b), how can we find the correct phylogeny? A first simplification is introduced by the fact that we only search for an unrooted tree (see Fig. 8.1c–e), which means a tree where we do not know the location of the last common ancestor of the four species (the specific point of the tree that is called the root, see Fig. 8.1a). In order to know the position of the root, we need to know in which direction the characters have evolved (this is called character polarisation). If, in certain cases, morphologists are able to polarize characters thanks to palaeontological data, the same approach is impossible with molecular data (or only possible under very strong assumptions on the evolutionary process (Gu and Li 1998)). The position of the root in a molecular phylogeny is thus normally determined by the addition of the sequence of an outgroup, about which we know that it is not part of the group under study. This allows us to polarize the evolution of characters, and thus to reconstruct the order of emergence of the ingroup species. For example, we can use birds to root the mammalian tree, or, in the case of Fig. 8.1, species number four can be used as an outgroup if we have additional external information supporting this fact.

By only considering unrooted trees, the number of possible trees is lowered: with four (n) species there are only three unrooted trees ($[2n-5] \times [2n-7] \times \ldots \times 3 \times 1$) but 15 rooted trees ($[2n-3] \times [2n-5] \times \ldots \times 5 \times 3$). In the case of Fig. 8.1, we have to find among the three possible unrooted trees the right one: tree X, which unites species 1 and 2, tree Y, which unites species 1 and 3, or tree Z, which unites species 2 and 3 (Fig. 8.1c). If we look at the first amino acid position, all species have a lysine (K), except species 4 that has a histidine

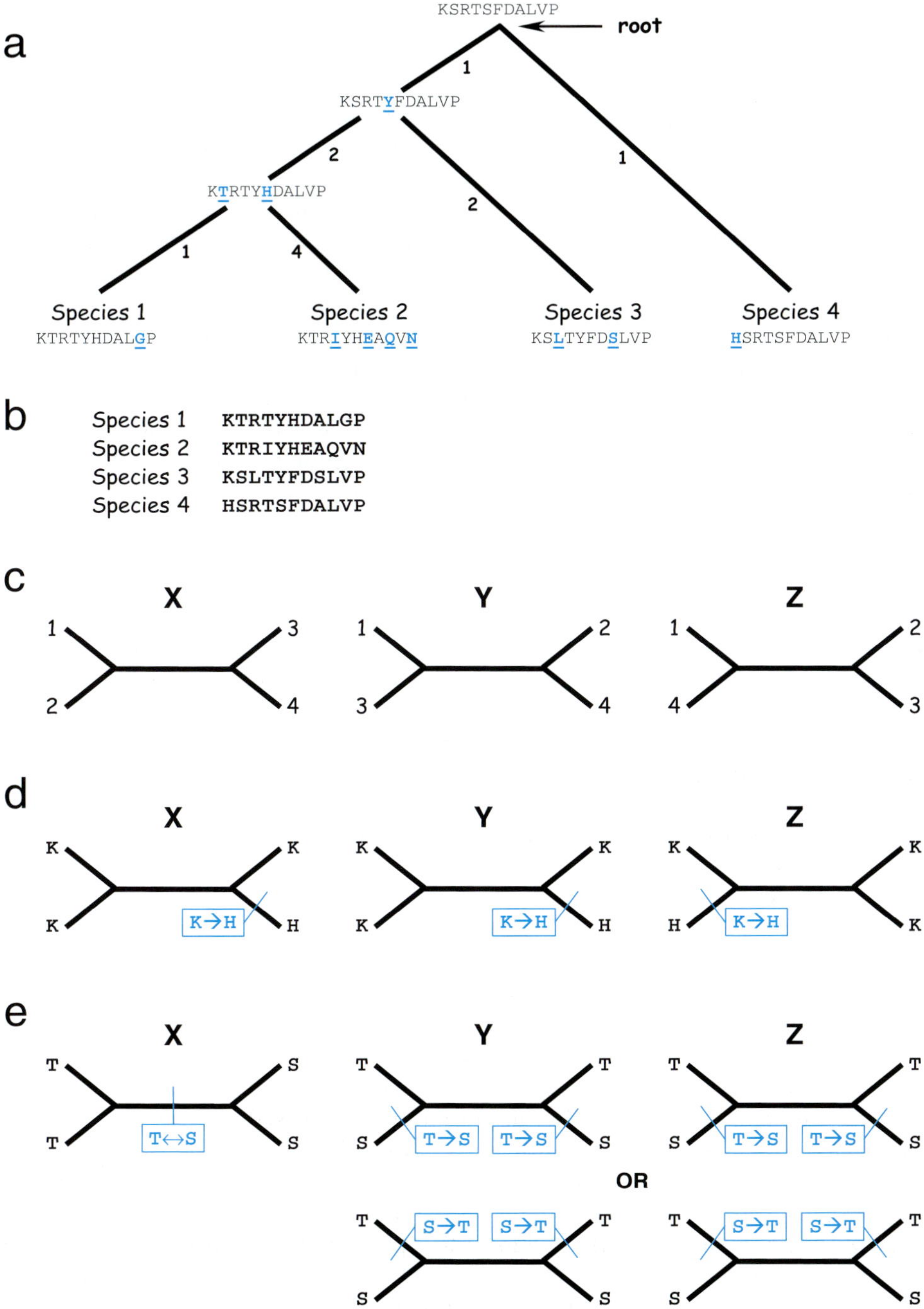

Fig. 8.1. Scheme illustrating the principles of phylogenetic reconstruction methods. An ideal case is presented, because each position changed once and only once

(H). To explain the actual distribution, it is sufficient to postulate that there was a nucleotide substitution leading to the change of amino acid K into H on the branch that leads to species 4; this change is independent of the topology (Fig. 8.1d). This position is therefore not informative, because it does not give any clue that allows a choice to be made between the three alternative tree topologies, but it has an impact on the branch lengths of the tree by adding an additional step to the branch of this species. It should be noted that other explanations are possible (e.g. a change from K to H in the internal branch and another one from H to K in the branch leading to species 2), but, according to the rule known as Occam's razor, only the most parsimonious interpretation is retained.

We will shortly discuss the additional difficulties introduced by the genetic code. The 20 different amino acids used in normal proteins are encoded by a triplet code (three nucleotides), with a total of 64 different codons (4^3, there are four different bases). The genetic code is redundant, since the number of codons is over three times higher than that of amino acids. Most mutations of nucleotides do not lead to a change of the encoded amino acid (synonymous mutations). For the replacement of an amino acid, there may be anything from a single to up to three nucleotide changes necessary. There are changes corresponding to a single nucleotide change (e.g., glycine, GGN versus alanine, GCN), or two changes (e.g., threonine, ACN versus valine, GTN), or even three changes (e.g., tryptophan, TGG versus histidine, CAC/T). However, empirical experience indicates that for the amino-acid-based phylogenetic reconstructions, the way in which substitutions are realized at the nucleotide level is not of great consequence. Furthermore, for the inference of distant evolutionary relationships based on proteins, amino acids are more appropriate than codons (nucleotides).

The second position of the sequences is informative (Fig. 8.1e), because a T (threonine) is present in the species 1 and 2 and an S (serine) is found in the species 3 and 4. Taking the tree X, it would be sufficient to postulate a single change between T and S in the internal branch to explain the distribution of characters for this position in the given species. We actually do not know whether the change is from T to S or from S to T, because it is not possible to polarize the evolution of the sequences. If we try to explain the current character distribution with the trees Y and Z, we would need two independent substitution events. For example for tree Y, we need to assume one T to S substitution in the branch leading to species 3 and one T to S substitution in the branch leading to species 4 (or two S to T substitutions in the branches leading to species 1 and 2). In conclusion, position 2 needs one substitution to fit the tree X and two substitutions in the trees Y and Z. According to the principle of economy of hypotheses (also known under the name of Occam's razor or parsimony), one has to choose the tree X. A phylogenetic method, named maximum parsimony, chooses the tree that needs the lowest number of changes (or steps) over all positions. In the case of the sequences of Fig. 8.1b, the tree X, which has a total length of eleven steps, is the most parsimonious and is therefore favoured over the trees Y and Z, which need 13 steps to explain the data. This tree corresponds to the true tree of Fig. 8.1a. In fact, there are only two informative characters (positions number two and six); they correspond to changes in the central branch of the unrooted tree. Another important point is that the correct tree is found despite quite different evolutionary speeds among the species (e.g., species 2 evolves four times faster than species 1).

8.2 The Universal Tree of Life According to Woese

Woese and Fox published an analysis about the oligonucleotide catalogues of the ribosomal RNA (rRNA) of the so-called 16S rRNA, which is part of the small subunit (30S) of the prokaryotic ribosome (S=Svedberg coefficient, refers to the sedimentation coefficient, is dependent of shape and weight) from a large ensemble of organisms (Woese and Fox 1977). The most important result of their work was the discovery that a group of poorly characterised prokaryotes (thermoacidophiles, methanogenes and extreme halophiles) were so distantly related to both Eukaryotes and Eubacteria, that they had to be placed in a third major group, called the Archaebacteria. This separation (Fig. 8.2) was later confirmed by the analyses of the rRNA sequences of the small subunit (SSU) of the ribosome: the Archaebacteria did represent an independent branch on the universal rRNA tree of Life (Woese 1987). The root was placed on the bacterial branch in the universal tree of life (Fig. 8.2) based on the analyses of anciently duplicated genes (Gogarten et al. 1989; Iwabe et al. 1989). This led to a strong rejection of the traditional dichotomy between Eukaryotes and prokaryotes and subsequently a new tripartite distribution of the biodiversity was established, assigning the extant organisms to one of three domains: Eukaryotes (Eucarya), Archaebacteria (Archaea), and Eubacteria (Bacteria) (Woese et al. 1990).

As discussed above, establishing the order of descent in a group of organisms starting from a common ancestor necessitates the use of an external group, which allows rooting the phylogeny. The fact that there is obviously no external group to all extant life forms has hindered for a long time the rooting of the tree of life. However, this problem was going to be solved by the use of anciently duplicated (paralogous) genes (Schwartz and Dayhoff 1978). If a pair of paralogous genes is universally distributed (which means that the two copies are found in all representatives of the three domains), we can deduce that the duplication, from which they originated, must have taken place in the last universal common ancestor (LUCA) or more likely before. In this context, one paralogous gene furnishes the external group for the phylogeny established by the second copy and vice versa. The first two identified pairs of universal paralogous sequences (paralogs) were the translation elongation factors of EF-Tu/1α and EF-G/2 (Iwabe et al. 1989) and the V- and F-type ATPases (Gogarten et al. 1989).

The four resulting phylogenies support the rooting along the bacterial branch of the tree of life, thereby proposing a sister-group relationship between Archaea

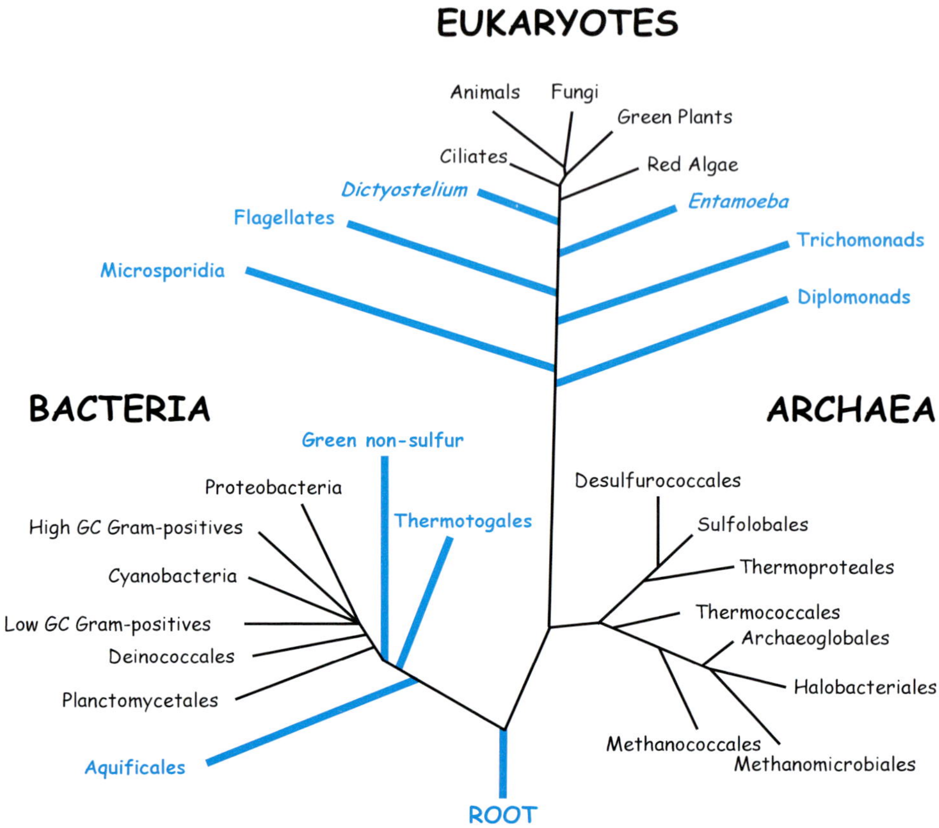

Fig. 8.2. The classical view of the universal tree of life. The topology was inspired by (Stetter 1996), and is principally based on the comparison of the rRNA sequences. Branches that could be affected by the long-branch attraction artefact (e.g., the position of the root on the bacterial branch or the early emergence of the hyperthermophilic taxa among the Bacteria) are indicated by *thick blue lines*

and Eukaryotes (Fig. 8.2). Unfortunately, the pairs of universal paralogs remain very rare. Until now only five additional anciently duplicated gene pairs have been found: the tRNA synthetases (tRS) for the amino acids valine and isoleucine (Brown and Doolittle 1995) and for tyrosine and tryptophane (Brown et al. 1997), an internal duplication in the subunit B of the carbamoyl-phosphate synthetase (Lawson et al. 1996), two components of the pathway directing nascent translation products into the endoplasmic reticulum (ER), SRP54/SRα (Gribaldo and Cammarano 1998), and the histidine biosynthetic genes *hisA*/*hisF* (Charlebois et al. 1997). All of the reciprocally rooted phylogenies based on these markers confirmed the bacterial rooting, with the only exception being the *hisA*/*hisF* pair. The rRNA phylogeny rooted between the Bacteria and the group Archaea/Eukaryotes became in consequence the universal tree of life shown in all the textbooks even for pupils (Fig. 8.2). Currently the classification of the life in three domains and the deduction that LUCA had a prokaryotic-type organisation is representing a paradigm that is largely accepted by the scientific community with only a few exceptions (Gupta 1998; Margulis 1996; Mayr 1998). Two other major evolutionary inferences were deduced from the tree of Fig. 8.2.

First, the phylogenies based on rRNA data suggest that the most deeply emerging branches in both prokaryotic domains correspond to hyperthermophilic organisms (Fig. 8.2). The members of the two known hyperthermophilic bacterial phyla – Aquificales and Thermotogales – are representing the most basal branches, followed by the moderate thermophilic bacteria, and finally the remaining species emerge (mostly mesophilic Bacteria). This observation linked to the traditional bacterial rooting of the universal tree of life and an early emergence of the hyperthermophilic Archaea was leading to the hypothesis that LUCA was a hyperthermophilic organism (Stetter 1996). Furthermore, the fact that the branches of the hyperthermophilic Bacteria are shorter than those of the mesophilic taxa was taken as an indication that a reconstruction artefact could not be responsible for their early branching (see below) and that the hyperthermophilic Bacteria are "living fossils", meaning that they had retained many ancestral features since LUCA. The idea of a hyperthermophilic LUCA was tightly linked to the hypothesis of a "hyperthermophilic Eden", which is suggesting that life originated and evolved in an extremely hot environment (more precisely a hydrothermal vent), and that the ancestral organisms remained hyperthermophilic until the appearance of the last common ancestor of the three domains (Gogarten-Boekels et al. 1995; Imai et al. 1999; Nisbet and Sleep 2001; Olsen et al. 1994; Pace 1991; Stetter 1996; Woese 1987).

Secondly, the phylogeny of Eukaryotes based on rRNA data (Fig. 8.2) can be divided into two parts, a basal part with several successively emerging amitochondriate lineages (such as, e.g. microsporidia, diplomonads and trichomonads), followed by mitochondriate groups (e.g. Euglenozoa or amoebae), and finally the remaining groups containing all late emerging lineages, that form a poorly resolved huge radiation, called the crown (Knoll 1992; Sogin 1991). The early appearance of amitochondriate protists in the rRNA tree had been interpreted as proof of the Archezoa hypothesis (Cavalier-Smith 1987), which suggested that several eukaryotic lineages originated before the mitochondrial endosymbiosis (microsporidia, diplomonads, trichomonads and Archaemoeba). Archezoa are often regarded as "living fossils", which furnish us information about the ancestral Eukaryotes (Knoll 1992; Sogin 1991).

The general idea of the evolutionary scenarios that may be deduced from this tree is that simple organisms preceded complex organisms (even if the definition of simplicity is not clear). The hypothesis that the "simple" Prokaryotes would be the ancestors of the "complex" Eukaryotes is a typical illustration. Furthermore, the simple Eukaryotes without mitochondria preceded the complex Eukaryotes with mitochondria. In fact, the idea that evolution is going from simple to complex is almost the direct inheritance of the scale of beings (scala naturae) from Aristotle. The latter was classifying the objects, living or inanimate, only in function of their complexity (from mineral to man), of course without assuming that the organisms have the capacity to change. The evolutionary theory was adapted to this philosophic frame simply by supposing those simple organisms are "living fossils", which ultimately gave rise to complex organisms. This apparent congruence between the modern results (ribosomal tree) and solidly rooted philosophical concepts (Aristotle) probably largely explains the fast and considerable acceptance of this new paradigm.

Nevertheless, this view of the universal tree of life is based on a very limited sampling of the genome (ca. 1000 nucleotides for the 16S rRNA, and ca. 1000 amino acids for the anciently duplicated paralogs). With the establishment of several whole genome sequences from the three domains, it rapidly became obvious that the trees based on alternative markers were largely contradicting the rRNA-based phylogeny and also each other, thus undermining the general consensus (Pennisi 1998). The reasons for these contradictions reside often in biological phenomena, like, e.g., lateral/horizontal gene transfers (LGT) or in the comparison of paralogous instead of orthologous genes (hidden paralogy) (Doolittle 1999c). However, mathematical features (tree-reconstruction artefacts) are more and more identified as an additional and important source of incorrect results in molecular phylogenetic reconstruction (Philippe and Laurent 1998).

8.3 Reconstruction Artefacts

8.3.1 Multiple Substitutions Generate Reconstruction Problems

After explaining the principles of phylogenetic reconstruction, we will now inspect the limitations of these methods. In Fig. 8.3a, the same phylogenetic tree as in Fig. 8.1a is shown, displaying the same number of substitutions; the only difference is that it is not exactly the same positions that have changed, as is indicated by the arrows. In the tree of Fig. 8.1a each position only changed once, whereas in Fig. 8.3 three positions (the second, the sixth and the eleventh) encountered two changes/substitutions. Let us see the influence of these multiple substitutions on the phylogenetic reconstruction methods. The second position was subject to a reversion (return to the previous state) that changed the T to S in the branch leading to the species 2. Therefore a position that was originally supporting the tree X (see Fig. 8.1e) is now noninformative. In the sixth position, which was also supporting tree X, there is a second substitution from F to L located on the branch leading to species 3 and it is also becoming noninformative, because we need two changes to explain the observed distribution for all three possible tree topologies (Fig. 8.3c). Even if the species 1 and 2 share the same amino acid (H), this does not necessarily mean that they are specifically related, a fact that was well explained by Willy Hennig (Hennig 1966). One has therefore to be critical towards the intuitive assumption that the more similar two organisms the more closely related they are.

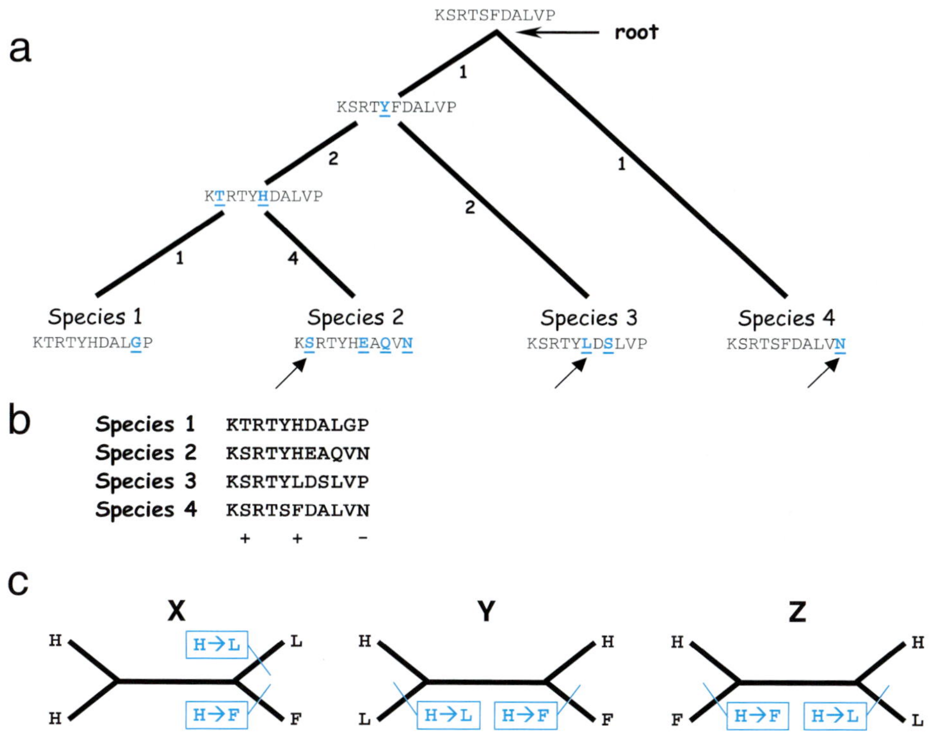

Fig. 8.3. Schema showing the limits of the molecular phylogenies. It is almost exactly the same tree as in Fig. 8.1, the only difference is that certain positions changed several times. These multiple substitutions generate noise that is excluding the reconstruction of the right tree

The eleventh position did experience a convergence on the branch leading to the species 2 and 4; this means the same amino acid, N, originated independently at the same position twice. Unfortunately, this position is now becoming informative, but in favour of the wrong tree (tree Y). In conclusion, the multiple substitutions destroyed the phylogenetic signal of two positions that support the right tree (indicated by a+ in Fig. 8.3b) and did create an erroneous informative position (indicated by a− in Fig. 8.3b). In this case, the most parsimonious tree based on the alignment is tree Y, thus leading to the inference of the wrong tree. Therefore the substitutions, which are the basis of phylogenetic information, are also responsible for its disappearance if they affect the same position several times.

8.3.2 Mutational Saturation Versus Resolving Power

As we saw before, positions that changed only once furnish reliable information for inferring bifurcations in the phylogenetic trees (e.g. the change from K to Y in site 1 of Fig. 8.4). If we have a sufficient number of these positions, we are able to reconstruct any given evolutionary history (Swofford et al. 1996). In practice, in order to resolve a phylogeny of about 50 species we will need more than 150 reliable positions (at least three changes for each node). Unfortunately, substitutions involve most of the time only a subsample of the total number of positions during history (Fitch and Markowitz 1970), generating more noise than signal for molecular phylogenetics (e.g. site 2, with 25 changes in Fig. 8.4). The positions that contain essential information (changes on the internal branches) often reach a state of mutational saturation, because they have lost all valuable information (site 2 in Fig. 8.4), e.g. the signal they contain is becoming random. Although the resolving power of molecular phylogenies is directly proportional to the length of the data set and to its evolutionary speed, it is diminishing if the

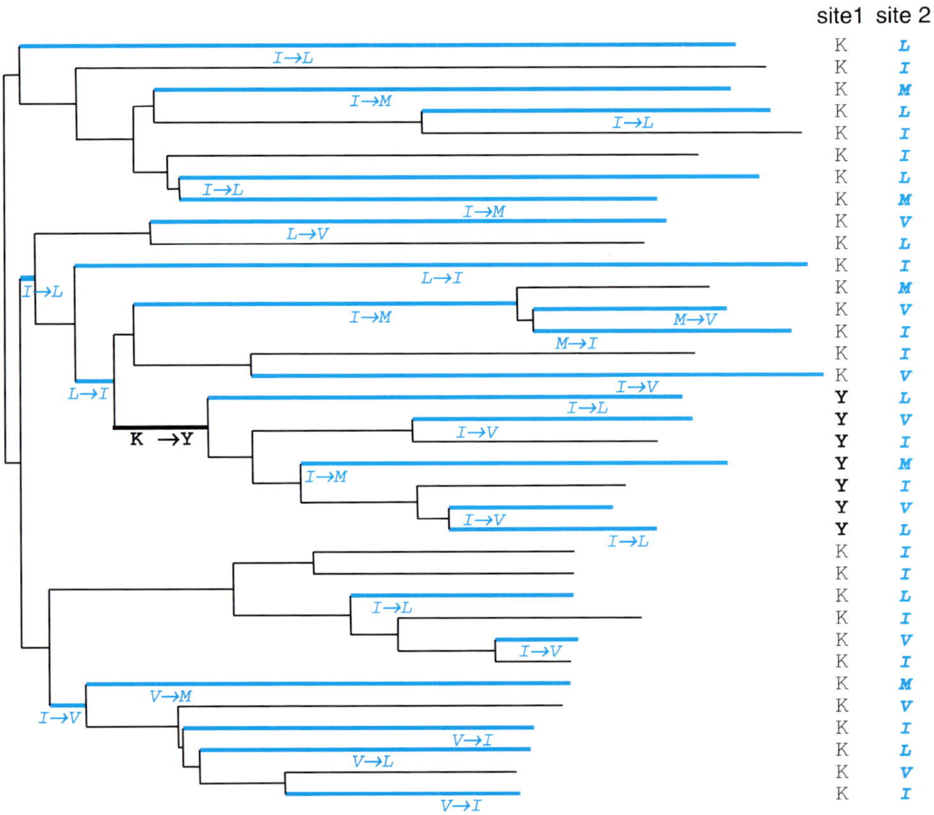

Fig. 8.4. Archetype of "good" and "bad" positions. Position 1 changed once (lysine versus tyrosine) and it is conserving a weak signal. In contrast, position 2 underwent multiple changes between several hydrophobic residues, which have completely intermingled the phylogenetic signal; more exactly it reached the point of mutational saturation, and in consequence it will essentially add more noise rather than phylogenetic information

speed of a given position is too high. In conclusion, there is the following major paradox: an informational overload is finally leading to the loss of the original phylogenetic signal.

Although an important number of the positions is highly conserved (and allows the alignment to be created), most of the other positions normally had changed much more than once. In order to demonstrate this fact, we analysed an alignment of 97 valyl/isoleucyl-tRNA synthetases (tRS) using the maximum parsimony (MP) method (Fig. 8.5). Five or less changes were inferred for only about 20% of the positions (68 of 344, with 15 constant sites). The information given by these positions consists of 144 changes (generally called steps). In contrast, for the 20% of the most variable positions (67 positions with more than 35 substitutions) a total of 2929 changes are inferred. Because the noisy positions are contributing tremendously more than the slowly evolving reliable ones to the choice of the best tree (2929 vs. 144 steps), it is of the utmost importance to develop methods that allow the efficient discrimination between signal and noise (see below).

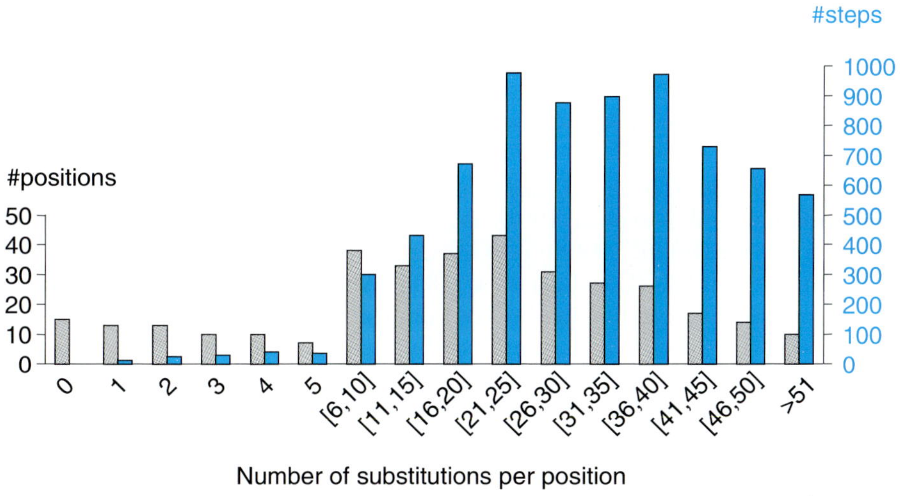

Fig. 8.5. Mutational saturation of the Ile/Val-tRNA synthetase sequences. On the X axis, the positions are regrouped according to their number of changes, calculated by the maximum parsimony method. On the Y axis, the number of positions that belong to the different classes is indicated on the *left* columns (in *light grey*) and the total number of steps that they are adding to the most parsimonious tree on the *right* columns (*blue*). The information added by the positions that only accepted a few changes (e.g., the positions with up to five changes furnished 144 steps), is largely outnumbered by those introduced by the positions that frequently changed (for example, the positions with more than 35 inferred changes furnished 2929 steps). The more noisy a position is, the more important is its contribution to the choice of the tree, thus leading us to the question of the validity of the bacterial rooting of the universal tree, observed for this gene (Brown and Doolittle 1995)

Saturation can also be identified by the presence of a plateau in a diagram, where the number of multiple substitutions calculated based on the tree is compared to the observed differences, for pairwise comparisons of sequences (Philippe et al. 1994). Mutational saturation represents a central issue of the reconstruction of ancient phylogenies, because all markers used for the inference of these trees are highly saturated (Philippe and Adoutte 1998; Philippe and Forterre 1999; Roger et al. 1999). Completely saturated data should not contain any signal and, in principle, should produce a star-like topology (without any resolution). Nevertheless, biases of multiple natures can lead in an erroneous way to resolved trees, because character states can be shared by chance between evolutionary distant species. This is linked to the fact that these biases are not correctly treated by the current tree-reconstruction methods. Three principal biases had been identified: (1) compositional bias, (2) variable rates of evolution and (3) heterotachy (also entitled covarion structure).

8.3.3 Compositional Bias

It was earlier shown that the G + C content of the rRNA was fundamentally heterogeneous among species (spreading between 30 and 70%) and could lead to an erroneous grouping of distantly related species that share a similar content (Loomis and Smith 1990; Weisburg et al. 1989). For example, thermophilic organisms have an elevated G + C content (Galtier and Lobry 1997); this produces a systematic bias. Modifications of the three principal tree-reconstruction methods were developed in order to take this problem into account (e.g. transversion-based parsimony (Woese et al. 1991), log-det distance (Lockhart et al. 1994) and nonstationary maximum likelihood (Galtier and Gouy 1995)). Despite some observable improvements (Lockhart et al. 1994; Woese et al. 1991), these methods did not considerably modify the inferred phylogenies. The corrections may even lead to less correct results, as exemplified by the enhanced statistical support for the early emergence of microsporidia, if the G + C content is taken into account (Galtier and Gouy 1995).

Furthermore, contrary to the first affirmations (Hasegawa and Hashimoto 1993), protein sequences can also be sensitive to the influence of compositional bias (Foster and Hickey 1999). For example, in a recent phylogeny based on a concatenation of ribosomal proteins, taxa with G + C rich genomes (e.g. *Mycobacterium* and *Deinococcus*) were artefactually put together (Brochier et al. 2002). Nevertheless, despite these few counterexamples, the impact of compositional biases is reasonably well treated by the current methods and cannot therefore be considered as a major problem, especially because it is quite easy to identify.

8.3.4 Long-branch Attraction

Another well-known bias, which can lead to errors in phylogenetic reconstructions, if the true phylogenetic signal is attenuated, is the one introduced by the variation of the evolutionary speed among species. For example, the evolutionary rates of the rRNA differ by a factor of 100 among planktonic foraminifers (Pawlowski et al. 1997). The artefact known as long-branch attraction (LBA) leads to the association of the longest branches, that correspond to the fastest evolving species, irrespective of their real phylogenetic affiliations (Felsenstein 1978). In the case of phylogenies rooted by a distantly related external group (outgroup), independently fast-evolving sequences of the studied taxa (ingroup) will emerge separately as the deepest branches, because they are attracted by the long branch of the distant outgroup (Philippe and Laurent 1998). Microsporidia that were proposed to be an early branching eukaryotic lineage (Vossbrinck et al. 1987), represent a spectacular example of LBA, because they are in reality highly evolved fungi (Keeling 2001b). The LBA artefact actually represents a major problem for molecular phylogenies, because it may affect the position of practically every deeply diverging lineage. In consequence, the position of many taxa, shown in bold in the universal tree of Fig. 8.2, should be considered as highly suspect, especially in the case of Eukaryotes, and at a lesser degree also for Bacteria.

Three main approaches were used to detect and/or reduce the impact of the LBA artefact: (1) enhanced or modified taxon sampling, (2) focusing on slowly evolving positions, and (3) better models of sequence evolution.

It was suggested very early during the development of molecular systematics that the addition of species reduces the impact of the LBA artefact, via breaking long branches (Hendy and Penny 1989). The eukaryotic phylogeny based on the elongation factor EF-1α furnishes a good example of this approach, even if the resolution of the tree is becoming limited concomitantly with the rising number of species (Moreira et al. 1999). Even if the theoretical basis is still under study (Poe and Swofford 1999), the practical approach is proving that this is an efficient method. Alternatively, if several representatives are available for a given group, it is important to do a more specific choice of taxa conserving only those that are more slowly evolving, rather than increasing naively the number of species. For example, the monophyly of the moulting animals (Ecdysozoa) was established by using a slowly evolving nematode species (Aguinaldo et al. 1997). The branch length of a given species taken from the root to the tip of a phylogenetic tree is roughly reflecting its evolutionary rate (e.g., short branches correspond to slowly evolving sequences).

The phylogenetic inference can also be improved by using the more slowly evolving positions (Olsen 1987), because they are less likely to contain noise (which means multiple substitutions, see Figs. 8.3 and 8.4) and will therefore more likely maintain the ancient phylogenetic signal (Felsenstein 2001). Our group recently developed a method, named SF (slow/fast) method, in order to detect these positions (Brinkmann and Philippe 1999). In short, the number of changes for each position is calculated by a maximum parsimony analysis independently for several predefined monophyletic groups. The evolutionary rate of a position is estimated as the sum of the number of substitutions observed in all groups; this estimate is therefore independent of the phylogenetic relationships among the groups. The number of substitutions at which the noise is exceeding the signal is not *a priori* known. Different data sets, containing all positions with less than n changes, are successively generated, covering the complete range of the relationships between signal and noise, and the evolution of the phylogenetic inference is also followed at the same time. If a rapidly evolving group is badly placed by the standard methods of phylogenetic reconstruction because of an LBA artefact, its true position should be found by using the most slowly evolving positions, and it will emerge progressively earlier when the more rapidly evolving positions are taken increasingly into account. Even if the reliable signal is rare and the resolution is often weak (see below), the SF method is efficient in detecting groups that are most likely incorrectly located because of LBA and may help to find the true affinity (Brinkmann and Philippe 1999; Philippe et al. 2000).

The LBA artefact is due to the fact that the observed number of substitutions is an underestimate of the real number, because of multiple substitutions that had taken place in the same position (Fig. 8.4); this is even more pronounced when a taxon is rapidly evolving (Olsen 1987). This underestimation happens if the model of sequence evolution used in the reconstruction process is not sufficiently reflecting the reality. The first model proposed for the detection of multiple substitutions assumes that all types of substitution are equivalent and that all sites have the same probability of change (Jukes and Cantor 1969). A progress was made by differentiating the types of substitutions (e.g., transitions vs. transversions) (Swofford et al. 1996). However, the most significant improvement was obtained when the unrealistic assumption of an equal substitution rate for all sites had been rejected (Yang 1996). For instance, the presence of invariant positions, although noninformative, violates the assumption of an equal substitution rate for all sites, and their identification and subsequent elimination improves the phylogenetic reconstruction (Lockhart et al. 1996). The support for the early emergence of the microsporidia in the phylogenies based on the elongation factors considerably decreased if the constant positions were removed from the analysis (Hirt et al. 1999). Actually, the most widely used method is based on the implementation of a model with variable rates between the positions (rate-across sites or RAS), modelling the heterogeneity of the substitution rates between the sites in the form of a continuous distribution of a Γ law type. This kind of approach, for example, helped to demonstrate that the microsporidia are highly derived fungi (Van de Peer et al. 2000).

8.3.5 Heterotachy

Models of the RAS type treat the variation of the substitution rate along the entire molecule, but they assume that the rate of a given position remains the same during its entire history (i.e. on all the branches of a phylogenetic tree). We named this hypothesis homotachy, for the same speed in Greek (Lopez et al.

2002). In fact homotachy has been shown to be unrealistic since the 1970s, with the pioneering work of Fitch (Fitch 1971), who demonstrated, for the cytochrome c, that the evolutionary rate of positions across the molecule are distributed differently between fungi and animals. A simple model of sequence evolution, the covarion model, was suggested to explain these observations. This model assumes that, at a given moment, only a fraction of the positions (the concomitantly variable codons, or covarions) is free to accept substitutions, with a constant probability for each of them (Fitch and Markowitz 1970). The rejection of the homotachy is therefore often described under the name of "covarion-like" phenomenon (Galtier 2001; Lockhart et al. 1996; Penny et al. 2001). Because large taxonomic samples are necessary for testing the hypothesis of homotachy (Lopez et al. 1999), the current massive increase of the data bases for molecular sequences permits a more thorough study of this question. We recently analysed about 2000 vertebrate cytochrome b sequences and demonstrated that 95% of the variable positions are significantly heterogeneous with respect to their substitution rates with time (Lopez et al. 2002). We used the term heterotachy instead of "covarion-like" to describe this major phenomenon, because the covarion model is only one among many other possible heterotachous models, and in addition it does not describe satisfactorily the evolution of the cytochrome b sequences (Lopez et al. 2002).

Heterotachy can generate reliable phylogenetic signals in sequence datasets (Lopez et al. 1999), in fossilizing ancient substitutions (e.g. a position that changed once several billions years ago became invariable rapidly after that change, thus excluding the possibility that multiple subsequent substitutions can erase this signal). Sometimes, heterotachy can represent an important source of trouble. It was shown that the presence of invariable positions in alignments can lead to an erroneous tree reconstruction (including the maximum likelihood methods, which are usually the most robust) (Lockhart et al. 1996). The analysis of the 16S rRNA and the elongation factor tufA of a number of non-photosynthetic and of oxygen-liberating photosynthetic prokaryotes (cyanobacteria and their endosymbiotic descendents chloroplasts) for example, has proved that the sharing of a similar ensemble of covarions (sites free to vary) can bias the estimate of the origin of chloroplasts (Lockhart et al. 1998). Because programs that take into account the phenomenon of heterotachy were not available, an easy method for improving the fit of the RAS model to the data is simply to eliminate the heterotachous positions from the alignment. This approach had shown, based on a concatenation of the 18S and 28S rRNA sequences, that a profoundly revised phylogeny of the Eukaryotes (compare Figs. 8.2 and 8.7) is not significantly rejected by these markers (Philippe and Germot 2000). The reconstruction of ancient phylogenies could especially benefit from the implementation of heterotachous models. The evolutionary processes of the heterotachous sequences could be responsible for the conservation of an ancient phylogenetic signal even if the mutational saturation would have almost totally eradicated all evolutionary messages (e.g., if a substitution did happen in an early emerging lineage and remained unchanged ever since) (Lopez et al. 1999; Penny et al. 2001). To model the heterotachy is very complex, but there is an increasing effort to achieve its implementation in the phylogenetic reconstruction methods (Galtier 2001; Huelsenbeck 2002; Penny et al. 2001).

8.3.6 Rare Genomic Events as an Alternative Approach?

In order to avoid the pitfalls described above, the approach called "hennigian" (the use of characters less sensitive towards saturation) could be powerful (Philippe and Laurent 1998). The comparison of rare genomic changes (e.g. insertions/deletions, or indels) could represent valuable markers supporting the monophyly of a group (Rokas and Holland 2000). Several promising results were obtained by using, for example, the integration of retroposons in the genome for mammals (Lum et al. 2000), the positions of introns for vertebrates (Venkatesh et al. 1999), and the order of mitochondrial genes for animals (Boore and Brown 1998). For the ancient phylogenies, this approach is less convincing. However, four indels had been used to support the opisthokont monophyly (animals and fungi) (Baldauf and Palmer 1993). One of these indels was also present in microsporidia, in agreement with the proximity of this group to the fungi (Hashimoto and Hasegawa 1996).

Small-sized indels could be particularly misleading as phylogenetic markers, because they are easily subject to convergence. In fact, they are usually found in regions of the protein (e.g. in loops exposed to the surface) that can easily accept the insertion or the deletion of segments representing several hydrophilic amino acids. For example, the reliability of an indel of two amino acids in the enolase, suggested as a support in favour of the early emergence of the trichomonads in the eukaryotic tree (Keeling and Palmer 2000), was seriously questioned (Bapteste and Philippe 2002; Hannaert et al. 2000). Assuming indels to have arisen several times independently by convergent evolution (and therefore being misleading) seems reasonable in the case of a few positions, but this problem should be less relevant in the case of large indels. But can they in consequence be considered to be perfect characters?

The eukaryotic gene coding for the valyl tRNA synthetase (ValRS), whose product is present in the cytoplasm and in the mitochondria of fungi, is considered to be probably of mitochondrial origin. In the phylogenies based on the ValRS, the eukaryotic sequences form a monophyletic group with the Proteobacteria (Brown and Doolittle 1995). This affiliation was further supported by the presence of a highly conserved, large insertion of 37 amino acids, shared by the eukaryotic and the γ-proteobacterial sequences, which was completely in agreement with the phylogeny based on this gene (Hashimoto et al. 1998). Nevertheless, because α-Proteobacteria are most closely related to the mitochondria, the mitochondrial origin of the ValRS was not completely convincing. A lateral gene transfer (LGT) of the ValRS gene from an archaebacterium to

Rickettsia, the first available sequence from an α-Proteobacteria, unfortunately did not allow this test to be performed (Woese et al. 2000). We decided to create an updated alignment of the ValRS consisting of 185 sequences, containing numerous α-Proteobacteria generated in the meantime. A phylogenetic analysis of the ValRS (Fig. 8.6) indicates that the α and the β/γ-Proteobacteria form two monophyletic groups, which are sister-groups, as expected. However, Eukaryotes did not show any specific affinity to the α-Proteobacteria, but rather emerge earlier, most likely because of a LBA artefact. Furthermore, the large insertion although present in all β/γ-Proteobacteria and in all Eukaryotes, is surprisingly absent in the α-Proteobacteria (see Fig. 8.6). At first sight, this observation seems to exclude the mitochondrial origin of the eukaryotic ValRS gene and would rather suggest a lateral gene transfer (LGT) from the β/γ-Proteobacteria to Eukaryotes. But, another well-conserved insertion of six amino acids is shared by a monophyletic group of four γ-Proteobacteria and by two representatives of the α subdivision (*Sinorhizobium* and *Agrobacterium*). These organisms are nevertheless solidly grouped with the other α-Proteobacteria, which do not have this insertion (Fig. 8.6). The same insertion, although longer and less conserved, is also present in one of the members of the high GC Gram-positives (*Mycobacterium*). In consequence, because the two large indels in the sequence of the ValRS are in conflict with each other and also in conflict with the phylogeny, they must be considered as ambiguous characters, and did not give any answer to the question of the origin of the eukaryotic ValRS.

Furthermore, rare genomic events are equally sensitive to LGT, as standard phylogenetic inference. For example, a large indel in the amino-terminal part of the HSP70 proteins (heat shock protein, 70 kDa) was for a long time considered as evidence to unite the Archaea and the Gram-positive bacteria, reinforced by the claim that some HSP70-based phylogenies seem to support this point of view (Gupta and Singh 1994). This indel was subsequently used for supporting a new hypothesis about a chimerical origin of the Eukaryotes (Gupta and Singh 1994). However, by using a larger taxonomic sampling, it was shown that (1) several Gram-negative bacteria did have a HSP70 gene with the same deletion that was supposed to be specific for the Gram-positive bacteria (Philippe et al. 1999) and (2) the majority of the archaeal species do not have a HSP70 gene, suggesting that the Archaea originally did not possess any HSP70 gene at all (Gribaldo et al. 1999). In fact, the taxonomic distribution of HSP70 genes in the complete genomes actually accessible (its absence in all Crenarchaeota, in the Pyrococci and in *Methanococcus*), strongly suggests a single LGT event with a Gram-positive bacterium, after the emergence of *Methanococcus* (see Fig. 2 in (Forterre et al. 2002)). In conclusion, rare genomic changes could be only useful markers in the reconstruction of phylogenies if they are used in addition to sequence data (Rokas and Holland 2000), but their direct use as an indicator of relatedness between species (Gupta and Johari 1998) should always be regarded very carefully.

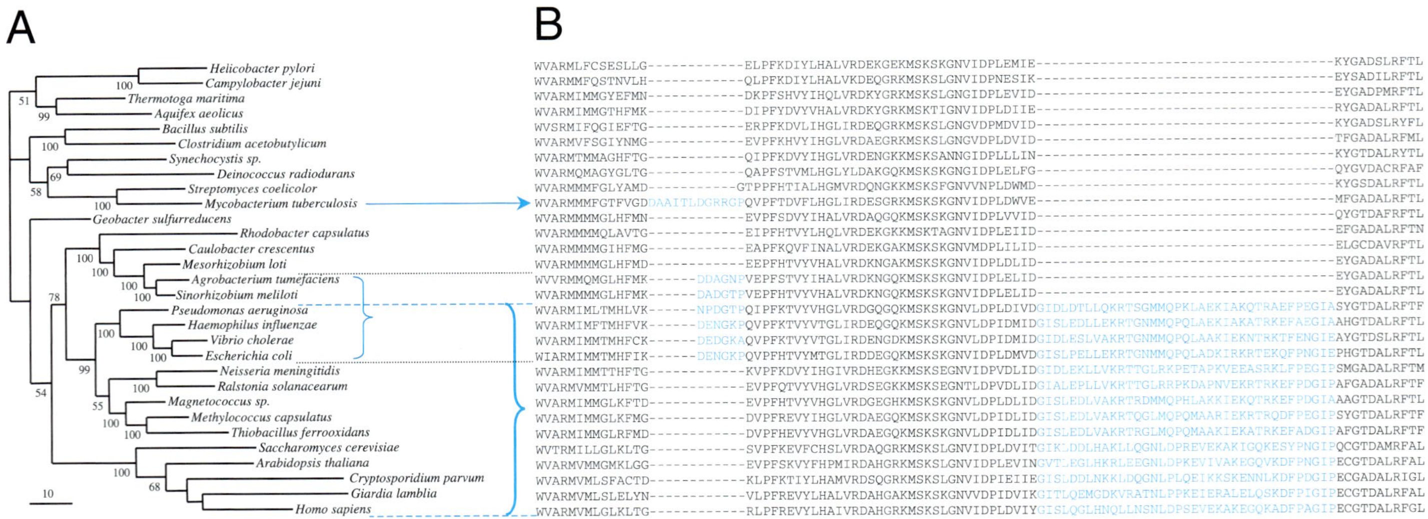

Fig. 8.6. Rare genomic events and the evolution of the Ile/Val-tRNA synthetase. (**A**) a representative sample of 30 sequences has been selected starting from a dataset containing 185 sequences of ValRS covering a huge taxonomic diversity. The tree has been inferred with a maximum likelihood method with the program protml and bootstrap values (shown *left* of the nodes) were calculated using the RELL method (Adachi and Hasegawa 1996). The proteobacteria of the α and β/γ subdivisions form two monophyletic groups, which are sistergroups, as expected. Surprisingly, Eukaryotes did not show any affinity for the α-Proteobacteria. (**B**) an extract of the ValRS alignment, which is focused on the region surrounding two highly conserved insertions. The large insertion of 37 residues has been proposed as a strong support in favour of the mitochondrial origin of the eukaryotic ValRS (Hashimoto et al. 1998), although this insertion is present in all β/γ-Proteobacteria and Eukaryotes, it is surprisingly absent from the α-Proteobacteria. Furthermore, the phylogenetic distribution of another well-conserved insertion of six amino acids is clearly in conflict with the phylogeny presented in (**A**). It is in fact shared by four γ-Proteobacteria and two β-Proteobacteria, which do not form a common group at all. In addition, the taxonomic distribution of this indel is also in conflict with the larger indel. Although, longer and less conserved, an insertion at the same position than the six-amino-acid insertion is also present in several high G + C Gram-positives (more precisely, present in *Mycobacterium*, but absent in *Streptomyces*)

8.4 Lateral Gene Transfer and the Quest for a Phylogeny of the Organisms

The sequencing of complete genomes has confirmed the high frequency of LGTs in prokaryotes (Koonin et al. 2001; Ochman et al. 2000) and of genes duplications in Eukaryotes (Lynch and Conery 2000). Therefore it is potentially dangerous to deduce the phylogeny of the organisms starting from trees that are only based on a single gene (Page and Charleston 1997). The presence of multiple copies of a homologous gene in a single genome opens the way to recombination. But, if a gene becomes involved in recombination events, the most appropriate mathematical representation of its history would be a network, rather than a tree (i.e. a connected graph without cycles). Difficult to identify, the intragene recombination was recently demonstrated through the incongruent taxonomic distribution of indels in the enolase and in the inositol monophosphate dehydrogenase (IMPDH), rendering the phylogenetic reconstruction of Eukaryotes based on these genes elusive (Bapteste and Philippe 2002; Keeling and Palmer 2001). In fact, recombination between paralogous or xenologous genes is probably more frequent than was previously thought (Archibald and Roger 2002), rendering the phylogenetic inference even more problematic (Posada and Crandall 2002).

The methods used to identify the LGTs (so-called indirect methods) are principally based on the non-normal nucleotide composition and the fact that the most similar sequences in the data banks did not belong to closely related species (similarity is normally estimated by a BLAST search). But these approaches are only poor indicators (Guindon and Perriere 2001; Koski and Golding 2001; Koski et al. 2001; Wang 2001). This is illustrated by the fact that four alternative methods, detected different groups of genes (Ragan 2001). The most likely explanation is that these indirect methods could potentially detect LGT of different ages (Lawrence and Ochman 2002; Ragan 2001). In consequence, albeit slow, the thorough phylogenetic approach represents the only direct detection method that should be used systematically (see below).

The underlying assumption of this method is that a phylogeny of organisms exists and can be represented in the form of a bifurcating tree. Because LGTs seem to be common and affect all genes (Koonin et al. 2001), it was suggested that "the history of life could not be correctly represented in the form of a tree" (Doolittle 1999c). Other authors have less radical opinions on the damaging impact of LGTs on the reconstruction of the evolutionary history. For example, it was suggested that LGTs could have played an important role only during a precellular stage of the evolution because it was hypothesised that "primitive entities were basically modular (loosely coupled) in construction", thus facilitating LGT events (Woese 2000). Alternatively, only a subdivision of the genetic materials could be affected by this phenomenon, excluding a veritable core of nontransferable genes (Jain et al. 1999). In any case, it seems that the LGTs likely represent an important evolutionary force in the diversification of the prokaryotes (Doolittle 1999b; Martin 1999; Ochman et al. 2000).

In order to examine the sheer existence of the phylogeny of organisms, several groups applied genomic approaches (Fitz-Gibbon and House 1999; Korbel et al. 2002; Lin and Gerstein 2000; Tekaia et al. 1999; Wolf et al. 2001). These methods are normally based on gene content (presence or absence) or the order of genes. The resulting phylogenies are more or less similar to the rRNA-based tree. For example, the monophyly of the three domains and of several other major groups (like animals, spirochaetes, Proteobacteria) is generally recovered. However, unexpected relationships, like that between *Thermoplasma* (Euryarchaeota) and the Crenarchaeotes (Korbel et al. 2002), demonstrate biases introduced in these approaches by frequent LGTs between phylogenetically distant taxa, but that are coexisting in the same environment (Ruepp et al. 2000). Another striking example is given by the hyperthermophilic bacteria, which probably acquired up to 24% of their genes from the Archaea (Aravind et al. 1998; Nelson et al. 1999). Furthermore, the convergent loss/gain of a comparable ensemble of genes due to similar physiological conditions, e.g. the adaptation to intracellular parasitism, can equally introduce biases. In conclusion, the methodology based on complete genomes should rather be considered as phenetic rather than as phylogenetic approaches (Doolittle 1999a; Wolf et al. 2001). Large-scale studies based on the concatenation of a great number of genes (Brown et al. 2001), despite their great potential for the inference of phylogenetic trees, can also be influenced by LBA artefacts, especially if they are applied to prokaryotes. In fact, if too many of the laterally transferred genes are included in the data sets, the results could correspond to a simple mean of the LGTs frequencies between the organisms, rather than being a correct indicator of the phylogeny of the organisms. Circular and therefore specious methodology could be applied in order to recover the tree in which one believes. For example, Brown and collaborators (2001) observed incongruities between their tree based on several fused genes and the one based on the rRNA. They obtained an increase of the congruence (e.g. the basal emergence of the hyperthermophilic bacteria) by eliminating the nine genes that did not recover the monophyly of the bacterial clade out of a total of 23. Such an approach based on only 14 genes does not represent convincing evidence in favour of the existence of a phylogeny of the organisms.

In order to avoid these limitations, our group recently developed a method that evaluates congruence among genes without a reference phylogeny (Brochier et al. 2002). Starting with 45 complete bacterial genomes, 59 markers implicated in translation (57 proteins and the 16S and 23S) were chosen, which were supposed to be excluded from LGTs (Jain et al. 1999), as well as 39 markers well known for being transferred (essentially tRNA synthetases (Woese et al. 2000), but also some ribosomal proteins (Brochier et al. 2000)). To estimate the congruence, each marker was described by a vector containing its likelihood values for an ensemble of representative topologies. These vectors were subsequently examined by a principal-component analysis (PCA). If two genes share the same evolutionary history, they furnish comparable support for the same topologies and will therefore group in the same area of the PCA. After eliminating stochastic effects, 46 of the 59 markers a priori supposed to be nontransferable, and six of the 39 a priori supposed transferable formed a compact cloud of points. In consequence, the identification of a genuine core of 52 genes, that belong a posteriori to the nontransferred fraction of the genome among the 45 analysed species, argues strongly in favour of a phylogeny of organisms for the Bacteria. Furthermore, the phylogeny based on these proteins is in good agreement with that based on the two concatenated rRNA genes (Brochier et al. 2002). The PCA method also allowed detection of the very likely cases of LGTs that were not reported earlier (e.g. for the translation elongation factor EF-G).

We applied the same approach to 14 archaeal species and obtained very similar results: a core of nontransferred genes could be detected and the phylogeny based on their concatenation gives a good approximation of the phylogeny of the species (e.g. obtained by the concatenated rRNAs) (Matte-Tailliez et al. 2002). Without minimising the importance of the LGTs in Prokaryotes, these results nevertheless demonstrate that a universal tree of organisms really exists and that its inference represents the primary challenge of the molecular phylogenies. We will now focus on the recent progress made to resolve this challenge.

8.5 A New Evaluation of the Universal Tree of Life

In light of the tree-reconstruction problems described above, especially the LBA artefact, it is legitimate to question whether the standard view of the ancient phylogenies is really sound. Especially, many of the basal branches (shown in bold on Fig. 8.2) could potentially be incorrect.

8.5.1 The Root of the Universal Tree of Life

The inference of the root of the universal tree is fundamentally the most difficult duty for a molecular phylogeneticist, because it corresponds to the most ancient evolutionary event that one can study. It is somewhat surprising that the vast majority of the researchers take the bacterial rooting for granted (Fig. 8.2), because this had been deduced from a few genes (less than ten), each of them having only a few alignable positions (ca. 100), and a very limited taxon sampling. This is even more striking because the methods used to reconstruct the trees were rather rudimentary, in general distance-based methods without correction for the variation of the evolutionary rates across sites. This approach is known to be very sensitive to LBA artefacts (Swofford et al. 1996), particularly with a very limited species sampling (Hillis 1996; Lecointre et al. 1993; Philippe and Douzery 1994). In a reanalysis with an improved taxon sampling, the majority of the phylogenies used for the inference of the root of the universal tree of life (Brown and Doolittle 1995; Brown et al. 1997; Gogarten et al. 1989; Gribaldo and Cammarano 1998; Iwabe et al. 1989; Lawson et al. 1996) indicated a much more complex evolutionary pattern (Philippe and Forterre

1999). For example, in the phylogeny based on the isoleucyl tRNA synthetase the two prokaryotic domains are so extremely intermingled that it is very difficult, even impossible, to know which sequences are the true representatives of the Archaea and of the Bacteria, respectively. More fundamentally, all genes revealed very high levels of mutational saturation (Philippe and Forterre 1999). In fact, these highly conserved genes contain by essence a certain number of invariant positions. But saturation is free to accumulate at variable positions, which furnish noise rather than phylogenetic information. In consequence, instead of informative positions (like the position 2 in Fig. 8.4), noisy positions (like the position 2 in Fig. 8.4) furnish the principal information to determine the location of the root, as illustrated by the case of the duplicated genes of the isoleucyl/valyl tRNA synthetases (Fig. 8.5). In that instance, the most slowly evolving positions (less than ten changes) furnish a negligible number of steps (442) in comparison to those (2032) generated by the rapidly evolving positions (more than 40 changes). Encountering such a high level of saturation, the classical methods of reconstruction have only a moderate performance and are susceptible to artefacts.

Furthermore, among the three domains of life, the bacterial branch is always the longest one for all of the genes used to root the universal trees of life (Philippe and Forterre 1999). Because this root is inferred by using an anciently duplicated paralogous gene, which *per se* represents a very long branch, a LBA artefact between these two long branches is predictable because of the mutational saturation. Therefore, the bacterial rooting could be simply the result of an artefact. Since branch length equals λt, with λ being the substitution rate and t the absolute time, the long bacterial branch can be schematically explained in two opposite ways: the origin of the bacterial domain is really more ancient (and t has a very elevated value) or the bacterial genes evolve more rapidly than their archaeal and eukaryotic homologs (and λ has a very high value). In the first case, the bias created by the LBA artefact is acting in favour of the true phylogeny, whereas in the second case it will hinder its discovery. A better way to distinguish between these two hypotheses (high value of t or λ) is to apply methods that focus on the slowly evolving positions, because they are less sensitive to the LBA artefact (Olsen 1987).

Such a method was developed by our group and applied to the paralogous pair EF1α/EF2 (Lopez et al. 1999). By using the more slowly evolving positions, the two opposite topologies (Bacteria, (Archaea, Eukaryotes)) – the current paradigm – and ((Bacteria, Archaea), Eukaryotes) – prokaryotic monophyly – could not be distinguished. This result is in agreement with an elevated λ (substitution rate) and not an elevated time t. The most concrete point of this analysis was the very long branch that separates EF1α and EF2, that renders it very difficult to avoid LBA artefact. We subsequently focused our attention on the SRP54/SRα sequences, because these two genes are by far the most similar pair among the anciently duplicated paralogs (Brinkmann and Philippe 1999). When the positions, containing the most reliable phylogenetic signal were extracted (positions with less than six changes), the Bacteria displayed a very long branch, but were nevertheless a sister-group of the Archaea (indicating a eukaryotic rooting of the universal tree). This fact constitutes a strong argument in favour of the hypothesis assuming a large λ. The utilisation of the positions with seven substitutions is provoking a shift from the topology ((B,A),E) to (B,(A,E)). Quite interestingly, this shift is not due to any signal in favour of the topology (B,(A,E)), but rather to a rejection of the ((B,A),E) by noisy char-

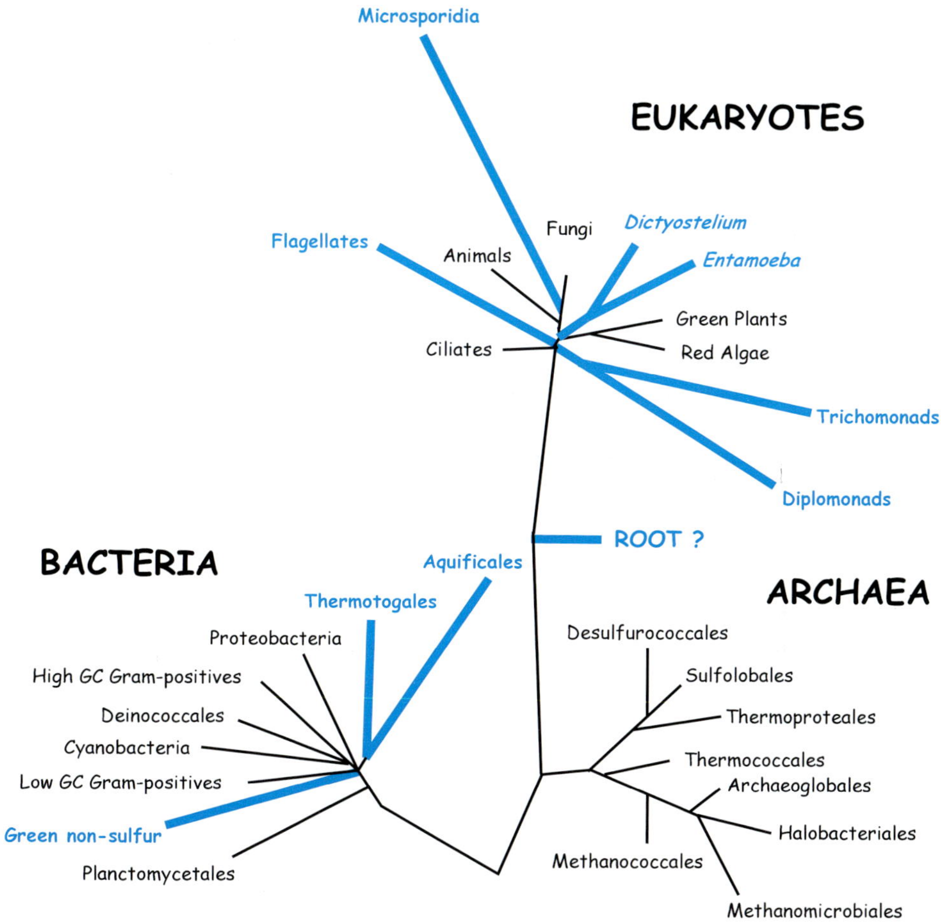

Fig. 8.7. A new vision on the universal tree of life. Taxa that are incorrectly placed in Fig. 8.2 because of the LBA artefact are replaced on this phylogeny, based either on the reanalysis of the rRNA (e.g., hyperthermophilic bacteria) and of the anciently duplicated genes (for example the root), or based on the analyses of new markers (for example, the grouping of *Dictyostelium* with *Entamoeba*). It should be noted that, while some of these revisions are robust (e.g. microdsporidia with fungi), several others still remain open (e.g. the root)

acters, because of a LBA artefact. In conclusion, the study of slowly evolving positions provides a serious proof that the rooting in the bacterial branch is due to an elevated λ(LBA) and not an elevated t (the true phylogeny). The statistical support for the eukaryotic rooting (the monophyly of Prokaryotes, Fig. 8.7) is weak, because the slowly evolving positions correspond to too small a number. The implications of the rooting in the eukaryotic branch have been discussed in detail elsewhere ((Forterre and Philippe 1999a; Forterre and Philippe 1999b); Penny and Poole 1999; Philippe and Forterre 1999), and will only be mentioned very briefly below. The most important consequence is that such a rooting prevents knowledge of whether LUCA was a Prokaryote or a Eukaryote. It seems therefore that the rooting of the universal tree of life should be considered as an open question.

8.5.2 Prokaryotic Phylogeny

The fact that the branches leading to the hyperthermophilic bacteria in the tree based on the rRNA are shorter than those leading to the mesophiles had been taken as an indication that a LBA artefact was not responsible for their early emergence, but rather that the hyperthermophilic bacteria were "living fossils". The observations about the prokaryotic phylogeny and the rooting in the eukaryotic branch also led to the hypothesis of thermoreduction (Forterre 1995), that propose that the prokaryotic organisation of the cells emerged originally as an adaptation of life to high temperatures.

The first branches of the bacterial phylogeny are potentially misplaced for the reason of the LBA, because the external group (Archaea and Eukaryotes) is very distant (Fig. 8.2). We recently applied the SF-method, which focuses on the most slowly evolving positions (Brinkmann and Philippe 1999), to a large dataset of the rRNA containing the representatives of the majority of the prokaryotic phyla and we demonstrated that the early emergence of hyperthermophilic bacteria (Aquificales and Thermotogales) is due to a LBA artefact (Brochier and Philippe 2002). This fact argues in favour of a secondary adaptation to high temperatures in these lineages. This deduction is in agreement with an estimate of the G + C content of the rRNA of LUCA (54–56%)(Galtier et al. 1999), a value that is incompatible with a growth temperature higher than 80°C. The complete genomes of *Aquifex* and *Thermotoga* showed that numerous genes found in these species (respectively 16% and 24%) have been acquired by LGTs from the Archaea (Aravind et al. 1998; Deckert et al. 1998; Nelson et al. 1999). Such a high flux of genes could have helped the adaptation to a life at high temperature in hyperthermophilic bacteria. An example of this type of genes is the reverse gyrase, which generates a positive superturn (supercoil) and very likely serves in the stabilisation of the DNA at high temperature. A detailed phylogenetic analysis of this gene supports the independent acquisitions of this gene by *Aquifex* and *Thermotoga* from Archaea (Forterre et al. 2000). These recent results therefore strongly argue against the hypothesis of a hyperthermophilic ancestor of Bacteria and advocate towards a nonhyperthermophile (in contrast, the ancestor of Archaea is most likely a hyperthermophile (Forterre et al. 2002)).

Surprisingly, the SF-method applied to the bacterial rRNA suggests, with a reasonable statistical support (bootstrap value of 75%), that the Planctomycetales may be the earliest emerging bacterial phylum (Brochier and Philippe 2002). This phylum is an important bacterial division where the members share several remarkable characteristics, like the absence of peptidoglycan in their cell wall and reproduction via budding (Fuerst 1995). Another particularly intriguing trait is the existence of an internal membrane, simple or double, around the bacterial chromosome in the genera *Gemmata* and *Pirellula*, respectively. Although this structure had been compared to the nuclear envelope of the Eukaryotes (Fuerst 1995), the evolutionary homology with the nuclear membrane has not been proven. Unfortunately, only a few studies were made on this group, despite their special characteristics; yet their implication in the anaerobic oxidation of ammonium has been recently demonstrated (Strous et al. 1999). If the early emergence of the Planctomycetales favoured in our reanalysis of the rRNA is confirmed with additional phylogenetic markers, the origin of Bacteria and the nature of LUCA should be seriously re-evaluated. With the exception of the basal branches, the bacterial phylogeny based on the rRNA has been globally confirmed by the analyses of complete genomes, such as the trees based on the gene order or on the gene content (Korbel et al. 2002; Wolf et al. 2001), or by the simultaneous analyses of numerous genes (Brochier and Philippe 2002; Brown et al. 2001; Daubin and Gouy 2001; Wolf et al. 2001). Surprisingly, the resolution between the major bacterial phyla has been slightly improved despite the use of very large datasets (Fig. 8.7). Only three new supergroups have been suggested by the phylogenetic analysis of concatenated genomic datasets; these are Chlamydiales/Spirochaetes, Thermotogales/Aquificales and high G + C Gram-positive bacteria/Deinococcales/Cyanobacteria (Brochier et al. 2002; Brown et al. 2001; Daubin and Gouy 2001; Wolf et al. 2001). However, it is not impossible that these affiliations could have been influenced by one of the biases discussed above; i.e. a very elevated evolutionary rate (due to parasitism), the amino-acid composition (linked to hyperthermophily) and the genomic G + C content, respectively.

In conclusion, since new data generated in numerous genome-sequencing projects are released rapidly, approaches based on large concatenations of genes represent the most promising perspective for resolving the prokaryotic phylogeny. It is, nevertheless, crucial to work on the methodological improvement of tree reconstruction, in order to extract the maximum information from this enormous quantity of sequences. One example of a methodological improvement will be given in the following section.

8.5.3 Eukaryotic Phylogeny

The Archaezoan hypothesis, proposing eukaryotic lineages that appeared before the mitochondrial endosymbiosis, has been criticised. In fact, it is possible to know whether a species does not have mitochondria because they have been lost, or because they were never present. Indeed, after the mitochondrial endosymbiosis, the essential part of the genetic inheritance of the α-Proteobacteria has been either lost or transferred to the nucleus (Lang et al. 1999). The proteins encoded by these genes are often imported into the mitochondria, but are also located in other compartments (cytoplasma, nucleus, etc.). Therefore, if a species is losing their mitochondria, or at least the oxidative respiration that constitutes its principal function, during the adaptation to an anaerobic environment, it will likely conserve some genes of mitochondrial origin. The discovery of genes of mitochondrial origin in all of the eukaryotic lineages without mitochondria (especially, the early emerging ones like the diplomonads and the microsporidia) strongly rejected the Archaezoan hypothesis (Embley and Hirt 1998). Furthermore, the early emergence of these lineages, which represented the second basis of this hypothesis, is highly suspicious, given the very long branch leading to the prokaryotic outgroup (Fig. 8.2). In fact, numerous independent analyses have shown that the basal branches in rRNA-based phylogenies are incorrectly placed because of the LBA artefact (for a review see (Philippe 2000)).

The challenge is now to find enough correct phylogenetic positions of all the lineages for which rRNA sequences are rapidly evolving. Analyses based on a single gene shed some light on these questions. For example, microsporidia have been grouped together with the fungi thanks to phylogenies based on tubulin (Edlind et al. 1996), the mitochondrial HSP70 (Germot et al. 1997), and the RNA polymerases II (Hirt et al. 1999). The trees based on actin (Keeling 2001a) and an indel found in polyubiquitin (Archibald et al. 2003) have suggested the grouping of foraminifers and Cercozoa, a group of highly heterogeneous protists. The principal drawback of phylogenies based on a single gene is a very limited resolution of the relationships between the principal eukaryotic lineages (Budin and Philippe 1998). In consequence, the necessity to increase the resolving power imposes the simultaneous use of several genes. The first attempts to reconstruct the eukaryotic phylogeny by using a large number of data were done starting in the mid-1990s (Kuma et al. 1995; Nikoh et al. 1994). Unfortunately, the taxon sampling was very limited (four or five species), rendering these phylogenetic inferences not very reliable. Several attempts have subsequently been made with a larger number of species, although this reduced the number of genes used, invariably leading to poorly resolved phylogenies (Budin and Philippe 1998; Germot and Philippe 1999).

One notable exception is a phylogeny based on concatenated mitochondrial genes (cob1, cox1, cox2, and cox3), which provided a strong statistical support for the monophyly of Opisthokonta (animal+fungi) and of Plantae (red algae and green plants) (Burger et al. 1999). The recent explosion of the sequence data for a much larger taxonomic variety, led to an increase of analyses based on multiple genes, in particular for angiosperms (for review see (Chase and Fay 2001)), the mammals (e.g. (Madsen et al. 2001; Murphy et al. 2001)), but also for the Eukaryotes ((Arisue et al. 2002a; Arisue et al. 2002b); Baldauf et al. 2000; Bapteste et al. 2002; Fast et al. 2002; Moreira et al. 2000).

Several recent results have clarified important questions of eukaryotic evolution (e.g. the acquisition and the loss of organelles). The analyses based on multi-gene fusions (Burger et al. 1999; Moreira et al. 2000) furnished significant support for a single primary endosymbiosis with a cyanobacterium at the origin of chloroplasts, because they indicated the monophyly of Plantae (green plants, red algae and glaucocystophytes). Furthermore, the grouping of the two principal phyla that harbour a chloroplast enveloped by four membranes, the alveolates and the stramenopiles (Arisue et al. 2002b; Baldauf et al. 2000; Bapteste et al. 2002; Fast et al. 2001) strongly suggested that secondary endosymbioses (the phagocytotic uptake of a photosynthetic Eukaryote by another Eukaryote) are much rarer than was previously thought. In fact, only two secondary endosymbiotic events could have taken place, in contrast to seven as was previously assumed (Cavalier-Smith 1999). Nevertheless, in order to definitely solve this question, additional data from other lineages of photosynthetic Eukaryotes (especially, haptophytes, cryptophytes, chlorarachniophytes and the euglenoids) are needed. The discovery of two new eukaryotic super-groups of organisms without mitochondria, *Entamoeba* and *Mastigamoeba* on the one hand (Arisue et al. 2002a; Bapteste et al. 2002), trichomonads and diplomonads on the other hand (Baldauf et al. 2000; Embley and Hirt 1998; Henze et al. 2001), also reduced the expected number of independent secondary losses of mitochondria. Anyhow, the independent losses of mitochondria and chloroplasts are likely much more frequent than their acquisition by endosymbiosis (see also below).

8.6 The Importance of an Evolution by Simplification and by Extinction

The idea that evolution is going from simple to complex, which is inherited from the philosophy of Aristotle (as already mentioned), deserves to be discussed here in more detail in light of the recent results from molecular phylogenies. First of all, it is certain that starting from the origin of life (and even before) up to the emergence of LUCA, a very important rise in complexity of living organisms happened. But, there is no proof that this rise in complexity arrived in a linear way and that there had not also been episodes of secondary simplification. There is often a tendency to extrapolate the phenomenon of a steady rise in complexity to the period after LUCA. But in particular, the often-committed error is to associate the simplicity of extant species with primitiveness, even often considering them as inferior (e.g., the lower animals to designate the invertebrates).

The concept of the “living fossil” is intimately mixed with this problem. Hence, the microsporidia, which are simple Eukaryotes (without flagella, Golgi apparatus and mitochondria) were often considered as primitive species, the “living fossils” from an era antedating the mitochondrial endosymbiosis. Unfortunately, the notion of a “living fossil” is highly arguable. In fact, all extant species are the descendants from a single common ancestor, LUCA, and did evolve exactly during the same period of time: they are therefore all approximately equally evolved (not taking into account different rates of evolution). To say that an extant species is a fossil implies the assumption that it is less evolved than the others. However, a major issue concerning this concept is that its application is highly subjective. For example, the coelacanth, probably the most famous example of a "living fossil", has a morphology that is globally similar to that of certain several hundred million years old fossils, but it is very likely that its physiology, and in consequence also its life style, is very different. Furthermore, it is obvious that the genome of the coelacanth is as evolved as the genome of all other vertebrates. One can propose an objective measure for the degree of evolution of an organism, and it would be the number of generations since the last common ancestor. This measure is perfectly reasonable because the influence of the natural selection rises with the number of generations (or of DNA replications). Based on these grounds, the less-evolved species would be those that have the longest generation times (e.g., the elephant, man or the turtle) and the most highly evolved would be those that have the smallest generation times (e.g., the Bacteria *Escherichia coli* or *Bacillus subtilis*). In fact, the sheer concept of evolved or primitive species is without any interest if one considers extant species and is only correct for palaeontologists. Let us go back to the problem of the evolution from simple to complex without further considering the useless complications introduced by the use of expressions like "primitive species" or "lower organisms".

The molecular phylogenies have indeed demonstrated that in numerous cases the species having numerous simple characteristics often emerge late in the trees. For example, the Bacteria (*Mycoplasma*) and the Archaea (*Thermoplasma*) without a cell wall arose among species with a cell wall, thus strongly suggesting secondary loss of cell walls. Eukaryotes without flagella, like red algae and the majority of fungi, emerge in between organisms that have well-developed flagella, suggesting that this complex organelle (composed of several hundred different proteins) was lost at least twice. More surprisingly, chloroplasts were lost in several independent lineages. In certain cases, the photosynthetic function was lost, but the structure was conserved, like for example in *Plasmodium falciparum*, the intracellular parasite that causes malaria (Waller et al. 1998). In other cases, like for example ciliates (Fast et al. 2001), diplomonads, trichomonads and the trypanosomes (Andersson and Roger 2002; Hannaert et al. 2003), it seems that the original chloroplast was completely lost and that only some genes of cyanobacterial origin were conserved in the nucleus. Finally, as already discussed before, mitochondria were lost several times. To be more precise, the respiratory function based on oxygen as well as the genome were lost, but a mitochondrial structure (limited by two membranes) is conserved. In particular, thanks to monoclonal antibodies directed against the chaperonin proteins of mitochondrial origin, it was possible to demonstrate the presence of highly reduced mitochondria in *Entamoeba histolytica* (Mai et al. 1999; Tovar et al. 1999), in the microsporidia

(Williams et al. 2002) and in diplomonads (Tovar et al. 2003). It is conceivable that the complete loss of mitochondrial structures is impossible, because the mitochondria are implicated in numerous, nonrespiratory functions central to eukaryotic cells, like the iron-sulphur protein maturation (Katinka et al. 2001; Tovar et al. 2003).

In contrast to what seems to indicate the first results (Fig. 8.2), molecular phylogenies have revealed that the evolution by secondary simplification represents a major process. They indeed simply allowed recovering a quite old result. In 1943, André Lwoff published a book exclusively dedicated to the importance of the evolution from the complex towards the simple, as its title suggests, "L'évolution physiologique. Etude des pertes de fonctions chez les microorganismes" (The physiological evolution: Studies of functional losses in microorganisms) (Lwoff 1943). The introduction of this book, which is still perfectly up to date and whose lecture we strongly recommend, explains very well the reasons of the psychological rejection that we have of regressive evolution. In particular, because of positivism, there exists a very strong association between evolution and progress on the one hand and between progress and complexity on the other hand. Therefore, an evolution by simplification is strongly opposite to the myth of progress. More recently, the palaeontologist Stephen Jay Gould renewed that discussion and added a very simple but also very convincing argument (Gould 1996). Let us assume that evolution corresponds to a random walk, going sometimes to the complex, sometimes to the simple. Since living organisms cannot exceed a minimal limit of simplicity (especially, if one does not consider parasitic species), the random walk will generate more and more complex organisms, even if the majority of species is relatively simple. Therefore, the existence of complex organism like animals and plants does not at all mean that there is a tendency of the evolution towards complexity, but can very well be explained by a stochastic process.

In conclusion, it is certain that in order to explain the evolution of life one has to take into account both the evolution from simple to complex and the evolution from complex to simple (Forterre and Philippe 1999a). In fact, it would be the best to completely re-evaluate the problem. The fundamental point is that the evolutionary steps have to be simple, in order that the innovations can be maintained by natural selection, either for simplifying or for making the organisms more complex. It is therefore useless to look for simple organisms, in order to inform us about the ancient evolution. In contrast, it is indispensable to study the evolutionary process and to develop evolutionary scenarios about the origin of the major groups of organisms, which suppose simple evolutionary stages. A provoking proposition inspired by this idea is that LUCA would have resembled more a Eukaryote than a prokaryote. The major argument employed is that Eukaryotes have much more RNA-based reactions than the prokaryotes (Forterre and Philippe 1999a; Poole et al. 1999). If one accepts the hypothesis about the RNA world for the origin of life (see also Maurel and Haenni Chap. 6, Part II), the replacement of RNA-based reactions by protein-based reactions is essentially an irreversible phenomenon. In this sense, the scarcity of catalytic RNA in prokaryotes supports their late emergence (or of their derived status). Even if this hypothesis remains controversial, one should remember that the picture that we have about the ancient evolution is under major reconstruction (compare Figs. 8.2 and 8.7).

8.7 Exobiology, a Procession of Extinctions?

We have not discussed a major problem that is limiting the efficiency of the methods based on the comparison of extant species. This is the question of extinctions. For example, if one takes a look at the mammals, the last common ancestor of the extant species has an age of about 130 million years, whereas the divergence between mammals and the other vertebrates happened about 310 million years ago. This indicates that, on a branch corresponding to 180 million years of evolution, only a single species has survived; these circumstances exclude the access of biologists to any living material that originated within this period. It is extremely difficult to understand the appearance of the mammalian characteristics only based on the extant species (like for example the bones that form the internal ear). In this special case, rich fossil data allowed us to elucidate numerous problems (like the progressive movement of the jaw bones towards the inner ear). Unfortunately, if one is interested in more ancient evolution, palaeontology is barely useful because of the generally bad fossilisation of micro-organisms. More precisely, the fossils usually provide only morphological information, which is of extremely limited taxonomic use, because of their mostly very simple rod-shaped or coccoid morphology. It is even difficult to recognize these objects as true fossils since minerals can mimic the simple morphologies of a large part of the micro-organisms, as exemplified by the debate about the archaean Cyanobacteria (Brasier et al. 2002; Garcia-Ruiz et al. 2003; Schopf et al. 2002). The existence of long internal branches, where not a single extant (e.g. surviving) species emerges (see the basis of Bacteria and Eukaryotes in Fig. 8.7), is therefore a major limitation on the inferences. It is possible that a better description of the terrestrial biodiversity, especially due to the molecular ecological methods, will allow several additional important lineages to be discovered. In fact, numerous species, even the majority, have escaped to the classical culture methods, in both prokaryotes (Pace 1997) and Eukaryotes (Lopez-Garcia et al. 2001). If one extrapolates the observations made based on the fossil record of multicellular organisms, it is expected that the vast majority of these species, especially those that could have witnessed the simple stages of ancient evolution have disappeared. These extinctions explain the difficulties that can arise if one studies the evolution of life since LUCA despite the knowledge of the complete sequence of more than one hundred genomes. The exobiology can therefore be of great help to us, by discovering other living systems. In light of the importance of the historical contingency (Gould 1996), these systems would very likely have evolved in different ways and would have suffered from completely independent extinction events. This would constitute, in some respect, a parade of extinctions that would have taken place on Earth.

Acknowledgement

We thank Bernard Barbier, Christophe Douady, Hervé Martin, and David Moreira for many useful comments and Muriel Gargaud for her kindness, her patience, and her never-ending interest in the subject. This work was supported by the Canada research chair program.

References

Adachi J, Hasegawa M (1996) MOLPHY version 2.3: programs for molecular phylogenetics based on maximum likelihood. *Comput. Sci. Monogr.* **28**:1–150

Aguinaldo AM, Turbeville JM, Linford LS, Rivera MC, Garey JR, Raff RA, Lake JA (1997) Evidence for a clade of nematodes, arthropods and other moulting animals. *Nature* **387**:489–493

Andersson JO, Roger AJ (2002) A cyanobacterial gene in nonphotosynthetic protists – an early chloroplast acquisition in Eukaryotes? *Curr. Biol.* **12**:115–119

Aravind L, Tatusov RL, Wolf YI, Walker DR, Koonin EV (1998) Evidence for massive gene exchange between archaeal and bacterial hyperthermophiles. *Trends Genet.* **14**:442–444

Archibald JM, Longet D, Pawlowski J, Keeling PJ (2003) A novel polyubiquitin structure in cercozoa and foraminifera: evidence for a new eukaryotic supergroup. *Mol. Biol. Evol.* **20**:62–66

Archibald JM, Roger AJ (2002) Gene duplication and gene conversion shape the evolution of archaeal chaperonins. *J. Mol. Biol.* **316**:1041–1050

Arisue N, Hashimoto T, Lee JA, Moore DV, Gordon P, Sensen CW, Gaasterland T, Hasegawa M, Müller M (2002a) The phylogenetic position of the pelobiont *Mastigamoeba balamuthi* based on sequences of rDNA and translation elongation factors EF-1a and EF-2. *J. Eukaryot. Microbiol.* **49**:1–10

Arisue N, Hasshimoto T, Yoshikawa H, Nakamura Y, Nakamura G, Nakamura F, Yano T-A, Hasegawa M (2002b) Phylogenetic position of *Blastocystis hominis* and of stramenopiles inferred from multiple molecular sequence data. *J. Eukaryot. Microbiol.* **49**:42–53

Baldauf SL, Palmer JD (1993) Animals and fungi are each other's closest relatives: congruent evidence from multiple proteins. *Proc. Natl. Acad. Sci. USA* **90**:11558–11562

Baldauf SL, Roger AJ, Wenk-Siefert I, Doolittle WF (2000) A kingdom-level phylogeny of Eukaryotes based on combined protein data. *Science* **290**:972–977

Bapteste E, Brinkmann H, Lee JA, Moore DV, Sensen CW, Gordon P, Durufle L, Gaasterland T, Lopez P, Muller M, Philippe H (2002) The analysis of 100 genes supports the grouping of three highly divergent amoebae: *Dictyostelium*, *Entamoeba*, and *Mastigamoeba*. *Proc. Natl. Acad. Sci. USA* **99**:1414–1419

Bapteste E, Philippe H (2002) The potential value of indels as phylogenetic markers: position of trichomonads as a case study. *Mol. Biol. Evol.* **19**:972–977

Boore JL, Brown WM (1998) Big trees from little genomes: mitochondrial gene order as a phylogenetic tool. *Curr. Opin. Genet. Dev.* **8**:668–674

Brasier MD, Green OR, Jephcoat AP, Kleppe AK, Van Kranendonk MJ, Lindsay JF, Steele A, Grassineau NV (2002) Questioning the evidence for Earth's oldest fossils. *Nature* **416**:76–81

Brinkmann H, Philippe H (1999) Archaea sister group of Bacteria? Indications from tree reconstruction artifacts in ancient phylogenies. *Mol. Biol. Evol.* **16**:817–825

Brochier C, Bapteste E, Moreira D, Philippe H (2002) Eubacterial phylogeny based on translational apparatus proteins. *Trends Genet.* **18**:1–5

Brochier C, Philippe H (2002) Phylogeny: a non-hyperthermophilic ancestor for bacteria. *Nature* **417**:244

Brochier C, Philippe H, Moreira D (2000) The evolutionary history of ribosomal protein RpS14: horizontal gene transfer at the heart of the ribosome. *Trends Genet.* **16**:529–533

Brown JR, Doolittle WF (1995) Root of the universal tree of life based on ancient aminoacyl-tRNA synthetase gene duplications. *Proc. Natl. Acad. Sci. USA* **92**:2441–2445

Brown JR, Douady CJ, Italia MJ, Marshall WE, Stanhope MJ (2001) Universal trees based on large combined protein sequence data sets. *Nat. Genet.* **28**:281–285

Brown JR, Robb FT, Weiss R, Doolittle WF (1997) Evidence for the early divergence of tryptophanyl- and tyrosyl-tRNA synthetases. *J. Mol. Evol.* **45**:9–16

Budin K, Philippe H (1998) New insights into the phylogeny of Eukaryotes based on ciliate Hsp70 sequences. *Mol. Biol. Evol.* **15**:943–956

Burger G, Saint-Louis D, Gray MW, Lang BF (1999) Complete sequence of the mitochondrial DNA of the red alga *Porphyra purpurea.* Cyanobacterial introns and shared ancestry of red and green algae. *Plant Cell* **11**:1675–1694

Castresana J (2000) Selection of conserved blocks from multiple alignments for their use in phylogenetic analysis. *Mol. Biol. Evol.* **17**:540–552

Cavalier-Smith T (1987) Eukaryotes with no mitochondria. *Nature* **326**:332–333

Cavalier-Smith T (1999) Principles of protein and lipid targeting in secondary symbiogenesis: euglenoid, dinoflagellate, and sporozoan plastid origins and the Eukaryote familly tree. *J. Eukaryot. Microbiol.* **46**:347–366

Charlebois RL, Sensen CW, Doolittle WF, Brown JR (1997) Evolutionary analysis of the hisCGABdFDEHI gene cluster from the archaeon *Sulfolobus solfataricus* P2. *J. Bacteriol.* **179**:4429–4432

Chase MW, Fay MF (2001) Ancient flowering plants: DNA sequences and angiosperm classification. *Genome Biol.* **2**:REVIEWS1012

Daubin V, Gouy M (2001) Bacterial molecular phylogeny using supertree approach. Genome Inform Ser Workshop *Genome Inform.* **12**:155–164

Deckert G, Warren PV, Gaasterland T, Young WG, Lenox AL, Graham DE, Overbeek R, Snead MA, Keller M, Aujay M, Huber R, Feldman RA, Short JM, Olsen GJ, Swanson RV (1998) The complete genome of the hyperthermophilic bacterium Aquifex aeolicus. *Nature* **392**:353–358

Doolittle WF (1999a) Lateral gene transfer, genome surveys, and the phylogeny of prokaryotes. Technical comment on M. Huynen et al. *Science* **286**:1443

Doolittle WF (1999b) Lateral genomics. *Trends Cell Biol.* **9**:M5–8

Doolittle WF (1999c) Phylogenetic classification and the universal tree. *Science* **284**: 2124–2129

Edlind TD, Li J, Visvesvara GS, Vodkin MH, McLaughlin GL, Katiyar SK (1996) Phylogenetic analysis of beta-tubulin sequences from amitochondrial protozoa. *Mol. Phylogenet. Evol.* **5**:359–367

Embley TM, Hirt RP (1998) Early branching Eukaryotes? *Curr. Opin. Genet. Dev.* **8**:624–629

Fast NM, Kissinger JC, Roos DS, Keeling PJ (2001) Nuclear-encoded, plastid-targeted genes suggest a single common origin for apicomplexan and dinoflagellate plastids. *Mol. Biol. Evol.* **18**:418–426

Fast NM, XUE L, Bingham S, Keeling PJ (2002) Re-examining alveolate evolution using multiple protein molecular phylogenies. *J. Eukaryot. Microbiol.* **49**:30–37

Felsenstein J (1978) Cases in which parsimony or compatibility methods will be positively misleading. *Syst. Zool.* **27**:401–410

Felsenstein J (2001) Taking Variation of Evolutionary Rates Between Sites into Account in Inferring Phylogenies. *J. Mol. Evol.* **53**:447–455

Fitch WM (1971) The nonidentity of invariable positions in the cytochromes c of different species. *Biochem. Genet.* **5**:231–241

Fitch WM, Markowitz E (1970) An improved method for determining codon variability in a gene and its application to the rate of fixation of mutations in evolution. *Biochem. Genet.* **4**:579–593

Fitz-Gibbon ST, House CH (1999) Whole genome-based phylogenetic analysis of free-living microorganisms. *Nucleic Acids Res.* **27**:4218–4222

Forterre P (1995) Thermoreduction, a hypothesis for the origin of prokaryotes. *C.R. Acad. Sci. III.* **318**:415–422

Forterre P, Bouthier De La Tour C, Philippe H, Duguet M (2000) Reverse gyrase from hyperthermophiles: probable transfer of a thermoadaptation trait from archaea to bacteria. *Trends Genet.* **16**:152–154

Forterre P, Brochier C, Philippe H (2002) Evolution of the Archaea. *Theor. Popul. Biol.* **61**:409–422

Forterre P, Philippe H (1999a) The last universal common ancestor (LUCA), simple or complex? *Biol. Bull.* **196**:373–375; discussion 375–377

Forterre P, Philippe H (1999b) Where is the root of the universal tree of life? *BioEssays* **21**:871–879

Foster PG, Hickey DA (1999) Compositional bias may affect both DNA-based and protein-based phylogenetic reconstructions. *J. Mol. Evol.* **48**:284–290

Fuerst JA (1995) The planctomycetes: emerging models for microbial ecology, evolution and cell biology. *Microbiology* **141**:1493–1506

Galtier N (2001) Maximum-likelihood phylogenetic analysis under a covarion-like model. *Mol. Biol. Evol.* **18**:866–873

Galtier N, Gouy M (1995) Inferring phylogenies from DNA sequences of unequal base compositions. *Proc. Natl. Acad. Sci. USA* **92**:11 317–11 321

Galtier N, Lobry JR (1997) Relationships between genomic G + C content, RNA secondary structures, and optimal growth temperature in prokaryotes. *J. Mol. Evol.* **44**:632–636

Galtier N, Tourasse N, Gouy M (1999) A nonhyperthermophilic common ancestor to extant life forms. *Science* **283**:220–221

Garcia-Ruiz JM, Hyde ST, Carnerup AM, Christy AG, Van Kranendonk MJ, Welham NJ (2003) Self-assembled silica-carbonate structures and detection of ancient microfossils. *Science* **302**:1194–1197

Germot A, Philippe H (1999) Critical analysis of eukaryotic phylogeny: a case study based on the HSP70 family. *J. Eukaryot. Microbiol.* **46**:116–124

Germot A, Philippe H, Le Guyader H (1997) Evidence for loss of mitochondria in Microsporidia from a mitochondrial- type HSP70 in *Nosema locustae. Mol. Biochem. Parasitol.* **87**:159–168

Gogarten JP, Kibak H, Dittrich P, Taiz L, Bowman EJ, Bowman BJ, Manolson MF, Poole RJ, Date T, Oshima T, Konishi J, Denda K, Yoshida M (1989) Evolution of the vacuolar H+-ATPase: implications for the origin of Eukaryotes. *Proc. Natl. Acad. Sci. USA* **86**:6661–6665

Gogarten-Boekels M, Hilario E, Gogarten JP (1995) The effects of heavy meteorite bombardment on the early evolution – the emergence of the three domains of life. *Orig. Life Evol. Biosph.* **25**:251–264

Gould SJ (1996) Full House: *The Spread of Excellence From Plato to Darwin.* Harmony Books

Gribaldo S, Cammarano P (1998) The root of the universal tree of life inferred from anciently duplicated genes encoding components of the protein-targeting machinery. *J. Mol. Evol.* **47**:508–516

Gribaldo S, Lumia V, Creti R, de Macario EC, Sanangelantoni A, Cammarano P (1999) Discontinuous occurrence of the hsp70 (dnaK) gene among Archaea and sequence features of HSP70 suggest a novel outlook on phylogenies inferred from this protein. *J. Bacteriol.* **181**:434–443

Gu X, Li WH (1998) Estimation of evolutionary distances under stationary and nonstationary models of nucleotide substitution. *Proc. Natl. Acad. Sci. USA* **95**:5899–5905

Guindon S, Perriere G (2001) Intragenomic base content variation is a potential source of biases when searching for horizontally transferred genes. *Mol. Biol. Evol.* **18**:1838–1840

Gupta RS (1998) What are Archaebacteria: life's third domain or monoderm prokaryotes related to Gram-positive bacteria? A new proposal for the classification of prokaryotic organisms. *Mol. Microbiol.* **229**:695–708

Gupta RS, Johari V (1998) Signature sequences in diverse proteins provide evidence of a close evolutionary relationship between the deinococcus-thermus group and cyanobacteria. *J. Mol. Evol.* **46**:716–720

Gupta RS, Singh B (1994) Phylogenetic analysis of 70 kD heat shock protein sequences suggests a chimeric origin for the eukaryotic cell nucleus. *Curr. Biol.* **4**:1104–1114

Hannaert V, Brinkmann H, Nowitzki U, LeeJ.A., Albert M-A, Sensen CW, Gaasterland T, Müller M, Michels P, Martin W (2000) Enolase from *Trypanosoma brucei*, from the amitochondriate protist *Mastigamoeba balamuthi*, and from the chloroplast and cytosol of *Euglena gracilis*: pieces in the evolutionary puzzle of the eukaryotic glycolytic pathway. *Mol. Biol. Evol.* **17**:989–1000

Hannaert V, Saavedra E, Duffieux F, Szikora JP, Rigden DJ, Michels PA, Opperdoes FR (2003) Plant-like traits associated with metabolism of Trypanosoma parasites. *Proc. Natl. Acad. Sci. USA* **100**:1067–1071

Hasegawa M, Hashimoto T (1993) Ribosomal RNA trees misleading? *Nature* **361**:23

Hashimoto T, Hasegawa M (1996) Origin and early evolution of Eukaryotes inferred from the amino acid sequences of translation elongation factors 1alpha/Tu and 2/G. *Adv. Biophys.* **32**:73–120

Hashimoto T, Sanchez LB, Shirakura T, Muller M, Hasegawa M (1998) Secondary absence of mitochondria in *Giardia lamblia* and *Trichomonas vaginalis* revealed by valyl-tRNA synthetase phylogeny. *Proc. Natl. Acad. Sci. USA* **95**:6860–6865

Hendy M, Penny D (1989) A framework for the quantitative study of evolutionary trees. *Syst. Zool.* **38**:297–309

Hennig W (1966) *Phylogenetic Systematics.* University of Illinois Press, Urbana

Henze K, Horner DS, Suguri S, Moore DV, Sanchez LB, Muller M, Embley TM (2001) Unique phylogenetic relationships of glucokinase and glucosephosphate isomerase of the amitochondriate Eukaryotes *Giardia intestinalis, Spironucleus barkhanus* and *Trichomonas vaginalis. Gene* **281**:123–131

Hickson RE, Simon C, Perrey SW (2000) The Performance of Several Multiple-Sequence Alignment Programs in Relation to Secondary-Structure Features for an rRNA Sequence. *Mol. Biol. Evol.* **17**:530–539

Hillis DM (1996) Inferring complex phylogenies. *Nature* **383**:130–131

Hirt RP, Logsdon JM, Jr., Healy B, Dorey MW, Doolittle WF, Embley TM (1999) Microsporidia are related to fungi: evidence from the largest subunit of RNA polymerase II and other proteins. *Proc. Natl. Acad. Sci. USA* **96**:580–585

Huelsenbeck JP (2002) Testing a covariotide model of DNA substitution. *Mol. Biol. Evol.* **19**:698–707

Imai E, Honda H, Hatori K, Brack A, Matsuno K (1999) Elongation of oligopeptides in a simulated submarine hydrothermal system. *Science* **283**:831–833

Iwabe N, Kuma K, Hasegawa M, Osawa S, Miyata T (1989) Evolutionary relationship of Archaebacteria, Eubacteria, and Eukaryotes inferred from phylogenetic trees of duplicated genes. *Proc. Natl. Acad. Sci. USA* **86**:9355–9359

Jain R, Rivera MC, Lake JA (1999) Horizontal gene transfer among genomes: The complexity hypothesis. *Proc. Natl. Acad. Sci. USA* **96**:3801–3806

Jukes TH, Cantor CR (1969) Evolution of protein molecules. In: Munro HN (ed.) *Mammalian Protein Metabolism.* Academic Press, New York, p 21–132

Katinka MD, Duprat S, Cornillot E, Metenier G, Thomarat F, Prensier G, Barbe V, Peyretaillade E, Brottier P, Wincker P, Delbac F, El Alaoui H, Peyret P, Saurin W, Gouy M, Weissenbach J, Vivares CP (2001) Genome sequence and gene compaction of the Eukaryote parasite *Encephalitozoon cuniculi. Nature* **414**:450–453

Keeling PJ (2001a) Foraminifera and Cercozoa are related in actin phylogeny: two orphans find a home? *Mol. Biol. Evol.* **18**:1551–1557

Keeling PJ (2001b) Parasites go the full monty. *Nature* **414**:401–402

Keeling PJ, Palmer JD (2000) Parabasalian flagellates are ancient Eukaryotes. *Nature* **405**:635–637

Keeling PJ, Palmer JD (2001) Lateral transfer at the gene and subgenic levels in the evolution of eukaryotic enolase. *Proc. Natl. Acad. Sci. USA* **98**:10 745–10 750

Knoll AH (1992) The early evolution of Eukaryotes: a geological perspective. *Science* **256**:622–627

Koonin EV, Makarova KS, Aravind L (2001) Horizontal gene transfer in prokaryotes: quantification and classification. *Annu. Rev. Microbiol.* **55**:709–742

Korbel JO, Snel B, Huynen MA, Bork P (2002) SHOT: a web server for the construction of genome phylogenies. *Trends Genet.* **18**:158–162

Koski LB, Golding GB (2001) The closest BLAST hit is often not the nearest neighbor. *J. Mol. Evol.* **52**:540–542

Koski LB, Morton RA, Golding GB (2001) Codon bias and base composition are poor indicators of horizontally transferred genes. *Mol. Biol. Evol.* **18**:404–412

Kuma K, Nikoh N, Iwabe N, Miyata T (1995) Phylogenetic position of Dictyostelium inferred from multiple protein data sets. *J. Mol. Evol.* **41**:238–246

Lang BF, Gray MW, Burger G (1999) Mitochondrial genome evolution and the origin of Eukaryotes. *Annu. Rev. Genet.* **33**:351–397

Lawrence JG, Ochman H (2002) Reconciling the many faces of lateral gene transfer. *Trends Microbiol.* **10**:1–4

Lawson FS, Charlebois RL, Dillon JA (1996) Phylogenetic analysis of carbamoylphosphate synthetase genes: complex evolutionary history includes an internal duplication within a gene which can root the tree of life. *Mol. Biol. Evol.* **13**:970–977

Lecointre G, Philippe H, Le HLV, Le Guyader H (1993) Species sampling has a major impact on phylogenetic inference. *Mol. Phylogenet. Evol.* **2**:205–224

Lin J, Gerstein M (2000) Whole-genome trees based on the occurrence of folds and orthologs: implications for comparing genomes on different levels. *Genome Res.* **10**:*808–818*

Lockhart P, Steel M, Hendy M, Penny D (1994) Recovering evolutionary trees under a more realistic model of sequence evolution. *Mol. Biol. Evol.* **11**:605–612

Lockhart PJ, Larkum AW, Steel M, Waddell PJ, Penny D (1996) Evolution of chlorophyll and bacteriochlorophyll: the problem of invariant sites in sequence analysis. *Proc. Natl. Acad. Sci. USA* **93**:1930–1934

Lockhart PJ, Steel MA, Barbrook AC, Huson D, Charleston MA, Howe CJ (1998) A covariotide model explains apparent phylogenetic structure of oxygenic photosynthetic lineages. *Mol. Biol. Evol.* **15**:1183–1188

Loomis WF, Smith DW (1990) Molecular phylogeny of *Dictyostelium discoideum* by protein sequence comparison. *Proc. Natl. Acad. Sci. USA* **87**:9093–9097

Lopez P, Casane D, Philippe H (2002) Heterotachy, an important process of protein evolution. *Mol. Biol. Evol.* **19**:1–7

Lopez P, Forterre P, Philippe H (1999) The root of the tree of life in the light of the covarion model. *J. Mol. Evol.* **49**:496–508

Lopez-Garcia P, Rodriguez-Valera F, Pedros-Alio C, Moreira D (2001) Unexpected diversity of small Eukaryotes in deep-sea Antarctic plankton. *Nature* **409**:603–607

Lum JK, Nikaido M, Shimamura M, Shimodaira H, Shedlock AM, Okada N, Hasegawa M (2000) Consistency of SINE insertion topology and flanking sequence tree: quantifying relationships among cetartiodactyls [In Process Citation]. *Mol. Biol. Evol.* **17**:1417–1424

Lwoff A (1943) *L'évolution Physiologique. Etude des Pertes de Fonctions Chez les Microorganismes.* Hermann et Cie, Paris

Lynch M, Conery JS (2000) The evolutionary fate and consequences of duplicate genes. *Science* **290**:1151–1155

Madsen O, Scally M, Douady CJ, Kao DJ, DeBry RW, Adkins R, Amrine HM, Stanhope MJ, de Jong WW, Springer MS (2001) Parallel adaptive radiations in two major clades of placental mammals. *Nature* **409**:610–614

Mai Z, Ghosh S, Frisardi M, Rosenthal B, Rogers R, Samuelson J (1999) Hsp60 is targeted to a cryptic mitochondrion-derived organelle ("crypton") in the microaerophilic protozoan parasite *Entamoeba histolytica*. *Mol. Cell. Biol.* **19**:2198–2205

Margulis L (1996) Archaeal-eubacterial mergers in the origin of Eukarya: phylogenetic classification of life. *Proc. Natl. Acad. Sci. USA* **93**:1071–1076

Martin W (1999) Mosaic bacterial chromosomes: a challenge en route to a tree of genomes. *Bioessays* **21**:99–104

Matte-Tailliez O, Brochier C, Forterre P, Philippe H (2002) Archaeal phylogeny based on ribosomal proteins. *Mol. Biol. Evol.* **19**:631–639

Mayr E (1998) Two empires or three? *Proc. Natl. Acad. Sci. USA* **95**:9720–9723

Moreira D, Le Guyader H, Philippe H (1999) Unusually high evolutionary rate of the elongation factor 1 alpha genes from the Ciliophora and its impact on the phylogeny of Eukaryotes. *Mol. Biol. Evol.* **16**:234–245

Moreira D, Le Guyader H, Philippe H (2000) The origin of red algae: implications for the evolution of chloroplasts. *Nature* **405**:69–72

Murphy WJ, Eizirik E, Johnson WE, Zhang YP, Ryder OA, O'Brien SJ (2001) Molecular phylogenetics and the origins of placental mammals. *Nature* **409**:614–618

Nelson KE, Clayton RA, Gill SR, Gwinn ML, Dodson RJ, Haft DH, Hickey EK, Peterson JD, Nelson WC, Ketchum KA, McDonald L, Utterback TR, Malek JA, Linher KD, Garrett MM, Stewart AM, Cotton MD, Pratt MS, Phillips CA, Richardson D, Heidelberg J, Sutton GG, Fleischmann RD, Eisen JA, Fraser CM et al. (1999) Evidence for lateral gene transfer between Archaea and bacteria from genome sequence of Thermotoga maritima. *Nature* **399**:323–329

Nikoh N, Hayase N, Iwabe N, Kuma K, Miyata T (1994) Phylogenetic relationship of the kingdoms Animalia, Plantae, and Fungi, inferred from 23 different protein species. *Mol. Biol. Evol.* **11**:762–768

Nisbet EG, Sleep NH (2001) The habitat and nature of early life. *Nature* **409**:1083–1091

Ochman H, Lawrence JG, Groisman EA (2000) Lateral gene transfer and the nature of bacterial innovation. *Nature* **405**:299–304

Olsen G (1987) Earliest phylogenetic branching: comparing rRNA-based evolutionary trees inferred with various techniques. *Cold Spring Harb. Symp. Quant. Biol.* **LII**:825–837

Olsen GJ, Woese CR, Overbeek R (1994) The winds of (evolutionary) change: breathing new life into microbiology. *J. Bacteriol.* **176**:1–6

Pace NR (1991) Origin of life – facing up to the physical setting. Cell **65**:531–533

Pace NR (1997) A molecular view of microbial diversity and the biosphere. *Science* **276**:734–740

Page RD, Charleston MA (1997) From gene to organismal phylogeny: reconciled trees and the gene tree/species tree problem. *Mol. Phylogenet. Evol.* **7**:231–240

Pawlowski J, Bolivar I, Fahrni JF, de Vargas C, Gouy M, Zaninetti L (1997) Extreme differences in rates of molecular evolution of foraminifera revealed by comparison of ribosomal DNA sequences and the fossil record. *Mol. Biol. Evol.* **14**:498–505

Pennisi E (1998) Genome data shake tree of life. *Science* **280**:672–674

Penny D, McComish BJ, Charleston MA, Hendy MD (2001) Mathematical elegance with biochemical realism: the covarion model of molecular evolution. *J. Mol. Evol.* **53**:711–723

Penny D, Poole A (1999) The nature of the last universal common ancestor. *Curr. Opin. Genet. Dev.* **9**:672–667

Philippe H (2000) Long branch attraction and protist phylogeny. *Protist* **51**:307–316

Philippe H, Adoutte A (1998) The molecular phylogeny of Eukaryota: solid facts and uncertainties. In: Coombs G, Vickerman K, Sleigh M, Warren A (eds) *Evolutionary Relationships among Protozoa*. Kluwer, Dordrecht, p 25–56

Philippe H, Budin K, Moreira D (1999) Horizontal transfers confuse the prokaryotic phylogeny based on the HSP70 protein family. *Molecular Microbiology* **31**:1007–1009

Philippe H, Douzery E (1994) The pitfalls of molecular phylogeny based on four species, as illustrated by the Cetacea/Artiodactyla relationships. *J. Mammal. Evol.* **2**:133–152
Philippe H, Forterre P (1999) The rooting of the universal tree of life is not reliable. *J. Mol. Evol.* **49**:509–523
Philippe H, Germot A (2000) Phylogeny of Eukaryotes based on ribosomal RNA: Long-branch attraction and models of sequence evolution. *Mol. Biol. Evol.* **17**:830–834
Philippe H, Laurent J (1998) How good are deep phylogenetic trees? *Curr. Opin. Genet. Dev.* **8**:616–623
Philippe H, Lopez P, Brinkmann H, Budin K, Germot A, Laurent J, Moreira D, Müller M, Le Guyader H (2000) Early branching or fast evolving Eukaryotes? An answer based on slowly evolving positions. *Philos. Trans. R. Soc. Lond. B Biol. Sci.* **267**:1213–1221
Philippe H, Sörhannus U, Baroin A, Perasso R, Gasse F, Adoutte A (1994) Comparison of molecular and paleontological data in diatoms suggests a major gap in the fossil record. *J. Evolut. Biol.* **7**:247–265
Poe S, Swofford DL (1999) Taxon sampling revisited. *Nature* **398**:299–300
Poole A, Jeffares D, Penny D (1999) Early evolution: prokaryotes, the new kids on the block. *BioEssays* **21**:880–889
Posada D, Crandall KA (2002) The effect of recombination on the accuracy of phylogeny estimation. *J. Mol. Evol.* **54**:396–402.
Ragan MA (2001) On surrogate methods for detecting lateral gene transfer. *FEMS Microbiol. Lett.* **201**:187–191
Roger AJ, Sandblom O, Doolittle WF, Philippe H (1999) An evaluation of elongation factor 1 alpha as a phylogenetic marker for Eukaryotes. *Mol. Biol. Evol.* **16**:218–233
Rokas A, Holland PWH (2000) Rare genomic changes as a tool for phylogenetics. *Trends Ecol. Evol.* **15**:454–459
Ruepp A, Graml W, Santos-Martinez ML, Koretke KK, Volker C, Mewes HW, Frishman D, Stocker S, Lupas AN, Baumeister W (2000) The genome sequence of the thermoacidophilic scavenger *Thermoplasma acidophilum*. *Nature* **407**:508–513
Schopf JW, Kudryavtsev AB, Agresti DG, Wdowiak TJ, Czaja AD (2002) Laser-Raman imagery of Earth's earliest fossils. *Nature* **416**:73–76
Schwartz RM, Dayhoff MO (1978) Origins of prokaryotes, Eukaryotes, mitochondria, and chloroplasts. *Science* **199**:395–403
Sogin ML (1991) Early evolution and the origin of Eukaryotes. *Curr. Opin. Genet. Dev.* **1**:457–463
Stetter KO (1996) Hyperthermophiles in the history of life. *Ciba Found. Symp.* **202**:1–10
Strous M, Fuerst JA, Kramer EH, Logemann S, Muyzer G, van de Pas-Schoonen KT, Webb R, Kuenen JG, Jetten MS (1999) Missing lithotroph identified as new planctomycete. *Nature* **400**:446–449
Swofford DL, Olsen GJ, Waddell PJ, Hillis DM (1996) Phylogenetic inference. In: Hillis DM, Moritz C, Mable BK (eds) *Molecular Systematics*. Sinauer Associates, Sunderland, p 407–514
Tekaia F, Lazcano A, Dujon B (1999) The genomic tree as revealed from whole proteome comparisons. *Genome Res.* **9**:550–557
Tovar J, Fischer A, Clark CG (1999) The mitosome, a novel organelle related to mitochondria in the amitochondrial parasite *Entamoeba histolytica*. *Mol. Microbiol.* **32**:1013–1021

Tovar J, Leon-Avila G, Sanchez LB, Sutak R, Tachezy J, Van Der Giezen M, Hernandez M, Muller M, Lucocq JM (2003) Mitochondrial remnant organelles of Giardia function in iron-sulphur protein maturation. *Nature* **426**:172–176

Van de Peer Y, Ben Ali A, Meyer A (2000) Microsporidia: accumulating molecular evidence that a group of amitochondriate and suspectedly primitive Eukaryotes are just curious fungi. *Gene* **246**:1–8

Venkatesh B, Ning Y, Brenner S (1999) Late changes in spliceosomal introns define clades in vertebrate evolution. *Proc. Natl. Acad. Sci. USA* **96**:10 267–10 271

Vossbrinck CR, Maddox JV, Friedman S, Debrunner-Vossbrinck BA, Woese CR (1987) Ribosomal RNA sequence suggests microsporidia are extremely ancient Eukaryotes. *Nature* **326**:411–414

Waller RF, Keeling PJ, Donald RG, Striepen B, Handman E, Lang-Unnasch N, Cowman AF, Besra GS, Roos DS, McFadden GI (1998) Nuclear-encoded proteins target to the plastid in Toxoplasma gondii and Plasmodium falciparum. *Proc. Natl. Acad. Sci. USA* **95**:12 352–12 357

Wang B (2001) Limitations of compositional approach to identifying horizontally transferred genes. *J. Mol. Evol.* **53**:244–250

Weisburg WG, Giovannoni SJ, Woese CR (1989) The Deinococcus-Thermus phylum and the effect of rRNA composition on phylogenetic tree construction. *Syst. Appl. Microbiol.* **11**:128–134

Williams BA, Hirt RP, Lucocq JM, Embley TM (2002) A mitochondrial remnant in the microsporidian *Trachipleistophora hominis*. *Nature* **418**:865–869

Woese CR (1987) Bacterial evolution. *Microbiol. Rev.* **51**:221–271

Woese CR (2000) Interpreting the universal phylogenetic tree. *Proc. Natl. Acad. Sci. USA* **97**:8392–8396

Woese CR, Achenbach L, Rouviere P, Mandelco L (1991) Archaeal phylogeny: reexamination of the phylogenetic position of Archaeoglobus fulgidus in light of certain composition-induced artifacts. *Syst. Appl. Microbiol.* **14**:364–371

Woese CR, Fox GE (1977) Phylogenetic structure of the prokaryotic domain: the primary kingdoms. *Proc. Natl. Acad. Sci. USA* **74**:5088–5090

Woese CR, Kandler O, Wheelis ML (1990) Towards a natural system of organisms: proposal for the domains Archaea, Bacteria, and Eucarya. *Proc. Natl. Acad. Sci. USA* **87**:4576–4579

Woese CR, Olsen GJ, Ibba M, Soll D (2000) Aminoacyl-tRNA synthetases, the genetic code, and the evolutionary process. *Microbiol. Mol. Biol. Rev.* **64**:202–236

Wolf YI, Rogozin IB, Grishin NV, Tatusov RL, Koonin EV (2001) Genome trees constructed using five different approaches suggest new major bacterial clades. *BMC Evol. Biol.* **1**:8

Yang Z (1996) Among-site rate variation and its impact on phylogenetic analyses. *Trends Ecol. Evol.* **11**:367–370

Zuckerkandl E, Pauling L (1965) Evolutionary divergence and convergence in proteins. In: Bryson V, Vogel HJ (eds) *Evolving Genes and Proteins*. Academic Press, New York, p 97–166

9 Extremophiles

Purificación López-García

9.1 Some Concepts About Extremophiles

Thought sterile for a long time, extreme environments are known today to host a variety of well-adapted organisms so-called "extremophiles" that are their natural inhabitants. Salt-loving organisms were among the first organisms studied from particularly enduring habitats. The first halophilic species was isolated from salted fish by Farlow as early as 1880, although it was properly described some years later (Farlow, 1880). In 1936, Eleazari Volcani isolated a number of microbial strains living at very high salt concentration, up to 30–34% (weight/volume) from the Dead Sea (Wilansky, 1936). He thus demonstrated the inappropriate designation of this water body and became a historical pioneer on extremophile research, since he devoted most of his scientific career to the study of halophilic microbes. However, salt-loving organisms, or halophiles, began to attract much more attention from microbiologists some years later, when it was discovered that most of them belonged to the third domain of life, the archaea, whose recognition dates back to the late 1970s (Woese and Fox, 1977). It was also during this decade that extreme thermophiles and hyperthermophiles growing well above 60 to 80 °C began to be isolated. In 1969, Thomas Brock isolated a thermophilic bacterium living at 60–80 °C, *Thermus aquaticus*, from a hot spring at Yellowstone National Park in the US (Brock and Freeze, 1969). A year later, it was the turn of *Sulfolobus acidocaldarius*, the first isolated hyperthermophilic archaeon, which was able to grow at temperatures up to 85 °C and at low pH (1–5) (Brock et al., 1972). It was a big surprise for biologists who, since Pasteur's times, had the idea in mind that no living being could survive above 80 °C (Madigan et al., 2002). These first discoveries triggered the search for extremophiles that, assisted by important technical developments in microbiology, have had since then an extraordinary impact on our understanding of microbial diversity and evolution in our planet.

The present review aims at summarising the general knowledge about extreme environments and extremophiles, what they are, how we study them, why they are so appealing to us, and what their link is to exo/astrobiology issues. This brief overview is logically very far from exhaustive; references have been kept to minimum numbers and often correspond to general revisions. The reader is therefore invited to consult more specialised work for deeper insights into the different areas.

9.1.1 What is an Extremophile?

An extremophile is an organism that lives optimally in an extreme environment. What is then an extreme environment? Intuitively, we immediately think of a habitat where conditions are harsh and hostile to us. This anthropocentric vision is too subjective, although relatively valid (we are indeed mesophiles). In more neutral terms, an environment can be defined as extreme when the physicochemical parameters characterising it approach the limits within which life is known to occur. This is a strict biological definition. In this sense, we should talk of "bioextremes". Thus, if we take the example of temperature, life expands between approximately $-20\,^{\circ}\mathrm{C}$ (some bacteria) and 110–120 °C (some archaea). In this interval organisms are usually divided into different categories depending on their minimal and maximal growth temperatures (Fig. 9.1). Any organism grows best at a temperature that is optimal, and growth rate decreases when temperature shifts up or down. Following the above definition, psychrophiles and hyperthermophiles, living close to temperature limits for life, are extremophiles.

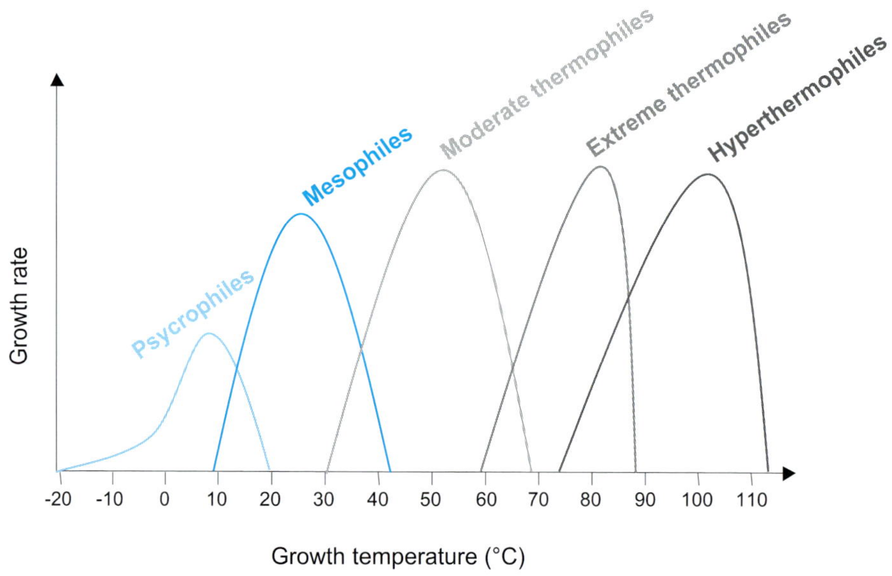

Fig. 9.1. Classification of organisms according to their growth temperature. Psycrophiles and hyperthermophiles are extremophiles

9.1.2 Some Extremophile Features

An extremophilic organism not only tolerates but also requires extreme conditions to live. It is capable of developing its whole life cycle in these conditions. On the contrary, an organism that is extremotolerant can endure extreme values of one or more physicochemical parameters, but its optimal growth occurs under milder conditions. Thus, halotolerant bacteria survive at high salt concentrations but grow better at lower salinity. Similarly, barotolerant organisms stand high pressures but live optimally at lower pressure. We need also to distinguish between life under extreme conditions and under stressing conditions. Stressing environments are often changing biotopes, such as intertidal zones, but they do not display limiting life conditions permanently. Here, organisms are adapted to tolerate change out of the optimal conditions. In fact, extreme environments are most frequently very stable at a biological scale (e.g. many geothermal areas, the Earth crust, the deep ocean, the Antarctic ice).

From a biological point of view, extreme environments are characterised by the predominance of microbial life (bacteria, archaea and, except for very high temperature environments, unicellular eukaryotes). Another important feature is that microbial diversity decreases as one gets closer to life-limiting conditions (increasingly fewer species, although their biomass can be eventually high). This is particularly well illustrated in the case of high temperature. Eukaryotic life is known to exist up to $\sim 60-62\,^{\circ}\mathrm{C}$ (some fungi and protists), but above those values only prokaryotes seem to occur. Thermophilic bacteria and archaea are diverse but, again, hyperthermophily (optimal growth above $80\,^{\circ}\mathrm{C}$, as originally defined by Stetter in 1989) appears limited among bacteria. A few hyperthermophilic bacterial lineages are known, the most hyperthermophilic bacteria being able to grow up to $95\,^{\circ}\mathrm{C}$. Only hyperthermophilic archaea are known to live at temperatures higher than $100\,^{\circ}\mathrm{C}$ (Stetter, 1996).

Extremophiles are normally pluriextremophiles, being adapted to live in habitats where various physicochemical parameters reach extreme values. For instance, many hot springs are acid or alkaline at the same time, and usually rich in metal content; several hypersaline lakes are very alkaline; the deep ocean is generally cold, oligotrophic (very low nutrient content), and exposed to high pressure; pressure and temperature increase progressively in the subsurface, etc. All these are biotopes for specifically adapted microbial communities.

9.1.3 Why Extremophiles are Interesting?

Since they began to be discovered, extremophiles have fascinated and attracted scientists from various disciplines:

- **Molecular biology.** Many different studies have been devoted to understand the molecular adaptations needed to work under conditions that would severely alter, denature or prevent function of nonextremophile's molecules (see below and Table 9.1).
- **Biotechnology.** The potential to use extremophiles or their molecules for industrial application greatly encouraged (and still does) extremophile research. In any organism, enzymes (proteins catalysing transformations of chemical

substrates into products) are logically adapted to work optimally at its natural growth conditions. Enzymes from extremophiles are therefore functional at high temperature, high or low pH, high salt concentration, etc. They are normally very stable and resistant to denaturing agents as well, which may be an additional interest for some industrial processes. Enzymes from organisms adapted to very diverse extreme conditions, also called "extremozymes" by some authors, offer a vast choice to industry due to their functionality in media where traditional enzymes cannot work (Deming, 1998; Horikoshi, 1999; Hough and Danson, 1999; Oren, 1999). Furthermore, many extremophiles display novel metabolic pathways, which increase the panorama of possible reactions for biotechnology.

- **Evolution and phylogeny.** On the one hand, it was partly due to the first studies on extremophiles that the third domain of life, the archaea, was discovered, with fundamental consequences for evolutionary biology (Woese and Fox, 1977; Woese et al., 1990). On the other hand, physicochemical conditions on early Earth could have been similar to those of some contemporary extreme environments. These could be then used as inspiring models to unravel ancient geochemical cycles, the origin of life and its first diversification (Nisbet and Sleep, 2001).
- **Exobiology.** The occurrence of extremophiles living under conditions that are similar to those found in other planets raises more realistic hopes about the possibility to discover extraterrestrial life.

9.2 Microbial Diversity

Methods to study microbial diversity have greatly evolved in the last thirty years. The application of molecular techniques led first to the discovery of archaea, within which we still find today most extremophily records, and later to the detection of an unexpected microbial diversity in very different environments, including extreme biotopes.

Classical microbiology is based on isolation and cultivation of micro-organisms. The objective is to obtain pure cultures (all cells belonging to the same microbial species or strain) that can be assayed for different metabolic or physiological properties in various culture media. However, in most cases, neither morphology nor phenotypic features are informative enough to affiliate micro-organisms to a given phylogenetic (evolutionary-related) family, that is, to achieve a natural or phylogenetic classification. To overcome this problem, we began to use the evolutionary information stored in nucleic acid (DNA and RNA) nucleotide sequences and in protein amino-acid sequences, thus applying the idea that Zuckerkandl and Pauling had published in 1965 (Zuckerkandl and Pauling, 1965). From all biological macromolecules, the RNA of the small ribosomal subunit (16S/18S rRNA) was initially chosen by Woese and colleagues in the 1970s by its properties of universality and reasonable conservation (Woese, 1987; Woese and Fox, 1977). Sequence comparison of rRNAs from many different organisms, including "rare" species coming from extreme environments, led to the recognition of three domains of cellular life on Earth, Archaea, Bacteria and Eukarya. Since then, 16S/18S rRNA has been the most widely used molecular marker, with thousands of sequences from cultivated and noncultivated species being deposited in public databases, and has further confirmed the tripartite division of cellular organisms on Earth (Fig. 9.2).

The application of molecular methods based on 16/18S rRNA gene amplification directly from environmental samples has been revolutionary for microbial diversity studies. This is particularly true for extreme environments for which classical culture methods allow the recovery of usually less than 1% of the actual diversity (Amann et al., 1995). This strategy permits, in principle, the identification of the different rRNA genes present in a given environment (phylotypes). rRNA genes can be amplified by the polymerase chain reaction (PCR) from DNA extracted directly from environmental samples, and then they can be sequenced and the sequences compared with the thousands of rRNA sequences available in databases. In this way, the new environmental sequences can be placed in phylogenetic trees where they represent organisms living in nature even when these have not been or cannot be isolated. The use of this approach has demonstrated that life is much more diversified than we ever thought with the discovery, among many other lineages, of whole organismal groups for which no cultivated member is known (Fig. 9.2) (Hugenholtz et al., 1998; Pace, 1997).

Classical and molecular approaches in microbiology are indeed complementary. Both of them have been, and continue to be, applied to the study of microbes living in extreme habitats leading to the discovery (still very partial) of an astonishing diversity of extremophiles.

9.3 Extreme Environments and Their Inhabitants

9.3.1 Extremophiles and Extremotolerants

It is beyond the scope of this review to decorticate in detail extremophiles and their natural environments. Table 9.1 summarises general features of extremophilic and extremotolerant organisms: their classification as a function of the physicochemical parameter considered, their most outstanding molecular adaptations, as well as some examples of species or groups of extremophiles and the habitats they colonise. Figures 9.3 to 9.9 illustrate different extreme environments. Temperature, pH, salinity (osmotic pressure), pressure, dryness (water availability) and exposure to radiation are parameters whose values usually configure the extreme nature of a particular environment (Table 9.1). Some authors also consider vacuum, gravity, and partial oxygen pressure (Rothschild and Mancinelli, 2001). Nevertheless, no organism is known to live under hypo- or hypergravity and, although many organisms are able to resist vacuum conditions, they do it in an inactive state (dormant or resistance forms), that is, they

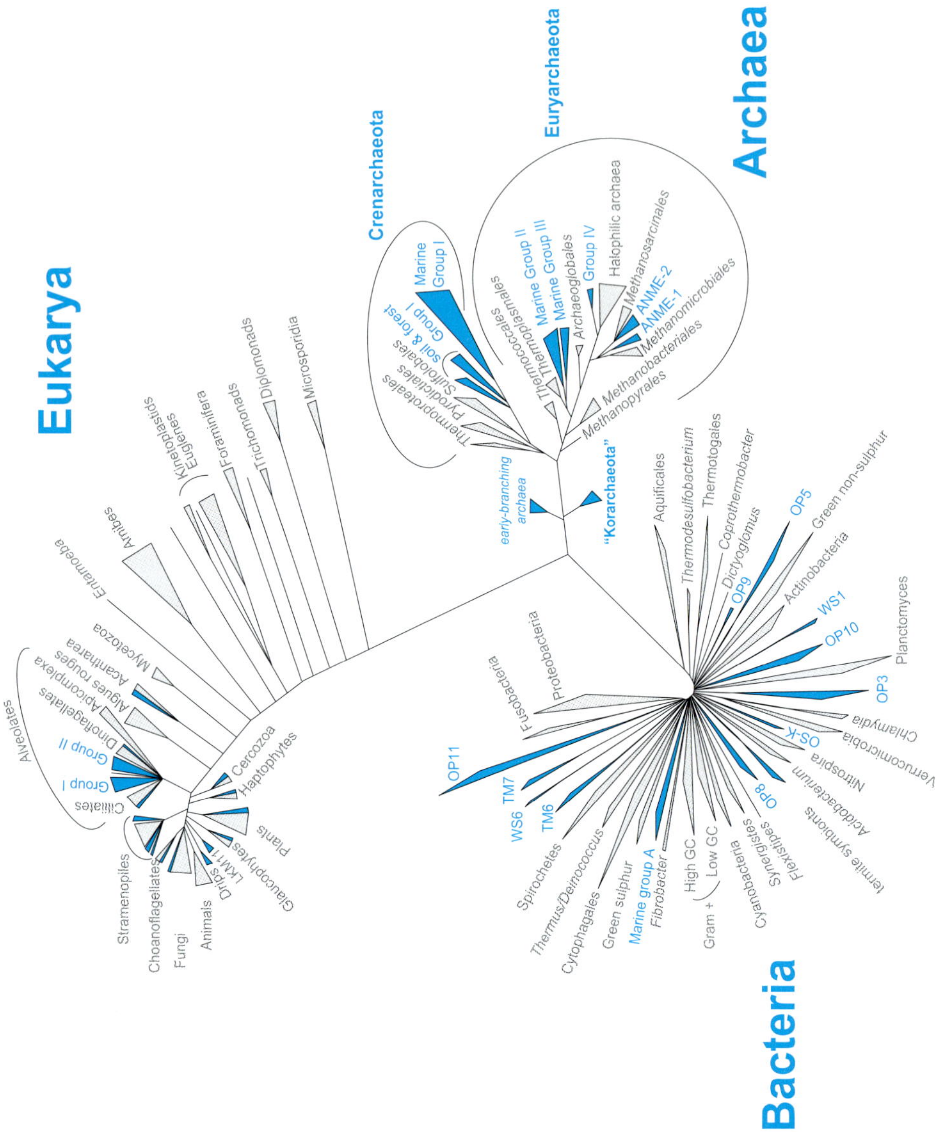

Fig. 9.2. Schematic diagram showing the tree of life based on 16S/18S rRNA sequences. Each *line* represents a sequence (in turn representing the corresponding organism); each *triangle* represents a group of closely related sequences. *Blue triangles* correspond to environmental lineages for which no cultivated members are known

Fig. 9.3. Sampling at a hot mud pool in a hydrothermal field at the Azores archipelago (Photo courtesy of P. Forterre and E. Marguet)

do not depend on those conditions for a living. Oxygen induces the formation of free radicals that can damage cells. Indeed, most organisms have mechanisms to neutralise oxygen-derived deleterious effects. In this sense, aerobic organisms like us could be considered extremophiles. However, a vast diversity of organisms belonging to the three life domains is able to live optimally at all oxygen partial pressures found on Earth: from complete absence to atmospheric levels and even more, for instance in oxygen-producing photosynthetic bacterial mats. This indicates that life has adapted to all possible terrestrial niches with regard to oxygen concentration, without it being a general limiting factor. Therefore, anaerobic organisms cannot be considered extremophiles by this feature.

For some of the parameters mentioned above, we find only extremotolerant organisms. Thus, there are radiotolerant micro-organisms, but none is "radiophile". In addition to the parameters listed in Table 9.1, oligotrophic conditions, or extremely low nutrient concentrations, which very often characterise some extreme environments, constitute a major limiting factor for life.

Table 9.1. Major features of extreme environments and associated biota

Physico-chemical parameter	Type of organism	Definition/ Optimal growth	Molecular adaptations
Temperature	Hyperthermophile	$> 80\,°C$	• DNA stability: existence of a specific enzyme, reverse gyrase, acting on DNA topology • Lipids more saturated, membrane lipids forming a single monolayer more impermeable to protons • Protein stability: increased salt and disulfide interactions, increased hydrophobicity and compaction, less subunits • K^+ or compatible solute-enriched cytoplasm
	Thermophile	$60-80\,°C$	• More saturated lipids • Protein stability: increased hydrophobicity
	Psychrophile	$< 5\,°C$	• More unsaturated lipids • Proteins: increased flexibility, less interactions among protein domains, high complementarity substrate–enzyme • Antifreezing molecules in cytoplasm
pH	Acidophile	$pH < 2-3$	• Large internal buffering capacity • External cell surface positively charged • Proton export by specific membrane pumps
	Alkaliphile	$pH > 9-10$	• Large internal buffering capacity • External cell surface negatively charged • Proton import by antiporters
Salinity	Halophile	High salt concentration ($\sim 2-5$ M NaCl)	• K^+ or compatible solute-enriched cytoplasm • Proteins enriched in negatively charged amino acids
Pressure	Barophile (Piezophile)	High pressure	• More unsaturated lipids • Specific adaptations in proteins
Desiccation	Xerophile	Conditions of anhydrobiose	• Increased internal osmolarity • DNA stability: protein binding and strong repair mechanisms
Radiation	Radio-tolerant	support high radiation doses (ionising, UV, etc.)	• Proteins stabilising DNA and strong repair systems
Metals	Metallo-tolerant	support high heavy metal concentrations	• Specific detoxification mechanisms • Selective metal accumulation (sequestration) inside or outside cells

Table 9.1. (*continued*)

Physico-chemical parameter	Examples of biotopes	Biodiversity/examples
Temperature	• Continental and deep-sea hydrothermal systems • Geysers • Solfataras • Deep continental and oceanic subsurface regions	• Archaea dominate: *Methanopyrus, Pyrolobus, Thermococcus, Methanothermus, Pyrodictium,* "korarchaeota", etc. • Bacteria: *Aquifex, Thermotoga* • **NO** eukaryotes
	• Hydrothermal systems • Mineral piles • Industrial wastewaters • Deep subsurface	• Archaea: *Thermoplasma, Methanobacterium,* etc. • Bacteria: *Thermus, Bacillus stearothermophilus* • Eukaryotes: some fungi
	• Deep ocean • Arctic and Antarctic ice • Sea ice (seasonal) • Snow and high mountain • Permafrost	• Archaea: *Cenarchaeum symbiosum,* many non-cultivable deep-sea and sea ice lineages • Bacteria: *Flavobacterium, Psychrobacter,* etc. • Eukaryotes: red algae, diatoms, etc.
pH	• Mining areas • Hot acid springs • Acid solfataras	• Archaea: *Picrophilus, Thermoplasma, Sulfolobus,* etc. • Bacteria: *Acidithiobacillus, Thiobacillus, Leptospirillum,* etc. • Eukaryotes: many fungi, the red algae *Cyanidium*
	• Soda lakes • Alkaline hot springs	• Archaea: *Natronobacterium, Natronococcus,* etc. • Bacteria: *Spirulina,* other cyanobacteria • Eukaryotes: several protists
Salinity	• Coastal salterns • Some soda lakes (Natron type) • Deep-sea brines • Evaporites, salt mines	• Archaea: *Halobacterium, Haloferax, Halococcus,* etc. • Bacteria: *Salinibacter, Halomonas,* etc. • Eukaryotes: *Artemia salina, Dunaliella,* some yeast
Pressure	• Deep sea • Deep subsurface	• Archaea: many noncultivable deep-sea lineages • Bacteria: *Shewanella, Colwelia,* etc. • Eukaryotes: fungi, abyssal fauna
Desiccation	• Hot and cold deserts • Many salterns	• Archaea: halophilic archaea and others • Bacteria: *Metallogenium, Pedomicrobium,* etc. • Eukaryotes: fungi, lichens, etc.
Radiation	• Radioactive waste • Natural radioactive mines • Deserts, solar salterns • High mountain	• Archaea: some *Thermococcus* spp., halophiles (UV) • Bacteria: *Deinococcus radiodurans*
Metal	• Mining regions • Metal-contaminated aquifers • Wastewaters	• Archaea: *Acidianus, Thermoplasma,* etc. • Bacteria: *Acidithiobacillus, Leptospirillum* • Eukaryotes: some fungi, algae and plants

Fig. 9.4. Square halophilic archaea as seen by scanning electron microscopy. Several halophilic archaea belonging to the genus *Haloarcula* living close to saturation acquire these unusual morphologies. (Photo courtesy of F. Rodriguez-Valera)

Fig. 9.5. The Rio Tinto (Spain). Its waters are highly acidic (pH 1.9–2.5) and saturated in ferric iron, responsible for the *red colour*. The river bed is densely colonised by acidophilic micro-organisms; filaments of green algae are shown to the *right*. (Photo courtesy of A.I. Lopez-Archilla)

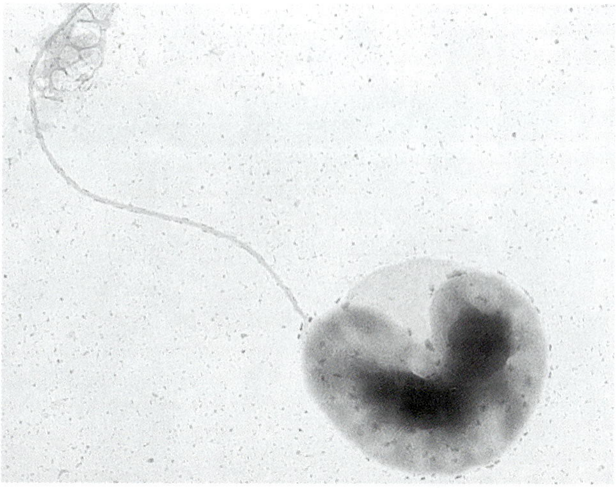

Fig. 9.6. Transmission electron micrograph of an acidophilic chemolitoautotrophic bacterium. (Photo courtesy of A.I. Lopez-Archilla)

Fig. 9.7. Endolithic community living in gypsum. The green layer is composed of thermophilic algae (50 °C), *Galdieria sulfuraria* (shown in *inset*). (Photo courtesy of A.I. Lopez-Archilla)

Fig. 9.8. Metallotolerant organisms living on copper sulfate crystals. (Photo courtesy of A.I. Lopez-Archilla)

Fig. 9.9. Deep subsurface cave (Pozo Alfredo, 800 m deep, Spain). Thermophilic archaea and bacteria have been identified in this subsurface environment. (Photo courtesy of A.I. Lopez-Archilla)

9.3.2 Phylogenetic Groups Best Adapted to Extreme Conditions

Some extremophiles correspond to groups of micro-organisms belonging to specific phylogenetic lineages that are particularly well adapted to specific life-limiting conditions and restricted types of environments. For instance only several archaeal and a few bacterial lineages are hyperthermophiles (Stetter, 1996). Conversely, we also find extremophiles broadly scattered in the phylogenetic tree that are adapted to the same extreme condition. This is the case for different psychrophiles or barophiles, for which we find members dispersed in the three domains of life.

There are also groups of organisms belonging to the same phylogenetic family that, without holding all extremophile records, have adapted to very diverse extreme or moderately extreme conditions. Among bacteria, the best-adapted group to various extreme conditions is possibly that of cyanobacteria. They often form microbial mats with other bacteria, from Antarctic ice to continental hot springs. Having been isolated from Yellowstone living at temperatures of $\sim 70\,^{\circ}\mathrm{C}$ they constitute, together with some green sulfur bacteria[1], the most thermophilic photosynthesisers known (Madigan et al., 2002; Paerl et al., 2000). They can live in both aerobic and anaerobic conditions, and they can also develop in hypersaline and alkaline lakes, they support high metal concentrations and they tolerate xerophilic conditions, forming endolithic communities in desertic regions. However, they do not appear to withstand acid conditions, and they are rarely found at pH values lower than 5–6.

Among eukaryotes, fungi, alone or in symbiosis with cyanobacteria or algae forming lichens, are certainly the most versatile and ecologically successful phylogenetic lineage. With the exception of hyperthermophily, although they represent the most thermophilic eukaryotes, living up to $60–64\,^{\circ}\mathrm{C}$, they adapt well to the rest of extreme environments. Most are aerobic, but many can be anaerobic. Fungi live in acidic and metal-enriched waters from mining regions, alkaline settings, hypersaline regions such as the Dead Sea, hot and cold deserts, the deep ocean (they have been even isolated from the Mariana Trench at $\sim 11\,000$m deep), etc. (Buchalo et al., 1998; Lopez-Archilla et al., 2001; Sterflinger, 1998; Takami et al., 1997).

Archaea are generally less versatile than bacteria and eukaryotes. In contrast, they are generally very skilful in adapting to different extreme conditions, holding frequently extremophily records. Thus, *Pyrolobus fumarii*, coming from a deep-sea vent from the Mid-Atlantic Ridge and living up to $113\,^{\circ}\mathrm{C}$, was the most hyperthermophilic organism known until very recently (Blöchl et al., 1997) but, in 2003, another archaeon growing up to $121\,^{\circ}\mathrm{C}$ (strain 121) was isolated from a Pacific deep-sea vent (Kashefi and Lovley, 2003). Similarly, some archaea are among the most halophilic and alkaliphilic micro-organisms

[1] Green sulfur bacteria, e.g. *Chlorobium* species, are strictly anaerobic and phototrophic, gaining energy from light (via a bacteriochlorophyll) and using a reduced sulfur species (instead of H_2O) as an electron acceptor.

known (Oren, 1994). In addition, we could mention methanogens as the most versatile among the archaea. They are strict anaerobes that occupy the entire thermal gradient from ∼ 110°C (e.g. *Methanopyrus kandlerii*) (Burggraf et al., 1991) to 0–2°C, living in cold marine sediments (Lanoil et al., 2001). Many methanogens are also acidophilic, alkaliphilic or halophilic (Madigan et al., 2002).

9.3.3 Resistance Forms and Longevity

How much and how long can micro-organisms resist exceedingly harsh conditions? These are basic questions with regard to the eventual possibility of transporting life forms (e.g. in meteorites) from one planet to another. When the environmental conditions become adverse, most micro-organisms get into latent states that allow them to survive until conditions become favourable again. We can distinguish different cellular states going from normal metabolic activity to death: growing state, resting state (cells do not reproduce), dormancy (absence of metabolism), mummified state (cells undergo already irreversible transformations) (Muliukin et al., 2002) and death. This progression is accompanied by changes at the internal cellular structure. Micro-organisms can survive over more or less long periods of time before death. Many of them, such as Gram positive bacteria or fungi, make spores, special resistance forms that allow them to survive and also to disseminate while assuring an effective protection. Some spores are extraordinarily resistant to extreme conditions and to time. Experiences of exposure to the outer space suggest that *Bacillus subtilis* (a Gram positive bacterium) spores would survive for years if they were just covered by a thin layer protecting them from ultraviolet radiation, such as polysaccharides or a layer of cells. However, they will not last more than a few hours in a single layer. In a similar way, halophilic archaea included in salt crystals resisted the exposure to outer space conditions during the two weeks that the experiment lasted (Rothschild and Mancinelli, 2001). This is a remarkable resistance because halophilic archaea do not form spores. Halophiles are to date the micro-organisms that hold the longevity record (Fish et al., 2002; Grant et al., 1998).

The most efficient ways of biological conservation are cryopreservation and desiccation. All microbiologists use low temperatures to conserve and store most prokaryotic strains, keeping them at −80°C or immersed in liquid nitrogen for years. To get a high survival rate, one has only to add a little glycerol to the culture medium in order to avoid the formation of big water crystals that would damage cells. Cryopreservation occurs also in nature, and reports of species retrieved from permafrost after several thousand years exist in the literature (Kochkina et al., 2001; Shi et al., 1997). Desiccation is also very efficient as it can be readily deduced from the long survival of halophilic archaea in salt crystals. Anecdotally, microbial strains from culture collections are frequently delivered lyophilised.

9.4 Extremophiles and Exobiology

In the above, we have taken a brief look at the different types of extreme environments on Earth, the organisms that occupy them, their adaptations and their long-term survival strategies. As can be realised, life is present on Earth in any place where physicochemical conditions are permissive. If one extends this idea to other planets, the hypothesis of extraterrestrial life becomes more than plausible. The only condition that appears necessary for life, as we know it, is the presence of liquid water (and possibly a carbon-based biochemistry as well).

In our solar system, the two candidates that appear more susceptible to host life or having hosted it in the past are Mars and Europa. Early in its history, the conditions on Mars resembled those of the Earth so that, even if Mars were sterile today, life could have existed on the red planet and therefore left fossil remains. An ocean of liquid water possibly existed that was subsequently lost by evaporation, at least partially, since some water persists frozen at the poles and possibly in the subsurface (Boynton et al., 2002; Forsythe and Zimbelman, 1995; Kerr, 2000). Europa, one of Jupiter's satellites, is fully covered by ice, but some data suggest that an underlying ocean of liquid water might exist and, perhaps, some hydrothermal activity (Carr et al., 1998; Kivelson et al., 2000).

In the following, we will briefly consider the kind of extremophiles and extreme habitats that have raised a special interest among exobiologists.

9.4.1 Hyperthermophiles

Since their discovery, hyperthermophiles have played a key role in generating new hypotheses on the origin of life (on Earth and, by extrapolation, to any other planet exhibiting analogous conditions to those of the early Earth).

Origin-of-life models based on the classical Oparin–Haldane theory propose that life arose at low temperature and was first heterotrophic (Lazcano and Miller, 1996). Life would have been generated in a kind of marine "primitive soup" where simple organic molecules would have been formed, and then accumulated, via prebiotic synthesis helped by ultraviolet radiation and electric discharges. Miller's famous experience in 1953 appeared to be a strong support for this hypothesis, since it demonstrated that amino acids could form in a mixture of reduced gases, supposedly representative of the primitive atmosphere, after exposure to electric discharges (Miller, 1953). However, subsequent experiences showed that amino-acid yield and variety strongly diminished if the starting gas mixture was less reducing and, therefore, in greater agreement with present-day ideas on the early atmosphere composition (Schlesinger and Miller, 1983). From these primordial simple organic ingredients, larger macromolecules would have been formed, RNA molecules, proteins and (somewhat later) DNA, from which primitive cells evolved (Fig. 9.10).

However, during the last decades of the twentieth century, and more particularly after hyperthermophiles were discovered, the idea of a hot origin of life expanded in the scientific community (Pace, 1991). Life would have been initially chemolithoautotrophic (inorganic sources of matter and energy) and would have been born at high temperature in environments similar to contemporary marine hydrothermal vents (Henley, 1996) (Fig. 9.10). The idea of a hot origin of life is sustained in three types of arguments:

- **Geological.** Different geological data suggest that the Earth was hotter than today at the time when life arose about 3.8 Ga ago, with extensive volcanism and hydrothermalism (see Martin's Chap. 4, Part I in this book). Data on oxygen isotopes in archaean sedimentary rocks appear to further sustain this view

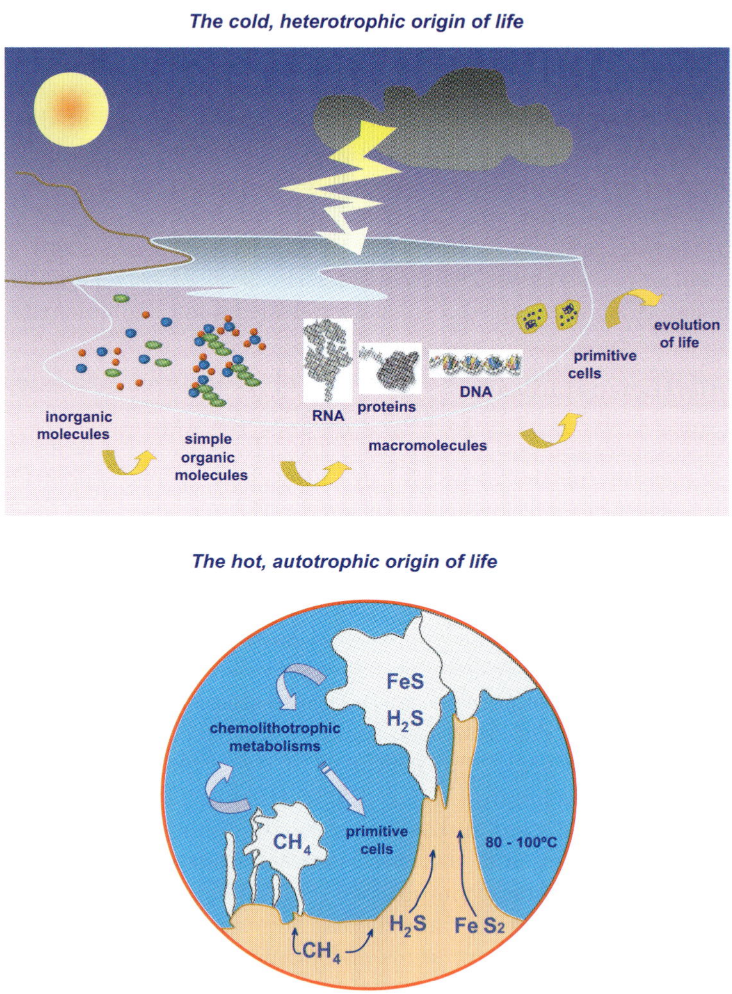

Fig. 9.10. Schematic diagram showing the two main global scenarios for the origin of life

(Knauth, 1992). In addition, the oldest unambiguous microbial fossils known today correspond most likely to thermophilic prokaryotes (Summons, 1999) identified in fossil hydrothermal settings from South Africa. The existence of the earliest putative cyanobacteria fossils in stromatolithe-like structures is very speculative and controversial (Brasier et al., 2002; Schopf et al., 2002). Geological data thus appear compatible with models in which life evolved from a thermophilic ancestor and the subsequent adaptation of microbial communities to increasingly cooler conditions (Nisbet and Sleep, 2001).

- **Phylogenetic.** Several authors have placed the root of the tree of life in the bacterial branch using a strategy based on the comparison of ancient duplicated genes (reviewed in Brown and Doolittle, 1997). If this position is correct, the lineages that appeared to be placed closer to the root were all hyperthermophilic, which supported the idea that the last common universal ancestor was itself a hyperthermophile (Barns et al., 1996; Stetter, 1996). This, of course, led to the idea that life had originated at high temperature, although some authors alternatively suggested that the origin of life took place at low temperature but that the Last Universal Common Ancestor (LUCA) was a hyperthermophile having survived the intensive meteorite bombardments Earth suffered early in its history (e.g. Lazcano and Miller, 1996). In any case, it is difficult to reconstruct ancient phylogenetic relationships among organisms because the evolutionary signal may have been lost with time, or it may be obscured by successive mutations and differences in evolutionary rates (see Brinkmann and Philippe's Chap. 8, Part II, in this book). Indeed, methods of phylogenetic reconstruction today allow the root of the tree of life to be placed safely neither in the bacterial nor in the eukaryotic branch. The phylogenetic argument should thus be taken with caution in this highly controversial debate (Brown et al., 2001; Forterre and Philippe, 1999).
- **Metabolic.** These are based on the fact that many hyperthermophiles are chemolithoautotrophs, displaying a large variety of possible redox reactions to obtain energy (Table 9.2) (Madigan et al., 2002). Furthermore, these reactions frequently involve hydrogen, iron and sulfur. This recalls the model proposed by Wächtershäusser who hypothesized that life appeared on pyrite-like surfaces and that the first organisms gained their energy from FeS reduction by H_2S generating H_2 and FeS_2 (Wächtershäusser, 1988). Most hyperthermophiles are also strict anaerobes (let us recall here that oxygen was not present in significant proportions in the early Earth atmosphere and ocean).

If life arose in a hot environment on Earth, it would be possible that it originated also on Mars, where initial conditions were similar, or in other planets with hydrothermal activity and analogous environmental settings.

9.4.2 Psychrophiles

Psychrophilic micro-organisms living at very low temperatures constitute another focus of research for exobiologists. Extremely cold environments such as

Table 9.2. Chemoautotrophic metabolisms in hyperthermophiles. Bacterial genera are labeled with an asterisk, the remaining genera ascribe to the Archaea (López-García)

Energy-yielding reactions	Representative genera
$H_2 + S^0 \rightarrow H_2S$	*Acidianus*, *Pyrodictium*, *Thermoproteus*
$H_2 + NO_3^- \rightarrow NO_2^- + H_2O\,(NO_2, N_2)$	*Pyrobaculum*, *Stygioglobus*, *Aquifex**, *Pyrodictium*, *Thermoproteus*
$4H_2 + NO_3^- + 2H^+ \rightarrow NH_4 + 3H_2O$	*Pyrolobus*
$2H_2 + O_2 \rightarrow 2H_2O$	*Acidianus*, *Sulfolobus*, *Pyrobaculum*, *Aquifex**
$2S^{\circ} + 3O_2 + 2H_2O \rightarrow 2H_2SO_4$	*Sulfolobus*, *Acidianus*
$4FeS_2 + 15O_2 + 2H_2O \rightarrow 2Fe_2(SO_4)_3 + 2H_2SO_4$	*Sulfolobus*
$10FeCO_3 + 2NO_3^- + 24H_2O \rightarrow 10Fe(OH)_3 + N_2 + 10HCO_3^- + 8H^+$	*Ferroglobus*
$4H_2 + SO_4^- + 2H^+ \rightarrow 4H_2O + H_2S$	*Archaeoglobus*
$4H_2O + CO_2 \rightarrow CH_4 + 2H_2O$	*Methanopyrus*, *Methanococcus*

the Antarctic or the permafrost are interesting as analogous environments to cold extraterrestrial regions, for example the hypothetical ocean underlying Europa's ice cover. They could also serve as a model of past fossil life preservation on ice. Subglacial Antarctic lakes, particularly lake Vostok (see J.R. Petit's Chap. 7, Part I, in this book), at a depth of more than 3750m under the continental ice cover and a pressure of more than 350 bars, may host microbial communities isolated for several million years (Siegert et al., 2001).

9.4.3 Halophiles and Evaporites

Halophilic micro-organisms can survive for long periods within salt crystals in salt mines or in evaporitic rocks (Fish et al., 2002; Grant et al., 1998). They can even be active in fluid inclusions (Rothschild et al., 1994). Evaporites are important in the fossil record since they tend to preserve well microbial structures and algal filaments. Recent Martian exploration revealed the existence of geological structures resembling evaporitic deposits in impact craters showing traces of water flow (Newsom et al., 2001). These are interesting sampling sites due to their potential to have conserved past life signatures.

9.4.4 The Deep Biosphere

The discovery of microbial life in Earth's subsurface has reinforced the idea that life can occur today in other planets of our solar system and even beyond. During the last several years evidence on the existence of a deep microbial biosphere based on chemolithotrophy has been accumulating (Stetter et al., 1993; Pedersen, 2000). Underground biotopes comprise continental and sea floor, deep buried sediments, oil and petroleum reservoirs, deep caves and deep aquifers. The ocean subseafloor close to midoceanic ridges is probably one of the most successfully colonised subterranean habitats, because the mixture of hydrothermal fluid and seawater that penetrates in the substratum is highly enriched in inorganic electron donors and acceptors for chemolithotrophic redox energy reactions (Summit and Baross, 2001). The oceanic crust would thus be the largest habitat on Earth, and its microbial inhabitants could constitute a considerable portion of the total biomass in our planet, even when the microbial density is generally scarce and the metabolic rates low due to overall oligotrophic conditions. The depth reached by biological alterations in subseafloor basaltic glass appears limited by temperatures permissive for life (Fisk et al., 1998; Furnes et al., 1999), which suggests that temperature is a major limiting factor for underground life.

Subterranean microbial communities are characterised by chemolithoautotrophic metabolisms in reducing and strict anaerobic conditions. The role of hydrogen appears to be primordial (Pedersen, 2000). A recent study showed that more than 95% of the population found in a hot deep aquifer were methanogenic archaea utilising H_2 and CO_2 to produce CH_4 (Chapelle et al., 2002). The implications of these studies for exobiology are numerous, since analogous subsurface environments are likely to exist in other planets that still maintain a geothermal activity (Weiss et al., 2000). In addition, these environments are stable and protected from ultraviolet irradiation even when they had a thin atmosphere (e.g. Mars).

9.5 Perspectives

During recent years, the discovery of extremophiles and the limits of life on Earth have allowed questions on the existence of extraterrestrial life to be formulated in a more rigorous way. We need now to continue exploring the micro-organisms colonising extreme environments, particularly those from less-accessible biotopes such as the deep subsurface that still remain very poorly known. The results obtained in this way should allow identification of potential habitats for life (as we know it) on other planets. One of the most important challenges today is the identification of markers of incontestable biogenic origin that could be afterwards detected in other planets, either *in situ* or by direct analysis after an eventual sample return. In the future, exobiology will certainly require a further interdisciplinary approach to answer solidly questions regarding the origin of life, to identify unambiguously biosignatures and, finally, to develop methodologies to detect them on Earth or on other planets.

References

General

Madigan M. T., Martinko J. M. and Parker J. (2002). *Brock Biology of Microorganisms*, 10th edn. New Jersey, Prentice-Hall, Inc.

Specialised

Amann R. I., Ludwig W. and Schleifer K. H. (1995). Phylogenetic identification and in situ detection of individual microbial cells without cultivation. *Microbiol. Rev.* **59**, 143–169.

Barns S. M., Delwiche C. F., Palmer J. D., Dawson S. C., Hershberger K. L. and Pace N. R. (1996). Phylogenetic perspective on microbial life in hydrothermal ecosystems, past and present. *Ciba Found Symp.* **202**, 24–32; discussion 32–39.

Blöchl E., Rachel R., Burgraff S., Hafenbradl D., Jannasch H. W. and Stetter K. O. (1997). *Pyrolobus fumarii*, gen. and sp. nov., represents a novel group of archaea, extending the upper temperature limit for life to 113 °C. *Extremophiles* **1**, 14–21.

Boynton W. V., Feldman W. C., Squyres S. W., Prettyman T. H., Bruckner J., Evans L. G., Reedy R. C., Starr R., Arnold J. R., Drake D. M., Englert P. A., Metzger A. E., Mitrofanov I., Trombka J. I., D'Uston C., Wanke H., Gasnault O., Hamara D. K., Janes D. M., Marcialis R. L., Maurice S., Mikheeva I., Taylor G. J., Tokar R. and Shinohara C. (2002). Distribution of hydrogen in the near surface of Mars: evidence for subsurface ice deposits. *Science* **297**, 81–85.

Brasier M. D., Green O. R., Jephcoat A. P., Kleppe A. K., Van Kranendonk M. J., Lindsay J. F., Steele A. and Grassineau N. V. (2002). Questioning the evidence for Earth's oldest fossils. *Nature* **416**, 76–81.

Brock T. D., Brock K. M., Belly R. T. and Weiss R. L. (1972). *Sulfolobus*: a new genus of sulfur-oxidizing bacteria living at low pH and high temperature. *Arch Mikrobiol.* **84**, 54–68.

Brock T. D. and Freeze H. (1969). *Thermus aquaticus gen. n.* and *sp. n.*, a nonsporulating extreme thermophile. *J. Bacteriol.* **98**, 289–297.

Brown J. R. and Doolittle W. F. (1997). Archaea and the prokaryote-to-eukaryote transition. *Microbiol. Mol. Biol. Rev.* **61**, 456–502.

Brown J. R., Douady C. J., Italia M. J., Marshall W. E. and Stanhope M. J. (2001). Universal trees based on large combined protein sequence data sets. *Nat. Genet.* **28**, 281–285.

Buchalo A. S., Nevo E., Wasser S. P., Oren A. and Molitoris H. P. (1998). Fungal life in the extremely hypersaline water of the Dead Sea: first records. *Proc. R. Soc. Lond. B Biol. Sci.* **265**, 1461–1465.

Burggraf S., Stetter K. O., Rouviere P. and Woese C. R. (1991). *Methanopyrus kandleri*: an archaeal methanogen unrelated to all other known methanogens. *Syst. Appl. Microbiol.* **14**, 346–351.

Carr M. H., Belton M. J., Chapman C. R., Davies M. E., Geissler P., Greenberg R., McEwen A. S., Tufts B. R., Greeley R., Sullivan R., Head J. W., Pappalardo R. T., Klaasen K. P., Johnson T. V., Kaufman J., Senske D., Moore J., Neukum G., Schubert G., Burns J. A., Thomas P. and Veverka J. (1998). Evidence for a subsurface ocean on Europa. *Nature* **391**, 363–365.

Chapelle F. H., K, O. N., Bradley P. M., Methe B. A., Ciufo S. A., Knobel L. L. and Lovley D. R. (2002). A hydrogen-based subsurface microbial community dominated by methanogens. *Nature* **415**, 312–315.

Deming J. W. (1998). Deep ocean environmental biotechnology. *Curr. Opin. Biotechnol.* **9**, 283–287.

Farlow, W. G. (1880). *On the nature of the peculiar reddening of salted codfish during the summer season.* U.S. Commission of Fish and Fisheries, pp. 969–974.

Fish S. A., Shepherd T. J., McGenity T. J. and Grant W. D. (2002). Recovery of 16S ribosomal RNA gene fragments from ancient halite. *Nature* **417**, 432–436.

Fisk M. R., Giovannoni S. J. and Thorseth I. H. (1998). Alteration of oceanic volcanic glass: textural evidence of microbial activity. *Science* **281**, 978–980.

Forsythe R. D. and Zimbelman J. R. (1995). A case for ancient evaporite basins on Mars. *J. Geophys. Res.* **100**, 5553–5563.

Forterre P. and Philippe H. (1999). Where is the root of the universal tree of life? *Bioessays* **21**, 871–879.

Furnes H., Muehlenbachs K., Tumyr O., Torsvik T. and Thorseth I. H. (1999). Depth of active bio-alteration in the ocean crust: Costa Rica Rift (Hole 504B). *Terra Nova* **11**, 228–233.

Grant W. D., Gemmell R. T. and McGenity T. J. (1998). Halobacteria: the evidence for longevity. *Extremophiles* **2**, 279–287.

Henley R. W. (1996). Chemical and physical context for life in terrestrial hydrothermal systems: chemical reactors for the early development of life and hydrothermal ecosystems. *Ciba Found. Symp.* **202**, 61–76; discussion 76–82.

Horikoshi K. (1999). Alkaliphiles: some applications of their products for biotechnology. *Microbiol Mol Biol Rev* **63**, 735–750, table of contents.

Hough D. W. and Danson M. J. (1999). Extremozymes. *Curr. Opin. Chem. Biol.* **3**, 39–46.

Hugenholtz P., Pitulle C., Hershberger K. L. and Pace N. R. (1998). Novel division level bacterial diversity in a Yellowstone hot spring. *J. Bacteriol.* **180**, 366–376.

Kashefi K. and Lovley D. R. (2003). Extending the upper temperature limit for life. *Science* **301**, 934.

Kerr R. A. (2000). Planetary science. Buried channels may have fed Mars ocean. *Science* **287**, 1727–1728.

Kivelson M. G., Khurana K. K., Russell C. T., Volwerk M., Walker R. J. and Zimmer C. (2000). Galileo magnetometer measurements: a stronger case for a subsurface ocean at Europa. *Science* **289**, 1340–1343.

Knauth L. P. (1992). Origin and diagenesis of cherts: An isotopic perspective. In *Isotopic signatures and sedimentary records*, pp. 123–152. Edited by N. Clauer and S. Chanduri. Berlin: Springer-Verlag.

Kochkina G. A., Ivanushkina N. E., Karasev S. G., Gavrish E., Gurina L. V., Evtushenko L. I., Spirina E. V., Vorob'eva E. A., Gilichinskii D. A. and Ozerskaia S. M. (2001). Micromycetes and actinobacteria under conditions of many years of natural cryopreservation. *Mikrobiologiia* **70**, 412–420.

Lanoil B. D., Sassen R., La Duc M. T., Sweet S. T. and Nealson K. H. (2001). Bacteria and archaea physically associated with gulf of Mexico gas hydrates. *Appl. Environ. Microbiol.* **67**, 5143–5153.

Lazcano A. and Miller S. L. (1996). The origin and early evolution of life: prebiotic chemistry, the pre-RNA world, and time. *Cell* **85**, 793–798.

Lopez-Archilla A. I., Marin I. and Amils R. (2001). Microbial community composition and ecology of an acidic aquatic environment: the tinto river, Spain. *Microb. Ecol.* **41**, 20–35.

Madigan M. T., Martinko J. M. and Parker J. (2002). *Brock Biology of Microorganisms*, 10th edn. New Jersey: Prentice-Hall, Inc.

Miller S. (1953). A production of amino acids undert possible primitive Earth conditions. *Science* **117**, 528–529.

Muliukin A. L., Sorokin V. V., Vorob'eva E. A., Suzina N. E., Duda V. I., Gal'chenko V. F. and El'-Registan G. I. (2002). Detection of microorganisms in the environment and the preliminary appraisal of their physiological state by X-ray microanalysis. *Mikrobiologiia* **71**, 836–848.

Newsom H. E., Hagerty J. J. and Thorsos I. E. (2001). Location and sampling of aqueous and hydrothermal deposits in Martian impact craters. *Astrobiology* **1**, 71–88.

Nisbet E. G. and Sleep N. H. (2001). The habitat and nature of early life. *Nature* **409**, 1083–1091.

Oren A. (1994). The ecology of extremely halophilic archaea. *FEMS Microbiol. Rev.* **13**, 415–440.

Oren A. (1999). Bioenergetic aspects of halophilism. *Microbiol. Mol. Biol. Rev.* **63**, 334–348.

Pace N. R. (1991). Origin of life–facing up to the physical setting. *Cell* **65**, 531–533.

Pace N. R. (1997). A molecular view of microbial diversity and the biosphere. *Science* **276**, 734–740.

Paerl H. W., Pinckney J. L. and Steppe T. F. (2000). Cyanobacterial-bacterial mat consortia: examining the functional unit of microbial survival and growth in extreme environments. *Environ. Microbiol.* **2**, 11–26.

Pedersen K. (2000). Exploration of deep intraterrestrial microbial life: current perspectives. *FEMS Microbiol. Lett.* **185**, 9–16.

Rothschild L. J., Giver L. J., White M. R. and Mancinelli R. L. (1994). Metabolic activity of microorganisms in gypsum-halite crusts. *J. Phycol.* **30**, 431–438.

Rothschild L. J. and Mancinelli R. L. (2001). Life in extreme environments. *Nature* **409**, 1092–1101.

Schlesinger G. and Miller S. (1983). Prebiotic synthesis in atmospheres containing CH_4, CO and CO_2. I. Amino acids. *J. Mol. Evol.* **19**, 376–382.

Schopf J. W., Kudryavtsev A. B., Agresti D. G., Wdowiak T. J. and Czaja A. D. (2002). Laser–Raman imagery of Earth's earliest fossils. *Nature* **416**, 73–76.

Shi T., Reeves R. H., Gilichinsky D. A. and Friedmann E. I. (1997). Characterization of viable bacteria from Siberian permafrost by 16S rDNA sequencing. *Microb. Ecol.* **33**, 169–179.

Siegert M. J., Ellis-Evans J. C., Tranter M., Mayer C., Petit J. R., Salamatin A. and Priscu J. C. (2001). Physical, chemical and biological processes in Lake Vostok and other Antarctic subglacial lakes. *Nature* **414**, 603–609.

Sterflinger K. (1998). Temperature and NaCl-tolerance of rock-inhabiting meristematic fungi. *Antonie Van Leeuwenhoek* **74**, 271–281.

Stetter K.O. (1989) Extremely thermophilic chemolithoautotrophic Archaebacteria. *In* H.G. Schlegel and B. Bowien (eds.), *Extremely Thermophilic Chemolithoautotrophic Archaebacteria.* Springer-Verlag, Berlin, pp. 167–176.

Stetter K. O., Huber R., Blöchl E., Kurr M., Eden R. D., Fiedler M., Cash H. and Vance I. (1993). Hyperthermophilic archaea are thriving in deep North Sea and Alaskan oil reservoirs. *Nature* **365**, 743–745.

Stetter K. O. (1996). Hyperthermophilic prokaryotes. *FEMS Microbiol. Rev.* **18**, 149–158.

Summit M. and Baross J. A. (2001). A novel microbial habitat in the mid-ocean ridge subseafloor. *Proc. Natl. Acad. Sci. USA* **98**, 2158–2163.

Summons R. (1999). Molecular probing of deep secrets. *Nature* **398**, 752–753.

Takami H., Inoue A., Fuji F. and Horikoshi K. (1997). Microbial flora in the deepest sea mud of the Mariana Trench. *FEMS Microbiol. Lett.* **152**, 279–285.

Wächtershäusser G. (1988). Pyrite formation, the first energy source for life: a hypothesis. *Syst. Appl. Microbiol.* **10**, 207–210.

Weiss B. P., Yung Y. L. and Nealson K. H. (2000). Atmospheric energy for subsurface life on Mars? *Proc. Natl. Acad. Sci. USA* **97**, 1395–1399.

Wilansky B. (1936). Life in the Dead Sea. *Nature* **138**, 467.

Woese C. R. (1987). Bacterial evolution. *Microbiol. Rev.* **51**, 221–271.

Woese C. R. and Fox G. E. (1977). Phylogenetic structure of the prokaryotic domain: the primary kingdoms. *Proc. Natl. Acad. Sci. USA* **74**, 5088–5090.

Woese C. R., Kandler O. and Wheelis M. L. (1990). Towards a natural system of organisms: proposal for the domains Archaea, Bacteria, and Eucarya. *Proc. Natl. Acad. Sci. USA* **87**, 4576–4579.

Zuckerkandl E. and Pauling L. (1965). Molecules as documents of evolutionary history. *J Theor. Biol.* **8**, 357–366.

Part III

Appendices

1 Earth Structure and Plate Tectonics: Basic Knowledge

Hervé Martin

1.1 Earth Internal Structure

The Zapoliarny borehole in Russia only reaches a depth of 12.5 km; samples of deeper parts of our planet were brought up by volcanoes, but their original depth never exceeds a few hundred kilometres. Consequently, most of our knowledge of the Earth's internal composition and structure is indirect and mostly due to studies of propagation rates of seismic waves, although gravimetric data have also been used. The structure of the Earth can be schematically described as consisting of concentric shells that are from the centre to the surface (Fig. 1.1).

1.1.1 Inner Core (from 6378 to 5155 km Depth)

It is solid and mainly consists of iron with smaller amounts of nickel; its density is about 12 and its temperature at least 6000 K.

Photo 1.1. TTG sampled at 11 047.9 m depth in the Zapoliarny borehole in Kola Peninsula (Russia). This is the deepest borehole in the world until today (Photo H. Martin)

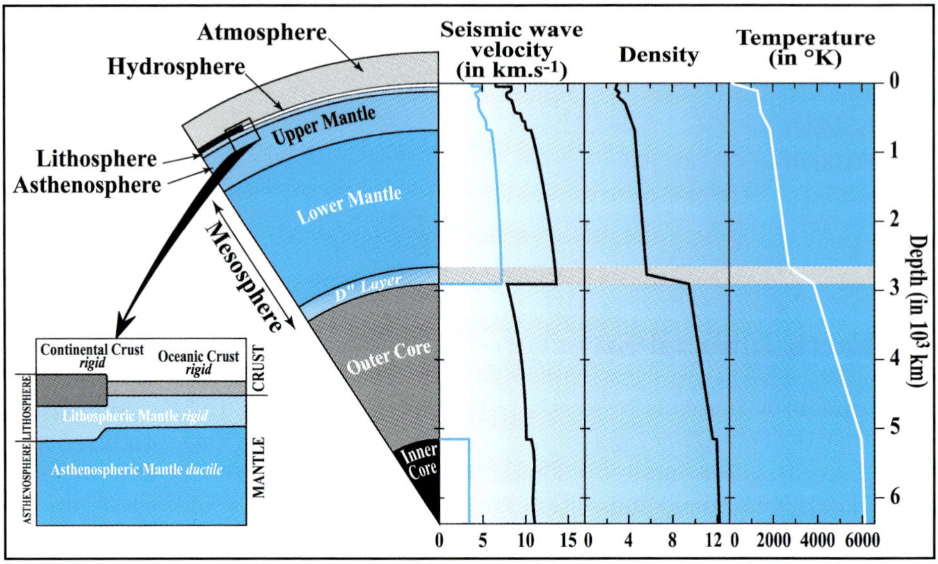

Fig. 1.1. Schematic cross section of the Earth showing its internal structure. On the right side are shown seismic wave velocity, density and temperature variations vs. depth. Shear seismic waves are in blue and compression waves in black. It must be noted that shear waves do not travel through the outer core, pointing to its liquid state. The inset on the left side illustrates the distinction that is made between mantle/crusts and asthenosphere/lithosphere pairs

1.1.2 Outer Core (from 5155 to 2891 km Depth)

It is separated from the inner core by the Lehman discontinuity at 5155 km.

Shear seismic waves (S waves) do not travel through the outer core (Fig. 1.1), which indicates that it is liquid. As for the inner core, it is composed of iron and nickel but also contains lighter element such as S, Si, O, and possibly K. From top to bottom its density ranges between 9.5 and 11.5 and its temperature between 3800 and 6000 K. As the Earth cooled it progressively crystallized at high pressure thus creating the solid inner core. During crystallization, light elements (Si, O, S, K) are rejected and remain in the liquid core.

The relative motions of the inner and outer cores as well as the convection in the outer core result in the creation of electromagnetic currents that are responsible for the Earth's magnetic field.

1.1.3 Lower Mantle (from 2891 to 670 km Depth)

It is separated from the outer core by the Gutenberg discontinuity at 2891 km.

Sometimes called the mesosphere, it consists of a silicate-rich solid, made up of perovskite $(Fe,Mg)SiO_3$ and magnesiowurstite $(Fe,Mg)O$. From top to bottom its density ranges between 4.5 and 5.5 and its temperature between 1900 and 2500 K. Velocities of shear seismic waves show that the lower mantle is solid, but on the geological timescale ($\sim 10^8$ years) it behaves as a fluid and is affected by convection.

The transition between the mantle and core consists of a 100–200 km thick heterogeneous layer called the D″ layer; through this zone temperature increases from 2500 to 3800 K. Through this thermal boundary layer, heat is only transferred by conduction and not by convection, consequently, this interface is the place where energy exchanges can take place, however, only a little matter exchange is suspected. Instabilities in this zone are considered as responsible for hot-spot magmatism. It has also been proposed that cold avalanches due to accumulation of subducted oceanic crust could create these instabilities.

1.1.4 Upper Mantle (from 670 km to 7 km Depth Under Oceans and 30 km Depth Under Continents)

It mainly consists of peridotite: a rock made up of olivine $(Fe,Mg)_2SiO_4$, orthopyroxene $(Fe,Mg)_2Si_2O_6$, clinopyroxene $Ca(Fe,Mg)Si_2O_6$ and an aluminium-rich mineral that is plagioclase $CaAl_2Si_2O_8$ (depth $< 30-40$ km), spinel $MgAl_2O_4$ ($30-40$ km $<$ depth $<$ 70 km) or garnet $Mg_3Al_2(SiO_4)_3$ (depth $>$ 70 km). Its density ranges from 3.3 to 4.5 and its temperature increases from 900–1000 K to 1900 K. It is also the place of active convection. Until now, scientific research has tried to determine if the whole mantle is affected by a single convection system or if upper and lower mantles convert independently.

The transition between lower and upper mantle is due to the transformation of olivine at low pressure into perovskite + magnesiowurstite at high pressure.

Between 100 and 200 km depth, seismic waves show a diminution of their propagation rate, this zone is called the low-velocity zone (LVZ). It is interpreted as being due to the diminution of peridotite rigidity but this zone remains solid as it allows transmission of shear seismic waves. (It must be noted that the notion of rigidity is relative and strongly depends on the timescale. For instance, the upper mantle can have an immediate rigid behaviour but is ductile when million years period is taken into account.)

1.1.5 Crusts (from 7 km Depth Under Oceans and 30 km Depth Under Continents to Surface)

It is separated from the upper mantle by the Mohorovicic (Moho) discontinuity.

There exist two different types of crusts:

Oceanic Crust

Its average composition is that of a basalt, it is thin ($\sim$ 7 km), has a density of 3.1 and its age never exceeds 180 Ma. It is generated in oceanic ridge systems by partial melting of the upper mantle and recycled into the mantle in subduction zones.

Continental Crust
It has an average thickness of $\sim$ 30 km but under mountains, it can be 70 km thick. It is highly heterogeneous with an average composition of granodiorite. When compared with mantle it has a relatively low density (2.75) such that it has high buoyancy and is almost not recycled into the mantle. This is why continental crust as old as 4.04 Ga can be found. Most of the new continental crust is generated in subduction-zone environments.

1.1.6 Hydrosphere

It consists of surface water: ocean, lakes, rivers, ice caps, clouds, etc. It strongly interacts with surface rocks.

1.1.7 Atmosphere

This is the gaseous envelope of the Earth, it can extend up to 500 km in altitude. This interface between the surface of the Earth and the extraterrestrial domain plays the role of energy filter.

Both the hydrosphere and atmosphere are the places where life developed.

1.1.8 Lithosphere and Asthenosphere

The distinction between core, mantle and crust is based on compositional differences. However, for the same chemical and mineralogical composition, depending, for instance, on temperature, rheological properties can be completely different. For instance, viscosity is strongly temperature dependent, it decreases when temperature increases. Consequently the uppermost part of the mantle that is cold has high viscosity and a rigid behaviour even at long timescale, whereas the deeper and hotter parts have a lower viscosity resulting in a ductile behaviour on the geological timescale ($\sim$ 100 Ma). Consequently, in the mantle–crusts system two rheological shells are distinguished:

Lithosphere
It consists of the crust plus the rigid part of the upper mantle, its average thickness is about 100 km under oceans and 150–200 km under continents. It has a rigid behaviour and is not affected by internal convection; it corresponds to a domain where heat is transferred by conduction.

Asthenosphere
It corresponds to the deeper and hotter parts of the mantle where peridotite has a ductile behaviour and where heat is dissipated by convection. It is generally considered that the lithosphere–asthenosphere limit corresponds to the 800 °C isotherm; it roughly begins with the LVZ. As it cools the asthenosphere is transformed into lithosphere.

1.2 Plate Tectonics

Since the 1960s, the concept of sea-floor spreading has been established. As the surface of the Earth can be considered as constant, the oceanic crust, which is progressively generated in rift systems, must return and be recycled into the mantle (subduction zone). Then plate tectonics was born and this concept progressively evolved and refined.

1.2.1 Plates on the Surface of the Earth

The surface of the Earth is divided into lithospheric plates made up of crust (oceanic or continental) and with the rigid part of the upper mantle; they move over the convecting asthenospheric mantle. Today 12 major plates are recognized on the surface of the Earth but several small (micro-plates) also exist (Table 1.1; Fig. 1.2).

Table 1.1. Name and surface of the main lithospheric plates

Plate name	Area (10^6 km^2)	Plate name	Area (10^6 km^2)
Pacific	103.28	South-American	43.62
African	78.00	Nazca	15.63
North American	75.89	Philippine	5.45
Eurasian	67.81	Arabian	5.01
Antarctic	60.92	Caribbean	3.32
Indo-Australian	59.07	Cocos	2.86

1.2.2 Margin Definitions

Some of these plates contain both oceanic and continental crust (i.e. Eurasian or African), others are exclusively oceanic (i.e. Pacific or Nazca). Most of the seismic and volcanic activity of our planet is concentrated at plate margins. Two kinds of continent–ocean limit must be distinguished:

- ***Passive margin:*** both oceanic and continental crust belong to the same plate (limit between Europe and East North Atlantic), there is no seismic or volcanic activity;
- ***Active margin:*** oceanic and continental crusts do not belong to the same plate; generally the oceanic crust disappears under continental crust giving rise to an important seismic and volcanic activity.

There exist 3 main kinds of boundary between lithospheric plates:

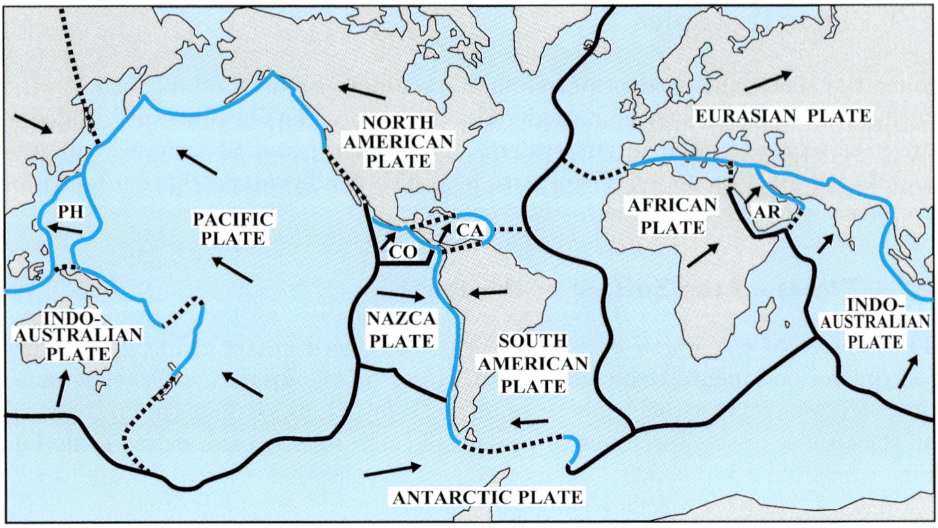

Fig. 1.2. Schematic map of the surface of the Earth showing the 12 main lithospheric plates. *Black heavy line* = divergent margin; *blue heavy line* = convergent margin; *heavy dotted line* = transform margin. Arrows show the main direction of plate motion. CO = Coco plate; CA = Caribbean plate; PH = Philippine plate; AR = Arabian plate

- ***Divergent margin:*** also called constructive margin, this is where new material resulting from upper-mantle melting is added to the oceanic lithosphere; the two plates pull away from each other. They correspond to the midoceanic ridges that form a 80 000 km long submarine relief.
- ***Convergent margin:*** also called destructive margin. This is where the oceanic crust plunges under another plate and sinks into the mantle (subduction). Subduction can occur under either continental or oceanic lithosphere. Because of its low density and of its correlated high buoyancy, continental crust is almost never subducted deep into the mantle. Consequently when both plates contain a continent, the two continents collide and form a mountain chain as, for instance, the subduction of the Indian plate under the Eurasian plate led to the collision of continents and gave rise to the Himalaya Mountains.
- ***Transform margin:*** also called conservative margin, there, crust is neither produced nor destroyed as the plates slide horizontally past each other. The San Andreas Fault perfectly illustrates this type of contact between Pacific and North-American plates.

1.2.3 Divergent Margin

Midoceanic ridge

The rate of oceanic spreading ranges between 2 and 17 cm yr^{-1}.

Divergent margins are located over the ascending branch of a convection cell (Fig. 1.3). Due to the low thermal diffusion rates, solid upper mantle peridotite ascents adiabatically. Figure 1.4 shows that during adiabatic ascent a dry ($H_2O \sim 0.1\%$) peridotite will crosscut its dry solidus temperature (temperature where dry peridotite begins to melt) at low pressure (in the domain of stability of spinel or even of plagioclase) thus giving rise to a basaltic magma whose crystallization generates new oceanic crust.

As this magmatism is generally submarine, ocean water penetrates into the fractures of the newly formed oceanic crust.

During its descent into the crust water progressively warms up, thus creating a convective water circulation into the oceanic crust. This water will not only exchange energy with basalts but it will also chemically interact with it. This is well exemplified by the black or white smokers that correspond to the escape of warm ($> 300\,^\circ C$) hydrothermal water rich in sulfurs dissolved from the basalts. Another important effect of water is that a basalt initially composed of anhydrous minerals (olivine $(Fe,Mg)_2SiO_4$, orthopyroxene $(Fe,Mg)_2Si_2O_6$, clinopyroxene $Ca(Fe,Mg)Si_2O_6$ and plagioclase $CaAl_2Si_2O_8$) will be transformed into a hydrous mineral-bearing rock (talc $(Fe,Mg)_3Si_4O_{10}(OH)_2$;

Morphology	Ridge	Cliff	Trench
Surface variation	Increase: Accretion	No	Decrease: Subduction
Cinematic	Divergent	Transcurrent	Convergent

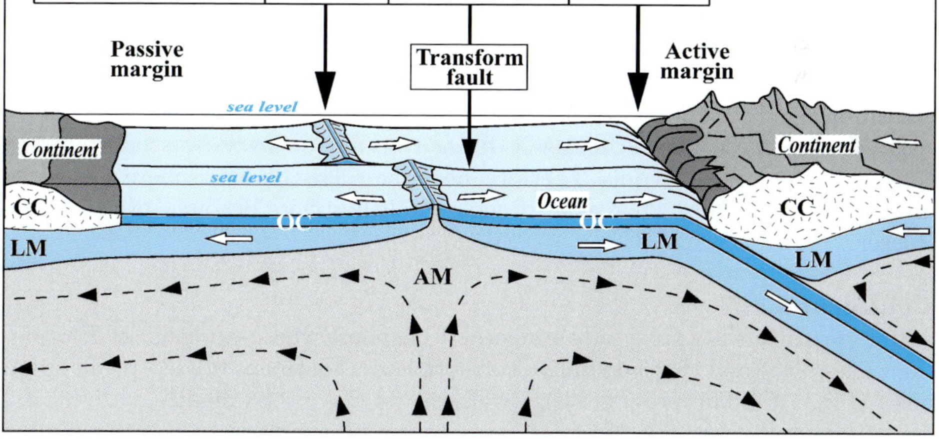

Fig. 1.3. Schematic block diagram showing different possible relationships between plates as well as the terminology generally used. CC = continental crust; OC = oceanic crust; LM = lithospheric mantle; AM = asthenospheric mantle. *Black arrows* and *dotted lines* show convective displacement into the upper mantle

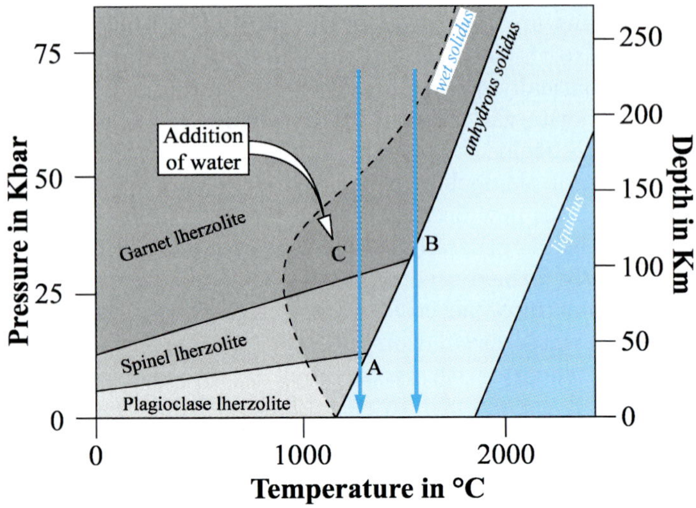

Fig. 1.4. Pressure vs. temperature diagram showing the domains of stability of garnet, spinel and plagioclase lherzolite of the lithospheric mantle. *Heavy lines* represent solidus and liquidus of dry mantle, *dotted line* is the solidus of hydrous mantle lherzolite. In ridge environments mantle melts by adiabatic decompression, the dry *solidus curve* is cross-cut at shallow depth (**A**). Melting in hot-spot environments is also due do adiabatic decompression but as mantle has a deeper origin it is hotter and melting takes place at greater depth (**B**). In subduction zones, mantle melts by addition of water due to dehydration of the subducted oceanic crust, water significantly lowers the solidus temperature (**C**)

serpentine $(Fe,Mg)_6Si_4O_{10}(OH)_8$, amphibole $Ca_2(Fe,Mg)_5Si_8O_{22}(OH)_2$, chlorite $Fe_3Mg_3AlSi_3O_{10}(OH)_8$). The oceanic crust that contained about 0.3% water when it emplaced will contain between 1.5 and 5% water after hydrothermal alteration.

In the course of time the newly formed oceanic crust moves far from the ridge and progressively cools. Therfore, near the ridge the lithosphere was warm and consequently thin whereas as it cools the lithosphere becomes progressively thicker.

Continental rift

A continental rift is a zone of continental lithosphere where extensional deformation takes place and that generally evolves towards continent break-up and ocean birth (Fig. 1.5). The process starts with the ascent of hot mantle material under continental crust. As this material is hot and buoyant, it causes the doming of the overlying plate. The warmed lithosphere transforms into asthenosphere. Doming of continental crust provokes its stretch and thinning due to its split into large blocks limited by faults, thus resulting in a collapsed valley called rift. As thinning proceeds, crust begins to be pulled apart, to extend and as in oceanic ridge, adiabatic decompression of mantle peridotite leads to its melting and to the genesis of basaltic magma. Progressive and successive injection of basalt in the central part of the rift contributes to pushing plates apart, the opened space being immediately filled with magma. This mechanism results in the genesis of oceanic crust, whose surface increases as continents separate (Fig. 1.5).

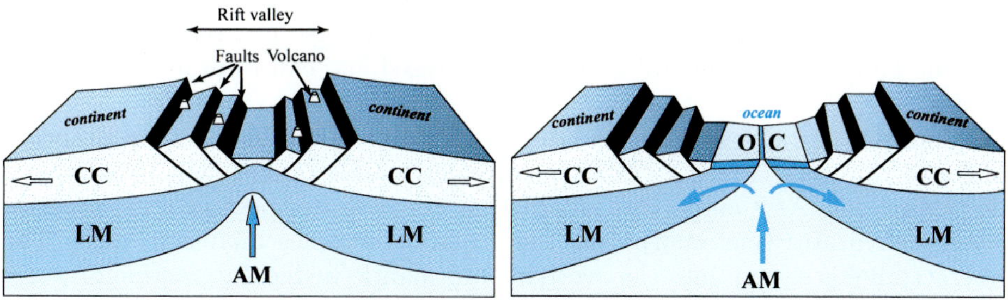

Fig. 1.5. Schematic diagrams showing how continental crust breaks (*left*) and how ocean birth occurs (*right*). CC = continental crust; OC = oceanic crust; LM = lithospheric mantle; AM = asthenospheric mantle

1.2.4 Convergent Margin

Subduction zone

To compensate for new material formed in ridges, old oceanic lithosphere has to be destroyed which is realised in subduction zones (Fig. 1.2). As it cools, the density of oceanic lithosphere increases and consequently, its buoyancy decreases. When its density becomes greater than that of the underlying asthenosphere, then oceanic lithosphere can spontaneously sink into the mantle. In this case as it is old, oceanic lithosphere is thick and subduction angle is high (Fig. 1.6).

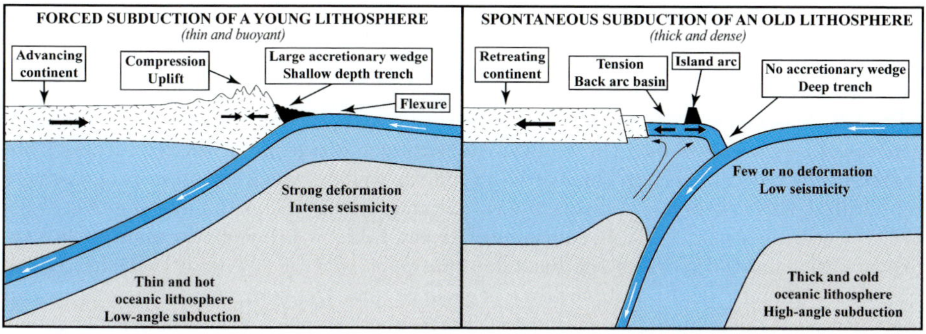

Fig. 1.6. Schematic cross sections showing the contrasted characteristics of both spontaneous and forced subduction

This geometry induces a deep trench and generally the associated seismicity is moderate (i.e. Marianna, Japan).

When strain constraints are too high young oceanic lithosphere can be forced to subduct, even if its density is lower than the asthenosphere one. In this situation (Fig. 1.5) the lithosphere is thin and angle of subduction is low (it can even be almost flat). As oceanic lithosphere resists subduction, seismic activity is generally very important. All sediments emplaced on the ocean floor do not enter into subduction and form an accretionary wedge in a relatively shallow trench. Thus, the continental crust is compressed and can form mountains (i.e. Andes).

Contrary to ridge systems, subduction corresponds to descending branches of convection cells (Fig. 1.2); consequently dry mantle cannot melt by adiabatic decompression. As it enters into subduction, the hydrated oceanic crust progressively dehydrates (metamorphism) such that it becomes unable to melt. The liberated fluids rise up into the overlying dry mantle wedge. Interaction of water with dry peridotite significantly lowers its solidus temperature (Fig. 1.6) such that it can melt. Fluids released by subducted oceanic crust not only lower mantle solidus temperature, but they also transport dissolved elements (U, K, Rb, etc.) that modify the mantle peridotite composition (metasomatism). The generated magmas are andesitic or granodioritic in composition, they form new (juvenile) continental crust. The descending residual oceanic crust accumulates at

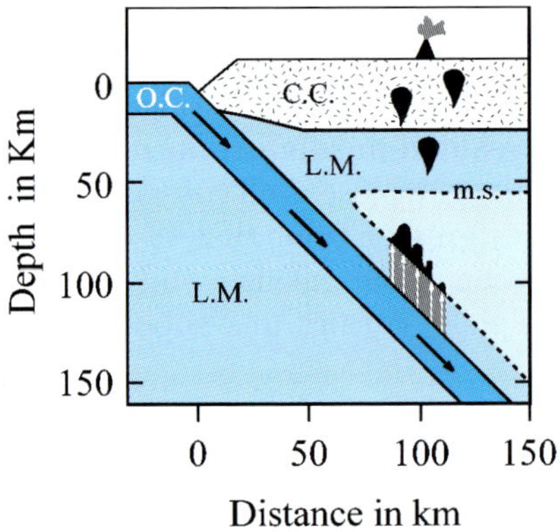

Fig. 1.7. Schematic cross section of a subduction zone showing the process of magma generation: as it enters into the mantle, the subducted oceanic crust dehydrates. Liberated fluids rise-up through the overlying mantle wedge, which becomes progressively rehydrated, this lowers its solidus temperature and induces its partial melting (Fig. 1.3). O.C. = oceanic crust; C.C. = continental crust; L.M. = lithospheric mantle; m.s. = lithospheric mantle hydrous solidus (the *pale blue part* of the mantle wedge corresponds to the part whose temperature is greater than the hydrous solidus one: it can melt if water is available); *black domains* correspond to place where magma is generated and emplaced; whereas *grey area* with *white arrows* represents place where fluids pass through

the 660-km seismic discontinuity. When stored oceanic crust exceeds a threshold it suddenly sinks into the mantle as a cold avalanche that reaches the mantle–core boundary (D″ layer). In some exceptional situations generally related to subduction of very young oceanic lithosphere, oceanic crust can melt thus generating magmas called adakites that also will contribute to form juvenile continental crust.

Collision zone

When the plate implied in subduction carries a continent, this latter, due to its high buoyancy and contrarily to oceanic lithosphere, does not significantly sink into the mantle; which results in continental collision (Fig. 1.8). The over-stacking of continental crust segments significantly thickens continental lithosphere and generates mountain chains. Part of the continental crust transported to depth, can melt and generates granitic magmas. As these latter are produced by melting of the continental crust they do not constitute new addition to this crust whose volume remains unchanged: this mechanism is called recycling.

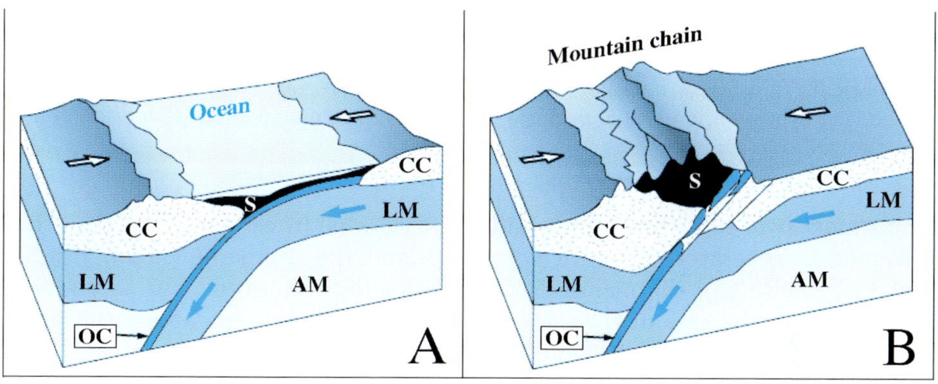

Fig. 1.8. Schematic cross sections showing how subduction and closure of an ocean (**A**) can evolve towards the collision of two continental crusts. This process, still active in the Alps and the Himalayas, generates mountain chains and thickens continental crust (**B**). O.C. = oceanic crust; S. = sediments; C.C. = continental crust; L.M. = lithospheric mantle; A.M. asthenospheric mantle

1.2.5 Hot Spots

In some places volcanism occurs in the middle of plates (Fig. 1.9) it is called hot-spot magmatism (Hawaii, La Réunion, etc.). These magmas are considered as being generated by the ascent of hot lower mantle (mesosphere) diapir, possibly coming from the core mantle boundary (D″ layer). The source of these diapirs appears as immobile on the scale of several tens of millions of years. Lithospheric

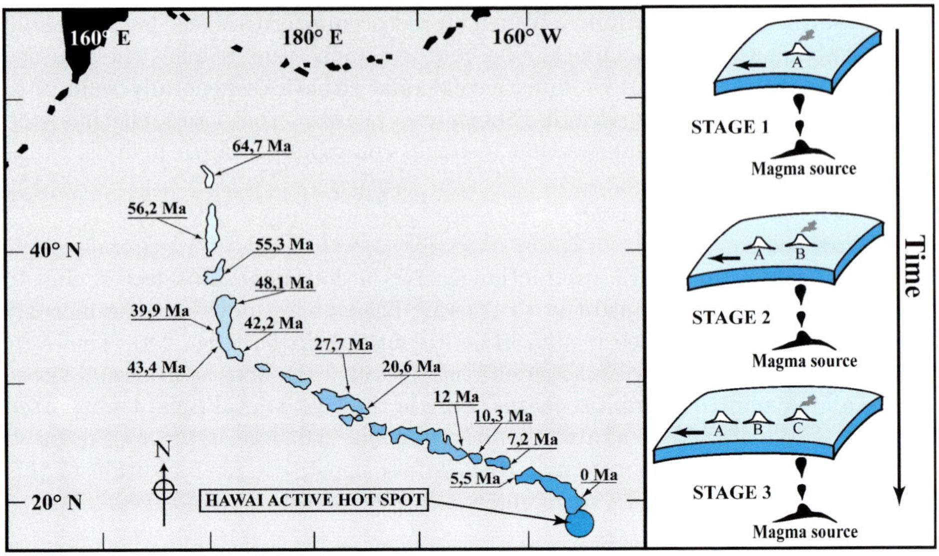

Fig. 1.9. Map of Hawaii-Emperor chain showing that the age of volcanoes regularly decreases towards Hawaii that remains the unique active edifice. *Inset*, on the *right*, schematically illustrates hot-spot island genesis. Magma source is immobile and deep seated, the lithospheric plate is perforated by ascending magma and its motion generates the volcanic chain

plates move over these fixed spots and are perforated by ascending magma, thus this motion generates chains of volcanic island (i.e. Emperor Chain, Hawaii; Fig. 1.9). Melting of the mantle is due to its adiabatic ascent, but as diapir is hotter than in the case of the midocean ridge, melting occurs at greater depth in asthenosphere between 100 and 200 km (Fig. 1.3).

Hot-spot activity can also generate large igneous provinces (LIP), which corresponds to the emplacement of huge amounts of basaltic magma either in continental (i.e. Deccan trapps) or oceanic environments (i.e. Ontong-Java, Kerguelen oceanic plateaus). Their formation requires that enormous volumes of hot mantle rise up until the base of the lithosphere. They are generally attributed to melting in the head of large starting mantle plumes (diapirs).

Recently, three great mesosphere domains (Hawaiian, Tristan and Icelandic) were recognized based on their hot-spot characteristics. They are called "mesoplates", their upper surface corresponds to the asthenosphere-mesosphere frontier and they are limited by the deep subduction zones that extend below 200 km.

1.2.6 Wilson Cycle

In 1968 Wilson developed the concept of cyclicity of oceanic spreading: Ocean begins its life by breaking a continent, it extends through a ridge system and disappears in subduction or collision; this evolution is known as the Wilson cycle. On the other hand, collision between two independent continents assembles them and leads to the formation of a single and greater new continent. Succession of subductions and continental collisions gives rise to supercontinents. The supercontinent cycle involves formation of a supercontinent from smaller continental blocks followed by fragmentation, new ocean opening and again assembly of a new supercontinent. Such a cycle lasts about 500 Ma. For instance, about 245 Ma ago, almost all present-day continents were joined and formed a supercontinent called Pangea. At 245 Ma the Atlantic Ocean began to open and India started to migrate from Madagascar towards Asia: a new supercycle started. Continent breaking continues today, for example in the Red Sea and in the East African rift.

The breakup of an existing supercontinent is due to hot-spot activity and also to the thermal shield effect due to continental crust that acts as a screen that makes internal heat release more difficult. In both cases, heat accumulates under continental crust and provokes its thickening that leads to its breakup.

1.2.7 Energy for Plate Tectonics

All the energy necessary to allow plate tectonics to work is the internal energy of the Earth. Today, this energy has been evaluated at 42 TW (42×10^{12} W). Its main source is the decay of long-lived radiogenic elements (^{235}U, ^{238}U, ^{232}Th and ^{40}K). They are mostly concentrated in the continental crust that produces about 20% of the heat of the Earth. They are less concentrated in the mantle, but due to its greater volume, it contributes about 55% of the heat budget. The core does not contain significant amounts of radiogenic elements. However, crystallization of the outer core produces latent heat ($\sim$ 10%). The last 15% of released heat corresponds to fossil accretion energy. When it formed, due to accretion energy, the Earth was hot, since 4.55 Ga it slowly cooled (200 to 250 °C since 4.0 Ga).

References

Condie, K.C., 1997. *Plate Tectonics and Crustal Evolution.* Pergamon, Oxford, 476 pp.
Fowler, C.M.R., 2004. *The Solid Earth: An Introduction to Global Geophysics.* Cambridge University Press, Cambridge, 500 pp.
Turcotte, D.L. and Schubert, G., 2002. 1.2 *Geodynamics.* Wiley, John and Sons, 472 pp.
Wilson, J.T. (ed.), 1976. *Continents Adrift and Continents Aground.* W.H. Freeman and Co., San Francisco, 230 pp.
Wilson, M., 1991. *Igneous Petrogenesis: a Global Tectonic Approach.* Harper Collins Academic, 466 pp.

2 Useful Astrobiological Data

2.1 Physical and Chemical Data

1. International System Units

	Name	Symbol	
Basic units			
Length	Metre	m	
Mass	Kilogram	kg	
Time	Second	s	
Electric current	Ampere	A	
Thermodynamic temperature	Kelvin	K	
Amount of substance	Mole	mol	
Luminous intensity	Candela	cd	
Additional units			
Plane angle	Radian	rad	
Solid angle	Steradian	sr	
Derived units			
Energy	Joule	J	$m^2\,kg\,s^{-2} \equiv N\,m$
Force	Newton	N	$m\,kg\,s^{-2} \equiv J\,m^{-1}$
Pressure	Pascal	Pa	$m^{-1}\,kg\,s^{-2} \equiv N\,m^{-2}$
Electrical charge	Coulomb	C	$s\,A$
Magnetic flux	Weber	Wb	$m^2\,kg\,s^{-2}\,A^{-1}$
Magnetic flux density	Tesla	T	$kg\,s^{-2}\,A^{-1} \equiv V\,s\,m^{-2}$
Magnetic field strength			$A\,m^{-1}$
Inductance	Henry	H	$m^2\,kg\,s^{-2}\,A^{-2}$
Power	Watt	W	$m^2\,kg\,s^{-2} \equiv J\,s^{-1}$
Electric potential difference	Volt	V	$m^2\,kg\,s^{-3}\,A \equiv J\,C^{-1}$
Frequency	Hertz	Hz	s^{-1}

2. Other Units

	Name	Symbol	
Length	Fermi	fm	10^{-15} m
	Angstrom	Å	10^{-10} m
	Micron	µm	10^{-6} m
Time	Minute	min, mn	60 s
	Hour	h	3600 s
	Day	d	86 400 s
	Year	a, y	3.156×10^{7} s
Volume	Litre	l	$10^{-3}\,\mathrm{m}^3 \equiv 1\,\mathrm{dm}^3$
Mass	Ton	t	10^3 kg
Temperature	Degree Celsius	°C	Temperature comparison:
			$0\,°\mathrm{C} = 273.15\,\mathrm{K} = 32\,°\mathrm{F}$
			$100\,°\mathrm{C} = 373.15\,\mathrm{K} = 212\,°\mathrm{F}$
			Temperature scale:
			$\mathrm{K} = °\mathrm{C} = 1.8\,°\mathrm{F}$
Force	Dyne (cgs system)	dyn	10^{-5} N
Energy	Erg (cgs system)	erg	10^{-7} J
	Calorie	cal	4.184 J
	Electron-volt	eV	1.6×10^{-19} J
	Kilowatt-hour		$3600 \times 10^3\,\mathrm{J} =$
			8.6042×10^5 cal
Pressure	Bar	bar	10^5 Pa
	Atmosphere	atm	1.01325×10^5 Pa
	mercury mm	torr	1.3332×10^2 Pa
Magnetic flux density	Gauss	gauss	10^{-4} T
Concentration	Molarity	M	$\mathrm{mol\,l^{-1}}$
Molar mass			g/mol
Angle	Degree	°	$1° = \pi/180$
			$= 1.74 \times 10^{-2}$ rad
	Minute	′	$1' = \pi/10\,800$
			$= 2.91 \times 10^{-4}$ rad
	Second	″	$1'' = \pi/648\,000$
			$= 4.85 \times 10^{-6}$ rad
			1 Radian $= 57.296\,°$
			$= 3.44 \times 10^{3}\,'$
			$= 2.06 \times 10^{5}\,''$

3. International System Prefixes

10^{-1}	deci	d	10^{1}	deca	da
10^{-2}	centi	c	10^{2}	hecto	h
10^{-3}	milli	m	10^{3}	kilo	k
10^{-6}	micro	μ	10^{6}	mega	M
10^{-9}	nano	n	10^{9}	giga	G
10^{-12}	pico	p	10^{12}	tera	T
10^{-15}	femto	f	10^{15}	peta	P
10^{-18}	atto	a	10^{18}	exa	E
10^{-21}	zepto	z	10^{21}	zetta	Z
10^{-24}	yocto	y	10^{24}	yotta	Y

4. Fundamental Physical Constants

Newtonian constant of gravitation	G	$6.67259(85) \times 10^{-11}\,\mathrm{N\,m^2\,kg^{-2}}$
Acceleration of free fall on Earth	g	$9.80665\,\mathrm{ms^{-1}}$
Speed of light in vacuum	c	$2.99792458 \times 10^{8}\,\mathrm{m\,s^{-1}}$
Planck constant	h	$6.6260755(40) \times 10^{-34}\,\mathrm{J\,s}$
Planck mass	$(hc/2\pi G)^{1/2}$	$2.17671(14) \times 10^{-8}\,\mathrm{kg}$
Planck length	$(hG/2\pi c^3)^{1/2}$	$1.61605(10) \times 10^{-35}\,\mathrm{m}$
Planck time	$(hG/2\pi c^5)^{1/2}$	$5.39056(34) \times 10^{-44}\,\mathrm{s}$
Boltzmann constant	k	$1.380658(12) \times 10^{-23}\,\mathrm{J\,K^{-1}}$
Black body constant	a	$7.564 \times 10^{-16}\,\mathrm{J\,m^{-3}\,K^{-4}}$
Stefan Boltzmann constant	σ	$5.67051(19) \times 10^{-8}\,\mathrm{W\,m^{-2}\,K^{-4}}$
Perfect gas constant	R	$8.314510(70)\,\mathrm{J\,K^{-1}mol^{-1}}$
Elementary charge	e	$1.602176462(63) \times 10^{-19}\,\mathrm{C}$
Electron rest mass	m_{e}	$9.1093897(54) \times 10^{-31}\,\mathrm{kg}$
Proton rest mass	m_{p}	$1.6726231(10) \times 10^{-27}\,\mathrm{kg}$
Neutron rest mass	m_{n}	$1.6749286(10) \times 10^{-27}\,\mathrm{kg}$
Proton/electron mass ratio		$1836.152701(37)$
Hydrogen atomic mass		$1.6735344 \times 10^{-27}\,\mathrm{kg}$
Electron rest mass energy equivalent (eV)	$m_{\mathrm{e}}c^2$	$0.511 \times 10^{6}\,\mathrm{eV}$

(continued)

Atomic mass unit	u	$1.6605402(10) \times 10^{-27}\,\mathrm{kg} \equiv$ 1/12 of ^{12}C mass
Avogadro constant	N	$6.0221367(36) \times 10^{23}\,\mathrm{mol}^{-1}$
Vacuum permittivity constant	$1/(\mu_0 c^2)$	$8.854187817\times$ $10^{-12}\mathrm{m}^{-3}\,\mathrm{kg}^{-1}\,\mathrm{s}^4\,\mathrm{A}^2$
Vacuum permeability constant	μ_0	$4\pi 10^{-7}\,\mathrm{m\,kg\,s}^{-2}\,\mathrm{A}^{-2}$
Ice melting temperature ($P = 1\,\mathrm{atm}$)	$0\,^{\circ}\mathrm{C}$	$273.150\,\mathrm{K}$
Water triple point (H_2O)		$T = 273.160\,\mathrm{K}$, $P = 611.73\,\mathrm{Pa}$ (6.11 mbar)
Bohr radius	a_0	$0.529177249(24) \times 10^{-10}\,\mathrm{m}$
Electron volt-energy relationship		$1\,\mathrm{eV} = 1.60217733(49) \times 10^{-19}\,\mathrm{J}$
Electron volt-wavelength relationship		$1\,\mathrm{eV} = 12\,398.428 \times 10^{-10}\,\mathrm{m}$
Electron volt-frequency relationship		$1\,\mathrm{eV} = 2.41798836(72) \times 10^{14}\,\mathrm{s}^{-1}$
Electron volt-temperature relationship		$1\,\mathrm{eV} = 11\,604.45\,\mathrm{K}$

5. Other Physical, Chemical and Astronomical Symbols and Abbreviations

c	Mass concentration
	Light speed
E_a	Activation energy
k	Reaction rate
R	"Rectus" configuration (chirality)
S	"Sinister" configuration (chirality)
L,D	Old nomenclature still used to define the absolute configuration of oses and amino acids
δ	Chemical shift (RMN)
	Fractional charge
λ	Wavelength
μ	Dipolar moment
	Micron
	Reduced mass
ν	Frequency
[ ]	Concentration ($\mathrm{mol\,l^{-1}}$)
$\leftrightarrows$	Chemical equilibrium
$\leftrightarrow$	Mesomery
UV	Ultraviolet
IR	Infrared
NMR	Nuclear Magnetic Resonance
a	Ellipse semi great axis
e	Ellipse eccentricity
i	Inclination
PSR	Pulsar
QSO	Quasar
γ	Photon
v	Neutrino
☉	Sun
☿	Mercury
♀	Venus
⊕	Earth
♂	Mars
♃	Jupiter
♄	Saturn
♅	Uranus
♆	Neptune
♇	Pluto

References: *Allen's: Astrophysical quantities*, Arthur N. Cox Editor, (2000); *Introduction aux éphémérides astronomiques*, EDP Sciences, (1998); *Handbook of Chemistry and Physics*, D.R. Lide, editor-in-Chief 82nd Edition (2001–2002) CRC Press.

6. Periodic Table of Elements

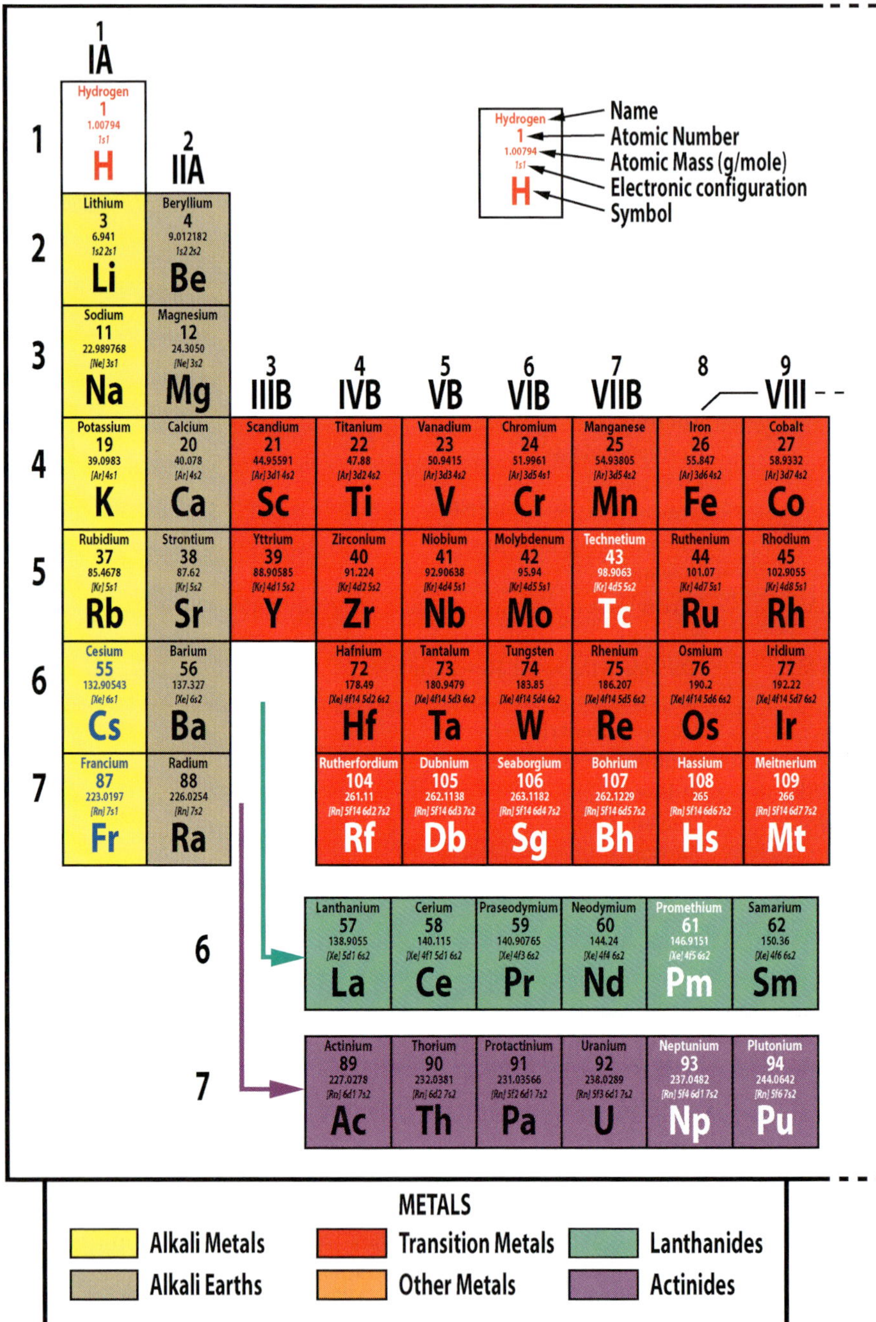

10	11 IB	12 IIB	13 IIIA	14 IVA	15 VA	16 VIA	17 VIIA	18 VIIIA
								Helium 2 4.002602 1s2 He
			Boron 5 10.811 1s2 2s2 2p1 B	Carbon 6 12.011 1s2 2s2 2p2 C	Nitrogen 7 14.00674 1s2 2s2 2p3 N	Oxygen 8 15.9994 1s2 2s2 2p4 O	Fluorine 9 18.9984032 1s2 2s2 2p5 F	Neon 10 20.1797 1s2 2s2 2p6 Ne
			Aluminium 13 26.981539 [Ne] 3s2 3p1 Al	Silicium 14 28.0855 [Ne] 3s2 3p2 Si	Phosphorus 15 30.973762 [Ne] 3s2 3p3 P	Sulfur 16 32.066 [Ne] 3s2 3p4 S	Chlorine 17 35.4527 [Ne] 3s2 3p5 Cl	Argon 18 39.948 [Ne] 3s2 3p6 Ar
Nickel 28 58.6934 [Ar] 3d8 4s2 Ni	Copper 29 63.546 [Ar] 3d10 4s1 Cu	Zinc 30 65.39 [Ar] 3d10 4s2 Zn	Gallium 31 69.723 [Ar] 3d10 4s2 4p1 Ga	Germanium 32 72.61 [Ar] 3d10 4s2 4p2 Ge	Arsenic 33 74.92159 [Ar] 3d10 4s2 4p3 As	Selenium 34 78.96 [Ar] 3d10 4s2 4p4 Se	Bromine 35 79.904 [Ar] 3d10 4s2 4p5 Br	Krypton 36 83.80 [Ar] 3d10 4s2 4p6 Kr
Palladium 46 106.42 [Kr] 4d10 Pd	Silver 47 107.8682 [Kr] 4d10 5s1 Ag	Cadmium 48 112.411 [Kr] 4d10 5s2 Cd	Indium 49 114.82 [Kr] 4d10 5s2 5p1 In	Tin 50 118.71 [Kr] 4d10 5s2 5p2 Sn	Antimony 51 121.757 [Kr] 4d10 5s2 5p3 Sb	Tellurium 52 127.60 [Kr] 4d10 5s2 5p4 Te	Iodine 53 126.90447 [Kr] 4d10 5s2 5p5 I	Xenon 54 131.29 [Kr] 4d10 5s2 5p6 Xe
Platinum 78 195.08 [Xe] 4f14 5d9 6s1 Pt	Gold 79 196.96654 [Xe] 4f14 5d10 6s1 Au	Mercury 80 200.59 [Xe] 4f14 5d10 6s2 Hg	Thallium 81 204.3833 [Xe] 4f14 5d10 6s2 6p1 Tl	Lead 82 207.2 [Xe] 4f14 5d10 6s2 6p2 Pb	Bismuth 83 208.98037 [Xe] 4f14 5d10 6s2 6p3 Bi	Polonium 84 208.9824 [Xe] 4f14 5d10 6s2 6p4 Po	Astatine 85 209.9871 [Xe] 4f14 5d10 6s2 6p5 At	Radon 86 222.0176 [Xe] 4f14 5d10 6s2 6p6 Rn

Europium 63 151.965 [Xe] 4f7 6s2 Eu	Gadolinium 64 157.25 [Xe] 4f7 5d1 6s2 Gd	Terbium 65 158.92534 [Xe] 4f9 6s2 Tb	Dysprosium 66 162.50 [Xe] 4f10 6s2 Dy	Holmium 67 164.93032 [Xe] 4f11 6s2 Ho	Erbium 68 167.26 [Xe] 4f12 6s2 Er	Thulium 69 168.93421 [Xe] 4f13 6s2 Tm	Ytterbium 70 173.04 [Xe] 4f14 6s2 Yb	Lutetium 71 174.967 [Xe] 4f14 5d1 6s2 Lu
Americium 95 243.0614 [Rn] 5f7 7s2 Am	Curium 96 247.0703 [Rn] 5f7 6d1 7s2 Cm	Berkelium 97 247.0703 [Rn] 5f9 7s2 Bk	Californium 98 251.0796 [Rn] 5f10 7s2 Cf	Einsteinium 99 252.0829 [Rn] 5f11 7s2 Es	Fermium 100 257.0951 [Rn] 5f12 7s2 Fm	Mendelevium 101 258.0986 [Rn] 5f13 7s2 Md	Nobelium 102 259.1009 [Rn] 5f14 7s2 No	Lawrencium 103 260.1053 [Rn] 5f14 6d1 7s2 Lr

NON-METALS

- Hydrogen
- Halogens
- Other Non-Metals
- Noble Gases

STANDARD CONDITIONS (P = 100 KPa; T = 298 K)

Gaz | Liquid | Solid | Synthetic

2.2 Astrophysical Data

1. Units and General Data

Distance units	
Astronomical unit (AU)	$1.4959787066 \times 10^{11}$ m
1 parsec (pc)	206 264.8 au
	3.0856776×10^{16} m
	3.2615638 light-years
1 light-year	9.460730×10^{15} m
	6.324×10^{4} AU
Time units	
1 sideral day	23 h 56 mn 04.098 s (average time)
1 tropic year	365.2422 average solar days
1 sideral year	365.2564 average solar days
Known universe	
Number of nucleons	$\sim 10^{80}$
Universe radius	$\sim 10^{26}$ m
Number of galaxies	$\sim 10^{11}$
Nebula recession speed	$\sim 3 \times 10^{-18}\,\mathrm{s}^{-1} (\sim 100\,(\mathrm{km/s})/\mathrm{Mpc})$
Our galaxy	
Number of stars	$\sim 1.6 \times 10^{11}$
Diameter	$\sim 10^{21}$ m
Mass	$\sim 8 \times 10^{41}$ kg
Sun	
Radius	6.95508×10^{8} m
Mass	1.9891×10^{30} kg
Sun luminosity	$3.845(8) \times 10^{26}$ W
Core temperature	1.557×10^{7} K
Core pressure	2.334×10^{16} Pa

(continued)

Photosphere temperature	5780 K
Corona temperature	2 to 3×10^6 K
Average mass density	$1.41 \times 10^3\,kg\,m^{-3}$
Earth/Moon	
Earth's mass	5.9742×10^{24} kg
Moon's mass	7.34×10^{22} kg
Earth's average radius	6371.23×10^3 m
Earth's equatorial radius	6378.136×10^3 m
Earth–Moon distance (semi-major axis)	3.844×10^8 m
Earth–Sun distance (semi-major axis)	1.496×10^{11} m

2. Compared Planetology

Object		Semi major axis of the orbit[1]		Revolution period (d: day, yr: year, terrestrial)	Eccentricity (e)
		AU	Million km		
Mercury	☿	0.3871	57.9092	87.969 d	0.2056
Venus	♀	0.7233	108.2089	224.701 d	0.0068
Earth	⊕	1.0000	149.65979	365.2564 d	0.0167
Moon		2.570×10^{-3}	0.3844	27.3217 d	0.0549
Mars	♂	1.5234	227.9364	686.98 d	0.0934
Jupiter	♃	5.2034	778.4120	11.862 yr	0.0484
Europe		4.4851×10^{-3}	0.671	3.551181 d	0.009
Saturn	♄	9.5371	1426.7254	29.457 yr	0.0542
Titan		8.167×10^{-3}	1.2218	15.945 d	0.0291
Uranus	♅	19.1913	2870.9722	84.019 yr	0.0472
Neptune	♆	30.0690	4498.2529	164.767 yr	0.0086
Pluto	♇	39.4817	5906.3762	247.688 yr	0.2488

[1] around the Sun or, in the case of satellites, around their planet

Object		Equatorial radius km	Mass ⊕ = 1	Average density $g\,cm^{-3}$	Surface gravity[1] $m\,s^{-2}$	Surface escape velocity[1] $km\,s^{-1}$
Sun	☉	695 508	332 946.0	1.41	274.0	617.5
Mercury	☿	2439.7	0.0553	5.43	3.72	4.25
Venus	♀	6051.8	0.8150	5.24	8.87	10.36
Earth	⊕	6378.136	1.000	5.515	9.81	11.18
Moon		1737.4	0.012300	3.34	1.62	2.38
Mars	♂	3397	0.1074	3.94	3.71	5.02
Jupiter	♃	71 492	317.82	1.33	23.12	59.54
Europe		1565	0.008026	3.04	1.8	2.0
Saturn	♄	60 268	95.161	0.70	8.96	35.49
Titan		2575	0.02226	1.90	0.14	2.7
Uranus	♅	25 559	14.371	1.30	8.69	21.29
Neptune	♆	24 764	17.147	1.76	11.0	23.71
Pluto	♇	1737.4	0.0022	1.1	0.81	1.27

[1] or at $P = 1$ bar for gaseous planets

Object		Mean obliquity (degree)	Visual geometric Albedo	Sideral rotation peroid (d:day, h; hour)
Sun	☉		–	25.7 d
Mercury	☿	0.0	0.11	58.6462 d
Venus	♀	177.3	0.65	243.0185 d [1]
Earth	⊕	23.45	0.37	23.9345 h
Moon		6.683	0.12	27.3217 d
Mars	♂	25.19	0.15	24.6230 h
Jupiter	♃	3.12	0.52	9.9249 h
Europe		–	0.67	2.5512 d
Saturn	♄	26.73	0.47	10.6562 h
Titan		–	0.22	15.9454 h
Uranus	♅	97.86	0.51	17.2400 h [1]
Neptune	♆	29.58	0.51	16.1100 h
Pluto	♇	119.61	0.3	153.29 h [1]

[1] retrograde rotation

3. Electromagnetic Spectrum

Wavelength	Frequency	Wave number	Equivalent energy	Equivalent temperature	Spectral Region		
1000 km	300 Hz				—		
					ULF		
100 km	3 kHz				—		
1 km	300 kHz						
10 m	30 MHz				—		Radio
					VHF		
1 m	300 MHz				—	— —	
					UHF		
10 cm	3 GHz				— —	centimetric	—
1 cm	30 GHz	$1\,\mathrm{cm}^{-1}$			—	— —	Microwave
						millimetric	
1 mm	300 GHz	$10\,\mathrm{cm}^{-1}$	1.2 meV	3 K	—	— —	— —
						submillimetric	
100 µm	3 THz	$100\,\mathrm{cm}^{-1}$	0.012 eV	30 K	—	— —	
					FIR		
40 µm					— —		
10 µm	30 THz	$1000\,\mathrm{cm}^{-1}$	0.12 eV	300 K	MIR	Infrared	
5 µm					— —		

(continued)

Wavelength	Frequency	Wave number	Equivalent energy	Equivalent temperature	Spectral Region		
1 µm	3×10^{14} Hz	10 000 cm^{-1}	1.24 eV	3000 K	NIR		
0.7 µm					——	——	
						Visible	
0.4 µm					——	——	
0.1 µm	3×10^{15} Hz		12.4 eV	3×10^{4} K		Ultraviolet	
10 nm	3×10^{16} Hz		124 eV	3×10^{5} K	——	——	——
1 nm	3×10^{17} Hz		1.24 keV	3×10^{6} K			
1 Å	3×10^{18} Hz		12.4 keV	3×10^{7} K		X-rays	High
10^{-11} m	3×10^{19} Hz		124 keV		——	——	Energies
10^{-12} m	3×10^{20} Hz		1.24 MeV			Gamma-rays	

Remarks:

- Wave numbers, energies and temperatures can be calculated for all existing wavelengths (here, they have been intentionally limited to the domains where they are generally used).
- Spectrum domain limits are no more than indicative and correspond to astronomical use.
- ULF: Ultra Low Frequency, VHF: Very High Frequency, UHF: Ultra High Frequency, FIR: Far Infrared, MIR: Mid Infrared, NIR: Near Infrared, UV: Ultraviolet
- Formula: $E = h\nu = hc/\lambda = \text{eV} = \text{kT}$

2.3 Geological Data

1. General Information

Whole Earth	
Earth's Age	4.55 Ga
Age of the oldest known terrestrial material (zircon from Jack Hills; Australia)	4.404 Ga
Age of the oldest known terrestrial rock (Acasta gneiss; Canada)	4.03 Ga
Equatorial radius	6378.136 km
Polar radius	6356.753 km
Ellipticity	0.00335281
Total surface	510×10^6 km^2
Ocean surface	357×10^6 km^2
Continent surface	153×10^6 km^2
Mass	5.9742×10^{24} kg
Mean density	5.515
Deeper oceanic trench (Marianas)	11 034 m
Higher mountain (Everest)	8863 m
Present day oceanic ridge length	~80 000 km
Solid inner core	
Depth	5155 km
Thickness	~1223 km
Average density	12
Volume	8.45×10^{10} km^3
Mass	1.18×10^{23} kg
Liquid outer core	
Depth	2891 km
Thickness	~2264 km
Average density	10
Volume	1.65×10^{11} km^3
Mass	1.65×10^{24} kg
Lower mantle (Mesosphere)	
Depth	660 km
Thickness	~2231 km
Average density	5
Volume	5.87×10^{11} km^3
Mass	2.93×10^{24} kg

(continued)

Upper mantle	
Depth	30 km
Thickness	~630 km
Average density	3.2
Volume	3.08×10^{11} km^3
Mass	0.986×10^{24} kg
Oceanic crust	
Average thickness	5 km
Average density	3.1
Age of the oldest known oceanic crust	0.165 Ga
Volume	2.0×10^{9} km^3
Mass	6.2×10^{21} kg
Continental crust	
Average thickness	30 km
Average density	2.75
Age of the oldest known continental crust	4.01 Ga
Volume	7×10^{9} km^3
Mass	1.9×10^{22} kg
Average surface magnetic flux density	5×10^{-5} T
Heat flux	42×10^{12} W
Range of lithospheric plate motion rate	1 to 10 cm yr^{-1}

2. Decay Constants (λ)

Parent nuclide	Daughter nuclide	Decay constant
^{228}Th	^{224}Ra	3.63×10^{-1} yr^{-1}
^{210}Pb	^{210}Bi	3.11×10^{-2} yr^{-1}
^{32}Si	^{32}P	2.1×10^{-3} yr^{-1}
^{226}Ra	^{222}Rn	4.33×10^{-4} yr^{-1}
^{14}C	^{14}N	1.245×10^{-4} yr^{-1}

(continued)

Parent nuclide	Daughter nuclide	Decay constant
^{231}Pa	^{227}Ac	$2.11 \times 10^{-5}\,yr^{-1}$
^{230}Th	^{226}Ra	$9.21 \times 10^{-6}\,yr^{-1}$
^{59}Ni	^{59}Co	$9.12 \times 10^{-6}\,yr^{-1}$
^{41}Ca	^{41}K	$6.93 \times 10^{-6}\,yr^{-1}$
^{81}Kr	^{81}Br	$3.03 \times 10^{-6}\,yr^{-1}$
^{234}U	^{230}Th	$2.83 \times 10^{-6}\,yr^{-1}$
^{36}Cl	^{36}Ar	$2.30 \times 10^{-6}\,yr^{-1}$
^{26}Al	^{26}Mg	$9.80 \times 10^{-7}\,yr^{-1}$
^{107}Pd	^{107}Ag	$6.5 \times 10^{-7}\,yr^{-1}$
^{60}Fe	^{60}Ni	$4.62 \times 10^{-7}\,yr^{-1}$
^{10}Be	^{10}B	$4.59 \times 10^{-7}\,yr^{-1}$
^{182}Hf	^{182}W	$7.7 \times 10^{-8}\,yr^{-1}$
^{129}I	^{129}Xe	$4.3 \times 10^{-8}\,yr^{-1}$
^{92}Nb	^{92}Zr	$1.93 \times 10^{-8}\,yr^{-1}$
^{53}Mn	^{53}Cr	$1.87 \times 10^{-8}\,yr^{-1}$
^{244}Pu	$^{131-136}Xe$	$8.66 \times 10^{-9}\,yr^{-1}$
^{235}U	^{207}Pb	$9.849 \times 10^{-10}\,yr^{-1}$
^{146}Sm	^{142}Nd	$6.73 \times 10^{-10}\,yr^{-1}$
^{40}K	^{40}Ar	$5.50 \times 10^{-10}\,yr^{-1}$
^{40}K	^{40}Ca	$4.96 \times 10^{-10}\,yr^{-1}$
^{238}U	^{206}Pb	$1.551 \times 10^{-10}\,yr^{-1}$
^{130}Te	^{130}Xe	$8.66 \times 10^{-22}\,yr^{-1}$
^{87}Rb	^{87}Sr	$1.42 \times 10^{-11}\,yr^{-1}$
^{187}Re	^{187}Os	$1.64 \times 10^{-11}\,yr^{-1}$
^{176}Lu	^{176}Hf	$1.93 \times 10^{-11}\,yr^{-1}$
^{232}Th	^{208}Pb	$4.95 \times 10^{-11}\,yr^{-1}$
^{40}K	^{40}Ar	$5.81 \times 10^{-11}\,yr^{-1}$
^{147}Sm	^{143}Nd	$6.54 \times 10^{-12}\,yr^{-1}$
^{138}La	^{138}Ce	$2.24 \times 10^{-12}\,yr^{-1}$
^{130}Te	^{130}Xe	$8.66 \times 10^{-22}\,yr^{-1}$

3. Temperature Range of Magma Emplacement

Granitic magma	700–800 °C
Basaltic magma	1250–1350 °C
Komatiitic magma	1650 °C

4. Average Compositions of Earth Shells

	N = Atomic number	M = Atomic mass (g/mol)	Unit	C1 Chondrite	Bulk Solid Earth
H	1	1.00794	%	2.02	0.0257
He	2	4.002602		ppm	56 *(nL/g)*
Li	3	6.941	ppm	1.49	1.08
Be	4	9.012182	ppm	0.0249	0.046
B	5	10.811	ppm	0.87	0.203
C	6	12.011	%	3.5	0.73
N	7	14.00674	ppm	3180	55
O	8	15.9994	%	46.4	30.5
F	9	18.998403	ppm	58.2	17.1
Ne	10	20.1797	ppm	203 *(pL/g)*	
Na	11	22.989768	%	0.51	0.18
Mg	12	24.3050	%	9.65	15.39
Al	13	26.981539	%	0.86	1.59
Si	14	28.0855	%	10.65	16.5
P	15	30.973762	ppm	1080	1101
S	16	32.066	%	5.4	0.634
Cl	17	35.4527	ppm	680	74.5
Ar	18	39.948	ppm	751 *(pL/g)*	
K	19	39.0983	%	0.0550	0.0164
Ca	20	40.078	%	0.925	1.71
Sc	21	44.95591	ppm	5.92	10.93
Ti	22	47.88	%	0.044	0.0813
V	23	50.9415	ppm	56	94.4
Cr	24	51.9961	ppm	2650	3720
Mn	25	54.93805	ppm	1920	1974
Fe	26	55.847	%	18.1	30.3
Co	27	58.9332	ppm	500	838
Ni	28	58.6934	ppm	10500	17 690
Cu	29	63.546	ppm	120	60.9
Zn	30	65.39	ppm	310	37.1
Ga	31	69.723	ppm	9.2	2.74

Table showing the compositions of the different Earth shells. Data are from Anders and Grevesse (1989) Taylor and Mc Lennan, Mc Donough et al. (1992), Mc Donough and Sun (1995), Rudnick and Fountain (1995), McDonough (1998), Rudnick and Gao (2003). Several other data are coming from the Geochemical Earth Reference Model (GERM) Internet site 'http://www.earthref.org).

(continued)

Core	Primitive Mantle	Upper Mantle	Oceanic Crust	Bulk Continental Crust	Ocean	Atmosphere
0.06	0.01	0.00025		0.14	10.8	0.000053
					0.0000069	5.2
0	1.6	1.57	10	16	0.18	
0	0.068	0.0442	0.5	1.9	0.000225	
0	0.3	0.07	4	11	4.39	
0.2	0.0120	0.0018		0.02	0.0028	0.0337
170	2	0.05		56	8.5	780900
0	44	44		45	85.7	20.95
0	25	9.8		553	1.3	
				0.00007	0.04	18
0	0.267	0.214	2.08	2.36	1.05	
0	22.8	22.8	4.64	3.20	0.135	
0	2.35	2.23	8.47	8.41	0.00000016	
6.4	21	21	23.1	26.77	0.0003	
3200	90	54	873	2000	0.07	
1.9	0.025	0.024		0.0404	0.0885	0.0001
200	17	2.55		244	19000	0.001
				1.2	0.6	9300
0	0.024	0.0024	0.125	0.99	0.038	
0	2.53	2.4	8.08	5.29	0.04	
0	16.2	15.4	38	21.9	0.0000006	
0	0.12	0.0928	0.9	0.54	0.0000001	
120	82	82	250	138	0.0019	
9000	2625	2625	270	135	0.0003	
4000	1045	1045	1000	852	0.0002	
85	6.26	6.26	8.16	5.6	0.00000056	
2500	105	105	47	26.6	0.00039	
52 000	1960	1960	135	59	0.0054	
125	30	29.1	86	27	0.0002	
0	55	55	85	72	0.0037	
0	4	3.8	17	16	0.00003	

(continued)

	N = Atomic number	M = Atomic mass (g/mol)	Unit	C1 Chondrite	Bulk Solid Earth
Ge	32	72.61	ppm	31	7.24
As	33	74.92159	ppm	1.85	1.67
Se	34	78.96	ppb	21 000	2560
Br	35	79.904	ppb	3570	300
Kr	36	83.80	ppb	8.7 *(pL/g)*	
Rb	37	85.4678	ppm	2.3	0.405
Sr	38	87.62	ppm	7.25	13.4
Y	39	88.90585	ppm	1.57	2.95
Zr	40	91.224	ppm	3.82	7.1
Nb	41	92.90638	ppm	0.240	0.45
Mo	42	95.94	ppm	0.9	1.66
Tc	43	98.9063	ppm		
Ru	44	101.07	ppb	710	1310
Rh	45	102.9055	ppb	130	242
Pd	46	106.42	ppb	550	1010
Ag	47	107.8682	ppb	200	54.4
Cd	48	112.411	ppb	710	76
In	49	114.82	ppb	80	7.5
Sn	50	118.71	ppb	1650	251
Sb	51	121.757	ppb	140	46.2
Te	52	127.60	ppb	2330	285.7
I	53	126.90447	ppb	450	50
Xe	54	131.29	ppb	8.6 *(pL/g)*	
Cs	55	132.90543	ppb	190	35
Ba	56	137.327	ppm	2.41	4.46
La	57	138.9055	ppm	0.237	0.437
Ce	58	140.115	ppm	0.613	1.148
Pr	59	140.90765	ppm	0.0928	0.171
Nd	60	144.24	ppm	0.457	0.844
Pm	61	146.9151	ppm		
Sm	62	150.36	ppm	0.148	0.274
Eu	63	151.965	ppm	0.0563	0.104
Gd	64	157.25	ppm	0.199	0.367
Tb	65	158.92534	ppm	0.0361	0.068
Dy	66	162.50	ppm	0.246	0.455
Ho	67	164.93032	ppm	0.546	0.102

(continued)

Core	Primitive Mantle	Upper Mantle	Oceanic Crust	Bulk Continental Crust	Ocean	Atmosphere
20	1.1	1.1	1.5	1.3	0.000051	
5	0.05	0.01	1	2.5	0.0026	
8000	75	71	160	130	0.00012	
700	50	5		880	65 000	
					0.29	1140
0	0.6	0.041	2.2	49	0.124	
0	19.9	12.93	130	320	8.1	
0	4.3	3.65	32	19	0.000013	
0	10.5	6.19	80	132	0.000026	
0	0.658	0.11	2.2	8	0.000015	
5	0.05	0.03	1	0.8	0.01	
4000	4.9	4.9	1	0.57	0.0007	
740	0.91	0.91	0.2	0.06		
3100	3.9	3.9	0.2	1.5		
150	8	7.8	26	56	0.28	
150	40	39.2	130	80	0.00011	
0	11	8.5	72	52	0.000115	
500	130	97.5	1400	1700	0.81	
130	5.5	1.1	17	200	0.33	
850	12	12	3	5		
130	10	1		710	64	
					0.066	86
65	21	504	30	2600	0.5	
0	6.6	0.45	25	456	0.021	
0	0.648	0.08	3.7	20	0.0000029	
0	1.675	0.538	11.5	43	0.0000028	
0	0.254	0.114	1.8	4.9	0.00000064	
0	1.25	0.738	10	20	0.0000028	
					–	
0	0.406	0.305	3.3	3.9	0.00000045	
0	0.154	0.119	1.3	1.1	0.0000013	
0	0.544	0.430	4.6	3.7	0.0000007	
0	0.099	0.080	0.87	0.6	0.00000014	
0	0.674	0.559	5.7	3.6	0.00000091	
0	0.149	0.127	1.3	0.77	0.00000022	

(continued)

	N = Atomic number	M = Atomic mass (g/mol)	Unit	C1 Chondrite	Bulk Solid Earth
Er	68	167.26	ppm	0.16	0.296
Tm	69	168.93421	ppm	0.0247	0.046
Yb	70	173.04	ppm	0.161	0.297
Lu	71	174.967	ppm	0.0246	0.046
Hf	72	178.49	ppm	0.103	0.191
Ta	73	180.9479	ppm	0.0136	0.025
W	74	183.85	ppm	0.093	0.170
Re	75	186.207	ppb	40	75.3
Os	76	190.2	ppb	490	900
Ir	77	192.22	ppb	455	835
Pt	78	195.08	ppb	1010	1866
Au	79	196.96654	ppb	140	164
Hg	80	200.59	ppb	300	23.1
Tl	81	204.3833	ppb	140	12.2
Pb	82	207.2	ppb	2470	232
Bi	83	208.98037	ppb	110	9.85
Po	84	208.9824			
At	85	209.9871			
Rn	86	222.0176			
Fr	87	223.0197			
Ra	88	226.0254			
Ac	89	227.0278			
Th	90	232.0381	ppm	0.029	0.054
Pa	91	231.03588			
U	92	238.0289	ppm	0.0074	0.0139

References

Anders, E. and Grevesse, N., 1989. Abundances of the elements: Meteoritic and solar. Geochimica Cosmochimica Acta, 53: 197–214.

McDonough, W.F., 1998. Earth's core. In: C.P. Marshall and R.W. Fairbridge (Editors), Encyclopedia of geochemistry. Kluwer Academic Publishers, Dordrecht, pp. 151–156.

McDonough, W.F. and Sun, S.-S., 1995. Composition of the Earth. Chemical Geology, 120: 223–253.

(continued)

Core	Primitive Mantle	Upper Mantle	Oceanic Crust	Bulk Continental Crust	Ocean	Atmosphere
0	0.438	0.381	3.7	2.1	0.00000087	
0	0.068	0.060	0.54	0.28	0.00000017	
0	0.441	0.392	5.1	1.9	0.00000082	
0	0.0675	0.061	0.56	0.30	0.00000015	
0	0.283	0.167	2.5	3.7	0.0000016	
0	0.037	0.006	0.3	0.7	0.0000025	
0.47	0.029	0.002	0.5	1	0.000001	
230	0.28	0.27	0.9	0.188	0.0084	
2750	3.43	3.43	0.004	0.041	0.0000017	
2550	3.18	3.18	0.02	0.037	0.00000115	
5700	7.07	7.07	2.3	1.5	0.000117	
500	0.98	0.96	0.23	1.3	0.011	
50	10	9.8	20	30	0.15	
30	3.5	3.5	12	500	0.0123	
4000	150	20	800	11 000	0.03	
25	2.5	0.5	7	18	0.017	
0	0.0795	0.06	0.22	5.6	0.0000004	
0	0.0203	0.002	0.1	1.3	0.0033	

McDonough, W.F., Sun, S.-S., Ringwood, A.E., Jagoutz, E. and Hofmann, A.W., 1992. Potassium, rubidium, and cesium in the Earth and Moon and the evolution of the mantle of the Earth. Geochimica et Cosmochimica Acta, 56(3): 1001–1012.

Rudnick, R.L. and Fountain, D.M., 1995. Nature and composition of the continental crust: a lower crustal perspective. Reviews in Geophysics, 33: 267–309.

Rudnick, R.L. and Gao, S., 2003. The Composition of the Continental Crust. In: R.L. Rudnick (Editor), The Crust. Treatise on Geochemistry. Elsevier-Pergamon, Oxford, pp. 1–64.

Taylor, S.R. and McLennan, S.M., 1985. The continental crust: its composition and evolution. Blackwell scientific Publications, Oxford, 312 pp.

5. Geological Time Scale and Live Evolution

Aeon	Era		System	Epoch	Ma	Life evolution (Ma)
PHANEROZOIC	Cenozoic	Quaternary	Quaternary	Holocene	Present – 0,01	Modern human beings (0,12)
				Pleistocene	1,75	Human evolution (2)
		Tertiary	Neogene	Pliocene	5,3	Australopithecus (4)
				Miocene	23,5	
			Paleogene	Oligocene	33,7	
				Eocene	53	Mammal's expansion (50)
				Paleocene	65	1st Primates (58)
	Mesozoic	Secondary	Cretaceous		135	Dinosaur's extinction (65)
			Jurassic		203	1st Mammals (180)
			Triassic		250	1st Dinosaurs (230)
	Paleozoic	Primary	Permian		295	
			Carboniferous		355	
			Devonian		410	1st Reptiles (360)
			Silurian		435	
			Ordovician		500	1st Amphibians (440) 1st Aerial plants (450)
			Cambrian		540	1st Vertebrates (fishes) (500)
PRECAMBRIAN	PROTEROZOIC				2500	1st Shelled Invertebrates (560) 1st Pluricellular organisms (1000) 1st Eukariotes (2000-2100)
	ARCHAEAN				4000	1st Cyanobacterias (2600-2700) 1st Bacterias (3500)
	HADEAN				4550	

Simplified geological time scale for Earth; main stages in life development are also shown. The International Stratigraphic Chart is available on UNESCO-IUGS Internet site: ftp://ftp.iugs.org/pub/iugs/iugs_intstratchart.pdf

2.4 Biochemical Data

1. The Main Prebiotic Precursors of Biomolecules (in brackets, the IUPAC nomenclature recommendation)

water (hydrogen oxyde)	H—O—H
acetaldehyde (ethanal)	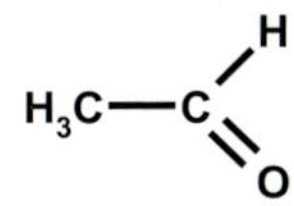
acetronitrile (methyl cyanide)	$H_3C—C\equiv N$
acetylene (ethyne)	$H—C\equiv C—H$
ammonia	NH_3
cyanhydric acid (hydrogen cyanide)	$H—C\equiv N$
cyanoacetylene (2-Propynenitrile)	$H—C\equiv C—C\equiv N$
ethylene (ethene)	$H_2C{=}CH_2$
formaldehyde (methanal)	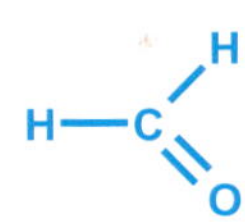
formic acid (methanoic acid)	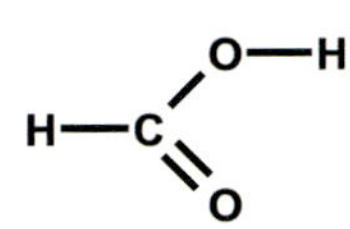
hydrogen sulfide	H—S—H
imidazole	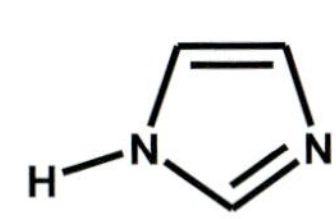
methane	CH_4
methanethiol	$H_3C—S—H$

2. Nucleic Bases

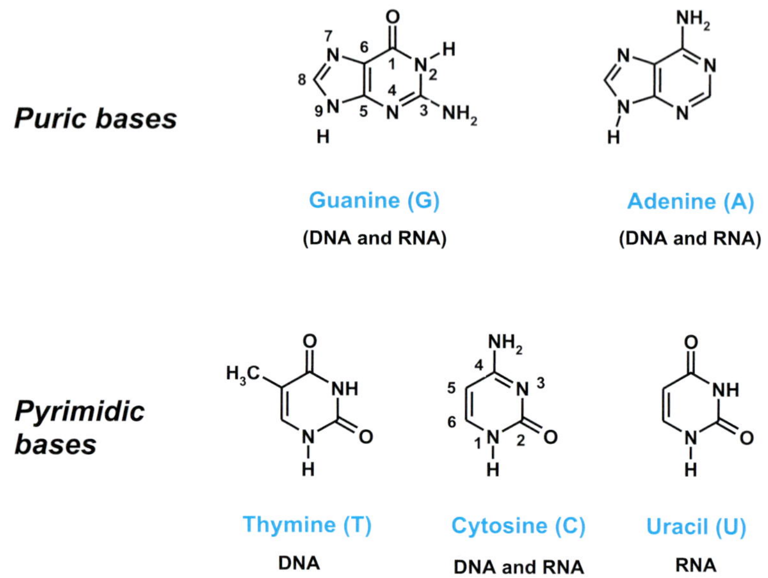

The five major nucleic bases (with atom numerotation)

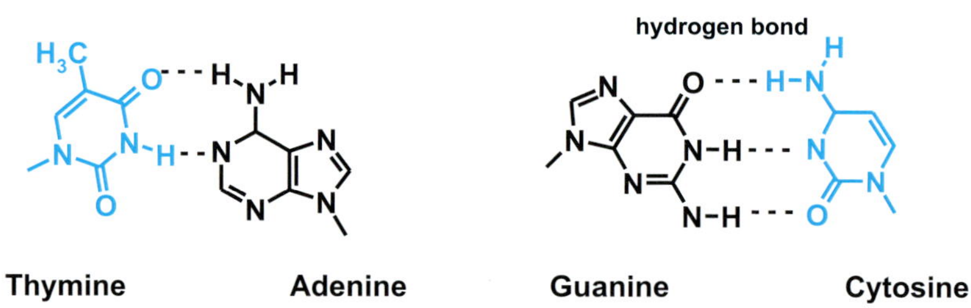

Watson-Crick base pairs

3. Nucleosides and Nucleotides

Nucleosides

Adenosine, 2'-deoxyadenosine
Guanosine, 2'-deoxyguanosine
Thymidine, 2'-deoxythymidine
Uridine
2'-deoxythymidine

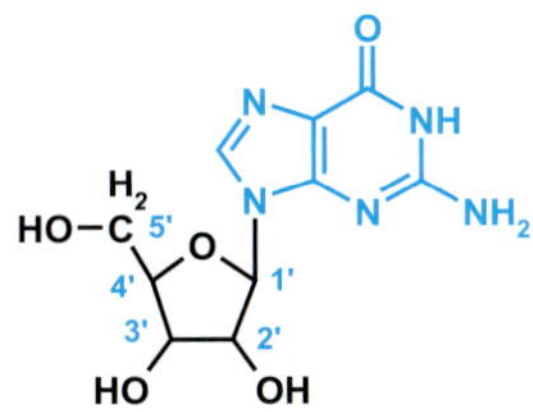

Guanosine

(RNA)

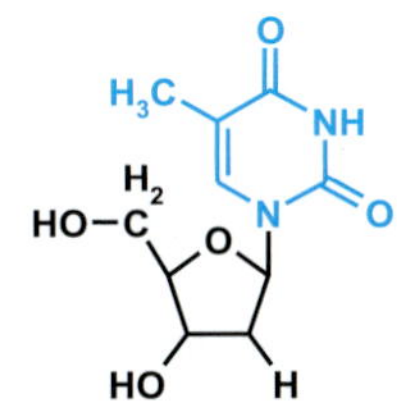

2'-deoxythymidine

(DNA)

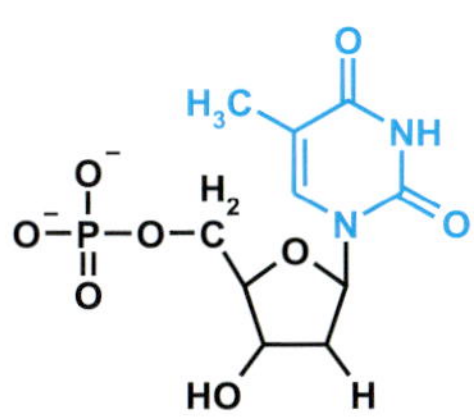

2'-deoxythymidine-5'monophosphate

(DNA)

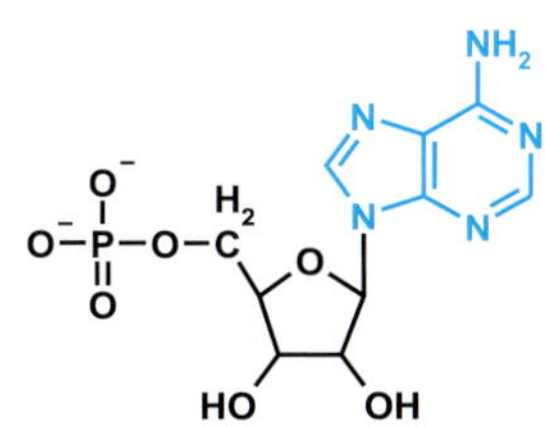

Adenosine-5'-monophosphate (AMP)

(RNA)

Some nucleotides

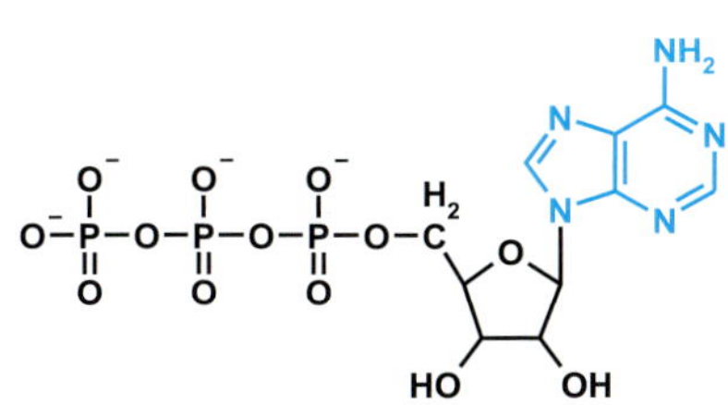

Adenosine-5'-triphosphate (ATP)

4. Amino Acids Genetically Coded

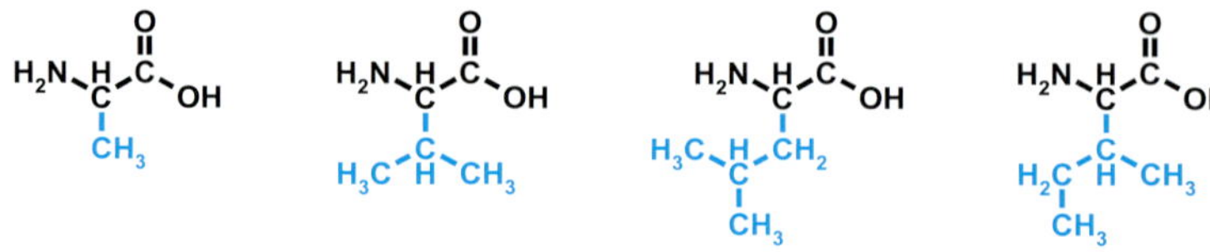

Alanine (Ala, A) | Valine (Val, V) | Leucine (Leu, L) | Isoleucine (Ile, I)

Hydrophobic aliphatic amino acids

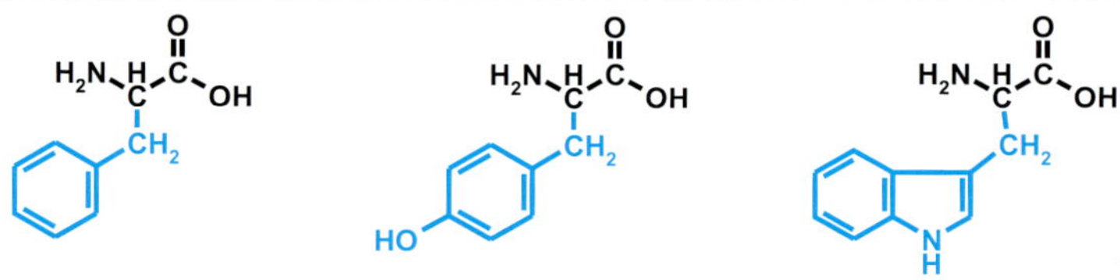

Phenylalanine (Phe, F) | Tyrosine (Tyr, Y) | Tryptophane (Trp, W)

Aromatic amino acids

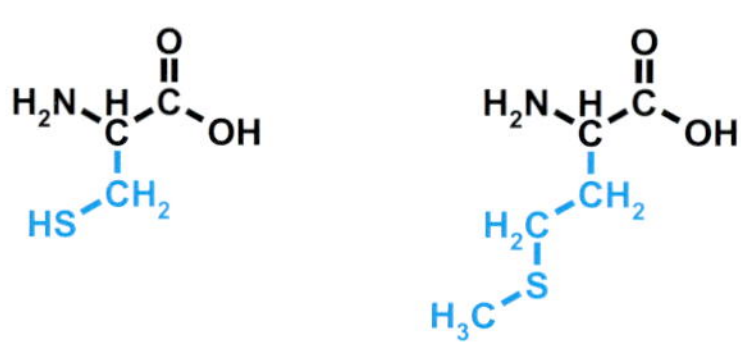

Cysteine (Cys, C) | Methionine (Met, M)

Sulfur containing amino acids

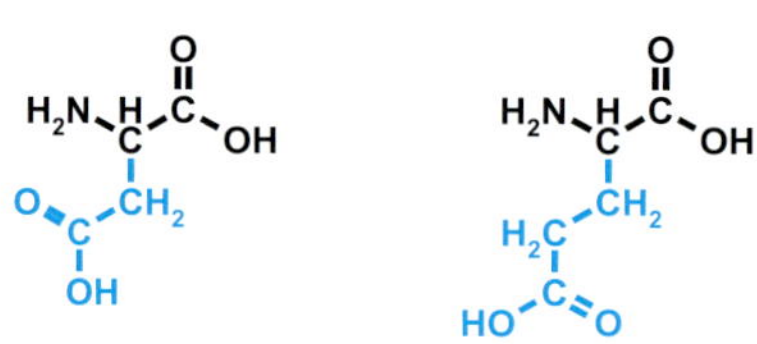

Aspartic acid (Asp, D) | Glutamic acid (Glu, E)

Acidic amino acids

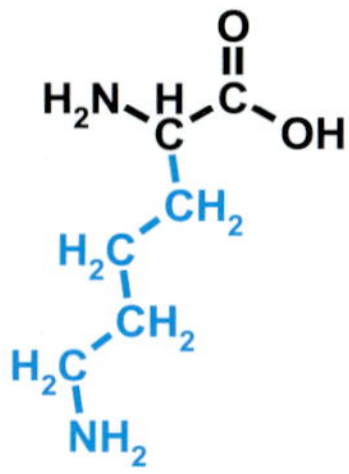

Lysine (Lys, K)

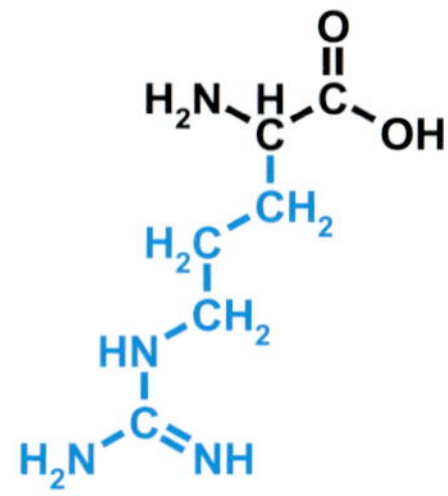

Arginine (Arg, R)

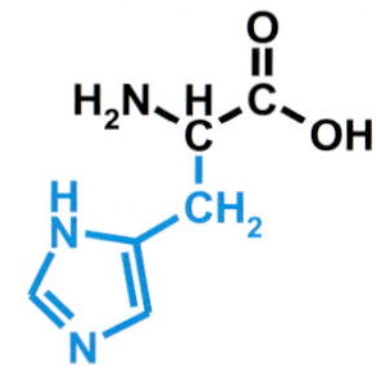

Histidine (His, H)

Basic amino acids

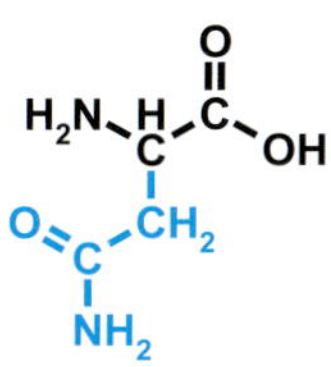

Asparagine (Asn, N)

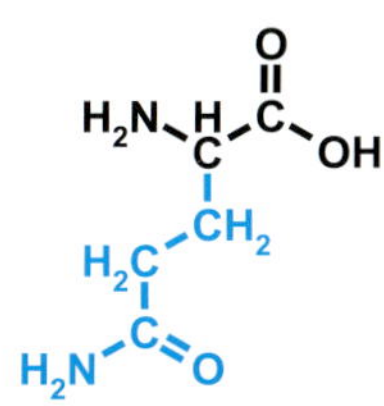

Glutamine (Gln, Q)

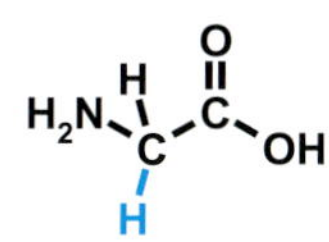

Glycine (Gly, G)

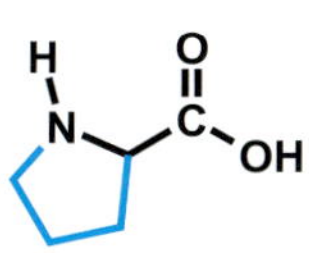

Proline (Pro, P)

N-alkyl amino acid

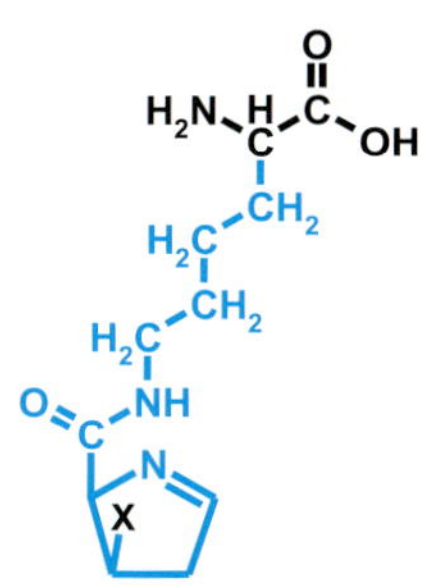

Pyrrolysine

X = NH_2, CH_3 or OH

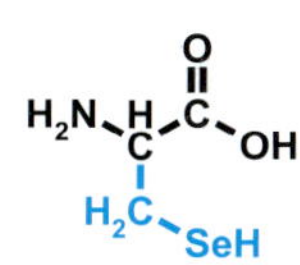

Selenocysteine

Rare amino acids

5. Peptide Bond Formation

leucine + aspartic acid ⟶ peptide + water

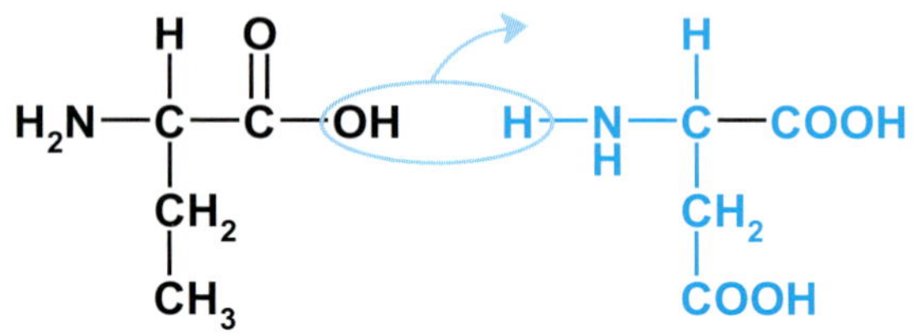

Leucine **Aspartic acid**

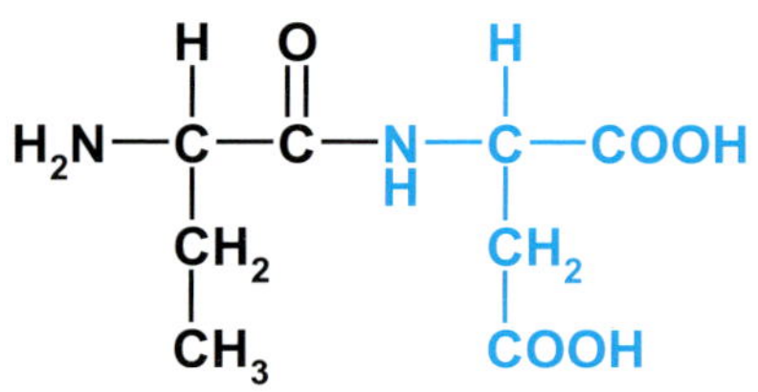

Dipeptide Leu-Asp

3 Glossary

As editors, we would like to thank Doctor D. Despois who initiated this glossary, for the book entitled "L'environnement de la Terre Primitive" published in 2001 by the Presses Universitaires de Bordeaux. Since this initial version, new terms have been added and today, the glossary contains more than 700 items. Considering the diversity of scientific fields that are of direct interest in astrobiology, any glossary devoted to astrobiology is necessarily incomplete. Nevertheless, the editors hope that the following glossary, even far from perfect, will help specialists in one field to obtain some pertinent information concerning other scientific fields without being obliged, at least in a first step, to refer to original papers or specialized books.

A

a (*astronomy*): Orbital parameter, semimajor axis of elliptical orbit.

Abiotic: In the absence of life.

Ablation (*astronomy*): Loss of matter by fusion or vaporization during the entrance of an object into the atmosphere.

Absolute magnitude: Magnitude of a stellar object when seen from a distance of 10 parsecs (32.6 light-years) from Earth.

Acasta: Region of Northern Territories (Canada) where the oldest relicts of continental crust were discovered. These gneisses were dated at 4.03 Ga.

Accretion (*astronomy*): Matter aggregation leading to the formation of larger objects (stars, planets, comets, asteroids). Traces of accretion of their parent body (asteroid) can be observed in some meteorites.

Accretion disk: Disk of matter around a star (or around a black hole) such that matter is attracted by the central object and contributes to its growth. Disks around protostars, newborn stars and T-Tauri stars are accretion disks.

Accretion rate (*astronomy*): Mass accreted per time unit (typically 10^{-5} to 10^{-8} solar mass per year).

Acetaldehyde: CH_3CHO, ethanal.

Acetic acid: CH_3COOH, ethanoic acid.

Acetonitrile: CH_3CN, cyanomethane, methyl cyanide.

Achiral: Not chiral.

Acidophile: Organism that "likes" acidic media, which needs an acidic medium.

Actinolite: Mineral. Inosilicate (double-chain silicate). [$Ca_2(Fe,Mg)_5$ Si_8O_{22} $(OH,F)_2$]. It belongs to the amphibole group.

Activation energy (E_a): Empirical parameter that permits, via an Arrhenius law, the temperature dependence of a reaction rate to be described. In the case of a reaction A + B → C, E_a can be described as the energy difference between the reactants A + B and the activated complex (AB^*) on the reaction pathway to C. It corresponds to an energy barrier.

Active site (*of an enzyme*): Part of an enzyme where substrate is specifically fixed, in a covalent or noncovalent way, and where catalysis takes place.

Activity (*solar or stellar*): All physical phenomena that are time dependent and that are related to star life (like stellar wind, solar prominence, sunspots). Their origin is mainly magnetic, they correspond to emissions of electromagnetic waves at different frequencies (from radiowaves to X-rays) and also to emissions of charged particles (protons, alpha particles and heavier particles). In the case of the Sun, the solar activity is cyclic. The shorter cycle is an eleven-year cycle.

Activity: In nonideal solution, activity plays the same role as mole fraction for ideal solution.

Adakite: Volcanic felsic rock generated in subduction zone, by partial melting of the subducted basaltic oceanic crust.

Adenine: Purine derivative that plays an important role in the living world as a component of nucleotides.

Adenosine: Nucleoside that results from condensation of adenine with ribose or desoxyribose.

ADP: Adenosine diphosphate (see also ATP).

Aerobian: Organism whose life requires free oxygen in its environment.

Aerobic respiration: Ensemble of reactions providing energy to a cell, oxygen being the ultimate oxidant of organic or inorganic electrons donors.

Aerosols: Liquid or solid submillimetric particles in suspension in a gas. Aerosols play an important role in atmospheric physics and chemistry.

AIB: See aminoisobutyric acid.

Akilia: Region of Greenland near Isua where are exposed sediments and volcanic rocks similar in both composition and age (3.865 Ga) to the Isua gneisses.

Alanine (*Ala*): Proteinic amino acid containing three carbon atoms.

Albedo: Fraction of the incident light that is reflected by a reflecting medium (i.e. atmosphere) or surface (i.e. ice cap). A total reflection corresponds to an albedo of 1.

Albite: Mineral. Tectosilicate (3D silicate). Sodic plagioclase feldspar $NaAlSi_3O_8$.

Alcohol: R–OH.

Aldehyde: R–CHO.

Aldose: Any monosaccharide that contains an aldehyde group (–CHO).

ALH84001: Martian meteorite found, in 1984, in the Alan Hills region (Antarctica). In 1996, the claim that it contains traces of metabolic activity and even, possibly, microfossils was the starting point of strong debates.

Allende: Large carbonaceous chondrite (meteorite) of 2 tons, of the C3/CV type and found in Mexico in 1969.

Allochtonous sediment: Sediment that formed in a place different from its present-day location (transported).

Alpha helix: A type of secondary helical structure frequently found in proteins. One helix step contains approximately 3.6 amino acid residues.

Alteration (*Weathering*): Modification of physical and chemical properties of rocks and minerals by atmospheric agents.

Amide: $R–CO–NH_2$, R–CO–NH–R', R–CO–NR'R" depending if the amide is primary, secondary or tertiary. The bond between the CO group and the N atom is generally called the amide bond but when it links two amino acid residues in a polypeptide or in a protein, it is called the peptide bond.

Amine: Derivatives of ammonia NH_3 in which one, two or three H atoms are substituted by an R group (a group containing only C and H atoms) to give primary, secondary or tertiary amines.

Amino acid (*AA*): Organic molecule containing a carboxylic acid function (COOH) and an amino function (generally, but not always, a NH_2 group). If the two functions are linked to the same carbon atom, the AA is an alpha amino acid. All proteinic AA are alpha amino acids.

Amino acids (*biological*): AA directly extracted from organisms, fossils, sedimentary rocks or obtained, after hydrolysis, from polypeptides or proteins found in the same sources.

Amino acids (*proteinic*): AA found as building blocks of proteins. All proteinic AA are homochiral and characterized by an L absolute configuration.

Amino isobutyric acid (*AIB*): Nonproteinic alpha amino, alpha methyl AA. Detected in chondrites.

Amino nitrile: Molecule containing a CN group and an amine function. It could have played a role in the prebiotic synthesis of amino acids.

Amitsôq: Region of Greenland where the oldest huge outcrops ($\sim 3000\,km^2$) of continental crust, dated at 3.822 Ga (gneiss) are exposed.

Amorphous: Solid state characterized by a lack of order at large distances.

Amphibole: Mineral family. Inosilicate (water-bearing double chain silicate), including actinolite, hornblende, glaucophane, etc.

Amphibolite: Rock generated by metamorphism of basalt. It mainly consists of amphibole and plagioclase feldspar crystals, sometimes associated with garnet.

Amphiphile: Molecule with a hydrophilic part and a lipophylic part.

Amplification (*of DNA*): Production, in relatively large quantity, of fragments of DNA by in vitro replication, starting from a very small initial sample (see: PCR).

Amu (atomic mass unit): Atomic mass unit such that the atomic mass of the ^{12}C (carbon isotope) has exactly a mass equal to 12.0000.

Anabolism: General term to designate a group of biochemical reactions involved in the biosynthesis of different components of living organisms.

Anaerobian: An organism that does not need free oxygen for his metabolism. In some cases, free oxygen is a poison for anaerobian organisms.

Anaerobic respiration: Ensemble of reactions providing energy to a cell, the ultimate oxidant being an inorganic molecule other than oxygen.

Anatexis: High-degree metamorphism where rock begins to undergo partial melting.

Andesite: Effusive mafic magmatic rock (volcanic); it mainly consists of sodic plagioclase feldspar + amphibole ± pyroxene crystals. These magmas are abundant in subduction zones. Diorite is its plutonic equivalent.

Anorthite: Mineral. Tectosilicate (3D silicate). Calcic plagioclase feldspar $CaAl_2Si_2O_8$.

Anticodon: Triplet of nucleotides of tRNA able to selectively recognize a triplet of nucleotides of the mRNA (codon).

Antisense: Strand of DNA that is transcripted into a mRNA (messenger RNA).

Aphelia: In the case of an object in elliptical motion around a star, the point that corresponds to the largest distance with respect to the star.

Apollo: Ensemble of asteroids whose orbits intersect Earth orbit.

Archaea: One of the three main domains of the living world. All the organisms of this domain are prokaryotes. Initially, these organisms were considered as the most primitive form of life but this is no longer accepted by most biologists. Most extremophiles such as hyperthermophiles and hyperacidophiles belong to the Archaea domain.

Archaean (*Aeon*): Period of time (Aeon) ranging from 4.0 to 2.5 Ga. Archaean aeon belongs to Precambrian. Unicellular life existed and possibly was already aerobe.

Archebacteria: See Archaea.

Arginine: Proteinic alpha amino acid containing six carbon atoms and a guanido group in the side chain.

Aspartic acid (*Asp*): Proteinic AA with an acidic side chain.

Asteroid belt: Ring-shaped belt between Mars and Jupiter where the majority of the asteroids of the Solar System are located.

Asteroid: Small object of the Solar System with a diameter less than 1000 km. Many of them are orbiting around the Sun, between Mars and Jupiter (asteroid belt).

Asthenosphere: Layer of the Earth mantle, located under the lithosphere and having a ductile behaviour. It is affected by convective movements. Depending on geothermal gradient, its upper limit varies between 0 km under midoceanic ridges and 250 km under continents.

Asymmetric: An atom is called asymmetric when it is surrounded by four ligands that are oriented in 3D space in such a way that they define an irregular tetrahedron. Such an atomic arrangement is chiral and can exist as two enantiomeric forms. These two enantiomers are called D or L depending on their absolute configuration. The D/L nomenclature is now replaced by the R/S nomenclature of Cahn, Ingold and Prelog but the D/L system is still accepted for amino acids and oses. The asymmetric carbon atom is a particular (but very important) example of an asymmetric atom.

Atmosphere: Gaseous envelope around a star, a planet or a satellite. In the absence of a rigid crust or ocean, the atmosphere is defined as the most external part of the object. *The primary atmosphere* of a young planet corresponds to the first gaseous envelope directly generated from protostellar nebula. *The primitive atmosphere* of the Earth corresponds to the atmosphere in which prebiotic chemistry may have occurred, when free oxygen was at very low concentrations. In the second part of the past century, this primitive atmosphere was considered as being highly reductive. Today, it is considered as mainly consisting of carbon dioxide, nitrogen and water vapour.

ATP synthetase: Enzyme involved in ATP synthesis.

ATP: Adenosine triphosphate. Molecule that plays an important role in the living world for energy transfers. Its hydrolysis in ADP and in inorganic phosphate is an exergonic reaction.

AU (*astronomical unit*): Average Earth–Sun distance corresponding to 149.6×10^6 km (or approximately 8 light-minutes or 100 Sun diameters).

Authigenic minerals: In a sedimentary environment, minerals generated by direct local precipitation of dissolved ions.

Autocatalysis: Chemical reaction such that a reaction product acts as a catalyst for its own synthesis.

Autochthonous sediment: Sediment that formed in the place where it is now.

Automaton (*chemical automaton*): As defined by A. Brack, chemical system able to promote its own synthesis.

Autotroph: Organism that is able to synthesize its own constituents from simple molecules like water and carbon dioxide.

B

Bacteria: One of the three domains of the living world. Organisms of this domain are all prokaryotes. They are the more abundant micro-organisms on Earth. One of their characteristics consists of the presence of a cell-wall containing muramic acid. Some of these organisms are extremophiles like *Aquifex* and *Thermoga*.

Bacteriochlorophylls: Pigments found in micro-organisms and containing a tetrapyrole acting as a ligand for an Mg cation (similarly to chlorophylls of green plants).

Barophile: Micro-organism that lives optimally (or can only grow) in high-pressure conditions such as deep-sea environments.

Basalt: Effusive mafic magmatic rock (volcanic); it mainly consists of plagioclase feldspar + pyroxenes ± olivine crystals. It results from 20 to 25% melting of the mantle. Gabbro is its plutonic equivalent.

Bases in nucleotides: See nucleic bases, purine bases and pyrimidine bases.

Benioff plane: In a subduction zone, interface between the subducted slab and the overlying mantle wedge.

Beta sheet: A secondary pleated structure frequently observed in proteins.

BIF (*banded iron formation*): Sedimentary rocks widespread in Archaean terrains and no longer generated today. They consist of alternation of black iron-rich (magnetite) and white amorphous silica-rich layers.

Bifurcation (thermodynamics): This term describes the behaviour of a system that, submitted to a very small variation in exchange conditions with its surroundings, jumps suddenly from one stationary state to another one. The characteristics of the new stationary state cannot be predicted on the basis of a complete knowledge of the initial stationary state.

Biofilm: Tiny film (few µm thick) consisting of colonies of micro-organisms fixed on an inorganic or organic surface.

Biogenic sediment: Sediment formed by precipitation of ions dissolved in water by the means of living beings (e.g., calcium ion in shells).

BIOPAN: Experimental module made by ESA to be fixed on a Russian satellite of the Photon type, the aim of the system is to expose samples or dosimeters to space conditions, such as vacuum, microgravity or radiation.

Biosignature: Observable considered as a piece of evidence for the presence of life.

Biosphere: Ensemble of species living on Earth.

Biosynthesis: Production of a chemical compound by a living organism, equivalent to anabolism.

Biotite: Mineral. Phyllosilicate (water-bearing sheet silicate). [$K(Fe,Mg)_3Si_3AlO_{10}$ $(OH)_2$]. It belongs to the mica group and is also called black mica.

Biotope: Smallest unit of habitat where all environment conditions and all type of organisms found within it are the same throughout.

Bipolar flow (*astronomy*): Flow of matter in two opposite directions, perpendicularly to the circumstellar disk associated to a newborn star. The components of the bipolar flow are molecules, atoms, ions and dust particles.

Birthline: Locus, in the Hertzsprung–Russell diagram where young stars become optically visible.

Black body (*radiation*): Radiation emitted by a body at a temperature T and such that the coupling between matter and radiation is perfect. Such a body is black. The total power emitted by unit area and the power emitted at a well-defined frequency depends only on the temperature (the T^4 dependence is given by the empirical Stefan law and by the theoretical Planck law). As a first approximation and on the basis of their emission properties, stars and planets can be described as black bodies.

Black smoker: Also called hydrothermal vent. Structure observed on the oceans floor generally associated with midocean ridges. There, hot hydrothermal fluid, rich in base metal sulfides, enters in contact with cold oceanic water. Polymetallic sulfides and calcium sulfate precipitate progressively building a columnar chimney around the vent.

Blast: Blast (basic local alignment search tool) represents a powerful and rapid way to compare new sequences to an already existing database, which may either contain nucleotides (Blastn) or proteins (Blastp). Since the BLAST algorithm establishes local as well as global alignments, regions of similarity embedded in otherwise unrelated proteins could be detected. The Blast program gives a value for each of the high-scoring results, together with the probability to find an identical score in a database of the same size by chance.

Blue algae: See cyanobacteria. Old name for some procaryotic unicellular that are not algae.

Bolometric light (*or bolometric magnitude*): Total radiation output by time unit of a stellar object.

Bootstrap: Statistical method with resampling, commonly used to measure the robustness and the reliability of phylogenic trees.

Branching ratio: In the case of a chemical or of a nuclear reaction, such that different reaction paths exist, the branching ratio is the relative rate constants or probabilities of occurrence per unit of time for each different path.

Breccias (geology): Sedimentary or magmatic rocks consisting of an accumulation of angular fragments in a sedimentary or magmatic matrix.

Bremsstrahlung: German word also used in French and English to describe the electromagnetic radiation emitted by high-speed particle (electron, proton) when deviated by a magnetic field.

Brown Dwarf: Space body born as a star, but whose mass is too small (< 80 Jupiter mass) to allow nuclear reaction: both core temperature and core pressure are insufficient to initiate hydrogen fusion.

C

C (*alpha*): Carbon atom linked to a chemical function we are interested in. More specifically, the carbon atom directly linked to the carboxylic function in amino acids. In alpha amino acids, the amino group is linked to the alpha carbon. The Greek letters alpha, beta, gamma ... are used to describe carbon atoms separated from the function by one, two, three ... other carbon atoms.

CAI (*Ca–Al Inclusions*): Ca-and Al-rich inclusions abundant in some chondrite meteorites.

Caldera: Circular km-sized structure due to the collapse of superficial formation induced by the emptying of an underlying magma chamber.

Carbonaceous chondrite: Chondrite with high carbon content. The famous Murchison meteorite is a carbonaceous chondrite.

Carbonate: $(CO_3)^{2-}$-bearing mineral e.g., Calcite = $CaCO_3$; Dolomite = $MgCa(CO_3)_2$.

Carbonation: In Ca-, Mg-, K-, Na- and Fe-bearing minerals, chemical reaction of alteration resulting in the formation of carbonates.

Carbonyl: –(CO)–, this chemical function is found in carboxylic acids, ketones, aldehydes, amides in the peptide bond and many other organic molecules.

Carboxylic acid: Organic molecule containing a COOH group.

C-asteroid: Asteroid containing carbon. C-asteroids are the parent-bodies of carbonaceous chondrites.

Catalysis: Chemical process such that a substance (catalyst) increases the reaction rate by changing reaction pathway but without being chemically modified during reaction. Enzymes are very efficient catalysts able to increase reaction rates by several orders of magnitude and also able to limit the number of secondary products and therefore to increase reaction selectivity.

CCD (*charge-coupled device*): Silicon photoelectronic imaging device containing numerous photosensors (often at least 1000×1000). The most used astronomic detector in the visible wavelength domain.

Cell: Complex system surrounded by a semipermeable membrane that can be considered as the basis unit of all living organisms.

Cenozoic or Caenozoic(*Era*): Period of time (Era) ranging from 65 Ma to today, it is also called Tertiary Era but in addition it also includes the Quaternary Era.

Chemical sediment: Sediment formed by direct precipitation of ions dissolved in water.

Chemolithoautotroph: Chemotroph that uses CO_2 as only source of carbon.

Chemolitotroph: Chemotroph that takes its energy from the oxidation of inorganic molecules. On Earth, the first living organisms could have been chemolitotrophs.

Chemoorganotroph: Chemotroph that takes its energy from oxidation of organic molecules.

Chemotroph: Organism that takes its free energy from the oxidation of chemicals.

Chicxulub: Large (180 km in diameter) impact crater, located in the Gulf of Mexico. It is assumed to be the result of the collision of a big (10 km in diameter) meteorite, 65 Ma ago. This impact is considered as the cause of the important biological crisis at the Cretaceous–Tertiary boundary that led to mass extinction of thousands of living species, such as ammonites, dinosaurs, etc.

Chirality: Property of an object (and therefore of a molecule) to be different from its mirror image in a plane mirror. A hand is an example of a chiral object (in Greek: *cheir* means hand). Any object is chiral or achiral and if it is chiral, it can exist as two enantiomorphous forms (called enantiomers for molecules).

Chlorophylls: Pigments of major importance in the oxygenic photosynthesis. The chemical structure of all chlorophylls is based on a porphyrin ring system chelating an Mg^{2+} cation. Chlorophylls are found in higher plants, algae and some micro-organisms.

Chloroplast: Subcellular structure that plays a fundamental role in photosynthesis in all photosynthetic eukaryotes. Chloroplasts have more probably an endosymbiotic origin.

Chondre: Small spherical aggregate of radiated silicate minerals (typically 1 mm in diameter) that is frequent in stony meteorites and especially in chondrites. Olivine is the main component of chondres.

Chondrite: Undifferentiated stony meteorite unmelted and frequently considered as a very primitive object. Chondrites have the same composition as the Sun except for volatile elements.

Chromatography: Preparative or analytical chromatography: experimental method based on the properties of all molecules to be absorbed more or less selectively by a solid phase (the stationary phase) and therefore to migrate at different rates when they are "pushed away" by a mobile phase that can be a ga s (gas chromatography or GC) or a liquid (LC or HPLC for high-performance LC).

Chromosome: Subcellular structure containing most of the genetic material of the cell. Prokaryotic cells generally contain only one chromosome made of a circular DNA molecule, while eukaryotic cells generally have several chromosomes, each of them containing a linear DNA molecule.

CIP (*for Cahn–Ingold–Prelog*): General nomenclature used in chemistry to describe chiral molecules and more generally stereoisomers. Following the International Union of Pure and Applied Chemistry (IUPAC), CIP nomenclature must replace all other nomenclatures including D, L Fisher nomenclature, except for amino acids and sugars for which the D/L nomenclature is still accepted.

Circular dichroism: The absorption coefficients of right and left circularly polarized light are different if the absorbing medium is chiral. By plotting the difference between the absorption coefficients as a function of the light wavelength, the curve obtained corresponds to the circular dichroism curve of the medium.

Circumstellar disk: Disk of gas and dust particles around a star.

Class [0, I, II, III] (*astronomy*): Classification of young stellar objects based on their electromagnetic emission in the microwave and infrared domains. It consists of an evolution sequence from the protostars (0 and I) to the T-Tauri stars (III).

Clast: Fragment of mineral or rock included in another rock.

Clay: Mineral family. Phyllosilicate (water-bearing sheet silicate), e.g., kaolinite, illite, smectite, montmorillonite.

CM matter: Pristine material of the Solar System, analogue to constitutive matter of CM carbonaceous meteorites (M = Mighei) and very abundant in micrometeorites.

Coacervat (*droplet*): Protein and polysaccharides containing emulsion. According to Oparin, model of protocells.

Codon: Triplet of nucleotides in a mRNA molecule that corresponds to a specific amino acid of a protein synthesized in a ribosome or that corresponds to a punctuation signal in protein synthesis.

Coenzyme: Small molecule that binds with an enzyme and that is necessary to its activity. ATP, CoA, NADH, biotine are examples of coenzymes. Coenzymes are often derived from vitamins.

Collapse (*astronomy*): Process that describes the formation of stars from dense cores. The process seems to be fast: less than 10^5 years.

Coma: Broadly spherical cloud of dust particles and gaseous molecules, atoms and ions surrounding cometary nucleus. It appears when a nucleus becomes active, generally when approaching the Sun.

Combustion: Used to describe any exothermal chemical reaction involving dioxygen and organic reactants. Sometimes used in astrophysics to describe the thermonuclear fusion reactions taking place in the stars and that, obviously, are also exothermal.

Comet: Small body of the Solar System with an average size of 1 to 100km, travelling generally on a strongly elliptic orbit around the Sun. Comets are constituted of ice and dust and are considered as the most primitive objects of the Solar System.

Cometary nucleus: Solid part of a comet (1–100km diameter), made of ice (H_2O, CH_3OH, CO...) and of dust.

Cometary tail: Part of an active comet; three different cometary tails are known depending on their composition (dust particles and molecules, neutral sodium atoms, ions).

Complex molecule (*astronomy*): Molecule containing more than 3 atoms.

Complexation: Chemical term used to describe the noncovalent interaction of a molecule or, more frequently, an inorganic ion with other molecules (called ligands) to give a supramolecular system described as a complex.

Condensation: Chemical reaction involving two molecules and leading to the formation of a new chemical bond between the two subunits but also to the elimination of a small molecule (generally a water molecule). The formation of bonds between the subunits of many biochemical polymers are condensation reactions (examples: polynucleotides, polypeptides, polysaccharides).

Configuration: Term used in stereochemistry. Stereoisomers (except if they are conformers) have different configurations. As an example, butene can be cis or trans and it corresponds to two different configurations. In the case of chiral molecules like the D- and L-valine, the stereoisomers are enantiomers and they have "opposite" configurations. It is important to make a clear distinction between the relative configurations of two enantiomers and the absolute configuration of each of them.

Conglomerate (*chemistry*): In the restricted case of crystallized chiral molecules, a crystalline state such that all molecules of the same chirality (homochiral) crystallize together giving a mixture of crystal that, themselves, are of opposite chirality.

Conglomerate (*geology*): Detrital sedimentary rock consisting of an accumulation of rounded fragments in a sedimentary matrix.

Continental margin: Submarine part of a continent making the boundary with the oceanic crust.

Continents: Emerged and associated shallow depth (< 300m) parts of the Earth's surface. Their average composition is that of a granitoid.

Continuum (*astronomy*): in emission (or absorption) spectroscopy, emission (or absorption) background of a spectrum extending over a large frequency domain. Frequently, lines are superimposed on the continuous spectrum. Blackbody radiation corresponds to a continuous spectrum.

Cool Early Earth: Period of time between Earth accretion (4.55 Ga) and the late heavy bombardment (4.0–3.8 Ga). This model considers that period of early Earth as quiescent with respect to meteoritic impacts, thus being potentially favourable for life development.

Core (*geology*): Central shell of a planet. On Earth the core mainly consists of iron with minor amounts of nickel and some traces of sulfur; it represents 16% of the volume but 33% of the mass of the planet. It is subdivided in a solid inner core from 5155 to 6378 km depth and a liquid outer core from 2891 to 5155 km depth.

COROT: French project for the search of extrasolar planets based on a 25-cm telescope and able to detect planets having a diameter equal to twice the Earth diameter and located at 0.5 AU from its star.

Cosmic rays: Highly accelerated ions coming from the Sun (solar wind, essentially protons) or coming from other and extrasolar sources (galactic cosmic rays).

Covalent bonds: Interatomic bonds such that two atoms share one, two or three electron pairs leading to the formation of a single, a double or a triple bond. A covalent bond is described as a polarized bond if the two bonded atoms have different electronegativities.

CPT (*theorem*): general theorem of physics that assumes that physical laws are unchanged if, simultaneously, space is reversed (parity operation), time is reversed (the sense of motion is reversed) and matter is replaced by antimatter. In many cases, but not all, the CP theorem alone is valid.

Craton: Huge block of old (often Precambrian in age) and very stable continental crust (see also shield).

Crust (*geology*): The more superficial shell of the solid Earth. Its lower limit with the mantle is called Morohovicic (Moho) discontinuity. Together with the rigid part of the upper mantle it forms the lithosphere. Two main crusts exist 1) oceanic crust, basaltic in composition and about 7 km thick, it constitutes the ocean floor; 2) continental crust is granitic in composition, with a thickness ranging between 30 and 80 km, it constitutes the continents.

Cryosphere: Part of the Earth surface made of ice.

Cumulate: Igneous rock generated by accumulation of crystals extracted from magma.

Cyanamid: NH_2–CO–CN.

Cyanhydric acid: HCN, hydrogen cyanide. Triatomic molecule that during prebiotic period could have played the role of starting material for purine synthesis.

Cyanoacetylene or better cyanoethyne: H–CC–CN.

Cyanobacteria: Micro-organism belonging to the Bacteria domain and able to perform oxygenic photosynthesis. In the past, these micro-organisms were improperly called "blue algae". The cyanobacteria could be the ancestors of chloroplasts.

Cyanogen: C_2N_2.

Cysteine (*Cys*): Protinic amino acid containing three C atoms and a –SH group in its side chain. In a proteinic chain, two cysteine residues can be linked together by a –S-S- bond (disulfide bond) often used to stabilize protein conformation. Two cysteines linked together by a disulfide bond is a cystine molecule.

Cytidine: The ribonucleoside of cytosine. The corresponding deoxyribonucleoside is called deoxycytidine.

Cytochrome c: One particular example of the large cytochrome family. Cytochrome c is a protein involved in the electron transfers associated to aerobic respiration. In the eukaryotic cells, cytochrome c is localized in the mitochondrias.

Cytoplasm: Whole content of a cell (protoplasm) except the nucleus whose content is called nucleoplasm.

Cytoplasmic membrane: also called plasmatic membrane or cell membrane.

Cytosine (*C*): One of the nucleic bases of pyrimidine type.

D

D/H ratio: D = deuterium; H = hydrogen. Due to their mass difference, molecules containing either H or D are able to fractionate. For instance, during vaporization, a molecule containing the light isotope (H) is more efficiently vaporized than its heavy equivalent (D). Chemical reactions, including biochemical ones are also able to fractionate these isotopes. Consequently, the D/H ratio can provide information on the biotic or abiotic origin of some organic molecules. The D/H ratio can be very high in chondritic organic matter due to D-enrichment during reactions taking place in the interstellar clouds.

Dalton (*Da*): Molecular mass unit equal to the sum of the atomic masses given in amu (atomic mass unit).

DAP: Abbreviation frequently used to design two different molecules, diamino propionic acid and diamino-pimelic acid.

Darwin: Research programme from the European Space Agency devoted to the search of extrasolar planets and the study of their atmosphere composition by spatial interferometry, in the infrared spectral region. Five independent telescopes of 1.5 m each constitute the basis of this very sophisticated system.

Daughter molecule (*in comet*): In the cometary coma, any molecule produced by photodissociation of a parent molecule coming from the nucleus.

Deamination: Reaction associated to the elimination of an amine group (NH_2, NHR or NRR′).

Decarboxylation: Reaction associated to the elimination of a CO_2 molecule.

Deccan (*Trapps*): Voluminous stacking of basaltic flows emplaced in Northwest India at the end of Cretaceous period. It constitutes evidence of an extremely important volcanic event contemporaneous with the extinction of dinosaurs.

Delta (*isotopic*) (δ): Difference between the isotopic ratio of a sample (R_e) and that of a standard (R_s). $\delta = 1000 \times (R_e - R_s)/R_s$, e.g., $\delta^{18}O$.

Denaturation: Change of the native conformation of a biopolymer. More specifically and in the case of proteins, denaturation can be induced by increasing temperature and/or pressure or by adding a chemical reagent (like urea). Denaturation can be reversible or irreversible. It is generally associated with a loss of enzymatic properties.

Dense clouds: See interstellar clouds.

Dense core (*astronomy*): Gravitationally bound substructure located inside a molecular cloud and surrounded by protoplanetary disks. Dense core collapsing leads to the formation of one or several stars.

Deoxyribose: Ose or monosaccharide having a structure identical to ribose except that the OH group in position $2'$ is replaced by an H atom. In living organisms, deoxyribose comes from ribose via a reduction reaction.

Depletion: Impoverishment of chemical element abundance when compared to a standard of reference composition.

Diagenesis: Chemical and/or biochemical transformation of sediment after its deposition. This process, which generally consists of cementation and compaction, transforms a running sediment into a compact rock.

Diapirism: Gravity driven magma or rock ascent in the Earth. Generally, low-density materials rise up into denser rocks.

Diastereoisomers (*or diastereomers*): Stereoisomers that are not enantiomers. Diastereisomers are characterized by physical and chemical properties that can be as different as observed for isomers having different connectivity (constitutional isomers).

Differentiation (*Earth*): Separation from a homogeneous body of several components whose physical and chemical properties are contrasted. In the case of Earth these components are core, mantle, crusts, hydrosphere and atmosphere.

Diffuse clouds: See interstellar clouds.

Dinitrogen: N_2, also frequently called nitrogen.

Dioxygen: O_2, also frequently called oxygen.

Disk (of second generation): Disk around a star resulting from the breaking off of solid bodies previously formed like planetesimals, asteroids or comets. The beta Pictoris disk is probably a secondary disk.

Dismutation: Reaction that from a single reactant gives two products one of them being more oxidized than the reactant and the other one more reduced.

A typical example is the transformation of an aldehyde into an alcohol and an acid.

Disulfide bond: Covalent bond between two S atoms. When the –SH groups of two cysteines react on each other in the presence of an oxidant, it gives a cystine molecule, i.e. a cysteine dimer linked by a –S-S-bond. Disulfide bridges stabilize the ternary and quaternary structures of proteins. Some irreversible denaturazing of proteins can be associated to the formation of disulfide bonds.

DNA: Desoxyribonucleic acid, long-chain polymer of desoxynucleotides. Support of the genetic code in most living cells. Frequently observed as a double helix made by two complementary strands.

Drake (*equation of*): Empirical formula (containing several adjustable parameters) that gives a rough estimation of the number of "intelligent civilizations" in the galaxy.

Dust (*interstellar, cometary*): Small solid particles (0.1–1 mm), generally made of silicates, metal ions and/or carbonaceous matter.

E

e (*astronomy*): Orbital parameter that measures the eccentricity of an elliptical orbit.

Eccentricity: Parameter (e) characterizing the shape of an orbit. e is equal to 0 for a circle, equal to 1 for a parabola, higher than 1 for a hyperbola and between 0 and 1 for an ellipse.

Ecliptic: Geometric plane of the Earth orbit. More precisely, average planar of Earth–Moon barycentre orbit.

Eclogite: High-degree metamorphosed basalt. It consists of an anhydrous rock made up of pyroxene and garnet.

Ecosystem: Community of organisms and their natural environment. Ecosystem = community + biotope.

Eddington: ESA project devoted to asteroseismology and search for extrasolar planets by the transits method. Kepler: similar project for exoplanetary science.

Electrophoresis: Analytical method used in chemistry and based on the difference between diffusion rates of ions when placed in an electric field. Initially the ions are adsorbed on a support or immersed into a viscous medium. Capillary electrophoresis is a technique adapted to the analysis of small samples.

Enantiomeric excess (*i.e. in per cent*): In the case of a mixture of two enantiomers whose respective concentrations are D and L with D greater than L, i.e. $= (D - L/D + L) \times 100$.

Enantiomers: Two stereoisomers that differ only due to their opposite chirality (like an idealized left hand and an idealized right hand).

Enantiomorph: Two objects that differ only due to their opposite chirality (like an idealized left hand and an idealized right hand).

Enantioselectivity: A reaction leading to products that are chiral is said to be enantioselective if it gives an excess of one enantiomer. Enantioselectivity is generally induced by a chiral reactant or a chiral catalyst (like an enzyme). Enantioselectivity is also used to describe a reaction such that starting from a mixture of enantiomers, one of them reacts faster than the other. In this case, the reactant must be chiral.

Enantiotopic (*chemistry*): A planar molecule like $H–CO–CH_3$ can be seen as a scalene triangle with summits of different colours. In the 2D space, such an object is chiral and the two faces are said to be enantiotopic. A chiral reagent is able to differentiate enantiotopic faces.

Endergonic: A chemical reaction or a physical change is endergonic if it requires a supply of free energy from its surroundings to succeed.

Endolithic (*biology*): Micro-organisms living in rocks.

Endosymbiosis: Process by which a eukaryotic cell lives in symbiosis with another cell (generally a prokaryotic cell) located in its cytoplasm. Chloroplasts and mitochondria are considered as vestiges of endosymbiotic prokaryotes.

Endothermic: A chemical reaction or a physical change is endothermic if it needs a supply of energy from its surroundings to succeed.

Enstatite: Mineral. Inosilicate (simple chain silicate). Mg-rich Pyroxene [$MgSiO_3$].

Envelope (*circumstellar*): In astronomy, cloud of dust particles and gas surrounding a newborn star (protostar).

Enzyme: In biochemistry, a molecule that catalyses a reaction. Most enzymes are proteins but some are polynucleotides (ribozymes).

Equator (*celestial*): Plane perpendicular to the rotation axis of the Earth and corresponding to an extension in space of the terrestrial equator.

Escherichia Coli: Bacteria found in the intestine and commonly used in experimental bacteriology. Its size is of the order of one micrometre. Its genome codes, approximately, for 3000 proteins.

Ester: Molecule that can be described as the result of condensation of an acid and an alcohol associated with elimination of a water molecule. (–O–CO–) is the ester bond.

Eubacteria: Sometimes used instead of bacteria, in order to point out the difference between Bacteria and Archaea. Archaeas themselves are sometimes called Archebacterias.

Eukaria: One of the three domains of life (together with bacteria and archaea). Eukaryotic cells are members of the Eukaria domain.

Eukaryote: Any organism from the Eukarya domain and characterized by a nucleus (containing the genetic material) separated from the cytoplasm by a membrane. Eukaryotes can be unicellular or pluricellular.

Europa: Satellite of Jupiter with a diameter of about 3100km (same size as the Moon). An ocean could exist beyond its icy surface.

Evaporite: Sedimentary rock generated by evaporation of huge volumes of water; chemical elements dissolved in water precipitate leading to the deposition of minerals such as halite (NaCl) or gypsum = ($CaSO_4$, $2H_2O$).

Exergonic: A chemical reaction or a physical change is exergonic when it provides free energy to its surroundings.

EXOCAM: Special reactor used to experimentally simulate exobiological processes.

Exon: Sequence of transcripted nucleotides that is present in natural RNA and that corresponds to a DNA sequence. In DNA, exons are separated by introns (intervening sequences). Exons and introns are respectively coding and noncoding sequences for proteins.

Exoplanet: Planet orbiting around a star other than the Sun. The number of already discovered exoplanets is greater than 100. (More information on this fast developing field is available on www.obspm.fr/planets.)

Exothermic: A chemical reaction or a physical change is exothermic if it provides energy to its surroundings.

Extinction (*interstellar, atmospheric*): Decrease of the light intensity of a star due to light diffusion or absorption by a medium (planetary atmosphere, interstellar cloud). In an interstellar cloud, visible magnitude can be reduced by a factor of 1, while the reduction can be as large as 100 in a protostar dense nucleus.

Extremophile: Micro-organism that optimally lives (or optimally grows) in "extreme" physicochemical environments (P, T, pH...).

F

Fatty acid: Carboxylic acid R–COOH where R is a long chain containing only C and H atoms.

Feldspar: Mineral family. Tectosilicate (3D silicate). Feldspars are subdivided into two main chemical groups: 1) Alkali feldspars ($NaAlSi_3O_8$ = Albite and $KAlSi_3O_8$ = Orthoclase); 2) Plagioclase feldspars ($NaAlSi_3O_8$ = Albite and $CaAl_2Si_2O_8$ = Anorthite). These minerals constitute 52% of the continental and oceanic crusts.

Fermentation: Biochemical process such that complex organic molecules (i.e. glucose) are transformed into low molecular mass molecules (i.e. ethanol) by cells, in anaerobic conditions. Fermentation corresponds to an oxidation process but the final electron acceptor is an organic molecule instead of oxygen. During fermentation as during respiration, ATP is produced but less efficiently.

Ferrihydrite: Iron hydroxide, $5Fe_2O_39H_2O$.

Fisher: German chemist who was the first to introduce the D/L nomenclature to differentiate enantiomers and to characterize their absolute configurations.

Fisher–Tropsch (*reaction of*): Reaction that gives hydrocarbons from a mixture of H_2 and CO. The FT reaction had and still has a great industrial importance but it could also have been important in prebiotic chemistry. The FT reaction requires metallic catalysts.

Flint: Rock mainly made of amorphous silica and having a biogenic origin. It frequently appears as nodules in chalk or limestone.

Fluid inclusion: 1 to 100 µm-sized cavities in minerals that contain fluids trapped during mineral crystallization.

Formaldehyde or methanal: The simplest aldehyde (H–CO–H).

Formation (*geology*): Group of terrains or rocks having the same characteristics.

Formic acid: HCOOH, methanoic acid.

Formose (*reaction*): Starting from formaldehyde in water solution at high pH, this reaction leads to the formation of a large variety of sugars. Its importance in prebiotic chemistry remains an open question. It is also called Butlerow reaction.

Fossil (*geology*): All kinds of trace of passed life (bone, shell, cast, biomolecule, track, footprint, etc.).

Frasil: Ice disks with a diameter of a few mm that are observed in water as soon as surfusion occurs. Frasil is common in Artic and Antarctic rivers but also below the huge ice platforms moving forward in the Antarctic Ocean.

Free-fall time: Time required for an object of mass m, initially at rest, to reach an object of mass M ($M > m$) under effect of gravitation alone. It gives a good approximation of the time required for an accretion disk to collapse during the protostar stage (typically 10^5 years).

Furanose: A "furanose ring" is a cyclic ose formed of 4 carbons and an oxygen atom.

G

Ga: Giga annum = one billion years (= Gyr). (Ga is more frequently used in geology, whereas Gyr is preferred in astronomy.)

Gabbro: Plutonic magmatic rock. It has a granular texture and mainly consists of pyroxenes and plagioclase (± olivine). Basalt is its effusive equivalent.

Gaia: Ambitious project of ESA to measure the position of one billion stars with a precision of one microarcsecond. Gaia is essentially devoted to the search for extraterrestrial planets.

Galactic "open" cluster (*astronomy*): Cluster that can contain from a dozen to a few thousand stars ("I" population) and are younger than globular clusters.

Garnet: Mineral. Nesosilicate (isolated SiO_4 tetrahedron). Its general composition is $Y_2^{+++}X_3^{++}(SiO_4)_3$ (Y = Al, Fe^{+++}, Cr and X = Ca, Mg, Fe^{++}, Mn). In the mantle, garnet is stable at high pressure (> 70 km depth) where it is the only aluminium-bearing phase.

GC (*gas chromatography*): Chromatographic method using gas as the moving phase. It can be used in analytical and preparative chemistry.

Gene: Segment of DNA, containing hundreds to thousands of nucleotides, found in a chromosome. A gene codes for a specific protein.

Genome: Ensemble of genes of an organism.

Genotype: Ensemble of the genetic characters of an organism.

Geographic pole: Point where the rotation axis (instantaneous) of a planet intersects the globe surface.

Geothermal gradient: Thermal gradient corresponding to the temperature increase with depth. In the Earth crust, geothermal gradient is $\sim 30°C\ km^{-1}$.

Giant planets: Large-size planets of low density, such as Jupiter, Saturn, Uranus and Neptune. One can distinguish two groups: gaseous giants, Jupiter and Saturn, mainly made up of gas (H_2, He) coming from the protosolar nebula and icy giants, Uranus and Neptune, rich in ice (H_2O, NH_3, and CH_4). They all have a core made of heavy elements and were formed in the outer part of the solar nebula, beyond the ice line.

Glaciation: Cold period in the Earth history characterized by the presence of a large cryosphere (ice). Frozen water accumulates and forms ice caps on the continents; consequently, it becomes unable to return to ocean whose level decreases. The main glaciations occurred during Precambrian, Early Cambrian, end of Tertiary and during Quaternary.

Glass inclusion: 1 to 100 µm-sized cavities in minerals that contain magma trapped during mineral crystallization.

Glass: Amorphous material. In volcanic rocks, it can result from the rapid cooling of the magma.

Globular cluster (*astronomy*): Large spherical cluster containing from a few thousand to several million old stars ("II" population).

Glutamic acid (*Glu*): Alpha AA with a side chain containing an acidic COOH function. Described as a hydrophilic AA.

Glutamine (*Glu*): Amino acid containing 5 C atoms with a NH_2 group in the side chain; it is considered as hydrophilic.

Glycan: Synonymous of polysaccharides.

Glyceraldehyde: ($HO–CH_2–CHOH–CHO$); the simplest aldose, containing only one chirotopic carbon atom. The two enantiomers play an historical role in stereochemistry because they are at the origin of the D, L nomenclature that describes absolute configuration (Fisher). Glyceraldehyde is the biochemical precursor of other oses.

Glycerol: ($HO–CH_2–CHOH–CH_2OH$); 1,2,3-propanetriol also called glycerine, component of many membrane phospholipids (which are esters of glycerol).

Glycine (*Gly*): The simplest amino acid and the only one that is achiral.

Glycolic acid: $HO–CH_2–COOH$.

Gneiss: Metamorphic rock made up of quartz, feldspars and micas. All mica crystals show the same orientation thus defining a surface of preferential cutting up called "foliation plane".

Gondwana (*Gondwanaland*): Palaeozoic super-continent formed by convergence and agglomeration of continents (Peninsular India, Madagascar, Africa, Australia, South America and Antarctica) due to plate tectonic activity. It was mainly located in the South hemisphere.

Gram +: Bacteria previously coloured during a Gram test and that does not lose the colour after a treatment with ethanol ("positive" response).

Granite: Plutonic magmatic rock. It has a granular texture and mainly consists of quartz, alkali feldspar and plagioclase feldspar; mica can be present whereas amphibole is rare. Rhyolite is its effusive equivalent.

Granitoid: Family of quartz-bearing plutonic magmatic rocks including granites, granodiorites, tonalites and trondhjemites.

Granodiorite: Plutonic magmatic rock. It is similar to granite but contains no more than 10% alkali feldspar.

Green bacteria: Micro-organisms of the Bacteria domain able to perform anoxic photosynthesis.

Greenhouse effect: Warming of a planet surface due to the trapping by the planet's atmosphere of the electromagnetic waves received and radiated by the planet.

Greenstone belts: Volcanic (basalts and komatiites) and volcano-sedimentary formation widespread in Archaean terrains. It generally presents an elongated shape ($\sim$ 100 km long and few tens of km wide). Its green colour is due to metamorphism of basalts and komatiites.

Guanine (*G*): Nucleic base with purine structure.

Gyr: Giga year = one billion years (= Ga).

H

Hadean (*Aeon*): Period of time (Aeon) ranging from 4.55 Ga (Earth formation) to 4.0 Ga (oldest known rock: Acasta gneisses). Hadean aeon belongs to Precambrian.

Harzburgite: Peridotite made up of olivine and orthopyroxene. It generally corresponds to residual mantle after lherzolite melting and extraction of basaltic magma.

HD 209458b: First exoplanet, whose previous detection by radial velocimetry method has been confirmed by the observation of a transit in front of its parent star (HD 209458).

Heavy element (*astronomy*): Any element other than hydrogen and helium.

Helium: Rare gas whose ^{3}He isotope is used in geology as a marker of a recent degassing process from the deep mantle.

Hertzsprung–Russel diagram (*HR diagram*): In astronomy, a two-dimensional diagram with star temperature (or spectral type) as abscissa and star luminosity (or absolute magnitude) as coordinate. Temperatures decrease from left to right and the spectral types sequence is OBAFGKM.

Heterocycle: Cyclic organic molecule containing heteroatoms (i.e. atoms other than C) as constituents of the cyclic structure.

Heterotrophous: Organism that uses reduced organic molecules as principal carbon source for its biosynthesis. Nowadays, these reduced organic molecules are generally produced by other organisms. At the beginning of life, these reduced molecules were probably found in the environment.

Histidine (*His*): Proteinic amino acid containing an imidazole group in its side chain. Histidine residues (hystidyl) are frequently found in active sites of enzymes.

HMT: ($C_6H_{12}N_4$); hexamethylenetetramine, could be a minor component of a comet nucleus.

Homeostasis: Property of a living organism to maintain unchanged some of its physicochemical characteristics even in the presence of a change in the environment. Homeostasis requires autoregulation.

Homoacetogens: Micro-organisms of the Bacteria domain producing acetate from H_2 and CO_2.

Homochirality: Of the same chirality. All proteinic amino acids are L, while ribose in RNA or ATP is always D. The origin of homochirality for the large majority of the chiral constituents of organisms remains an active research subject.

Homology (*biology*): Two structures in two different species are homologous (and therefore comparable) irrespective of their forms, if they are connected in the same way to identical structures.

Hornblende: Mineral. Inosilicate (double-chain silicate). $[Na_{0-1}Ca_2(Fe^{++},Mg)_{3-5}(Al, Fe^{+++})_{0-2}\ Si_{8-6}Al_{0-2}O_{22}\ (OH,F)_2]$. It belongs to the amphibole group.

Hot Jupiter: Jupiter massive-like exoplanet, orbiting close to a star. Most of the extrasolar planets so far discovered belong to this type.

Hot spot: see mantle plume.

HPLC (*high-performance liquid chromatography*): Very efficient liquid-phase chromatography performed under high pressure up to 100–400 bars.

Hydrocarbons: Molecules containing only C and H atoms. If the hydrocarbon contains an aromatic system, it is called aromatic. If the hydrocarbon contains only tetracoordinated C atoms, the hydrocarbon is called aliphatic. Some hydrocarbons result from the polymerization of isoprene (2-methylbutene); they are called isoprenoid hydrocarbons. Latex contains isoprenoid hydrocarbons.

Hydrogen bond: Intermolecular but sometimes intramolecular low energy bond (about 20 kJ/mol) involving generally an H atom linked to an electronegative atom like O, N, S and an atom bearing nonbonding electron pairs such as O or N. The H bond implies an H donor and an H acceptor.

Hydrogen cyanide: H–CN.

Hydrogen sulfide: H_2S.

Hydrolysis: Cleavage of a molecule due to reaction with H_2O.

Hydrothermal vent: See black smoker.

Hydroxy acid: Carboxylic acid containing also an alcohol function. Glycolic acid is the simplest hydroxy acid.

Hydroxyl (*group*): –OH or alcohol group.

Hypoxanthine (*6-hydroxypurine*): Purine base and biological precursor of adenine and guanine.

I

Ice shelves: Ice platforms generated by glaciers whose ice progresses from Antarctica land over ocean. Ross shelf is about 1000 km wide and extents over 600 km into the ocean. They are the source of very large tabular icebergs.

Ices (*astronomy*): Solid form (crystalline or amorphous) of volatile molecules like water, carbon dioxide or ammonia.

IDP: Interplanetary dust particle.

Igneous rocks: Magmatic rocks = due to magma crystallization.

Impactite: Heterogeneous breccia generated at depth by the impact of a large-size stellar body ($>$ 10 000 tons). It is made up of fragments of the rock substrate included in a vitreous matrix.

Inclination (1) (*astronomy*): Angle between the orbital plane of a solar object and the ecliptic plane in degrees (always lower than 90°).

Inclination (2) (*astronomy*): Angle between the orbital plane of an interstellar object and the "sky plane" (plane perpendicular to the "line of sight", i.e. the straight line joining the observer and the stellar object).

Indels: Acronym for insertions/deletions. Phylogenetic analysis on several DNA or protein sequences requires sequences with same length. During the process of alignment, alignment gaps (indels) must be introduced in sequences that have undergone deletions or insertions.

Interplanetary dust: Small grains left behind by asteroids and comets, and dispersed in a cloud including the whole Solar System.

Interstellar cloud: Cloud of gas (98%) and dust (2%). The gas is mainly H (diffuse cloud) or H_2 (molecular cloud). Molecular clouds are called dense clouds ($n(H_2) > 10^3$ molecules cm^{-3}) or dark clouds if dense and cold (10–20 K).

Intertidal environment: Environment between high and low tide. Also called tide range.

Intron: Noncoding sequence of nucleotides that separates exons. Introns are removed during the maturation processes of the three types of RNA by splicing.

Iodine (*I*): Halogen element. As iodine is incorporated in living organisms, including marine organisms, it provides information on interactions between living world and marine sediments.

Ion–molecule (*reaction*): Reaction between two gaseous reactants and initiated by ionizing cosmic rays, X-rays or UV radiation. They are important in interstellar clouds and in planetary ionospheres.

Iridium (*Ir*): Element belonging to platinum element family. Its concentration in Earth crust is extremely low. A local Ir enrichment as at the Cretaceous–Cainozoic (K/T) boundary is interpreted as a strong argument in favour of meteoritic impact or of deep originated volcanic activity.

Isochron (geology): Rectangular diagram plotting isotopic ratio of a disintegration system (abscissa = parental isotope; ordinate = daughter isotope) (e.g., $^{87}Sr/^{86}Sr$ versus $^{87}Rb/^{86}Sr$). In this diagram, cogenetic rocks of the same age plot along a straight line whose slope is proportional to age. This method of age determination is widely used in geology.

Isocyanhydric acid: H–N=C.

Isocyanic acid: HN=C=O.

Isoleucine (*Ile*): Proteinic amino acid containing six carbon atoms and described as hydrophobic. Ile is considered as one of the prebiotic AA.

Isotopic ratio: Concentration ratio of two isotopes or concentration ratio of two isotopomers of a molecule (like H_2O and D_2O). Isotopic ratio can provide information on the age of a sample (when used in isochron calculation) as well as on its origin and source.

Isovaline: Hydrophobic nonproteinaceous amino acid. This constitutional isomer of valine contains five C atoms.

Isua: Region of Greenland where the oldest sediments so far recognized 3.865 Ga (gneiss) are exposed. They contain carbon whose origin could be biogenic. (See also Akilia.)

J

Jeans escape: Process leading to the escape of atomic or molecular species from a planet atmosphere. It happens when the thermal agitation rate is greater than the escape rate. The lighter elements or molecules (like H, H_2 or He) escape faster than the heavier ones.

Jovian planets: Other name for giant planets.

Jupiter: The fifth and largest planet (1400 times the Earth volume, 320 times the Earth mass) of the Solar System. Jupiter is 5.2 AU away from the Sun. Its gaseous envelope mainly made of H_2 and He, surrounds a core of ice and rocks (10–20 Earth mass).

Juvenile gases: Gases produced by or trapped inside the Earth and that reach the surface of Earth for the first time. ^{3}He is an example of juvenile gas detected in submarine geothermal fluids.

K

K/T (Strata): Few centimetre-thick sedimentary layer located at the Cretaceous–Cainozoic boundary. Its iridium-enrichment is interpreted as due to a giant meteoritic impact.

Kepler: Spacecraft NASA mission devoted to detection of Earth-type exoplanets (equipped with a 1-m telescope).

Keplerian rotation: Orbital motion that follows Kepler's laws.

Kerogen: Insoluble organic matter found in terrestrial sediments and in some types of meteorites like carbonaceous chondrites.

Kilo base or kilo base pair (kb): Unit used to measure the number of nucleotides in a gene or a genome: 1000 base pairs of DNA or 1000 bases of RNA.

Komatiite: Ultramafic high-Mg lava. It contains olivine and pyroxene; minerals that sometimes can have needle or dendritic shapes (spinifex texture). Komatiites were abundant before 2.5 Ga and extremely rare after.

Kuiper Belt, or Edgeworth–Kuiper: A large ring-shaped reservoir of comets beyond Neptune at about 30 astronomical units (AU) from the Sun.

L

L/D (*ratio*): For a chiral molecule, the ratio between the L and the D enantiomer concentrations.

Lactic acid: HO–CH(CH_3)–COOH.

Lagrange points: The five points determining the equilibrium position of a body of negligible mass in the plane of revolution of two bodies (ex: star-planet couple) in orbit around their common-gravity centre.

Late Heavy Bombardment: Heavy bombardment of the Moon (and certainly also of the Earth and other telluric planets) that happened between 4 and 3.8 Ga ago. It could correspond to either the end of a long period of bombardment by asteroids, meteorites and comets or to a short-term cataclysmic phenomenon.

Laurasia: Palaeozoic supercontinent formed by convergence and agglomeration of continents (Europe, North America and Asia) due to plate tectonic activity. It was mainly located in the northern hemisphere. Continent resulting as Pangaea broke into two parts at the end of Palaeozoic.

Lava: Magma emplaced as flow at Earth surface.

Leucine (*Leu*): Proteinic amino acid containing six C atoms and considered as one of the prebiotic AA.

Lherzolite: Peridotite made up of olivine and pyroxenes (ortho- and clino-pyroxenes) as well as of an Al-bearing mineral. Its melting generates basaltic magmas leaving a harzburgite residue.

Ligand: Any atom or group of atoms bonded to the atom we are interested in. As an example, the four ligands of a chirotopic (asymmetric) carbon atom are necessarily different.

Light-year: Measure of distance used in astronomy, it corresponds to the distance that light runs in one year (0.946×10^{16} m).

Liquidus: Line that, in composition vs. temperature or pressure vs. temperature diagrams, separates the domain where crystals and liquid coexist from the field where only liquid exists.

Lithosphere: External rigid shell of the Earth. Its definition is based on the rheological behaviour of rocks. It includes crusts (continental and oceanic) as well as the upper rigid part of the mantle. Its thickness varies between 0 and 250 km and more or less corresponds to the 800°C isotherm.

LUCA: Last universal common ancestor. Hypothetical micro-organism that stood at the root of all lines leading to the present-day living beings. Appeared after a long evolution, it cannot be considered as a primitive form of life.

Lysine (*Lys*): Proteinic amino acid containing six C atoms with an amino group in its side chain and that, therefore, is basic and hydrophilic. Lysyl residues are frequently found in the active site of enzymes.

M

Ma: Mega annum = Mega year = one million years (= Myr).

Mag: See Magnitude.

Magma: Molten rock that can be completely liquid or consists of a mixture of liquid and crystals. It is produced by high-temperature ($> 650\,^\circ$C for granite; $> 1200\,^\circ$C for basalt) melting of pre-existing rocks. Mantle melting generates basalts, whereas oceanic crust fusion rather generates adakites or TTG and continental crust gives rise to granites.

Magnetic anomaly: Difference between the measured and the theoretical value of the magnetic field intensity of Earth.

Magnetic pole: Point where the magnetic dipole axis of a planet intersects the globe surface.

Magnetite: Mineral: Iron oxide [$Fe^{++}Fe^{+++}{}_2O_4$]. Its ferromagnetic properties make it able to record past Earth magnetic field characteristics. It can also exist in some bacteria called "magnetotactic".

Magnitude (Mag) (*astronomy*): Measure of brightness of a stellar object on a logarithmic scale. The difference between two successive magnitudes is a factor 2.512. Mag $= -2.5 \log_{10} (I/I_0)$; The less bright a star, the greater is its magnitude. The magnitude is calculated on a chosen spectral interval (visible, IR) or on the total spectrum (bolometric magnitude).

Major half-axis: For an elliptic orbit, half of the distance aphelia–perihelia.

Mantle plume: Ascending column of hot mantle assumed to be generated near the mantle-core boundary or at the upper/lower mantle boundary, (= hot spot). Close to the surface, this column can melt giving rise to oceanic island magmatism (i.e. Hawaii; La Réunion, etc.).

Mantle: In a planet, the mantle is the shell comprised between the crust and the core. On Earth it represents 82% of the volume and 2/3 of the mass, it is divided into upper mantle (until $\sim$ 700 km depth) and lower mantle (until 2900 km depth).

Mars Express: ESA space mission towards Mars.

Mass loss rate (*astronomy*): Mass ejected per time unit by a star during its formation. Ejection takes place through stellar winds and bipolar jets (typically 10^{-5} to 10^{-8} solar mass per year).

Maturation (*genetics*): Transformation step of mRNAs leading to their functional form. It occurs by splicing "introns sequences".

Megaton: Unit of energy equivalent to the energy released by 10^9 kg of TNT (trinitrotoluene). Corresponds to 4.2×10^{15} J.

Mercaptans: Other name of thiols: sulfur analogues of alcohols.

Mesophase: Matter state exhibiting characteristics of two phases. Liquid crystals have the fluidity of liquids but are characterized by an order at short range similar to what is observed in crystals.

Mesotartric acid: HOOC–CHOH–CHOH–COOH, alpha-beta-dihydroxy-succinic acid, molecule containing two asymmetric carbon of opposite chirality. This molecule is achiral by internal compensation.

Mesozoic (*Era*): Period of time (Era) ranging from 250 Ma to 65 Ma, it is also called Secondary Era.

Metabolism: All the reactions taking place in a cell or in an organism. Metabolism is divided into two subclasses: anabolism and catabolism. Most metabolic reactions are catalyzed by proteinic enzymes.

Metamorphism: Solid-state transformation of a rock due to change in pressure and/or temperature conditions. New mineral assemblage, stable in new P–T conditions will appear. New minerals crystallise perpendicular to oriented pressure thus defining a new planar structure called foliation. Most often metamorphism corresponds to dehydration of the rock.

Metasediments: Sediments transformed by metamorphism.

Metasomatism: Change in rock composition due to fluid circulation. For instance, in a subduction zone, the fluids (mainly water) released by dehydration of the subducted slab, up-rise through the mantle wedge. These fluids, which also contain dissolved elements, not only rehydrate the mantle peridotite, but also modify its composition.

Meteor Crater: Impact crater in Arizona. It is about 1.2 km in diameter and 170 m deep. The impact, which took place 50 000 years ago, was due to an iron meteorite of about 25 m in diameter. Impacts of this kind generally occur every 25 000 year.

Meteorite: Extraterrestrial object, fragment of an asteroid, of a planet (like Mars) or of the Moon that falls on the Earth's surface.

Methanogen: Archeobacteria producing methane, CH_4, from CO_2 and H_2. Some methanogens are hyperthermophilic.

Methionine (*Met*): Proteinic amino acid containing five carbon atoms with a $-SCH_3$ group in its side chain.

Methylalanine: Synonymous of alpha aminoisobutyric acid (alpha-AIB).

MGS (*Mars Global Surveyor*): American probe that has carried out a complete cartography of Mars (from September 1997).

MHD: Magnetohydrodynamic.

Mica: Mineral family. Phyllosilicate (water-bearing sheet silicate): biotite = black mica [$K(Fe,Mg)_3Si_3AlO_{10}\,(OH)_2$]; muscovite = white mica [$KAl_2Si_3AlO_{10}\,(OH)_2$].

Micrometeorite: Very tiny meteorite (< 1 mm). The 50–400 μm fraction is the most abundant found on Earth. Micrometeorites constitute more than 99% of the extraterrestrial material able to reach the Earth surface (major impacts excepted).

Micro-organism: Organism invisible without a microscope. Includes prokaryotes and unicellular eukaryotes (i.e. yeasts).

Microspheres: Spherical clusters of organic molecules found in Precambrian rocks or produced in laboratory from amino acids polymers (proteinoids). Today, the Fox proteinoids microspheres are not still considered as plausible models of primitive cells.

Microsporidia: Parasitic unicellular eukaryotes that have been shown to be highly derived fungi from the fungi. Microsporidia were thought for some time to be primitive.

Migmatite: High-temperature metamorphic rock affected by partial melting.

Milankovitch (*theory of*): Theory connecting the Earth climate variations to astronomic variations such as changes of Earth's orbit or obliquity with time.

Miller–Urey (*experiment of*): One of the first experimental simulations of what was considered as atmospheric prebiotic chemistry (1953). Synthesis of a large variety of organic molecules including few amino acids from a very simple mixture containing reduced small molecules (H_2, CH_4, NH_3 and H_2O) submitted to an electric discharge.

Mineral: Solid material defined by both its chemical composition and crystalline structure.

Minimal protosolar nebula: Minimal mass of gas, necessary for the formation of the Solar System planets (= mass of all planets + H + He $\approx$ 0.04 solar mass).

Minor planet: Asteroid or planetoid.

Mitochondria: Organelles in the cytoplasm of all eukaryotic cells where ATP synthesis takes place during aerobic respiration. Mitochondria have their own DNA and could have an endosymbiotic origin.

Mitosis: Nucleus division, cell-division step including cytokinesis.

MM: Micrometeorites.

Moho (*Mohorovicic*): Discontinuity in seismic waves that marks the crust/ mantle boundary.

Mole: Amount of substance of a system which contains as many elementary entities as they are atoms in 0.012kg of carbon -12.

Molecular clouds: See interstellar clouds.

Molecular flow: See bipolar flow.

Monophyly: Term that describes a taxonomic group sharing a single ancestor and all its descendants (i.e. the mammals).

Monosaccharides: See oses.

Montmorillonite: Mineral. Phyllosilicate (water-bearing sheet silicate). Clay mineral belonging to the smectite group.

MORB (*mid-ocean ridge basalt*): Basalt generated in midoceanic ridge systems where oceanic crust is created. Most of ocean floor has a MORB composition.

m-RNA: Messenger RNA. Obtained by transcription of a DNA segment and able to orient the synthesis of a specific protein in the ribosome.

MS (*mass spectrometry*): Analytical method involving a preliminary ionization of atoms or ionization and fragmentation of molecules followed by measurement of atomic or molecular masses. These measurements can be carried out from precise study of ion trajectories or time of flight in an electric and/or magnetic field.

MSR (*Mars Sample Return*): NASA-CNES space mission project for the return of Martian soil samples (~ 1 kg) extracted by automatic probes. Launch expected between 2009 and 2014, and samples return three to five years later.

Murchison: Carbonaceous chondrite (CM) that fell in Australia in 1969. Fragments recovered immediately after the fall were (and still are) subjected to many analyses, mainly chemical. More than 500 organic compounds were identified, including amino acids and nucleic bases.

Muscovite: Mineral. Phyllosilicate (water-bearing sheet silicate). [$KAl_2Si_3AlO_{10}(OH)_2$]. It belongs to the mica group and is also called white mica.

Mutation: Any change of the genetic material, transmitted to the descendants.

My: Mega year = one million years (= Ma).

Mycoplasma: The simplest and the smallest known micro-organisms, they live as parasites in animal or vegetal cells. They have long been considered as possible analogues of the first cells; now considered as Gram+ bacteria that lack their rigid cell wall and evolved by reduction.

N

Nanobacteria: Hypothetical bacteria, whose size could be around few nanometres, smaller than any known bacteria. Their existence is very much debated.

N-carbamoyl-amino acids: A molecule showing many similarities with amino-acids except that one of the H atoms of the amino group is substituted by the carbamoyl polyatomic group ($-CO-NH_2$). N-carbamoyl-amino acids could have been prebiotic precursors of some amino acids.

Neutral (mutations) (*genetics*): Term coming from the neutral theory of molecular evolution proposed by Kimura. A neutral mutation is a mutation leading to sequences selectively and functionally equivalent. They are said to be neutral with regard to evolution.

NGST (*Next-Generation Space Telescope*): NASA project of space telescope (4 m), it must succeed the HST (Hubble Space Telescope).

Nitrification: Microbial oxidation, autotrophic or heterotrophic, of ammonium to nitrate.

Nitrile: R–CN where the CN group is the cyano group (cyano as prefix but nitrile as suffix).

Nonreducing atmosphere: Atmosphere of CO_2, N_2, H_2O where hydrogen is absent or in low quantity, either in the form of free H_2 or hydrogen-containing compounds, such as methane or ammonia. Also named oxidized or neutral atmosphere according to its composition.

Nonsense (*codon*): When a codon (triplet of nucleotides) does not specify an amino acid but corresponds to a termination codon (Term.). In the "universal code", these codons are UAA, UAG and UGA.

Normative (*rock composition*): Rock mineralogical composition recalculated from its chemical composition.

Nuclear pores: Complex structures, highly specialized, embedded in the nuclear membrane. They allow the transfer of macromolecules between nucleoplasm and cytoplasm.

Nucleic acid: Long-chain polymeric molecule obtained by condensation of nucleotides. DNA and RNAs are nucleic acids.

Nucleic base: Linked to ribose or desoxyribose by a hemi-acetal bond, it gives nucleosides. Nucleic bases are purine bases or pyrimidine bases. Nucleosides together with a phosphate group are the subunits of nucleotides. By condensation, nucleotides give the polynucleotides (including DNA and all RNAs).

Nucleides: Constituents of atom nucleus, i.e. protons and neutrons.

Nucleon: Common name for proton or neutron.

Nucleoplasm: Protoplasm within the nucleus of eukaryotes.

Nucleotide: Molecule made by condensation of a base (purine or pyrimidine), an ose and a phosphate group linked to the ose. Nucleotides are ribonucleotides when the ose is ribose or deoxyribonucleotide when the ose is deoxyribose. DNA is a polydeoxyribonucleotide while RNAs are polyribonucleotides. The symbol of a nucleotide is determined by the base (A for adenine, C for cytosine, G for guanine, T for thymine, U for uracil).

Nucleus (*Biology*): Eukaryote cell substructure that contains the chromosomes.

O

Obduction: Mechanism leading to the thrusting of oceanic lithosphere onto continental crust.

Obliquity: Angle between the ecliptic and the celestial equator, actually, 23.3 degrees for the Earth. This angle is $> 90°$ if the planet has a retrograde rotation.

Oceanic rift: Central depression in a midocean ridge, this is the place where oceanic plates are created.

Oligomer: Small polymer, generally containing less than 25 monomeric units.

Oligomerization: Polymerization involving a small number of monomers.

Oligopeptides: Small polypeptide (less than 25 AA residues even if the definition is not so strict).

Olivine: Mineral. Nesosilicate (isolated SiO_4 tetrahedron). $[(Fe, Mg)_2SiO_4]$. This mineral which belongs to the peridot family, is silica-poor and is one of the main components of the terrestrial upper mantle. It is also common in meteorites.

Oort cloud: Huge spherical collection of comets, orbiting the Sun between 10 000 and 100 000 AU.

Ophiolite: Part of oceanic lithosphere tectonically emplaced (obducted) on a continental margin.

Optical activity: Orientation change of the linearly polarized plane of light after its passage through a chiral medium.

Optical rotatory dispersion: Change of the optical power of a chiral medium with the wavelength of the linearly polarized light.

Orbital Migration: Change, in the course of time, of a planet–star distance; this hypothesis is proposed in order to explain the presence of massive planets close to their star.

Organic molecule: Until the beginning of the 19th century, organic molecules were molecules extracted from plants or animals. Today, any molecule containing carbon atoms is called organic. Carbonates, CO and CO_2 are borderline cases. Organic molecules generally contain C atoms with an oxidation number lower than 4.

Organometallic: Organic molecule containing one or more metallic atoms bonded by covalent bonds or by coordination to the organic moiety of the molecules. Metallic salts of organic acids are not considered as organometallic compounds.

Orgueil: Large CI carbonaceous chondrite (very primitive), without chondres, which fell in France in 1864, near Montauban.

Orogenesis: Mountain-chain genesis.

Orthoclase: Mineral. Tectosilicate (3D silicate). Alkali feldspar $KAlSi_3O_8$.

Oses (*or saccharides*): Large group of molecules of primeval importance in the living world. Some of them are monomers like glucose ($C_6(H_2O)_6$) or ribose ($C_5(H_2O)_5$). Some of them are dimers like lactose or saccharose; some of them are polymers (polysaccharides) like starch or cellulose.

Outer membrane: membrane surrounding the plasma membrane in Gram negative bacteria (i.e. *Escherichia coli*).

P

P4: Certification for laboratories accredited to analyze high-risk infectious agents. Such laboratories must be protected against any risk of contamination by viruses and micro-organisms, from inside to outside and from outside to inside. These two conditions must absolutely be realised for extraterrestrial sample analysis.

PAH (*polycyclic aromatic hydrocarbons*): Organic molecules like naphthalene or anthracene containing several fused aromatic rings.

Paleosoil: Fossil soil. These formations are able to have recorded O_2 and CO_2 concentration of the primitive terrestrial atmosphere.

Paleozoic or Palaeozoic(*Era*): Period of time (Era) ranging from 540 Ma to 250 Ma, it is also called Primary Era.

Pangaea: Supercontinent that existed at the end of Paleozoic era (–225 Ma) and that later broke into two parts: Laurasia (N) and Gondwana (S).

Paralogous: Paralogous genes originate in gene-duplication events, in contrast to the standard orthologous genes, which originate via speciation events.

Parent molecule (*in comet*): Molecule present in the cometary nucleus.

Parity (*violation of*): Characterizes any physical property that changes when space is inverted or, in other words, which is not the same in the mirror world. Parity violation is observed in several phenomena related to the weak intranuclear interactions like the beta radioactivity. Parity violation is at the origin of the very small energy difference between enantiomers (PEVD for parity-violation energy difference).

Parsec (*secpar*): Unit of astronomical length of 3×10^{18} cm (about 200 000 AU or 3.26 light-years); it is based on the distance from the Earth at which the stellar parallax is 1 second of arc. Average distance between stars in the Sun vicinity is around 1 parsec.

Pathfinder: American probe that landed on Mars on July 4th, 1997, formally named the Mars Environmental Survey (MESUR). It contained a rover "Sojourner" that explored Ares Vallis for several months. For instance, an "alpha proton X-ray spectrometer" was used to analyse soil and rock samples in order to determine their mineralogy.

PCR (*polymerase chain reaction*): Experimental method that, by successive molecular duplications, leads to a dramatic increase of a small initial amount of DNA. This result is obtained with an enzyme called DNA-polymerase isolated from thermophile bacteria.

Peptide: Polymer obtained by condensation of amino acids. In a peptide (or polypeptide), the number of AA residues is generally lower than 60. With a higher number of residues, the polymer generally adopts a well-defined conformation and is called a protein.

Peridot: Mineral family, olivine is a peridot.

Peridotite: Rock made up of olivine and pyroxenes (ortho- and clino-pyroxenes) as well as of an Al-bearing mineral (spinel at low pressure and garnet at high pressure. Earth mantle is made up of peridotite.

Perihelia: In the case of an object in elliptical motion around a star, the point that corresponds to the shortest distance with respect to the star.

Permafost: In arctic regions, permanently frozen soil or subsoil.

Petrogenesis: Mechanism(s) of formation of rocks.

pH: Measure of the acidity of an aqueous solution. $pH = -\log_{10}(H^+)$ where (H^+) is the molar concentration of hydroxonium ion in solution. $pH = 7$ corresponds to the neutrality while a solution with $pH > 7$ is basic and a solution with $pH < 7$ is acidic.

Phanerozoic (*Aeon*): Period of time (Aeon) ranging from 0.54 Ga to today; it followed Precambrian and was characterized by metazoa development.

Phenetic: Taxonomic system for living beings based on overall or observable similarities (phenotype) rather than on their phylogenetic or evolutionary relationships (genotype).

Phenotype: The observable characteristics of an organism, i.e. the outward, physical manifestation of an organism.

Phenylalanine (*Phe*): Amino acid containing an aromatic phenyl group in its side chain. Phenylalanine contains nine amino acid residues.

Phosphoric acid: H_3PO_4, dihydrogen phosphate. Molecule that plays an important role in the living world: nucleotides are esters of phosphoric acid.

Photochemistry: Chemistry involving energy supply coming from "light" (from IR to UV). When electromagnetic frequency is greater (X-rays, beta-rays or gamma rays), the term radiochemistry is generally used. Specific processes like photoactivation, photoionisation, photodissociation (including the production of free radicals), and photolysis are various aspects of photochemistry.

Photosynthesis: Synthesis using photons as energy supply. Photosynthesis can be performed at laboratory or industrial level as a subdomain of photochemistry. In biochemistry, the term "photosynthesis" describes the different biosynthetical pathways leading to the synthesis of molecules under the influence of light. The photons are absorbed by cell pigments and their energy is converted into chemical energy stored in chemical bonds of complex molecules, the starting material for the synthesis of these complex molecules being small molecules like CO_2 and H_2O. It is important to make a clear distinction between the aerobic photosynthesis (also called oxygenic photosynthesis) and the anaerobic photosynthesis. In the first case, water is the reductive chemical species and O_2 is a by-product of the reaction. It must be kept in mind that atmospheric dioxygen as well as oxygen atoms of many oxidized molecules on Earth surface have a biosynthetical origin. Green plants, algae and cyanobacteria are able to perform aerobic photosynthesis; their pigments are chlorophylls. The anoxygenic photosynthesis is not based on H_2O but on reductive species like H_2S. Finally, some halophile archaeas contain a pigment called bacteriorhodopsin and are able to use light energy supply for ATP synthesis. So, light energy is converted into chemical energy or better, in chemical free energy.

Phototroph: Organism whose energy source is light (photosynthesis).

Phylogenesis: History of the evolution of a group of organisms.

Phylogenic tree: Schematic representation describing the evolutive relationships between organisms. It gives an image of the evolution pattern.

Phylotype: Environmental sequence, representing an organism. Sequence of a clone obtained from environment and representing an organism.

Phylum (*pl. phyla*): Group of organisms evolutionary connected at high taxonomic rank.

Pillow lava: Lava extruded under water (ocean, lake) that produces its typical rounded pillow shape. Pillow lava form the upper part of oceanic crust.

PIXE (*proton-induced X-ray emission*): Device that allows the detection and identification of many elements (metals) in proteins.

Plagioclase: Mineral. Tectosilicate (3D silicate). Calco-sodic feldspar whose composition ranges between a sodic ($NaAlSi_3O_8$ = Albite) and a calcic ($CaAl_2Si_2O_8$ = Anorthite) poles. They represent about 40% of the Earth crust minerals.

Planet: Body formed in circumstellar disks by accretion of planetesimals and may be of gas.

Planetesimals: Small solid bodies ($\sim$ 1–100 km) formed in the protosolar nebula, probably similar to asteroids and comets. Their collision and accretion built the planets.

Plankton: The whole organisms living in water and drifting along ocean and lake currents. It includes zooplankton (small animals) and phytoplankton (plants). Fishes and sea mammals, able to swim independently of current flow, constitute the nekton. Neston refers to organisms drifting at the air/water interface, whereas benthos designates organisms living in/on the aquatic ground.

Plasma membrane (*or cell membrane*): Semipermeable membrane that surrounds contemporary cells. It consists of a double layer of amphiphilic molecules (hydrophobic tail with hydrophilic head), mainly phospholipids, with proteins embedded in it. It may also contain some molecules, such as cholesterol or triterpenes, able to rigidify the whole.

Plasmid: Extrachromosomic DNA found in bacteria and yeasts, able to replicate independently.

Plasts: Organites found in phototrophic Eukaryotes and containing photosynthetic pigments such as chlorophylls.

Plate (*lithospheric*): Piece of the rigid lithosphere that moves over the ductile asthenosphere affected by convection.

Plate tectonic: Theory that describes and explains the rigid lithospheric plate motion.

Plutonic: Magmatic rock, resulting of the slow cooling and crystallisation of magma at depth. Its texture is granular (big crystals), e.g., granite.

PMS Star: Newborn star such that the internal temperature is still too low ($< 10^7$ K) to initiate nuclear fusion of hydrogen into helium (i.e. T-Tauri stars) and characterized by a low mass.

PNA (*peptide nucleic acids*): Synthetic analogues of nucleic acids such that the ose-phosphate strand is replaced by a polypeptide backbone to which the bases are linked.

Polarized light: It is important to differentiate two limit cases. A linearly polarized light (or plane polarized light) is an electromagnetic radiation, whatever is its frequency, such that the electric vector and thus also the magnetic vector oscillate in a plane. If this radiation travels through a chiral medium, the plane of polarization is deviated (optical rotation). A circularly polarized light is an electromagnetic radiation such that its electric vector and thus also its magnetic vector describe, in space, a helix around the propagation direction. This helix can be right or left corresponding to the two possible circularly polarized lights of a definite frequency. The linearly polarized light corresponds to the superposition of two circularly polarized lights of the same frequency.

Pole motion: Geographic pole motion at the Earth surface. This motion has low amplitude, since the pole only moves a few metres.

Polymerization: Chemical reaction such that molecules called monomers are covalently linked together and form long-chain molecules, with or without branching. The polymerization can be the result of an addition of polymers like in polyethylene or polystyrene (industrial polymers) but the polymerization can also be the result of condensation reactions involving at each step, the elimination of a small molecule (generally water). Nylon is an industrial condensation polymer and most biological macromolecules such as polypeptides, polynucleotides and polysaccharides are condensation polymers.

Polynucleotide: Polymer resulting from condensation of nucleotides.

Polypeptide: Polymer resulting from condensation of amino acids.

Polysaccharide: Polymer resulting from condensation of oses (in the past, oses were also called monosaccharides).

POM (*polyoxymethylene*): Polymer of formaldehyde.

Population I stars: Stars enriched in heavy elements (O, C, N...). They have been formed from the interstellar gas enriched by the previous generation(s) of stars formed over the billions of years of lifetime of the Galaxy. They are predominantly born inside the galactic disk. The Sun belongs to population I stars.

Massive, hot stars are necessarily young and are therefore always population I stars.

Population II stars: Stars poor in heavy elements (O, C, N...). They have been formed from low-metallicity gas that has not been enriched by successive generations of stars. They are believed to be born in the early ages of the Galaxy. They are actually distributed predominantly in the halo of the Galaxy thus confirming that they are probable remnants of the infancy of our Galaxy.

Poynting–Robertson (effect): Effect of stellar light on a small orbiting particle. This causes the particle to fall slowly towards the star. Small particles (below 1 cm) are more affected because the effect varies as the reciprocal of particle size.

ppb (*part per billion*): Relative concentration in mass = nanogram/gram.

ppbv (*part per billion in volume*): Relative concentration in volume = nanolitre/litre).

ppm (*part per million*): Relative concentration in mass = microgram/gram.

ppmv (*part per million in volume*): Relative concentration in volume = microlitre/litre).

Prebiotic chemistry: All chemical reactions that have contributed to the emergence of life.

Precambrian: Group of aeons ranging from Earth genesis (4.55 Ga) until the beginning of Palaeozoic era (0.54 Ga). It includes: Hadean, Achaean and Proterozoic aeons.

Precession (*of equinoxes*): Motion of Earth's rotation axis with respect to the celestial sphere due to the other bodies of the Solar System and more particularly to the Moon. The terrestrial pole describes a circle on the celestial sphere in 26 000 years.

Primary structure (*of a protein*): Sequence of the amino acid residues in the proteinic chain.

Primitive Earth: The young Earth from its formation until 2.5 Ga.

Primitive nebula: See protosolar nebula.

Prion: Protein able to induce a pathological state because its conformation is modified.

p-RNA: Synthetic RNA molecule in which the sugar is a pyranose instead of a furanose.

Prokaryote: Micro-organism in which chromosomes are not separated from the cytoplasm by a membrane. Bacteria and Archaea are prokaryotes.

Proline (*Pro*): Amino acid containing six C atoms with a unique characteristic. Its side chain links together the alpha atom and the N atom to give a five-membered ring containing one N atom and four C atoms; the amino group is no longer a $-NH_2$ group but a –NH- group. Proline is frequently observed in protein secondary structures called beta turns.

Proper motion (*astronomy*): Apparent angular movement of a star on the celestial sphere during a year (perpendicular to the line of sight). Proper motion analysis can lead to detection of planets in orbit around this star.

Propionaldehyde: CH_3–CH_2–CHO or propanal.

Proteins: Long-chain biological polymers obtained by condensation of amino acids. The degree of polymerization (number of residues) ranges between 60 and 4000. Some proteins form aggregates. For example, haemoglobin is a tetramer containing two proteinic chains of one type and two proteins chains of another type; in each chain, a heme molecule containing a ferrous cation is settled in without being covalently bonded. Structural proteins are components of muscles or flagella; most enzymes are proteins and proteins are also carriers of other molecules like dioxygen or carbon dioxide. The activity of proteins is extremely dependent on their molecular conformation.

Proteobacteria: Group of bacteria including *Escherichia coli* and purple bacteria (photosynthetic bacteria). Mithochondria are relics of proteobacteria.

Proterozoic (*Aeon*): Period of time (Aeon) ranging from 2.5 to 0.54 Ga. This aeon belongs to Precambrian and follows Achaean. Apparition of oxygen in atmosphere and of metazoa.

Proton motive force: Free-energy difference (measured in volts) associated with proton translocation across a membrane. It depends on the electrical membrane potential and on the pH difference between the two reservoirs separated by the membrane. The proton motive force provides energy required for ATP synthesis.

Protoplanetary disks: Disk around a newborn star where accretion of planets is supposed to take place.

Protosolar nebula: Rotating disk of gas, dust and ice, from which the Solar System is originated.

Protostar: several similar definitions exist. 1) newborn star such that half of its luminosity is due to accretion. 2) Body involved in an accretion process that will bring it on the main sequence. 3) Collapsing interstellar cloud. 4) Young object that is not yet optically visible. Protostars are rare because their time life is short (10^4 to 10^5 years): "Protostars are the Holy Grail of IR and submillimetric astronomy".

Psychrophilic organism: Organism that lives optimally at a temperature lower than 10 °C. Some psychrophilic organisms live at temperatures lower than 0 °C.

Pulsar: Small neutron star in very fast rotation (one rotation in less than 1 s) emitting a highly focalized radiation, circularly polarized and detected as very regular pulses. Pulsars are remnants of supernovae.

Purine bases: Guanine and adenine are examples of purine bases because their molecular skeleton corresponds to purine. These bases are found in DNA as in RNAs.

Purple bacteria: Micro-organisms of the bacteria domain able to perform anoxic photosynthesis.

PVED (*parity-violation energy difference*): Energy difference between enantiomers due to parity violation at the level of the weak forces.

Pyranose: A "pyranose ring" is a cyclic ose formed of 5 carbons and an oxygen atom. It is the more stable form of oses.

Pyrimidine bases: Thymine, cytosine and uracil are examples of pyrimidic bases because their molecular skeleton corresponds to pyrimidine. Cytosine is found in DNA as in the RNAs while thymine is specific of DNA and uracil is specific of RNAs.

Pyrite: Mineral. Sulfide [FeS_2].

Pyrolysis: Thermal degradation of a molecule.

Pyroxene: Mineral family. Inosilicate (simple chain silicate). Divided in two families: 1) Orthopyroxene [$(Fe,Mg)_2Si_2O_6$] (e.g., enstatite) and 2) Clinopyroxenes [$(Ca,Fe,Mg)_2$ Si_2O_6] (e.g., augite, diopside).

Q

Q: Orbital parameter of a planet orbiting a star; distance between planet aphelia and star.

q: Orbital parameter; distance between planet perihelia and star.

Quartz: Mineral. Tectosilicate (3D silicate). It crystallizes in hexagonal or rhombohedric systems. In magma it characterizes silica sursaturation. The rhombohedric crystals are chiral: quartz can therefore exist as D- or L-quartz depending on the helicity of the –O-Si-O-Si- chain. A chiral quartz crystal can induce an enantioselectivity during a chemical reaction between achiral reactants via a catalytic effect.

Quaternary structure (*of a protein*): In the case of protein-forming supramolecular aggregates like haemoglobins, the quaternary structure corresponds to the arrangements in space of the subunits.

R

R: solar radius (0.69×10^6 km) or 1/200 AU; approximately 10 times the Jupiter radius and 100 times the Earth radius.

Racemate (*crystal*): Crystalline form of a chiral substance such that each unit cell of the crystal contains an equal number of molecules of opposite chirality.

Racemic (*mixture*): Mixture of enantiomers containing an equal number of the two enantiomers. Such a mixture is described as achiral by external compensation.

Racemisation: Diminution of the initial enantiomeric excess of a homochiral ensemble of molecules or of a nonracemic mixture of enantiomers. It corresponds to an equilibration reaction: the system evolves spontaneously towards a state characterized by higher entropy and therefore, lower free energy. The highest entropy and lower free energy corresponds to the racemic mixture. Racemisation can be a very slow process: this is why enantiomers of amino acids, sugars and many other components of living species can be separated in many cases.

Radial velocity: Star velocity component parallel to the view line. It causes the frequency shift observed in spectral emission lines (Doppler effect). Periodical change in radial velocity can be an indirect proof that a planet orbits around the star (reflex motion).

Radiation pressure: Pressure applied by electromagnetic waves on any atom, molecule or particle.

Radioactivity (*long-lived species*): Radioactive elements with long period (10^9 to 10^{11} years); they are still present in the Solar System.

Radioactivity (*short-lived species*): Radioactive elements with short period ($< 10^7$ years). Nowadays they have totally disappeared in the Solar System.

Raman (*spectroscopy*): Physical method used for molecular structural analysis. Raman spectroscopy is based on inelastic diffusion of visible or UV light and gives information about the vibration modes of diffusing molecules. Raman spectroscopy must be considered as a complementary method with respect to IR spectroscopy.

Rare Earth Elements (*REE*): Chemical elements with very similar chemical properties. This family (lanthanides) ranges from lanthanum ($Z = 57$) to lutetium ($Z = 71$). In geochemistry, they are commonly used as geological tracers of magmatic processes. Indeed they are poorly sensitive to weathering or metamorphism, but on the contrary they are excellent markers of magmatic processes such as melting or crystallization.

Rare gases (*He, Ne, Kr, Ar, Xe, Rn*): Monatomic gases corresponding to the (VIIIA) period of Mendeleyev periodic table (see "Astrobiological data"). Their isotopes can be used to trace some geological events of Earth differentiation (e.g., atmosphere and ocean formation).

Red giant (*astronomy*): Old star (spectral type K or M) still performing fusion of hydrogen but already having a helium core.

Reducing atmosphere: Atmosphere with high hydrogen content. Carbon, oxygen and nitrogen mainly exist as CH_4, H_2O and NH_3.

Reduction potential: Measure of the tendency for a molecule or an ion to give an electron to an electron acceptor that, itself, can be a molecule, an ion or an electrode. A conventional reduction potential scale for molecules in water allows determination, a priori, of what chemical species will be the electron acceptor and what will be the electron donor when they are mixed together at a well-defined concentration.

Refractory: Substance which remains solid in all temperature conditions available in a particular body of the Solar System (ex. dust particles in a comet). If not refractory a substance is said to be volatile.

Region (*HII region*): Interstellar cloud such that H exists essentially as H^+. Ionization is due to intense UV radiation coming from OB stars. HII is the old name used by spectroscopists to describe lines coming from H^+ recombination.

Replication: Biochemical process by which a DNA strand or a RNA molecule is copied into the complementary molecule. Replication is different from transcription and from translation.

Retrotransposons: DNA sequences able to move from one site to another along the chromosome. Retroposons belong to a transposons family that requires RNA molecule as intermediate. The more frequent retrotransposons are the Alu sequences; around one million of such sequences have been identified in the human genome.

Ribose: Aldopentose of major importance in the living world; part of RNA nucleotides.

Ribosomes: Intracellular structures containing many rRNA (ribosomal RNA) molecules together with a complex of 60–80 proteins. Synthesis of proteins takes place in ribosomes by condensation of amino acids; information about the correct sequence is given by a mRNA (messenger RNA), while each amino acid is linked to a specific tRNA (transfer RNA).

Ribozyme: RNA molecule acting as an enzyme.

Ridge (*geology*): Submarine mountain chain located at divergent lithospheric plate margins. On the Earth the total length of ridges is 60 000km.

Rift: Rift valley limited by faults = graben.

RNA world: Often considered as an early hypothetical stage of life evolution, based on a life without protein and such that RNAs would have played a double role: catalysis and support of genetic information. This theory is mainly based on the discovery of catalytic RNAs (ribozymes) and on an increasing knowledge about the importance of RNAs in contemporary life. For other scientists, the RNA world is the stage of evolution that preceded the DNA emergence.

RNA: Ribonucleic acid, a class of nucleic acids containing ribose as a building block of its nucleotides. RNAs themselves are divided into subgroups like messenger-RNA (m-RNA), transfer-RNA (t-RNA), ribosomal-RNA (r-RNA).

Rock: Solid material made up of mineral assemblage. It constitutes telluric planets, asteroids and probably the core of giant planets.

Rocky planets: Other name for telluric planets.

Rodinia: Super continent that formed at about 750Ma ago.

Root (*genetics*): for a particular genetic tree, the last common ancestor of all the organisms of the tree.

Rosetta: ESA mission launched at the beginning of 2004; it will reach the P/Cheryumov-Gerasimenko comet after a 10-year trip. An orbiter will follow the comet during one year, and a lander will perform "in situ" analyses on comet nucleus surface.

Rotatory power: Deviation of the polarization plane of a radiation of well-defined frequency by a chiral medium.

rRNA: Ribosomal RNA.

r-RNA: Ribosomal RNA. r-RNAs are the major components of ribosomes. Some of them have catalytic properties.

Rubisco (*or RuBisCo*): Ribulose-1,5 biphosphate carboxylase-oxygenase; enzyme that catalyses CO_2 metabolism. It is the most abundant enzyme on Earth.

Runaway greenhouse effect: Amplification of a greenhouse effect due to vaporization of molecules able to absorb infrared radiation emitted or received by the planet and that, themselves, contribute to greenhouse effect (i.e. Venus).

Runaway growth (*astronomy*): Increasing rate of planetary accretion with planet growth, it leads to an increasing size difference between the small and large bodies.

Runoff channels: Kind of channels observed on Mars and that seems to be originated by large water flows over a long period of time.

S

Sagduction: Gravity-driven rock deformation. When high-density rocks (e.g., komatiites) emplace over low-density rocks (e.g., TTG), they create an inverse density gradient that results in the vertical sinking of dense rocks in the lighter ones. Sagduction was widespread before 2.5 Ga.

Sarcosine: $(CH_3)NH–CH_2–COOH$, nonproteinic N-methyl amino acid.

Schist: Fine-grain sedimentary rock characterized by a cleavage (slate). It results from fine sediment (i.e. mud) metamorphism.

Secondary structure (*of a protein*): Spatial arrangement of the main chain of a protein. Alpha-helices, beta-sheets, beta-turns are examples of secondary structures.

Sedimentary rock: Rock generated on Earth surface. It can consist of the accumulation of rock particles (detrital) or of organic matter (oil, petrol, coal). It can also be produced by physicochemical or biogenic precipitation of dissolved ions. Detrital particles and ions derived of alteration and erosion of pre-existing rocks.

Selenocysteine: Frequently described as the 21st proteinic amino acid because sometimes coded by the genetic code. It has the structure of the cysteine with a Se atom replacing the S atom.

Semi-minor axis: Orbital parameter. Small axis of orbital parameter.

Sense: DNA strand that is not copied into mRNA. Therefore, the sequence of the sense strand corresponds to the mRNA sequence that, itself, corresponds to the transcription of the antisense strand. Sense and antisense DNA sequences are strictly complementary.

Sequence (*genetics*): Series of directly linked nucleotides in a DNA or a RNA strand; series of directly linked amino acids residues in proteins.

Sequence (main sequence) (*astronomy*): Stage in star evolution when it performs hydrogen fusion in its core. In a HR diagram, main sequence stars draw a straight line.

Sequencing: Experimental determination of a DNA, RNA or protein sequence.

Serine (*Ser*): Proteinic amino acid with a $-CH_2-OH$ side chain. Being hydrophilic, serine is generally present in the external part of the skin and can be used as the proof for human contamination of meteoritic samples. Serine is produced in very small amounts during simulation experiments considered as experimental models of interstellar chemistry.

Serpentine: Mineral family. Phyllosilicate (water-bearing sheet silicate) (i.e. Antigorite [$Si_4O_{10}Mg_6(OH)_8$]). They are generated by olivine (and sometimes pyroxene) alteration. They play an important role in internal water cycle.

SETH (*search for extraterrestrial homochirality*): Search for proof of enantiomeric excesses in extraterrestrial objects.

SETI (*search for extraterrestrial intelligence*): Search for electromagnetic signals (essentially in the radiowave domain) that should be intentionally emitted by some living organisms and coming from sources located outside the Solar System.

Shield (*geology*): Huge block of old (often Precambrian in age) and very stable continental crust, e.g., Baltic shield (includes Finland, Sweden, Norway, and Western part of Russia).

Shocked quartz: Quartz crystal whose structure contains defaults characteristic of the high pressures realised by meteorite impacts.

Siderite: Mineral, iron carbonate [$FeCO_3$].

Siderophile: Elements frequently associated with iron (like Au, Pt, Pd, Ni. . .).

SIDP (*stratospheric IDP*): IDP collected in the stratosphere by captors placed on airplanes.

Silicate: Wide mineral family of silicium oxides. The structure is based on a tetrahedron $(SiO_4)^{4-}$.

Site (*active site*): For an enzyme, specific locus where the substrate is fixed, ready to react with the reactant.

SL (*astronomy*): Solar luminosity (3.826×10^{24} W) or 3.826×10^{33} erg s^{-1}.

SM: Solar mass (2×10^{30} kg).

Small bodies: Comets and asteroids.

SMOW (*standard mean ocean water*): Standard reference sample for H and O isotopic abundance measurements.

SNC: Family of about 30 meteorites that, based on several experimental observations, are considered as having a Martian origin. SNC is the abbreviation for Shergotty, Nakhla and Chassigny, three meteorites of this family.

Solar constant: Total energy delivered each second by the Sun and measured in W m^{-2}, the surface being placed at 1AU of the Sun and perpendicular to the light rays. The value of the solar constant is equal to 1360 W m^{-2}.

Solar type star: Star of G type, similar to the Sun.

Solidus: Line that, in composition vs. temperature or pressure vs. temperature diagrams, separates the domain where crystals and liquid coexist from the domain where only crystals (solid) exists.

Spallation: Atomic nuclei breaking due to the collision between two atomic nuclei with energy greater than the coulombic barrier; it leads to the formation of new elements, stable or radioactive.

Spectral type: Star-classification procedure based on electromagnetic spectrum that, itself, depends on star surface temperature. The OBAFGKM classification ranges from surface temperature of about 50 000 K (O type) to 3500 K (M type = Sun).

Spinel: Mineral. Oxide [$MgAl_2O_4$]. Stable at low pressure (depth $<$ 70 km). In the mantle, at shallow depth, spinel is the only aluminium-bearing phase.

Spore (*biology*): a) Resistant, dormant, encapsulated body formed by certain bacteria in response to adverse environmental conditions. b) Small, usually single-cell body, highly resistant to desiccation and heat and able to develop into a new organism.

Star: Celestial object, generally spherical, where thermonuclear reactions take place (e.g., the Sun) or will take place in the future (e.g. the PMS stars) or has taken place in the past (neutron star).

Stellar cluster: Group of a few hundreds to several millions of stars. The smallest groups are named associations. Most of the stars are formed in open clusters.

Stereoisomers (*or stereomers*): Isomers (molecules with identical atomic composition, i.e. the same number of the same atoms) such that the atoms are identically interconnected by covalent bonds. If two stereoisomers are different in the same way that a left hand is different from a right hand, the stereoisomers are enantiomers. In all other cases, stereoisomers are called diastereoisomers (or diastereomers).

Stereoselectivity: When a chemical reaction leads to the formation of stereoisomers and when these latter are not exactly produced in the same amount, the reaction is called stereoselective. Similarly, when two stereoisomers react at a different rate with a reactant, the reaction is stereoselective. In the chemical literature, some authors have introduced subtle differences between stereoselectivity and stereospecificity.

Steric effect: One of the multiple "effects" introduced by organic chemists to explain the relative stability or the relative reactivity inside a group of molecules. The steric effect takes into account the size of each atom or groups of atoms. For instance, the steric effect of a $-C(CH_3)_3$ group is larger than the steric effect of a CH_3 group. The steric effect can be explained on the basis of repulsive term in the Van der Waals forces.

Stony meteorites: Mainly made up of silicates, they can contain from 0 to 30% of metal grains and several per cent of sulfides. They can be differentiated (achondrite) or undifferentiated (chondrite) even if CI undifferentiated chondrites do not contain chondrules.

Stop codon (*genetics*): Codon that does not code for amino acids but that indicates the translation end. For the Universal code, these codons are UAA, UAG and UGA (synonym: termination codons).

Stratopause: Atmospheric boundary between stratosphere and mesosphere.

Stratosphere: Atmospheric layer located above the troposphere and below the mesosphere, between 9–17 km and 50 km. In the stratosphere, temperature slightly increases with altitude that prevents it from convective movements.

Strecker (*synthesis*): Synthetic method producing amino acids from aldehyde, HCN, NH_3 and water. Frequently considered as important prebiotic reaction.

Stromatolite: Sedimentary structure consisting of laminated carbonate or silicate rocks and formed by the trapping, binding, or precipitation of sediment by colonies of micro-organisms (bacteria and cyanobacteria).

Strong force: One of the four fundamental forces in physics that contribute to the stability of the atomic nucleus.

Subduction: Plate-tectonic mechanism where an oceanic plate sinks under an other lithospheric plate (generally a continental plate, but sometimes also an oceanic plate).

Sublimation: Direct phase change from solid to gas state.

Succinic acid: $HOOC-CH_2-CH_2-COOH$.

Sun: Star belonging to the main sequence (in H-R diagram); it is 4.56 Ga old. Sun–Earth distance corresponds to 8 light-minutes or 1 astronomic unit (1 AU).

Supernova: Exploding star that, before explosion, was either a binary star (type I) or a massive star (type II). After explosion, the remnant becomes a neutron star.

Symbiosis: Prolonged association between two (or more) organisms that may benefit each other. In the case of endosymbiosis, one organism lives within another one; their relationship is irreversible and implies a complete interdependence, such as the two become a single functional organism. Mitochondria and chloroplasts are remnants of the endosymbiosis of photosynthetic bacteria.

Synchrotron radiation: Electromagnetic waves covering a large frequency domain and emitted under vacuum by high-velocity electrons or ions, when their trajectory is altered, as by a magnetic field. Synchrotron radiation is naturally polarized.

Synonymous (*codons*): Different codons that specify the same amino acid; e.g., AAA and AAG specify lysine.

T

T Tauri star: Newborn low-mass star (lower than two solar masses) that starts to become optically observable. Classical T-Tauri are very young (less than 1 Ma old) and are not yet on the main sequence (PMS stars). They are surrounded by a disk, their luminosity is variable with an excess of IR with respect to UV. T-Tauri stars we observe today are probably similar to the young Sun.

Taxonomy: Science that classifies living species. Similar species belong to the same taxon.

Tectonic: Study of structure (and deformation) of rocks and Earth crust.

Tektites: Natural glasses formed at very high temperature ($> 2000\,^\circ$C) by meteoritic impacts creating craters greater than 10 km in diameter.

Telluric planets: Small and dense rocky planets (density: 3 to 5.5 g cm^{-3}). These planets (namely Mercury, Venus, Earth and Mars) are silicate rich and were formed in the inner part of the protosolar nebula, beyond the dust line.

Terraforming: Voluntary transformation of a planetary atmosphere in order to allow colonization by plants and animals (including man).

Tertiary structure (*of a protein*): Arrangement of the side chains of the residues in space, overall conformation of a protein.

Theia: Name given by some authors to Mars-sized object, which, after impacting the young Earth, led to the Moon formation. This cataclysmic event took place 4.5 to 4.45 years ago.

Thermolysis: Thermal degradation of a molecule into smaller fragments (atoms, radicals, molecules and, rarely, ions).

Thermonuclear reaction: Reaction leading to formation of heavier nuclei by fusion of lighter nuclei. Thermonuclear reactions require high-T and high-P conditions. In natural conditions, these reactions occur spontaneously in cores of main sequence stars or of heavier stars like giant stars. It occurs also during supernovae explosions.

Thermophile: Organism living optimally at high temperatures. They can be divided into "moderate thermophiles", living optimally between 40 and 60 °C, extreme thermophiles, between 60 and 80 °C, and hyperthermophiles, living optimally at temperatures higher than 80 °C (and up to $\sim$ 120 °C).

Thin section (*geology*): Rock slice generally 30-µm thick. At such a thickness most minerals are transparent and can be observed by transmitted light with a polarizing microscope.

Thiocyanate: R–S–CN.

Thioester: R–S–CO–R′.

Thiol: R–SH.

Tholins: Solid mixture of complex organic molecules obtained by irradiation of reduced gases like CH_4, NH_3. Could be present on Titan.

Threonine (*Thr*): Proteinic amino acid containing four C atoms and a –OH group in its side chain. Threonine is described as a hydrophilic amino acid.

Thymine (*T*): Purinic nucleic base purine and specific of DNA.

Titan: The biggest satellite of Saturn ($\sim$ 5000 km in diameter; which is approximately the same size as Mercury). In the Solar System, this is the only satellite that possesses a dense atmosphere. Very probably, organic reactions are taking place in its multicomponent hydrocarbon-bearing atmosphere.

Titus–Bode law: Empirical law giving approximately the planet–Sun distance d as a function of the planet ranking n ($d = (4 + 3.2^{(n-2)})/10$). Mercury is an exception to this law ($d = 0.4$ AU).

Tonalite: Plutonic magmatic rock (granitoid), made up of quartz and calcic plagioclase feldspar; biotite and sometimes amphibole are minor mineral phases. Tonalite does not contain alkali feldspar. Dacite is its effusive equivalent.

TPF (*Terrestrial Planet Finder*): NASA project with a similar goal as the ESA Darwin project, i.e. discovery of extrasolar terrestrial planets.

Transcription: Synthesis of an m-RNA as a copy of an antisens DNA single strand.

Transduction: Transfer of genetic material from one bacteria to another through viral infection.

Transfer of genes (*horizontal transfer*): Transfer of a gene from one organism to another that does not belong to the same species. Such transfer can occur through viral infection or by direct inclusion of genetic material present in the external medium. Horizontal transfer is different from vertical transfer from parents to children.

Transferase: Enzyme that catalyses transfer of a chemical group from one substrate to another.

Transform fault: Boundary between two lithospheric plates that slide without any crust creation or destruction.

Transit: Motion of a planet in front of the disk of its star.

Translation: Sequence enzymatic reactions such that the genetic information coded into a messenger RNA (m-RNA) leads to the synthesis of a specific protein.

Triple point: In a P–T phase diagram of a pure compound, it corresponds to the unique P, T value where the three phases (gas, liquid, solid) coexist at equilibrium. For water, $P = 6.11\,\mathrm{mbar}$ and $T = 273.16\,\mathrm{K}$.

t-RNA synthetase: Enzyme that catalyzes, in a very specific way, bond formation between an amino acid and its t-RNA.

t-RNA: Transfer RNA. Polymer containing 70 to 80 ribonucleotides and specific of each amino acid to which it is linked. Able to recognize a triplet of nucleotides of m-RNA (codon) by a specific molecular-recognition process involving a triplet of nucleotides of the t-RNA (anti-codon). The t-RNA's play a fundamental role for proteinic synthesis.

Trojans: Family of asteroids located at the Lagrange point on the Jupiter orbit. Their position together with the Sun and Jupiter positions determines an equilateral triangle.

Trondhjemite: Plutonic magmatic rock (granitoid), made up of quartz and sodic plagioclase feldspar; biotite is a minor mineral phase. Tonalite and trondhjemite are similar rocks except that in tonalite plagioclase is calcic, whereas it is sodic in trondhjemite.

Tropopause: Atmospheric boundary between troposphere and stratosphere.

Troposphere: Lowest part of Earth's atmosphere, as the temperature at its base is greater than at its top, it is the place of active convection. On Earth, the troposphere thickness ranges between 9 km (pole) and 17 km (equator). Most meteorological phenomena take place in troposphere.

Tryptophane (*Trp*): Proteinic amino acid containing eleven C atoms; tryptophane contains a heterocycle in its side chain. It is described as an aromatic amino acid.

TTG: Tonalite, Trondhjemite, Granodiorite. Rock association typical of the continental crust generated during the first half of Earth's history.

Tunguska event: Explosion that took place in 1908 (June 30th) in Siberia and devastated $2000\,\mathrm{km}^2$ of forest. It was probably due to the impact of a comet or of an asteroid of a size of a few tens of metres.

Tunnel effect: Description of tunnel effect requires the use of quantum mechanics because it is a direct consequence of the wave properties of particles. When a system A gives another system B, while the internal energy of A is lower than the energy barrier (activation energy) required to cross the barrier from A to B, it can be said that A gives B by passing "through the barrier" by a tunnel effect.

Tyrosine (*Tyr*): Proteinic amino acid containing nine C atoms. Its side chain contains a phenolic group.

U

Uracil (*U*): Nucleic base belonging to pyrimidine family and specific of RNAs.

Urea (H_2N–CO–NH_2): First organic molecule that has been synthesised from a mixture of inorganic molecules (Wohler).

UV radiation: Electromagnetic radiation characterized by wavelengths ranging from 0.01 to 0.4 micrometres (energies from 124 to 3.1 eV).

V

Valine (Val): Hydrophobic proteinic amino acid containing five carbon atoms.

Van der Waals forces (*VdW*): Interatomic forces acting between nonbonded atoms at the intramolecular level but also at the intermolecular levels. Repulsive VdW forces are responsible for the less than infinite compressibility of matter and for the fact that atoms have sizes. Attractive VdW forces are responsible for matter cohesion. VdW forces play a major role in biochemistry: together with H bonds and electrostatic interactions, they determine the preferred conformations of molecules and they contribute to molecular-recognition phenomena.

Vernal point: Sun location on the celestial sphere at the vernal equinox (spring equinox). It is the origin of coordinates in the equatorial system.

Vertical tectonic: See sagduction.

Viking: NASA mission to Mars that started in 1976. The two landers (Chryse Planitia and Utopia Planitia) performed a series of very ambitious experiments to detect the presence of life on Mars. Unfortunately, many results were ambiguous.

Virus: System containing DNA or RNA surrounded by a proteinic envelope (capside). When introduced in a living cell, a virus is able to replicate its genetic material by using the host-cell machinery.

VLBI (*Very Long Baseline Interferometry*): Technique that allows a very accurate determination (50 microarcsec) of the position of astronomical sources of radiowaves. This method is based on interferometry measurements using very distant radiotelescopes (large base) located on the same continent or on different continents or even on Earth and on a satellite.

VLT (*Very Large Telescope*): Group of four large telescopes (4 to 8 metres) and several smaller telescopes, able to work as interferometers and located in Chile. VLT is managed by ESO.

Volatile: see refractory.

Volcanoclastic sediment: Material due to sedimentation of volcanic products (i.e. ashes) in the sea or in a lake.

W

Wall (*of a cell*): Extracellular membrane. In bacteria, cell-wall structure is complex: the walls of gram-positive and gram-negative bacteria are different.

Water triple point: See triple point.

Watson–Crick: Canonical model of DNA (double helix) involving the pairing of two polynucleotide strands via H bonds between A and T or G and C. RNA is generally single-stranded but Watson–Crick pairing can occur locally within a single strand. When happens, it involves A–U and G–C pairing.

Weak bonds: Intermolecular or intramolecular bonds involving nonbonded atoms (atoms not bonded by covalent, coordination or electrostatic bonds). H bonds are well-known examples of weak bonds but Van der Waals forces and electrostatic interactions also contribute to weak bonds. The weak bonds play a fundamental role in the living world: they determine the conformation of molecules and more particularly the conformation(s) of biopolymers; they are responsible for the specificity of molecular recognition. The intermolecular association due to weak bonds is generally reversible.

Weak force: One of the four fundamental forces of physics. Parity is violated for this force. The coupling between weak forces and electromagnetic forces is at the origin of the very small energy difference between enantiomers (PVED for parity-violation energy difference).

Weathering: See alteration.

Wind (*solar or stellar*): Flow of ionized matter ejected at high velocity (around 400 km/s) by a star. Solar wind mainly contains protons.

Wobble (*genetics*): Describes imprecision in base pairing between codons and anticodons. It always involves position 3 of codon, mainly when the base is U.

X

Xenolith: Inclusion or enclave of foreign rock or mineral (xenocrystal), in a magmatic rock.

Y

Young sun paradox: Apparent contradiction between the lower brightness of the young Sun, (70% of the present-day intensity in the visible spectral range) and the early presence of liquid water (–4.4 Ga) on Earth. A strong greenhouse effect due to high concentrations of atmospheric CO_2 could account for this apparent paradox.

YSO (*Young Stellar Object*): Star that has not yet completed the process of star formation. YSO includes objects ranging from dense cores, (that can be detected in the submillimetric IR frequency range) to pre-main sequence stars (T Tauri, Herbig AeBe) and HII regions.

YSO: Young stellar object.

Z

Z (*astronomy*): Abundance of "heavy elements" i.e. all elements except H and He.

ZAMS (***zero-age main sequence***): Ensemble of newborn stars in which H fusion has just started.

Zircon: Mineral. Nesosilicate (isolated SiO_4 tetrahedron). [$ZrSiO_4$]. This mineral which also contains traces of Th and U is extremely resistant to weathering and alteration. This is why it is commonly used to determine rock ages. The oldest zircon crystals so far dated gave an age of 4.4 Ga. They represent the oldest known terrestrial material.

Zodiacal light: Diffuse faint light observed in a clear sky close to the ecliptic. It is due to the diffusion of the solar radiation by the electrons and the interplanetary dusts. Also used for any light diffused by dust particles in a planetary system.

4 Authors (Photos and addresses)

ALEKHINA Irina
Senior Researcher
Division Molecular and Radiation Biophysics,
Petersburg Nuclear Physics Institute
Leningrad region, Russia
E-mail: `alekhina@omrb.pnpi.spb.ru`

Molecular biology: Ancient DNA, molecular microbe diversity, extreme environment, ice cores and subglacial lake environments

BARBIER Bernard
CNRS Researcher
Center for Molecular Biophysics, Orléans, France
E-mail: `barbier@cnrs-orleans.fr`

Physico-chemistry/analytical chemistry: Interstellar chemistry, prebiotic chemistry, evolution, origin of the genetic code

BIBRING Jean-Pierre
Professor
Institut d'Astrophysique Spatiale,
Paris Sud University,
Orsay, France
E-mail: `bibring@ias.u-psud.fr`

Planetology: Formation and evolution of the Solar System

BOITEAU Laurent
CNRS Researcher
Chemistry Department
University of Montpellier 2, France
E-mail: laurent.boiteau@univ-montp2.fr

Chemistry, physical organic chemistry: Origins of life, origins of amino acids and peptides

BRACK André
CNRS Research Director Emeritus
Center for molecular biophysics, Orléans, France
E-mail: brack@cnrs-orleans.fr

Chemistry: Origins of life, prebiotic peptides, life on Mars

BRINKMANN Henner
Research assistant
Biochemistry Department, University of Montréal, Montréal, Québec, Canada
E-mail: Henner.Brinkmann@umontreal.ca

Molecular biology: Evolution and molecular phylogeny, origin and early evolution of life, plastid and mitochondrial endosymbioses, evolution of the main lineages of the three domains

BULAT Sergey
Senior Researcher Group Leader
Division Molecular and Radiation Biophysics,
Petersburg Nuclear Physics Institute,
Leningrad region, Russia
email: bulat@omrb.pnpi.spb.ru

Molecular biology: Ancient DNA, microbe diversity, phylogenetics, extreme environment, icy (subglacial lake) and deep sub-terrestrial environments

COLL Patrice
Assistant Professor
Laboratoire Interuniversitaire
des Systèmes Atmosphériques,
University Paris 12 and Paris 7, Créteil, France
E-mail: `pcoll@lisa.univ-paris12.fr`

Chemistry/Planetology: Astro/Exobiology, extraterrestrial organic chemistry and physics (Titan, Mars,...), theoritical and experimental developments

COMMEYRAS Auguste
Professor Emeritus
Chemistry Department,
University of Montpellier 2, France
E-mail: `acommeyras@univ-montp2.fr`

Chemistry, physical organic chemistry: Origins of life, origins of amino acids and peptide

COTTIN Hervé
Associate Professor
Laboratoire Interuniversitaire
des Systèmes Atmosphériques,
University Paris 12 and Paris 7, Créteil, France
E-mail: `cottin@lisa.univ-paris12.fr`

Astrochemistry: Molecular complexity in space, organic chemistry in comets and ISM. Laboratory simulations and modelling

CRONIN John
Professor Emeritus,
Arizona State university, Tempe, Arizona (USA)
E-mail: `jcronin@asu.edu`

Chemistry: Origin of life, evolution of organic matter, organic chemistry of meteorites

DESPOIS Didier
CNRS Researcher
Observatoire Aquitain des Sciences de l'Univers, Bordeaux, France
E-mail: despois@obs.u-bordeaux1.fr

Astrophysics/exobiology: Comets, radioastronomy, stars and solar system formation

FORTERRE Patrick
Professor
Institut de Génétique et Microbiologie, University Paris Sud, Orsay, France
Head of the Department of fundamental and medical microbiology at the Pasteur Institut, Paris
E-mail: forterre@igmors.u-psud.fr
forterre@pasteur.fr

Biology/exobiology: Hyperthermophiles, archaea, phylogenomics, universal tree of life, LUCA, origin of DNA

GARGAUD Muriel
CNRS Researcher
Observatoire Aquitain des Sciences de l'Univers, Bordeaux, France
E-mail: muriel@obs.u-bordeaux1.fr

Astrophysics/astrobiology: Interstellar medium physico-chemistry, origins of life

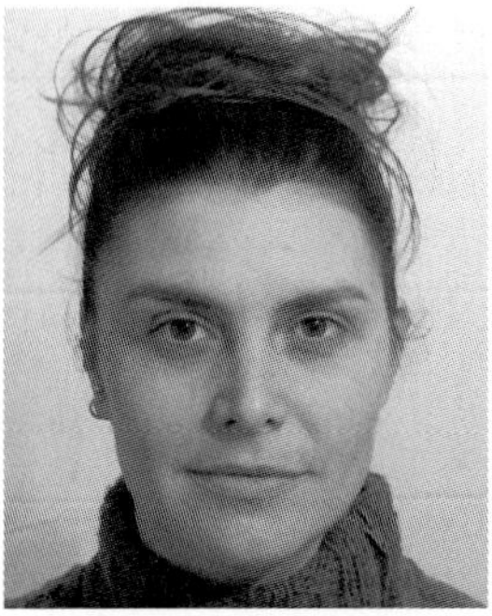

GRIBALDO Simonetta
Post-doc researcher,
Institut de Génétique et Microbiologie,
University Paris-Sud, Orsay, France
E-mail: simonetta.gribaldo@igmors.u-psud.fr

Biology: Early evolution, molecular phylogenetics, comparative genomics

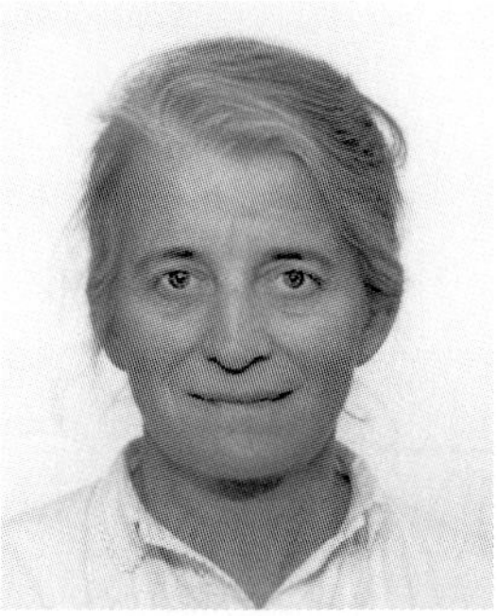

HAENNI Anne-Lise
CNRS Research Director Emeritus
Institut Jacques-Monod,
University Paris 6 and Paris 7, Paris, France
E-mail: haenni@ijm.jussieu.fr

Biochemistry: RNA viruses, protein biosynthesis

LEGER Alain
CNRS Research Director
Institut d'Astrophysique Spatiale,
University of Paris, Orsay, France
E-mail: Alain.Leger@ias.u-psud.fr

Astronomy/Astrobiology: Telluric extrasolar planets, Ocean-Planets Darwin mission (PI of initial proposal), COROT mission

LOPEZ-GARCIA Purificación
CNRS Researcher
Unité d'Ecologie, Systématique et Evolution,
University Paris-Sud, Orsay, France
E-mail: puri.lopez@ese.u-psud.fr

Microbiology: Microbial diversity, phylogeny and evolution, extreme environments

MARTIN Hervé
Professor
Laboratoire Magmas et Volcans,
Blaise Pascal University, Clermont-Ferrand, France
E-mail: `H.Martin@opgc.univ-bpclermont.fr`

Geochemistry: Geological and geochemical evolution of the primitive Earth. Subduction zone magmatism

MAUREL Marie-Christine
Professor
Biochemistry of Evolution
and Molecular Adaptability
Institut Jacques-Monod,
University Paris 6, Paris, France
E-mail: `maurel@ijm.jussieu.fr`

Biology-Biochemistry: Molecular evolution, origins of life, etiology, activity and persistence of RNA

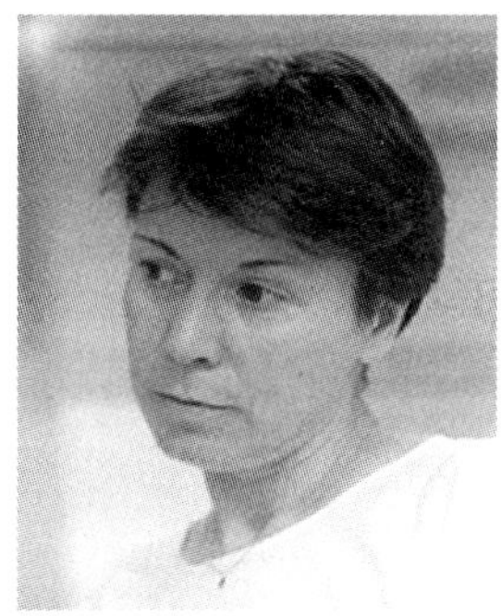

MONTMERLE Thierry
Director of Laboratoire d'Astrophysique de Grenoble
Joseph Fourier University, Grenoble, France
E-mail: `montmerle@obs.ujf-grenoble.fr`

Astrophysics: High-energy phenomena in the interstellar medium, formation and early evolution of stars and planets, irradiation effects by hard radiation and energetic particles in star-forming regions, stellar magnetic fields

MORBIDELLI Alessandro
CNRS Researcher
Observatoire de la Côte d'Azur, Nice, France
E-mail: morby@obs-nice.fr

Astrophysics: Dynamics of asteroids and trans-Neptunian objects, origin and evolution of near-Earth objects, primordial sculpting of the Solar System

NAVARRO-GONZALES Rafael
Research Professor
Universidad National Autonoma de Mexico
Ciudad Universitaeia, Mexico City
E-mail: navarro@nuclecu.unam.mx

Atmospheric Chemistry and Astrobiology: Lightning chemistry in the atmosphere of Earth and of other planets and satellites: impact in the origin and evolution of the biosphere. Study of terrestrial analogs of Mars and Europa for the identification of biosignatures required for the search of extraterrestrial life.

NAKATANI Yoichi
CNRS Research Director Emeritus
Louis Pasteur University, Strasbourg, France
E-mail: nakatani@chimie.u-strasbg.fr

Chemistry: Membrane chemistry, biochemistry and biophysics

OLLIVIER Marc
Assistant astronomer
Institut d'Astrophysique Spatiale,
University Paris-Sud, Orsay, France
E-mail: marc.ollivier@ias.u-psud.fr

Astrophysics: Remote detection and analysis of extrasolar planets, exobiology, high angular resolution technics, high precision photometry

OURISSON Guy
Emeritus Professor,
Louis Pasteur University, Strasbourg, France
Member of the French Academy of Sciences
E-mail: ourisson@chimie.u-strasbg.fr

Organic chemistry: Natural substances, biological and fossil membranes, origin of life

PETIT Jean-Marc
CNRS Researcher
Observatory of Besançon,
University of Franche-Comté, Besançon, France
E-mail: petit@obs-besancon.fr

Astronomy: Formation and evolution of the Solar System; Dynamical and collisional evolution of small bodies

PETIT Jean Robert
CNRS Research Director
Laboratoire de Glaciologie et de Géophysique de l'Environnement, Grenoble, France
E-mail: petit@lgge.obs.ujf-grenoble.fr

Paleoclimatogy: Ice geochemistry, geophysics, polar ice cores, subglacial lake environments

PHILIPPE Hervé
Professor
Biochemistry department, University of Montréal, Montréal, Québec, Canada
E-mail: herve.philippe@umontreal.ca

Biology: Bio-informatics, universal tree of life, molecular phylogeny, comparative genomics model of sequence evolution

PINTI Daniele L.
Professor
GEOTOP-UQAM-McGill,
Montréal, QC, Canada
E-mail: pinti.daniele@uqam.ca

Isotope geochemistry: Studies of the elemental and isotopic composition of early atmosphere; biogeochemistry of nitrogen in early Archean

RAULIN François
Professor
Laboratoire Interuniversitaire
des Systèmes Atmosphériques,
University Paris 12 and Paris 7, Créteil, France
E-mail: raulin@lisa.univ-paris12.fr

Chemistry/Planetology: Exo/astrobiology, extraterrestrial organic chemistry (Titan, Giant planets, Comets, Mars, ...), experimental, theoretical and observational approaches

REISSE Jacques
Professor Emeritus,
Free University of Brussels, Belgium
Member on the Belgian Academy of Sciences
E-mail: jreisse@ulb.ac.be

Chemistry: Stereochemistry and chirality, NMR, sonochemistry, organic cosmochemistry, study of the liquid phase

SELSIS Franck
CNRS Researcher
Centre de recherchen en astrophysique
Ecole Normale Superieure, Lyon, France
E-mail: selsis@ens-lyon.fr

Planetary Sciences and Astrobiology: Formation and evolution of planetary atmospheres, atmospheres of exoplanets: detection and characterization

SOTIN Christophe
Professor
Laboratoire de Planétologie et Géodynamique,
Nantes University, Nantes, France
E-mail: sotin@chimie.univ-nantes.fr

Planetary Geology and Geophysics: Structure and evolution of planetary interiors, infrared remote sensing of planetary surfaces, mineralogy, icy satellites of the giant planets

VANDENABEELE-TRAMBOUZE Odile
CNRS Researcher
Chemistry Department,
University of Montpellier 2, France
E-mail: o.trambouze@univ-montp2.fr

Analytical chemistry: Development of analytical methodologies for trace analysis of organic compounds in meteorites and complex matrices

WESTALL Frances
CNRS Research Director
Centre de Biophysique Moléculaire, Orléans, France
E-mail: westall@cnrs-orleans.fr

Geology: Early Earth, early life, extraterrestrial life

5 Index

Printing: Krips bv, Meppel
Binding: Stürtz, Würzburg